GENERAL CHEMISTRY

Readings from

**SCIENTIFIC
AMERICAN**

GENERAL CHEMISTRY

with introductions by

James B. Ifft
University of Redlands

John E. Hearst
University of California, Berkeley

W. H. Freeman and Company
San Francisco

Library of Congress Cataloging in Publication Data

Ifft, James B.   1935–   comp.
  General chemistry; readings from Scientific
American.

  1. Chemistry—Addresses, essays, lectures.
I. Hearst, John E., 1935–   joint comp.
II. Scientific American.   III. Title.
QD39.I35   540   73–13624
ISBN 0–7167–0886–8
ISBN 0–7167–0885–X (pbk.)

Most of the SCIENTIFIC AMERICAN articles in
General Chemistry are available as separate Offprints.
For a complete list of more than 950 articles now
available as Offprints, write to W. H. Freeman and
Company, 660 Market Street, San Francisco,
California 94104.

Printed in the United States of America

9 8 7 6 5 4 3 2 1

This reader is designed to be a supplement to an introductory course in general chemistry. General chemistry courses are undergoing major changes at present, and this is reflected in the content of the reader. There are two basic reasons for these changes. First, college education has become the privilege of a far greater number of high-school graduates than were able to enter college a mere quarter-century ago. Furthermore, at the University of California, Berkeley, for example, about forty percent of these undergraduates enroll in a general chemistry course. A very small fraction of these students are destined to major in physical science or engineering. Therefore, we interpret this increase in enrollment as an indication that science is being studied because it is intrinsically interesting. This brings us to the second reason for the changes that are occurring in general chemistry courses, for there is no legitimate reason why this intrinsic interest should not be exploited and encouraged in the process of teaching chemistry. Because so many students today arrive in college with an interest in themselves and in their society, examples of chemistry from biology or ecology are stimulating to them. Evolutionary changes in curricula are not new, of course, for the curricula in universities have always been designed to be sensitive to the interests of that society to which the universities belong.

Scientific American is an excellent source for literature about the changing interests in science, for its success as a magazine depends on attracting an audience more general than the professional scientists whose opinions and interests dominate the professional science literature. Nevertheless, when we first conceived the idea of this reader, we were concerned that the relation between the subject matter of the typical general chemistry course and past *Scientific American* articles would not be sufficiently apparent to make the reader possible. We were pleasantly surprised to find a wealth of pertinent material that required no prior knowledge of the subject, material written in sufficiently simple language to provide source material for general chemistry. That so much of the material is related to biology as well is not a matter of our design so much as a reflection of the main interests of this period in scientific research.

Our reader has eight sections. The first section contains two articles, one by Linus Pauling entitled "Chemistry" and the other by George Wald entitled "The Origin of Life." These are designed to provide the definitions for chemistry and the limits of phenomena which are normally called chemistry. The second section deals with the atomic hypothesis, atomic structure, the periodic table and the elements, and the chemical bond. Section III deals with molecular structure from a very biological point of view, but nevertheless provides examples of molecular geometry, electronic excitation by light, chemical change induced by light absorption, biological specificity resulting from complex and detailed molecular structure, and the forces between molecules. The last of these examples is provided by a study of the cell membrane, which is an example of the behavior of polar and nonpolar chemical groups in water. Section IV provides a computer simulation of the motion of molecules in gases, liquids, and solids, and discusses crystal structure, electrical conductivity, transistors, and lattice dislocation. Ice provides an example of the nature of the hydrogen bond and an introduction to the notion of disorder or entropy. Section V deals with thermodynamics, kinetics, equilibrium, and oxidation-reduction. Section VI provides a description of several analytical techniques, including X-ray crystallography, infrared spectroscopy, gas chromatography, and mass spectrometry. Section VII goes further into organic chemistry than most general chemistry courses do, but we feel students will find the articles stimulating and understandable. The last section provides myriad examples of chemistry in life. These include hydrogen bonding, optical rotation, chemical synthesis, molecular structure, weak acids and bases, oxidation-reduction, amino acids, proteins, carbohydrates, and nucleic acids. In addition to learning about protein structure and function, and the genetic code, students can learn about centrifugation, sequencing methods, molecular evolution, mutation, and what has become the central dogma of biology. The section is divided into two parts. The first, articles 28 through 31, deals with proteins. The second, articles 32 through 36, encompasses a discussion of nucleic acids.

Every article has been selected for its clarity and applicability to the material generally considered in a general chemistry course.

Have we gone too far into biology? We think not. Biological systems have been closely associated with chemistry for at least a century. Remember that organic chemistry originated in studies of the chemistry of biological extracts. The ultimate place for molecular biology in the undergraduate curriculum may be uncertain, but it is now biologists and chemists who must be reminded that what is molecular is chemical. They must also be reminded that without the tools of chemistry—X-ray crystallography, chromatography, ultracentrifugation, optical spectroscopy, isotope labelling and, lately, nuclear magnetic resonance—knowledge of molecular biology would not exist. Why not, then, use these most exciting applications of the techniques of chemistry to assist in the teaching of chemistry?

Are examples from molecular biology too complex to be clear to beginning students? Again, we think not. These examples do, moreover, provide students with an opportunity for establishing a relationship between scientific thought, their own thoughts about self-worth, and their attitudes toward life. The central dogma of molecular biology is as provocative, therefore, as Darwinism. It is good education.

We are pleased to acknowledge the contribution Linus Pauling has made to our book. His article written in 1950 provides an initial overview of the then-current state of chemistry and gives his predictions about the future successes to be expected in chemistry. In an epilogue beginning on page 419, written for this book, he reviews those earlier comments and provides several predictions about what the future promises beyond 1973. We express our appreciation to Professor Pauling for the model he provided us as an outstanding teacher and for the insights he has shared with us and our readers in his epilogue.

October 1973

James B. Ifft
John E. Hearst

CONTENTS

VII ORGANIC CHEMISTRY

VIII THE CHEMISTRY OF LIFE

Note on cross-references: References to articles included in this book are noted by the title of the article and the page on which it begins; references to articles that are available as Offprints, but are not included here, are noted by the article's title and Offprint number; references to articles published by SCIENTIFIC AMERICAN, but which are not available as Offprints, are noted by the title of the article and the month and year of its publication.

I

CHEMISTRY:
A PERSPECTIVE IN TIME

I
CHEMISTRY:
A PERSPECTIVE
IN TIME

The chemical properties of matter have been studied for centuries, yet chemistry as we think of it today is a young and fast-changing science. The first chemistry of the kind we consider important today began less than two centuries ago with the work of Lavoisier, which demonstrated the role of oxygen in combustion. The significant advances since then have all depended upon quantitative reasoning. In a way, Lavoisier and the chemists of his day were the beneficiaries of the first precise scientific instruments. But even for a century after Lavoisier, chemists' work consisted mainly of the analysis and synthesis of hundreds of small inorganic and organic molecules. It is doubtful if any of the nineteenth-century chemists could have even conceived that by the 1970's some major contributions of chemists would be the development of the transistor, plastics, and wonder drugs; that chemists would be creating a crisis with chemical pesticides and then devising new and safer ones; or that they would be dealing with environmental issues of energy production, birth control, and food production, or leading the investigations that have resulted in a detailed understanding of the structures of giant biopolymers and how these function.

Every science is dominated by a few great men and women who have made significant contributions to that science, given perspective to the efforts to extend its capabilities, and written with clarity and insight into the nature of their discipline.

Two such men, Linus Pauling and George Wald, have written the first two articles in this reader. Two decades ago each independently stepped back from his experimental work long enough to survey the status of his field. Both provide an excellent historical perspective and give an overview of the content of their disciplines as they seemed in 1950 (Pauling) and 1954 (Wald). Articles by these two scientists have been chosen not only because of the brilliance of their work and the breadth of their understanding of their disciplines, but also because they represent, between them, the full range of the field of chemistry as we view it today. Pauling's early contributions were primarily in the field of quantum mechanics. His early work on the nature of the chemical bond provided the foundation for much of our understanding of the properties of matter. More recently he has made significant contributions to biology, George Wald's discipline. Wald has elucidated the molecular mechanism of vision and has written extensively in the field of biochemistry.

Each man suggested, in the articles now reprinted in this reader, areas in which rapid progress might be made in the future. Because we are now a part of that future, it is exciting to look back at the perspective of these leading scientists.

In "Chemistry," Linus Pauling briefly traces the origin of chemistry from the pioneering work of Lavoisier up to the middle of the twentieth century. A comparison of the material described in this 1950 article with some of the content of the articles elsewhere in the reader is of interest. For example, a comparison of the periodic table used by Pauling in his article with that used by Frieden in "The Chemical Elements of Life," the fourth article, indicates that one of the elements, argon, had acquired a new symbol, Ar, rather than A, and that five additional elements had been discovered. Pauling mentions the importance of catalysts in the chemical laboratory and in the chemical industry. The articles by Faller, Haensel and Burwell, and Phillips, numbers 15, 16, and 31, indicate that not only has much progress been made in understanding the mechanism of catalysis by simple metallic surfaces, but that the mechanism by which the most intricate of all catalysts, enzymes, function has been determined for one enzyme —lysozyme. Pauling points out that the bases of thermodynamics were essentially complete early in this century. The articles by Hubbert and Summers, numbers 13 and 14, demonstrate a new thrust for thermodynamics in particular and science in general—the application of scientific knowledge to the solution of critical environmental problems, specifically, our energy resources.

In 1950, a large number of polymers of considerable practical use had been synthesized and an understanding of the structures of simple biopolymers was at hand. Article 29, by Merrifield, demonstrates that today chemists can synthesize not only Bakelite but one of the most complex polymers on Earth as well, a protein. We now understand not only the structure of methane but also the geometry of the genetic material, DNA. Our understanding of the electronic structure of molecules has deepened appreciably with the advent of high-speed computers, which have permitted accurate calculations of larger molecules. The elegant and informative diagrams in Wahl's article, 5, testify to the power of these new methods. In his article Pauling mentions the development of new analytical methods such as chromatography. Section VI, on instrumental methods of analysis, provides a number of elegant examples of how several analytical techniques— X-ray crystallography, infrared spectroscopy, gas chromatography, and mass spectrometry—have been developed or extended to yield the most detailed information available on molecular structure.

Pauling made a number of predictions in 1950 about where chemistry might be in the year 2000. We are now halfway through this time span. It is interesting, therefore, to see how far along we are in achieving the goals Pauling hoped would be met.

The first prediction was that we would be able to determine the rate of any chemical reaction from a fundamental understanding of the forces between atoms. In general, this goal has not been met yet. Although Phillips's article, number 31, indicates that at least for one enzyme, we do have a detailed understanding of how this particular catalyst, lysozyme, acts to sever a covalent bond, this understanding has not permitted a calculation of the rate of catalysis. Studies currently in progress using molecular beams should soon provide the background information needed to make predictions of reaction rates possible for very simple systems. Nor is it possible at present to synthesize catalysts to order. Haensel and Burwell point out in article 16 that there are innumerable reactions for which no catalysts are available.

Pauling predicted that a large number of inorganic polymers would be synthesized during this half century. The development of large polyphosphate molecules with interesting properties indicates that this is being done.

The prediction that the structure of proteins and nucleic acids would be determined has also been realized. It is reasonable to assume that the structure of genes will be known before the next quarter-century passes.

These are a truly remarkable set of predictions by a truly remarkable scientist.

George Wald wrote "The Origin of Life" in 1954, just four years after Pauling wrote his article. A biologist at Harvard, Wald has been attracted by the chemical aspects of biology, and devoted much of his effort toward elucidating the molecular mechanism of vision. The versatility required for the solution of such a complex problem is illustrated by Wald's outstanding work, some of which is described in article 9, "Molecular Isomers in Vision," by Hubbard and Kropf. Wald's studies required knowledge of molecular spectroscopy and molecular structure as well as the physiology of the eye. Thus Wald has worked where chemistry and biology join, and his ideas and interests are representative of this interface.

In his article of twenty years ago, Wald was most interested in the chemical evolution of the organic molecules from which life is structured. At the time, the primary structures of carbohydrates, fats, proteins, and nucleic acids were known. No sequence information and no forms of secondary structures are mentioned in Wald's article, but the alpha helix was a recent discovery (see article 28, "Proteins," by P. Doty), and the sequences of the two chains of insulin had just been determined by Sanger and his co-workers (see "The Insulin Molecule" by E. O. P. Thompson, *Scientific American* Offprint 42). Wald's perspective was wider: how had the elementary molecules of life come into being?

The Russian biochemist A. I. Oparin published *The Origin of Life* in 1936; it was among the first references about a period of chemical evolution in the Earth's history. Nearer to the time of Wald's article, the first laboratory demonstration of the feasibility of the spontaneous creation of the organic molecules of life was made by S. L. Miller, who showed that a mixture of water vapor (H_2O), methane (CH_4), ammonia (NH_3), and hydrogen (H_2), when circulated past an electric spark, resulted in the formation of glycine and alanine, two amino acids. Now nearly all the amino acids have been formed, through minor modifications of this experiment, from the chemicals believed present in the early atmosphere of the Earth (see article 23, by Eglinton and Calvin, "Chemical Fossils," for other current investigations of chemical evolution).

Wald discusses the probability of unlikely events in relation to the formation of organic molecules and provides the reader with an elegant introduction to the basis of the second law of thermodynamics (which is an important topic in article 12; see "Ice," by L. K. Runnels). Wald argues that with sufficient time unlikely events can become probable, and that processes such as chemical evolution will occur. He discusses the competition between the forces that aggregate molecules into larger structures and forces of disruption, alluding to the constant motion of molecules and therefore the dynamics of evolutionary processes (see article 10, by B. J. Alder and T. E. Wainwright, "Molecular Motions").

Wald notes that structure is as important as composition, and that knowledge of three-dimensional structure can provide insights into the origin of life; it is interesting to see how other authors in this book have explored these ideas (see article 30, by R. E. Dickerson, "The Structure and History of an Ancient Protein," and article 31, by D. C. Phillips, "The Three-dimensional Structure of an Enzyme Molecule"). Wald uses collagen fibrils as an example of how structures can result from the forces of "integration" or spontaneous aggregation (see article 28, by P. Doty, "Proteins," in which the structure of collagen is discussed). He also alludes to hydrophobic interactions in a discussion of the aggregation of lecithins and cephalins into myelin figures (such interactions are described more fully in article 8, by

C. F. Fox, "The Structure of Cell Membranes").

The last part of Wald's article contains a remarkable discussion of the energy requirements of life. Wald is convinced that life originated as a fermentation process that used the energy stored in the organic molecules formed by chemical evolution. Before the depletion of these organic molecules, photosynthesis evolved, and cells became independent of the stored organic molecules for their energy. Instead, they synthesized their own organic molecules by using the energy of the sun's light and carbon dioxide, CO_2, in the atmosphere. Photosynthesis produced molecular oxygen, O_2, in the atmosphere, and finally respiration evolved as a far more efficient process for the production of energy (see article 19, by P. Cloud and A. Gibor, "The Oxygen Cycle").

The molecular mechanism of energy production and transfer in the living cell remains one of the exciting topics of study in contemporary chemistry (as Cloud and Gibor show in "The Oxygen Cycle" and as R. E. Dickerson shows in "The Structure and History of an Ancient Protein," article 30). The mechanism of vision is being studied extensively. The cell membrane has become the focus of much of the present interest in neurobiology (see article 8, by C. F. Fox, "The Structure of Cell Membranes").

And so, like Linus Pauling, George Wald was able, twenty years ago, to give a perspective of science that was remarkably accurate in predicting present interests.

1

CHEMISTRY

LINUS PAULING
September 1950

Aided by the new ideas of physics, the chemists have welded a huge body of facts into a unified system. Many of their fundamental advances have quickly become part of technology

The half-century we are just completing has seen the evolution of chemistry from a vast but largely formless body of empirical knowledge into a coordinated science. This transformation resulted mainly from the development of atomic physics. After the discovery of the electron and of the atomic nucleus, physicists made rapid progress in obtaining a detailed understanding of the electronic structure of atoms and simple molecules, culminating in the discovery of quantum mechanics. The new ideas about electrons and atomic nuclei were soon introduced into chemistry, leading to the formulation of a powerful structural theory which has welded most of the great mass of chemical facts into a unified system. At the same time great steps forward have been taken through the application of new physical techniques to chemical problems, and also through the continued effective use of the techniques of chemistry itself.

Chemistry is a young science. The chemical revolution took place only a little more than 150 years ago, when Antoine Laurent Lavoisier first clearly explained the role of oxygen in combustion and the nature of elementary and compound substances. Before Lavoisier chemical operations had been carried out according to recipes, and chemical reactions had been discovered only by haphazard trial. His new approach led to the rapid collection of a great amount of information about inorganic and organic substances. In 1828 Friedrich Wöhler achieved the first synthesis of an organic substance of animal origin (urea) from inorganic materials, and in the following decades many thousands of new substances were synthesized and their properties investigated. In 1852 Sir Edward Frankland formulated the theory of valence, and in 1858 Friedrich August Kekulé perceived that carbon has four valences; this insight gave great impetus to organic chemistry. Louis Pasteur's discovery of optical activity (the property, possessed by tartaric acid and many

other substances, of rotating the plane of polarization of polarized light) and its explanation by means of the theory of the tetrahedral carbon atom by Jacob van't Hoff and Joseph LeBel effectively completed the classical structural theory of organic chemistry. Guided by this theory, and making use of many special techniques of analysis and synthesis, the organic chemist then investigated great numbers of natural substances and new substances made in the laboratory. Many of them were found to be valuable as dyes, as medicines, in foods and for special industrial purposes, and an immense organic chemical industry was developed, largely based on coal tar.

DURING the first half of the 20th century organic chemistry has advanced along an extension of this road. The theory of the structure of organic molecules has become more precise and more useful through the incorporation in it of the theory of resonance and the

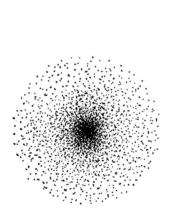

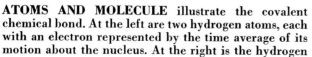

ATOMS AND MOLECULE illustrate the covalent chemical bond. At the left are two hydrogen atoms, each with an electron represented by the time average of its motion about the nucleus. At the right is the hydrogen molecule, in which two nuclei share two electrons. The drawings in this article are by Roger Hayward and are from Pauling's books *General Chemistry* and *College Chemistry*, published by W. H. Freeman and Company.

electronic theory of valence in general, and also of information about distances between atoms, bond angles and other features of molecular structure determined by spectroscopic analysis or by the diffraction of X-rays or electron waves. Many new synthetic procedures and analytical methods have been discovered. One of the most valuable new methods is the chromatographic technique for separating pure constituents from mixtures of substances, which was invented by Michael Tswett in 1906. In industrial chemistry we have seen an important shift: petroleum has to a great extent replaced coal tar as the raw material for the preparation of organic compounds.

Another striking aspect of organic chemistry in our century is the part played by special catalysts, both in the laboratory and in the industrial plant. This 20th-century development, which had its first success in the production of catalysts for the conversion of atmospheric nitrogen into ammonia and nitric acid, has risen to immense importance, especially in the manufacture of valuable products from petroleum. Great progress has resulted also from the expansion of the field of effective endeavor of the organic chemist to include giant molecules — molecules containing thousands or millions of atoms. In the 19th century the organic chemist could work confidently and effectively only with relatively simple substances. Then during the first decades of the 20th century he made effective headway in analyzing the structure and properties of macromolecular natural materials such as cotton, rubber and wood. Armed with knowledge obtained in this way, he essayed to synthesize new fibers, new elastomers, new plastics, and he succeeded not only in obtaining satisfactory substitutes for many natural materials but also in making many materials with far superior properties.

In inorganic chemistry there was a period of inactivity in the first part of our half-century, and then came a rebirth. The last two decades have seen the completion of the periodic table as far as atomic number 98, with the addition of technetium (43), promethium (61), astatine (85), francium (87), neptunium (93), plutonium (94), americium (95), curium (96), berkelium (97) and californium (98). Spectroscopic and diffraction studies of the structure of molecules and crystals have provided a penetrating insight into the nature of the interactions between atoms and molecules. This has been combined with quantum mechanics to yield a broad and powerful electronic theory of valence and chemical combination, permitting the correlation of the structure and properties of inorganic substances as well as of organic substances. The theory of resonance of molecules

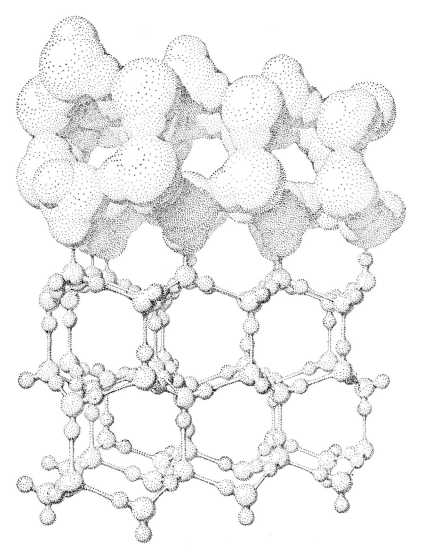

CRYSTAL STRUCTURE of ice is based upon the geometry of the water molecule. The molecules at the top of this drawing are shown with their atoms and interatomic distances in the correct proportion. The atoms at the bottom have been reduced in size to clarify the structure of the crystal.

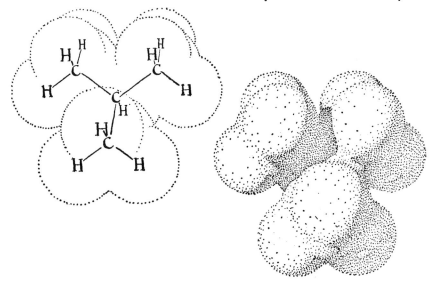

MOLECULAR STRUCTURE of the hydrocarbon isobutane is one of many of such structures worked out by organic chemists. At the right the atoms and the interatomic distances of the molecule are shown in their correct proportions. At the left is a diagram showing the skeletal geometry of the molecule.

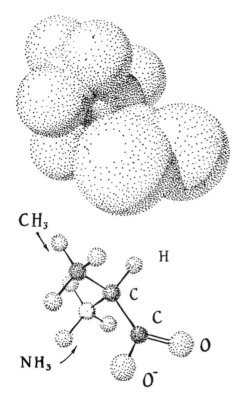
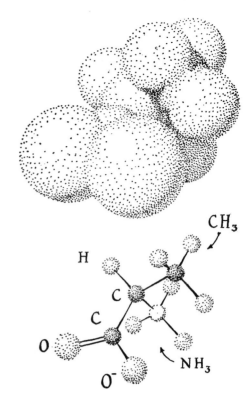

THE AMINO ACID alanine is one of the 20-odd amino acids that make up proteins. Like all the other amino acids except one, alanine has two stereoisomers or molecular mirror images: *d*-alanine (*left*) and *l*-alanine (*right*). Curiously only amino acids of the *l* configuration are found in proteins. Although the structure of the amino acids is well known, their arrangement in proteins is one of the fundamental problems of chemistry.

	H 1				He 2		

O	I	II	III	IV	V	VI	VII	O
He 2	Li 3	Be 4	B 5	C 6	N 7	O 8	F 9	Ne 10
Ne 10	Na 11	Mg 12	Al 13	Si 14	P 15	S 16	Cl 17	A 18

O	I	II	III	IVa	Va	VIa	VIIa	VIII			Ib	IIb	IIIb	IVb	V	VI	VII	O
A 18	K 19	Ca 20	Sc 21	Ti 22	V 23	Cr 24	Mn 25	Fe 26	Co 27	Ni 28	Cu 29	Zn 30	Ga 31	Ge 32	As 33	Se 34	Br 35	Kr 36
Kr 36	Rb 37	Sr 38	Y 39	Zr 40	Nb 41	Mo 42	Tc 43	Ru 44	Rh 45	Pd 46	Ag 47	Cd 48	In 49	Sn 50	Sb 51	Te 52	I 53	Xe 54
Xe 54	Cs 55	Ba 56	La* 57	Hf 72	Ta 73	W 74	Re 75	Os 76	Ir 77	Pt 78	Au 79	Hg 80	Tl 81	Pb 82	Bi 83	Po 84	At 85	Rn 86
Rn 86	Fr 87	Ra 88	Ac♦ 89	Th 90	Pa 91	U 92	Np 93	Pu 94										

✱ Rare-earth metals	Ce 58	Pr 59	Nd 60	Pm 61	Sm 62	Eu 63	Gd 64	Tb 65	Dy 66	Ho 67	Er 68	Tm 69	Yb 70	Lu 71
♦ Uranium metals	Th 90	Pa 91	U 92	Np 93	Pu 94	Am 95	Cm 96	Bk 97	Cf 98					

THE PERIODIC TABLE of the chemical elements has in recent years been filled in and enlarged. Four elements have been inserted into the former table: technetium (Tc), promethium (Pm), astatine (At) and francium (Fr). The table has also been extended beyond uranium (U) by neptunium (Np), plutonium (Pu), americium (Am), curium (Cm), berkelium (Bk) and californium (Cf). Atomic number of each element is beneath symbol.

among two or more valence-bond structures has found valuable applications in both inorganic and organic chemistry.

IN RECENT years practical inorganic chemistry has developed rapidly. Many new and important compounds of fluorine and silicon (fluorocarbons, silicones) have been made. The manufacture of plutonium and the controlled release of atomic energy have been accompanied by extensive chemical studies of uranium and the transuranium elements, of the rare-earth metals and of the elements formed as fission products.

Chemical thermodynamics—the study of the chemical effects of energy and temperature—is essentially a 20th-century development. It is true that the first and second laws of thermodynamics, dealing with heat transfer and entropy, had been formulated by 1851, and that Josiah Willard Gibbs had published his masterful series of papers on the application of thermodynamics to chemical phenomena in the period between 1873 and 1878. But the impact of this work on chemistry was not felt until after 1900. At the turn of the century Walter Nernst discovered the third law of thermodynamics, relating to the behavior of substances at low temperatures, and many chemists, among whom Gilbert Newton Lewis deserves special mention, labored to collect thermodynamic data and to weld them into a practical system. Quantum statistical mechanics has shown how the knowledge of interatomic distances and force constants obtained from spectroscopic and diffraction studies can be used in the application of chemical thermodynamics to practical problems. In the design of industrial plants the modern chemist — especially the petroleum chemist—may depend on thermodynamic information obtained by calculation from interatomic distances in molecules.

Information about the thermodynamic properties of substances, especially the absolute entropy, for application of the third law, often is obtainable only by measurements made down to very low temperatures. Early in the present century Kamerlingh Onnes, extending the pioneer work of Sir James Dewar, obtained temperatures slightly below 1 degree Kelvin by the evaporation of liquid helium. For some time it seemed impossible to achieve a closer approach to absolute zero; then William F. Giauque suggested in 1924 and later put into practice a new method—cooling by demagnetization. With this technique he and other investigators succeeded in reaching temperatures as low as about .001 degree K.

Thus chemical thermodynamics has rapidly developed to the point where it is possible for tables of the thermodynamic functions of chemical substances to be constructed. With the aid of these tables a reliable prediction can be made as to whether any chemical reaction involving these substances can be made to take place or is thermodynamically impossible. This prediction, however, does not satisfy the chemist; he wants also to know whether the reaction will proceed rapidly enough to provide a satisfactory yield of the product in the available time. The study of the speed of chemical reactions is another important branch of physical chemistry. In this field some progress has been made, but the goal of the formulation of a complete theory of reaction kinetics, analogous to the now essentially completed system of chemical thermodynamics, seems to lie far ahead.

WHAT will the next 50 years bring? How much greater understanding and mastery of chemical substances than we now possess will the chemist of the year 2000 have? We may hope that he will have obtained such penetrating

Pauling

knowledge of the forces between atoms and molecules that he will be able to predict the rate of any chemical reaction with reasonable reliability. In order to do this he will have to find out how catalysts work in accelerating chemical reactions. At the present time no one knows why a particular catalyst is effective for a particular reaction; the preparation of catalysts is essentially an empirical art. Perhaps in the next half-century chemists will succeed in preparing catalysts to order. In addition the chemist of the future may well be able to make use of new aids to cause desired chemical reactions to take place. One of these aids might be high-energy rays—alpha particles, electrons, positrons, gamma rays —made available by the uranium pile. As new materials capable of withstanding very high temperatures and pressures are developed, new chemical reactions can be made to occur. And the development of a greater understanding of the relation between the molecular structure and the chemical and physical properties of substances should permit predictions

to be made as to the types of new substances that need to be synthesized for various special purposes.

The recent successful development of valuable new compounds of silicon and fluorine suggests that other elements too may be put to additional uses. The chemistry of very large inorganic molecules has been neglected. We may look forward particularly to progress in the study and use of compounds of elements that have a strong tendency to be present in large molecules, notably phosphorus, vanadium, molybdenum, wolfram and tantalum.

The metals constitute a great class of substances that deserves more thorough study by chemists. Organic and ordinary inorganic materials have been assiduously investigated year after year, but metals and alloys, including intermetallic compounds, have been neglected. The coming half-century should see the development of a sound theory of the structural chemistry of metallic substances. Metallography will thereby become a science, and the straightforward formulation of new alloys with special properties and valuable uses will become possible.

IN organic chemistry there exists a field with equally broad room for progress: investigation of the structure of physiologically active substances, especially vitamins and drugs, and the synthesis of new ones. This work has been handicapped by the lack of a theory of the molecular structural basis of physiological activity. The next half-century should witness the development of such a theory. This would involve also the solution of the problem of the structure of proteins, nucleic acids and other macromolecular constituents of living organisms, including enzymes and ultimately genes. When the mechanism of drug action has been elucidated, it will be possible for chemists to make greater and greater contributions to the problem of good health and the control of physical and mental disease. Instead of synthesizing great numbers of substances at random, the chemist will be able to plot the molecular structure of the most likely substance for each use and synthesize it for trial.

BIBLICAL ACCOUNT of the origin of life is part of the Creation, here illustrated in a 16th-century Bible printed in Lyons. On the first day (*die primo*) God created heaven and the earth. On the second day (*die secundo*) He separated the firmament and the waters. On the third day (*die tertio*) He made the dry land and plants. On the fourth day (*die quarto*) He made the sun, the moon and the stars. On the fifth day (*die quinto*) He made the birds and the fishes. On the sixth day (*die sexto*) He made the land animals and man. In this account there is no theological conflict with spontaneous generation. According to *Genesis* God, rather than creating the animals and plants directly, bade the earth and waters bring them forth. One theological view is that they retain this capacity.

2

THE ORIGIN OF LIFE

GEORGE WALD

August 1954

How did living matter first arise on the earth? As natural scientists learn more about nature they are returning to a hypothesis their predecessors gave up almost a century ago: spontaneous generation

About a century ago the question, How did life begin?, which has interested men throughout their history, reached an impasse. Up to that time two answers had been offered: one that life had been created supernaturally, the other that it arises continually from the nonliving. The first explanation lay outside science; the second was now shown to be untenable. For a time scientists felt some discomfort in having no answer at all. Then they stopped asking the question.

Recently ways have been found again to consider the origin of life as a scientific problem—as an event within the order of nature. In part this is the result of new information. But a theory never rises of itself, however rich and secure the facts. It is an act of creation. Our present ideas in this realm were first brought together in a clear and defensible argument by the Russian biochemist A. I. Oparin in a book called *The Origin of Life*, published in 1936. Much can be added now to Oparin's discussion, yet it provides the foundation upon which all of us who are interested in this subject have built.

The attempt to understand how life originated raises a wide variety of scientific questions, which lead in many and diverse directions and should end by casting light into many obscure corners. At the center of the enterprise lies the hope not only of explaining a great past event—important as that should be—but of showing that the explanation is workable. If we can indeed come to understand how a living organism arises from the nonliving, we should be able to construct one—only of the simplest description, to be sure, but still recognizably alive. This is so remote a possibility now that one scarcely dares to acknowledge it; but it is there nevertheless.

One answer to the problem of how life originated is that it was created. This is an understandable confusion of nature with technology. Men are used to making things; it is a ready thought that those things not made by men were made by a superhuman being. Most of the cultures we know contain mythical accounts of a supernatural creation of life. Our own tradition provides such an account in the opening chapters of *Genesis*. There we are told that beginning on the third day of the Creation, God brought forth living creatures—first plants, then fishes and birds, then land animals and finally man.

Spontaneous Generation

The more rational elements of society, however, tended to take a more naturalistic view of the matter. One had only to accept the evidence of one's senses to know that life arises regularly from the nonliving: worms from mud, maggots from decaying meat, mice from refuse of various kinds. This is the view that came to be called spontaneous generation. Few scientists doubted it. Aristotle, Newton, William Harvey, Descartes, van Helmont, all accepted spontaneous generation without serious question. Indeed, even the theologians—witness the English Jesuit John Turberville Needham—could subscribe to this view, for *Genesis* tells us, not that God created plants and most animals directly, but that He bade the earth and waters to bring them forth; since this directive was never rescinded, there is nothing heretical in believing that the process has continued.

But step by step, in a great controversy that spread over two centuries, this belief was whittled away until nothing remained of it. First the Italian Francesco Redi showed in the 17th century that meat placed under a screen, so that flies cannot lay their eggs on it, never develops maggots. Then in the following century the Italian abbé Lazzaro Spallanzani showed that a nutritive broth, sealed off from the air while boiling, never develops microorganisms, and hence never rots. Needham objected that by too much boiling Spallanzani had rendered the broth, and still more the air above it, incompatible with life. Spallanzani could defend his broth; when he broke the seal of his flasks, allowing new air to rush in, the broth promptly began to rot. He could find no way, however, to show that the air in the sealed flask had not been vitiated. This problem finally was solved by Louis Pasteur in 1860, with a simple modification of Spallanzani's experiment. Pasteur too used a flask containing boiling broth, but instead of sealing off the neck he drew it out in a long, S-shaped curve with its end open to the air. While molecules of air could pass back and forth freely, the heavier particles of dust, bacteria and molds in the atmosphere were trapped on the walls of the curved neck and only rarely reached the broth. In such a flask the broth seldom was contaminated; usually it remained clear and sterile indefinitely.

This was only one of Pasteur's experiments. It is no easy matter to deal with so deeply ingrained and common-sense a belief as that in spontaneous generation. One can ask for nothing better in such a pass than a noisy and stubborn opponent, and this Pasteur had in the naturalist Félix Pouchet, whose arguments before the French Academy of Sciences drove Pasteur to more and more rigorous experiments. When he had finished, nothing remained of the belief in spontaneous generation.

We tell this story to beginning students of biology as though it represents a triumph of reason over mysticism. In fact it is very nearly the opposite. The reasonable view was to believe in spontaneous generation; the only alternative, to believe in a single, primary act of supernatural creation. There is no third position. For this reason many scientists a century ago chose to regard the belief in spontaneous generation as a "philosophical necessity." It is a symptom of the philosophical poverty of our time that this necessity is no longer appreciated. Most modern biologists, having reviewed with satisfaction the downfall of the spontaneous generation hypothesis, yet unwilling to accept the alternative belief in special creation, are left with nothing.

I think a scientist has no choice but to approach the origin of life through a hypothesis of spontaneous generation. What the controversy reviewed above showed to be untenable is only the belief that living organisms arise spontaneously under present conditions. We have now to face a somewhat different problem: how organisms may have arisen spontaneously under different conditions in some former period, granted that they do so no longer.

The Task

To make an organism demands the right substances in the right proportions and in the right arrangement. We do not think that anything more is needed—but that is problem enough.

The substances are water, certain salts—as it happens, those found in the ocean—and carbon compounds. The latter are called *organic* compounds because they scarcely occur except as products of living organisms.

Organic compounds consist for the most part of four types of atoms: carbon, oxygen, nitrogen and hydrogen. These four atoms together constitute about 99 per cent of living material, for hydrogen and oxygen also form water. The organic compounds found in organisms fall mainly into four great classes: carbohydrates, fats, proteins and nucleic acids. The illustrations on this and the next three pages give some notion of their composition and degrees of complexity. The fats are simplest, each consisting of three fatty acids joined to glycerol. The starches and glycogens are made of sugar units strung together to form long straight and branched chains. In general only one type of sugar appears in a single starch or glycogen; these molecules are large, but still relatively simple. The principal function of carbohydrates and fats in the organism is to serve as fuel—as a source of energy.

The nucleic acids introduce a further level of complexity. They are very large structures, composed of aggregates of at least four types of unit—the nucleotides—brought together in a great variety of proportions and sequences. An almost endless variety of different nucleic acids is possible, and specific differences among them are believed to be of the highest importance. Indeed, these structures are thought by many to be the main constituents of the genes, the bearers of hereditary constitution.

Variety and specificity, however, are most characteristic of the proteins, which include the largest and most complex molecules known. The units of which their structure is built are about 25 different amino acids. These are strung together in chains hundreds to thousands of units long, in different proportions, in all types of sequence, and with the greatest variety of branching and folding. A virtually infinite number of different proteins is possible. Organisms seem to exploit this potentiality, for no two species of living organism, animal or plant, possess the same proteins.

Organic molecules therefore form a large and formidable array, endless in variety and of the most bewildering complexity. One cannot think of having organisms without them. This is precisely the trouble, for to understand how organisms originated we must first of all explain how such complicated molecules could come into being. And that is only the beginning. To make an organism requires not only a tremendous variety of these substances, in adequate amounts and proper proportions, but also just the right arrangement of them. Structure here is as important as composition—and what a complication of structure! The most complex machine man has devised—say an electronic brain—is child's play compared with the simplest of living organisms. The especially trying thing is that complexity here involves such small dimensions. It is on the molecular level; it consists of a detailed fitting of molecule to molecule such as no chemist can attempt.

The Possible and Impossible

One has only to contemplate the magnitude of this task to concede that the spontaneous generation of a living organism is impossible. Yet here we are—as a result, I believe, of spontaneous generation. It will help to digress for a mo-

CARBOHYDRATES comprise one of the four principal kinds of carbon compound found in living matter. This structural formula represents part of a characteristic carbohydrate. It is a polysaccharide consisting of six-carbon sugar units, three of which are shown.

ment to ask what one means by "impossible."

With every event one can associate a probability—the chance that it will occur. This is always a fraction, the proportion of times the event occurs in a large number of trials. Sometimes the probability is apparent even without trial. A coin has two faces; the probability of tossing a head is therefore 1/2. A die has six faces; the probability of throwing a ·deuce is 1/6. When one has no means of estimating the probability beforehand, it must be determined by counting the fraction of successes in a large number of trials.

Our everyday concept of what is impossible, possible or certain derives from our experience: the number of trials that may be encompassed within the space of a human lifetime, or at most within recorded human history. In this colloquial, practical sense I concede the spontaneous origin of life to be "impossible." It is impossible as we judge events in the scale of human experience.

We shall see that this is not a very meaningful concession. For one thing, the time with which our problem is concerned is geological time,.and the whole extent of human history is trivial in the balance. We shall have more to say of this later.

But even within the bounds of our own time there is a serious flaw in our judgment of what is possible. It sounds impressive to say that an event has never been observed in the whole of human history. We should tend to regard such an event as at least "practically" impossible, whatever probability is assigned to it on abstract grounds. When we look a little further into such a statement, however, it proves to be almost meaningless. For men are apt to reject reports of very improbable occurrences. Persons of good

judgment think it safer to distrust the alleged observer of such an event than to believe him. The result is that events which are merely very extraordinary acquire the reputation of never having occurred at all. Thus the highly improbable is made to appear impossible.

To give an example: Every physicist knows that there is a very small probability, which is easily computed, that the table upon which I am writing will suddenly and spontaneously rise into the air. The event requires no more than that the molecules of which the table is composed, ordinarily in random motion in all directions, should happen by chance to move in the same direction. Every physicist concedes this possibility; but try telling one that you have seen it happen. Recently I asked a friend, a Nobel laureate in physics, what he would say if I told him that. He laughed and said that he would regard it as more probable that I was mistaken than that the event had actually occurred.

We see therefore that it does not mean much to say that a very improbable event has never been observed. There is a conspiracy to suppress such observations, not among scientists alone, but among all judicious persons, who have learned to be skeptical even of what they see, let alone of what they are told. If one group is more skeptical than others, it is perhaps lawyers, who have the harshest experience of the unreliability of human evidence. Least skeptical of all are the scientists, who, cautious as they are, know very well what strange things are possible.

A final aspect of our problem is very important. When we consider the spontaneous origin of a living organism, this is not an event that need happen again and again. It is perhaps enough for it to happen once. The probability with

which we are concerned is of a special kind; it is the probability that an event occur *at least once*. To this type of probability a fundamentally important thing happens as one increases the number of trials. However improbable the event in a single trial, it becomes increasingly probable as the trials are multiplied. Eventually the event becomes virtually inevitable. For instance, the chance that a coin will not fall head up in a single toss is 1/2. The chance that no head will appear in a series of tosses is $1/2 \times 1/2 \times 1/2$ . . . as many times over as the number of tosses. In 10 tosses the chance that no head will appear is therefore 1/2 multiplied by itself 10 times, or 1/1,000. Consequently the chance that a head will appear at least once in 10 tosses is 999/1,000. Ten trials have converted what started as a modest probability to a near certainty.

The same effect can be achieved with any probability, however small, by multiplying sufficiently the number of trials. Consider a reasonably improbable event, the chance of which is 1/1,000. The chance that this will not occur in one trial is 999/1,000. The chance that it won't occur in 1,000 trials is 999/1,000 multiplied together 1,000 times. This fraction comes out to be 37/100. The chance that it will happen at least once in 1,000 trials is therefore one minus this number—63/100—a little better than three chances out of five. One thousand trials have transformed this from a highly improbable to a highly probable event. In 10,000 trials the chance that this event will occur at least once comes out to be 19,999/20,000. It is now almost inevitable.

It makes no important change in the argument if we assess the probability that an event occur at least two, three, four or some other small number of

FATS are a second kind of carbon compound found in living matter. This formula represents the whole molecule of palmitin, one of the commonest fats. The molecule consists of glycerol (*11 atoms at the far left*) and fatty acids (*hydrocarbon chains at the right*).

times rather than at least once. It simply means that more trials are needed to achieve any degree of certainty we wish. Otherwise everything is the same.

In such a problem as the spontaneous origin of life we have no way of assessing probabilities beforehand, or even of deciding what we mean by a trial. The origin of a living organism is undoubtedly a stepwise phenomenon, each step with its own probability and its own conditions of trial. Of one thing we can be sure, however: whatever constitutes a trial, more such trials occur the longer the interval of time.

The important point is that since the origin of life belongs in the category of at-least-once phenomena, time is on its side. However improbable we regard this event, or any of the steps which it involves, given enough time it will almost certainly happen at least once. And for life as we know it, with its capacity for growth and reproduction, once may be enough.

Time is in fact the hero of the plot. The time with which we have to deal is of the order of two billion years. What we regard as impossible on the basis of human experience is meaningless here. Given so much time, the "impossible" becomes possible, the possible probable, and the probable virtually certain. One has only to wait: time itself performs the miracles.

Organic Molecules

This brings the argument back to its first stage: the origin of organic compounds. Until a century and a quarter ago the only known source of these substances was the stuff of living organisms. Students of chemistry are usually told that when, in 1828, Friedrich Wöhler synthesized the first organic compound, urea, he proved that organic compounds do not require living organisms to make

them. Of course it showed nothing of the kind. Organic chemists are alive; Wöhler merely showed that they can make organic compounds externally as well as internally. It is still true that with almost negligible exceptions all the organic matter we know is the product of living organisms.

The almost negligible exceptions, however, are very important for our argument. It is now recognized that a constant, slow production of organic molecules occurs without the agency of living things. Certain geological phenomena yield simple organic compounds. So, for example, volcanic eruptions bring metal carbides to the surface of the earth, where they react with water vapor to yield simple compounds of carbon and hydrogen. The familiar type of such a reaction is the process used in old-style bicycle lamps in which acetylene is made by mixing iron carbide with water.

Recently Harold Urey, Nobel laureate in chemistry, has become interested in the degree to which electrical discharges in the upper atmosphere may promote the formation of organic compounds. One of his students, S. L. Miller, performed the simple experiment of circulating a mixture of water vapor, methane (CH_4), ammonia (NH_3) and hydrogen—all gases believed to have been present in the early atmosphere of the earth—continuously for a week over an electric spark. The circulation was maintained by boiling the water in one limb of the apparatus and condensing it in the other. At the end of the week the water was analyzed by the delicate method of paper chromatography. It was found to have acquired a mixture of amino acids! Glycine and alanine, the simplest amino acids and the most prevalent in proteins, were definitely identified in the solution, and there were indications it contained aspartic acid and two others. The yield was surprisingly

high. This amazing result changes at a stroke our ideas of the probability of the spontaneous formation of amino acids.

A final consideration, however, seems to me more important than all the special processes to which one might appeal for organic syntheses in inanimate nature.

It has already been said that to have organic molecules one ordinarily needs organisms. The synthesis of organic substances, like almost everything else that happens in organisms, is governed by the special class of proteins called enzymes—the organic catalysts which greatly accelerate chemical reactions in the body. Since an enzyme is not used up but is returned at the end of the process, a small amount of enzyme can promote an enormous transformation of material.

Enzymes play such a dominant role in the chemistry of life that it is exceedingly difficult to imagine the synthesis of living material without their help. This poses a dilemma, for enzymes themselves are proteins, and hence among the most complex organic components of the cell. One is asking, in effect, for an apparatus which is the unique property of cells in order to form the first cell.

This is not, however, an insuperable difficulty. An enzyme, after all, is only a catalyst; it can do no more than change the *rate* of a chemical reaction. It cannot make anything happen that would not have happened, though more slowly, in its absence. Every process that is catalyzed by an enzyme, and every product of such a process, would occur without the enzyme. The only difference is one of rate.

Once again the essence of the argument is time. What takes only a few moments in the presence of an enzyme or other catalyst may take days, months or years in its absence; but given time, the end result is the same.

NUCLEIC ACIDS are a third kind of carbon compound. This is part of desoxyribonucleic acid, the backbone of which is five-carbon sugars alternating with phosphoric acid. The letter R is any one of four nitrogenous bases, two purines and two pyrimidines.

Indeed, this great difficulty in conceiving of the spontaneous generation of organic compounds has its positive side. In a sense, organisms demonstrate to us what organic reactions and products are *possible*. We can be certain that, given time, all these things must occur. Every substance that has ever been found in an organism displays thereby the finite probability of its occurrence. Hence, given time, it should arise spontaneously. One has only to wait.

It will be objected at once that this is just what one cannot do. Everyone knows that these substances are highly perishable. Granted that, within long spaces of time, now a sugar molecule, now a fat, now even a protein might form spontaneously, each of these molecules should have only a transitory existence. How are they ever to accumulate; and, unless they do so, how form an organism?

We must turn the question around. What, in our experience, is known to destroy organic compounds? Primarily two agencies: decay and the attack of oxygen. But decay is the work of living organisms, and we are talking of a time before life existed. As for oxygen, this introduces a further and fundamental section of our argument.

It is generally conceded at present that the early atmosphere of our planet contained virtually no free oxygen. Almost all the earth's oxygen was bound in the form of water and metal oxides. If this were not so, it would be very difficult to imagine how organic matter could accumulate over the long stretches of time that alone might make possible the spontaneous origin of life. This is a crucial point, therefore, and the statement that the early atmosphere of the planet was virtually oxygen-free comes forward so opportunely as to raise a suspicion of special pleading. I have for this reason taken care to consult a number of

geologists and astronomers on this point, and am relieved to find that it is well defended. I gather that there is a widespread though not universal consensus that this condition did exist. Apparently something similar was true also for another common component of our atmosphere—carbon dioxide. It is believed that most of the carbon on the earth during its early geological history existed as the element or in metal carbides and hydrocarbons; very little was combined with oxygen.

This situation is not without its irony. We tend usually to think that the environment plays the tune to which the organism must dance. The environment is given; the organism's problem is to adapt to it or die. It has become apparent lately, however, that some of the most important features of the physical environment are themselves the work of living organisms. Two such features have just been named. The atmosphere of our planet seems to have contained no oxygen until organisms placed it there by the process of plant photosynthesis. It is estimated that at present all the oxygen of our atmosphere is renewed by photosynthesis once in every 2,000 years, and that all the carbon dioxide passes through the process of photosynthesis once in every 300 years. In the scale of geological time, these intervals are very small indeed. We are left with the realization that all the oxygen and carbon dioxide of our planet are the products of living organisms, and have passed through living organisms over and over again.

Forces of Dissolution

In the early history of our planet, when there were no organisms or any free oxygen, organic compounds should have been stable over very long periods. This is the crucial difference between

the period before life existed and our own. If one were to specify a single reason why the spontaneous generation of living organisms was possible once and is so no longer, this is the reason.

We must still reckon, however, with another destructive force which is disposed of less easily. This can be called spontaneous dissolution—the counterpart of spontaneous generation. We have noted that any process catalyzed by an enzyme can occur in time without the enzyme. The trouble is that the processes which synthesize an organic substance are reversible: any chemical reaction which an enzyme may catalyze will go backward as well as forward. We have spoken as though one has only to wait to achieve syntheses of all kinds; it is truer to say that what one achieves by waiting is *equilibria* of all kinds—equilibria in which the synthesis and dissolution of substances come into balance.

In the vast majority of the processes in which we are interested the point of equilibrium lies far over toward the side of dissolution. That is to say, spontaneous dissolution is much more probable, and hence proceeds much more rapidly, than spontaneous synthesis. For example, the spontaneous union, step by step, of amino acid units to form a protein has a certain small probability, and hence might occur over a long stretch of time. But the dissolution of the protein or of an intermediate product into its component amino acids is much more probable, and hence will go ever so much more rapidly. The situation we must face is that of patient Penelope waiting for Odysseus, yet much worse: each night she undid the weaving of the preceding day, but here a night could readily undo the work of a year or a century.

How do present-day organisms manage to synthesize organic compounds against the forces of dissolution? They do so by a continuous expenditure of

PROTEINS are a fourth kind of carbon compound found in living matter. This formula represents part of a polypeptide chain, the backbone of a protein molecule. The chain is made up of amino acids. Here the letter R represents the side chains of these acids.

FILAMENTS OF COLLAGEN, a protein which is usually found in long fibrils, were dispersed by placing them in dilute acetic acid. This electron micrograph, which enlarges the filaments 75,000 times, was made by Jerome Gross of the Harvard Medical School.

energy. Indeed, living organisms commonly do better than oppose the forces of dissolution; they grow in spite of them. They do so, however, only at enormous expense to their surroundings. They need a constant supply of material and energy merely to maintain themselves, and much more of both to grow and reproduce. A living organism is an intricate machine for performing exactly this function. When, for want of fuel or through some internal failure in its mechanism, an organism stops actively synthesizing itself in opposition to the processes which continuously decompose it, it dies and rapidly disintegrates.

What we ask here is to synthesize organic molecules without such a machine. I believe this to be the most stubborn problem that confronts us—the weakest link at present in our argument. I do not think it by any means disastrous, but it calls for phenomena and forces some of which are as yet only partly understood and some probably still to be discovered.

Forces of Integration

At present we can make only a beginning with this problem. We know that it is possible on occasion to protect molecules from dissolution by precipitation or by attachment to other molecules. A wide variety of such precipitation and "trapping" reactions is used in modern chemistry and biochemistry to promote syntheses. Some molecules appear to acquire a degree of resistance to disintegration simply through their size. So, for example, the larger molecules composed of amino acids—polypeptides and proteins—seem to display much less tendency to disintegrate into their units than do smaller compounds of two or three amino acids.

Again, many organic molecules display still another type of integrating force—a spontaneous impulse toward structure formation. Certain types of fatty molecules—lecithins and cephalins—spin themselves out in water to form highly oriented and well-shaped structures—the so-called myelin figures. Proteins sometimes orient even in solution, and also may aggregate in the solid state in highly organized formations. Such spontaneous architectonic tendencies are still largely unexplored, particularly as they may occur in complex mixtures of substances, and they involve forces the strength of which has not yet been estimated.

What we are saying is that possibilities exist for opposing *intra*molecular dissolution by *inter*molecular aggregations of various kinds. The equilibrium between union and disunion of the amino acids that make up a protein is all to the advantage of disunion, but the aggregation of the protein with itself or other molecules might swing the equilibrium in the opposite direction: perhaps by removing the protein from access to the water which would be required to disintegrate it or by providing some particularly stable type of molecular association.

In such a scheme the protein appears only as a transient intermediate, an unstable way-station, which can either fall back to a mixture of its constituent amino acids or enter into the formation of a complex structural aggregate: amino acids $\leftrightarrows$ protein $\rightarrow$ aggregate.

Such molecular aggregates, of various degrees of material and architectural complexity, are indispensable intermediates between molecules and organisms. We have no need to try to imagine the spontaneous formation of an organism by one grand collision of its component molecules. The whole process must be gradual. The molecules form aggregates, small and large. The aggregates add further molecules, thus growing in size and complexity. Aggregates of various kinds interact with one another to form still larger and more complex structures. In this way we imagine the ascent, not by jumps or master strokes, but gradually, piecemeal, to the first living organisms.

First Organisms

Where may this have happened? It is easiest to suppose that life first arose in the sea. Here were the necessary salts and the water. The latter is not only the principal component of organisms, but prior to their formation provided a medium which could dissolve molecules of the widest variety and ceaselessly mix and circulate them. It is this constant mixture and collision of organic molecules of every sort that constituted in large part the "trials" of our earlier discussion of probabilities.

The sea in fact gradually turned into a dilute broth, sterile and oxygen-free. In this broth molecules came together in increasing number and variety, sometimes merely to collide and separate, sometimes to react with one another to produce new combinations, sometimes to aggregate into multimolecular formations of increasing size and complexity.

What brought order into such complexes? For order is as essential here as composition. To form an organism, molecules must enter into intricate designs and connections; they must eventually form a self-repairing, self-constructing dynamic machine. For a time this problem of molecular arrangement seemed to present an almost insuperable obstacle in the way of imagining a spontaneous origin of life, or indeed the laboratory

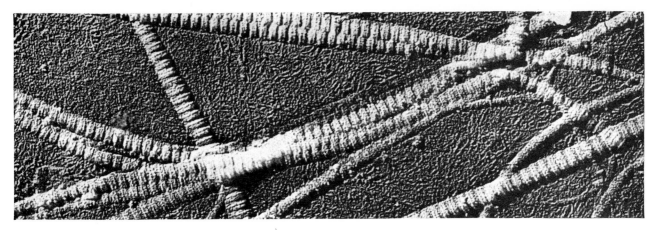

FIBRILS OF COLLAGEN formed spontaneously out of filaments such as those shown on the opposite page when 1 per cent of sodium chloride was added to the dilute acetic acid. These long fibrils are identical in appearance with those of collagen before dispersion.

synthesis of a living organism. It is still a large and mysterious problem, but it no longer seems insuperable. The change in view has come about because we now realize that it is not altogether necessary to *bring* order into this situation; a great deal of order is implicit in the molecules themselves.

The epitome of molecular order is a crystal. In a perfect crystal the molecules display complete regularity of position and orientation in all planes of space. At the other extreme are fluids—liquids or gases—in which the molecules are in ceaseless motion and in wholly random orientations and positions.

Lately it has become clear that very little of a living cell is truly fluid. Most of it consists of molecules which have taken up various degrees of orientation with regard to one another. That is, most of the cell represents various degrees of approach to crystallinity—often, however, with very important differences from the crystals most familiar to us. Much of the cell's crystallinity involves molecules which are still in solution—so-called liquid crystals—and much of the dynamic, plastic quality of cellular structure, the capacity for constant change of shape and interchange of material, derives from this condition. Our familiar crystals, furthermore, involve only one or a very few types of molecule, while in the cell a great variety of different molecules come together in some degree of regular spacing and orientation—*i.e.*, some degree of crystallinity. We are dealing in the cell with highly mixed crystals and near-crystals, solid and liquid. The laboratory study of this type of formation has scarcely begun. Its further exploration is of the highest importance for our problem.

In a fluid such as water the molecules are in very rapid motion. Any molecules dissolved in such a medium are under a constant barrage of collisions with water molecules. This keeps small and moderately sized molecules in a constant turmoil; they are knocked about at random, colliding again and again, never holding any position or orientation for more than an instant. The larger a molecule is relative to water, the less it is disturbed by such collisions. Many protein and nucleic acid molecules are so large that even in solution their motions are very sluggish, and since they carry large numbers of electric charges distributed about their surfaces, they tend even in solution to align with respect to one another. It is so that they tend to form liquid crystals.

We have spoken above of architectonic tendencies even among some of the relatively small molecules: the lecithins and cephalins. Such molecules are insoluble in water yet possess special groups which have a high affinity for water. As a result they tend to form surface layers, in which their water-seeking groups project into the water phase, while their water-repelling portions project into the air, or into an oil phase, or unite to form an oil phase. The result is that quite spontaneously such molecules, when exposed to water, take up highly oriented positions to form surface membranes, myelin figures and other quasi-crystalline structures.

Recently several particularly striking examples have been reported of the spontaneous production of familiar types of biological structure by protein molecules. Cartilage and muscle offer some of the most intricate and regular patterns of structure to be found in organisms. A fiber from either type of tissue presents under the electron microscope a beautiful pattern of cross striations of various widths and densities, very regularly spaced. The proteins that form these structures can be coaxed into free solution and stirred into completely random orientation. Yet on precipitating, under proper conditions, the molecules realign with regard to one another to regenerate with extraordinary fidelity the original patterns of the tissues [*see illustration above*].

We have therefore a genuine basis for the view that the molecules of our oceanic broth will not only come together spontaneously to form aggregates but in doing so will spontaneously achieve various types and degrees of order. This greatly simplifies our problem. What it means is that, given the right molecules, one does not have to do everything for them; they do a great deal for themselves.

Oparin has made the ingenious suggestion that natural selection, which Darwin proposed to be the driving force of organic evolution, begins to operate at this level. He suggests that as the molecules come together to form colloidal aggregates, the latter begin to compete with one another for material. Some aggregates, by virtue of especially favorable composition or internal arrangement, acquire new molecules more rapidly than others. They eventually emerge as the dominant types. Oparin suggests further that considerations of optimal size enter at this level. A growing colloidal particle may reach a point at which it becomes unstable and breaks down into smaller particles, each of which grows and redivides. All these phenomena lie within the bounds of known processes in nonliving systems.

The Sources of Energy

We suppose that all these forces and factors, and others perhaps yet to be revealed, together give us eventually the first living organism. That achieved, how does the organism continue to live?

We have already noted that a living organism is a dynamic structure. It is the site of a continuous influx and outflow of matter and energy. This is the very sign of life, its cessation the best evidence of death. What is the primal organism to use as food, and how derive the energy it needs to maintain itself and grow?

For the primal organism, generated under the conditions we have described, only one answer is possible. Having arisen in an oceanic broth of organic molecules, its only recourse is to live upon them. There is only one way of doing that in the absence of oxygen. It is called fermentation: the process by which organisms derive energy by breaking organic molecules and re-arranging their parts. The most familiar example of such a process is the fermentation of sugar by yeast, which yields alcohol as one of the products. Animal cells also ferment sugar, not to alcohol but to lactic acid. These are two examples from a host of known fermentations.

The yeast fermentation has the following over-all equation: $C_6H_{12}O_6 \rightarrow 2\ CO_2 + 2\ C_2H_5OH +$ energy. The result of fragmenting 180 grams of sugar into 88 grams of carbon dioxide and 92 grams of alcohol is to make available about 20,000 calories of energy for the use of the cell. The energy is all that the cell derives by this transaction; the carbon dioxide and alcohol are waste products which must be got rid of somehow if the cell is to survive.

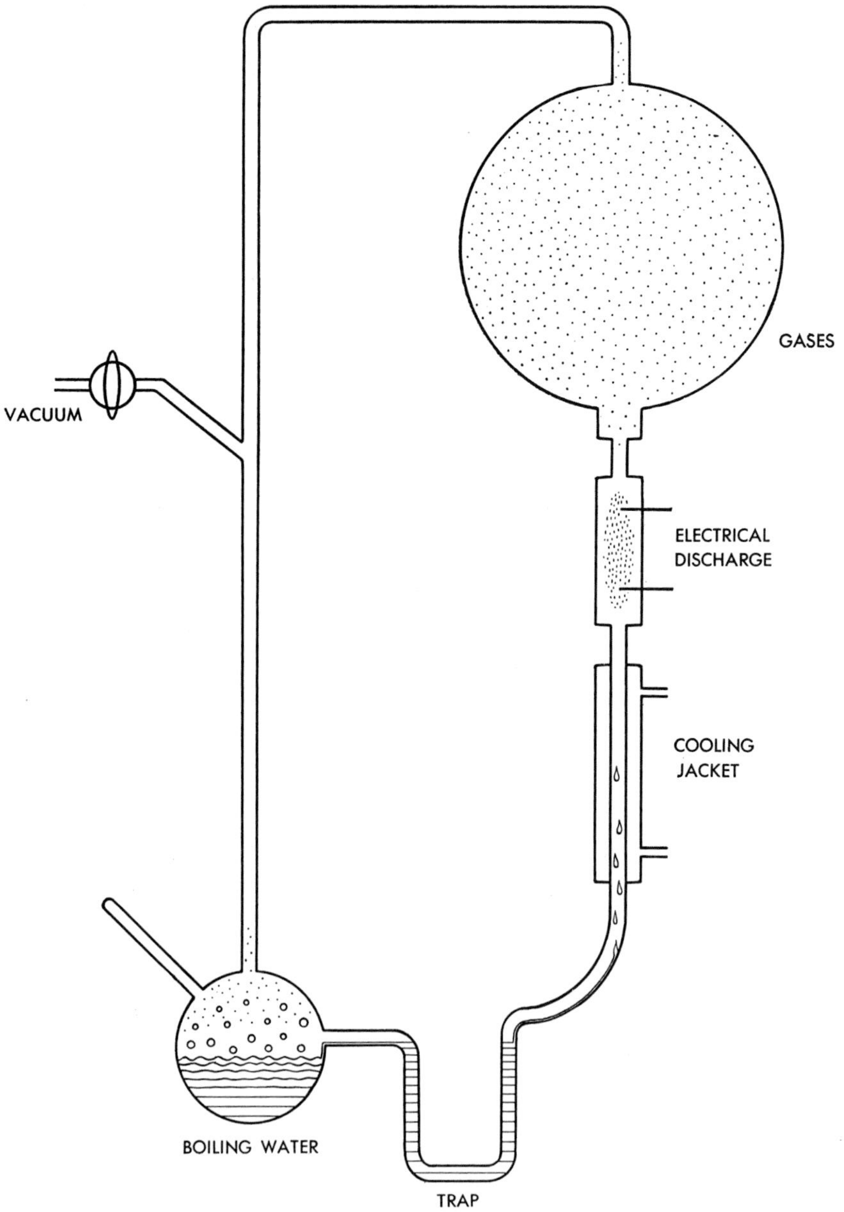

The cell, having arisen in a broth of organic compounds accumulated over the ages, must consume these molecules by fermentation in order to acquire the energy it needs to live, grow and reproduce. In doing so, it and its descendants are living on borrowed time. They are consuming their heritage, just as we in our time have nearly consumed our heritage of coal and oil. Eventually such a process must come to an end, and with that life also should have ended. It would have been necessary to start the entire development again.

Fortunately, however, the waste product carbon dioxide saved this situation. This gas entered the ocean and the atmosphere in ever-increasing quantity. Some time before the cell exhausted the supply of organic molecules, it succeeded in inventing the process of photosynthesis. This enabled it, with the energy of sunlight, to make its own organic molecules: first sugar from carbon dioxide and water, then, with ammonia and nitrates as sources of nitrogen, the entire array of organic compounds which it requires. The sugar synthesis equation is: $6\ CO_2 + 6\ H_2O +$ sunlight $\rightarrow C_6H_{12}O_6 + 6\ O_2$. Here 264 grams of carbon dioxide plus 108 grams of water plus about 700,000 calories of sunlight yield 180 grams of sugar and 192 grams of oxygen.

This is an enormous step forward. Living organisms no longer needed to depend upon the accumulation of organic matter from past ages; they could make their own. With the energy of sunlight they could accomplish the fundamental organic syntheses that provide their substance, and by fermentation they could produce what energy they needed.

Fermentation, however, is an extraordinarily inefficient source of energy. It leaves most of the energy potential of organic compounds unexploited; consequently huge amounts of organic material must be fermented to provide a modicum of energy. It produces also various poisonous waste products—alcohol, lactic acid, acetic acid, formic acid and so on. In the sea such products are readily washed away, but if organisms were ever to penetrate to the air and land, these products must prove a serious embarrassment.

One of the by-products of photosynthesis, however, is oxygen. Once this was available, organisms could invent a new way to acquire energy, many times as efficient as fermentation. This is the

EXPERIMENT of S. L. Miller made amino acids by circulating methane (CH_4), ammonia (NH_3), water vapor (H_2O) and hydrogen (H_2) past an electrical discharge. The amino acids collected at the bottom of apparatus and were detected by paper chromatography.

process of cold combustion called respiration: $C_6H_{12}O_6 + 6\ O_2 \rightarrow 6\ CO_2 + 6\ H_2O +$ energy. The burning of 180 grams of sugar in cellular respiration yields about 700,000 calories, as compared with the approximately 20,000 calories produced by fermentation of the same quantity of sugar. This process of combustion extracts all the energy that can possibly be derived from the molecules which it consumes. With this process at its disposal, the cell can meet its energy requirements with a minimum expenditure of substance. It is a further advantage that the products of respiration—water and carbon dioxide—are innocuous and easily disposed of in any environment.

Life's Capital

It is difficult to overestimate the degree to which the invention of cellular respiration released the forces of living organisms. No organism that relies wholly upon fermentation has ever amounted to much. Even after the advent of photosynthesis, organisms could have led only a marginal existence. They could indeed produce their own organic materials, but only in quantities sufficient to survive. Fermentation is so profligate a way of life that photosynthesis could do little more than keep up with it. Respiration used the material of organisms with such enormously greater efficiency as for the first time to leave something over. Coupled with fermentation, photosynthesis made organisms self-sustaining; coupled with respiration, it provided a surplus. To use an economic analogy, photosynthesis brought organisms to the subsistence level; respiration provided them with capital. It is mainly this capital that they invested in the great enterprise of organic evolution.

The entry of oxygen into the atmosphere also liberated organisms in another sense. The sun's radiation contains ultraviolet components which no living cell can tolerate. We are sometimes told that if this radiation were to reach the earth's surface, life must cease. That is not quite true. Water absorbs ultraviolet radiation very effectively, and one must conclude that as long as these rays penetrated in quantity to the surface of the earth, life had to remain under water. With the appearance of oxygen, however, a layer of ozone formed high in the atmosphere and absorbed this radiation. Now organisms could for the first time emerge from the water and begin to populate the earth and air. Oxygen provided not only the means of obtaining adequate energy for evolution but the protective blanket of ozone which alone made possible terrestrial life.

This is really the end of our story. Yet not quite the end. Our entire concern in this argument has been to bring the origin of life within the compass of natural phenomena. It is of the essence of such phenomena to be repetitive, and hence, given time, to be inevitable.

This is by far our most significant conclusion—that life, as an orderly natural event on such a planet as ours, was inevitable. The same can be said of the whole of organic evolution. All of it lies within the order of nature, and apart from details all of it was inevitable.

Astronomers have reason to believe that a planet such as ours—of about the earth's size and temperature, and about as well-lighted—is a rare event in the universe. Indeed, filled as our story is with improbable phenomena, one of the least probable is to have had such a body as the earth to begin with. Yet though this probability is small, the universe is so large that it is conservatively estimated at least 100,000 planets like the earth exist in our galaxy alone. Some 100 million galaxies lie within the range of our most powerful telescopes, so that throughout observable space we can count apparently on the existence of at least 10 million million planets like our own.

What it means to bring the origin of life within the realm of natural phenomena is to imply that in all these places life probably exists—life as we know it. Indeed, I am convinced that there can be no way of composing and constructing living organisms which is fundamentally different from the one we know—though this is another argument, and must await another occasion. Wherever life is possible, given time, it should arise. It should then ramify into a wide array of forms, differing in detail from those we now observe (as did earlier organisms on the earth) yet including many which should look familiar to us—perhaps even men.

We are not alone in the universe, and do not bear alone the whole burden of life and what comes of it. Life is a cosmic event—so far as we know the most complex state of organization that matter has achieved in our cosmos. It has come many times, in many places—places closed off from us by impenetrable distances, probably never to be crossed even with a signal. As men we can attempt to understand it, and even somewhat to control and guide its local manifestations. On this planet that is our home, we have every reason to wish it well. Yet should we fail, all is not lost. Our kind will try again elsewhere.

II

ATOMS AND THE CHEMICAL BOND

II

ATOMS AND THE CHEMICAL BOND

INTRODUCTION

The existence of atoms has been hypothesized from the time of Democritus, but it has been only from the beginning of the nineteenth century that chemists have found theories about atoms indispensable to the advance of chemistry. The atomic hypothesis is the foundation of modern science. The evidence that matter is composed of atoms is overwhelming; indeed, it is now possible to provide a picture of atoms in crystalline arrays on the surface of metals. Article 3, by Erwin Müller, "Atoms Visualized," presents such a "picture." Although the field emission microscope that Müller developed was not important in establishing the atomic hypothesis, it is one of a number of modern techniques that prove the existence of atoms. In its original design the field emission microscope produced images that were limited in resolution because they were formed by electrons which were emitted from a metal needle tip and which impinged on a fluorescent screen. Müller points out that the poor resolution was an intrinsic limitation resulting from the long wavelength of the emitted electrons. The microscope was therefore redesigned so that the images on the screen would be formed by helium ions. Helium atoms introduced near the surface of the positive needle tip lose an electron to the tip and, as positive helium ions, are repelled toward the screen. Because helium ions are 7,300 times heavier than electrons, after acceleration in the electrical field of the field emission microscope the helium ions have 85 times more momentum than electrons accelerated to the same kinetic energy, and thus produce a sharper image of the atoms in the needle tip from which they stream.

The arguments used by Müller here are examples of wave-particle duality, the de Broglie equation, and the diffraction or interference of waves. The de Broglie equation states that all particles have wave-like motions and that the greater the momentum of the particle, the smaller is its wavelength. The helium ion has a wavelength 1/85 that of an electron of equal kinetic energy. The smaller wavelength of helium ions results in a sharper diffraction pattern and therefore greater resolution.

There are 105 different kinds of atoms, called elements. "The Chemical Elements of Life" by Earl Frieden introduces the elements and the periodic table. The elemental abundances are provided for the universe, the Earth's crust, seawater, and the human body (see also article 17, by F. MacIntyre, "Why the Sea Is Salt"; A. J. Deutsch, "The Abundance of the Elements," *Scientific American*, October 1950; and W. A. Fowler, "The Origin of the Elements," [Offprint 210]). What is appealing about Frieden's use of the periodic table is that he emphasizes those elements that are known to be essential for life. He enables the reader to think about elements as being parts of living organisms as well as parts of inanimate objects.

The list of elements essential for life is still not fully known because it is hard to establish the requirements of living organisms for trace amounts of an element. Frieden sets the number at 24 elements. Furthermore, in some cases these basic elements must be present in specific molecules. The essential nutritional requirements of man,

for instance, begin with amino acids—nitrogen-containing molecules
—of which the following eight,

L-Lysine	L-Valine
L-Tryptophan	L-Methionine
L-Phenylalanine	L-Leucine
L-Threonine	L-Isoleucine

are all required to the extent of 0.2 to 1.2 grams per day. Other classes
of compounds required for life are the vitamins and the essential
fatty acids.

It is useful to continue to think in terms of the most basic characteristics of elements or atoms, of which hydrogen is the simplest and
lightest. The hydrogen atom consists of one proton and one electron.
If a sample of hydrogen atoms is heated to a very high temperature,
the hydrogen atoms will emit electromagnetic radiation. Examination
of this radiation with a simple prism reveals that light of discrete
wavelengths and frequencies is emitted. Such an emission spectrum
for hydrogen atoms in the ultraviolet is shown in Figure II.1.

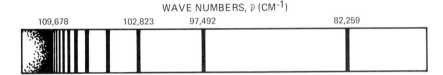

Figure II.1   Ultraviolet light emitted by hydrogen atoms.

The numbers called wave numbers above the lines in the spectrum
are the reciprocals of the wavelengths of the light. They are proportional to the energy of a photon, the unit of light, having this wavelength. Several other groups of lines are found for H atoms in other
regions of the emission spectrum, which, if shown here, would extend
toward the right. All the wave numbers of the lines in the H atom
emission spectrum are related to simple integers through the Rydberg
equation

$$\tilde{\nu} \; (\text{cm}^{-1}) = 109{,}678 \left(\frac{1}{n_2{}^2} - \frac{1}{n_1{}^2} \right)$$

where n_1 and n_2 are integral numbers, and $n_1 > n_2$.

These line spectra can be explained by the postulate that an electron in an H atom is in a definite energy state and that when a photon
is emitted with a definite energy, the electron drops to a lower energy

Table II.1   Properties of the quantum numbers

Quantum number	Name	What they describe	Possible values
n	Principal	Size, energy, number of nodal surfaces	$n = 1, 2, 3, 4, \ldots$
l	Azimuthal	Shape, total angular momentum	$l = 0, 1, 2, \ldots, n-2, n-1$
m	Magnetic	Direction, angular momentum along the z axis	$m = 0, \pm 1, \pm 2, \ldots, \pm l$
s	Spin	Direction of electron spin	$s = \pm\tfrac{1}{2}$

state also with a definite energy. The energy levels for hydrogen are shown in Figure II.2. The ultraviolet spectrum in Figure II.1 shows the light emitted when a hydrogen electron drops to the lowest energy level from the higher energy levels.

In its most stable state, the electron in the hydrogen atom in Figure II.2 occupies the orbital represented by the dash at $n = 1$, $l = 0$. At

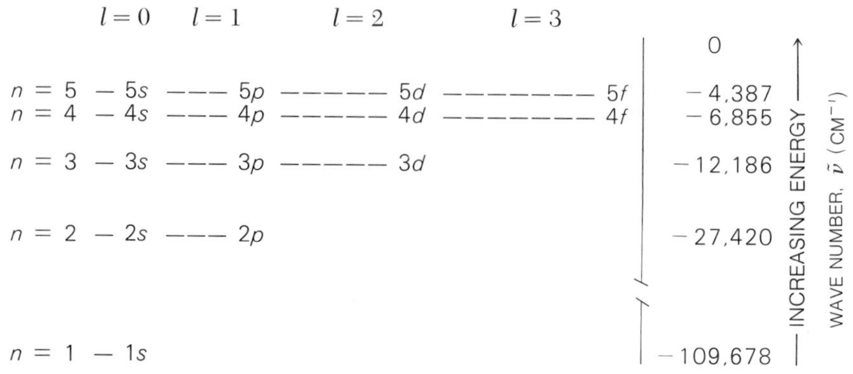

Figure II.2 Energy level diagram for the hydrogen atom (wave number scale discontinuous).

higher energies, the electron would occupy other orbitals. But these configurations would be less stable.

The energy levels of the hydrogen atom, Figure II.2, are characterized by four quantum numbers, which are represented by the letters n, l, m, and s. Properties of the quantum numbers are shown in Table II.1, on the previous page.

The energy of the hydrogen atom depends only on the principal quantum number n. For the values of n there can be many states in which electron distributions take different shapes and directions. This is demonstrated by Figure II.3, which shows the electron distributions for many of these states. At the top of Figure II.2 a second notation was used in which $l = 0$ is called an s state, $l = 1$ a p state, $l = 2$ a d state, and $l = 3$ an f state. These quantum states, shown for the H atom, are applicable qualitatively to atoms having more electrons. The energies of the quantum states in atoms of higher atomic numbers are different from those in the H atom, and are different from each other, but the general shapes of the electron distributions are retained. The more complex energy-level diagrams for the many-electron atoms are the basis for the ordering of elements in the periodic table.

By using spectroscopy, it has been possible to determine just what the universe is made of, for each element has its characteristic absorption and emission spectra, which enable astronomers to determine the elemental composition of stars. These spectra also prove the existence of discrete energy levels in atoms. Each electron in an atom is in a definite energy state called a quantum state and each such state is identified by a set of quantum numbers. For each such state a wave function or probability density function indicates the probability that an electron is located in a unit volume element at some position relative to the nucleus.

The Schrödinger wave equation, discussed in "Chemistry by Computer," makes it possible to compute the probability density functions for the electrons in atoms and simple molecules. Use of the computer can provide some information about probability density functions not readily obtained by experiment. In "Chemistry by Computer," Wahl has computed the probability density functions for all the electrons in the first ten elements in the periodic table. The contour

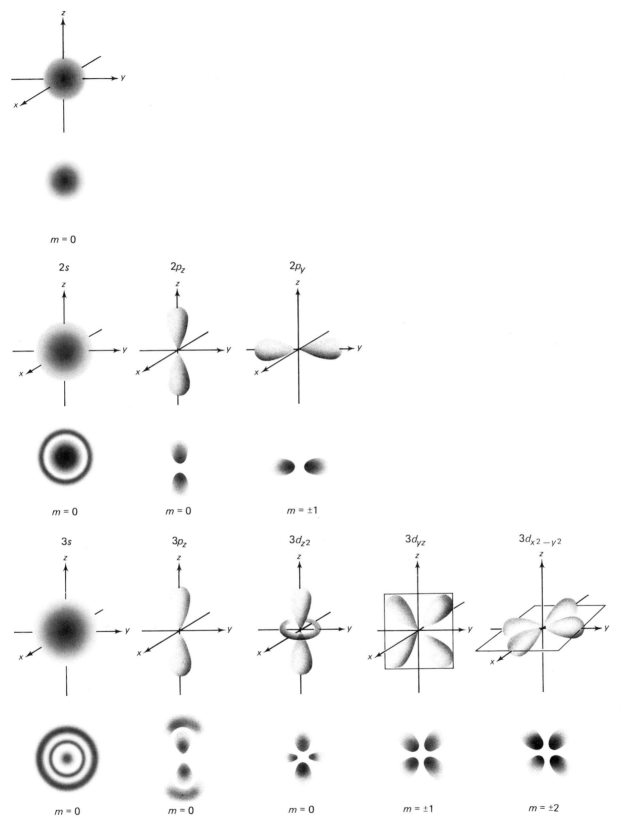

Figure II.3 Electron distributions for many of the energy states of the hydrogen atom. The top member of each pair of diagrams is a three-dimensional representation of an atomic orbital, and above each is a quantum identification. The bottom member of each pair is a cross-section through the most revealing plane of symmetry in that orbital and represents the approximate electron density in that plane. The lobes of the other $2p$ orbital (the $2p_x$, not shown) are directed along the x axis; in other regards the cross-sectional representation of this orbital is equivalent to those of the $2p_y$ and $2p_z$ orbitals. The lobes of the other $3p$ orbitals (the $3p_x$ and $3p_y$) are directed along the x and y axes, respectively. Their cross-sections are equivalent in shape but different in orientation, just like the $2p$ orbitals. Two $3d$ orbitals are not shown; the $3d_{xy}$ orbital has four lobes directed along perpendicular axes oriented in the xy plane $45°$ between x and y; similarly, the $3d_{xz}$ orbital has four lobes oriented on perpendicular axes in the xz plane $45°$ between x and z. Each indicated $\pm$ value of m corresponds to two orbitals with equivalent cross-sections oriented differently in space. For example, the $3d_{xy}$, $3d_{xz}$, $3d_{yz}$, and $3d_{x^2-y^2}$ orbitals have identical cross-sections, all of which are associated with m values -2, -1, $+1$, and $+2$, though not, in this figure, in specific ways.

diagrams presented in this article are nearly exact solutions to the Schrödinger equation and are a precise method of presenting the atomic orbitals or energy states. In Wahl's article the formation of diatomic molecules and a chemical bond are represented by a figure in which two independent atoms — shown as contour-line drawings — are brought together to form molecular orbitals in which the probability density functions are such that electrons can no longer be identified with one atom or the other. Some of these molecular states are bonding states, labelled *g*, and result in a net attraction between atoms. Some states are antibonding, labelled *u*, and result in a net repulsion between atoms. The sum of the electrons in the *g* states minus the number of electrons in the *u* states must be greater than zero or no net chemical bond exists. Ionic bonding is associated with the transfer of an electron from one atom to another in bond formation, as shown for the molecule of LiF.

Molecules have definite energy states just as atoms do. Electronic excitation generally requires large amounts of energy and is produced by the absorption of electromagnetic radiation in the ultraviolet or the visible ranges. ("Molecular Isomers in Vision" by Hubbard and Kropf discusses the effects of light absorption in these regions of the spectrum.) Wahl discusses molecular vibrations, a process associated with absorption of lower-energy radiation in the infrared (see also Crawford's article, "Chemical Analysis by Infrared").

Finally, Wahl presents some truly remarkable representations of simple chemical reactions which clearly demonstrate that chemical reactions are the redistribution of electrons. To describe the geometry of molecules accurately requires not one, but many different representations. The many representations of molecules are presented in Figure II.4; all the diagrams depict the geometry of the water molecule, each with slightly different emphasis.

GRAPHIC FORMULA

ELECTRON CONTOUR DIAGRAM

BALL-AND-STICK MODEL

SPACE-FILLING MODEL

Figure II.4 Various representations of molecular structure. The graphic formula for a molecule of H_2O is shown at the left. Other ways the molecule can be represented are shown in the other diagrams.

3

ATOMS VISUALIZED

ERWIN W. MÜLLER
June 1957

*The field ion microscope makes pictures of atoms in a
metal crystal by accelerating positive ions from a fine
needle of the metal to a fluorescent screen*

The picture on the cover of Scientific American, June 1957, was
a photograph of atoms. Nonsense,
you may say; atoms are much too small
to be visible in any microscope, and,
besides, they could not possibly be colored, for they are a great deal smaller
than a wavelength of light. I have to

confess that the colors were a trick of
photography, but the fact remains: it was
an actual photograph of individual
atoms—tungsten atoms on a needle tip a
thousand times finer than the tip of an
ordinary pin. The picture was produced
with a new microscope which has made
it possible for the first time to photo-

graph atoms as the building stones of a
piece of matter. With this tool we can
watch the fascinating changes in a metal
crystal as atoms are added or torn out
of the lattice.

The instrument is an improved version of the field emission microscope
which I introduced in 1936 and de-

ATOMS IN A CRYSTAL of tungsten appear as small luminous
spots in this photograph of the fluorescent screen of a field ion
microscope at Pennsylvania State University. The atoms, in the tip
of a tungsten needle, are enlarged some two million diameters.

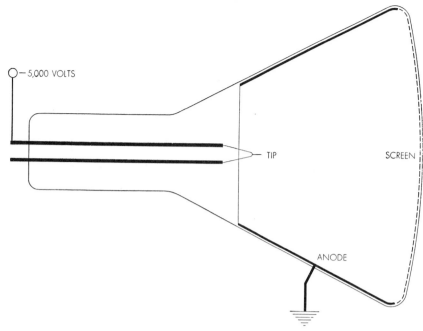

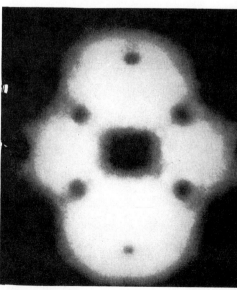

FIELD EMISSION MICROSCOPE, the forerunner of the field ion microscope, is schematically depicted in cross section. A fuzzy image of the atoms in the tip of an extremely fine needle (center) is formed on a fluorescent screen by a stream of electrons from the tip.

TUNGSTEN NEEDLE of the field emission microscope (left) is compared with the tip of an ordinary pin in this photomicrograph.

scribed in an article in this magazine several years ago ["A New Microscope," by Erwin W. Müller; SCIENTIFIC AMERICAN, May, 1952]. In principle the field emission microscope is a very simple device, resembling a television tube. It has a fluorescent screen and a fine metal needle which corresponds to the electron gun of the cathode-ray tube. By means of a high electric voltage we strip particles off the rounded needle tip and fire them at the screen, where they

form a picture of the atomic structure of the tip surface. The needle is very slender indeed: its round tip has a radius of only 1,000 Angstrom units (one 100,-000th of a centimeter). The picture magnifies the tip surface a million times.

This magnification is sufficient to make individual atoms visible. In the original version of the microscope, however, the image is not sharp enough to separate the atoms. Its resolution (*i.e.*, the narrowest distance at which

one object can be distinguished from an adjacent one) is about 20 Angstroms, whereas the atoms on the needle tip are only from three to 10 Angstroms apart. Consequently the original microscope shows only fuzzy images of a few large, prominent atoms.

The resolution obtainable with that first version of the field emission microscope is inherently limited by the fact that the particles emitted from the needle tip are electrons. Unfortunately elec-

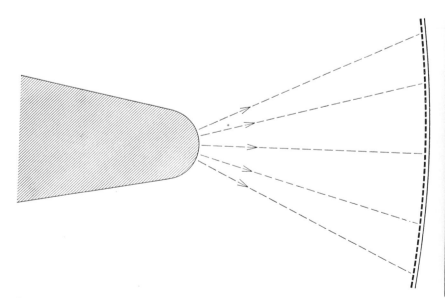

ELECTRONS which form the image of the field emission microscope are simply emitted by the rounded tip of the needle, here enormously enlarged and moved close to the screen.

IMAGE of field emission microscope shows symmetry of crystal planes, but no atoms.

trons are apt to stray a little instead of traveling straight to the screen. That is, they spread out like shot from a shotgun. Moreover the electrons, behaving as waves, produce a diffraction effect. The result is that they give a fuzzy image of the point source on the needle tip from which they came.

A little thought about physical principles suggests a way to overcome these limitations. Suppose we use heavier particles (*e.g.*, ionized atoms) instead of electrons. Since the heavier particles have a much shorter wavelength, there will be less wave interference and therefore less diffraction. More important, we can reduce the shotgun spread of the heavier particles by cooling the needle tip to a very low temperature—a stratagem which does not work with electrons.

I therefore decided to modify the microscope to employ positive ions (atoms with an electron removed) as the projectiles. Now you cannot tear atoms out of a metal as easily as you can electrons. Very well, then, let us bring gas atoms up to the needle tip, ionize them there, and thus provide ourselves with a supply of projectiles. The gas we shall use is helium. We put a little helium in the microscope tube—not much, because we want the gas to be thin enough so that the ions will not collide with atoms on their way to the screen. We apply a high positive voltage to the needle tip, creating an electric field amounting to 400 million volts per centimeter at the curved surface of the tip. Now when a helium atom comes very close to the tip, one of its electrons jumps to the tip, and then the positively ionized helium atom is immediately propelled by the high positive voltage on the needle tip toward the screen at the other end of the tube. In this manner thousands of helium atoms are ionized at protrusions on the needle-tip surface (*e.g.*, steps in the crystal lattice), and they stream to the screen, reproducing there an image of each atom in the rows forming the lattice steps [*see photographs on pages 30 and 31*].

The reason the tip must be cooled is that many helium atoms reach it without being ionized; at ordinary temperatures these bounce off obliquely in random directions. If we cool the tip to somewhere near absolute zero, the gas atoms will stick to it for a while and lose their kinetic energy. As they are reevaporated, they hop about on the tip surface [*see diagram p. 32*]. They cannot become ionized unless they hop at least five Angstroms away from the surface. If the temperature of the needle tip is kept at about 21 degrees above absolute zero, five Angstroms will be the average hopping height of the helium atoms, and they are most likely to reach this height above protrusions (*e.g.*, lattice steps) on the tip surface, where the electrical field is exceptionally strong.

The principal features of the microscope are depicted in the accompanying diagram [*below*]. The tip is cooled by means of liquid hydrogen, enclosed within an outer mantle of liquid nitrogen. Voltages up to 30,000 volts can be applied to the tip; to prevent electrical discharges within the tube which would break down the voltage, the glass walls of the tube are covered with a conduc-

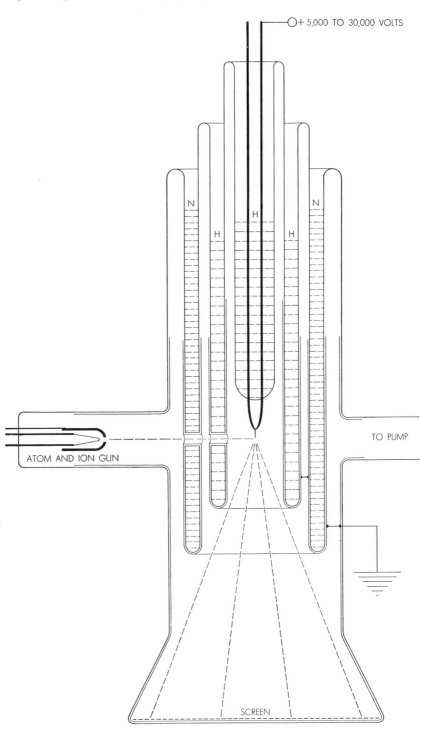

FIELD ION MICROSCOPE is a modification of the field emission microscope. Where the needle of the field electron microscope is charged negatively, the needle of the field ion microscope is charged positively. The positive ions which form its image are made from a thin gas of helium atoms introduced into the microscope. The needle is cooled by an arrangement of vessels containing liquid nitrogen (N) and hydrogen (H). The crystal lattice of the needle may be experimentally damaged by bombarding it with the gun at left.

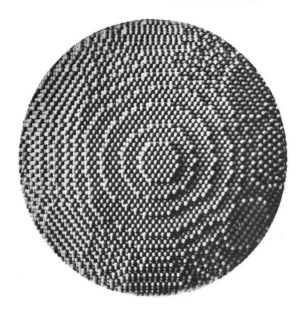

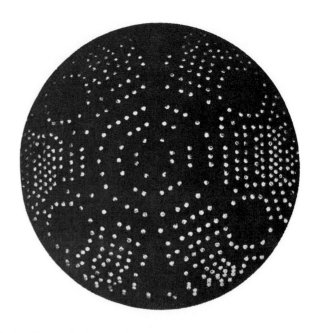

MODEL OF TUNGSTEN NEEDLE TIP (radius: 130 Angstrom units) at left is made of cork balls. The balls at the corners of planes of atoms in the crystal lattice are covered with luminescent paint; the photograph in center, made in the dark, shows their loca-

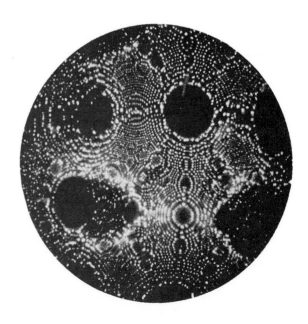

RHENIUM NEEDLE TIP is shown in these field ion microscope pictures. The three pictures show successive stages in the evapora-tion of rhenium atoms from the surface of the needle under the influence of the electric field; as the evaporation proceeds, the

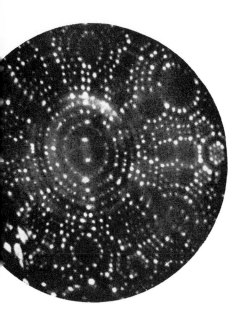

tion. At right is a field ion picture of a tungsten needle tip (radius: 160 Angstroms).

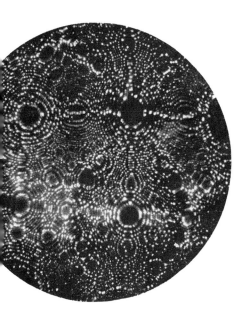

configuration of the crystal lattice changes. The tip is enlarged some 750,000 diameters.

tive coating which serves as a ground electrode. Helium atoms (or electrons or metal atoms) are introduced into the evacuated tube by a gun firing them at the needle tip from the side of the tube. The tip itself may be as small as 70 Angstroms in radius, giving a magnification of 10 million on the screen (the magnification is determined by the distance to the screen divided by the radius of the tip). But a tip of this size sends so few ions to the screen that the image is extremely faint. For most experiments we use a tip of 1,000 Angstroms or 500 Angstroms radius, the latter yielding a magnification of two million. These tips produce images bright enough to be recorded by sensitive film with an exposure time of one minute.

The preparation of such fine needle tips is, needless to say, a major problem. They are made by extremely careful and tricky etching of a fine wire. After a tip of the desired size has been etched, it must be polished, because its surface is rough (on the atomic scale) and is covered with a layer of oxidized metal, adsorbed oxygen or other contaminants. To finish the tip we insert it in the microscope and apply a very strong electric field, amounting sometimes to more than 500 million volts per centimeter. The field evaporates off the contaminants and the rough top layer of metal atoms, producing a smooth surface of regular lattice steps. As the field cleans and smooths the tip, we can watch the peeling process on the screen.

Once the tip has been polished, it stays clean indefinitely, so long as the strong electric field used for observing in the microscope is turned on. This is a remarkable situation, as any physicist or chemist will recognize. Even at the highest vacuum achievable in the laboratory today, a surface will become covered with a layer of gas molecules within an hour. In the ion emission microscope, thanks to the high electric field, not a single contaminating atom or molecule is deposited on the needle tip. We can study ideal surfaces and experimentally faulted ones in their pristine state.

Why were the atoms colored in the cover picture mentioned earlier? The colors were simply a device to highlight atomic alterations of the tip surface. If we want to see the results of treatments of the tip which remove or add a few atoms or shift atoms around, it is helpful to label them with markers. Our problem is like that of an astronomer comparing two photographs of a section of the sky to find a new or variable star

which should be present in the second but not in the first. He must hunt for it among thousands of stars on his plate. Similarly a few new atoms are difficult to find among the thousands of atoms in a photograph of the needle tip. To mark the changes, we developed a technique employing superimposed photographs in different colors. Let us say we wish to submit the needle tip to a treatment which will deposit a few atoms on the tip and remove a few. We photograph the black-and-white picture of the tip made before the treatment in green light, and the post-treatment picture in red light. Now we superimpose the red print on the green and make a color photograph of the composite picture with a conventional camera. Wherever the positions of the atoms are unchanged, the atoms will look yellow—the sum of red and green. Where we see green dots, these must mark atoms which were present in the first picture but were removed before the second. Red dots, on the other hand, mark atoms recorded only in the second picture—*i.e.*, new atoms added to the tip surface by the experimental treatment.

By this means we not only can see atomic changes on the surface of the metallic tip but also can determine some properties of the atoms involved. For instance, the strength of the voltage used to evaporate atoms from the tip tells us what the binding energy of the green (*i.e.*, removed) atoms was.

The technique now permits us to study metals at the atomic level with a precision never before possible. Bombarding a sample with fast ions or neutrons from a nuclear reactor, we can see the effect of the impact of a single particle on the structure of the crystal lattice. We can investigate the atomic details of dislocations or faults in the metal and can observe the results of treatments such as annealing. We may even be able to unveil some of the mysteries of the action of catalysts.

So far we have obtained good images only of metals such as tungsten, rhenium, tantalum and molybdenum and some of their highly refractory alloys—carbides and nitrides. The technique at present is limited to the very hard metals. But they may serve as models to yield information about the softer metals. It has always been the belief of physicists that the behavior of "simple cases" such as an ideally perfect surface can give insight into the ordinary—and more complex—systems with which we have to deal in everyday technology.

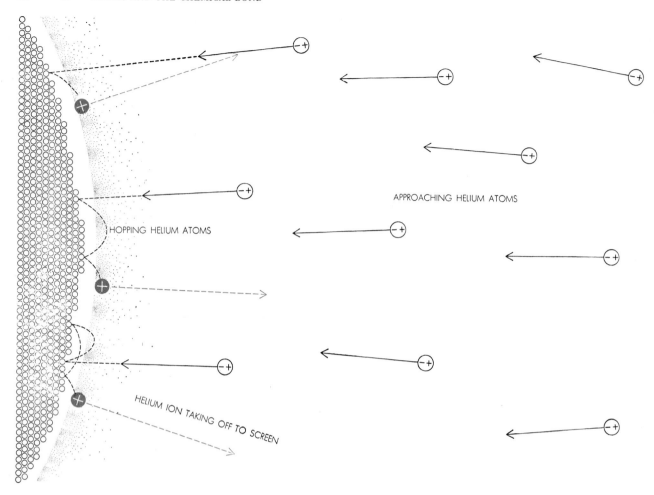

HOPPING HELIUM ATOMS

APPROACHING HELIUM ATOMS

HELIUM ION TAKING OFF TO SCREEN

CREATION OF IONS in the field ion microscope is outlined in this supermicroscopic diagram. The atoms in the crystal lattice of the needle tip are at left. Neutral helium atoms (− +) introduced into the microscope land on the needle surface and momentarily stick to it. When they depart from the surface, they encounter re-gions above steps in the lattice where the electric field is excep-tionally strong (*heavier stippling*). At a distance of five Angstrom units the helium atoms lose an electron to the positively charged needle surface and become positive ions (*color*). They are repelled by the positive needle and travel to the screen of the microscope.

4

THE CHEMICAL ELEMENTS OF LIFE

EARL FRIEDEN
July 1972

*Until recently it was believed that living matter
incorporated 20 of the natural elements. Now it has
been shown that a role is played by four others:
fluorine, silicon, tin and vanadium*

How many of the 90 naturally occurring elements are essential to life? After more than a century of increasingly refined investigation, the question still cannot be answered with certainty. Only a year or so ago the best answer would have been 20. Since then four more elements have been shown to be essential for the growth of young animals: fluorine, silicon, tin and vanadium. Nickel may soon be added to the list. In many cases the exact role played by these and other trace elements remains unknown or unclear. These gaps in knowledge could be critical during a period when the biosphere is being increasingly contaminated by synthetic chemicals and subjected to a potentially harmful redistribution of salts and metal ions. In addition, new and exotic chemical forms of metals (such as methyl mercury) are being discovered, and a complex series of competitive and synergistic relations among mineral salts has been encountered. We are led to the realization that we are ignorant of many basic facts about how our chemical milieu affects our biological fate.

Biologists and chemists have long been fascinated by the way evolution has selected certain elements as the building blocks of living organisms and has ignored others. The composition of the earth and its atmosphere obviously sets a limit on what elements are available. The earth itself is hardly a chip off the universe. The solar system, like the universe, seems to be 99 percent hydrogen and helium. In the earth's crust helium is essentially nonexistent (except in a few rare deposits) and hydrogen atoms constitute only about .22 percent of the total. Eight elements provide more than 98 percent of the atoms in the earth's crust: oxygen (47 percent), silicon (28 percent), aluminum (7.9 percent), iron (4.5 percent), calcium (3.5 percent), so-dium (2.5 percent), potassium (2.5 percent) and magnesium (2.2 percent). Of these eight elements only five are among the 11 that account for more than 99.9 percent of the atoms in the human body. Not surprisingly nine of the 11 are also the nine most abundant elements in seawater [*see illustration on page 34*].

Two elements, hydrogen and oxygen, account for 88.5 percent of the atoms in the human body; hydrogen supplies 63 percent of the total and oxygen 25.5 percent. Carbon accounts for another 9.5 percent and nitrogen 1.4 percent. The remaining 20 elements now thought to be essential for mammalian life account for less than .7 percent of the body's atoms.

The Background of Selection

Three characteristics of the biosphere or of the elements themselves appear to have played a major part in directing the chemistry of living forms. First and foremost there is the ubiquity of water, the solvent base of all life on the earth. Water is a unique compound; its stability and boiling point are both unusually high for a molecule of its simple composition. Many of the other compounds essential for life derive their usefulness from their response to water: whether they are soluble or insoluble, whether or not (if they are soluble) they carry an electric charge in solution and, not least, what effect they have on the viscosity of water.

The second directing force involves the chemical properties of carbon, which evolution selected over silicon as the central building block for constructing giant molecules. Silicon is 146 times more plentiful than carbon in the earth's crust and exhibits many of the same properties. Silicon is directly below carbon in the periodic table of the elements; like carbon, it has the capacity to gain four electrons and form four covalent bonds.

The crucial difference that led to the preference for carbon compounds over silicon compounds seems traceable to two chemical features: the unusual stability of carbon dioxide, which is readily soluble in water and always monomeric (it remains a single molecule), and the almost unique ability of carbon to form long chains and stable rings with five or six members. This versatility of the carbon atom is responsible for the millions of organic compounds found on the earth.

Silicon, in contrast, is insoluble in water and forms only relatively short chains with itself. It can enter into longer chains, however, by forming alternating bonds with oxygen, creating the compounds known as silicones (–Si–O–Si–O–Si–). Carbon-to-carbon bonds are more stable than silicon-to-silicon bonds, but not so stable as to be virtually immutable, as the silicon-oxygen polymers are. Nevertheless, silicon has recently been shown to be essential in a way as yet unknown for normal bone development and full growth in chicks.

The third force influencing the evolutionary selection of the elements essential for life is related to an atom's size and charge density. Obviously the heavy synthetic elements from neptunium (atomic number 93) to lawrencium (No. 103), along with two lighter synthetic elements, technetium (No. 43) and promethium (No. 61), were never available in nature. (The atomic number expresses the number of protons in the nucleus of an atom or the number of electrons around the nucleus.) The eight heavy elements in another group (Nos. 84 and 85 and Nos. 87 through 92) are too radioactive to be useful in living structures. Six more elements are inert gases

COMPOSITION OF UNIVERSE		COMPOSITION OF EARTH'S CRUST		COMPOSITION OF SEAWATER		COMPOSITION OF HUMAN BODY	
PERCENT OF TOTAL NUMBER OF ATOMS							
H	91	O	47	H	66	H	63
He	9.1	Si	28	O	33	O	25.5
O	.057	Al	7.9	Cl	.33	C	9.5
N	.042	Fe	4.5	Na	.28	N	1.4
C	.021	Ca	3.5	Mg	.033	Ca	.31
Si	.003	Na	2.5	S	.017	P	.22
Ne	.003	K	2.5	Ca	.006	Cl	.03
Mg	.002	Mg	2.2	K	.006	K	.06
Fe	.002	Ti	.46	C	.0014	S	.05
S	.001	H	.22	Br	.0005	Na	.03
		C	.19			Mg	.01
ALL OTHERS <.01		ALL OTHERS <.1		ALL OTHERS <.1		ALL OTHERS <.01	

Al	ALUMINUM	C	CARBON	Fe	IRON	O	OXYGEN	S	SULFUR
B	BORON	Cl	CHLORINE	Mg	MAGNESIUM	K	POTASSIUM	Ti	TITANIUM
Br	BROMINE	He	HELIUM	Ne	NEON	Si	SILICON		
Ca	CALCIUM	H	HYDROGEN	N	NITROGEN	Na	SODIUM		

CHEMICAL SELECTIVITY OF EVOLUTION can be demonstrated by comparing the composition of the human body with the approximate composition of seawater, the earth's crust and the universe at large. The percentages are based on the total number of atoms in each case; because of rounding the totals do not exactly equal 100. Elements in the colored boxes in the last column appear in one or more columns at the left. Thus one sees that phosphorus, the sixth most plentiful element in the body, is a rare element in inanimate nature. Carbon, the third most plentiful element, is also very scarce elsewhere.

with virtually no useful chemical reactivities: helium, neon, argon, krypton, xenon and radon. On various plausible grounds one can exclude another 24 elements, or a total of 38 natural elements, as being clearly unsatisfactory for incorporation in living organisms because of their relative unavailability (particularly the elements in the lanthanide and actinide series) or their high toxicity (for example mercury and lead). This leaves 52 of the 90 natural elements as being potentially useful.

Only three of the 24 elements known to be essential for animal life have an atomic number above 34. All three are needed only in trace amounts: molybdenum (No. 42), tin (No. 50) and iodine (No. 53). The four most abundant atoms in living organisms—hydrogen, carbon, oxygen and nitrogen—have atomic numbers of 1, 6, 7 and 8. Their preponderance seems attributable to their being the smallest and lightest elements that can achieve stable electronic configura-

tions by adding one to four electrons. The ability to add electrons by sharing them with other atoms is the first step in forming chemical bonds leading to stable molecules. The seven next most abundant elements in living organisms all have atomic numbers below 21. In the order of their abundance in mammals they are calcium (No. 20), phosphorus (No. 15), potassium (No. 19), sulfur (No. 16), sodium (No. 11), magnesium (No. 12) and chlorine (No. 17). The remaining 10 elements known to be present in either plants or animals are needed only in traces. With the exception of fluorine (No. 9) and silicon (No. 14), the remaining eight occupy positions between No. 23 and No. 34 in the periodic table [see illustration at right]. It is interesting that this interval embraces three elements for which evolution has evidently found no role: gallium, germanium and arsenic. None of the metals with properties similar to those of gallium (such as aluminum and indium) has

proved to be useful to living organisms. On the other hand, since silicon and tin, two elements with chemical activities similar to those of germanium, have just joined the list of essential elements, it seems possible that germanium too, in spite of its rarity, will turn out to have an essential role. Arsenic, of course, is a well-known poison.

Functions of Essential Elements

Some useful generalizations can be made about the role of the various elements. Six elements—carbon, nitrogen, hydrogen, oxygen, phosphorus and sulfur—make up the molecular building blocks of living matter: amino acids, sugars, fatty acids, purines, pyrimidines and nucleotides. These molecules not only have independent biochemical roles but also are the respective constituents of the following large molecules: proteins, glycogen, starch, lipids and nucleic acids. Several of the 20 amino acids contain sulfur in addition to carbon, hydrogen and oxygen. Phosphorus plays an important role in the nucleotides such as

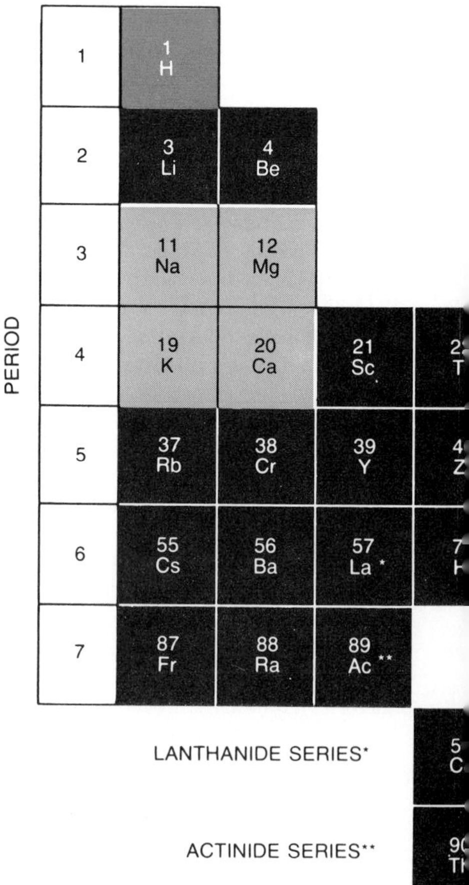

LANTHANIDE SERIES*

ACTINIDE SERIES**

ESSENTIAL LIFE ELEMENTS, 24 by the latest count, are clustered in the upper half of the periodic table. The elements are ar-

adenosine triphosphate (ATP), which is central to the energetics of the cell. ATP includes components that are also one of the four nucleotides needed to form the double helix of deoxyribonucleic acid (DNA), which incorporates the genetic blueprint of all plants and animals. Both sulfur and phosphorus are present in many of the small accessory molecules called coenzymes. In bony animals phosphorus and calcium help to create strong supporting structures.

The electrochemical properties of living matter depend critically on elements or combinations of elements that either gain or lose electrons when they are dissolved in water, thus forming ions. The principal cations (electron-deficient, or positively charged, ions) are provided by four metals: sodium, potassium, calcium and magnesium. The principal anions (ions with a negative charge because they have surplus electrons) are provided by the chloride ion and by sulfur and phosphorus in the form of sulfate ions and phosphate ions. These seven ions maintain the electrical neutrality of body fluids and cells and also play a part in

maintaining the proper liquid volume of the blood and other fluid systems. Whereas the cell membrane serves as a physical barrier to the exchange of large molecules, it allows small molecules to pass freely. The electrochemical functions of the anions and cations serve to maintain the appropriate relation of osmotic pressure and charge distribution on the two sides of the cell membrane.

One of the striking features of the ion distribution is the specificity of these different ions. Cells are rich in potassium and magnesium, and the surrounding plasma is rich in sodium and calcium. It seems likely that the distribution of ions in the plasma of higher animals reflects the oceanic origin of their evolutionary antecedents. One would like to know how primitive cells learned to exclude the sodium and calcium ions in which they were bathed and to develop an internal milieu enriched in potassium and magnesium.

The third and last group of essential elements consists of the trace elements. The fact that they are required in extremely minute quantities in no way di-

minishes their great importance. In this sense they are comparable to the vitamins. We now know that the great majority of the trace elements, represented by metallic ions, serve chiefly as key components of essential enzyme systems or of proteins with vital functions (such as hemoglobin and myoglobin, which respectively transports oxygen in the blood and stores oxygen in muscle). The heaviest essential element, iodine, is an essential constituent of the thyroid hormones thyroxine and triiodothyronine, although its precise role in hormonal activity is still not understood.

The Trace Elements

To demonstrate that a particular element is essential to life becomes increasingly difficult as one lowers the threshold of the amount of a substance recognizable as a "trace." It has been known for more than 100 years, for example, that iron and iodine are essential to man. In a rapidly developing period of biochemistry between 1928 and 1935 four more elements, all metals, were shown to be

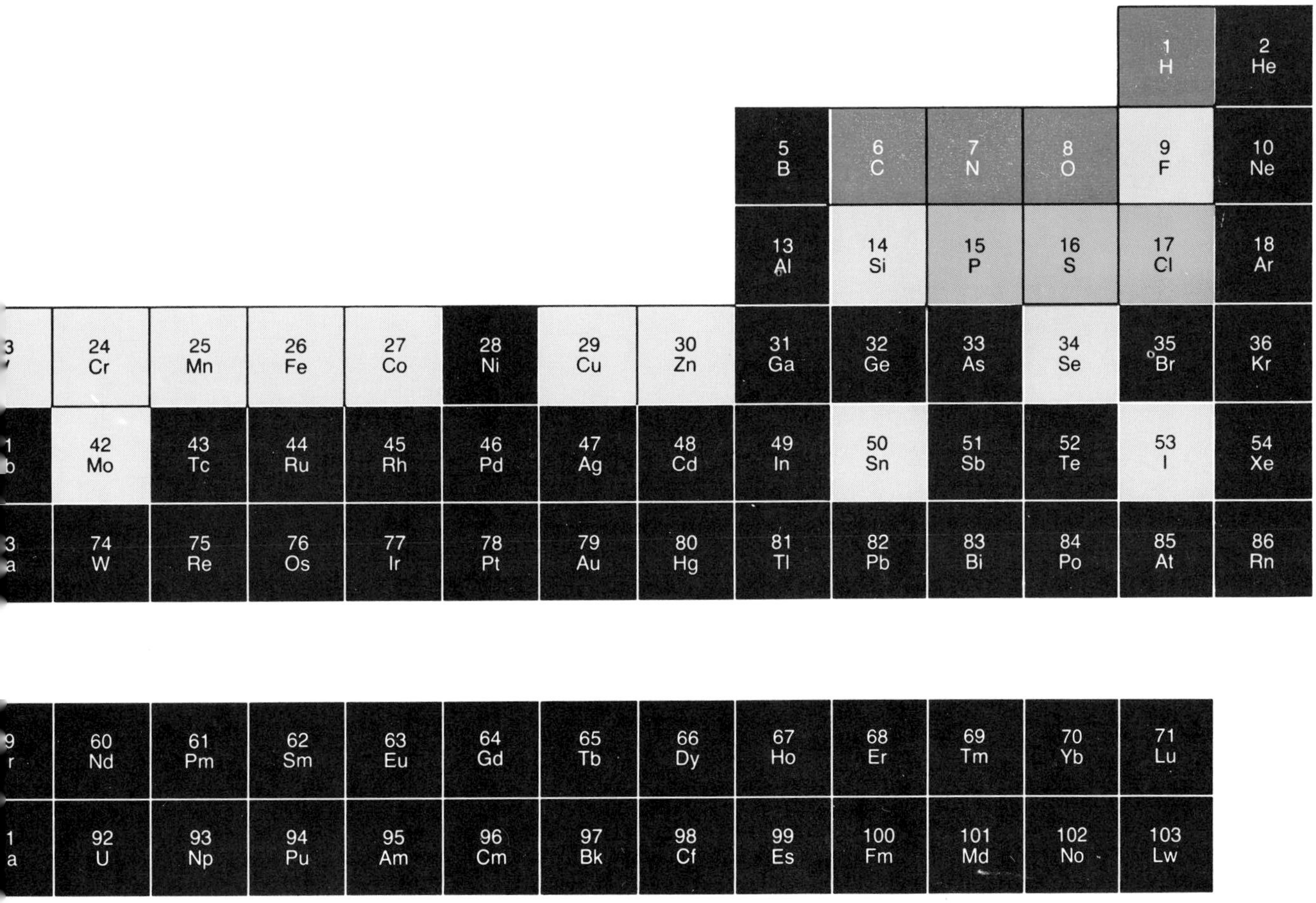

ranged according to their atomic number, which is equivalent to the number of protons in the atom's nucleus. The four most abundant elements that are found in living organisms (hydrogen, oxygen, carbon and nitrogen) are indicated by dark color. The seven next most common elements are in lighter color. The 13 elements that are shown in lightest color are needed only in traces.

ELEMENT	SYMBOL	ATOMIC NUMBER	COMMENTS
HYDROGEN	H	1	Required for water and organic compounds.
HELIUM	He	2	Inert and unused.
LITHIUM	Li	3	Probably unused.
BERYLLIUM	Be	4	Probably unused; toxic.
BORON	B	5	Essential in some plants; function unknown.
CARBON	C	6	Required for organic compounds.
NITROGEN	N	7	Required for many organic compounds.
OXYGEN	O	8	Required for water and organic compounds.
FLUORINE	F	9	Growth factor in rats; possible constituent of teeth and bone.
NEON	Ne	10	Inert and unused.
SODIUM	Na	11	Principal extracellular cation.
MAGNESIUM	Mg	12	Required for activity of many enzymes; in chlorophyll.
ALUMINUM	Al	13	Essentiality under study.
SILICON	Si	14	Possible structural unit of diatoms; recently shown to be essential in chicks.
PHOSPHORUS	P	15	Essential for biochemical synthesis and energy transfer.
SULFUR	S	16	Required for proteins and other biological compounds.
CHLORINE	Cl	17	Principal cellular and extracellular anion.
ARGON	A	18	Inert and unused.
POTASSIUM	K	19	Principal cellular cation.
CALCIUM	Ca	20	Major component of bone; required for some enzymes.
SCANDIUM	Sc	21	Probably unused.
TITANIUM	Ti	22	Probably unused.
VANADIUM	V	23	Essential in lower plants, certain marine animals and rats.
CHROMIUM	Cr	24	Essential in higher animals; related to action of insulin.
MANGANESE	Mn	25	Required for activity of several enzymes.
IRON	Fe	26	Most important transition metal ion; essential for hemoglobin and many enzymes.
COBALT	Co	27	Required for activity of several enzymes; in vitamin B_{12}.
NICKEL	Ni	28	Essentiality under study.
COPPER	Cu	29	Essential in oxidative and other enzymes and hemocyanin.
ZINC	Zn	30	Required for activity of many enzymes.
GALLIUM	Ga	31	Probably unused.
GERMANIUM	Ge	32	Probably unused.
ARSENIC	As	33	Probably unused; toxic.
SELENIUM	Se	34	Essential for liver function.
MOLYBDENUM	Mo	42	Required for activity of several enzymes.
TIN	Sn	50	Essential in rats; function unknown.
IODINE	I	53	Essential constituent of the thyroid hormones.

SOME TWO-THIRDS OF LIGHTEST ELEMENTS, or 21 out of the first 34 elements in the periodic table, are now known to be essential for animal life. These 21 plus molybdenum (No. 42), tin (No. 50) and iodine (No. 53) constitute the total list of the 24 essential elements, which are here enclosed in colored boxes. It is possible that still other light elements will turn out to be essential. The most likely candidates are aluminum, nickel and germanium. The element boron already appears to be essential for some plants.

essential: copper, manganese, zinc and cobalt. The demonstration can be credited chiefly to a group of investigators at the University of Wisconsin led by C. A. Elvehjem, E. B. Hart and W. R. Todd. At that time it seemed that these four metals might be the last of the essential trace elements. In the next 30 years, however, three more elements were shown to be essential: chromium, selenium and molybdenum. Fluorine, silicon, tin and vanadium have been added since 1970.

The essentiality of five of these last seven elements was discovered through the careful, painstaking efforts of Klaus Schwarz and his associates, initially located at the National Institutes of Health and now based at the Veterans Administration Hospital in Long Beach, Calif. For the past 15 years Schwarz's group has made a systematic study of the trace-element requirements of rats and other small animals. The animals are maintained from birth in a completely isolated sterile environment [see illustration on page 39].

The apparatus is constructed entirely of plastics to eliminate the stray contaminants contained in metal, glass and rubber. Although even plastics may contain some trace elements, they are so tightly bound in the structural lattice of the material that they cannot be leached out or be picked up by an animal even through contact. A typical isolator system houses 32 animals in individual acrylic cages. Highly efficient air filters remove all trace substances that might be present in the dust in the air. Thus the animals' only access to essential nutrients is through their diet. They receive chemically pure amino acids instead of natural proteins, and all other dietary ingredients are screened for metal contaminants.

Since the standards of purity employed in these experiments far exceed those for reagents normally regarded as analytically pure, Schwarz and his co-workers have had to develop many new analytical chemical methods. The most difficult problem turned out to be the purification of salt mixtures. Even the purest commercial reagents were contaminated with traces of metal ions. It was also found that trace elements could be passed from mothers to their offspring. To minimize this source of contamination animals are weaned as quickly as possible, usually from 18 to 20 days after birth.

With these precautions Schwarz and his colleagues have within the past several years been able to produce a new

deficiency disease in rats. The animals grow poorly, lose hair and muscle tone, develop shaggy fur and exhibit other detrimental changes [see illustration on page 40]. When standard laboratory food is given these animals, they regain their normal appearance. At first it was thought that all the symptoms were caused by the lack of one particular trace element. Eventually four different elements had to be supplied to complete the highly purified diets the animals had been receiving. The four elements proved to be fluorine, silicon, tin and vanadium. A convenient source of these

elements is yeast ash or liver preparations from a healthy animal. The animals on the deficiency diet grew less than half as fast as those on a normal or supplemented diet. Growth alone, however, may not tell the entire story. There is some evidence that even the addition of the four elements may not reverse the loss of hair and skin changes resulting from the deficiency diet.

Functions of Trace Elements

The addition of tin and vanadium to the list of essential trace metals brings

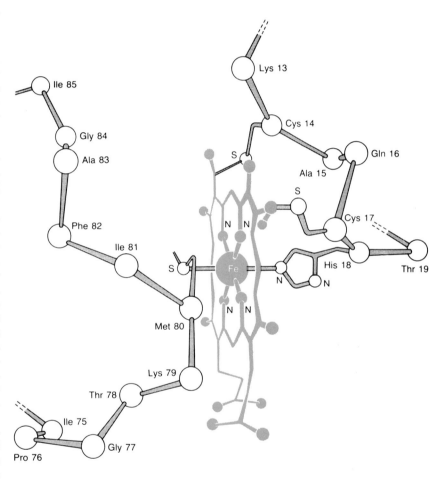

Ala	ALANINE	His	HISTIDINE	Phe	PHENYLALANINE
Cys	CYSTEINE	Ile	ISOLEUCINE	Pro	PROLINE
Gln	GLUTAMINE	Lys	LYSINE	Thr	THREONINE
Gly	GLYCINE	Met	METHIONINE		

THE METALLOENZYME CYTOCHROME C is typical of metal-protein complexes in which trace metals play a crucial role. Cytochrome c belongs to a family of enzymes that extract energy from food molecules. It consists of a protein chain of 104 amino acid units attached to a heme group (color), a rosette of atoms with an atom of iron at the center. This simplified molecular diagram shows only the heme group and several of the amino acid units closest to it. The iron atom has six coordination sites enabling it to form six bonds with neighboring atoms. Four bonds connect to nitrogen atoms in the heme group itself, and the remaining two bonds link up with amino acid units in the protein chain (histidine at site No. 18 and methionine at site No. 80). The illustration is based on the work of Richard E. Dickerson of the California Institute of Technology, in whose laboratory the complete structure of horse-heart cytochrome c was recently determined.

METAL	ENZYME	BIOLOGICAL FUNCTION
IRON	FERREDOXIN	Photosynthesis
	SUCCINATE DEHYDROGENASE	Aerobic oxidation of carbohydrates
IRON IN HEME	ALDEHYDE OXIDASE	Aldehyde oxidation
	CYTOCHROMES	Electron transfer
	CATALASE	Protection against hydrogen peroxide
	[HEMOGLOBIN]	Oxygen transport
COPPER	CERULOPLASMIN	Iron utilization
	CYTOCHROME OXIDASE	Principal terminal oxidase
	LYSINE OXIDASE	Elasticity of aortic walls
	TYROSINASE	Skin pigmentation
	PLASTOCYANIN	Photosynthesis
	[HEMOCYANIN]	Oxygen transport in invertebrates
ZINC	CARBONIC ANHYDRASE	CO_2 formation; regulation of acidity
	CARBOXYPEPTIDASE	Protein digestion
	ALCOHOL DEHYDROGENASE	Alcohol metabolism
MANGANESE	ARGINASE	Urea formation
	PYRUVATE CARBOXYLASE	Pyruvate metabolism
COBALT	RIBONUCLEOTIDE REDUCTASE	DNA biosynthesis
	GLUTAMATE MUTASE	Amino acid metabolism
MOLYBDENUM	XANTHINE OXIDASE	Purine metabolism
	NITRATE REDUCTASE	Nitrate utilization
CALCIUM	LIPASES	Lipid digestion
MAGNESIUM	HEXOKINASE	Phosphate transfer

WIDE VARIETY OF METALLOENZYMES is required for the successful functioning of living organisms. Some of the most important are given in this list. The giant oxygen-transporting molecules hemoglobin and hemocyanin are included in the list (in brackets) even though they are not strictly enzymes, that is, they do not act as biological catalysts.

to 10 the total number of trace metals needed by animals and plants. What role do these metals play? For six of the eight trace metals recognized from earlier studies (that is, for iron, zinc, copper, cobalt, manganese and molybdenum) we are reasonably sure of the answer. The six are constituents of a wide range of enzymes that participate in a variety of metabolic processes [see illustration above].

In addition to its role in hemoglobin and myoglobin, iron appears in succinate dehydrogenase, one of the enzymes needed for the utilization of energy from sugars and starches. Enzymes incorporating zinc help to control the formation of carbon dioxide and the digestion of proteins. Copper is present in more than a dozen enzymes, whose roles range from the utilization of iron to the pigmentation of the skin. Cobalt appears in enzymes involved in the synthesis of DNA and the metabolism of amino acids. Enzymes incorporating manga-

nese are involved in the formation of urea and the metabolism of pyruvate. Enzymes incorporating molybdenum participate in purine metabolism and the utilization of nitrogen.

These six metals belong to a group known as transition elements. They owe their uniqueness to their ability to form strong complexes with ligands, or molecular groups, of the type present in the side chains of proteins. Enzymes in which transition metals are tightly incorporated are called metalloenzymes, since the metal is usually embedded deep inside the structure of the protein. If the metal atom is removed, the protein usually loses its capacity to function as an enzyme. There is also a group of enzymes in which the metal ion is more loosely associated with the protein but is nonetheless essential for the enzyme's activity. Enzymes in this group are known as metal-ion-activated enzymes. In either group the role of the metal ion may be to maintain the proper confor-

mation of the protein, to bind the substrate (the molecule acted on) to the protein or to donate or accept electrons in reactions where the substrate is reduced or oxidized.

In 1968 the complete three-dimensional structure of the first metalloenzyme, cytochrome c, was published [see the article "The Structure and History of an Ancient Protein," by Richard E. Dickerson, beginning on page 326]. Cytochrome c, a red enzyme containing iron, is universally present in plants and animals. It is one of a series of enzymes, all called cytochromes, that extract energy from food molecules by the stepwise addition of oxygen.

The complete amino acid sequence of cytochrome c obtained from the human heart was determined some 10 years ago by a group led by Emil L. Smith of the University of California at Los Angeles and by Emanuel Margoliash of Northwestern University. The iron atom is partially complexed with an intricate organic molecule, protoporphyrin, to form a heme group similar to that in hemoglobin. Of the iron atom's six coordination sites, four are attached to the heme group through nitrogen atoms. The other two sites form bonds with the protein chain; one bond is through a nitrogen atom in the side chain of a histidine unit at site No. 18 in the protein sequence and the other bond is through a sulfur atom in the side chain of a methionine unit at site No. 80 [see illustration on preceding page].

Although the cytochrome c molecule is complicated, it is one of the simplest of the metalloenzymes. Cytochrome oxidase, probably the single most important enzyme in most cells, since it is responsible for transferring electrons to oxygen to form water, is far more complicated. Each molecule contains about 12 times as many atoms as cytochrome c, including two copper atoms and two heme groups, both of which participate in transferring the electrons.

More complicated yet is cysteamine oxygenase, which catalyzes the addition of oxygen to a molecule of cysteamine; it contains one atom each of three different metals: iron, copper and zinc. There are many other combinations of metal ions and unique molecular assemblies. An extreme example is xanthine oxidase, which contains eight iron atoms, two molybdenum atoms and two molecules incorporating riboflavin (one of the B vitamins) in a giant molecule more than 25 times the size of cytochrome c.

The metal-containing proteins of another group, the metalloproteins, closely resemble the metalloenzymes except that they lack an obvious catalytic function. Hemoglobin itself is an example. Others are hemocyanin, the copper-containing blue protein that carries oxygen in many invertebrates, metallothionein, a protein involved in the absorption and storage of zinc, and transferrin, a protein that transports iron in the bloodstream. There may be many more such compounds still unrecognized because their function has escaped detection.

The Newest Essential Elements

Much remains to be learned about the specific biochemical role of the most recently discovered essential elements. In 1957 Schwarz and Calvin M. Foltz, working at the National Institutes of Health, showed that selenium helped to prevent several serious deficiency diseases in different animals, including liver necrosis and muscular dystrophy. Rats were protected against death from liver necrosis by a diet containing one-tenth of a part per million of selenium. Comparably low doses reversed the white muscle disease observed in cattle and sheep that happen to graze in areas where selenium is scarce.

In April a group at the University of Wisconsin under J. T. Rotruck reported a direct biochemical role for selenium.

Oxidative damage to red blood cells was detected in rats kept on a selenium-deficient diet. This damage was related to reduced activity of an enzyme, glutathione peroxidase, that helps to protect hemoglobin against the injurious oxidative effects of hydrogen peroxide. The enzyme uses hydrogen peroxide to catalyze the oxidation of glutathione, thus keeping hydrogen peroxide from oxidizing the reduced state of iron in hemoglobin. Oxidized glutathione can readily be converted to reduced glutathione by a variety of intracellular mechanisms. There is some reason to believe glutathione peroxidase may even contain some form of selenium acting as an integral part of the functional enzyme molecule.

The physiological importance of chromium was established in 1959 by Schwarz and Walter Mertz. They found that chromium deficiency is characterized by impaired growth and reduced life-span, corneal lesions and a defect in sugar metabolism. When the diet is deficient in chromium, glucose is removed from the bloodstream only half as fast as it is normally. In rats the deficiency is relieved by a single administration of 20 micrograms of certain trivalent chromic salts. It now appears that the chromium ion works in conjunction with insulin, and that in at least some cases

diabetes may reflect faulty chromium metabolism.

After developing the all-plastic trace-element isolator described above, Schwarz, David B. Milne and Elizabeth Vineyard discovered that tin, not previously suspected as being essential, was necessary for normal growth. Without one or two parts per million of tin in their diet, rats grow at only about two-thirds the normal rate.

The next element shown to be essential in mammals by the Schwarz group was vanadium, an element that had been detected earlier in certain marine invertebrates but whose essentiality had not been demonstrated. On a diet in which vanadium is totally excluded rats suffer a retardation of about 30 percent in growth rate. Schwarz and Milne found that normal growth is restored by adding one-tenth of a part per million of vanadium to the diet. At higher concentrations vanadium is known to have several biological effects, but its essential role in trace amounts remains to be established. A high dose of vanadium blocks the synthesis of cholesterol and reduces the amount of phospholipid and cholesterol in the blood. Vanadium also promotes the mineralization of teeth and is effective as a catalyst in the oxidation of many biological substances.

The third element most recently iden-

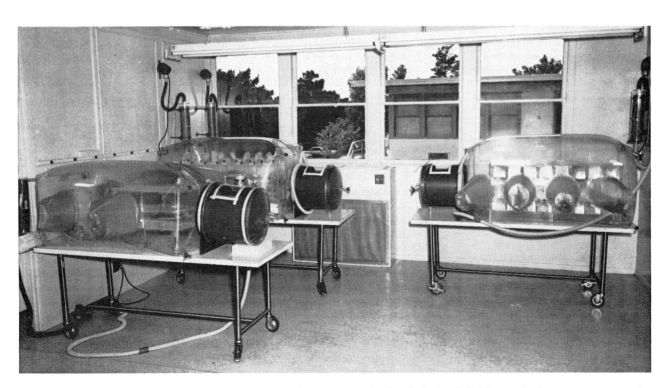

NUTRITIONAL NEEDS OF SMALL ANIMALS are studied in a trace-element isolator, a modification of the apparatus originally conceived to maintain animals in a germ-free environment. To prevent unwanted introduction of trace elements the isolator is built completely of plastics. It holds 32 animals in separate cages, individually supplied with food of precisely known composition. The system was designed by Klaus Schwarz and J. Cecil Smith of the Veterans Administration Hospital in Long Beach, Calif.

tified as being essential is fluorine. Even with tin and vanadium added to highly purified diets containing all other elements known to be essential, the animals in Schwarz's plastic cages still failed to grow at a normal rate. When up to half a part per million of potassium fluoride was added to the diet, the animals showed a 20 to 30 percent weight gain in four weeks. Although it had appeared that a trace amount of fluorine was essential for building sound teeth, Schwarz's study showed that fluorine's biochemical role was more fundamental than that. In any case fluoridated water provides more than enough fluorine to maintain a normal growth rate.

Although there were earlier clues that silicon might be an essential life element, firm proof of its essentiality, at least in

young chicks, was reported only three months ago. Edith M. Carlisle of the School of Public Health at the University of California at Los Angeles finds that chicks kept on a silicon-free diet for only one or two weeks exhibit poor development of feathers and skeleton, including markedly thin leg bones. The addition of 30 parts per million of silicon to the diet increases the chicks' growth more than 35 percent and makes possible normal feathering and skeletal development. Considering that silicon is not only the second most abundant element in the earth's crust but is also similar to carbon in many of its chemical properties, it is hard to see how evolution could have totally excluded it from an essential biochemical role.

Nickel, nearly always associated with

iron in natural substances, is another element receiving close attention. Also a transition element, it is particularly difficult to remove from the food used in special diets. Nickel seems to influence the growth of wing and tail feathers in chicks but more consistent data are needed to establish its essentiality. One incidental result of Schwarz's work has been the discovery of a previously unrecognized organic compound, which will undoubtedly prove to be a new vitamin.

Synergism and Antagonism

The interaction of the various essential metals can be extremely complicated. The absence of one metal in the diet can profoundly influence, either positively or negatively, the utilization of another metal that may be present. For example, it has been known for nearly 50 years that copper is essential for the proper metabolism of iron. An animal deprived of copper but not iron develops anemia because the biosynthetic machinery fails to incorporate iron in hemoglobin molecules. It has only recently been found in our laboratories at Florida State University that ceruloplasmin, the copper-containing protein of the blood, is a direct molecular link between the two metals. Ceruloplasmin promotes the release of iron from animal liver so that the iron-binding protein of the serum, transferrin, can complex with iron and transfer it to the developing red blood cells for direct utilization in the biosynthesis of hemoglobin. This represents a synergistic relation between copper and iron.

As an example of antagonism between elements one can cite the instance of copper and zinc. The ability of sheep or cattle to absorb copper is greatly reduced if too much zinc or molybdenum is present in their diet. Evidently either of the two metals can displace copper in an absorption process that probably involves competition for sites on a metal-binding protein in the intestines and liver.

The recent discoveries present many fresh challenges to biochemists. One can expect the discovery of previously unsuspected metalloenzymes containing vanadium, tin, chromium and selenium. New compounds or enzyme systems requiring fluorine and silicon may also be uncovered. The multiple and complex interdependencies of the elements suggest many hitherto unrecognized and important facts about the role and interrelations of metal ions in nutrition and in health and disease.

TRACE-ELEMENT DEFICIENCY developed when the rat at the top of this photograph was kept in the trace-element isolator for 20 days and fed a diet from which fluorine, tin and vanadium had been carefully excluded. The healthy animal at the bottom was fed the same diet but was kept under ordinary conditions. It was evidently able to obtain the necessary trace amounts of fluorine, tin and vanadium from dust and other contaminants.

5

CHEMISTRY BY COMPUTER

ARNOLD C. WAHL

April 1970

*Computed quantum-mechanical models of the electronic
structure of atoms and molecules can provide a reliable
and comprehenisve alternative to the traditional
experimental approach to chemistry*

Traditionally chemistry has been a predominantly experimental discipline, consisting for the most part of direct measurements of the properties of matter and laboratory analyses of the reactions by which chemical substances are transformed into other chemical substances. The great contribution of modern quantum theory to chemistry has been to enrich the language of the chemist and to emphasize and explain the atomic and molecular basis of chemical phenomena. Assuming that one starts with a fundamentally sound theoretical model of the structure of the individual atoms or molecules, and of the nature of the forces between them, the laws of basic electrostatics, classical physics, quantum mechanics and statistical mechanics in principle provide a means for computing the macroscopic outcome of a chemical experiment—without ever performing the experiment! This possibility is particularly intriguing when the answers one could theoretically obtain in this way might not be accessible experimentally, as for example with highly corrosive substances or substances that exist either for a very short time or only at very high temperatures.

Since the introduction of the fundamental wave equation of quantum mechanics by Erwin Schrödinger in 1926, much of the work of quantum chemists has been focused on its solution for specific chemical systems; in other words, · on the problem of constructing adequate mathematical models of atomic and molecular structure in order to obtain reliable nonexperimental information about chemical processes and to unify existing information. The major difficulty encountered in this effort arises from the intractable multidimensional differential equations that the Schrödinger equation demands as the only proper way of describing such a complex system of subatomic particles.

One of the most useful approaches to this problem has been the "electron orbital" picture of atoms and molecules. The orbital picture for atoms was a direct outgrowth of Schrödinger's solution of his wave equation for the hydrogen atom; during the period from 1927 to 1932 this picture was extended to many-electron atoms by William and Douglas R. Hartree, Vladimir Alexandrovitch Fock and James C. Slater and to molecules by Robert S. Mulliken, Felix Hund and J. E. Leonard-Jones. The essence of this theory is that the electrons of a given atom or molecule occupy a set of distinct orbitals. Each orbital is characterized by a set of "quantum numbers," denoting various properties of the electrons in that orbital (for instance their spin, angular momentum and the probability of finding the electrons in various regions of space). One can build up an orbital picture that represents the approximate structure of the entire atom or molecule as a product of these individual orbitals.

Until recently the quantitative accuracy of this orbital model for molecules was rather poor, and it found its greatest use as a conceptual and interpretive tool. The advent of large electronic computers, however, has made it possible to perform the vast amount of algebra and arithmetic required to determine molecular orbitals and to make some necessary improvements beyond the orbital picture, providing in many cases reliable

FIRST 10 ATOMS in the periodic table of the elements are depicted on the following page in the form of sets of computer-generated diagrams showing the spatial distribution of the atoms' electrons in terms of contour lines of equal charge density. Each atom is represented by one diagram showing its total electron density and one or more diagrams showing the electron density of its constituent "orbitals." Under the name and symbol of each atom at left is the orbital configuration of that atom's complement of electrons; for example, in the orbital picture the configuration of the eight electrons in the oxygen atom is designated $1s^2 2s^2 2p^4$, which means that two electrons occupy the oxygen atom's $1s$, or lowest-energy, orbital, two electrons occupy the $2s$, or next-higher, orbital, and four electrons occupy the $2p$ orbitals (for which three different orientations, corresponding to alignment of the orbital's long axis along the x, y or z directions respectively, are possible). The electron densities of the orbitals were computed directly from each atom's orbital "wave function," a complex mathematical expression that is obtained by approximately solving the Schrödinger wave equation for the atom. This orbital picture yields a good approximation of the electronic structure of atoms and molecules. (The $2p$ orbitals must be spherically averaged when added to the total electron density, since the electrons can occupy different combinations of the $2p$ orbitals with equal probability.) The total and orbital energies are given below the respective diagrams in hartrees (an atomic unit of energy equal to 27.7 electron volts). The unit of charge density is one electron per cubic bohr, where one bohr (the atomic unit of distance) equals 5.29×10^{-9} centimeter; thus all diagrams are at a scale that represents a magnification of about a half-billion diameters. The highest contour value plotted corresponds to a density of one electron per cubic bohr and the value of the succeeding contours decreases by a factor of two down to 4.9×10^{-4} electron per cubic bohr. All plots are in a plane passing through the nucleus of the atom. This illustration and the ones on pages 45, 46 and 48 are based on a series of wall charts prepared by the author and his colleagues and published by McGraw-Hill, Inc.

ATOMS

ATOMIC ORBITALS

TOTAL | 1s | 2s

2p

2p_x | 2p_y | 2p_z

HYDROGEN (H)
1s

−.5 | −.5

HELIUM (He)
1s^2

−2.861680 | −.91795

0   5   10
DISTANCE (BOHRS)

LITHIUM (Li)
1s^{2}2s

−7.236414 | −2.47774 | −.19632

BERYLLIUM (Be)
1s^{2}2s^2

−14.57302 | −4.73266 | −.30927

BORON (B)
1s^{2}2s^{2}2p

−24.52906 | −7.69533 | −.49470 | −.30983

CARBON (C)
1s^{2}2s^{2}2p^2

−37.63133 | −11.32550 | −.70562 | −.43333 | −.43333

NITROGEN (N)
1s^{2}2s^{2}2p^3

−54.40093 | −15.62898 | −.94528 | −.56754 | −.56754 | −.56754

OXYGEN (O)
1s^{2}2s^{2}2p^4

−74.80939 | −20.66860 | −1.24427 | −.63186 | −.63186 | −.63186

FLUORINE (F)
1s^{2}2s^{2}2p^5

−99.40933 | −26.38265 | −1.57245 | −.72994 | −.72994 | −.72994

NEON (Ne)
1s^{2}2s^{2}2p^6

−128.5471 | −32.77233 | −1.93031 | −.85034 | −.85034 | −.85034

and comprehensive chemical information independently of experiment. To give some idea of the computational power modern computers have made available to quantum chemists, a calculation of the orbital picture of a simple diatomic, or two-atom, molecule that 20 years ago would have required about 15 man-years of labor consumes only about 20 seconds on the most capable computer available today.

So far the chemistry amenable to such computer modeling is concerned only with small systems, but they are by no means chemically uninteresting systems. What is more, the success achieved in these limited cases portends an increasingly important role for "computational chemistry" in the future. Potentially this new approach presents the chemist with an opportunity to view in unprecedented detail and with arbitrary magnification or time scale the various stages of such chemical processes as the formation, excitation, ionization, vibration or collision of molecules. What is perhaps most exciting about this new tool is that it is not merely an information-retrieval or information-extrapolation system but rather a true information-creation system; through an inductive mathematical process beginning with the Schrödinger equation it creates information in usable form where there was none before. Thus its greatest value will be as an independent check on the answers provided by experimental chemistry, as an independent means of predicting answers currently inaccessible to experimental chemistry and as a conceptual framework for empirical information.

The Theoretical Approach

In formulating his wave equation Schrödinger cast in rigorous mathematical form the earlier body of quantum-mechanical knowledge. His equation, which can be written in a variety of forms, in general gives the total energy of any system of particles in terms of that system's characteristic "wave function." When the Schrödinger equation is solved for a particular system, it turns out that the wave function obtained must satisfy a set of distinct conditions. These imply that the energy of the system is itself "quantized"; in other words, the system can exist only in certain definite energy states. Another implication of the Schrödinger equation is that the positions of the individual particles that constitute the system cannot be located exactly in space; all that can be obtained is a mathematical expression giving the probability of finding any particle in various regions of space. Thus one way of looking at the structure of a quantum-mechanical system such as an atom or a molecule is as though it were an electron "cloud" of varying density surrounding a central nucleus (or group of nuclei). The exact shape and value of the charge density of the cloud are the quantities that are specified directly by the characteristic wave function of the system.

For example, in the case of the simplest atom, namely the hydrogen atom, which is made up of one nucleus (a single proton) and one electron, the solution of the Schrödinger equation yields a series of exact wave functions corresponding to the various allowed energy states of the atom. These are known as hydrogenic orbitals. As soon, however, as one goes to the slightly more complicated system of the helium atom, which has two electrons instead of one, an exact solution of the Schrödinger equation is impossible. Instead one must resort to special approximation procedures that approach the exact result very closely.

Beyond three- or four-electron systems, even these approximate solutions of the Schrödinger equation call for awesome amounts of computation, since the exact total wave function of such systems is a function of all $3N$ dimensions of the N electrons. Hence some more generalized simplifying approximation must be made in order to solve the Schrödinger equation for the larger systems. The most successful approximation of this type is the orbital model, in which the $3N$-dimensional wave function is expressed in a similar-looking but actually much simpler form: as the product of N three-dimensional wave functions, each of which describes the orbital for a single electron. This method of computing approximate atomic structures by means of their atomic orbitals was perfected largely through the efforts of Hartree and Fock; the resulting total wave functions are sometimes called Hartree-Fock

MAGNITUDE OF THE PROBLEM involved in performing the computations necessary to extend the orbital model to molecules is suggested by this schematic illustration of a water molecule. In order to solve the Schrödinger equation for such a quantum-mechanical system one must take into account (among other things) the potential energy represented by the interactions between all the particles in the system; in the water molecule these interactions include three nucleus-nucleus repulsions, 30 electron-nucleus attractions and 45 electron-electron repulsions. Typical interparticle distances over which such interactions must be calculated and averaged within the electron clouds are indicated for five of the total of 13 particles in the system (all three nuclei but only two of the 10 electrons are shown).

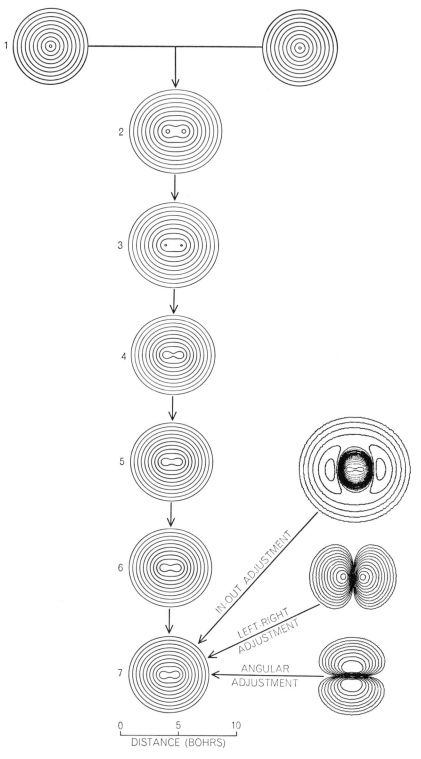

TYPICAL PROCEDURE followed in developing a mathematical model of a simple dia-tomic molecule can be visualized in stages. First the charge density for two isolated hydro-gen atoms is computed (*1*). Then the two hydrogen atoms are brought together as if they were classical (non-quantum-mechanical) spherical clouds of electronic charge (*2*). Next quantum mechanics is invoked and electron exchange is allowed to take place between the two atoms (*3*). This now produces a valid molecular orbital wave function, but it is not yet the best possible one. The shape of the molecular orbital is therefore changed in a variety of ways (*4, 5, 6*) until the energy associated with it reaches a minimum, ultimately giving one of the best approximate solutions of the Schrödinger equation for the molecule (*6*). Further adjustments must be added to the best molecular orbital picture in order to allow the electrons to avoid each other more effectively in an in-out, left-right and angular sense (*lower right*); the resulting "electron-correlated" picture (*7*) is not visibly altered, but the predicted binding energy and the description of the bonding process are greatly improved.

atomic orbitals. Hartree-Fock atomic or-bital pictures of the first 10 atoms of the periodic table are shown in the illustra-tion on page 42 in terms of contour lines of equal electron density.

The Hartree-Fock atomic orbital mod-el provides a comprehensive framework for discussing the internal structure of individual atoms and in fact can be viewed as the quantum-mechanical basis and analogue of the periodic table. In particular this model has been concep-tually invaluable in sorting out the in-credible intricacies of atomic spectra, in effect answering the question of what is really happening when the energy states of an atom change. In addition the or-bital model yields many useful chemical properties of atoms, such as their size and their approximate ionization poten-tial (the amount of energy required to strip an atom of its electrons). Nonethe-less, inherent defects in the orbital model for atoms prevent the accurate predic-tion of such properties as the affinity of an atom for an extra electron, the energy differences between atomic states and the total atomic energies.

The Molecular Orbital Model

The extension of the orbital model to molecules was a formidable com-putational task, primarily because the mathematics involved in the evaluation of the interactions among electrons in the molecule is much more difficult than in the case of the atom. (The spherical symmetry of the atom is a great advan-tage here.) The first step in simplifying the Schrödinger equation for molecules was in effect to separate out the motion of the nuclei, that is, to consider the nu-clei to be at rest and to solve the Schrö-dinger equation for each electron mov-ing with respect to this fixed framework. The motion of the nuclei can then be studied as if it were taking place on a potential-energy surface determined by the total electronic energy of the system. Max Born and J. Robert Oppenheimer were able to show quite early that this is usually a valid approximation, owing to the fact that the comparatively massive nuclei move much more slowly than the lighter electrons; as a result the electron-ic wave function can adjust itself instan-taneously to the position of the nuclei at any given time.

How does one actually go about de-veloping a mathematical model of a sim-ple diatomic molecule, say the hydrogen molecule (H_2)? It is helpful to visualize the job in stages [*see illustration at left*]. First we may compute precisely the charge density for two isolated hydrogen atoms, displaying the result in the form of an atomic orbital picture composed of contour lines of equal electron density. We then bring the two hydrogen atoms together as if they were classical (non-quantum-mechanical) spherical clouds of electronic charge. This would be a valid picture if it were not for the fact that according to the laws of quantum mechanics electrons are indistinguishable and therefore cannot be rigidly assigned to one hydrogen atom or the other; rather they must be allowed

to "exchange," or mix freely, between the two atoms.

Next we invoke quantum mechanics and allow electron exchange to take place. This now produces a valid molecular orbital wave function, but it is not yet the best possible one. In order to obtain the best possible molecular orbital wave function (the Hartree-Fock function) we must vary the form of the molecular orbital picture until the energy associated with it reaches a minimum. A convenient way of doing this in forming the molecular orbital is to include in ad-

dition to the simple 1s, or lowest-energy, hydrogenic function higher functions such as the 2s, 2p, 3d and 4f functions. When allowed to combine with the 1s wave function, these additions can change the shape of the molecular orbital in a variety of ways and ultimately give the best solution of the Schrödinger equation for the molecule. We have displayed such best molecular orbitals in terms of contour diagrams of electronic charge density for a variety of diatomic molecules [*see illustrations below and on page 46*].

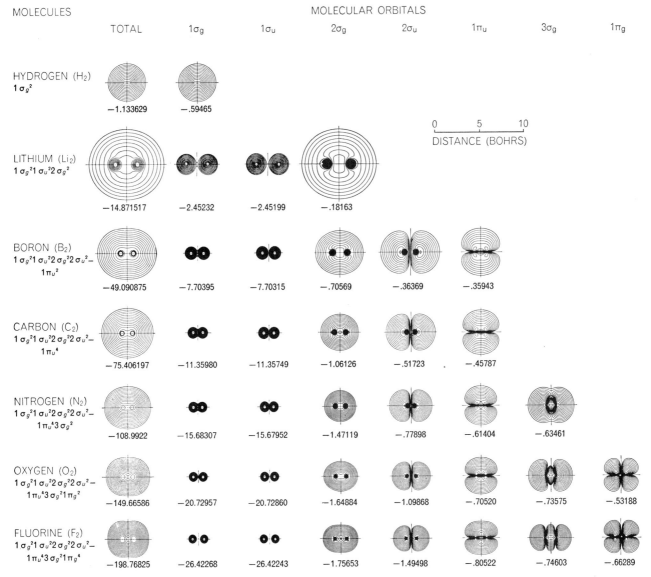

MOLECULES		MOLECULAR ORBITALS						
	TOTAL	1σ_g	1σ_u	2σ_g	2σ_u	1π_u	3σ_g	1π_g
HYDROGEN (H$_2$) 1σ_g^2	−1.133629	−.59465						
LITHIUM (Li$_2$) 1$\sigma_g^2$1$\sigma_u^2$2σ_g^2	−14.871517	−2.45232	−2.45199	−.18163				
BORON (B$_2$) 1$\sigma_g^2$1$\sigma_u^2$2$\sigma_g^2$2σ_u^2− 1π_u^2	−49.090875	−7.70395	−7.70315	−.70569	−.36369	−.35943		
CARBON (C$_2$) 1$\sigma_g^2$1$\sigma_u^2$2$\sigma_g^2$2σ_u^2− 1π_u^4	−75.406197	−11.35980	−11.35749	−1.06126	−.51723	−.45787		
NITROGEN (N$_2$) 1$\sigma_g^2$1$\sigma_u^2$2$\sigma_g^2$2σ_u^2− 1$\pi_u^4$3σ_g^2	−108.9922	−15.68307	−15.67952	−1.47119	−.77898	−.61404	−.63461	
OXYGEN (O$_2$) 1$\sigma_g^2$1$\sigma_u^2$2$\sigma_g^2$2σ_u^2− 1$\pi_u^4$3$\sigma_g^2$1π_g^2	−149.66586	−20.72957	−20.72860	−1.64884	−1.09868	−.70520	−.73575	−.53188
FLUORINE (F$_2$) 1$\sigma_g^2$1$\sigma_u^2$2$\sigma_g^2$2σ_u^2− 1$\pi_u^4$3$\sigma_g^2$1π_g^4	−198.76825	−26.42268	−26.42243	−1.75653	−1.49498	−.80522	−.74603	−.66289

DISTANCE (BOHRS): 0 — 5 — 10

SEVEN COVALENT DIATOMIC MOLECULES (that is, molecules that share their outermost pairs of electrons) are represented here by contour diagrams of their total and orbital electron densities; the molecules shown are those that result from combining in homonuclear pairs seven of the first 10 atoms in the periodic table. The electronic configurations of the molecules (that is, the number of electrons in each of a given molecule's set of orbitals) are given at left under the names and symbols of the respective molecules. The subscript letters *g* and *u* indicate whether a molecule's "inver-

sion symmetry" (its symmetry across a central plane) is either even (in German *gerade*) or odd (*ungerade*). As in the case of the atomic orbitals on page 42, orbital energy is given below each diagram in hartrees, and charge density is plotted in electrons per cubic bohr. The diatomic molecules of helium (He$_2$), beryllium (Be$_2$) and neon (Ne$_2$), which are members of this homonuclear series, are not bound in their ground, or lowest-energy, state and hence are not displayed here. In this illustration and the one on page 46 three additional, low-value contour lines have been plotted.

The molecular orbital model has proved to be a great aid to the chemist in talking and thinking about chemical bonding and molecular energy states. It also yields by computation many molecular properties. Among these are molecular charge density, bond lengths and angles (which determine the size and shape of the molecule), ionization potentials, magnetic moments and vibrational frequencies.

One of the main inadequacies of the orbital model is its inability to quantitatively predict chemical-bond strengths and energy differences between various states of a molecule and its ions. This defect results from the fact that the error in the orbital approximation is a sensitive function of both the number of electrons in a given chemical system and the way they are distributed in space. Thus when a molecule forms from two atoms, the error in the orbital approximation of the molecule is larger than the error in the orbital approximation of the two separated atoms. Moreover, when a molecule or an atom gains or loses an electron, or changes its electronic state, the error of the orbital approximation also changes. This error, which arises from the fact that in the orbital picture electrons can only avoid one another on the average and are not able to correlate these motions instantaneously (as the Schrödinger equation demands), is accordingly called the correlation error.

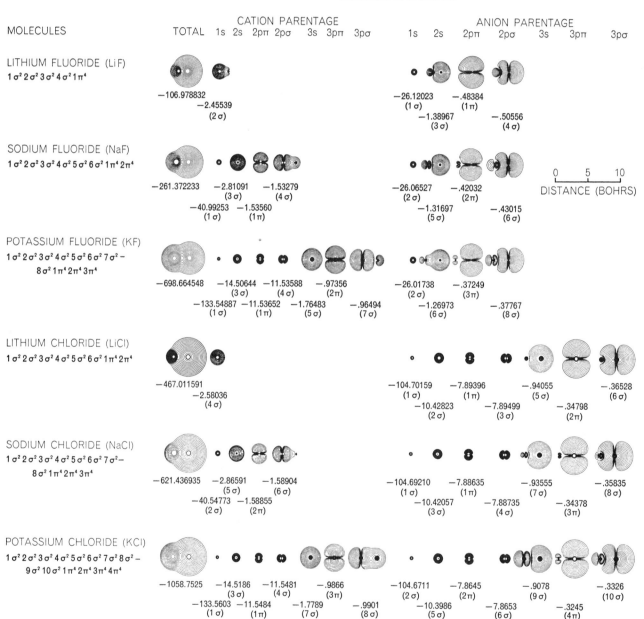

SIX IONIC DIATOMIC MOLECULES (that is, molecules formed by the attraction between positively charged ions, called cations, and negatively charged ions, called anions) are shown in these contour diagrams of electron density. The ionic systems displayed are known as alkali halides. The molecular orbital structures are arranged according to their separated ion parentage; the set arising from the cation is at left and the set arising from the anion is at right. Again total and orbital energies are given in hartrees, and charge densities are plotted in electrons per cubic bohr. The names, symbols and electronic configurations of the molecules are indicated at left; in this case, however, the corresponding molecular orbitals are identified in parentheses directly under each orbital energy.

Significant progress has been made by quantum chemists in modifying the orbital picture to allow electron correlation. This is customarily done either by abandoning the orbital picture completely (a great conceptual sacrifice) and allowing the wave function to depend on the distances between all electrons or by extending the orbital picture so that it enables the electrons to spend a fraction of their time in a variety of different orbital configurations instead of just one; this enables the electrons to avoid one another more effectively. When these correlation terms are added to the best molecular orbital picture so as to allow the electrons to avoid one another in an in-out, left-right and angular sense, the charge density is not significantly affected, but the binding energy and the description of the bonding process are greatly improved. Some effective correlating orbitals are displayed in the illustration on page 44.

Computing Molecular Properties

The orbital picture enables us to look at a chemical process such as molecular excitation (the promotion of electrons into higher energy states) or ionization (the removal of electrons from a molecule) in terms of changes in the electronic charge distribution [see illustration on page 52]. In order to have the energy-level spacing or the ionization potentials correct, however, one must go beyond the molecular orbital model and include electron correlation. The correlated orbital model generally gives an error of several electron volts in both ionization and excitation energies. This error is sometimes tolerable, but often a much higher precision is needed to be of genuine aid to the experimentalist in interpreting atomic or molecular spectra.

Molecular properties associated with the vibration of molecules are customarily studied mathematically by solving the Schrödinger equation to obtain the value of the molecular energy for various fixed positions of the nuclei in the molecule. This procedure determines a "potential well," or energy curve, in which the nuclei are considered to vibrate [see illustration on page 49]. This approach is quite good in most cases when the change of molecular energy can be computed accurately. Since the molecular orbital model gives the change in molecular energy with nuclear position fairly well near the equilibrium nuclear positions of the molecules, the vibrational frequency can be determined within about 10 percent; for higher precision

one must go beyond the orbital picture. (A molecule can absorb or emit even smaller quanta of energy through changes in its rotation.)

An important class of molecular properties that can be studied by means of the orbital approach involves only the individual coordinates of each electron; accordingly such properties are known as "one electron" properties. These properties are evaluated by summing over all space the contribution of all electrons to the total electronic charge distribution. A typical example is the electric-dipole moment of a diatomic molecule, which is a measure of the imbalance between the electronic and the nuclear charge distribution. Other one-electron properties are evaluated in a similar manner and are obtainable from the orbital picture, usually to within a precision better than 20 percent. For molecular properties that depend on the simultaneous position of two electrons or on a fine detail of the wave function, such as its behavior at the nucleus, correlation effects are important and must be included.

The concept of the chemical bond has been indispensable in discussing the mechanism by which elements and compounds are held together. Theoretically this concept states simply that a given chemical compound is stable if its energy is lower than that of the competing reactions; thus a chemical bond is said to exist between two atoms if there exists an equilibrium internuclear separation such that any deviation from it leads to a higher energy for the system. Quantum chemists, however, have sought for many years to produce a more tangible characterization of the chemical bond, preferably in terms of the electronic charge distribution in the molecule itself. One way to analyze the chemical bond is to study the difference in electronic charge density and its associated energy between a molecule and its constituent atoms. Most workers agree that the chemical bond can be associated with an overall contraction of the charge cloud, leading to an increase in charge density between the nuclei (see top illustration on page 51). Nonetheless, there is still some ambiguity as to how exactly this change in charge density is associated with the various energy variations. Another way of looking at the chemical bond is in terms of the force exerted on the nuclei by the charge cloud. This approach has been applied recently to a variety of molecules. Accurate computer chemistry provides an unprecedented opportunity to probe the details of chemical bonding.

As our theoretical models are further refined and as our mathematical apparatus and computers increase in efficiency over the next few years we expect to be able to observe an entire simple chemical reaction on the computer. Thus when we write $2H_2 + O_2 \rightleftharpoons 2H_2O$, we should like to see a continuous process in terms of accurate electronic charge densities [see bottom illustration on page 51]. We would then be able to explore fully the changes in the energy and electron distribution as the hydrogen and oxygen molecules approach each other from different angles and pass through a series of intermediate steps to yield the stable water molecule. It is perhaps in this respect of being able to explore slowly and visually a complicated chemical process

FORMATION OF DIATOMIC MOLECULES can be visualized by arranging sequences of computer-generated electron-density diagrams depicting pairs of atoms at successively closer internuclear distances. The three examples on the following page show two isolated hydrogen atoms coming together to form the covalent-bonded H_2 system (left), a neutral lithium atom and a neutral fluorine atom joining to form the ionic-bonded LiF system (center) and two helium atoms being forced together to form the unstable He_2 system (right). Below each sequence of diagrams is a corresponding graph giving the changes that take place in the total energy of the system during the successive stages of molecular formation. In the H_2 sequence the initial delocalization of the electrons is evidenced by the disappearance of the innermost contour (stage d). This is followed by an overall contraction of the electron-density cloud as the formation process continues with a gradual buildup of charge density between and around the two nuclei. The equilibrium, or minimum-potential-energy, configuration of the molecule is reached at stage g, which corresponds to the bottom of a "potential well" in the energy graph below. In the LiF sequence a drastic change in charge density takes place at an internuclear distance of 13.9 bohrs (stage b) when the system changes from a neutral one to a more stable ionic one. The comparatively undistorted Li^+ ion and F^- ion then come together to the equilibrium LiF separation (stage g). In the He_2 sequence the two helium atoms repel each other, in contrast to the stable H_2 system in which the two atoms attract each other. The electronic charge density in the He_2 system is pushed out from between the two nuclei because as noble-gas atoms their outermost electron shells are filled and the type of electron-sharing that was possible between the unfilled shells of the hydrogen atoms cannot take place.

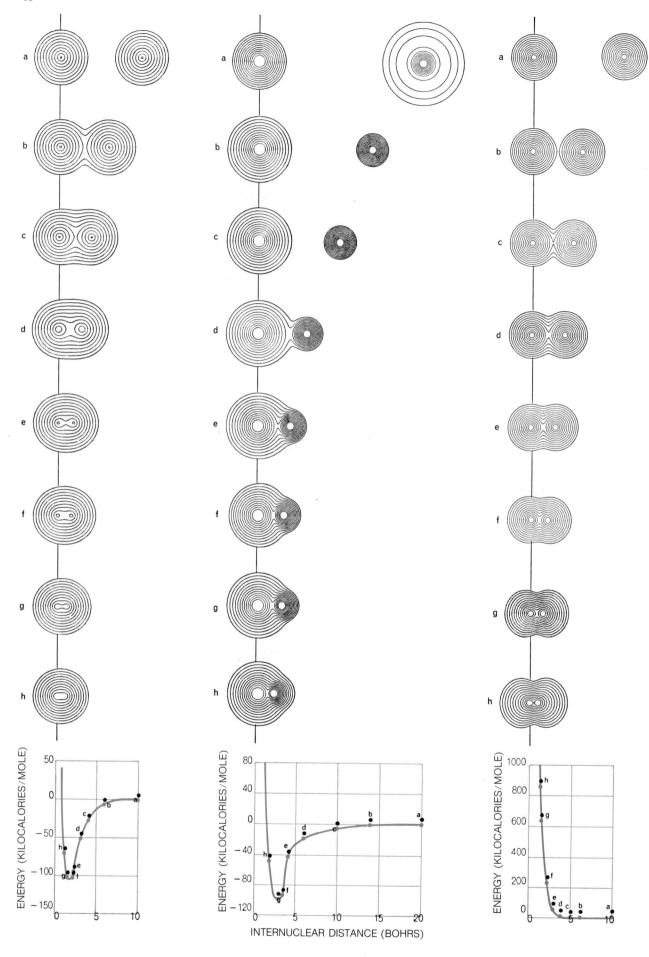

48

in terms of both its quantitative features and its conceptual features that such computer "experiments" offer their most informative potential. Such computer studies of chemical reactions, which are now in their early stages, also promise by extensive interplay with parallel experimental results to characterize and define such often vague concepts as the transition state of a molecule. In addition fine details of the nature of the chemical-reaction surface (that is, the energy changes associated with the approach of reacting atoms from all possible directions) are accessible to the computer experiment, whereas the conventional physical experiment often sees these effects only as the average over time or space of a continuous chemical process, a result that cannot usually be translated unambiguously into a detailed reaction surface.

On to Bulk Properties

As I have mentioned, one of the most compelling reasons for being able to evaluate the properties and interactions between individual atoms and molecules is that highly sophisticated statistical theories have been developed that allow the calculation of the equation of state and transport properties of a macroscopic system from the characteristic atomic and molecular wave functions. Such theories are now quite accurate for dilute gases and are becoming increasingly precise for dense gases, liquids and solids. We know, for example, that for a "perfect" gas, namely one made up of particles with no volume and with no forces between particles, the equation of state is $PV = RT$, where P is the gas pressure, V the gas volume, R a constant and T the temperature. This equation is quite accurate for real gases provided that they are quite dilute; however, even then the deviations from the ideal case are important, and one must modify the perfect-gas equation of state to take into account the fact that the atoms or molecules of the gas occupy volume and have forces operating between them. One of the most successful modifications of this type is called the virial equation of state, which is written $PV/RT = 1 + B(T)/V + C(T)/V^2 + D(T)/V^3 \ldots$ The coefficients $B(T)$, $C(T)$ and $D(T)$, which depend on temperature, are respectively called the second, third and fourth virial coefficients. Statistical mechanics can be used to evaluate these virial coefficients directly in terms of the forces between atoms or molecules. Thus computer calculations of interatomic and intermolecu-

lar forces, when coupled with statistical mechanics, can lead directly to an improved equation of state.

Another important class of bulk chemical properties, called the transport properties, includes viscosity (transport of momentum through a fluid), diffusion (transport of mass) and thermal conductivity (transport of thermal energy). These properties are directly related to the size and shape of individual atoms and molecules and to the forces between them. They are hence susceptible to investigation by theoretical computations.

A New Chemical Instrument

What I have been describing are typical mathematical models of molecular structure and simple chemical processes. These concepts and models have been developed and refined by chemists over the past 50 years. It is only in the past 20 years, however, that such models have been computationally developed, and it is only in even more recent years that they have been comprehensively programmed on large-memory, high-speed digital computers to be operative for many different molecules. When applied cautiously, they are now capable

of yielding useful chemical information, and thus they provide an independent and competitive investigative technique, particularly when an experiment is very difficult or even dangerous. Accordingly, if the experimental worker faced with a chemical problem is to be encouraged to ask the question "Should I measure the answer or should I compute the answer?" some attention must be given to making it possible for the nonspecialized worker to utilize sophisticated computer technology.

Here I should like to discuss BISON, a computing tool we have developed in our laboratory that addresses itself to this problem. BISON is most properly viewed as a new chemical instrument to which the chemist can easily turn for certain types of chemical information. BISON has been designed with four basic principles in mind: (1) responsiveness to questions posed in the chemist's natural language; (2) ability to estimate the reliability of the computed answers; (3) foolproof operation with a great deal of procedural experience programmed into the system; (4) heavy use of computer graphics.

In BISON the first two requirements are accomplished by means of an INTER-

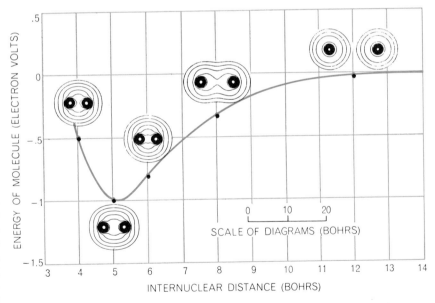

MOLECULAR VIBRATION, another way in which a molecule can absorb or emit energy in discrete quanta, is responsible for certain chemical properties that can be studied mathematically by solving the Schrödinger equation to obtain the value of the molecular energy for various fixed positions of the nuclei. This procedure results in a potential-energy curve in which the nuclei are considered to vibrate. The shape of this potential well determines the frequency of the molecular vibrations. The molecule can emit or absorb vibrational energy in multiples of this fundamental vibrational frequency. (The changes in energy that occur during molecular vibration are usually much smaller than those associated with electronic excitation.) As these contour diagrams of a diatomic lithium molecule (Li_2) show, the electron density changes quite smoothly as the nuclei vibrate in the potential well. If the molecule absorbs enough vibrational energy through radiation or collisions with other molecules, atoms or surfaces, it "dissociates," or breaks up into its constituent atoms.

VIEW module, which converses by teletypewriter with the investigator in a conversational FORTRAN language to define his request clearly and to inform him of the time involved and the precision in the computed answer. An essential feature of INTERVIEW is that it converses in the chemist's natural language and not in a highly complex or specialized input code. Since INTERVIEW is based on a key-word vocabulary, the chemist's questions and answers can be stated in any way as long as a few key words natural to the chemical problem are included. A typical conversation between the chemist and BISON's INTERVIEW module might go like this:

Chemist: I should like the dipole moment of the lowest-lying doublet π state of the calcium oxide (CaO) molecule.

BISON: At which internuclear distance?

Chemist: The equilibrium separation.

BISON: I shall require four hours of machine time. The probable error in the computed dipole moment will be approximately 20 percent. Should I proceed?

The third requirement—foolproof operation—is fulfilled by a PROCEDURE EXECUTIVE module. Once the chemist has conversed with BISON through the INTERVIEW module and placed his "order," the system "takes off" on its own without further inconvenience to the chemist. The PROCEDURE EXECUTIVE module supervises a sequence of calculations necessary in order to answer the chemist's question.

Currently BISON is based on the Hartree-Fock molecular orbital model of diatomic molecules, and PROCEDURE EXECUTIVE, by supervising the "workhorse," or unit-operation, program, can yield reliable charge densities, molecular sizes

and shapes, ionization potentials, dipole moments and other related molecular properties. For certain classes of diatomic molecules it also yields useful potential curves, vibrational frequencies, binding energies and transition probabilities.

BISON will generate and plot on film, paper or a cathode ray screen contour diagrams of the orbital and total density of a molecule or a chemical process. All the diagrams used in this article were drawn by BISON. This display feature is particularly fascinating when it is used to watch a molecule form, or pass through a series of energy states, or encounter another molecule or an atom, or ultimately engage in a chemical reaction. The display feature will soon be operative, using precomputed and stored charge densities interpolated between computed points so that they can be quickly manipulated in "real time." Cur-

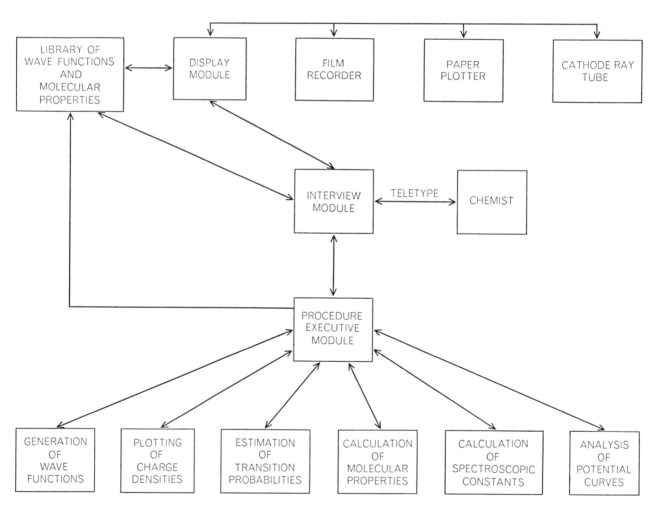

BISON, a new computing instrument developed by the author and his colleagues at the Argonne National Laboratory, is designed to make it possible for a chemist who is not a computer specialist to obtain certain types of chemical information with the aid of rigorous mathematical models and sophisticated computer technology. At present the investigator converses with BISON by teletypewriter through an INTERVIEW module in the chemist's natural language. Relying on a library of previous experience, INTERVIEW informs him of the time required to fill his request and the precision of the computed answer. If the chemist decides to place his "order," a PROCEDURE EXECUTIVE module then "takes off" on its own, supervising a sequence of calculations necessary to answer the chemist's question. The electronic charge densities of the atoms or molecules involved in a chemical process can be produced and displayed on film, paper or the screen of a cathode ray tube. All the diagrams used to illustrate this article were produced by BISON.

SODIUM-LITHIUM MOLECULE   SODIUM ATOM   LITHIUM ATOM
(NaLi)   (Na)   (Li)

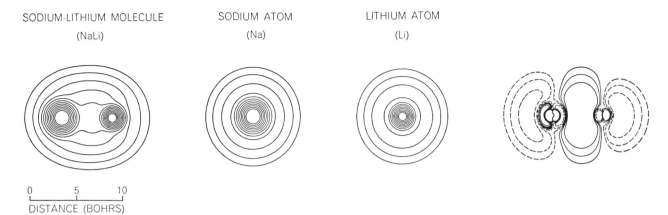

0   5   10
DISTANCE (BOHRS)

CONCEPT OF CHEMICAL BOND can be studied by observing the changes that take place in the computed electronic charge density as the molecule forms. For example, these diagrams show the electron density of a stable sodium-lithium molecule (*left*), the electron density of its separated constituent atoms (*center*) and the "difference density" (*right*), that is, the difference between the density of the stable molecule and the density of the undistorted atoms brought together to the equilibrium molecular position. The solid contour lines at right denote regions of positive difference density (regions where the charge density has increased); the broken lines denote regions of negative difference density (regions where the charge density has decreased).

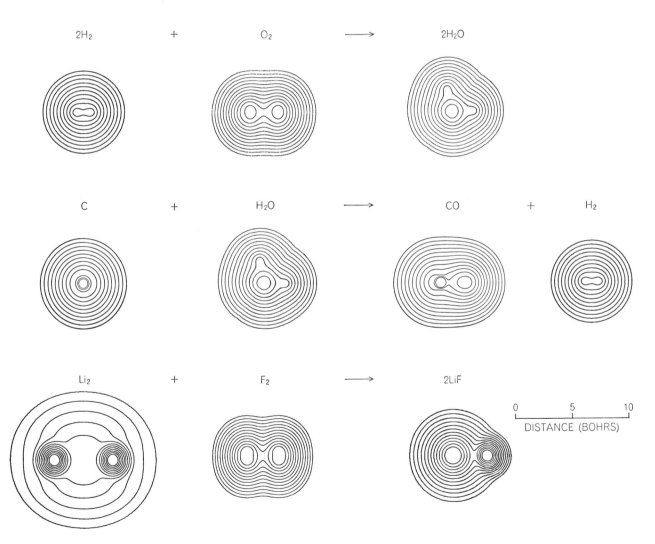

2H$_2$ +   O$_2$ $\longrightarrow$ 2H$_2$O

C   +   H$_2$O $\longrightarrow$ CO   +   H$_2$

Li$_2$ +   F$_2$ $\longrightarrow$ 2LiF

0   5   10
DISTANCE (BOHRS)

THREE CHEMICAL REACTIONS are portrayed here in terms of changes in computed electron-density diagrams. The reaction at top shows two hydrogen molecules (H$_2$) combining with an oxygen molecule (O$_2$) to form two water molecules (H$_2$O). The reaction at middle shows a carbon atom (C) and a water molecule (H$_2$O) combining to form a carbon monoxide molecule (CO) and a hydrogen molecule (H$_2$). The reaction at bottom shows a lithium molecule (Li$_2$) and a fluorine molecule (F$_2$) combining to form two lithium fluoride molecules (LiF). At present even such relatively simple reactions present a considerable challenge to computer chemistry, but it appears reasonable to expect that this general approach will be extended to much more complicated reactions.

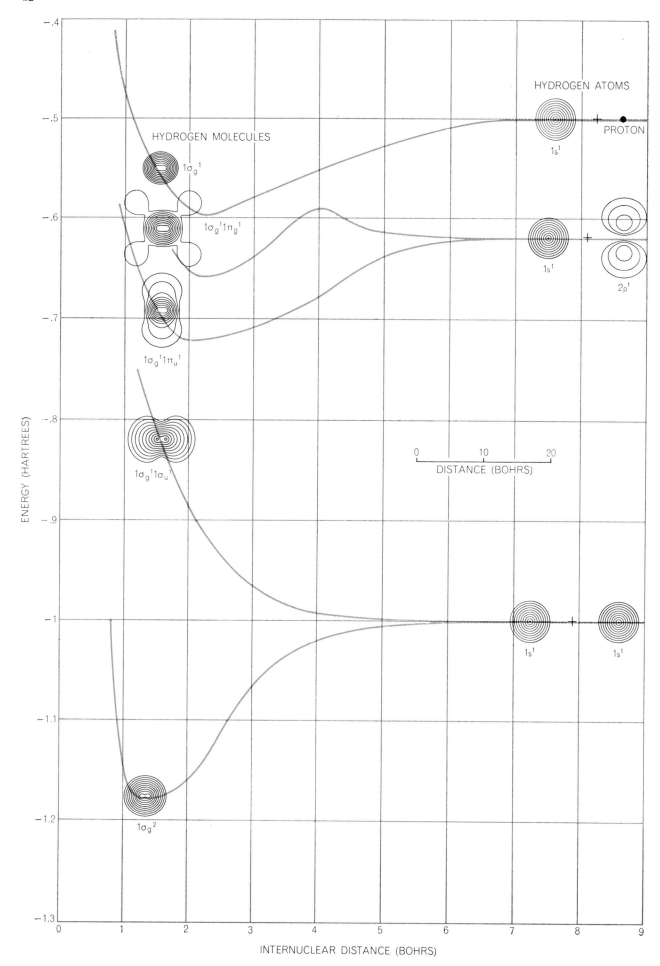

HYDROGEN ATOMS

PROTON

$1s^1$

HYDROGEN MOLECULES

$1\sigma_g^{\ 1}$

$1\sigma_g^{\ 1}1\pi_g^{\ 1}$

$1s^1$

$2p^1$

$1\sigma_g^{\ 1}1\pi_u^{\ 1}$

$1\sigma_g^{\ 1}1\sigma_u^{\ 1}$

0   10   20
DISTANCE (BOHRS)

$1s^1$ $1s^1$

$1\sigma_g^{\ 2}$

ENERGY (HARTREES)

−.4

−.5

−.6

−.7

−.8

−.9

−1

−1.1

−1.2

−1.3

0   1   2   3   4   5   6   7   8   9

INTERNUCLEAR DISTANCE (BOHRS)

rently the calculation required for this purpose is too time-consuming except for the most trivial molecules.

It is important to stress that BISON is at present an information-creation system. Starting from first principles, BISON obtains by computation the answer to the chemist's questions. Thus it is an independent source of information, not a system that simply retrieves information or interpolates or extrapolates it from large tables of stored data. The plan is to include in BISON such a retrieval capacity and to have BISON upgrade its own store of information as it answers new questions.

Future plans for BISON include the extension of the system to simple polyatomic molecules for the detailed study of chemical reactions, the inclusion of other molecular properties and the development of a transport-property module that will evaluate bulk physical properties from computed atomic and molecular properties. The incorporation of sophisticated computing technology into systems that fulfill "instrument" requirements is certainly becoming an important new development in the physical sciences. As theoretical techniques in all disciplines become more comprehensive and powerful, and as anticipated improvements in computer speed, sophistication, organization and accessibility through widely distributed terminals become a reality, more and more experimental workers will undoubtedly turn to the computer experiment as an alternative to the traditional one.

←————

EXCITATION AND IONIZATION of a hydrogen molecule are viewed according to the orbital model as changes in the density and shape of the molecule's electron cloud. When such a molecule is excited to a higher energy state by absorbing energy from outside in discrete quanta (*corresponding to the vertical distance between the colored energy curves*), its charge cloud typically becomes more extended and more complicated. (When the molecule emits energy, the reverse changes occur.) When enough energy is absorbed, the molecule's outer electron is ejected, creating a positive molecular ion of hydrogen (*top left*). The curves indicate how the total energy of the system changes as the two hydrogen atoms (*right*) come together in various ways. The electronic configurations next to the diagrams indicate the atomic or molecular orbital that is in each case the dominant contributor to the total wave function. At higher excited states of the molecule the area enclosed by the highest-value, innermost contour first becomes smaller and then disappears as charge density is removed from this region.

III

MOLECULAR STRUCTURE
AND BIOLOGICAL SPECIFICITY

III

MOLECULAR STRUCTURE AND BIOLOGICAL SPECIFICITY

INTRODUCTION

Molecular structure determines most of the commonly observed properties of matter. For example, melting points, boiling points, solubilities in various liquids, absorption of electromagnetic radiation, and chemical reactivities, in principle are all predictable from a detailed knowledge of molecular structure. In practice, the theoretical chemist finds such predictions far from easy. There are, however, many generalizations related to molecular structure which provide chemists with intuitions that make qualitative predictions possible.

For example, the melting points and boiling points of a chemically similar family of nonpolar compounds increase with increasing molecular weight. The hydrocarbons that are isolated from petroleum provide an example of such a trend: natural gas, which is burned in stoves and hot water heaters, is primarily methane (CH_4), with molecular weight 16, but liquid gasoline is a mixture of hydrocarbons ranging from hexane (C_6H_{14}) to nonane (C_9H_{20}), with molecular weights ranging from 86 to 128. The more viscous fuel oils or the kerosene used in diesel engines and for heating homes have molecular weights from 142 to 268; the lubricating oils and solid greases have still higher molecular weights.

The attractive force between nonpolar molecules that causes this liquefaction and solidification is called the London force after the physicist F. London. The London force is always attractive. It originates in the concerted motions of electrons in neighboring atoms that are not chemically bonded to each other. The attractive London forces are very short-range, and for the force to be significant, the molecules must be very near one another, say within one atomic radius of one another.

A molecule is said to have a dipole moment if the center of positive electrical charge in the molecule does not coincide with the center of negative electrical charge. A typical example is the hydrochloric acid molecule, HCl. When a hydrogen atom is isolated in space, its electron distribution is spherically symmetrical about its nucleus and it has no dipole moment. The same is true of an isolated chlorine atom. When the HCl molecule forms, much of the electron density originally around the hydrogen nucleus is attracted to the more electronegative chlorine atom, creating a dipole moment in which the negative end of the dipole is at the chlorine atom and the positive end is at the hydrogen atom. A molecule with a large dipole moment is called a polar molecule. Such polar molecules are attracted to each other not only by London forces but also by dipole-dipole and dipole-induced dipole interactions. The sum of these three types of intermolecular forces is called the van der Waals force or interaction, and it results in energies that are between 0.1 kcal/mole and 1.0 kcal/mole for small molecules, far less than the energies of chemical bonds. Because of these additional modes of interaction, polar molecules generally have higher melting and boiling points than nonpolar molecules of the same molecular weight. An example

is the hydrocarbon isobutane,
$$CH_3-\overset{\overset{\displaystyle CH_3}{|}}{\underset{\underset{\displaystyle H}{|}}{C}}-CH_3,$$
which has a molecular weight of 58 and a boiling point of $-10°C$. Acetone,
$$CH_3-\overset{\overset{\displaystyle O}{\|}}{C}-CH_3,$$
a ketone with a very polar $C=O$ bond has the same molecular weight but a boiling point of $57°C$, $67°C$ higher.

There are other forces which contribute to intermolecular interactions. One of these is the hydrogen bond, which is a bond between a hydrogen atom covalently bonded to an electronegative atom such as N, O, or F, and a nonbonding pair of electrons on another electronegative atom. The most common hydrogen bonds are listed below.

Electron pair acceptors	Electron pair donors
O—H	⋯ O
O—H	⋯ N
N—H	⋯ O
N—H	⋯ N

All of these hydrogen bonds have energies between 3 and 6 kcal/mole, about one tenth that of covalent bonds but 10 times that of van der Waals interactions. Hydrogen bonds account for the very high melting point and boiling point of water, which would have properties similar to methane if water were not polar and if hydrogen bonds did not exist. You will find some details about hydrogen bonds and ice structure in article 12 (section IV), "Ice," by L. K. Runnels.

Molecules which have an electrical charge are called ions. This charge affects the properties of the molecule, inasmuch as ions attract other ions of opposite charge. Such interactions are called ionic interactions or charge-charge interactions.

The solubility of substances in liquids is related to the chemical properties of both the substance and the liquid. For example, salts or ionic compounds are soluble in polar solvents but insoluble in nonpolar solvents. Most salts are therefore soluble in water. Generally, polar substances are miscible or soluble in each other, and nonpolar substances, such as benzene and gasoline, are soluble in one another, but polar and nonpolar substances are not miscible. There are, of course, many intermediate substances, such as ethanol,
$$H-\overset{\overset{\displaystyle H}{|}}{\underset{\underset{\displaystyle H}{|}}{C}}-\overset{\overset{\displaystyle H}{|}}{\underset{\underset{\displaystyle H}{|}}{C}}-OH,$$
which is partly polar because of the OH group and partly nonpolar because of the CH_3CH_2 group. Ethanol dissolves well in water and in gasoline.

Hydrogen bonding is important for solubility; although acetone, $CH_3-CO-CH_3$, cannot form a hydrogen bond to itself (it does not contain a hydrogen covalently bonded to an electronegative atom), it has nonbonding electrons which can be donated to a hydrogen bond to water. It therefore dissolves well in water.

The absorption of visible light by organic molecules is related to another structural property of molecules. Light in this region of the electromagnetic spectrum is generally not energetic enough to break covalent bonds, but it is sufficient to excite molecules to higher electronic energy levels. Only those molecules which have a spacing between their ground electronic state and an excited state equal to the

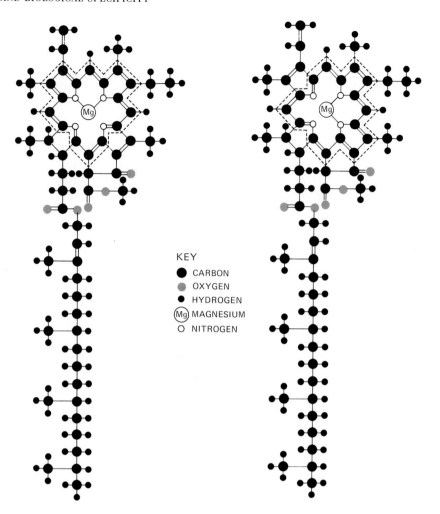

KEY
● CARBON
● OXYGEN
• HYDROGEN
(Mg) MAGNESIUM
O NITROGEN

Figure III.1   The chlorophyll molecule owes its photobiological activity to the rigid and intricate porphyrin structure at the end of the long carbon chain. The arrangement of the bonds in the porphyrin structure may resonate among both configurations diagrammed here. These and other possible configurations of the bonds help to make it possible for the chlorophyll molecule to trap and store energy which is conveyed to it by light quanta. [Adapted from "Life and Light," by George Wald, *Scientific American*, October 1959. Copyright © 1959 by Scientific American, Inc. All rights reserved.]

energy of a photon of visible light will absorb visible light. In the world of organic molecules, the only molecules having the ability to absorb visible light are those with delocalized electrons, having "conjugated" structures. Such molecules contain alternating carbon-carbon double bonds and carbon-carbon single bonds. An example of a conjugated structure is the chlorophyll molecule, which is responsible for the absorption of the sun's energy in all green plants. The ring of carbon atoms at the top of this structure is called a porphyrin ring and is a closed ring of alternating C=C double bonds and C—C single bonds. Another example is provided by the retinal (retinene) molecule, which is responsible for light absorption in the human retina.

"The Stereochemical Theory of Odor" by Amoore, Johnston, and Rubin, article 6, provides one of the simplest possible examples of intermolecular interactions. While most other biological specificities are far more sensitive to the detailed molecular structure and electron charge distribution in a molecule, the human olfactory area appears sensitive, instead, primarily to the three-dimensional shape of the odorous molecules, at least according to this theory, and to a lesser

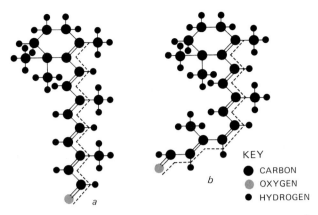

Figure III.2   The retinal molecule is the active agent in the pigments of vision. Upon absorbing the energy of light the geometry of the molecule changes from the so-called *trans* arrangement (*a*) to the *cis* arrangment (*b*). This change in structure triggers the process of vision. [Adapted from "Life and Light," by George Wald, *Scientific American*, October 1959, Copyright © 1959 by Scientific American, Inc. All rights reserved.]

extent are sensitive to the electronic properties of the molecule.

There are several requirements for substances to be odorous. They must be volatile and capable, therefore, of reaching the olfactory area as a vapor. The odorous molecules must have some solubility in water so they can penetrate the film of water covering the nerve endings in the nose. They must also have some solubility in lipids (fatty substances or nonpolar substances) so they can penetrate the surface membrane of the nerve cells (the mechanism by which this takes place is described in article 8, "The Structure of Cell Membranes").

Amoore, Johnston, and Rubin divide odors into seven primary classes: camphoraceous, musky, floral, pepperminty, ethereal, pungent, and putrid. Of these, the first five depend only on molecular shape, although all odorous molecules must satisfy the above-stated conditions of volatility and solubility. Only pungent and putrid odors are related to more subtle chemical properties of the molecules. Pungent molecules are electrophilic, meaning they have an attraction for electrons. This does not mean the molecule is a positive ion, as might be thought from the article, but that a site on the molecule has a strong affinity for electrons. The starred carbon atom in the structure for formic acid is such an electrophilic center because, as we stated earlier, a carbon-oxygen double bond, $\diagdown$C$=$O, is very polar.

Formic acid, showing the electrophilic center at the carbon atom.

Putrid molecules are nucleophilic, meaning they have an affinity for nuclei. This does not mean they are negative ions, as might be thought from the article, but that they contain a nucleophilic center which has a strong attraction for nuclei. The sulfur atom in hydrogen sulfide is such a nucleophilic center because of the polarity of this molecule and the nonbonding pairs of electrons on the sulfur.

Hydrogen sulfide, showing the nucleophilic center at the sulfur atom.

Pepperminty molecules are another minor exception to the geometric model of odor, for this odor is produced by wedge-shaped molecules which must be able to form a hydrogen bond at the point of the wedge.

"Pheromones," article 7, by Wilson, provides more spectacular examples of the chemical specificity of a class of compounds made and excreted by animals to regulate their communication and their societies, and to map their external environments. Insect chemoreceptors can detect a few hundred molecules of a specific pheromone per cubic centimeter and reject the signal of many other chemically similar molecules with amazing accuracy. The author estimates that ants may excrete as few as 10 pheromones that regulate their foraging for food, their sex lives, their defenses from intruders, and their disposal of dead carcasses. The chemical steps used in the animal for the synthesis and specific detection of these complex compounds are largely unknown. An investigation of the structures of the sex attractants shown in the article will indicate how sophisticated these detection mechanisms must be.

The cell membrane described in article 8, "The Structure of Cell Membranes," by Fox, provides another example of how living systems take advantage of the diversity of chemical properties and structures. The principles associated with membrane formation are similar to those associated with the action of soaps and detergents.

For example, a typical soap is sodium laurate

$$CH_3CH_2CH_2CH_2CH_2CH_2CH_2CH_2CH_2CH_2CH_2C \underset{O^-Na^+}{\overset{O}{<}}$$

The long carbon chain in this molecule is nonpolar and does not dissolve well in water. This end of the molecule is said to be "hydrophobic," or water hating. The polar end of the molecule, $-C \underset{O^-Na^+}{\overset{O}{<}}$ a salt dissolving well in water, is said to be "hydrophilic," or water loving, and it therefore gives the soap its solubility in water. Since the hydrophobic tails would rather interact with themselves than with

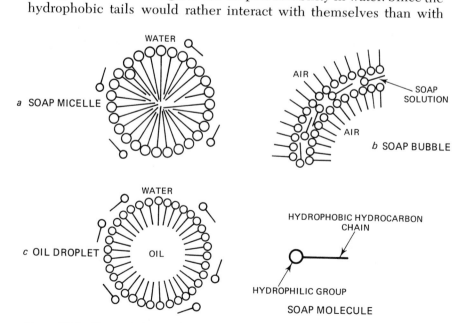

Figure III.3 How membrane formation is useful in soaps and detergents. *a*, Soap micelle; *b*, Section of soap bubble; *c*, Oil droplet in soap solution.

water, special structures form in soap solutions called micelles. In a micelle, all the hydrophobic tails of many soap molecules aggregate and interact with one another and all polar groups interact with the water. The interactions of a soap micelle are shown in Figure III.3, together with a soap bubble and the chemical mechanism for dissolving grease — a hydrophobic substance — in a soap solution.

The membrane-forming molecules in living cells are called phospholipids. They contain a polar phosphate group and generally two fatty acid chains similar to those in the soap molecule shown earlier. The hydrocarbon chains (hydrophobic) aggregate as in a micelle, but in phospholipids they form a membrane or lipid bilayer where the polar phosphate groups are on both exterior sides of the membrane and the hydrophobic chains are in the interior of the membrane. Such membranes form spontaneously in the laboratory, and are an excellent example of the solubility principles discussed above.

Fox's article shows electron micrographs of natural membranes that also contain proteins. These proteins must also have special solubility properties. They must be hydrophobic where they are in the interior of the membrane and hydrophilic where they come in contact with the exterior of the membrane. You will learn more about these properties of proteins from Section VIII of this reader. The detailed arrangement of proteins in membranes is still unknown. It is not known, for example, if proteins generally emerge from both sides of the membrane, are totally contained within the membrane, or if their arrangement is some combination of these models. Fox's article also proposes a mechanism for membrane transport involving the rotation of proteins in the membrane, but Fox's proposal is entirely speculative at this time and has not been proven.

"Molecular Isomers in Vision," article 9, by Hubbard and Kropf, provides a brief history of the development of stereochemistry. A century ago the notion that molecules have three-dimensional structures was introduced by Jacobus Henricus van't Hoff and was received with a great deal of skepticism. This idea is today one of the central principles of chemistry from which, as we have seen, many of the generalizations of the properties of matter emerge.

Hubbard and Kropf trace the development of the ideas of tetrahedral symmetry about carbon atoms, as in methane, CH_4, and the restricted rotation about carbon-carbon double bonds (cis-trans isomers). This then leads to the notion of delocalized electrons, conjugation, and π electrons. The joining of these two seemingly separate ideas — the light absorption associated with the delocalized electrons in retinal, and the cis-trans isomerization of the retinal molecule — provides a mechanism for one of the major steps in vision.

6

THE STEREOCHEMICAL THEORY OF ODOR

JOHN E. AMOORE, JAMES W. JOHNSTON, JR., AND MARTIN RUBIN
February 1964

There is evidence that the sense of smell is based on the geometry of molecules. Seven primary odors are distinguished, each of them by an appropriately shaped receptor at the olfactory nerve endings

A rose is a rose and a skunk is a skunk, and the nose easily tells the difference. But it is not so easy to describe or explain this difference. We know surprisingly little about the sense of smell, in spite of its important influence on our daily lives and the voluminous literature of research on the subject. One is hard put to describe an odor except by comparing it to a more familiar one. We have no yardstick for measuring the strength of odors, as we measure sound in decibels and light in lumens. And we have had no satisfactory general theory to explain how the nose and brain detect, identify and recognize an odor. More than 30 different theories have been suggested by investigators in various disciplines, but none of them has passed the test of experiments designed to determine their validity.

The sense of smell obviously is a chemical sense, and its sensitivity is pro-verbial; to a chemist the ability of the nose to sort out and characterize substances is almost beyond belief. It deals with complex compounds that might take a chemist months to analyze in the laboratory; the nose identifies them instantly, even in an amount so small (as little as a ten-millionth of a gram) that the most sensitive modern laboratory instruments often cannot detect the substance, let alone analyze and label it.

Two thousand years ago the poet Lucretius suggested a simple explanation of the sense of smell. He speculated that the "palate" contained minute pores of various sizes and shapes. Every odorous substance, he said, gave off tiny "molecules" of a particular shape, and the odor was perceived when these molecules entered pores in the palate. Presumably the identification of each odor depended on which pores the molecules fitted.

It now appears that Lucretius' guess was essentially correct. Within the past few years new evidence has shown rather convincingly that the geometry of molecules is indeed the main determinant of odor, and a theory of the olfactory process has been developed in modern terms. This article will discuss the stereochemical theory and the experiments that have tested it.

The nose is always on the alert for odors. The stream of air drawn in through the nostrils is warmed and filtered as it passes the three baffle-shaped turbinate bones in the upper part of the nose; when an odor is detected, more of the air is vigorously sniffed upward to two clefts that contain the smelling organs [see illustration on opposite page]. These organs consist of two patches of yellowish tissue, each about one square inch in area. Embedded in the tissue are two types of nerve fiber whose endings receive and detect the odorous molecules. The chief type is represented by the fibers of the olfactory nerve; at the end of each of these fibers is an olfactory cell bearing a cluster of hairlike filaments that act as receptors. The other type of fiber is a long, slender ending of the trigeminal nerve, which is sensitive to certain kinds of molecules. On being stimulated by odorous molecules, the olfactory nerve endings send signals to the olfactory bulb and thence to the higher brain centers where the signals are integrated and interpreted in terms of the character and intensity of the odor.

From the nature of this system it is obvious at once that to be smelled at all a material must have certain basic properties. In the first place, it must be volatile. A substance such as onion soup, for example, is highly odorous because it continuously gives off vapor that can reach the nose (unless the soup is im-

PRIMARY ODOR	CHEMICAL EXAMPLE	FAMILIAR SUBSTANCE
CAMPHORACEOUS	CAMPHOR	MOTH REPELLENT
MUSKY	PENTADECANOLACTONE	ANGELICA ROOT OIL
FLORAL	PHENYLETHYL METHYL ETHYL CARBINOL	ROSES
PEPPERMINTY	MENTHONE	MINT CANDY
ETHEREAL	ETHYLENE DICHLORIDE	DRY-CLEANING FLUID
PUNGENT	FORMIC ACID	VINEGAR
PUTRID	BUTYL MERCAPTAN	BAD EGG

PRIMARY ODORS identified by the authors are listed, together with chemical and more familiar examples. Each of the primary odors is detected by a different receptor in the nose. Most odors are composed of several of these primaries combined in various proportions.

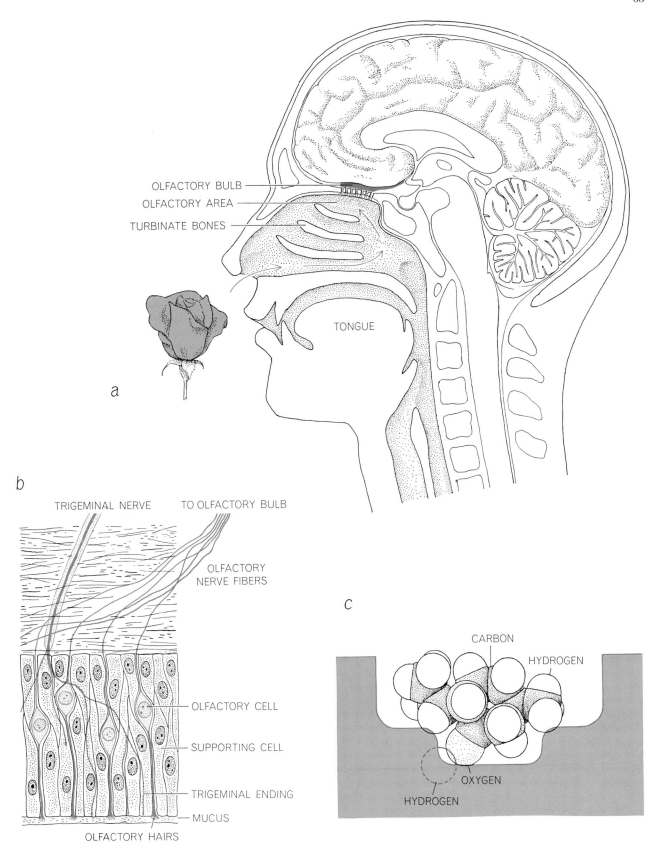

OLFACTORY BULB
OLFACTORY AREA
TURBINATE BONES

TONGUE

a

b

TRIGEMINAL NERVE   TO OLFACTORY BULB

OLFACTORY
NERVE FIBERS

c

CARBON   HYDROGEN

OLFACTORY CELL

SUPPORTING CELL

TRIGEMINAL ENDING

OXYGEN

MUCUS

HYDROGEN

OLFACTORY HAIRS

ANATOMY of the sense of smell is traced in these drawings. Air carrying odorous molecules is sniffed up past the three baffle-shaped turbinate bones to the olfactory area (*a*), patches of epithelium in which are embedded the endings of large numbers of olfactory nerves (*color*). A microscopic section of the olfactory epithelium (*b*) shows the olfactory nerve cells and their hairlike endings, trigeminal endings and supporting cells. According to the stereochemical theory different olfactory nerve cells are stimulated by different molecules on the basis of the size and shape or the charge of the molecule; these properties determine which of various pits and slots on the olfactory endings it will fit. A molecule of *l*-menthone is shown fitted into the "pepperminty" cavity (*c*).

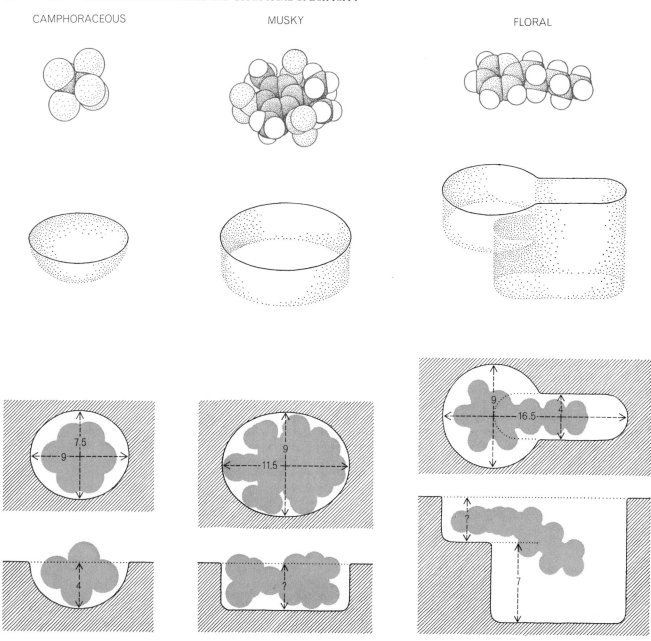

CAMPHORACEOUS   MUSKY   FLORAL

OLFACTORY RECEPTOR SITES are shown for each of the primary odors, together with molecules representative of each odor. The shapes of the first five sites are shown in perspective and (with the molecules silhouetted in them) from above and the side;

prisoned in a sealed can). On the other hand, at ordinary temperatures a substance such as iron is completely odorless because it does not evaporate molecules into the air.

The second requirement for an odorous substance is that it should be soluble in water, even if only to an almost infinitesimal extent. If it is completely insoluble, it will be barred from reaching the nerve endings by the watery film that covers their surfaces. Another common property of odorous materials is solubility in lipids (fatty substances); this enables them to penetrate the nerve endings through the lipid layer that

forms part of the surface membrane of every cell.

Beyond these elementary properties the characteristics of odorous materials have been vague and confusing. Over the years chemists empirically synthesized a wealth of odorous compounds, both for perfumers and for their own studies of odor, but instead of clarifying the properties responsible for odor these compounds seemed merely to add to the confusion. A few general principles were discovered. For instance, it was found that adding a branch to a straight chain of carbon atoms in a perfume molecule markedly increased the po-

tency of the perfume. Strong odor also seemed to be associated with chains of four to eight carbon atoms in the molecules of certain alcohols and aldehydes. The more chemists analyzed the chemical structure of odorous substances, however, the more puzzles emerged. From the standpoint of chemical composition and structure the substances showed some remarkable inconsistencies.

Curiously enough, the inconsistencies themselves began to show a pattern. As an example, two optical isomers—molecules identical in every respect except that one is the mirror image of the other —may have different odors. As another

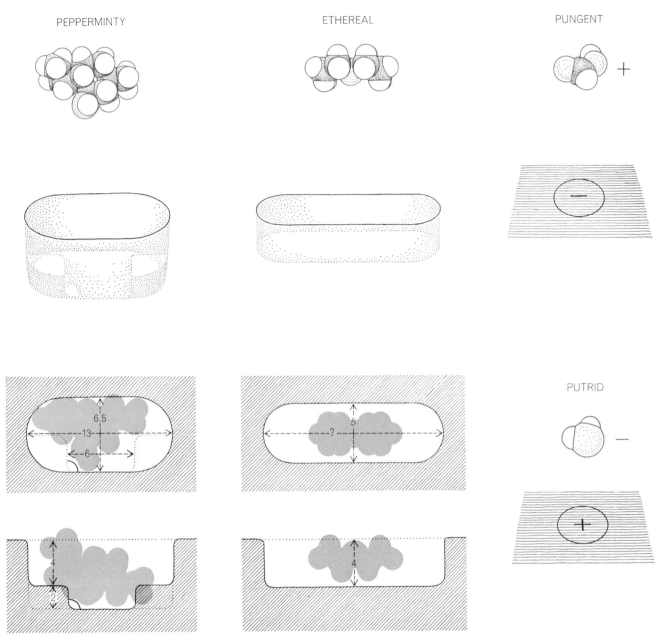

PEPPERMINTY   ETHEREAL   PUNGENT

PUTRID

known dimensions are given in angstrom units. The molecules are (*left to right*) hexachloroethane, xylene musk, alpha-amylpyridine, *l*-menthol and diethyl ether. Pungent (formic acid) and putrid (hydrogen sulfide) molecules fit because of charge, not shape.

example, in a compound whose molecules contain a small six-carbon-atom benzene ring, shifting the position of a group of atoms attached to the ring may sharply change the odor of the compound, whereas in a compound whose molecules contain a large ring of 14 to 19 members the atoms can be rearranged considerably without altering the odor much. Chemists were led by these facts to speculate on the possibility that the primary factor determining the odor of a substance might be the over-all geometric shape of the molecule rather than any details of its composition or structure.

In 1949 R. W. Moncrieff in Scotland gave form to these ideas by proposing a hypothesis strongly reminiscent of the 2,000-year-old guess of Lucretius. Moncrieff suggested that the olfactory system is composed of receptor cells of a few different types, each representing a distinct "primary" odor, and that odorous molecules produce their effects by fitting closely into "receptor sites" on these cells. His hypothesis is an application of the "lock and key" concept that has proved fruitful in explaining the interaction of enzymes with their substrates, of antibodies with antigens and of deoxyribonucleic acid with the "messenger" ribonucleic acid that presides at the synthesis of protein.

To translate Moncrieff's hypothesis into a practical approach for investigating olfaction, two specific questions had to be answered. What are the "primary odors"? And what is the shape of the receptor site for each one? To try to find answers to these questions, one of us (Amoore, then at the University of Oxford) made an extensive search of the literature of organic chemistry, looking for clues in the chemical characteristics of odorous compounds. His search resulted in the conclusion that there were seven primary odors, and in 1952 his findings were summed up in a stereochemical theory of olfaction that identified the seven odors and gave a detailed description of the size, shape and chemical affinities of the seven corresponding receptor sites.

To identify the primary odors Amoore started with the descriptions of 600 organic compounds noted in the literature as odorous. If the receptor-site hypothesis was correct, the primary odors should be recognized much more frequently than mixed odors made up of two or more primaries. And indeed, in the chemists' descriptions certain odors turned up much more commonly than others. For instance, the descriptions mentioned more than 100 compounds as having a camphor-like odor, whereas only about half a dozen were put in the category characterized by the odor of cedarwood. This suggested that in all likelihood the camphor odor was a primary one. By this test of frequency, and from other considerations, it was possible to select seven odors that stand out as probable primaries. They are: camphoraceous, musky, floral, pepperminty, ethereal (ether-like), pungent and putrid.

From these seven primaries every known odor could be made by mixing them in certain proportions. In this respect the primary odors are like the three primary colors (red, green and blue) and the four primary tastes (sweet, salt, sour and bitter).

To match the seven primary odors there must be seven different kinds of olfactory receptors in the nose. We can picture the receptor sites as ultramicroscopic slots or hollows in the nerve-fiber membrane, each of a distinctive shape and size. Presumably each will accept a molecule of the appropriate configuration, just as a socket takes a plug. Some molecules may be able to fit into two different sockets—broadside into a wide receptor or end on into a narrow one. In such cases the substance, with its molecules occupying both types of receptor, may indicate a complex odor to the brain.

The next problem was to learn the shapes of the seven receptor sites. This was begun by examining the structural formulas of the camphoraceous compounds and constructing models of their molecules. Thanks to the techniques of modern stereochemistry, which explore the structure of molecules with the aid of X-ray diffraction, infrared spectroscopy, the electron-beam probe and other means, it is possible to build a three-dimensional model of the molecule of any chemical compound once its structural formula is known. There are rules for building these models; also available are building blocks (sets of atomic units) on a scale 100 million times actual size.

As the models of the camphoraceous molecules took form, it soon became clear that they all had about the same shape: they were roughly spherical. Not only that, it turned out that when the models were translated into molecular dimensions, all the molecules also had about the same diameter: approximately seven angstrom units. (An angstrom unit is a ten-millionth of a millimeter.) This meant that the receptor site for camphoraceous molecules must be a hemispherical bowl about seven angstroms in diameter. Many of the camphoraceous molecules are rigid spheres that would inevitably fit into such a bowl; the others are slightly flexible and could easily shape themselves to the bowl.

When other models were built, shapes and sizes of the molecules representing the other primary odors were found [see illustration on preceding two pages]. The musky odor is accounted for by molecules with the shape of a disk about 10 angstroms in diameter. The pleasant floral odor is caused by molecules that have the shape of a disk with a flexible tail attached—a shape somewhat like a kite. The cool pepperminty odor is produced by molecules with the shape of a wedge, and with an electrically polarized group of atoms, capable of forming a hydrogen bond, near the point of the wedge. The ethereal odor is due to rod-shaped or other thin molecules. In each of these cases the receptor site in the nerve endings presumably has a shape and size corresponding to those of the molecule.

The pungent and putrid odors seem to be exceptions to the Lucretian scheme of shape-matching. The molecules responsible for these odors are of indifferent shapes and sizes; what matters in their case is the electric charge of the molecule. The pungent class of odors is produced by compounds whose molecules, because of a deficiency of electrons, have a positive charge and a strong affinity for electrons; they are called electrophilic. Putrid odors, on the other hand, are caused by molecules

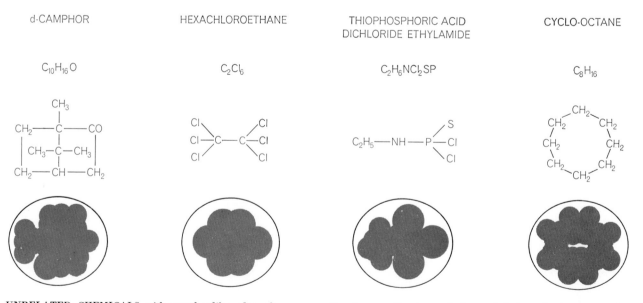

d-CAMPHOR	HEXACHLOROETHANE	THIOPHOSPHORIC ACID DICHLORIDE ETHYLAMIDE	CYCLO-OCTANE
$C_{10}H_{16}O$	C_2Cl_6	$C_2H_6NCl_2SP$	C_8H_{16}

UNRELATED CHEMICALS with camphor-like odors show no resemblance in empirical formulas and little in structural formulas. Yet, because the size and shape of their molecules are similar, they all fit the bowl-shaped receptor for camphoraceous molecules.

that have an excess of electrons and are called nucleophilic, because they are strongly attracted by the nuclei of adjacent atoms.

A theory is useful only if it can be tested in some way by experiment. One of the virtues of the stereochemical theory is that it suggests some very specific and unambiguous tests. It has been subjected to six severe tests of its accuracy so far and has passed each of them decisively.

To start with, it is at once obvious that from the shape of a molecule we should be able to predict its odor. Suppose, then, that we synthesize molecules of certain shapes and see whether or not they produce the odors predicted for them.

Consider a molecule consisting of three chains attached to a single carbon atom, with the central atom's fourth bond occupied only by a hydrogen atom [see top illustration at right]. This molecule might fit into a kite-shaped site (floral odor), a wedge-shaped site (pepperminty) or, by means of one of its chains, a rod-shaped site (ethereal). The theory predicts that the molecule should therefore have a fruity odor composed of these three primaries. Now suppose we substitute the comparatively bulky methyl group (CH_3) in place of the small hydrogen atom at the fourth bond of the carbon atom. The introduction of a fourth branch will prevent the molecule from fitting so easily into a kite-shaped or wedge-shaped site, but one of the branches should still be able to occupy a rod-shaped site. As a result, the theory predicts, the ether smell should now predominate.

Another of us (Rubin) duly synthesized the two structures in his laboratory at the Georgetown University School of Medicine. The third author (Johnston), also working at the Georgetown School of Medicine, then submitted the products to a panel of trained smellers. He used an instrument called the olfactometer, which by means of valves and controlled air streams delivers carefully measured concentrations of odors, singly or mixed, to the observer. The amount of odorous vapor delivered was measured by gas chromatography. A pair of olfactometers was used, one for each of the two compounds under test, and the observer was asked to sniff alternately from each.

The results verified the predictions. The panel reported that Compound A had a fruity (actually grapelike) odor, and that Compound B, with the methyl

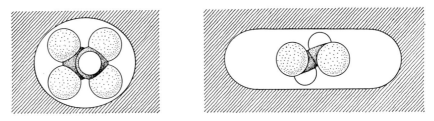

CHANGE IN SHAPE of a molecule changed its odor. The molecule at left smelled fruity because it fitted into three sites. When it was modified (right) by the substitution of a methyl group for a hydrogen, it smelled somewhat ethereal. Presumably the methyl branch made it fit two of the original sites less well but allowed it still to fit the ethereal slot.

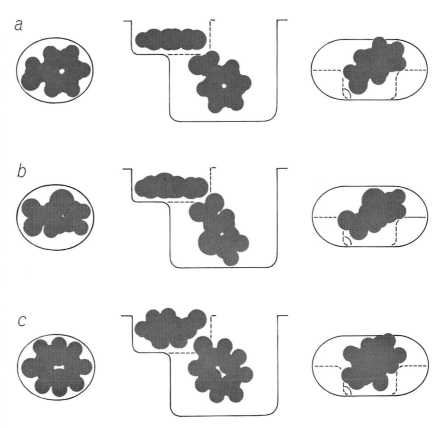

SINGLE CHEMICAL has more than one primary odor if its molecule can fit more than one site. Acetylenetetrabromide, for example, is described as smelling both camphoraceous and ethereal. It turns out that its molecule can fit either site, depending on how it lies.

COMPLEX ODORS are made up of several primaries. Three molecules with an almond odor are illustrated: benzaldehyde (a), alpha-nitrothiophen (b) and cyclo-octanone (c). Each of them fits (left to right) camphoraceous, floral (with two molecules) and pepperminty sites.

group substituted for the hydrogen atom, had a pronounced tinge of the ether-like odor. This experiment, and the theory behind it, make understandable the earlier finding that the odor of certain benzene-ring compounds changes sharply when the position of a group of atoms is shifted. The change in odor is due to the change in the over-all shape of the molecule.

A second test suggested itself. Could a complex odor found in nature be matched by putting together a combination of primary odors? Taking the odor of cedarwood oil as a test case, Amoore found that chemicals known to possess this odor had molecular shapes that would fit into the receptor sites for the camphoraceous, musky, floral and pepperminty odors. Johnston proceeded to try various combinations of these four primaries to duplicate the cedarwood odor. He tested each mixture on eight trained observers, who compared the synthetic odor with that of cedarwood oil. After 86 attempts he was able to produce a blend that closely matched the natural cedarwood odor. With the same four primaries he also succeeded in synthesizing a close match for the odor of sandalwood oil.

The next two tests had to do with the identification of pure (that is, primary) odors. If the theory was correct, a molecule that would fit only into a receptor site of a particular shape and size, and no other, should represent a primary odor in pure form. Molecules of the same shape and size should smell very much alike; those of a different primary shape should smell very different. Human subjects were tested on this point. Presented with the odors from a pair of different substances whose molecules nonetheless had the same primary shape (for example, that of the floral odor), the subjects judged the two odors to be highly similar to each other. When the pair of

substances presented had the pure molecular traits of different categories (for instance, the kite shape of the floral odor and the nucleophilic charge characteristic of putrid compounds), the subjects found the odors extremely dissimilar.

Johnston went on to make the same sort of test with honeybees. He set up an experiment designed to test their ability to discriminate between two odors, one of which was "right" (associated with sugar sirup) and the other "wrong" (associated with an electric shock). The pair of odors might be in the same primary group or in different primary groups (for example, floral and pepperminty). At pairs of scented vials on a table near the hive, the bees were first conditioned to the fact that one odor of a pair was right and the other was wrong. Then the sirup bait in the vials was replaced with distilled water and freshly deodorized scent vials were substituted for those used during the training period. The visits of the marked bees to the respective vials in search of sirup were counted. It could be assumed that they would tend to visit the odor to which they had been favorably conditioned and to avoid the one that had been associated with electric shock, provided that they could distinguish between the two.

So tested, the honeybees clearly showed that they had difficulty in detecting a difference between two scents within the same primary group (say pepperminty) but were able to distinguish easily between different primaries (pepperminty and floral). In the latter case they almost invariably chose the correct scent without delay. These experiments indicate that the olfactory system of the honeybee, like that of human beings, is based on the stereochemical principle, although the bee's smelling organ is different; it smells not with a nose but with antennae. Apparently the receptor sites on the antennae are differentiated by shape in the same way as those in the human nose.

A fifth test was made with human observers trained in odor discrimination. Suppose they were presented with a number of substances that were very different chemically but whose molecules had about the same over-all shape. Would all these dissimilar compounds smell alike? Five compounds were used for the test. They belonged to three different chemical families differing radically from one another in the internal structure of their molecules but in all five cases had the disk shape characteristic of the molecules of musky-odored substances. The observers, exposed to

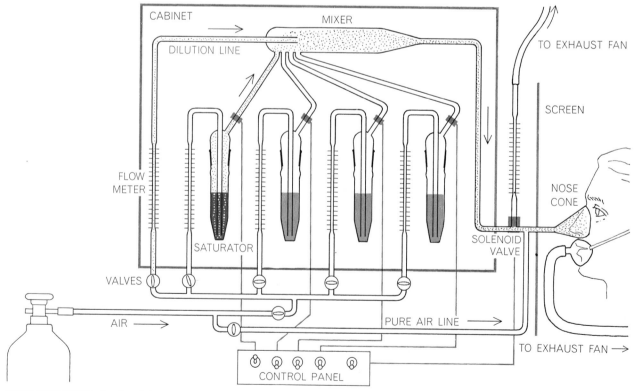

OLFACTOMETER developed by one of the authors (Johnston) mixes odors in precise proportions and delivers them to a nose cone for sampling. This schematic diagram shows the main elements. Air bubbles through a liquid in one of the saturators, picks up odorous molecules and is then diluted with pure air or mixed with air carrying other odors. The experimenter controls the solenoid valves.

the vapors of these five chemicals among many others by means of the olfactometer, did indeed pick out and identify all five as musky. By the odor test, however, they were often unable to distinguish these five quite different chemicals from one another.

Basically all this evidence in favor of the stereochemical theory was more or less indirect. One would like some sort of direct proof of the actual existence of differentiated receptor sites in the smelling organ. Recently R. C. Gesteland, then at the Massachusetts Institute of Technology, searched for such evidence. He devised a way to tap the electric impulses from single olfactory-nerve cells by means of microelectrodes. Applying his electrodes to the olfactory organ of the frog, Gesteland presented various odors to the organ and tapped the olfactory cells one by one to see if they responded with electric impulses. He found that different cells responded selectively to different odors, and his exploration indicated that the frog has about eight such different receptors. What is more, five of these receivers correspond closely to five of the odors (camphoraceous, musky, ethereal, pungent and putrid) identified as primary in the stereochemical theory! This finding, then, can be taken as a sixth and independent confirmation of the theory.

Equipped now with a tested basic theory to guide further research, we can hope for much faster progress in the science of osmics (smell) than has been possible heretofore. This may lead to unexpected benefits for mankind. For man the sense of smell may perhaps have become less essential as a life-and-death organ than it is for lower animals, but we still depend on this sense much more than we realize. One can gain some appreciation of the importance of smell to man by reflecting on how tasteless food becomes when the nose is blocked by a head cold and on how unpleasantly we are affected by a bad odor in drinking water or a closed room. Control of odor is fundamental in our large perfume, tobacco and deodorant industries. No doubt odor also affects our lives in many subtle ways of which we are not aware.

The accelerated research for which the way is now open should make it possible to analyze in fine detail the complex flavors in our food and drink, to get rid of obnoxious odors, to develop new fragrances and eventually to synthesize any odor we wish, whether to defeat pests or to delight the human nose.

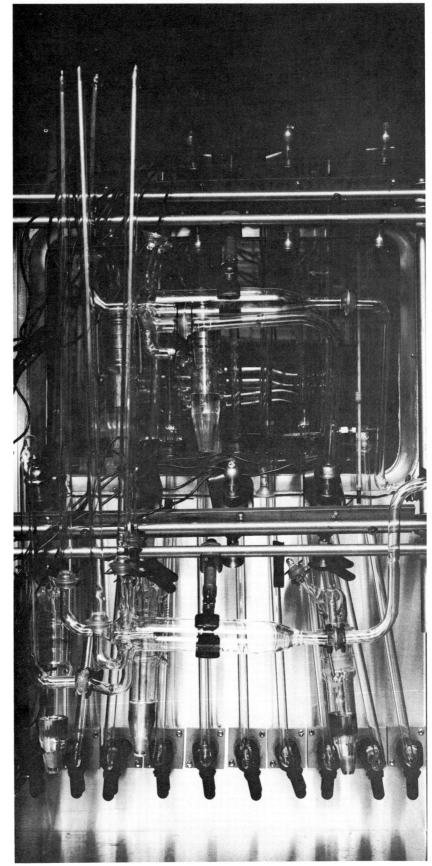

CONSTANT-TEMPERATURE CABINET maintains the olfactometer parts at 77 degrees Fahrenheit. The photograph shows the interior of the cabinet, containing two units of the type diagramed on the opposite page. Several of the saturators are visible, as are two mixers (*horizontal glass vessels*), each of them connected by tubing to a nose cone at right.

7

PHEROMONES

EDWARD O. WILSON
May 1963

A pheromone is a substance secreted by an animal that influences the behavior of other animals of the same species. Recent studies indicate that such chemical communication is surprisingly common

It is conceivable that somewhere on other worlds civilizations exist that communicate entirely by the exchange of chemical substances that are smelled or tasted. Unlikely as this may seem, the theoretical possibility cannot be ruled out. It is not difficult to design, on paper at least, a chemical communication system that can transmit a large amount of information with rather good efficiency. The notion of such a communication system is of course strange because our outlook is shaped so strongly by our own peculiar auditory and visual conventions. This limitation of outlook is found even among students of animal behavior; they have favored species whose communication methods are similar to our own and therefore more accessible to analysis. It is becoming increasingly clear, however, that chemical systems provide the dominant means of communication in many animal species, perhaps even in most. In the past several years animal behaviorists and organic chemists, working together, have made a start at deciphering some of these systems and have discovered a number of surprising new biological phenomena.

In earlier literature on the subject, chemicals used in communication were usually referred to as "ectohormones." Since 1959 the less awkward and etymologically more accurate term "pheromones" has been widely adopted. It is used to describe substances exchanged among members of the same animal species. Unlike true hormones, which are secreted internally to regulate the organism's own physiology, or internal environment, pheromones are secreted externally and help to regulate the organism's external environment by influencing other animals. The mode of influence can take either of two general forms. If the pheromone produces a more or less immediate and reversible change

in the behavior of the recipient, it is said to have a "releaser" effect. In this case the chemical substance seems to act directly on the recipient's central nervous system. If the principal function of the pheromone is to trigger a chain of physiological events in the recipient, it has what we have recently labeled a "primer" effect. The physiological changes, in turn, equip the organism with a new behavioral repertory, the components of which are thenceforth evoked by appropriate stimuli. In termites, for example, the reproductive and soldier castes prevent other termites from developing into

their own castes by secreting substances that are ingested and act through the *corpus allatum,* an endocrine gland controlling differentiation [see "The Termite and the Cell," by Martin Lüscher; SCIENTIFIC AMERICAN, May, 1953].

These indirect primer pheromones do not always act by physiological inhibition. They can have the opposite effect. Adult males of the migratory locust *Schistocerca gregaria* secrete a volatile substance from their skin surface that accelerates the growth of young locusts. When the nymphs detect this substance with their antennae, their hind legs,

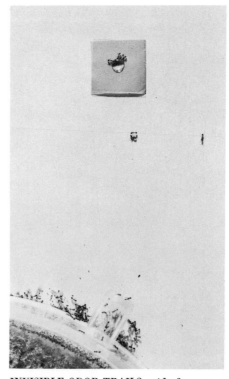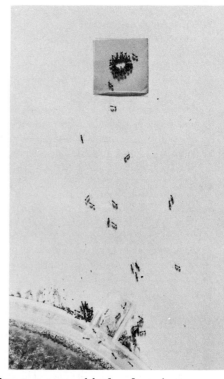

INVISIBLE ODOR TRAILS guide fire ant workers to a source of food: a drop of sugar solution. The trails consist of a pheromone laid down by workers returning to their nest after finding a source of food. Sometimes the chemical message is reinforced by the touching of antennae if a returning worker meets a wandering fellow along the way. This is hap-

some of their mouth parts and the antennae themselves vibrate. The secretion, in conjunction with tactile and visual signals, plays an important role in the formation of migratory locust swarms.

A striking feature of some primer pheromones is that they cause important physiological change without an immediate accompanying behavioral response, at least none that can be said to be peculiar to the pheromone. Beginning in 1955 with the work of S. van der Lee and L. M. Boot in the Netherlands, mammalian endocrinologists have discovered several unexpected effects on the female mouse that are produced by odors of other members of the same species. These changes are not marked by any immediate distinctive behavioral patterns. In the "Lee-Boot effect" females placed in groups of four show an increase in the percentage of pseudopregnancies. A completely normal reproductive pattern can be restored by removing the olfactory bulbs of the mice or by housing the mice separately. When more and more female mice are forced to live together, their oestrous cycles become highly irregular and in most of the mice the cycle stops completely for long periods. Recently W. K. Whitten of the Australian National University has discovered that the odor of a male mouse can initiate and

synchronize the oestrous cycles of female mice. The male odor also reduces the frequency of reproductive abnormalities arising when female mice are forced to live under crowded conditions.

A still more surprising primer effect has been found by Helen Bruce of the National Institute for Medical Research in London. She observed that the odor of a strange male mouse will block the pregnancy of a newly impregnated female mouse. The odor of the original stud male, of course, leaves pregnancy undisturbed. The mouse reproductive pheromones have not yet been identified chemically, and their mode of action is only partly understood. There is evidence that the odor of the strange male suppresses the secretion of the hormone prolactin, with the result that the *corpus luteum* (a ductless ovarian gland) fails to develop and normal oestrus is restored. The pheromones are probably part of the complex set of control mechanisms that regulate the population density of animals [see "Population Density and Social Pathology," by John B. Calhoun; SCIENTIFIC AMERICAN Offprint 506].

Pheromones that produce a simple releaser effect—a single specific response mediated directly by the central nervous system—are widespread in the

animal kingdom and serve a great many functions. Sex attractants constitute a large and important category. The chemical structures of six attractants are shown on page 77. Although two of the six—the mammalian scents muskone and civetone—have been known for some 40 years and are generally assumed to serve a sexual function, their exact role has never been rigorously established by experiments with living animals. In fact, mammals seem to employ musklike compounds, alone or in combination with other substances, to serve several functions: to mark home ranges, to assist in territorial defense and to identify the sexes.

The nature and role of the four insect sex attractants are much better understood. The identification of each represents a technical feat of considerable magnitude. To obtain 12 milligrams of esters of bombykol, the sex attractant of the female silkworm moth, Adolf F. J. Butenandt and his associates at the Max Planck Institute of Biochemistry in Munich had to extract material from 250,000 moths. Martin Jacobson, Morton Beroza and William Jones of the U.S. Department of Agriculture processed 500,000 female gypsy moths to get 20 milligrams of the gypsy-moth attractant gyplure. Each moth yielded only about .01 microgram (millionth of a gram) of

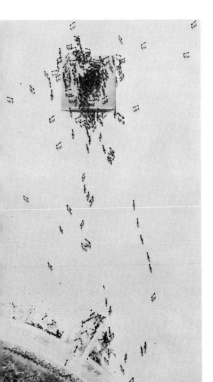

pening in the photograph at the far left. A few foraging workers have just found the sugar drop and a returning trail-layer is communicating the news to another ant. In the next two pictures the trail has been completed and workers stream from the nest in increasing numbers. In the fourth picture unrewarded workers return to the nest without laying trails and outward-bound traffic wanes. In the last picture most of the trails have evaporated completely and only a few stragglers remain at the site, eating the last bits of food.

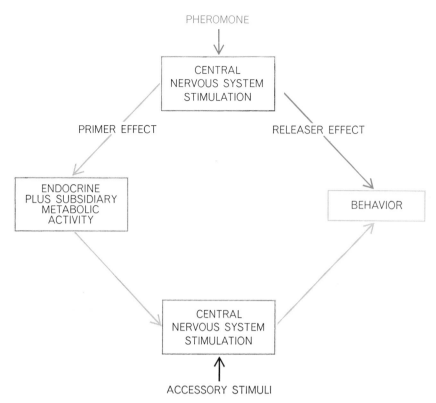

PHEROMONES INFLUENCE BEHAVIOR directly or indirectly, as shown in this schematic diagram. If a pheromone stimulates the recipient's central nervous system into producing an immediate change in behavior, it is said to have a "releaser" effect. If it alters a set of long-term physiological conditions so that the recipient's behavior can subsequently be influenced by specific accessory stimuli, the pheromone is said to have a "primer" effect.

gyplure, or less than a millionth of its body weight. Bombykol and gyplure were obtained by killing the insects and subjecting crude extracts of material to chromatography, the separation technique in which compounds move at different rates through a column packed with a suitable adsorbent substance. Another technique has been more recently developed by Robert T. Yamamoto of the U.S. Department of Agriculture, in collaboration with Jacobson and Beroza, to harvest the equally elusive sex attractant of the American cockroach. Virgin females were housed in metal cans and air was continuously drawn through the cans and passed through chilled containers to condense any vaporized materials. In this manner the equivalent of 10,000 females were "milked" over a nine-month period to yield 12.2 milligrams of what was considered to be the pure attractant.

The power of the insect attractants is almost unbelievable. If some 10,000 molecules of the most active form of bombykol are allowed to diffuse from a source one centimeter from the antennae of a male silkworm moth, a characteristic sexual response is obtained in most cases. If volatility and diffusion rate

are taken into account, it can be estimated that the threshold concentration is no more than a few hundred molecules per cubic centimeter, and the actual number required to stimulate the male is probably even smaller. From this one can calculate that .01 microgram of gyplure, the minimum average content of a single female moth, would be theoretically adequate, if distributed with maximum efficiency, to excite more than a billion male moths.

In nature the female uses her powerful pheromone to advertise her presence over a large area with a minimum expenditure of energy. With the aid of published data from field experiments and newly contrived mathematical models of the diffusion process, William H. Bossert, one of my associates in the Biological Laboratories at Harvard University, and I have deduced the shape and size of the ellipsoidal space within which male moths can be attracted under natural conditions [see bottom illustration on opposite page]. When a moderate wind is blowing, the active space has a long axis of thousands of meters and a transverse axis parallel to the ground of more than 200 meters at the widest point. The 19th-century

French naturalist Jean Henri Fabre, speculating on sex attraction in insects, could not bring himself to believe that the female moth could communicate over such great distances by odor alone, since "one might as well expect to tint a lake with a drop of carmine." We now know that Fabre's conclusion was wrong but that his analogy was exact: to the male moth's powerful chemoreceptors the lake is indeed tinted.

One must now ask how the male moth, smelling the faintly tinted air, knows which way to fly to find the source of the tinting. He cannot simply fly in the direction of increasing scent; it can be shown mathematically that the attractant is distributed almost uniformly after it has drifted more than a few meters from the female. Recent experiments by Ilse Schwinck of the University of Munich have revealed what is probably the alternative procedure used. When male moths are activated by the pheromone, they simply fly upwind and thus inevitably move toward the female. If by accident they pass out of the active zone, they either abandon the search or fly about at random until they pick up the scent again. Eventually, as they approach the female, there is a slight increase in the concentration of the chemical attractant and this can serve as a guide for the remaining distance.

If one is looking for the most highly developed chemical communication systems in nature, it is reasonable to study the behavior of the social insects, particularly the social wasps, bees, termites and ants, all of which communicate mostly in the dark interiors of their nests and are known to have advanced chemoreceptive powers. In recent years experimental techniques have been developed to separate and identify the pheromones of these insects, and rapid progress has been made in deciphering the hitherto intractable codes, particularly those of the ants. The most successful procedure has been to dissect out single glandular reservoirs and see what effect their contents have on the behavior of the worker caste, which is the most numerous and presumably the most in need of continuing guidance. Other pheromones, not present in distinct reservoirs, are identified in chromatographic fractions of crude extracts.

Ants of all castes are constructed with an exceptionally well-developed exocrine glandular system. Many of the most prominent of these glands, whose function has long been a mystery to entomologists, have now been identified as the source of pheromones [see illustra-

tion on page 75]. The analysis of the gland-pheromone complex has led to the beginnings of a new and deeper understanding of how ant societies are organized.

Consider the chemical trail. According to the traditional view, trail secretions served as only a limited guide for worker ants and had to be augmented by other kinds of signals exchanged inside the nest. Now it is known that the trail substance is extraordinarily versatile. In the fire ant (*Solenopsis saevissima*), for instance, it functions both to activate and to guide foraging workers in search of food and new nest sites. It also contributes as one of the alarm signals emitted by workers in distress. The trail of the fire ant consists of a substance secreted in minute amounts by Dufour's gland; the substance leaves the ant's body by way of the extruded sting, which is touched intermittently to the ground much like a moving pen dispensing ink. The trail pheromone, which has not yet been chemically identified, acts primarily to attract the fire ant workers. Upon encountering the attractant the workers move automatically up the gradient to the source of emission. When the substance is drawn out in a line, the workers run along the direction of the line away from the nest. This simple response brings them to the food source or new nest site from which the trail is laid. In our laboratory we have extracted the pheromone from the Dufour's glands of freshly killed workers and have used it to create artificial trails. Groups of workers will follow these trails away from the nest and along arbitrary routes (including circles leading back to the nest) for considerable periods of time. When the pheromone is presented to whole colonies in massive doses, a large portion of the colony, including the queen, can be drawn out in a close simulation of the emigration process.

The trail substance is rather volatile, and a natural trail laid by one worker diffuses to below the threshold concentration within two minutes. Consequently outward-bound workers are able to follow it only for the distance they can travel in this time, which is about 40 centimeters. Although this strictly limits the distance over which the ants can communicate, it provides at least two important compensatory advantages. The more obvious advantage is that old, useless trails do not linger to confuse the hunting workers. In addition, the intensity of the trail laid by many workers provides a sensitive index of the amount of food at a given site and the rate of its depletion. As workers move to and from

ANTENNAE OF GYPSY MOTHS differ radically in structure according to their function. In the male (*left*) they are broad and finely divided to detect minute quantities of sex attractant released by the female (*right*). The antennae of the female are much less developed.

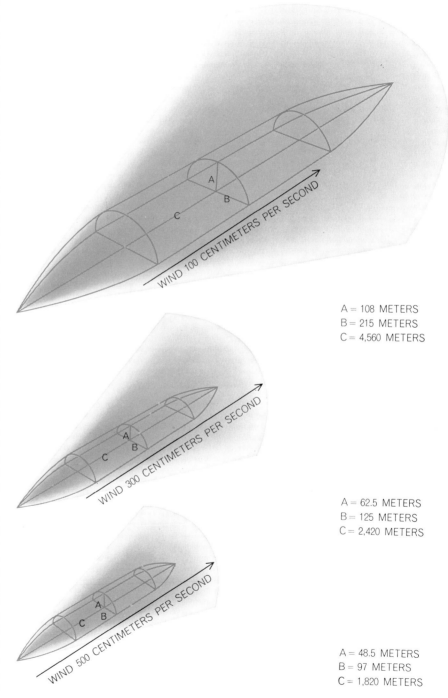

A = 108 METERS
B = 215 METERS
C = 4,560 METERS

A = 62.5 METERS
B = 125 METERS
C = 2,420 METERS

A = 48.5 METERS
B = 97 METERS
C = 1,820 METERS

ACTIVE SPACE of gyplure, the gypsy moth sex attractant, is the space within which this pheromone is sufficiently dense to attract males to a single, continuously emitting female. The actual dimensions, deduced from linear measurements and general gas-diffusion models, are given at right. Height (*A*) and width (*B*) are exaggerated in the drawing. As wind shifts from moderate to strong, increased turbulence contracts the active space.

FIRE ANT WORKER lays an odor trail by exuding a pheromone along its extended sting. The sting is touched to the ground periodically, breaking the trail into a series of streaks.

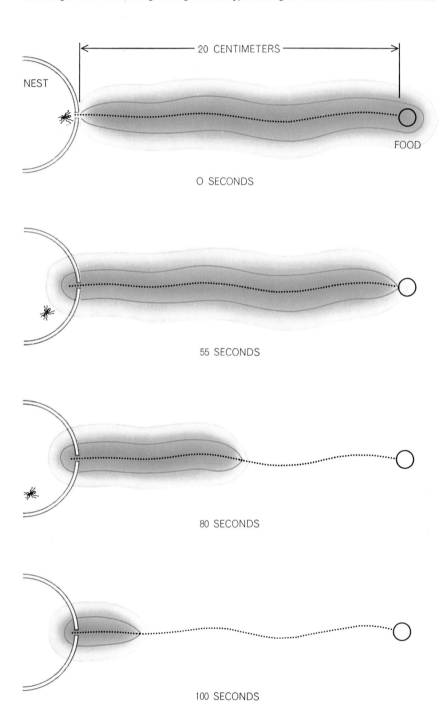

ACTIVE SPACE OF ANT TRAIL, within which the pheromone is dense enough to be perceived by other workers, is narrow and nearly constant in shape with the maximum gradient situated near its outer surface. The rapidity with which the trail evaporates is indicated.

the food finds (consisting mostly of dead insects and sugar sources) they continuously add their own secretions to the trail produced by the original discoverers of the food. Only if an ant is rewarded by food does it lay a trail on its trip back to the nest; therefore the more food encountered at the end of the trail, the more workers that can be rewarded and the heavier the trail. The heavier the trail, the more workers that are drawn from the nest and arrive at the end of the trail. As the food is consumed, the number of workers laying trail substance drops, and the old trail fades by evaporation and diffusion, gradually constricting the outward flow of workers.

The fire ant odor trail shows other evidences of being efficiently designed. The active space within which the pheromone is dense enough to be perceived by workers remains narrow and nearly constant in shape over most of the length of the trail. It has been further deduced from diffusion models that the maximum gradient must be situated near the outer surface of the active space. Thus workers are informed of the space boundary in a highly efficient way. Together these features ensure that the following workers keep in close formation with a minimum chance of losing the trail.

The fire ant trail is one of the few animal communication systems whose information content can be measured with fair precision. Unlike many communicating animals, the ants have a distinct goal in space—the food find or nest site—the direction and distance of which must both be communicated. It is possible by a simple technique to measure how close trail-followers come to the trail end, and, by making use of a standard equation from information theory, one can translate the accuracy of their response into the "bits" of information received. A similar procedure can be applied (as first suggested by the British biologist J. B. S. Haldane) to the "waggle dance" of the honeybee, a radically different form of communication system from the ant trail [see "Dialects in the Language of the Bees," by Karl von Frisch; SCIENTIFIC AMERICAN Offprint 130]. Surprisingly, it turns out that the two systems, although of wholly different evolutionary origin, transmit about the same amount of information with reference to distance (two bits) and direction (four bits in the honeybee, and four or possibly five in the ant). Four bits of information will direct an ant or a bee into one of 16 equally probable sectors of a circle and two bits will identify one of four equally probable distances. It is conceivable that these information values represent the maximum that can be achieved with the insect brain and sensory apparatus.

Not all kinds of ants lay chemical trails. Among those that do, however, the pheromones are highly species-specific in their action. In experiments in which artificial trails extracted from one species were directed to living colonies of other species, the results have almost always been negative, even among related species. It is as if each species had its own private language. As a result there is little or no confusion when the trails of two or more species cross.

Another important class of ant pheromone is composed of alarm substances. A simple backyard experiment will show that if a worker ant is disturbed by a clean instrument, it will, for a short time, excite other workers with whom it comes in contact. Until recently most students of ant behavior thought that

the alarm was spread by touch, that one worker simply jostled another in its excitement or drummed on its neighbor with its antennae in some peculiar way. Now it is known that disturbed workers discharge chemicals, stored in special glandular reservoirs, that can produce all the characteristic alarm responses solely by themselves. The chemical structure of four alarm substances is shown on page 79. Nothing could illustrate more clearly the wide differences between the human perceptual world and that of chemically communicating animals. To the human nose the alarm substances are mild or even pleasant, but to the ant they represent an urgent tocsin that can propel a colony into violent and instant action.

As in the case of the trail substances, the employment of the alarm substances appears to be ideally designed for the purpose it serves. When the contents of the mandibular glands of a worker of the harvesting ant (*Pogonomyrmex badius*)

are discharged into still air, the volatile material forms a rapidly expanding sphere, which attains a radius of about six centimeters in 13 seconds. Then it contracts until the signal fades out completely some 35 seconds after the moment of discharge. The outer shell of the active space contains a low concentration of pheromone, which is actually attractive to harvester workers. This serves to draw them toward the point of disturbance. The central region of the active space, however, contains a concentration high enough to evoke the characteristic frenzy of alarm. The "alarm sphere" expands to a radius of about three centimeters in eight seconds and, as might be expected, fades out more quickly than the "attraction sphere."

The advantage to the ants of an alarm signal that is both local and short-lived becomes obvious when a *Pogonomyrmex* colony is observed under natural conditions. The ant nest is subject to almost innumerable minor disturbances. If the

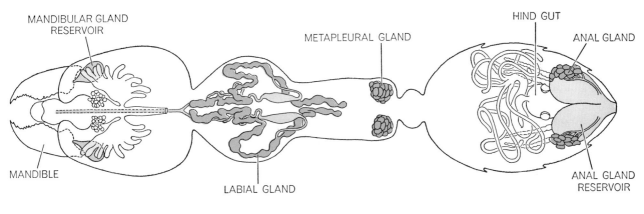

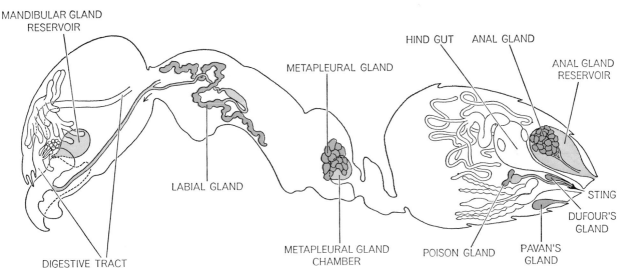

EXOCRINE GLANDULAR SYSTEM of a worker ant (*shown here in top and side cutaway views*) is specially adapted for the production of chemical communication substances. Some pheromones are stored in reservoirs and released in bursts only when needed; others are secreted continuously. Depending on the species, trail substances are produced by Dufour's gland, Pavan's gland or the poison glands; alarm substances are produced by the anal and mandibular glands. The glandular sources of other pheromones are unknown.

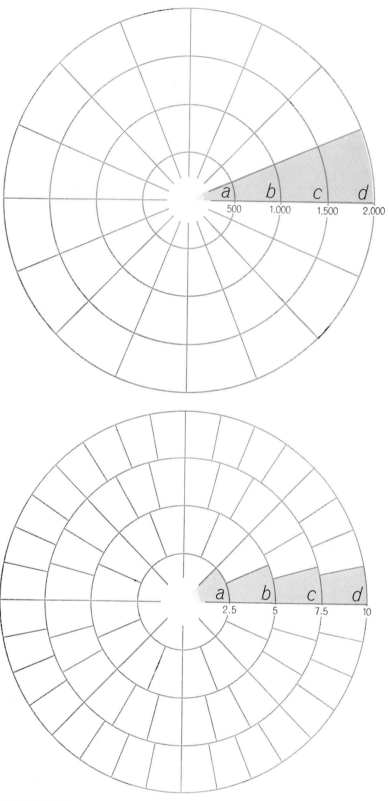

FORAGING INFORMATION conveyed by two different insect communication systems can be represented on two similar "compass" diagrams. The honeybee "waggle dance" (*top*) transmits about four bits of information with respect to direction, enabling a honeybee worker to pinpoint a target within one of 16 equally probable angular sectors. The number of "bits" in this case remains independent of distance, given in meters. The pheromone system used by trail-laying fire ants (*bottom*) is superior in that the amount of directional information increases with distance, given in centimeters. At distances *c* and *d*, the probable sector in which the target lies is smaller for ants than for bees. (For ants, directional information actually increases gradually and not by jumps.) Both insects transmit two bits of distance information, specifying one of four equally probable distance ranges.

alarm spheres generated by individual ant workers were much wider and more durable, the colony would be kept in ceaseless and futile turmoil. As it is, local disturbances such as intrusions by foreign insects are dealt with quickly and efficiently by small groups of workers, and the excitement soon dies away.

The trail and alarm substances are only part of the ants' chemical vocabulary. There is evidence for the existence of other secretions that induce gathering and settling of workers, acts of grooming, food exchange, and other operations fundamental to the care of the queen and immature ants. Even dead ants produce a pheromone of sorts. An ant that has just died will be groomed by other workers as if it were still alive. Its complete immobility and crumpled posture by themselves cause no new response. But in a day or two chemical decomposition products accumulate and stimulate the workers to bear the corpse to the refuse pile outside the nest. Only a few decomposition products trigger this funereal response; they include certain long-chain fatty acids and their esters. When other objects, including living workers, are experimentally daubed with these substances, they are dutifully carried to the refuse pile. After being dumped on the refuse the "living dead" scramble to their feet and promptly return to the nest, only to be carried out again. The hapless creatures are thrown back on the refuse pile time and again until most of the scent of death has been worn off their bodies by the ritual.

Our observation of ant colonies over long periods has led us to believe that as few as 10 pheromones, transmitted singly or in simple combinations, might suffice for the total organization of ant society. The task of separating and characterizing these substances, as well as judging the roles of other kinds of stimuli such as sound, is a job largely for the future.

Even in animal species where other kinds of communication devices are prominently developed, deeper investigation usually reveals the existence of pheromonal communication as well. I have mentioned the auxiliary roles of primer pheromones in the lives of mice and migratory locusts. A more striking example is the communication system of the honeybee. The insect is celebrated for its employment of the "round" and "waggle" dances (augmented, perhaps, by auditory signals) to designate the location of food and new nest sites. It is not so widely known that chemical signals play equally important roles in other aspects of honeybee life. The mother queen regulates the reproductive cycle of the colony by secreting from her mandibular glands a substance recently identified as 9-ketodecanoic acid. When this pheromone is ingested by the worker bees, it inhibits development of their ovaries and also their ability to manufacture the royal cells in which new queens are reared. The same pheromone serves as a sex attractant in the queen's nuptial flights.

Under certain conditions, including the discovery of new food sources, worker bees release geraniol, a pleasant-smelling alcohol, from the abdominal Nassanoff glands. As the geraniol diffuses through the air it attracts other workers and so supplements information contained in the waggle dance. When a worker stings an intruder, it discharges, in addition to the venom, tiny amounts of a secretion from clusters of unicellular glands located next to the basal plates of the sting. This secretion is responsible for the tendency, well known to bee-keepers, of angry swarms of workers to sting at the same spot. One component, which acts as a simple attractant, has been identified as isoamyl acetate, a compound that has a banana-like odor. It is possible that the stinging response is evoked by at least one unidentified alarm substance secreted along with the attractant.

Knowledge of pheromones has advanced to the point where one can make some tentative generalizations about their chemistry. In the first place, there appear to be good reasons why sex attractants should be compounds that contain between 10 and 17 carbon atoms and that have molecular weights between about 180 and 300—the range actually observed in attractants so far identified. (For comparison, the weight of a single carbon atom is 12.) Only compounds of roughly this size or greater can meet the two known requirements of a sex attractant: narrow specificity, so that only members of one species will respond to it, and high potency. Compounds that contain fewer than five or so carbon atoms and that have a molecular weight of less than about 100 cannot be assembled in enough different ways to provide a distinctive molecule for all the insects that want to advertise their presence.

It also seems to be a rule, at least with insects, that attraction potency increases with molecular weight. In one series of esters tested on flies, for instance, a doubling of molecular weight resulted in as much as a thousandfold increase in efficiency. On the other hand, the molecule cannot be too large and complex or it will be prohibitively difficult for the insect to synthesize. An equally important limitation on size is

BOMBYKOL (SILKWORM MOTH)

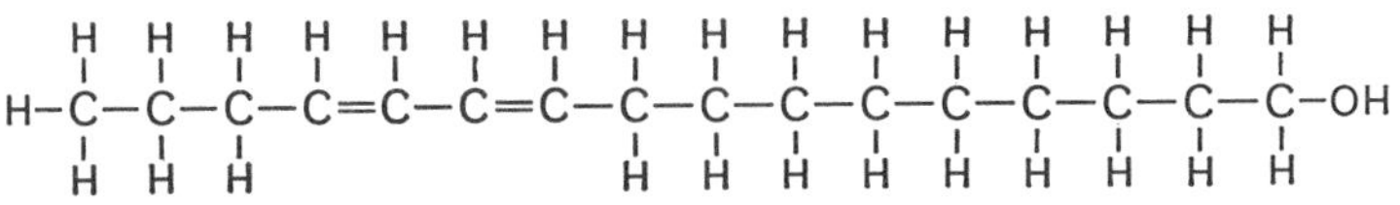

GYPLURE (GYPSY MOTH)

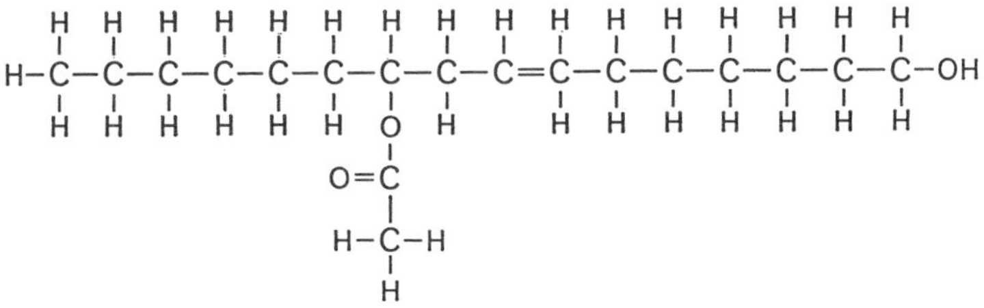

2,2-DIMETHYL-3-ISOPROPYLIDENECYCLOPROPYL PROPIONATE (AMERICAN COCKROACH)

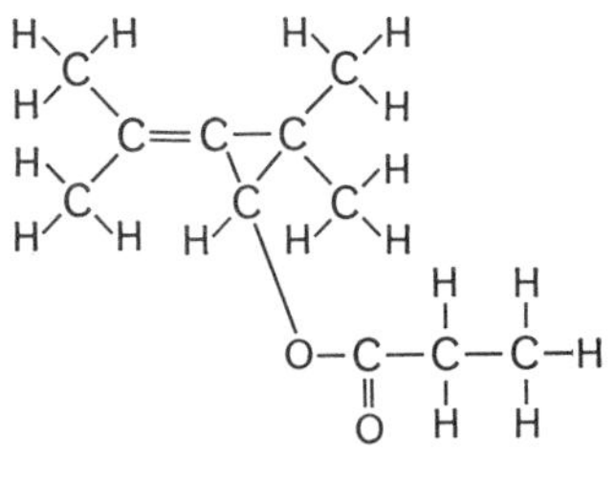

HONEYBEE QUEEN SUBSTANCE

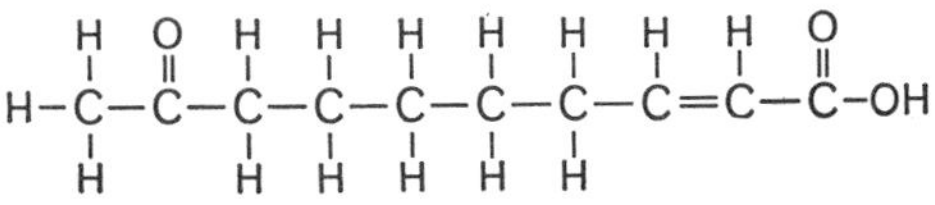

CIVETONE (CIVET)

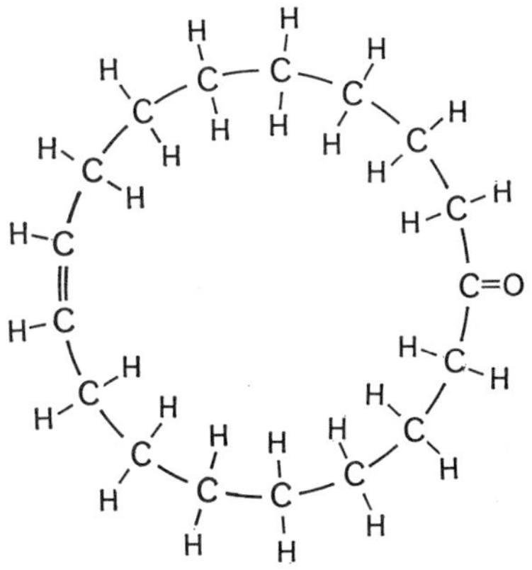

MUSKONE (MUSK DEER)

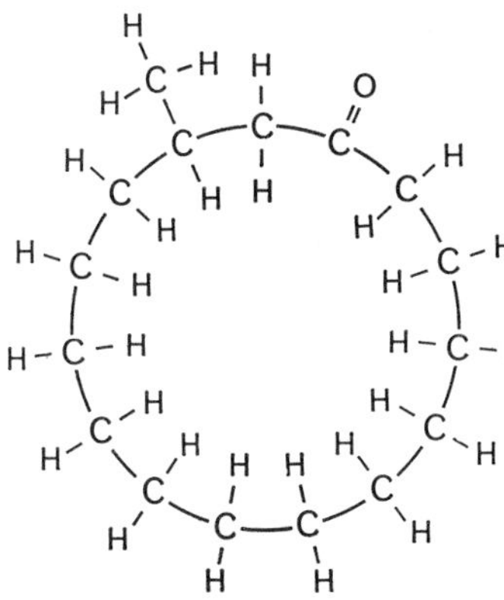

SIX SEX PHEROMONES include the identified sex attractants of four insect species as well as two mammalian musks generally believed to be sex attractants. The high molecular weight of most sex pheromones accounts for their narrow specificity and high potency.

the fact that volatility—and, as a result, diffusibility—declines with increasing molecular weight.

One can also predict from first principles that the molecular weight of alarm substances will tend to be less than those of the sex attractants. Among the ants there is little specificity; each species responds strongly to the alarm substances of other species. Furthermore, an alarm substance, which is used primarily within the confines of the nest, does not need the stimulative potency of a sex attractant, which must carry its message for long distances. For these reasons small molecules will suffice for alarm purposes. Of seven alarm substances known in the social insects, six have 10 or fewer carbon atoms and one (dendrolasin) has 15. It will be interesting to see if future discoveries bear out these early generalizations.

Do human pheromones exist? Primer pheromones might be difficult to detect, since they can affect the endocrine system without producing overt specific behavioral responses. About all that can be said at present is that striking sexual differences have been observed in the ability of humans to smell certain

ARTIFICIAL TRAIL can be laid down by drawing a line (*colored curve in frame at top left*) with a stick that has been treated with the contents of a single Dufour's gland. In the remaining three frames, workers are attracted from the nest, follow the artificial route in close formation and mill about in confusion at its arbitrary terminus. Such a trail is not renewed by the unrewarded workers.

DENDROLASIN (*LASIUS FULIGINOSUS*)

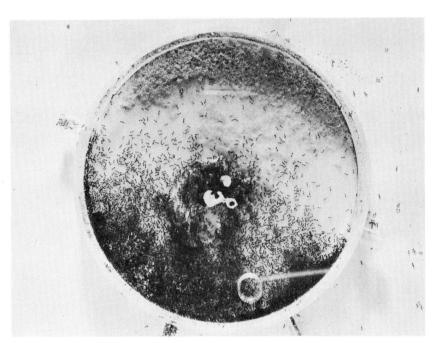

CITRAL (*ATTA SEXDENS*)

CITRONELLAL (*ACANTHOMYOPS CLAVIGER*)

2-HEPTANONE (*IRIDOMYRMEX PRUINOSUS*)

FOUR ALARM PHEROMONES, given off by the workers of the ant species indicated, have so far been identified. Disturbing stimuli trigger the release of these substances from various glandular reservoirs.

ably some pheromone "languages" will be found to have a syntax. It may be found, in other words, that pheromones can be combined in mixtures to form new meanings for the animals employing them. One would also like to know if some animals can modulate the intensity or pulse frequency of pheromone emission to create new messages. The solution of these and other interesting problems will require new techniques in analytical organic chemistry combined with ever more perceptive studies of animal behavior.

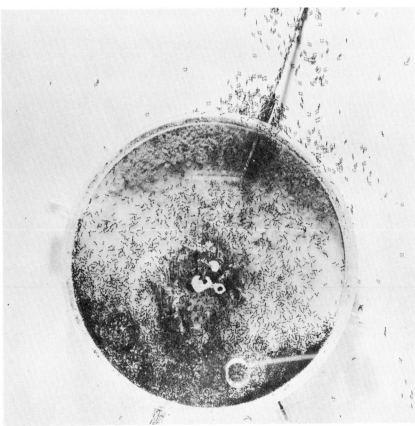

MASSIVE DOSE of trail pheromone causes the migration of a large portion of a fire ant colony from one side of a nest to another. The pheromone is administered on a stick that has been dipped in a solution extracted from the Dufour's glands of freshly killed workers.

substances. The French biologist J. Le-Magnen has reported that the odor of Exaltolide, the synthetic lactone of 14-hydroxytetradecanoic acid, is perceived clearly only by sexually mature females and is perceived most sharply at about the time of ovulation. Males and young girls were found to be relatively insensitive, but a male subject became more sensitive following an injection of estrogen. Exaltolide is used commercially as a perfume fixative. LeMagnen also reported that the ability of his subjects to detect the odor of certain steroids paralleled that of their ability to smell Exaltolide. These observations hardly represent a case for the existence of human pheromones, but they do suggest that the relation of odors to human physiology can bear further examination.

It is apparent that knowledge of chemical communication is still at an early stage. Students of the subject are in the position of linguists who have learned the meaning of a few words of a nearly indecipherable language. There is almost certainly a large chemical vocabulary still to be discovered. Conceiv-

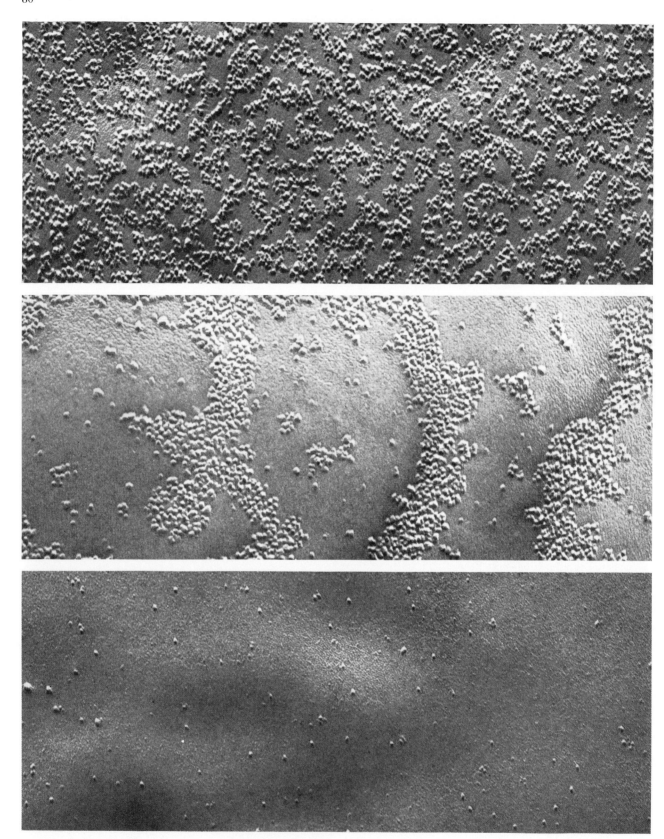

EVIDENCE FOR PROTEINS within the bilayer structure of cell membranes is provided by freeze-etch electron microscopy. A suspension of membranes in water is frozen and then fractured with a sharp blade. The fracture will often split a membrane in the middle along a plane parallel to the surface. After platinum and carbon vapors are deposited along the fracture surface the specimen can be studied in the electron microscope. The micrograph at the top shows many particles 50 to 85 angstroms in diameter embedded in a fractured membrane from rabbit red blood cells. The other two views show how the number of particles is greatly reduced if the membrane is first treated with a proteolytic enzyme that digests 45 percent (*middle*) or 70 percent (*bottom*) of the original membrane protein. The missing particles have presumably been digested by the enzyme. The membrane preparations are enlarged some 95,000 diameters in these micrographs made by L. H. Engstrom in Daniel Branton's laboratory at the University of California at Berkeley.

8

THE STRUCTURE OF CELL MEMBRANES

C. FRED FOX

February 1972

The thin, sturdy envelope of the living cell consists of lipid, phosphate and protein. The proteins act as both gatekeepers and active carriers, determining what passes through the membrane

Every living cell is enclosed by a membrane that serves not only as a sturdy envelope inside which the cell can function but also as a discriminating portal, enabling nutrients and other essential agents to enter and waste products to leave. Called the cytoplasmic membrane, it can also "pump" substances from one side to the other against a "head," that is, it can extract a substance that is in dilute solution on one side and transport it to the opposite side, where the concentration of the substance is many times higher. Thus the cytoplasmic membrane selectively regulates the flux of nutrients and ions between the cell and its external milieu.

The cells of higher organisms have in addition to a cytoplasmic membrane a number of internal membranes that isolate the structures termed organelles, which play various specialized roles. For example, the mitochondria oxidize foodstuffs and provide fuel for the other activities of the cell, and the chloroplasts conduct photosynthesis. Single-cell organisms such as bacteria have only a cytoplasmic membrane, but its structural diversity is sufficient for it to serve some or all of the functions carried out by the membranes of organelles in higher cells. It is clear that any model formulated to describe the structure of membranes must be able to account for an extraordinary range of functions.

Membranes are composed almost entirely of two classes of molecules: proteins and lipids. The proteins serve as enzymes, or biological catalysts, and provide the membrane with its distinctive functional properties. The lipids provide the gross structural properties of the membrane. The simplest lipids found in nature, such as fats and waxes, are insoluble in water. The lipids found in membranes are amphipathic, meaning that one end of the molecule is hydrophobic, or insoluble in water, and the other end hydrophilic, or water-soluble. The hydrophilic region is described as being polar because it is capable of carrying an ionic (electric) charge; the hydrophobic region is nonpolar.

In most membrane lipids the nonpolar region consists of the hydrocarbon chains of fatty acids: hydrocarbon molecules with a carboxyl group (COOH) at one end. In a typical membrane lipid two fatty-acid molecules are chemically bonded through their carboxyl ends to a backbone of glycerol. The glycerol backbone, in turn, is attached to a polar-head group consisting of phosphate and other groups, which often carry an ionic charge [*see illustration on next page*]. Phosphate-containing lipids of this type are called phospholipids.

When a suspension of phospholipids in water is subjected to high-energy sound under suitable conditions, the phospholipid molecules cluster together to form closed vesicles: small saclike structures called liposomes. The arrangement of phospholipids in the walls of both liposomes and biological membranes has recently been deduced with the help of X-ray diffraction, which can reveal the distance between repeating groups of atoms. An X-ray diffraction analysis by M. F. Wilkins and his associates at King's College in London indicates that two parallel arrays of the polar-head groups of lipids are separated by a distance of approximately 40 angstroms and that the fatty-acid tails are stacked parallel to one another in arrays of 50 or more phospholipid molecules.

The X-ray data suggest a structure for liposomes and membranes in which the phospholipids are arranged in two parallel layers [*see illustrations on page 83*]. The polar heads are arrayed externally on the bilayer surfaces, and the fatty-acid tails are pointed inward, perpendicular to the plane of the membrane surface. This model of phospholipid structure in membranes is identical with one proposed by James F. Danielli and Hugh Davson in the mid-1930's, when no precise structural data were available. It is also the minimum-energy configuration for a thin film composed of amphipathic molecules, because it maximizes the interaction of the polar groups with water.

Unlike lipids, proteins do not form orderly arrays in membranes, and thus their arrangement cannot be assessed by X-ray diffraction. The absence of order is not surprising. Each particular kind of membrane incorporates a variety of protein molecules that differ widely in molecular weight and in relative numbers; a membrane can incorporate from 10 to 100 times more molecules of one type of protein than of another.

Since little can be learned about the disposition of membrane proteins from a general structural analysis, investigators have chosen instead to study the orientation of one or a few species of the proteins in membranes. In the Danielli-Davson model the proteins are assumed to be entirely external to the lipid bilayer, being attached either to one side of the membrane or to the other. Although information obtained from X-ray diffraction and high-resolution electron microscopy indicates that this is probably true for the bulk of the membrane protein, biochemical studies show that the Danielli-Davson concept is an oversimplification. The evidence for alternative locations has been provided chiefly by Marc Bretscher of the Medical Research Council laboratories in Cambridge and by Theodore L. Steck, G. Franklin and Donald F. H. Wallach of the Harvard Medical School. Their results suggest that certain proteins penetrate the lipid bilayer and that others extend all the way through it.

Bretscher has labeled a major protein of the cytoplasmic membrane of human red blood cells with a radioactive substance that forms chemical bonds with the protein but is unable to penetrate the membrane surface. The protein was labeled in two ways [see illustration on pages 84 and 85]. First, intact red blood cells were exposed to the label so that it became attached only to the portion of the protein that is exposed on the outer surface of the membrane. Second, red blood cells were broken up before the radioactive label was added. Under these conditions the label could attach itself to parts of the protein exposed on the internal surface of the membrane as well as to parts on the external surface.

The two batches of membrane, labeled under the two different conditions, were treated separately to isolate the protein. The purified protein from the two separate samples was degraded into definable fragments by treatment with a proteolytic enzyme: an enzyme that cleaves links in the chain of amino acid units that constitutes a protein. A sample from each batch of fragments was now placed on the corner of a square of filter paper for "fingerprinting" analysis. In this technique the fragments are separated by chromatography in one direction on the paper and by electrophoresis in a direction at right angles to the first. In the chromatographic step each type of fragment is separated from the others because it has a characteristic rate of travel across the paper with respect to the rate at which a solvent travels. In the electrophoretic step the fragments are further separated because they have characteristic rates of travel in an imposed electric field.

Once a separation had been achieved the filter paper was laid on a piece of X-ray film so that the radioactively labeled fragments could reveal themselves by exposing the film. When films from the two batches of fragments were developed, they clearly showed that more labeled fragments were present when both the internal and the external surface of the cell membrane had been exposed to the radioactive label than when the outer surface alone had been exposed. This provides strong evidence that the portion of the protein that gives rise to the additional labeled fragments is on the inner surface of the membrane.

Steck and his colleagues obtained similar results with a procedure in which they prepared two types of closed-membrane vesicle, using as a starting material the membranes from red blood cells. In one type of vesicle preparation (right-side-out vesicles) the outer membrane surface is exposed to the external aqueous environment. In the other type of preparation (inside-out vesicles) the inner surface of the membrane is exposed to the external aqueous environment. When the two types of vesicle are treated with a proteolytic enzyme, only those proteins exposed to the external aqueous environment should be degraded. Steck found that some proteins are susceptible to digestion in both the right-side-out and inside-out vesicles, indicating that these proteins are exposed on both membrane surfaces. Other proteins are susceptible to proteolytic digestion in right-side-out vesicles but not in inside-out vesicles. Such proteins are evidently located exclusively on only one side of the membrane. This information lends credence to the concept of sidedness in membranes. Such sidedness had been suspected for many years because the inner and outer surfaces of cellular membranes are thought to have different biological functions. The development of a technique for preparing vesicles with right-side-out and inside-out configurations should be extremely useful in determining on which side of the membrane a given species of protein resides and thus functions.

Daniel Branton and his associates at the University of California at Berkeley have developed and exploited the technique of freeze-etch electron microscopy to study the internal anatomy of membranes. In freeze-etch microscopy a suspension of membranes in water is frozen rapidly and fractured with a sharp blade. Wherever the membrane surface runs parallel to the plane of fracture much of the membrane will be split along the middle of the lipid bilayer. A thin film of platinum and carbon is then evaporated onto the surface of the fracture. This makes it possible to examine the anatomy of structures in the fracture plane by electron microscopy.

The electron micrographs of the fractured membrane reveal many particles, approximately 50 to 85 angstroms in diameter, on the inner surface of the lipid bilayer. These particles are not observed if the membrane samples are first treated with proteolytic enzymes, indicating that the particles probably consist of protein [see illustration on page 80]. From quantitative estimates of the number of particles revealed by freeze-etching, Branton and his colleagues have suggested that between 10 and 20 percent of the internal volume of many biological membranes is protein.

Somewhere between a fifth and a quarter of all the protein in a cell is physically associated with membranes. Most of the other proteins are dissolved in the aqueous internal environment of the cell. In order to dissolve membrane proteins in aqueous solvents detergents must be added to promote their dispersion. One might therefore expect membrane proteins to differ considerably from soluble proteins in chemical composition. This, however, is not the case.

The amino acids of which proteins are composed can be classified into two groups: polar and nonpolar. S. A. Rosenberg and Guido Guidotti of Harvard University analyzed the amino acid composition of proteins from a number of membranes and found that they contain about the same percentage of polar and nonpolar amino acids as one finds in the soluble proteins of the common colon

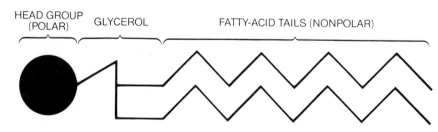

HEAD GROUP (POLAR)   GLYCEROL   FATTY-ACID TAILS (NONPOLAR)

TYPICAL MEMBRANE LIPID is a complex molecular structure, one end of which is hydrophilic, or water-soluble, and the other end hydrophobic. Such a substance is termed amphipathic. The hydrophilic, or polar, region consists of phosphate and other constituents attached to a unit of glycerol. The polar-head group, when in contact with water, often carries an electric charge. The glycerol component forms a bridge to the hydrocarbon tails of two fatty acids that constitute the nonpolar region of the lipid. In this highly schematic diagram the zigzag lines represent hydrocarbon chains; each angle is occupied by a carbon atom and two associated hydrogen atoms. The terminal carbon of each chain is bound to three hydrogen atoms. Phosphate-containing amphipathic lipids are called phospholipids.

bacterium *Escherichia coli.* Thus differences in amino acid composition cannot account for the water-insolubility of membrane proteins.

Studies conducted by L. Spatz and Philipp Strittmatter of the University of Connecticut indicate that the most likely explanation for the water-insolubility of membrane proteins is the arrangement of their amino acids. Spatz and Strittmatter subjected membranes of rabbit liver cells to a mild treatment with a proteolytic enzyme. The treatment released the biologically active portion of the membrane protein: cytochrome b_5. In a separate procedure they solubilized and purified the intact cytochrome b_5 and treated it with the proteolytic enzyme. This treatment also released the water-soluble, biologically active portion of the molecule, together with a number of small degradation products that were insoluble in aqueous solution. The biologically active portion of the molecule, whether obtained from the membrane or from the purified protein, was found to be rich in polar amino acids. The protein fragments that were insoluble in water, on the other hand, were rich in nonpolar amino acids. These observations suggest that many membrane proteins may be amphipathic, having a nonpolar region that is embedded in the part of the membrane containing the nonpolar fatty-acid tails of the phospholipids and a polar region that is exposed on the membrane surface.

We are now ready to ask: How do substances pass through membranes? The nonpolar fatty-acid-tail region of a phospholipid bilayer is physically incompatible with small water-soluble substances, such as metal ions, sugars and amino acids, and thus acts as a barrier through which they cannot flow freely. If one measures the rate at which blood sugar (glucose) passes through the phospholipid-bilayer walls of liposomes, one finds that it is far too low to account for the rate at which glucose penetrates biological membranes. Information of this kind has given rise to the concept that entities termed carriers must be present in biological membranes to facilitate the passage of metal ions and small polar molecules through the barrier presented by the phospholipid bilayer.

Experiments with biological membranes indicate that the hypothetical carriers are highly selective. For example, a carrier that facilitates the transport of glucose through a membrane plays no role in the transport of amino acids or other sugars. An interesting experimental

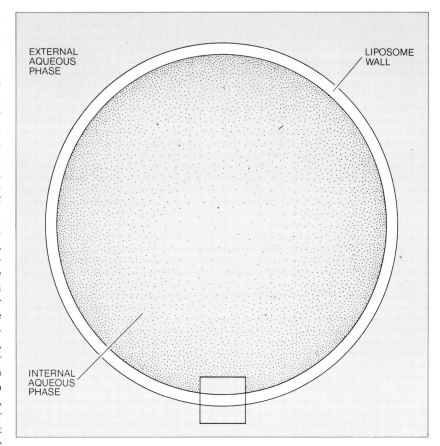

ARTIFICIAL MEMBRANE-ENCLOSED SAC, known as a liposome, is created by subjecting an aqueous suspension of phospholipids to high-energy sound waves. X-ray diffraction shows that the phospholipids in the liposome assume an orderly arrangement resembling what is found in the membranes of actual cells. Area inside the square is enlarged below.

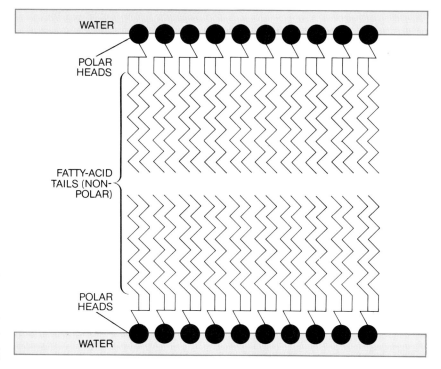

CROSS SECTION OF LIPOSOME WALL shows how the membrane is formed from two layers of lipid molecules. The polar heads of amphipathic lipids face toward the aqueous solution on each side while the nonpolar fatty-acid tails face inward toward one another.

system for measuring selective ion transport was developed by A. D. Bangham, M. M. Standish and J. C. Watkins of the Agricultural Research Council in Cambridge, England, and by J. B. Chappell and A. R. Crofts of the University of Cambridge. As a model carrier they used valinomycin, a nonpolar, fat-soluble antibiotic consisting of a short chain of amino acids (actually 12); such short chains are termed polypeptides to distinguish them from true proteins, which are much larger. Valinomycin combines with phospholipid-bilayer membranes and makes

them permeable to potassium ions but not to sodium ions.

The change in permeability is conveniently studied by measuring the change in electrical resistance across a phospholipid bilayer between two chambers containing a potassium salt in aqueous solution. The experiment is performed by introducing a sample of phospholipid into a small hole between the two chambers. The lipid spontaneously thins out until the chambers are separated by only a thin membrane consisting of a phospholipid bilayer. Electrodes

are then placed in the two chambers to measure the resistance across the membrane.

The resistance across a phospholipid bilayer in the absence of valinomycin is several orders of magnitude higher than the resistance across a typical biological membrane: 10 million ohms centimeter squared compared with between 10 and 10,000. This indicates that phospholipid-bilayer membranes are essentially impermeable to small hydrophilic ions. If a small amount of valinomycin (10^{-7} gram per milliliter of salt solution) is intro-

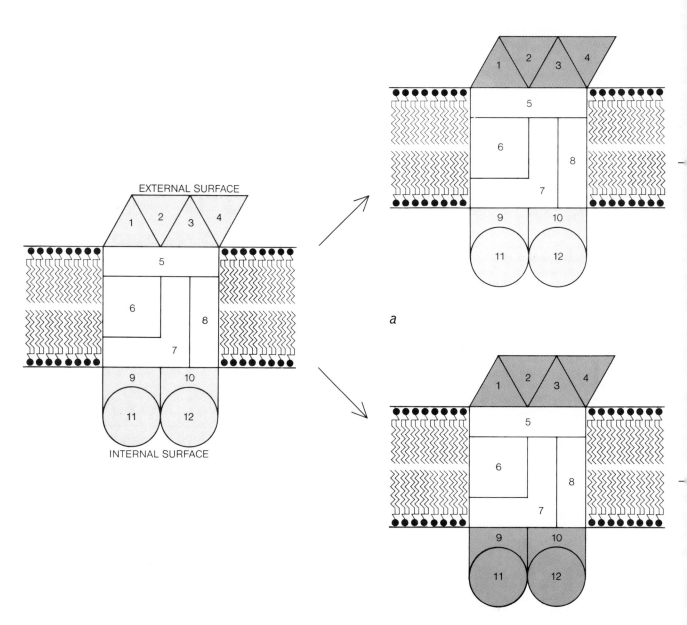

LOCATION OF PROTEINS IN MEMBRANES can be inferred by attaching radioactive labels to the proteins. These diagrams depict an experiment in which a major protein in the membrane of red blood cells was labeled (*a*). When intact cells (*top sequence*) are exposed to the radioactive substance, only the portion of the protein on the outside wall picks up the label (*color*). When the cells are broken before labeling (*bottom sequence*), the radioactive la-

bel is able to reach portions of the protein that are exposed to the internal as well as to the external surfaces of the membrane. This can be demonstrated by isolating and purifying the protein labeled under the two conditions. The protein is then broken up into defined fragments (*numbered shapes*) by treating it with a proteolytic enzyme (*b*). Portions of the two batches of fragments are spotted on the corners of filter paper for "fingerprinting" (*c*). This is a

duced into the chambers containing the potassium solution, the resistance falls by five orders of magnitude and the permeability of the phospholipid bilayer to potassium ions rises by a like amount. The permeability of the experimental membrane now essentially duplicates the permeability of biological membranes.

If the experiment is repeated with a sodium chloride solution in the chambers, one finds that the addition of valinomycin causes only a slight change in resistance. Hence valinomycin meets two of the most important criteria for a bio-

logical carrier: it enhances permeability and it is highly selective for the transported substance. The question that now arises is: How does valinomycin work?

First of all, valinomycin is nonpolar. Thus it is physically compatible with and can dissolve in the portion of the bilayer that contains the nonpolar fatty-acid tails. Second, valinomycin can evidently diffuse between the two surfaces of the bilayer. S. Krasne, George Eisenman and G. Szabo of the University of California at Los Angeles have shown that the enhancement of potassium-ion transport by

valinomycin is interrupted when the bilayer is "frozen" by lowering the temperature. Third, valinomycin must bind potassium ions in such a way that the ionic charge is shielded from the nonpolar region of the membrane. Finally, valinomycin itself must have a selective binding capacity for potassium ions in preference to sodium or other ions.

With valinomycin as a model for carrier-mediated transport, one can postulate three essential steps: recognition of the ion, diffusion of the ion through the membrane, and its release on the other

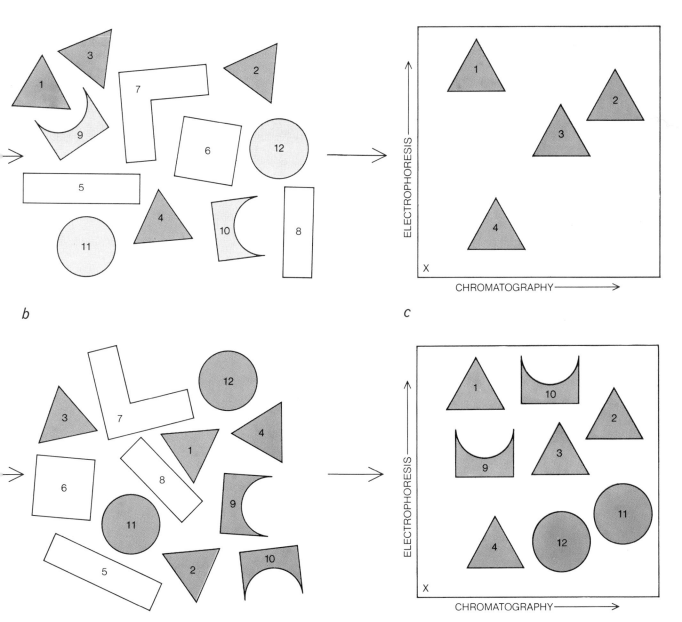

technique that combines chromatography with electrophoresis. By chromatography alone protein fragments would migrate at different rates depending primarily on their solubility in the solvent system. Electrophoresis involves establishing an electric-potential gradient along one axis of the filter paper. Since various fragments have different densities of electric charge they are further separated. A piece of X-ray film is then placed over each sheet of filter

paper. Radiation from the labeled fragments exposes the film and reveals where the various fragments have come to rest. A comparison of the X-ray films produced in the parallel experiments shows that more protein fragments are labeled when the red blood cells are broken before labeling and that the additional fragments (9, 10, 11, 12) must represent portions of the original protein that extend through the membrane and penetrate the inner surface.

side. In the first step some part of the valinomycin molecule, embedded in the membrane, "recognizes" the potassium ion as it approaches the surface of the membrane and captures it. In the second step the complex consisting of valinomycin and the potassium ion diffuses through the membrane. Finally, on reaching the opposite surface of the membrane the potassium ion is dissociated from the complex and is released.

The argument to this point can be summarized in a few words. The fundamental structure of biological membranes is a phospholipid bilayer, the phospholipid bilayer is a permeability barrier and carriers are needed to breach it. In addition, the membrane barrier must often be breached in a directional way. In a normally functioning cell hundreds of kinds of small molecule must be present at a higher concentration inside the cell than outside, and many other small molecules must be present at a lower concentration inside the cell than outside. For example, the concentration of potassium ions in human cells is more than 100 times greater than it is in the blood that bathes them. For sodium ions the concentrations are almost exactly reversed. The maintenance of these differences in concentration is absolutely es-

sential; even slight changes can result in death.

Although the model system based on valinomycin provides considerable insight into the function and selectivity of carriers, it sheds no light on the transport mechanism that can pump a substance from a low concentration on one side of the membrane to a higher concentration on the other. Our understanding of concentrative transport (or, as it is usually termed, active transport) owes much to the pioneering effort of Georges Cohen, Howard Rickenberg, Jacques Monod and their associates at the Pasteur Institute in Paris. The Pasteur group studied the transport of milk sugar (lactose) through the cell membrane of the bacterium *Escherichia coli.* Genetic experiments suggested that the carrier for lactose transport was a protein. Studies of the rate of transport revealed that the transport process behaves like a reaction catalyzed by an enzyme, giving further support to the idea that the carrier is a protein. The Pasteur group also found that the lactose-transport system is capable of active transport, producing a lactose concentration 500 times greater inside the cell than outside. The active-transport process depends on the expenditure of metabolic energy; poisons that block energy metabolism destroy

the ability of the cell to concentrate lactose.

A model that accounts for many (but not all) of the properties of the active-transport system that are typified by the lactose system postulates the existence of a carrier protein that can change its shape. The protein is visualized as resembling a revolving door in the membrane wall [*see illustration on opposite page*]. The "door" contains a slot that fits the target substance to be transported. The slot normally faces the cell's external environment. When the target substance enters the slot, the protein changes shape and is thereby enabled to rotate so that the slot faces into the cell. When the target substance has been discharged into the cell, the protein remains with its slot facing inward until the cell expends energy to rotate the protein so that the slot again faces outward.

Working with Eugene P. Kennedy at the Harvard Medical School in 1965, I succeeded in identifying the lactose-transport carrier. We found, as we had expected, that it is a protein with an enzyme-like ability to bind lactose. Since then a number of other transport carriers have been identified, and all turn out to be proteins. The lactose carrier resides in the membrane and is hydrophobic; thus it is physically compatible

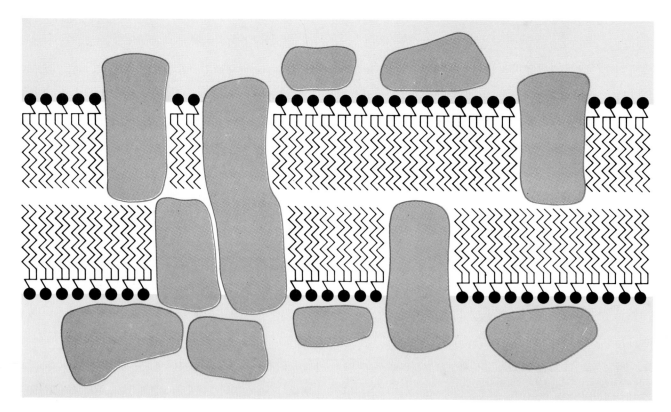

ANATOMY OF BIOLOGICAL MEMBRANE is suggested in this schematic diagram. Phospholipid molecules stacked side by side and back to back provide the basic structure. The gray shapes represent protein molecules. In some cases several proteins (for example the five at the left) are bound into a single functional complex. Proteins can occupy all possible positions with respect to the phospholipid bilayer: they can be entirely outside or inside, they can penetrate either surface or they can extend through the membrane.

with the nonpolar-lipid phase of the membrane.

In 1970 Ron Kaback and his associates at the Roche Institute of Molecular Biology observed that the energy that drives the active transport of lactose and dozens of other low-molecular-weight substances in *E. coli* is directly coupled to the biological oxidation of metabolic intermediates such as D-lactic acid and succinic acid. How energy derived from the oxidation of D-lactic acid can be used to drive active transport is one of the more interesting unsolved problems in membrane biology.

Since transport carriers must be mobile within the membrane in order to move substances from one surface to the other, one might guess that the region of the membrane containing the fatty-acid tails should not have a rigid crystalline structure. X-ray diffraction studies indicate that the fatty acids of membranes in fact do have a "liquid crystalline" structure at physiological temperature, that is, around 37 degrees Celsius. In other words, the fatty acids are not aligned in a rigid crystalline lattice. The techniques of electron paramagnetic resonance and nuclear magnetic resonance can be used to study the flexibility of the fatty-acid side chains in membranes. Several investigators, notably Harden M. McConnell and his associates at Stanford University, have concluded that the fatty acids of membranes are quasi-fluid in character.

Membranes incorporate two classes of fatty acids: saturated molecules, in which all the available carbon bonds carry hydrogen atoms, and unsaturated molecules, in which two or more pairs of hydrogen atoms are absent (with the result that two or more pairs of carbon atoms have double bonds). The fluid character of membranes is largely determined by the structure and relative proportion of the unsaturated fatty acids. In phospholipids consisting only of saturated fatty acids the fatty-acid tails are aligned in a rigidly stacked crystalline array at physiological temperatures. In phospholipids consisting of both saturated and unsaturated fatty acids the fatty acids are packed in a less orderly fashion and thus are more fluid. The double bonds of unsaturated fatty acids give rise to a structural deformation that interrupts the ordered stacking necessary for the formation of a rigid crystalline structure [*see illustration on next page*].

My colleagues and I at the University of Chicago (and later at the University of California at Los Angeles) and Peter Overath and his associates at the Uni-

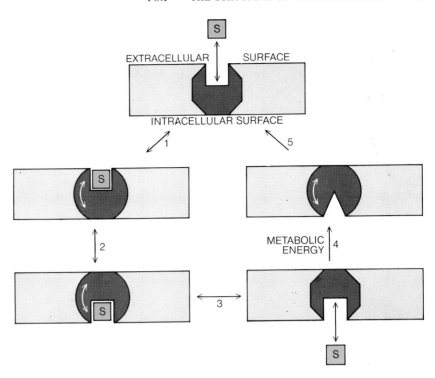

MECHANISM OF "ACTIVE" TRANSPORT may involve a carrier protein (*dark gray*) with the properties of a revolving door. A carrier protein can capture a substance, S, that exists outside the membrane in dilute solution and transport it to the inside of the cell, where the concentration of S is greater than it is outside. When S is bound to the protein, the protein changes shape (*1*), thus enabling it to rotate (*2*). When S becomes detached and enters the cell (*3*), the protein returns to its immobile form. Metabolic energy must be expended (*4*) to alter the protein's shape so that it can rotate and again present its binding site to the cell exterior (*5*). Other protein carriers have the capacity to transport substances from low concentration inside the cell to solutions of higher concentration outside the cell.

versity of Cologne have varied the fatty-acid composition of biological membranes to study the effects of fatty-acid structure on transport. When the membrane lipids are rich in unsaturated fatty acids, transport can proceed at rates up to 20 times faster than it does when the membrane lipids are low in unsaturated fatty acids. These experiments show that normal membrane function depends on the fluidity of the fatty acids.

The temperature at which cells live and grow can have a pronounced effect on the amount of unsaturated fatty acid in their membranes. Bacteria grown at a low temperature have membranes with a greater proportion of unsaturated fatty acid than those grown at a higher temperature. This adjustment in fatty-acid composition is necessary if the membranes are to function normally at low temperature. A similar adjustment can take place in higher organisms. For example, there is a temperature gradient in the legs of the reindeer; the highest temperature is near the body, the lowest is near the hooves. To compensate for this temperature gradient the cells near the hooves have membranes whose lipids are enriched in unsaturated fatty acids.

Although, as we have seen, phospholipids can spontaneously form bilayer films in water, this process only provides a physical rationale as to why the predominant structure in membranes is a phospholipid bilayer. The events leading to the assembly of a biological membrane are far more complex. The cells of higher organisms contain a number of unique membrane structures. They differ widely in lipid composition, and each type of membrane has its own unique complement of proteins. The diversity in protein composition and in the location of proteins within membranes explains the functional diversity of different types of membrane. Rarely does a single species of protein exist in more than one type of membrane.

Since all membrane proteins are synthesized at approximately the same cellular location, what is it that determines that one type of protein will be incorporated only into the cytoplasmic membrane and that another type will turn up only in a mitochondrial membrane? At present this question can be answered only by conjecture tinctured with a few facts. Two general hypotheses for membrane assembly can be offered. One pos-

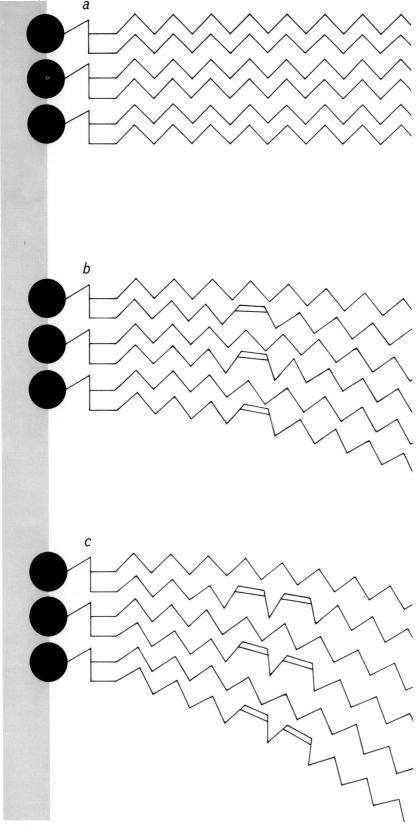

VARIATION IN FATTY-ACID COMPOSITION can disrupt the orderly stacking of phospholipids in a biological membrane. In a lipid layer composed entirely of saturated fatty acids (*a*) the fatty-acid chains contain only single bonds between carbon atoms and thus nest together to form rigid structures. In a lipid layer containing unsaturated fatty acids with one double bond (*b*) the double bonds introduce a deformation that interferes with orderly stacking and makes the fatty-acid region somewhat fluid. When fatty acids with two double bonds are present (*c*), the deformation and the consequent fluidity are greater still.

sibility is that new pieces of membrane are made from scratch by a self-assembly mechanism in which all the components of a new piece of membrane come together spontaneously. This new piece could then be inserted into an existing membrane. A second possibility is that newly made proteins are simply inserted at random into a preexisting membrane.

Recent studies in my laboratory at the University of California at Los Angeles and in the laboratories of Philip Siekevitz and George E. Palade at Rockefeller University support the second hypothesis. That is all well and good, but what determines why a given protein is incorporated only into a given kind of membrane? Although this must be answered by conjecture, it is known that many proteins are specifically bound to other proteins in the same membrane. Such protein-protein interactions are not uncommon; many of the functional entities in membranes are complexes of several proteins. Thus the proteins in a membrane may provide a template that is recognized by a newly synthesized protein and that helps to insert the newly synthesized protein into the membrane. In this way old membrane could act as a template for the assembly of new membrane. This might explain why different membranes incorporate different proteins.

Why, then, do different membranes have different lipid compositions? The answers to this question are even more obscure. In general lipids are synthesized within the membrane; the enzymes that catalyze the synthesis are part of the membrane. Some lipids, however, are made in one membrane and then shuttled to another membrane that has no inherent capacity to synthesize them. Since there is an interchange of lipids between various membranes, it seems unlikely that the variations in lipid composition in different membranes can be explained by dissimilarities in the synthetic capacity of a given membrane for a given type of lipid. There are at least two possible ways of accounting for differences in lipid composition. One possibility is that different membranes may destroy different lipids at different rates; another is that the proteins of one species of membrane may selectively bind one type of lipid, whereas the proteins of another species of membrane may bind a different type of lipid. It is obvious from this discussion that concrete evidence on the subject of membrane assembly is scant but that the problems are well defined.

9

MOLECULAR ISOMERS IN VISION

RUTH HUBBARD AND ALLEN KROPF

June 1967

Certain organic compounds can exist in two or more forms that have the same chemical composition but different molecular architecture. One of them is the basis for vision throughout the animal kingdom

Molecular biology, which today is so often associated with very large molecules such as the nucleic acids and proteins, actually embraces the entire effort to describe the structure and function of living organisms in molecular terms. We are coming to see how the manifold activities of the living cell depend on interactions among molecules of thousands of different sizes and shapes, and we can speculate on how evolutionary processes have selected each molecule for its particular functional properties. The significance of precise molecular architecture has become a central theme of molecular biology.

One of the more recent observations is that biological molecules are not static structures but, in a number of well-established cases, change shape in response to outside influences. As an example, the molecule of hemoglobin has one shape when it is carrying oxygen from the lungs to cells elsewhere in the body and a slightly different shape when it is returning to the lungs without oxygen [see "The Hemoglobin Molecule," by M. F. Perutz; SCIENTIFIC AMERICAN Offprint 196]. A somewhat similar changeability in the molecule of lysozyme, which breaks down the walls of certain bacterial cells, is described by David C. Phillips of the University of Oxford in SCIENTIFIC AMERICAN Offprint 1055. In this article we shall describe some of the simplest changes in shape that can take place in much smaller organic molecules and show how change of this type provides the basis of vision throughout the animal kingdom.

A "Childish Fantasy"

The notion that molecules of the same atomic composition might have different spatial arrangements is less than 100 years old. It dates back to a paper titled "*Sur les formules de structure dans l'espace,*" written in 1874 by Jacobus Henricus van't Hoff, then an obscure chemist at the Veterinary College of Utrecht. At that time it was still respectable to doubt the existence of atoms; to speak of the three-dimensional arrangement of atoms in molecules was a speculative leap of great audacity. Van't Hoff's paper provoked Hermann Kolbe, one of the most eminent organic chemists of his day, to publish a withering denunciation.

"Not long ago," Kolbe wrote in 1877, "I expressed the view that the lack of general education and of thorough training in chemistry of quite a few professors of chemistry was one of the causes of the deterioration of chemical research in Germany. . . . Will anyone to whom my worries may seem exaggerated please read, if he can, a recent memoir by a Herr van't Hoff on 'The Arrangements of Atoms in Space,' a document crammed to the hilt with the outpourings of a childish fantasy. This Dr. J. H. van't Hoff, employed by the Veterinary College at Utrecht, has, so it seems, no taste for accurate chemical research. He finds it more convenient to mount his Pegasus (evidently taken from the stables of the Veterinary College) and to announce how, on his daring flight to Mount Parnassus, he saw the atoms arranged in space."

Van't Hoff's "childish fantasy" was put forth independently by the French chemist Jules Achille le Bel and was soon championed by a number of leading chemists. In spite of Kolbe's opinion, evidence in support of the three-dimensional configuration of molecules rapidly accumulated. In 1900 van't Hoff was named the first recipient of the Nobel prize in chemistry.

Even before van't Hoff's paper of 1874 chemists had begun using the concept of the valence bond, commonly represented by a line connecting two atoms. It was not unnatural, therefore, to associate the valence bond concept with the idea that atoms were arranged precisely in space. The simplest hydrocarbon, methane (CH_4), would then be represented as a regular tetrahedron with a hydrogen atom at each vertex joined by a single valence bond to a carbon atom at the center of the structure [see illustration on page 91].

The valence bond remained an elusive concept, however, until G. N. Lewis postulated in 1916 that a common type of bond—the covalent bond—was formed when two atoms shared two electrons. "When two atoms of hydrogen join to form the diatomic molecule," he wrote, "each furnishes one electron of the pair which constitutes the bond. Representing each valence electron by a dot, we may therefore write as the graphical formula of hydrogen H : H." He visualized this bond to be "that 'hook and eye,' which is part of the creed of the organic chemist." To explain why electrons should tend to pair in this manner, Lewis could offer nothing beyond an intuitive principle that he called "the rule of two."

The rule of two entered the physicist's description of the atom when Wolfgang Pauli put forward the exclusion principle in 1923. This states that electrons in atoms and molecules are found in "orbitals" that can accommodate at most two electrons. Since electrons can be regarded as minuscule spinning negative charges, and thus as tiny electromagnets, the two electrons in each orbital must be spinning in opposite directions.

Let us return, however, to some of the chemical observations that gave rise to van't Hoff's ideas of stereochemistry late in the 19th century. Chemists were confronted by a series of puzzling observations best exemplified by two simple compounds: maleic acid and fumaric acid [see illustrations on page 92]. Here were two distinct, chemically pure substances, each with four atoms of carbon, four of hydrogen and four of oxygen ($C_4H_4O_4$). It was known, moreover, that the connections between atoms in the two molecules were exactly the same and that the two central carbon atoms in each molecule were connected by a double bond. Yet the two compounds were indisputably different. Whereas crystals of maleic acid melted at 128 degrees centigrade, crystals of fumaric acid did not melt until heated to about 290 degrees C. Furthermore, maleic acid was

about 100 times more soluble in water and 10 times stronger as an acid than fumaric acid. When maleic acid was heated in a vacuum, it gave off water vapor and became a new substance, maleic anhydride, which readily recombined with water and reverted to maleic acid. Fumaric acid underwent no such reaction. On the other hand, if either compound was heated in the presence of hydrogen, it was transformed into the identical compound, succinic acid ($C_4H_6O_4$), which contains two more hydrogen atoms per molecule than maleic or fumaric acid.

It was known that the four carbon atoms in maleic and fumaric acids form a chain. The only way to explain the differences between the compounds is to assume that the two halves of a molecule

that are connected by a double bond are not free to rotate with respect to each other. Thus the form of the $C_4H_4O_4$ molecule in which the two terminal COOH groups lie on the same side of the double bond (maleic acid) is not identical with the form in which the COOH groups lie on opposite sides (fumaric acid). Molecules that assume distinct shapes in this way are called geometrical or cis-trans isomers of one another. "Cis" is from the Latin meaning on the same side; "trans" means on opposite sides. Therefore maleic acid is the cis isomer of the $C_4H_4O_4$ molecule and fumaric acid is the trans isomer. When the double carbon-carbon bond of either isomer is reduced to a single bond by the addition of two more hydrogen atoms, the two halves of the molecule are free

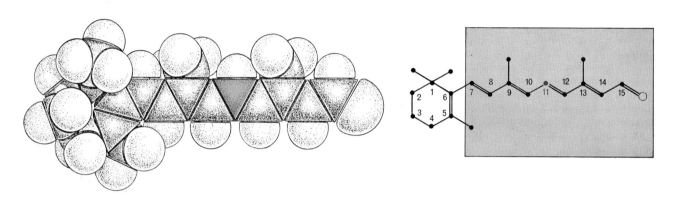

ALL-*TRANS* RETINAL

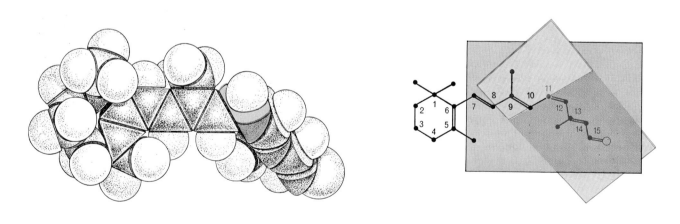

11-*CIS* RETINAL

FUNDAMENTAL MOLECULE OF VISION is retinal ($C_{20}H_{28}O$), also known as retinene, which combines with proteins called opsins to form visual pigments. Because the nine-member carbon chain in retinal contains an alternating sequence of single and double bonds, it can assume a variety of bent forms. Each distinct form is termed an isomer. Two isomers of retinal are depicted here. In the models (*left*) carbon atoms are dark, except carbon No. 11, which is shown in color; hydrogen atoms are light. The large atom attached to carbon No. 15 is oxygen. In the structural formulas (*right*) hydrogen atoms are omitted. The parts of each isomer that lie in a plane are marked by background panels. When tightly bound to opsin, retinal is in the bent and twisted form known as 11-*cis*. When struck by light, it straightens out into the all-*trans* configuration. This simple photochemical event provides the basis for vision.

to rotate with respect to each other and a single compound results: succinic acid.

Van't Hoff proposed that when two carbon atoms are joined by a single bond they can be regarded as the centers of two tetrahedrons that meet apex to apex, thus allowing the two bodies to rotate freely. To represent a double bond, he visualized the two tetrahedrons as being joined edge to edge so that they were no longer free to rotate. Apart from minor modifications his proposals have stood up extremely well.

Electrons in Orbitals

Van't Hoff's explanation, of course, was a purely formal one and provided no real insight into *why* a double bond prevents the parts of a molecule it joins from rotating with respect to each other. This was not understood for another 50 years, when the development of wave mechanics by Erwin Schrödinger set the stage for one of the most productive periods in theoretical chemistry. With Schrödinger's wave equation to guide them, chemists and physicists could compute the orbitals around atoms where pairs of electrons could be found. The valence bonds, which chemists had been drawing as lines for almost a century, now took on physical reality in the form of pairs of electrons confined to orbitals that were generally located in the regions where the valence lines had been drawn.

The first molecule to be analyzed successfully by the new wave mechanics was hydrogen (H_2). Walter Heitler and Fritz London applied Schrödinger's prescription and obtained the first profound insight into the nature of chemical bonding. Their results define the region in space most likely to be frequented by the pair of electrons associated with the two hydrogen atoms in the hydrogen molecule. The region resembles a peanut, each end of which contains a proton, or hydrogen nucleus [*see top illustration on page 93*].

The phrase "most likely to be frequented" must be used because, as Max Born convincingly argued in the late 1920's, the best one can do in the new era of quantum mechanics is to calculate the probability of finding electrons in certain regions; all hope of placing them in fixed orbits must be abandoned. The new methods were quickly applied to many kinds of molecule, including some with double bonds.

One of the most fruitful methods for describing doubly bonded molecules—the molecular-orbital method—was devised by Robert S. Mulliken of the Uni-

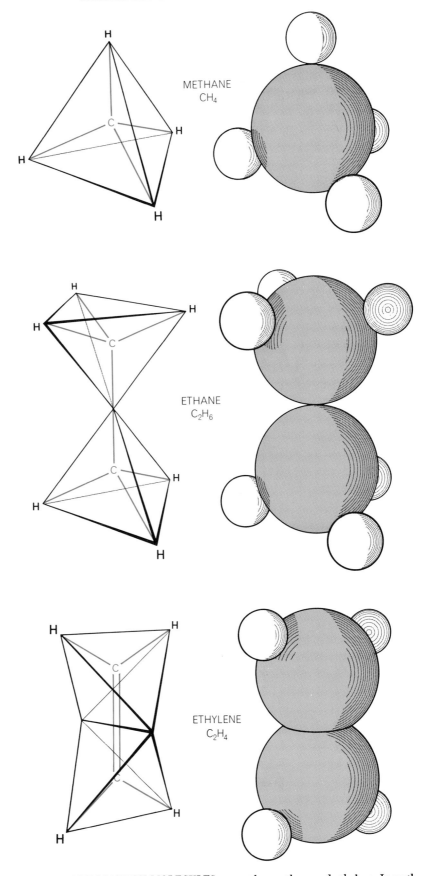

SIMPLEST HYDROCARBON MOLECULES are methane, ethane and ethylene. In methane the carbon atom lies at the center of a tetrahedron that has a hydrogen atom at each apex. The models at right show relative diameters of carbon and hydrogen. Ethane can be visualized as two tetrahedrons joined apex to apex. Ethylene, the simplest hydrocarbon that has a carbon-carbon double bond, can be visualized as two tetrahedrons joined edge to edge. The C=C bond in ethylene is about 15 percent shorter than the C—C bond in ethane.

versity of Chicago, who in 1966 received the Nobel prize in chemistry. In Mulliken's concept a double bond can be visualized as three peanut-shaped regions [see bottom illustration, opposite page]. The central peanut, the "sigma" orbital, encloses the nuclei of the two adjacent atoms, as in the hydrogen molecule. The other two peanuts, which jointly form the "pi" orbital, lie along each side of the sigma orbital. The implication of this model is that in forming the sigma orbital the two electrons occupy a common volume, whereas in forming the pi orbital both electrons tend to occupy the two separate volumes simultaneously.

One can say that the sigma bond connects the atoms like an axle that joins two wheels but leaves them free to rotate separately. The pi bond ties the

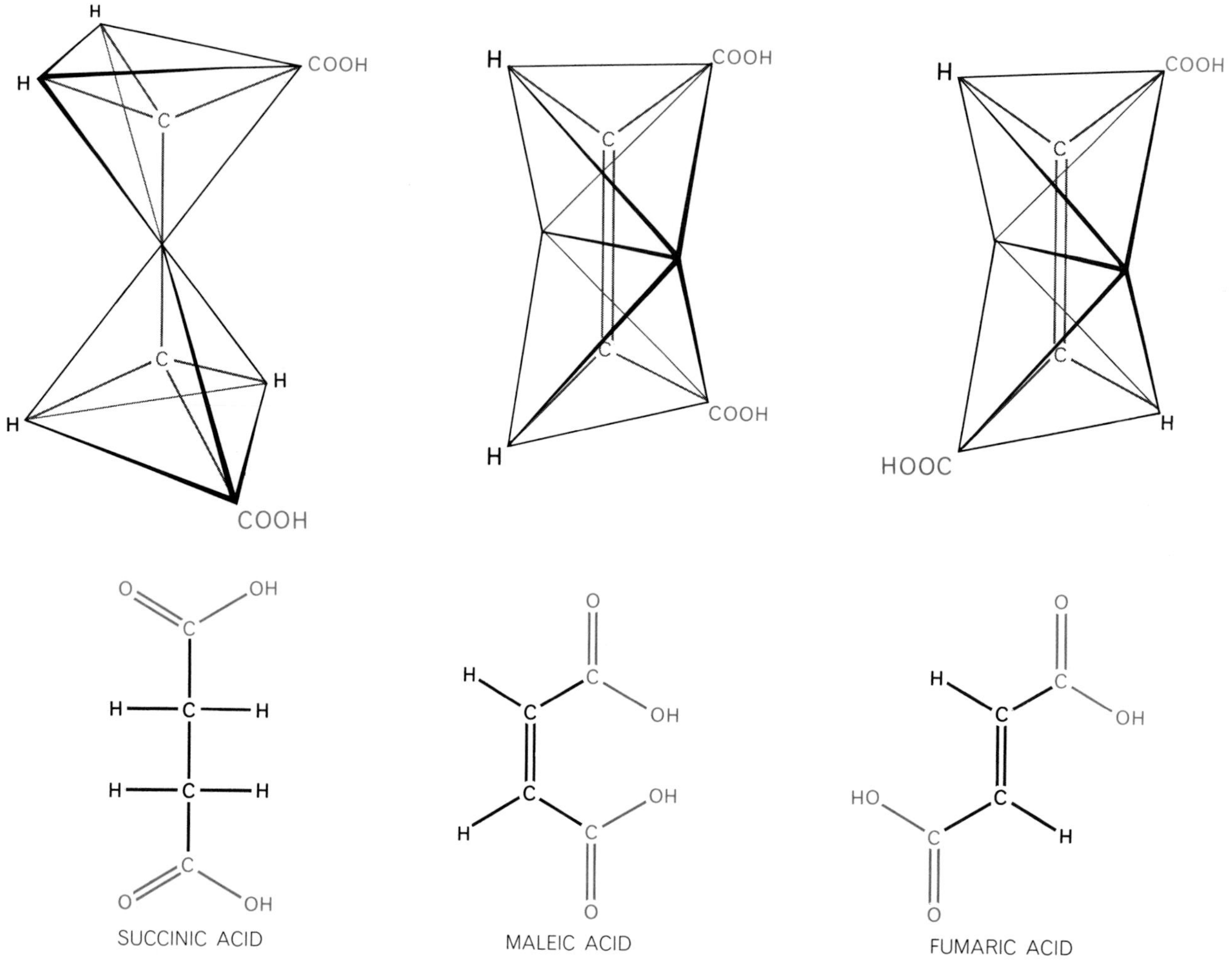

SUCCINIC ACID

MALEIC ACID

FUMARIC ACID

MOLECULAR PUZZLE was presented to chemists of the 19th century by maleic acid and fumaric acid, which have the same formula, $C_4H_4O_4$, and can be converted to succinic acid by the addition of two hydrogen atoms. Nevertheless, maleic and fumaric acids have very different properties. In 1874 Jacobus Henricus van't Hoff suggested that the central pair of carbon atoms in the three acids could be visualized as occupying the center of tetrahedrons that were joined edge to edge in the case of maleic and fumaric acids and apex to apex in the case of succinic acid. Thus the spatial relations of the two carboxyl (COOH) groups would be rigidly fixed in maleic and fumaric acids but not in succinic acid, because in the latter molecule the tetrahedrons would be free to rotate.

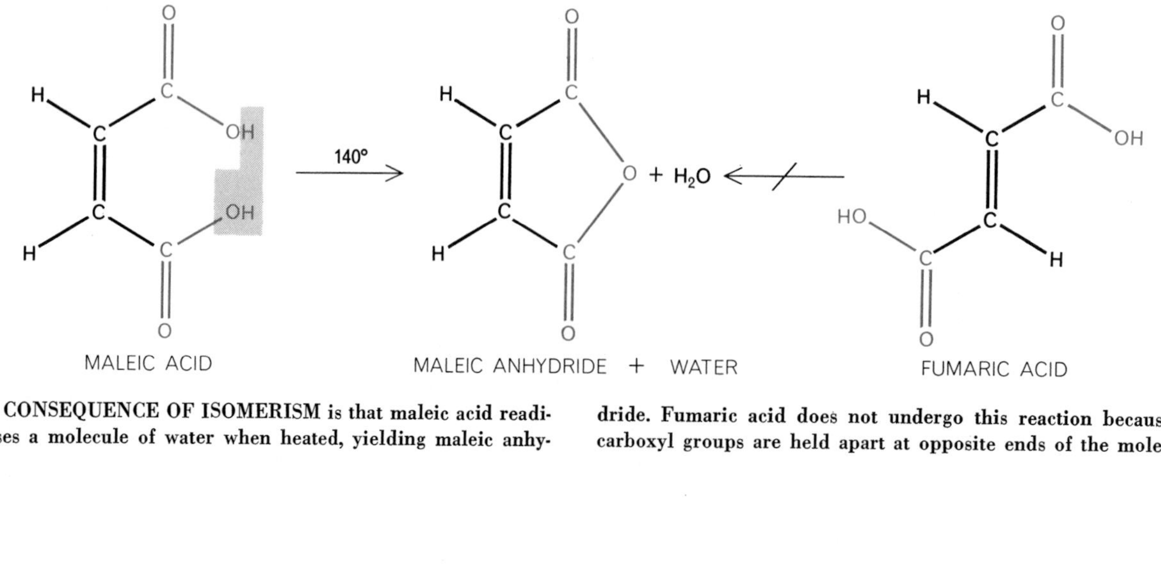

MALEIC ACID

MALEIC ANHYDRIDE + WATER

FUMARIC ACID

ONE CONSEQUENCE OF ISOMERISM is that maleic acid readily loses a molecule of water when heated, yielding maleic anhydride. Fumaric acid does not undergo this reaction because its carboxyl groups are held apart at opposite ends of the molecule.

two wheels together so that they must rotate as a unit. It also forces the two halves of the molecule to lie in the same plane, exactly as if two tetrahedrons were cemented edge to edge. In this way the molecular-orbital description of bonding provides a quantum-mechanical explanation of *cis-trans* isomerism.

The single bond joining two atoms, such as the carbon atoms of the two methyl groups in ethane (CH_3-CH_3), is a sigma bond, which leaves the groups attached to the two carbons free to rotate with respect to each other. Actually the two methyl groups in ethane are known to have a preferred configuration, so that they are not completely freewheeling. Nonetheless, at ordinary temperatures enough energy is available to make 360-degree rotations so frequent that derivatives of ethane (in which one hydrogen in each methyl group is replaced by a different kind of atom) do not form *cis-trans* isomers. There are exceptions, however, if the groups of atoms that replace hydrogen are so bulky that they collide and prevent rotation. In general, therefore, *cis-trans* isomerism is confined to molecules incorporating double bonds.

Electrons Delocalized

So much for molecules that have one double bond. What is the situation when a molecule has two or more double bonds? Specifically, what stereochemical behavior can be expected when single and double bonds alternate to form what is called a conjugated system?

The simplest conjugated system is found in 1,3-butadiene, a major ingredient in the manufacture of synthetic rubber, which can be written C_4H_6 or $CH_2=CH-CH=CH_2$. The designation "1,3" indicates that the double bonds originate at the first and third carbon atoms. Some of the properties of the biologically more interesting conjugated molecules are exhibited by butadiene. From the foregoing discussion one might expect that the second and third carbon atoms in the molecule would be free to rotate around the sigma bond connecting them. In actuality the rotation is not free: all the atoms in butadiene tend to lie in a plane.

It can also be shown that the energy content of each double bond in butadiene differs significantly from the energy content of the one double bond in the closely related compound 1-butene ($CH_2=CH-CH_2-CH_3$). The energy released in changing the two double bonds

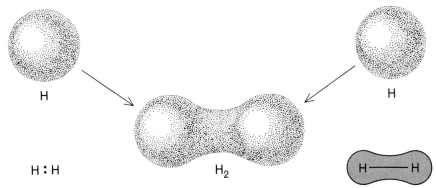

HYDROGEN MOLECULE is formed when two atoms of hydrogen (H) are joined by a chemical bond. The bond is created by the pairing of two electrons, one from each atom, which must have opposite magnetic properties if the atoms are to attract each other. The position of the electrons as they orbit around the hydrogen nuclei cannot be precisely known but can be represented by an "orbital," a fuzzy region in which the electrons spend most of their time. Known as a sigma orbital, it can be stylized as at lower right. The formula for the hydrogen molecule can be written as at lower left; the dots indicate electrons.

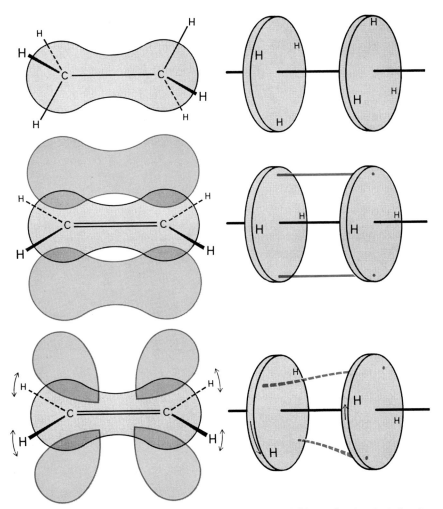

MOLECULAR ORBITALS help to explain why molecules held together by single bonds differ from molecules with double bonds. For example, the two carbon atoms in ethane are joined by two electrons in a sigma orbital, similar to the orbital in the hydrogen molecule. The two ends of the molecule, like wheels joined by a simple axle, are able to rotate. The two carbon atoms in ethylene (*middle*) are joined by two additional electrons in a "pi" orbital (*color*), as well as by two electrons in a sigma orbital. The four hydrogen atoms in ethylene are held in a plane perpendicular to the plane of the orbitals. The effect is as if two wheels were held together by two rigid rods in addition to an axle. When ethylene is in an "excited" state (*bottom*), one of the pi electrons occupies the four-lobed orbital. This lessens the rigidity of the double bond and gives it more of the character of a single bond.

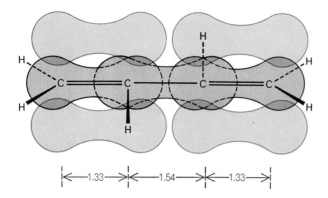

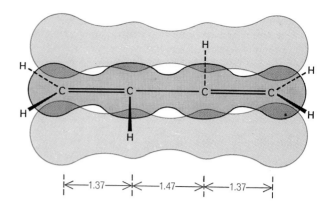

DELOCALIZED ORBITALS are found in "conjugated" systems: molecules in which single and double bonds alternate. The simplest conjugated molecule is 1,3-butadiene (C_4H_6). If its pi orbitals (*color*) were simply confined to the double bonds, as in ethylene, its orbital structure and carbon-carbon distances (in angstrom units) would be as shown at the left. Even in the lowest energy state, however, the pi electrons tend to spread across the entire molecule (*right*). As a result double bonds are lengthened and the single bond is shortened, making each type of bond more like the other. As a consequence the entire molecule is planar, or flat.

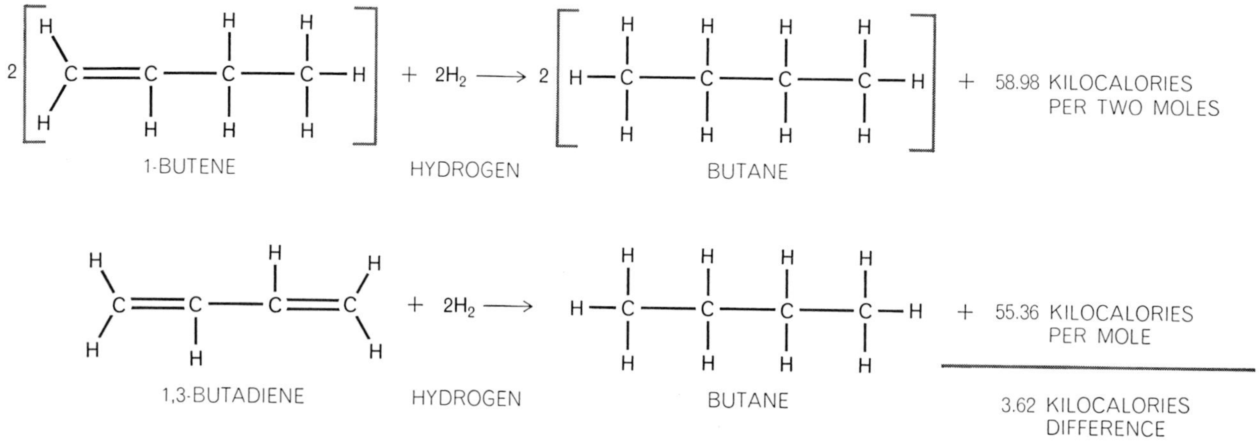

2 [1-BUTENE] + 2H₂ ⟶ 2 [BUTANE] + 58.98 KILOCALORIES PER TWO MOLES

1,3-BUTADIENE + 2H₂ ⟶ BUTANE + 55.36 KILOCALORIES PER MOLE

3.62 KILOCALORIES DIFFERENCE

EVIDENCE FOR BOND MODIFICATION in a conjugated molecule can be obtained by measuring the energy released when double bonds are converted to single bonds by adding hydrogen. Hydrogenation of the double bond in 1-butene, which is not a conjugated molecule, yields 29.49 kilocalories for every mole of reactant. A mole is a weight in grams equal to the molecular weight of a substance: 56 for butene and 54 for butadiene. Hydrogenation of two moles of butene, hence the hydrogenation of twice as many double bonds, would therefore yield 58.98 kilocalories. Hydrogenation of the same number of double bonds in butadiene (present in a single mole) yields only 55.36 kilocalories. The difference is 3.62 kilocalories per mole for the two bonds, or 1.81 kilocalories for each double bond. The lesser energy in the butadiene double bonds indicates that they are more stable than the double bond in 1-butene.

of butadiene into the single bonds of butane (CH_3–CH_2–CH_2–CH_3) is about 55,400 calories per mole of butane formed. (A mole is a weight in grams equal to the molecular weight of the molecule: 58 for butane, 54 for butadiene.) The energy released in converting 1-butene, which has only one double bond, into butane is about 29,500 calories per mole. When expressed in terms of equivalent numbers of double bonds hydrogenated, the latter reaction yields some 1,800 calories more than the former [*see lower illustration above*]. The greater energy release means that the double bond in 1-butene is more reactive than either of those in 1,3-butadiene.

The added stability of the bonds in 1,3-butadiene was not unexpected; the same kind of result had been obtained for benzene, whose famous ring structure is formed by six carbon atoms connected alternately by single and double bonds. One can picture the extra energy of stabilization as arising from the tendency of electrons in the pi orbitals to leak out and become delocalized. Indeed, the phenomenon is called delocalization. The pi orbitals spread over larger portions of conjugated molecules than one might have thought, so that the properties of delocalized systems can no longer be described in terms of the properties of the double and single bonds as they are usually drawn. In order to represent the pi orbitals of 1,3-butadiene more accurately one must stretch them across all four carbon atoms of the molecule [*see upper illustration above*]. The stretching helps to explain why butadiene is not completely free to rotate around the central single bond: the bond has some

of the characteristics of a double bond.

The altered character of butadiene's central carbon-carbon bond has been confirmed by X-ray-diffraction studies of butadiene. Whereas the usual carbon-carbon bond lengths are about 1.54 angstrom units for a single bond and 1.33 angstroms for a double bond, the length of the central single bond in 1,3-butadiene is only 1.47 angstroms. (An angstrom is 10^{-8} centimeter.) Linus Pauling, who did much to clarify the nature of the chemical bond, has estimated that the observed shortening of the central carbon-carbon bond of butadiene implies that it has about 15 percent of the double-bond character. One consequence of this is that the molecular configuration of butadiene tends to remain planar, or flat.

The tendency toward planarity in

conjugated systems was clearly demonstrated by the Scottish X-ray crystallographer J. M. Robertson and his colleagues in the mid-1930's. They compared the configurations of dibenzyl and *trans* stilbene, both of which contain two benzene rings joined by two carbon atoms [*see illustration at right*]. The difference between the two molecules is that in dibenzyl the two carbons are joined by a single bond, whereas in stilbene they are joined by a double bond. Robertson showed that the rings in dibenzyl are essentially at right angles to the connecting carbon-carbon bridge. In the *trans* form of stilbene the rings and the bridge lie in a plane, and the single bonds that join the rings to the two carbons in the bridge are foreshortened from the normal single-bond length to about 1.44 angstroms. In the *cis* form of stilbene the two rings cannot lie in a plane because they bump into each other.

Light-sensitive Molecules

We turn now to *cis-trans* isomerism in the family of molecules we have worked with most directly, the carotenoids and their near relatives, retinal (also known as retinene) and vitamin A. These molecules are built up from units of isoprene, which is like 1,3-butadiene in every respect but one: at the second carbon of isoprene a methyl group (CH_3) replaces the hydrogen atom present in butadiene. Natural rubber is polyisoprene, a long conjugated chain of carbon atoms with a methyl group attached to every fourth carbon.

The compound known as beta-carotene, which is responsible for the color of carrots, consists of an 18-carbon conjugated chain terminated at both ends by a six-member carbon ring, each of which adds another double bond to the conjugated system. The molecule has 40 carbon atoms in all and is presumably assembled from eight isoprene units [*see illustration on page 96*].

Until about 15 years ago the *cis-trans* isomers of the carotenoids entered biology in only one rather trivial way: in determining the color of tomatoes. Laszlo T. Zechmeister and his collaborators at the California Institute of Technology found in the early 1940's that normal red tomatoes contain the carotenoid lycopene in the all-*trans* configuration. (Lycopene differs from beta-carotene only in that the six carbon atoms at each end of the molecule do not close to form rings.) The yellow mutant known as the tangerine tomato contains

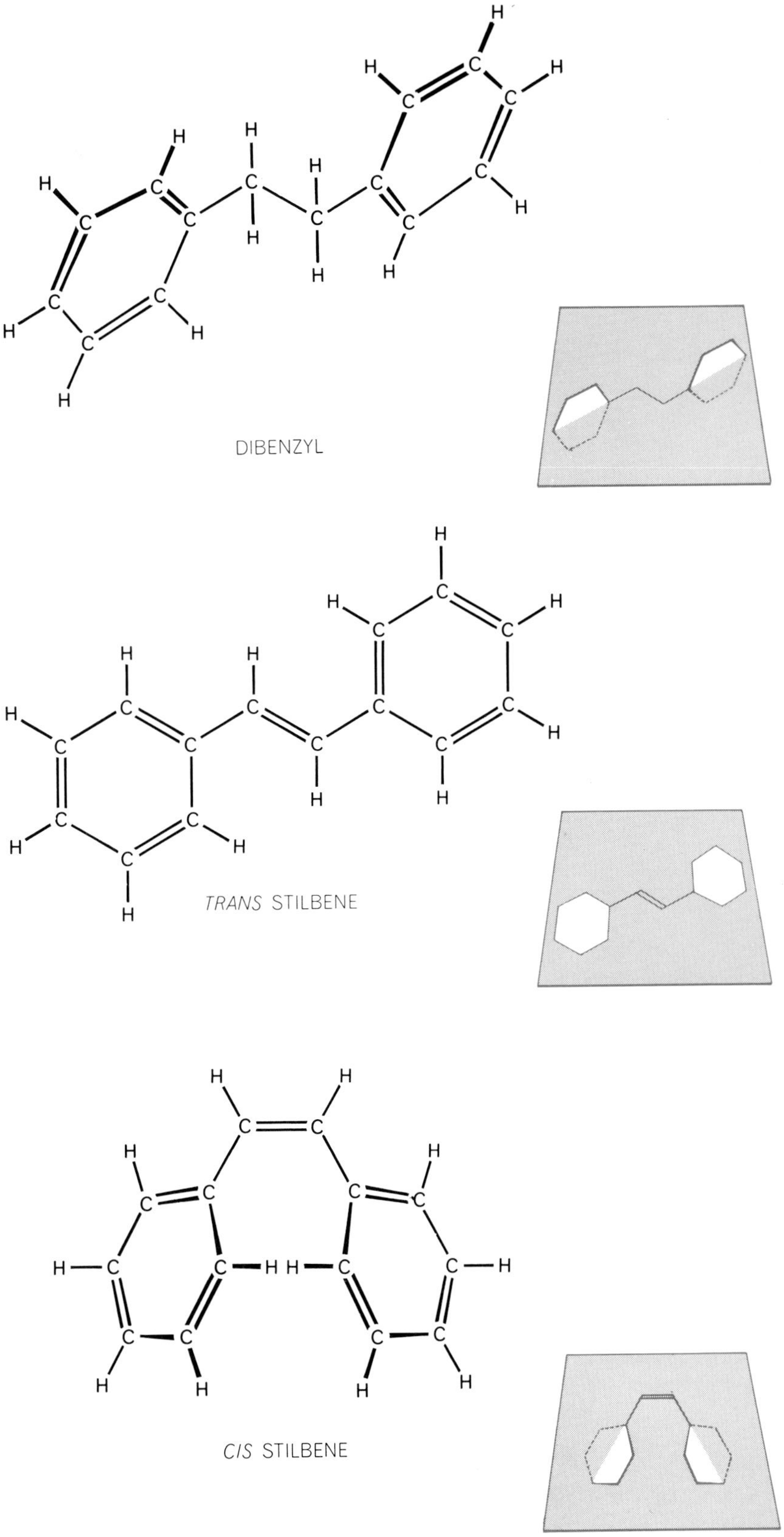

DIBENZYL

TRANS STILBENE

CIS STILBENE

PREFERENCE FOR FLATNESS in conjugated systems is exhibited by molecules of dibenzyl and two isomers of stilbene. The latter have a double bond in the carbon-carbon bridge linking the two benzene rings, whereas dibenzyl has a single bond. In dibenzyl the two rings are practically at right angles to the plane of the bridge. In *trans* stilbene all the atoms lie essentially in a plane. In *cis* stilbene, as can be demonstrated with molecular models, the two rings interfere with each other and thus cannot lie flat. The twisting of the rings has been established by X-ray studies of a related compound, *cis* azobenzene, in which the two rings are joined through a doubly bonded nitrogen (N=N) bridge.

ALL-*TRANS* CAROTENE

ALL-*TRANS* LYCOPENE

TWO NATURAL CAROTENOIDS are examples of highly conjugated systems. Like other carotenoids, they are built up from units of isoprene (C_5H_8), also known as 2-methyl butadiene. In these diagrams hydrogen atoms are omitted so that the carbon skeletons can be seen more clearly. Both molecules contain 40 carbon atoms and are symmetrical around the central carbon-carbon double bond, numbered 15–15'. Beta-carotene gives carrots their characteristic orange color. *Trans* lycopene is responsible for the red color of tomatoes.

a yellow *cis* isomer of lycopene called prolycopene. Zechmeister, who contributed more than anyone else to our present understanding of carotenoid chemistry, liked to demonstrate how a yellow solution of prolycopene, extracted from tangerine tomatoes, could be converted into a brilliant orange solution of all-*trans* lycopene simply by adding a trace of iodine in the presence of a strong light.

The discovery that the *cis-trans* isomerism of a carotenoid plays a crucial role in biology was made in the laboratory of George Wald of Harvard University, where it was found that the *cis-trans* isomerism of retinal is intrinsic to the way in which visual pigments react to light. The discovery of these pigments is usually attributed to the German physiologist Franz Boll.

The Chemistry of Vision

In 1877, the same year that Kolbe was ridiculing van't Hoff's work on stereochemistry, Boll noted that a frog's retina, when removed from the eye, was initially bright red but bleached as he watched it, becoming first yellow and finally colorless. Subsequently Boll observed that in a live frog the red color of the retina could be bleached by a strong light and would slowly return if the animal was put in a dark chamber. Recognizing that the bleachable substance must somehow be connected with the frog's ability to perceive light, Boll named it "erythropsin" or "Sehrot" (visual red). Before long Willy Kühne of Heidelberg found the red pigment in the retinal rod cells of many animals and renamed it "rhodopsin" or "Sehpurpur" (visual purple), which it has been called

ever since. Kühne also named the yellow product of bleaching "Sehgelb" (visual yellow) and the white product "Sehweiss" (visual white).

The chemistry of the rhodopsin system remained largely descriptive until 1933, when Wald, then a postdoctoral fellow working in Otto Warburg's laboratory in Berlin and Paul Karrer's laboratory in Zurich, demonstrated that the eye contains vitamin A. Wald showed that the vitamin appears when rhodopsin is bleached by light—the physiological process known as light adaptation—and disappears when rhodopsin is resynthesized during dark adaptation [see "Night Blindness," by John E. Dowling; SCIENTIFIC AMERICAN Offprint 1053]. He found that rhodopsin consists of a colorless protein (later named opsin) that carries as its chromophore, or color bearer, an unknown yellow carotenoid that he called retinene. Wald went on to show that the bleaching of rhodopsin to visual yellow corresponds to the liberation of retinene from its attachment to opsin, and that the fading of visual yellow to visual white represents the conversion of retinene to vitamin A. During dark adaptation rhodopsin is resynthesized from these precursors.

The chemical relation between retinene and vitamin A was elucidated in 1944 by R. A. Morton of the University of Liverpool. He showed that retinene is formed when vitamin A, an alcohol, is converted into an aldehyde, a change that involves the removal of two atoms of hydrogen from the terminal carbon atom of the molecule. As a result of Morton's finding the name retinene was recently changed to retinal.

In 1952 one of us (Hubbard), then a graduate student in Wald's laboratory, demonstrated that only the 11-*cis* isomer of retinal can serve as the chromophore of rhodopsin. This has since been confirmed for all the visual pigments whose chromophores have been examined. These pigments found in both the rod and cone cells of the eye contain various opsins, which combine either with retinal (strictly speaking retinal₁) or with a slightly modified form of retinal known as retinal₂. One other isomer of retinal, the 9-*cis* isomer, also combines with opsins to form light-sensitive pigments, but they are readily distinguishable from the visual pigments in their properties and have never been found to occur naturally. They have been called isopigments.

In 1959 we showed that the only thing light does in vision is to change the shape of the retinal chromophore by isomerizing it from the 11-*cis* to the all-*trans* configuration [see *illustration on page 90*]. Everything else—further chemical changes, nerve excitation, perception of light, behavioral responses—are consequences of this single photochemical act.

The change in the shape of the chromophore alters its relation to opsin and ushers in a sequence of changes in the mutual interactions of the chromophore and opsin, which is observed as a sequence of color changes. In vertebrates the all-*trans* isomer of retinal and opsin are incompatible and come apart. In some invertebrates, such as the squid, the octopus and the lobster, a metastable state is reached in which the all-*trans* chromophore remains bound to opsin.

Until the structure of opsin is established there is no way to know just how 11-*cis* retinal is bound to the opsin molecule. In the 1950's F. D. Collins, G. A. J. Pitt and others in Morton's laboratory showed that in cattle rhodopsin the aldehyde (C=O) group of 11-*cis* retinal forms what is called a Schiff's base with an amino (NH_2) group in the opsin molecule. Recently Deric Bownds in Wald's laboratory has found that the amino group belongs to lysine, one of the amino acid units in the opsin molecule, and has identified the amino acids in its immediate vicinity. There is little doubt that 11-*cis* retinal also has secondary points of attachment to opsin; otherwise it would be hard to explain why only the 11-*cis* isomer serves as the chromophore in visual pigments. Light changes the shape of the chromophore and thus alters its spatial relation to opsin. This leads, in turn, to changes in the shape of the opsin molecule [see lower illustration on opposite page]. The details of these changes, however, are still obscure.

How Molecules Twist

Let us examine somewhat more closely the various isomers of retinal. The six known isomers are illustrated at the left: the all-trans isomer and five cis isomers of one kind or another. Experiments with models, together with other evidence, show that four of the six isomers are essentially planar. The two that are not are the 11-cis isomer and the 11,13-dicis isomer. In these isomers there is considerable steric hindrance, or intramolecular crowding, between the hydrogen atom on carbon No. 10 (C_{10}) and the methyl group attached to C_{13}. Thus the double bond that joins C_{11} and C_{12} cannot be rotated by 180 degrees from the trans to a planar cis configuration. In the 11-cis isomers the tail of the molecule from C_{11} through C_{15} is therefore twisted out of the plane formed by the rest of the molecule.

This twisted geometry introduces two configurations, called enantiomers, that are mirror images of each other; if the molecule could be viewed from the ring end, one form would be twisted to the left and the other to the right. It is possible that opsin may combine selectively with only one enantiomer.

As Pauling had predicted in the 1930's, the steric hindrance that necessitates the twist in the 11-cis isomer makes it less stable than the all-trans or the 9-cis and 13-cis forms. We have recently found, for example, that the 11-cis form contains about 1,500 calories more "free energy" per mole than the trans form. One has to put in about 25,000 calories per mole, however, to rotate the molecule from one form to the other. This amount of energy, which is much more than a molecule is likely to acquire through chance collisions with its neighbors, is known as the activation energy: the energy required to surmount the barrier that separates the cis and trans states.

This raises an important point. How can two parts of a molecule be rotated around a double bond? The interconversion of cis and trans isomeric forms to another requires gross departures from flatness. How can this be accomplished?

Here we must introduce the concept of the excited state. One can think of molecules as existing in two kinds of state: a "ground," or stable, state of relatively low internal energy and various less stable states of higher energy—

the excited states. Molecules are raised from the ground state into one or another excited state by a sudden influx of energy, which can be in the form of heat or light. They return to the ground state by giving up their excess energy, usually as heat but occasionally as light, as in fluorescence or phosphorescence.

The orbital diagrams we have described apply to molecules in the ground state. When molecules are in an excited state, their electrons have more energy and therefore occupy different orbitals. Quantum-mechanical calculations show that an excited pi electron divides its time between the two ends of a double bond [see bottom illustration on page 93]. The net effect is to make the double bond in an excited molecule more like a single bond and less like a double bond. In a conjugated molecule, in which pi electrons are already delocalized, the changes in bond character are not uniform throughout the conjugated system but depend on the nature of the excitation and the structure of the molecule that is excited.

When one tries to isomerize carotenoids in the laboratory, it is usually helpful to add catalysts such as bromine or iodine. (The reader will recall that Zechmeister used iodine in his demonstrations.) Heat and light favor the existence of excited states. Bromine and iodine probably function by dissociating into atomic bromine and iodine, a process that is also favored by light. A bromine or iodine atom adds fleetingly to the double bond and converts it into a single bond, which is then momentarily free to rotate until the bromine or iodine atom has departed. The actual lifetime of the singly bonded form can be very brief indeed: the time required for one rotation around a carbon-carbon single bond is only about 10^{-12} second.

The Sensitivity of Eyes

It may seem remarkable that all animal visual systems so far studied depend on the photoisomerization of retinal for light detection. Three main branches of the animal kingdom—the mollusks, the arthropods and the vertebrates—have evolved types of eyes that differ profoundly in their anatomy. It seems that various anatomical (that is, optical) arrangements will do; apparently the photochemistry, once it had evolved, was universally accepted. Presumably the visual pigments of all animals must within narrow limits be equally sensitive to light, otherwise the more light-sensitive animals would eventually

SIX ISOMERS OF RETINAL are represented in skeleton form below the structure of all-trans vitamin A. Hydrogen atoms are omitted, except for the H in the hydroxyl group of vitamin A. If that H and one other on the final carbon are removed, all-trans retinal results. This isomer and 11-cis retinal, which combines with opsin to form rhodopsin, are the isomers involved in vision.

replace those whose eyes were less sensitive.

How sensitive to light is the animal retina? In a series of experiments conducted about 1940 Selig Hecht and his collaborators at Columbia University showed that the dark-adapted human eye will detect a very brief flash of light when only five quanta of light are absorbed by five rod cells. From this Hecht concluded that a single quantum is enough to trigger the discharge of a dark-adapted rod cell in the retina.

It is therefore essential that the quantum efficiency of the initial photochemical event be close to unity. In other words, virtually every quantum of light absorbed by a molecule of rhodopsin must isomerize the 11-*cis* chromophore to the all-*trans* configuration. It was shown many years ago by the British workers H. J. A. Dartnall, C. F. Goodeve and R. J. Lythgoe that an absorbed quantum has about a 60 percent chance of bleaching frog rhodopsin. One of us (Kropf) has found a similar quantum efficiency for the isomerization of the 11-*cis* retinal chromophore of cattle rhodopsin. Our work also shows that 11-*cis* retinal is more photosensitive than either the 9-*cis* or the all-*trans* isomers when they are attached as chromophores to opsin, and this may be the reason why the geometrically hindered and therefore comparatively unstable 11-*cis* isomer has evolved into the chromophore of the visual pigments.

We have also recently measured the quantum efficiency of the photoisomerization of retinal and several closely related carotenoids in solution. Retinal

RHODOPSIN

11-*CIS* RETINAL + OPSIN ⇌ ALL-*TRANS* RETINAL + OPSIN

11-*CIS* VITAMIN A + OPSIN ⇌ ALL-*TRANS* VITAMIN A + OPSIN

PHOTOCHEMICAL EVENTS IN VISION involve the protein opsin and isomers of retinal and its derivative, vitamin A. Opsin joined to 11-*cis* retinal forms rhodopsin. When struck by light, the 11-*cis* chromophore is converted to an all-*trans* configuration and subsequently all-*trans* retinal becomes detached from opsin. With the addition of two hydrogen atoms, all-*trans* retinal is converted to all-*trans* vitamin A. Within the eye this isomer must be converted to 11-*cis* vitamin A, thence to 11-*cis* retinal, which recombines with opsin to form rhodopsin.

turns out to be considerably more photosensitive than any of them and nearly as photosensitive as rhodopsin.

Although all animal eyes seem to employ 11-*cis* retinal as their light-sensitive agent, there are slight variations in the opsins that combine with retinal, just as there are variations in other proteins, such as hemoglobin, from species to species. Within the next few years we may learn the complete amino acid sequence of one of the opsins, and thereafter we should be able to compare such sequences for two or more species. It may be many years, however, before X-ray crystallographers have established the complete three-dimensional structure of an opsin molecule and are able to describe the site that binds it to retinal. One can conjecture that the binding site will be quite similar in the various opsins, even those from animals of different phyla, but there may be surprises in store. Whatever the precise details, it is clear that evolution has produced a remarkably efficient system for translating the absorption of light into the language of biochemistry—a language whose vocabulary and syntax are built on the various ways proteins interact with one another and with smaller molecules in their environment.

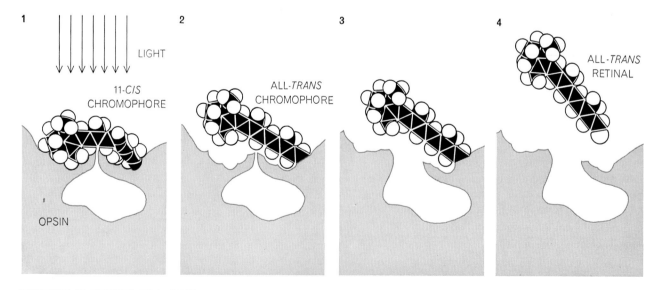

MOLECULAR EVENTS IN VISION can be inferred from the known changes in the configuration of 11-*cis* retinal after the absorption of light. In these schematic diagrams the twisted isomer is shown attached to its binding site in the much larger protein molecule of opsin (1). After absorbing light the 11-*cis* chromophore straightens into the all-*trans* isomer (2). Presumably a change in the shape of opsin (3) facilitates the release of all-*trans* retinal (4). The configuration of the binding site in opsin is not yet known.

IV

GASES, LIQUIDS, AND SOLIDS

IV

GASES, LIQUIDS, AND SOLIDS

INTRODUCTION

The science of chemistry is often associated with the study of chemical reactions. Hydrogen-oxygen explosions, the electrolysis of brine, or the combustion of a hydrocarbon represent some of the things which chemists might study. Indeed, the kinetics, thermodynamics, and equilibrium properties of chemical reactions do constitute a major part of chemists' studies. However, many chemists are also engaged in studies of the states of matter.

Have you ever wondered why the states of matter are limited to the three given in the title for this section? There is no *a priori* reason why this should be so. But if we exclude exotic matter such as plasmas and certain liquids near 0 K, our experience indicates that there are only the three in the title. They may be characterized by how they fill containers. Solids have a definite size and shape, liquids have a definite size but their shape conforms to the vessel which they occupy, and gases have neither a definite size nor a definite shape but expand to fill their restraining vessel.

A description of each state of matter is needed to understand how matter behaves. For instance, how do the properties of a system vary as the temperature and pressure vary? Such a description is called an equation of state. The only state of matter for which a satisfactory equation of state is available is the gaseous state. You undoubtedly are familiar with the equation of state for an ideal gas, $PV = nRT$. A number of empirical equations are available for real gases. A great deal of work has been done in trying to provide an equation of state for liquids. A recent attempt is summarized in the *Journal of Chemical Education*, **45**, 2 (1968). However, no generally applicable equation has yet been found.

Theoretical investigations of solids have centered primarily on an understanding of the electrical properties of solids. The band theory of metals provided the necessary theoretical understanding for development of the solid-state diode and transistor. As you know, these devices and the microminiaturization which followed have revolutionized the entire field of electronics and space exploration. As you will see in Mott's article, 11, "The Solid State," this theory can readily explain why sodium is a conductor and carbon is an insulator.

Another topic of interest to chemists is that of phase transitions. Under what conditions is one state of matter converted into another? What are the requirements for the coexistence of two phases? Can all three phases be present in a system at equilibrium? Phase diagrams are particularly helpful in answering these questions. The diagram for the most common liquid, water, is given in Figure IV.1.

The phase diagram for any substance provides interesting information about that substance. The water diagram, for instance, explains why ice-skating is possible. The slope of the line between the solid and the liquid is negative. Thus, the high pressure under the skate blade lowers the freezing point of water. The slight amount of liquid water thus produced provides the necessary lubricant for the skates to glide over the ice.

The liquid-gas curve in the water phase diagram provides an explanation of the effect of altitude on cooking times for food. As the

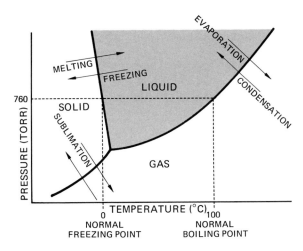

Figure IV.1   Phase diagram for water.

pressure decreases with increasing elevation, the temperature at which the vapor pressure equals that of the atmosphere also decreases and water boils at a lower temperature. The backpacker at the top of Pikes Peak must take the decreased pressure at that altitude into account when he boils a potato for his dinner.

The solid-gas curve of the diagram shows the vapor pressure of ice. It explains, for example, how clothes hung on a line to dry in a Minnesota winter when the temperature is 20°F can become thoroughly dry. The water sublimes.

The first article in this section, "Molecular Motions," reports the results of theoretical studies of the properties of solids, liquids, and gases and of the transitions between them. Two methods for calculating molecular distributions are reported. The Monte Carlo method consists, conceptually, of placing several balls in a box, specifying the x, y, and z coordinates of each ball, and then permitting one ball at a time to move to a new location, with new coordinates, subject only to the restraint that two balls can't occupy the same position at the same time. This is a conceptual method because, in fact, the balls and their locations are designated by computer-generated numbers. The method derives its name from the random way in which the moves are generated. This technique is excellent for providing a simulation of the placement of atoms and molecules within a system. The only limitation of the method is that it provides a statistical description only of the equilibrium properties of the system, such as the distributions of molecules when energy or pressure are constant, but it does not reproduce molecular motions.

The second method explored by Alder and Wainwright gives us a vivid feeling for how gases, liquids, and solids behave. In this dynamical method, the hypothetical particles are assigned both a position and a velocity. Fairly simple assumptions are made about how the particles, which represent molecules, interact when they approach each other. Then the computer simulates the motion of the particles and their tracks are recorded on the face of a cathode-ray tube which is linked to the computer doing the calculations. The figures on page 110 provide excellent representations of how solids and liquids differ. The two figures on page 111 demonstrate the dynamics of a liquid phase in contact with a bubble of gas. The blackness of the bubble at the top displays the low density of gases relative to that of liquids.

The next two articles in this section deal with the solid state. This would appear to be appropriate in view of the tremendous advances in our understanding of the properties of the solid state that developed in the 1960's, a decade that scientists already refer to as the solid-state decade.

Mott's article bears the title "Solid State" and gives a wide-ranging description of the solid state. Most of the discussion centers on crystalline solids. In these, the atoms are stacked more regularly than are the atoms in amorphous solids. Mott points out that most of our understanding of solids has come from X-ray diffraction (see Bragg's article about X-ray crystallography in section VI).

Fourteen crystal lattice systems describe all crystalline solids. These are shown in Mott's article. Most metals have hexagonal closest-packed or cubic closest-packed (face-centered cubic) structures. These two structures represent the closest packing of spheres of identical size. They differ only in that the third layer of one structural type is rotated 60° relative to the second. A significant number of metals also crystallize in the body-centered cubic structure.

A useful way to think of crystalline solids is to consider what species occupy the lattice sites in the fourteen lattice systems and what forces hold the lattice together. Table IV.1 extends the concept of metal lattices described above to three other classes.

Despite the fact that these chemical forces routinely align millions upon millions of atoms in a perfectly regular array, errors do occur that give rise to defects in the crystal lattices. Mott provides diagrams both of the crystal lattices and of actual metal surfaces displaying several types of lattice defects. Such defects are the source of such disparate behavior as the malleability of metals and the blue color of some salt crystals.

A special type of artificially induced defect produces one of the most useful types of solids and is related to the subject of the rest of Mott's article. The introduction of trace amounts of P or Al, for example, into very pure crystals of germanium or silicon provides the basis for manufacturing diodes and transistors. The band theory of metals explains the operation of these devices. Mott provides energy band diagrams to explain the difference between metals, semiconductors, and insulators.

A final comment regarding Mott's article is that graphite, which Mott shows in one diagram, is the ultimate extension of the conjugated

Table IV.1   Types of solids

	Molecular	Ionic	Metallic	Covalent
Particles that occupy lattice sites	Molecules or noble-gas atoms	Ions	Positive ions	Atoms
Type of bonding	van der Waals Hydrogen bonding	Electrostatic	Delocalized electrons	Shared electrons
Properties	Low melting points	High melting points	Moderately low to high melting points	Very high melting points
	Soft	Hard and brittle	Hard and soft	Very hard
	Poor electrical conductors	Poor electrical conductors	Good electrical conductors	Poor electrical conductors
Examples	H_2O, CO_2, H_2, Ar, and almost all organic compounds	NaCl, KBr, $LiNO_3$	Na, Cu, Ni	C (diamond), SiC (carborundum), SiS_2

systems described by Hubbard and Kropf in article 9, "Molecular Isomers in Vision," and by Breslow in article 25, "The Nature of Aromatic Molecules." A conjugated system, you may recall, is one having a series of alternating single and double bonds. Each layer of the graphite structure contains an almost endless sequence of alternating single and double bonds in hexagonal arrays.

The final article in this section deals with one of the most common and one of the most peculiar solids — ice. We can begin to understand some of the peculiar qualities of ice by visualizing an elementary model by which transition from the gaseous to the liquid to the solid state is shown by the different densities of atoms or molecules. This model predicts that the density of liquids is less than that of solids,

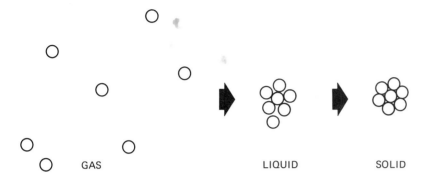

GAS   LIQUID   SOLID

Figure IV.2   General model showing distribution of atoms or molecules in a gas, a liquid, and a solid.

and, indeed, this behavior is observed in almost all cases. But a notable exception is water. Ice is less dense than liquid water. This is a most fortunate anomaly for ice skaters and winter fishermen.

This peculiarity is fairly well understood in terms of the unique hydrogen-bonding properties of water. A hydrogen bond is formed when a hydrogen atom bonded to an oxygen atom is attracted to an unshared electron pair on another oxygen atom, as we discussed in the introduction to section III. Water is ideally suited for a maximum number of hydrogen bonds because it possesses two hydrogen atoms bonded to an oxygen atom that has two unshared electron pairs. Runnels's article deals with the precise location of all of the atoms in the hydrogen-bonded ice structure.

The exact positions of one-third of the atoms in ice have been known for about thirty-five years. The oxygen atoms have a high enough electron density to deflect X rays. Thus crystallographers have been able to determine that the oxygen atoms form the symmetrical, interlocking hexagonal structure shown in Runnels's article. The problem is that the hydrogen atoms do not deflect X rays strongly enough to reveal their positions. Each molecule has six possible configurations in the ice lattice and scientists are still not certain which of these orientations exist. However, the creative use of an entirely different technique has shown that the actual number of different ways to orient the molecules is much less than six.

The technique is to use thermodynamics, one of the topics discussed in the next section. You probably recall from your study of thermodynamics that one of the important state functions is entropy. It is given the symbol S and is a measure of the randomness of a system. The greater the randomness a system possesses, the less ordered it is. Another way of saying that a system has a high degree of randomness is to say that it has available a large number of possible configurations. The diagrams showing the melting of a solid, the mixing of two gases, and the helix-coil transition of a protein (shown in Doty's article, 28, "Proteins") are all excellent examples of changes of state within a system so that they have a larger number of possible configurations and hence higher entropies.

The Austrian scientist Ludwig Boltzmann discovered a quantitative relation between entropy and the number of configurations (w) that a system possesses. His equation is

$$S = k \ln w$$

The proportionality constant, k, is called Boltzmann's constant and has the value 1.38×10^{-16} erg degree^{-1} molecule^{-1}. Runnels shows how this relation was employed to demonstrate that the number of configurations available to each molecule is slightly above 1.50, a value calculated by Linus Pauling several decades ago.

10

MOLECULAR MOTIONS

B. J. ALDER AND THOMAS E. WAINWRIGHT
October 1959

*One of the aims of molecular physics is to account
for the bulk properties of matter in terms of the
behavior of its particles. High-speed computers are
helping physicists realize this goal*

During the 19th century, as evidence in favor of the atomic theory mounted, an ancient hope of science began to bear fruit. As long ago as the first century B.C. Lucretius had proposed not only that matter is composed of tiny particles called atoms, but also that the behavior of these particles is the key to understanding the properties of bulk matter. For centuries this idea remained simply an interesting hypothesis. Then Isaac Newton set forth laws of motion from which the behavior of atoms might be calculated. At the same time a number of investigators were making quantitative observations on the gross properties of matter. The stage was set for an attempt to realize Lucretius' dream.

The earliest tries were mostly unsuccessful. But in 1739 Daniel Bernoulli succeeded in proving that the product of the pressure and the volume of a gas is proportional to the average kinetic

PHYSICAL MODEL of a molecular system consists of gelatin balls suspended in a tank of liquid. The tank is shaken to generate typical patterns of separations between balls. In the experiment depicted here separations were measured for the seven black balls.

energy of its atoms, provided that the atoms do not interact. His result is in fact valid for gases at very low density, where interactions between atoms or molecules are rare. It was not until the 19th century, however, that the program got under way on a grand scale, with the work of such men as Rudolf Clausius, James Clerk Maxwell and Ludwig Boltzmann. The task they faced was immense. Their predecessors had dealt with such problems as computing the orderly motions of a few planets under the gravitational attraction of the sun. These men were concerned with millions of particles, colliding with one another and darting about in all directions. To follow the trajectory of any individual atom in a piece of matter it would be necessary to know the position at all times of every other particle close enough to exert a force on it. The total force on the atom could then be computed from instant to instant, and, assuming that Newton's laws applied, its motion could be calculated.

In practice such a detailed calculation was hopelessly complicated for a dozen particles, let alone millions. Fortunately it was not necessary. The bulk properties of matter depend upon the average behavior of many atoms, and not upon the detailed motion of each. Therefore statistical methods could be used and a new branch of physics known as statistical mechanics grew up. The statistical approach fulfilled many of its founders' hopes. In some problems, however,

even this type of mechanics bogged down in cumbersome mathematics.

Today we are in a position to overcome some of the practical difficulties. Using high-speed computers we can perform calculations that were hitherto impossible. With the help of these machines we are moving a little closer to the ideal of understanding the properties of matter in terms of the mechanical behavior of its constituent particles.

In many applications of statistical mechanics it is possible to proceed with no knowledge of the velocities of individual atoms. We ask only how the atoms are distributed on the average in space. From this distribution alone many of the properties of the material can be calculated.

To appreciate what is meant by the spatial distributions, imagine that we have a microscope powerful enough to see the individual atoms or molecules in a sample of matter, and a camera fast enough to stop them in their rapid flight. A stereoscopic picture made with this arrangement shows how the particles are distributed at a given instant. We examine the particles in the snapshot in turn, measuring the distances between each one and all the others. The information is summarized on a graph, with distance of separation plotted along the horizontal axis, and the number of pairs at each distance along the vertical [see *illustration below*].

Suppose we make a number of snap-

shots in rapid succession, say 1,000 in a second. We plot the distances of separation in each picture and then obtain the average of all the curves. This graph represents the average distribution of pair separations for that second. If the system is in equilibrium, then the average distribution remains constant for all time.

Between each pair of molecules in a piece of matter there is a force that depends on the distance between them. Assuming that we know how the force varies with distance, we can, for example, use our average plot to compute the forces exerted on a typical molecule by its neighbors. This gives us the pressure of the system.

How do we actually find the distances of separation in a system of invisible particles darting about erratically at tremendous speeds? In some cases we can obtain the information experimentally. A beam of X-rays sent through a solid or a liquid is diffracted in a way that depends on the spacing between the particles of the substance. From the size and intensity of spots in the diffraction pattern we can deduce the distances of separation and the number of pairs at each separation. For a crystalline solid the pair-separation graph shows a number of distinct peaks and valleys, as in the illustration on the opposite page. The peaks indicate that the molecules tend to lie preferentially at certain specific distances from one another. This is what we should expect if the molecules are arranged in an orderly grid or lattice. As the temperature of the solid is lowered, the peaks in the plot become sharper, showing that the molecules vibrate less widely about their central positions in the lattice. With increasing temperature, on the other hand, the peaks grow broader. If the material is heated above its melting temperature, the peaks are still present at distances as small as a few molecular diameters. They disappear at large distances because the lattice structure has disintegrated, and there is no longer an ordering force between molecules at longer ranges.

From the point of view of statistical mechanics we should like to be able to find the distribution of distance between molecules theoretically and to explain how physical conditions such as temperature and density influence the distribution. This would enable us, for example, to predict the pressure or energy of a substance from its temperature and density.

One way of approaching the problem

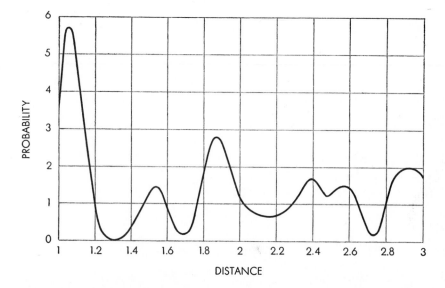

DISTANCE-OF-SEPARATION PLOT for a crystalline solid is characterized by peaks and valleys. Peaks represent preferred distances of separation between the molecules of the crystal. Horizontal axis measures distance in molecular diameters. Vertical axis measures the relative probability that pairs of particles will lie at each distance. From this curve it is possible to calculate the actual numbers of pairs that are separated by each distance.

is to construct a mental model of a system of particles, usually assuming a simplified law of force between them, and to devise some means of "shaking" the system; that is, moving the particles about at random. From time to time we stop shaking and make a "snapshot," recording the distances of separation of all the pairs of particles. Then the snapshots are averaged.

It should be understood that the random moves of the particles in the model are not the same as the motions that carry molecules from one place to another in real matter. The moves are simply a device for generating possible spatial arrangements. Because the moves are artificial, the order in which the resulting configurations occur is also unrealistic. In a system in equilibrium, however, the order of snapshots does not matter. The average plot is the same regardless of the sequence in which the individual curves are considered.

A relatively easy model to deal with is one in which the particles are taken to be hard spheres, like marbles. There is no attractive force between them, and they repel each other only when they collide. Simple though it is, this model duplicates some of the properties of real systems surprisingly well, as we shall see later on.

To construct a hard-sphere system we choose imaginary marbles of a suitable size and place them in an imaginary box (usually a cubical one for the sake of convenience). There is not only no force of attraction between marbles, but also no force of gravity; each sphere stays just where it is put. The size of the box depends on the density assigned to our hypothetical material. If we have chosen a low enough density, the process is straightforward. The marbles can be put in at random, with little chance that any will overlap. Such an overlap is of course disallowed. Whenever it occurs, we must remove the last marble put in and try placing it elsewhere. The same thing is true after shaking: snapshots in which marbles overlap are ruled out. At sufficiently low densities, corresponding to the gaseous state in real materials, the frequency of overlaps is negligible. It is not too difficult to generate mathematically as many distance-of-separation patterns as we wish.

As the density is increased, however, that is, as the average distance between marbles gets smaller, it becomes harder to put in the marbles at random without running into interference. With each succeeding particle we must try

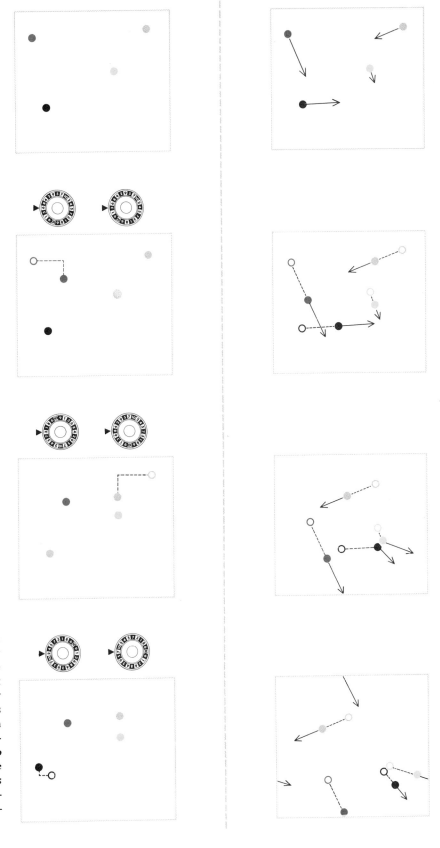

COMPUTER TECHNIQUES for calculating molecular distributions in three dimensions are illustrated schematically in these two-dimensional drawings. Monte Carlo method (*left*) moves particles according to a random-number table (represented by roulette wheels). Molecular-dynamical method (*right*) computes the actual trajectories. Open dots represent old positions of particles; solid dots, new positions. The particles are considered as being able to pass through the walls of the imaginary container and to re-enter on the opposite side.

more and more random positions before finding an empty spot. Eventually no more marbles can go in. The box is not completely full, but it is inefficiently packed. By rearranging the marbles already in it, we can make room for more. At this density matter begins to act like a liquid. In a liquid the spheres can rearrange themselves only if numbers of them move cooperatively. This situation makes the mathematics of shaking more cumbersome.

At still higher density, corresponding to that of a crystalline solid, the problem becomes simpler again. If the marbles are packed in an ordered array, and are touching one another, the box is completely full and shaking has no effect. The marbles cannot move at all. Now when the density is decreased slightly (by making the box bigger or the marbles smaller), the particles can rattle around their positions in the lattice, but cannot escape. This approximates the

situation in a real crystal. If the marbles are shrunk still further, they will eventually be able to escape from the cages formed by their neighbors and trade places. The lattice disappears and the solid melts. But each particle can escape from its lattice position only if its neighbors move cooperatively in such a way as to leave it a wide enough path. Again the mathematics becomes harder to carry out.

One way around the computational

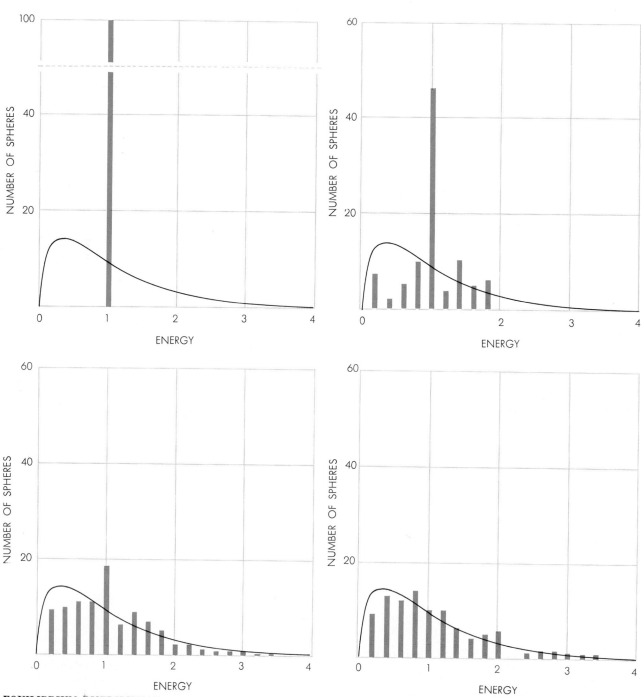

EQUILIBRIUM DISTRIBUTION of kinetic energies is rapidly attained by a system of 100 hard spheres, each having an energy of one unit (*top left*). Equilibrium pattern is represented by curve; numbers of particles at various energies, by bars. Drawings at top right, bottom left and bottom right show distribution after 50, 100 and 200 collisions, or an average of one, two and four per particle.

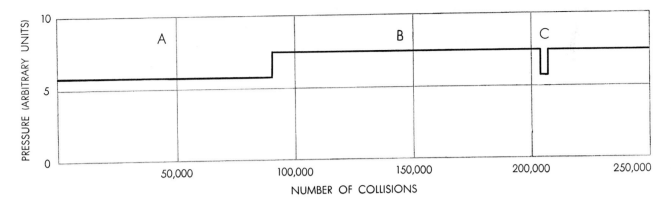

PRESSURE-JUMPS in a hard-sphere system signal changes in phase between the fluid and solid state. Region A represents the low-pressure, solid phase, which lasted for about 100,000 collisions. Then the pressure increased abruptly to that of a fluid (*region* B), continuing in this state for another 100,000 collisions. In region C it dropped briefly back to the solid value for another 3,000 collisions.

problem is to make a physical model and actually shake it. Some years ago Joel Hildebrand and his colleagues at the University of California performed just such an experiment. They suspended gelatin balls in a tank of liquid whose density was equal to that of the gelatin, so that the gravitational force was canceled by buoyancy. Placing the tank on a vibrating tray, they made a series of photographs from which the average distances of separation between the balls could be plotted. Their curve resembled plots made from X-ray studies of liquids.

No physical model, however, is altogether satisfactory. The chief problem is the difficulty of assuring that the shaking is really random.

With the help of a fast computer the experiment can be "performed" much more neatly by purely mathematical means on an ideal system. This has been done by a group at the Los Alamos Scientific Laboratory. Each particle is represented by a set of three numbers, specifying the three-dimensional coordinates of its center (x, y and z). Having located all the particles by feeding the appropriate sets of numbers into the machine, we make our first "photograph," recording the distances of separation between all pairs. Now we proceed to "shake" the particles one by one, using the so-called Monte Carlo method. We choose a particle at random and displace it, that is, change its x, y and z coordinates. The amount of each change is decided by picking one of a series of random numbers generated by the machine. If the move is a legal one—that is, if the displaced particle does not overlap any of the others—we make a second snapshot, again recording all distances of separation. Then we displace a second particle, and so on.

Whenever a move turns out to be illegal, we return the displaced particle to its former position and use the corresponding distribution of distances a second time in the averaging. This procedure gives effect to the comparative probability of each distribution. The more often a random change in a distribution leads to an impossible configuration, the more probable is that distribution.

Monte Carlo calculations would of course be impossible to carry out by hand. Machines are quick and tireless, but they too have limitations. The largest existing computers can handle no more than about 500 particles in a reasonable calculating time. Machines soon to be available will have capacities for some 10,000. Even the latter figure is infinitesimal compared with the number of molecules in any weighable sample of matter. Yet it is remarkable how closely a system of just a few dozen particles can approximate the properties of real matter. One stratagem that helps make this possible is the proper choice of "boundary conditions."

In any sizable piece of matter the overwhelming majority of molecules are inside, surrounded by other molecules and interacting with them. Only a tiny fraction is at the surface or boundary at any time. However, in our hypothetical system of a few dozen or a few hundred particles a substantial proportion will always be near the walls of their imaginary container. A random displacement is thus likely to bring them into contact with the boundary rather than with another particle. To avoid this atypical situation the computer is programmed to allow the particle to pass freely through the walls of the box. Whenever a particle moves out through one wall, it is immediately brought back in through the opposite wall. In effect we have put the opposite boundaries together, just as a string of beads is brought together by tying its ends. Now every particle is surrounded by neighbors and interacts

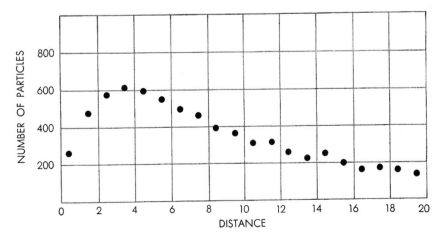

DIFFUSION OF PARTICLES in a specific time is plotted from molecular-dynamical calculation. Horizontal coordinate measures square of distance (in arbitrary units); vertical coordinate, numbers of particles that have moved various distances from starting point.

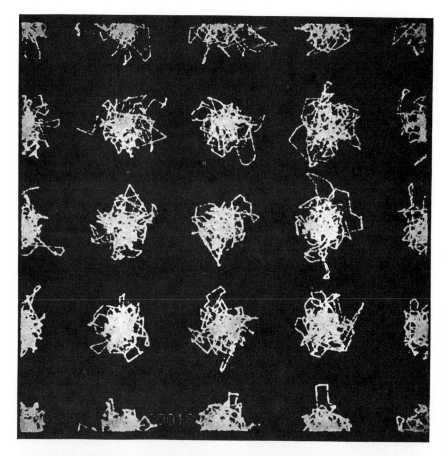

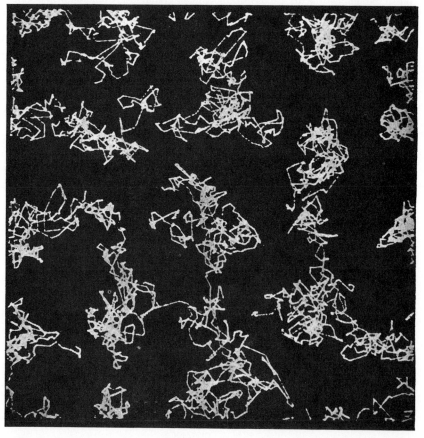

PATHS OF PARTICLES in molecular-dynamical calculation appear as bright lines on the face of a cathode-ray tube hooked to the computer. Each cluster in upper photograph represents two hard spheres, one behind the other. In solid state (*top*) particles can move only around well-defined positions; in fluid (*bottom*) they travel from one position to another.

with them instead of with the wall.

High-speed computers and the Monte Carlo method have enabled us to overcome many of the mathematical difficulties of statistical mechanics. Liquids and melting solids have been studied with the hard-sphere model and the pressure determined from the average distribution of pair separations.

Furthermore, it has been possible to consider more realistic intermolecular forces. For instance, two particles may be assumed to attract each other when they are closer than a certain distance, but repel each other when they actually collide. The shaking process now becomes more complicated. In the hard-sphere case one move is as good as another so long as it does not bring two particles into the same space. But when particles interact at a distance, some moves become more probable than others. Thus two particles within the range of attraction are more likely to approach each other than to separate. In carrying out the Monte Carlo calculations we assign probabilities reflecting the force that tends to encourage or inhibit each displacement: a given type of move might be allowed only every third time it came up.

With these refinements it is possible to take into account more of the properties of a real physical system, such as temperature. When attractive forces are considered, the likelihood of a particular change in the distribution of molecules depends on the temperature. Hence by varying the assigned probabilities we can play the Monte Carlo game at various temperatures. The distribution-of-distance plots can then be used to calculate the pressure of the system at different densities and temperatures, and the results compared with measurements on actual samples.

In this way the pressure of argon has been successfully calculated over a range of temperatures and densities that is wide enough to include the solid, liquid and gas phases. Moreover, the computations can easily be extended into regions of very high temperature and pressure, which are difficult to attain in the laboratory.

As has already been pointed out, the statistical mathematics of the Monte Carlo method produces accurate distributions of intermolecular distances. But it does not reproduce molecular motions. Hence it is restricted in its application to equilibrium properties such as pressure and energy. At the Lawrence Radiation Laboratory of the University of California we have undertaken a program to remove this deficiency. With the help of automatic computers we have been able to calculate the actual trajectories of a rather large number of individual particles. By means of such calculations it is possible to make theoretical determinations of properties that depend specifically upon details of molecular dynamics.

It turns out that a computer can follow the detailed motions of about as many particles as it can handle in the Monte Carlo method. We begin the calculation by assigning to each particle not only a position but also a velocity (*i.e.*, a speed and a direction of motion). Thereafter no element of chance is involved. The subsequent spatial patterns are predetermined. The machine calculates the path of every particle until two of them come close enough to exert a force on each other. Using Newton's laws of motion, the computer determines the effect of the forces on the two particles, and calculates their new velocities after the collision. Then it continues to compute all the trajectories until a second pair interacts, and so on. The same boundary conditions can be used as in the Monte Carlo method.

Of course we must make some assumption about the force between particles. The assumptions we make are not entirely realistic, but are designed to split a complex problem into simpler parts. Thus the mathematical complications arising out of realistic intermolecular forces can be isolated from those that are due simply to the large number of particles.

The two types of force that have been studied so far are the hard-sphere case and the so-called square-well interaction. In the latter there is no force between the particles until they approach to a certain distance. At this distance they abruptly attract each other. At shorter distances the attraction disappears, but if the particles then come close enough to touch each other they act like hard spheres and bounce apart. Since these particles move with constant velocity everywhere except when separated from one another by either the "attracting distance" or the "repelling distance," their motions are comparatively easy to compute. More realistic forces that vary smoothly with the distance of separation might be considered, but at the expense of more lengthy calculations.

The dynamical method is not restricted in its application to systems at equilibrium. It can, for example, be used to find how quickly the system reaches equilibrium. In one case a system of hard spheres was started out in the highly

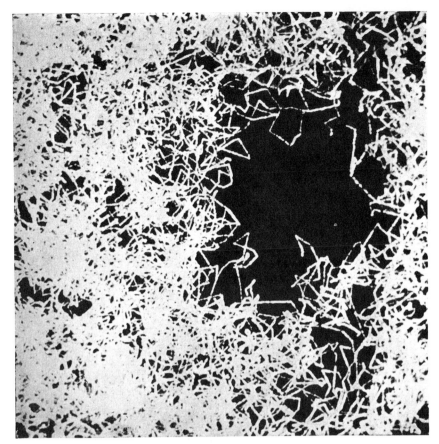

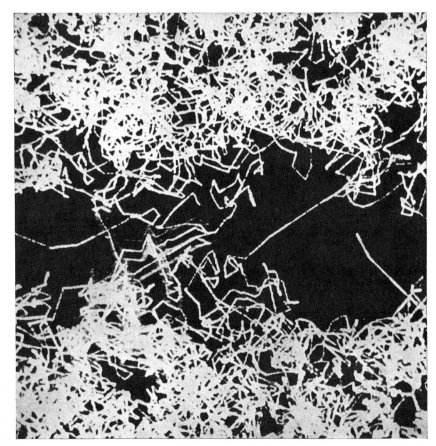

LIQUID-GAS SEPARATION is illustrated by molecular-dynamical calculations on particles with square-well interaction. Dark area in upper photograph represents a gaseous bubble, surrounded by particles whose motions characterize a liquid. Lower photograph shows the system at a later time, when some particles have vaporized and passed through the bubble.

atypical condition in which all the particles had the same kinetic energy. The statistically expected distribution of energies was attained after very few collisions [see illustration on page 108].

Once the system reaches equilibrium, distance-of-separation snapshots can be taken just as in the Monte Carlo method. These snapshots yield the same information as those achieved by statistical techniques. For example, they might be used to calculate the pressure. The dynamical method furnishes an additional way of arriving at the answer. If the system is considered to have reflecting walls, the pressure can also be calculated by considering the impacts of the molecules on the walls. Pressures in hard-sphere systems have been determined by both the Monte Carlo and the dynamical methods with very close agreement.

Because the configurations calculated by the dynamical method are in the correct temporal order, they enable us to study the time-dependent behavior of the system. This is important even after equilibrium has been attained. One example of time-dependent behavior is diffusion. At equilibrium the average total force on a molecule is zero, because the forces exerted on it by its neighbors are as likely to come from one side as another. Nevertheless any specific particle is likely to experience a succession of forces that leave it, after an interval of time, with a net displacement from its original position. In other words, it will diffuse. By noting the positions of the particles in successive snapshots, we can establish their rate of diffusion.

Albert Einstein's theory of diffusion predicts that the average of the squares

of the net displacements of the particles is proportional to the time during which diffusion takes place. Molecular-dynamical studies have shown that the theory is true for hard-sphere fluids and have also determined the constant of proportionality for a wide range of fluid densities.

So far the most extensive study carried out by the dynamical method has been the calculation of the pressure of a hard-sphere system over a large range of densities. The main purpose was to shed some light on a long-debated question: Can a hard-sphere material have a sharp freezing point? Does the change from the disorganized configuration of the molecules in a liquid to the ordered lattice of a crystal occur gradually or suddenly, if the molecules are assumed to have no attractive forces?

If hard spheres are in an orderly arrangement at a particular density, the system will have a lower pressure than if they are arranged at random, as in a liquid. Systems with as few as eight and as many as 256 hard-sphere particles have been studied. All exhibit sharp jumps in pressure when the density is held fixed somewhere between that of a crystal and that of a liquid [see illustration on page 109]. These pressure-jumps are associated with abrupt changes in the system from the orderly, crystalline phase to the random, fluid phase and vice versa. The size of the jump depends on the number of particles; the smaller the number, the larger the jump.

In all cases the system changed as a whole from one phase to the other: it was either entirely crystalline or entirely fluid. If a larger number of particles was used, we might expect to see a two-phase equilibrium, that is, part of the

system crystalline and part of the system fluid. This would correspond to the physical situation in which ice and water, for example, can exist side by side in equilibrium at the freezing point.

In order to demonstrate the phase changes more vividly, a display system similar to a television picture-tube was hooked up to the computer. Each particle in the system was then represented by a dot on the face of the tube. By focusing a camera on the screen and leaving its shutter open, it was possible to record the trajectories of the moving dots on film. The photographs on page 110 show the imaginary material in both the crystalline and the fluid phases.

If we want to study the liquid-gas phase transition, we must consider molecules with attractive forces. Such a study has been made with 32-particle systems with square-well interactions [see photographs on page 111].

We have mentioned only some of the molecular-dynamical calculations that have been carried out thus far. They represent an ideal example of the application of automatic computers to scientific research. A knowledge of the physical laws is assumed. The straightforward but tedious mathematical calculations are carried out by machines that are specifically designed for that purpose.

Perhaps it is not too much to expect that the information obtained by means of computing machines may play a role analogous to that of laboratory experiments in the development of theory. When we have built up a sufficiently large body of numerical computations, we may be able to discern generalizations that are not apparent to us now.

11

THE SOLID STATE

SIR NEVILL MOTT
September 1967

*Materials are solids, and solids are divided into two
general categories: crystalline, in which the atoms are
stacked in more or less regular arrays, and amorphous,
in which they are not*

If you take a paper clip and bend it, it stays bent; it doesn't spring back and it doesn't break. The metal of which the clip is made is said to be ductile. Some other materials are not ductile at all. If you try to bend a glass rod (unless you are holding it in a flame), it will simply break. It is said to be brittle. In this respect, as in many others, glass behaves quite differently from a metal. The difference must lie either in the particular atoms of which metals and glass are made up or in the way they are put together—probably both. There are of course many other differences between metals and glass. Metals conduct electricity and are therefore used for electrical transmission lines; glass hardly conducts electricity at all and can serve as an insulator. Glass is transparent and is used in windows, whereas a sheet of metal more than a millionth of an inch thick is quite opaque.

Students of such matters naturally want to understand the reasons for these differences in behavior. They want to study in detail the mechanical, electrical and optical properties of every kind of solid, and many other properties as well. Moreover, they want to provide the basis for choosing materials with desired properties in every branch of technology. During the past 20 years studies of this kind have been called solid-state physics, or sometimes, since the subject includes a great deal of chemistry, just "solid state." It is a major branch of science that has revealed new and previously unsuspected properties in materials. An example is the properties of semiconductors, knowledge of which has given rise to a flood of technological devices such as the transistor. Indeed, solid-state physics has become one of the most important branches of technology. Today engineers freely use expressions such

as "valence band" and "conduction band," which are terms of quantum mechanics as it is applied to solids. In solid state perhaps more than anywhere else quantum mechanics has ceased to be restricted to pure science and has become a working tool of technology.

Of course, solids were the subject of experimental investigation long before quantum mechanics was invented. I shall begin with the fact—known since the earliest studies of electric currents—that metals conduct electricity well and most other materials do not. With the discovery of the electron at the beginning of this century and the realization that it was a universal constituent of matter, it was assumed that in metals some or all of the atoms had lost an electron and that in insulators such as glass they had not. The electrons in a metal were thus free to move about and conduct electricity, whereas the electrons in an insulator were not.

Why this happened in metals had to await the discovery of quantum mechanics, and even now the answer is not quite clear. It has been known for some time, however, how to find the number of free electrons in a metal. The simplest way is based on the Hall effect: in the vicinity of a magnet the electrons carrying a current in a wire are pushed

sideways, so that a voltage—the Hall voltage—is set up across the wire. This voltage can be measured, and since it depends only on the speed with which each electron is moving down the wire, whereas the current depends both on the speed and on how many electrons there are, the measurement of both Hall voltage and current enables us to estimate the number of free electrons in a wire. It turns out that in a good conductor such as copper each atom has lost just about one electron. There must be in the metal a very dense gas of electrons, more than 10^{22} of them in a cubic centimeter.

The next question is: How are the atoms themselves arranged? Since the introduction of X-ray crystallography by William Bragg and his son Lawrence in 1911, this has been known for the simpler materials. Solids can be divided into two classes: crystalline and amorphous. In the crystalline group, which is the largest and includes the metals and most minerals, the atoms are arranged in a regular way; in many metals (for instance copper and nickel) they are packed together just as one would pack tennis balls into a box if one wanted to squash in as many as possible. In other metals (for instance iron) the structure is called body-centered cubic; there are four atoms at the corners of a

(overleaf)

ETCH PITS IN CADMIUM SULFIDE, a semiconductor widely used in photocells, are shaped like hexagons, reflecting the geometric pattern of the atoms that compose the material. In this photomicrograph the etched surface of a single crystal is viewed along the hexagonal axis. The colors arise from the interference of incident and reflected light. The concentric hexagons provide a contour map of the pyramidal pits and can be used to calculate their depth. The pit with most concentric bands (*top*) is the deepest. It is depressed about 10,000 atomic layers and measures about 100,000 atoms across. Such pits are formed by treatment of the material with hydrochloric acid. The pits mark defects in the crystalline structure of the material. The photomicrograph was made by Carl E. Bleil and Harry W. Sturner of the General Motors Research Laboratories. The magnification is approximately 800 diameters.

cube and one in the center. The arrangement of atoms in all crystalline solids falls into 14 such categories [*see illustration on following page*].

The commonest of the amorphous group of solids is glass. Its atoms are put together in a more disordered way than those of a metal [*see lower illustration on page 117*]. The structure of an amorphous material is much more difficult to discover than that of a crystalline solid, and considerable effort is now being made to learn more about the arrangement of atoms in such materials.

The crystalline structure of a metal such as iron presents to the eye a formidable array of atoms. How does an electric current manage to flow through such a material? One would think that no electron could get farther than from one atom to another without a collision, and that the electrons must percolate through the crystal the way hailstones sift through the leaves and branches of a tree and fall on someone taking shelter below. If this were the case, it would mean that the rate of drift of an electron gas, and therefore the current for a given voltage, would depend little on whether the arrangement of the atoms was regular or haphazard. That is far from being the truth. One of the most marked characteristics of metals is that they conduct much better at low temperatures than at high ones. For example, the amount of energy wasted by resistance in an electric cable is about 10 percent less in a typical Temperate Zone winter than in summer. At the very low temperatures obtainable in cryogenic laboratories an electron can go straight through millions of planes of atoms without being deflected from its path.

Electrical resistance occurs only if the atoms are *not* in a regular array. One such irregularity arises as the temperature increases; the atoms then start to swing around their average positions, each one being displaced up to as much as 10 percent from its normal position. Other evidence is provided by the fact that most metals conduct worse when they melt, and also by the fact that an alloy such as brass (a mixture of copper and zinc) conducts much worse than pure copper. A really good conductor is one in which atoms all of the same kind are arranged in a perfect crystalline array.

This was completely incomprehensible before the invention of quantum mechanics. Between 1924 and 1926 Erwin Schrödinger, Werner Heisenberg and Max Born showed how to set about explaining a host of phenomena that had

formerly been mysterious, and in the five years that followed the foundations were laid for the understanding of solids and of much else besides. One learned to say, when asking anything about electrons in atoms, molecules or solids: If you want to know what an electron does, forget about it and pretend there is a wave there. Calculate where the wave goes, and there you will find electrons.

It is a well-known property of waves that they can go through a regular array of obstacles of any kind. At first this seems surprising. It is easier to grasp that waves do not go through an irregular array of obstacles. That is why the headlights of a car cannot penetrate very far into a fog; the droplets of water scatter the light out of the headlight beams. If the droplets were arranged in some regular way, as the atoms in a crystal are, this would not happen; the light would go straight through.

This property of waves, which could be proved by quite simple mathematics and had been known long before quantum mechanics, showed in principle why good conductors of electricity have to be pure, crystalline and cold. Another property of waves enabled us to understand in the early 1930's why some materials were insulators and some conductors. The explanation was first given by A. H. Wilson of the University of Cambridge. The argument did not, as one might expect, seek to tie the electrons to the atoms in nonconducting materials. For both kinds of materials Wilson started by thinking of the electrons as being free to pass through the crystal as waves. The theory went on to show, however, that in some materials there cannot be any current because there will always be just as many electrons moving one way as the other.

The argument is a sophisticated one; it is based primarily on Wolfgang Pauli's exclusion principle, which says that no two electrons can ever move on exactly the same path with exactly the same speed. It is this principle that gives rise to the electron shells of atoms; there it is expressed by saying that no two electrons can have the same quantum number. Applied to metals, the exclusion principle means that electrons in a metal will have velocities lying between zero and some maximum velocity. For an insulator it happens that the limiting velocity has a value that is extremely awkward in view of the mathematical relation that exists between an electron's velocity and its wavelength. This relation is given by Louis de Broglie's formula in which wavelength equals Planck's

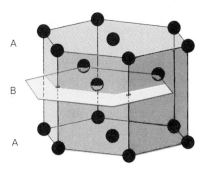

HEXAGONAL CLOSE-PACKED structure, a lattice arrangement common to many metals, is built of tightly nested layers of atoms. Three layers of hexagons provide the 17 atoms that form a crystal unit. Atoms in the layers labeled *A* fall directly over one another. Three atoms in layer *B* nest between.

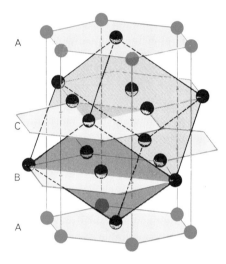

ALTERNATIVE CLOSE-PACKED structure can be built from layers of hexagons stacked in the sequence *ABCA*. In this arrangement 12 atoms can be selected that form a face-centered cube with the same packing density as the hexagonal close-packed structure. Metals commonly crystallize in either the hexagonal or the face-centered configuration. A less common form of metal crystal is the body-centered-cubic one.

constant divided by the mass times the velocity of the particle ($\lambda = h/mv$). The

FOURTEEN CRYSTAL SYSTEMS on the following page encompass all crystalline solids. The number of ways in which atomic arrangements can be repeated to form a solid is limited to 14 by the geometries of space division. Any one of these arrangements, when repeated in space, forms the lattice structure characteristic of a crystalline material. For example, cadmium sulfide, the crystal shown in color on page 114, has a lattice formed of hexagonal units. The shadows indicate the tilt away from the vertical.

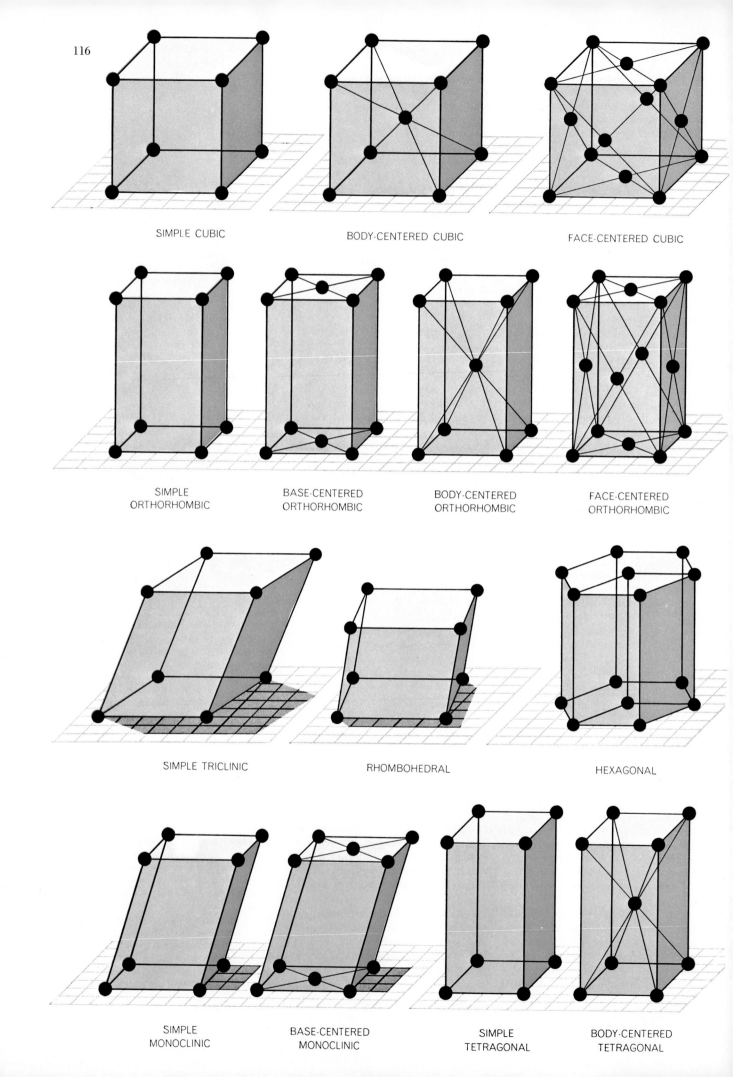

116

SIMPLE CUBIC

BODY-CENTERED CUBIC

FACE-CENTERED CUBIC

SIMPLE
ORTHORHOMBIC

BASE-CENTERED
ORTHORHOMBIC

BODY-CENTERED
ORTHORHOMBIC

FACE-CENTERED
ORTHORHOMBIC

SIMPLE TRICLINIC

RHOMBOHEDRAL

HEXAGONAL

SIMPLE
MONOCLINIC

BASE-CENTERED
MONOCLINIC

SIMPLE
TETRAGONAL

BODY-CENTERED
TETRAGONAL

wavelength of electrons in an insulator is awkward because it has a dimension that just fits into the distance between the atoms of which the material is composed. Under these conditions the wave will become a standing wave, and such a wave describes a situation in which the movement of electrons in one direction is exactly offset by the movement of other electrons in the reverse direction.

In many crystalline materials one finds this situation in which there can be no electric current. To overcome it a considerable amount of energy is required; an electron must be hit rather hard to put it in a position where its movement is not balanced by the movement of another electron. The needed energy is gained, of course, if the material is heated to a temperature that is high enough. All solids will conduct electricity to a certain extent if they are hot; they can also be made to conduct by energetic radiation such as ultraviolet or X rays.

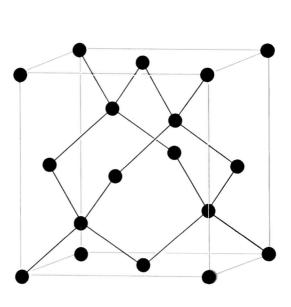
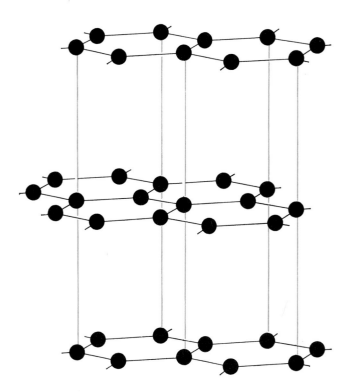

TWO FORMS OF CARBON exhibit markedly different properties owing to their different crystal structure. Diamond (left) consists of pairs of carbon atoms in a face-centered-cubic array. Each carbon atom is bound to four others. This tightly joined lattice contributes to diamond's hardness. In graphite (right), a soft material, carbon atoms are arranged in layers that are bound by weaker forces.

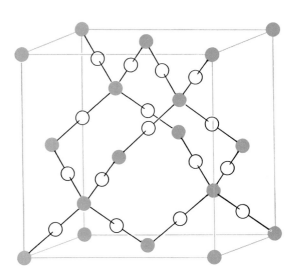
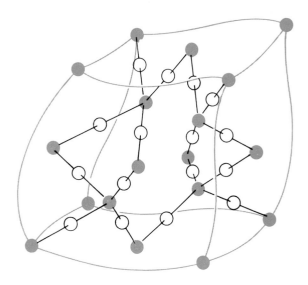

TWO FORMS OF SILICA demonstrate how molecules of the same composition can exist either as a crystal or as a glass. Cristobalite (left), a high-temperature form of quartz (SiO$_2$), is similar to diamond in that it has a face-centered-cubic structure. Silicon atoms (color) occupy the sites filled by carbon atoms in the diamond lattice. In addition an oxygen atom sits between every two silicon atoms. A conceivable glass structure (right) resembles a cristobalite structure that has been distorted. Also the three rings, each containing six silicon atoms, found in the cristobalite cell have been reconnected to form two rings with four silicons and one with eight.

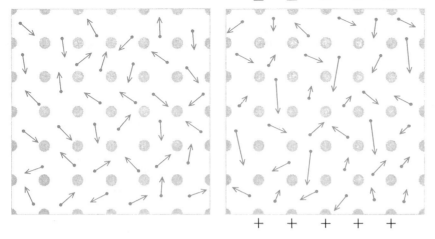

ELECTRICAL CONDUCTIVITY in a metal can be thought of as the movement of valence, or free, electrons (*color*) in a preferred direction. In the absence of an external electric field (*left*) the movement of any one electron is offset by the movement of another in the opposite direction. Within an electric field electrons move toward the positive plate (*right*).

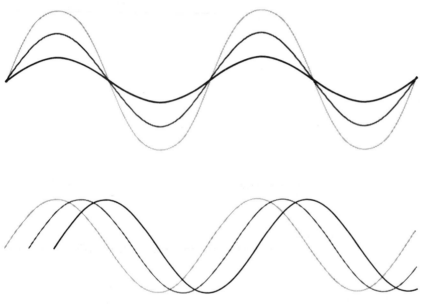

QUANTUM VIEW OF CONDUCTIVITY substitutes a wave for the motion of electrons. In an insulator (*top*) electron velocities correspond to a standing wave, one that does not move in any particular direction. In conductors electron velocities correspond to a running wave (*bottom*). The lines in the drawings show the form of the waves at successive moments.

No material can be a good insulator while it is exposed to X rays.

The theory I have just described makes use of difficult concepts of quantum mechanics, but the mathematics of it is not very complicated. It is the kind of theory a physics student learns in his final undergraduate year or first graduate one. The same cannot be said of the theoretical work that is currently being done in an effort to correct a grave omission in the theory. In the electron gas of a solid material the electrons are moving about all the time and bouncing off one another. To describe this bouncing mathematically is a formidable problem; it is very difficult to solve problems involving even three interacting bodies. Many-body theory is a subject of intensive research in many leading laboratories; it may be that when it is further along the metal-insulator problem will look rather different.

Meanwhile the simple model I have described has proved perfectly adequate for understanding the part of solid-state physics that is most important to technology, namely the semiconductors. These are materials that will carry an electric current but only a small one compared with a metal. Basically semicon-

ductors should be classed as nonmetals; when they are pure and at low temperatures, they do not conduct electric current. One makes them conduct by adding electrons to them. The simplest way to do this is to dissolve in the crystal of a semiconductor traces of some chemically different material, each atom of which easily gives up an electron. Germanium, a common raw material of transistors, becomes quite a good conductor when very small quantities of phosphorus (one part in a million) are added to it. The germanium atom has four outer electrons it can easily lose; the phosphorus atom has five. It is the extra electron of phosphorus that does the trick. When it is free of the atom, it can move about quite easily, like an electron in a metal, and its motion is not offset by the motion of any other electron. Germanium that has been "doped" with phosphorus is a semiconductor of the *n* type, the *n* standing for the negative charge contributed by the additional electrons.

The technological importance of semiconductors arises mainly from the fact that the contact between a semiconductor and a metal (or between two semiconductors) acts as a rectifier. This means that the material will pass electric current much more easily in one direction than in the other. A crude form of semiconductor rectifier was used in the earliest radio receivers; it was supplanted by the vacuum tube. The replacement of the vacuum tube by the transistor represents the return of the semiconductor rectifier in refined form.

In the *n* type of semiconductor I have described free electrons will flow from the semiconductor to a metal but not in the reverse direction. This is hardly surprising. A cold metal does not emit electrons; it must be heated until it glows to do so in a vacuum tube. In the semiconductor, however, the extra electrons donated by foreign atoms can move into the metal without undue difficulty. The barrier (more exactly the change in potential energy) that keeps electrons from going from the metal to the semiconductor helps them to go the other way.

The important properties of semiconductors, then, depend on the presence in the crystal of minute quantities of some impurity. This brings me to an important point about solids. They are never quite pure and their crystal lattices are rarely quite perfect; they have what we call defects. Such defects determine many significant properties of materials, particularly the mechanical ones.

One can obtain materials with impurities present in less than one part in

10 million, but even in such materials the impurity atoms will be only 10^{-5} centimeter apart, a distance shorter than a wavelength of light. Thus quite a small speck of even a very pure material will have plenty of impurities in it. Usually the impurity atoms simply replace an atom of the surrounding crystal. This is what happens when germanium is doped with phosphorus; four of the five outer electrons of a phosphorus atom participate in bonds with germanium atoms, so that the phosphorus atom is taken into the structure of the crystal.

Other forms of impurity can impart color to a crystal, and indeed the systematic study of colored rock salt and other salts such as potassium chloride by R. W. Pohl and his colleagues at the University of Göttingen before World War II makes Pohl one of the founders of solid-state physics. It takes only a trace of blue ink to color a glass of water, and in the same way traces of potassium in an initially transparent potassium chloride crystal will make the crystal dark blue [see "The Optical Properties of Materials," by Ali Javan; SCIENTIFIC AMERICAN, Sept. 1967]. The potassium is added by heating the crystal in alkali vapor. How is the extra potassium accommodated? It turns out that some of the sites that ought to be occupied by chlorine are empty, so that adding potassium makes the crystal expand by a readily calculable amount; this has been verified by experiment. But potassium chloride is a member of the class of ionic crystals, and one cannot have such a crystal in which a large number of chlorine sites are empty. It is not an atom that normally occupies any one of the sites but an ion, in this case a chlorine atom with an extra electron stuck to it. A crystal with all those vacant sites would have an enormous electric charge. What happens is that each of the vacant sites has an electron in it. It is these electrons that give the crystal its color, by absorbing certain wavelengths of light. Vacancies with an electron in them were named by Pohl F centers, from the German word *Farbe* (color).

It is not only impurities that make a crystal deviate from perfection. In metals at high temperatures, for instance, a number of sites—perhaps one in a mil-

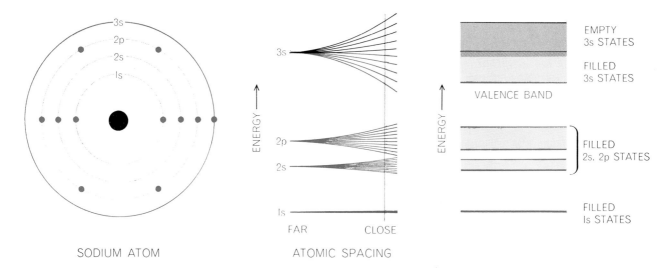

THEORY OF ELECTRICAL CONDUCTIVITY involves the behavior of an atom's outermost shell of electrons, the valence electrons. Sodium, for example, has 11 electrons arranged in four shells (*left*). The three inner shells are filled, but the valence shell could hold another electron. The electrons in each shell occupy specific energy levels (*middle*). As atoms are brought close together and begin to influence one another, the electrons are forced to have slightly different energies, since only two electrons can occupy precisely the same quantum state. The vertical line indicates the spacing of atoms in a crystal of sodium. Within a crystal containing some 10^{20} or more atoms the energy levels become densely filled bands (*right*). Since sodium has only one valence electron, only half of the energy states in the valence band are filled. As a result negligible energy is required to raise a valence electron to an empty state, where it is free to move inside the crystal and to conduct electricity. This is not true of nonconductors (*see illustration below*).

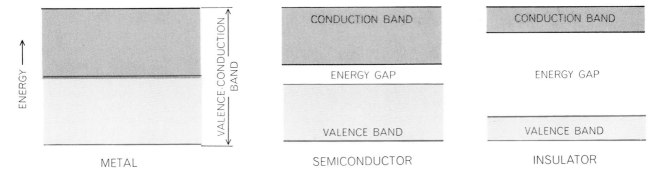

ELECTRICAL CONDUCTIVITY of a solid depends on the spacing and state of occupancy of the energy bands within its crystals. Many metals (*left*) resemble sodium in having a valence band that is only half-filled with electrons and therefore can readily act as a conduction band. Other metals have more complicated band structures but the net result is the same. In semiconductors (*middle*) there is a small energy gap between a filled valence band and the first permissible conduction band. It is not too difficult, however, for some of the electrons to acquire the energy needed to jump across the gap. In an insulator (*right*) the gap is not easily bridged.

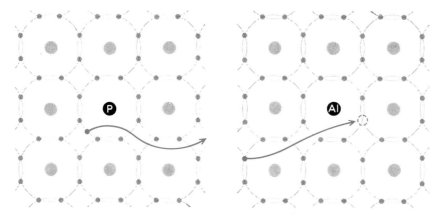

IMPURITY ATOM in the structure of germanium, a semiconductor, increases its conductivity. Each atom in a germanium crystal shares four valence electrons (*color*) with adjacent atoms. A phosphorus atom ("*P*" *at left*) has five valence electrons. The extra electron cannot fit into the regular structure. It is in a high-energy position like the free electrons of a metal. An impurity of aluminum ("*Al*" *at right*) also enhances germanium's conductivity. Aluminum, with only three valence electrons to contribute to the germanium structure, creates a vacant site, or electron hole, into which a nearby electron can move.

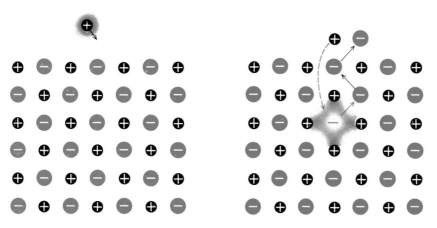

COLOR CENTER arises when electrons are trapped in certain crystals. The transparent crystal of potassium chloride (*left*) turns blue when potassium is added to it. A potassium atom (*black arrow*) attaches itself to the surface of the crystal and subsequently loses an electron (*broken arrow at right*). The electron trades places with a negatively charged chloride ion that has migrated outward to pair with the newly arrived potassium. The electron, which is held by adjacent positive ions, absorbs light, producing the color change.

lion—will be empty. Such sites are called vacancies. They arise because at high temperatures the atoms are vibrating, and now and then a vibration is so vigorous that a vacancy is produced at the surface; then it can move inward. Of course, vacancies will also move to the surface and disappear, and eventually a balance is set up, the number formed and the number disappearing being equal.

Vacancies move around in a crystal, much as molecules move in a gas but much more slowly. It is believed that when one metal mixes with another, which is what happens when two materials are welded together, atoms change places by jumping into vacancies, so that vacancies play an extremely important part in all the arts of metallurgy. It would be interesting if we could see vacancies, but that is beyond the power even of present-day electron microscopes. What we can see is little clusters of vacancies. When a metal is quenched (cooled quickly), any two vacancies that accidentally meet will stick together; the vibrations of the atoms are not vigorous enough to make them move apart. Little cavities form that have curious shapes varying from one material to another, some of which are shown in the illustrations on page 122.

The technique by which these pictures are made is transmission electron microscopy, and it turned out to be one of the most important techniques of solid-state physics. The electron microscope has the advantage that one can see objects much smaller than the wavelengths of light, which is of course not possible with the light microscope. Moreover, electrons will penetrate thin specimens of metals.

One of the great successes of this thin-film electron microscopy was the observation of another form of defect, namely the dislocation. This brings me back to the question of why a metal paper clip bends and a glass rod breaks. I remember discussing many years ago with Lawrence Bragg, the codiscoverer of X-ray crystallography, the possibility that this had something to do with the presence in metals of free electrons. It doesn't, except that metals usually have rather simple crystal structures; their atoms, having lost their outer electrons, don't form bonds with some specific number of atoms. It turns out that materials with simple crystal structures are often ductile and those with complicated ones are rarely so. Those that are amorphous are never ductile unless they are so hot that their atoms can change places quite easily. The key to the understanding of such behavior is the dislocation.

This concept was introduced in 1934 by Geoffrey I. Taylor of the University of Cambridge. The question Taylor asked himself was: When a crystalline substance is deformed, do the atoms all slip over one another together? Does the crystal pass suddenly from the undeformed state to the deformed one? For various reasons he thought that it would not, that it would instead deform through the motion of a kind of wrinkle in its structure—a dislocation in the regular array of its rows of atoms.

One of the reasons that led Taylor to put forward this hypothesis is that very pure metals are normally much softer than alloys and impure materials. The dislocation model makes it clear why impurities make a metal harder: the impurities collect in the dislocation and keep it from moving. In complicated crystal structures the dislocation itself is a complicated structure and cannot move easily. In glasses one cannot have dislocations at all.

Taylor's hypothesis explained facts known since the Bronze Age, but no one had actually seen dislocations in motion until 10 years ago. At that time some of my colleagues in the Cavendish Laboratory were examining thin metal films by transmission electron microscopy, and somewhat to their surprise they found that dislocation lines were visible. Moreover, the screen of the electron microscope showed the little lines darting forward; doubtless the metal film was buckling as the electron beam heated it up. I well remember our excitement the day one of our graduate students came into my office and said: "Prof, come and see some moving dislocations!"

Thus far I have described some of the main properties of solids that are important to the solid-state worker, particularly the electrical and mechanical properties. I have emphasized the great difference between the properties of one solid and another. There is another way of looking at these differences, namely the mechanisms by which the atoms or molecules of various materials stick together.

Perhaps the simplest of all solids are the frozen inert gases: helium, neon, argon and so on. These chemically unreactive gases do not form molecules; at any rate, if two atoms stick together, they stick so weakly that they will quickly be knocked apart by collisions with other atoms in the gas. But if you cool an inert gas, it will first liquefy and then solidify, and this shows that there is some kind of weak attraction between even the most inert atoms. The force between such atoms is called the van der Waals attraction. It is explained by quantum mechanics, but the explanation is not simple. In general it can be said that the negatively charged cloud of electrons surrounding the positively charged atomic nucleus can slightly shift its position, so that the center of negative charge does not quite coincide with the center of positive charge. As a result of this electrical imbalance a weak force is established that can attract other atoms.

The chemically inert atoms are those in which the electrons form what is called a closed shell; such shells are found in atoms with two electrons (he-

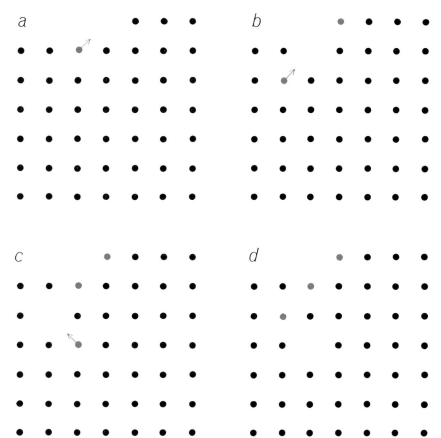

VACANT LATTICE SITES are created by shifting atoms. In a hot metal atoms can vibrate so vigorously that some at the surface leave their positions. From the row of atoms below one will move into the vacant site, called a vacancy. As a result of the sequence of events ("a" through "d") a vacancy moves inward. When two different metals are welded together, their atoms probably change places by jumping into each other's vacant sites.

lium), 10 (neon), 18 (argon) and so on. Certain ions have this property too. If an electron is removed from a sodium atom, 10 electrons are left and so the remaining negatively charged ion is inert like neon. By the same token, if one adds an electron to chlorine, which has 17 electrons, one gets 18 electrons and a positively charged inert ion. An important class of solids is the ionic salts, of which sodium

chloride is the best known; it is made up of inert positive and negative ions and holds together simply because the positive and negative charges attract each other.

There are solids made of atoms that are not chemically inert. Most atoms consist of an inert shell and a number of electrons in addition that can help the atom stick firmly to another atom. Car-

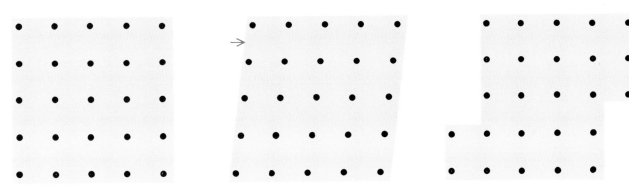

DEFORMATION OF METAL, in which dots represent atoms, takes place in successive steps: before deformation (left), following an elastic deformation such that the atoms will spring back (center) and after permanent bending (right). The atoms do not jump to their final positions all at once but move through the wrinkling process called a dislocation. The malleability of materials depends on the presence of dislocations in the crystal structure.

bon, silicon and germanium have four electrons outside an inert shell, and each of them can take part in forming a bond to another atom of the same kind. This kind of bonding is called "homopolar" or "covalent." The covalent bond involves the sharing of electrons between pairs of atoms.

These divisions are not hard and fast. Most of the minerals that make up the rocks of the earth's surface fall in neither one nor another. They are compounds, and the different atoms are to some extent charged; therefore they stick together partly like ionic substances such as salt. But there is a lot of electron-sharing too, and the simple classifications are not always useful.

Then there are molecular crystals. Hydrogen is the simplest example, although solid hydrogen can be obtained only at very low temperatures. The hydrogen atom has only one electron and can form a very strong bond with one other hydrogen atom, resulting in the molecule H_2. In solid or liquid hydrogen the molecules stick weakly together because only the van der Waals force holds them. Another example is water or ice, in which the H_2O molecules stick through a mechanism (the hydrogen bond) that seems to be halfway between ionic and covalent. Many organic materials are of this kind—wood and cotton, for instance. These are made up of polymer molecules, which have the form of long chains; covalent bonds link the atoms in each chain, and something like van der Waals bonds attract adjacent chains to each other [see "The Nature of Polymeric Materials," by Herman F. Mark; SCIENTIFIC AMERICAN, Sept. 1967].

Finally there are the metals. Here the outer electrons have left the atoms and can contribute to a current. The matter is discussed in the article "The Nature of Metals," by A. H. Cottrell, SCIENTIFIC AMERICAN, Sept. 1967. Here we need only say that to obtain an adequate theoretical description of cohesion in metals is complicated; the electrons are free, they repel one another but they are attracted by the ions. Of course attraction must win—otherwise no solid metal could exist. Detailed calculations have been carried out for only a few metals, although research in this area is very active, particularly with respect to accounting for the crystal structures observed in alloys. What one might say is that it is not surprising that one can form so many alloys; metal atoms are not particular about what other metal atoms they stick to. If all the electrons come off in any case, strong cohesion exists whatever the strength of the charge is on the ion and however many electrons there are.

Reflecting on solid-state science in 1967, one perceives the following main lines of advance. In theoretical physics and in fundamental physics generally the many-body problem—the interaction of all the electrons in a metal—continues to be of great interest and abounds in unanswered questions. Allied to this subject is our recent understanding of superconductivity, the complete disappearance of electrical resistance at very low temperatures. Then the study of surfaces and of the interface between metals and semiconductors has moved into the center of the picture, partly because of its extreme importance for electronic devices and partly because new techniques for investigating them have been introduced. Finally, many solid-state physicists are looking over their shoulder at biology. Since solid-state science deals mainly with the movement of charge and energy through relatively simple solids, it ought to have a lot to say about these processes in the vastly more complicated living tissues.

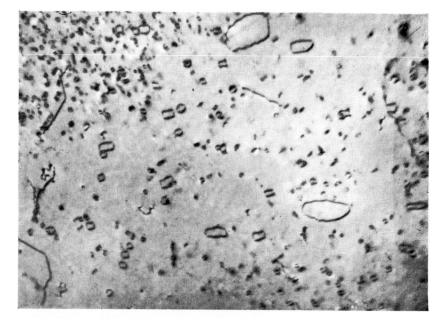

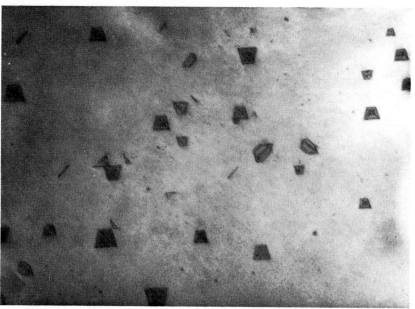

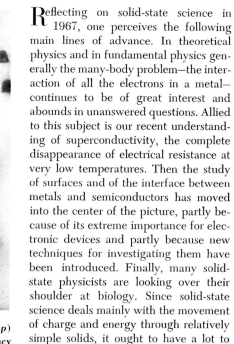

VACANCY CLUSTERS can assume different shapes in different metals. In aluminum (*top*) they are like disks; in gold (*bottom*) they are shaped like tetrahedrons. A single vacancy is too small to be visible. These electron micrographs were made by John Silcox of the University of Cambridge. The metals had been rapidly cooled. Magnification is 30,000 diameters.

12

ICE

L. K. RUNNELS
December 1966

*At the molecular level the seemingly rigid perfection
of a crystal of ice is disrupted by an astonishingly busy
traffic of molecules and migrating lattice faults*

Ice is a substance that has fascinated many of the foremost scientists of modern times. Chemists, physicists, mathematicians and even biologists have been drawn to the investigation of its somewhat mysterious crystal structure and its peculiar properties. Their exploration of the problems presented by ice have yielded a better understanding of the hydrogen bond (present in many important molecules, including proteins) and of the activity that goes on within a seemingly rigid crystal. The investigation of ice is a good example of how much can be learned from close study of a most commonplace object.

Let us begin by examining the structure of a single water molecule—say a molecule of water vapor in the air. The easiest way to visualize the molecule is to place it in an imaginary cube with the nucleus of the oxygen atom at the center of the cube and the two hydrogen nuclei at opposite corners of the cube's base [*see illustration on this page*]. Actually the shape of the figure is not exactly cubic—the "cube" should be slightly distorted—because the angle of attachment of the hydrogen atoms to the oxygen is 104.5 degrees, whereas in a perfect cube the lines from the corners to the center form an angle of 109.5 degrees. For our purpose, however, the cube figure is accurate enough. Of the oxygen atom's eight electrons two are held near the oxygen nucleus while another pair joins with two electrons from the hydrogen atoms in binding them to the oxygen; this leaves two "arms" of unshared electrons that point from the oxygen nucleus to other corners of the cube, as the illustration indicates.

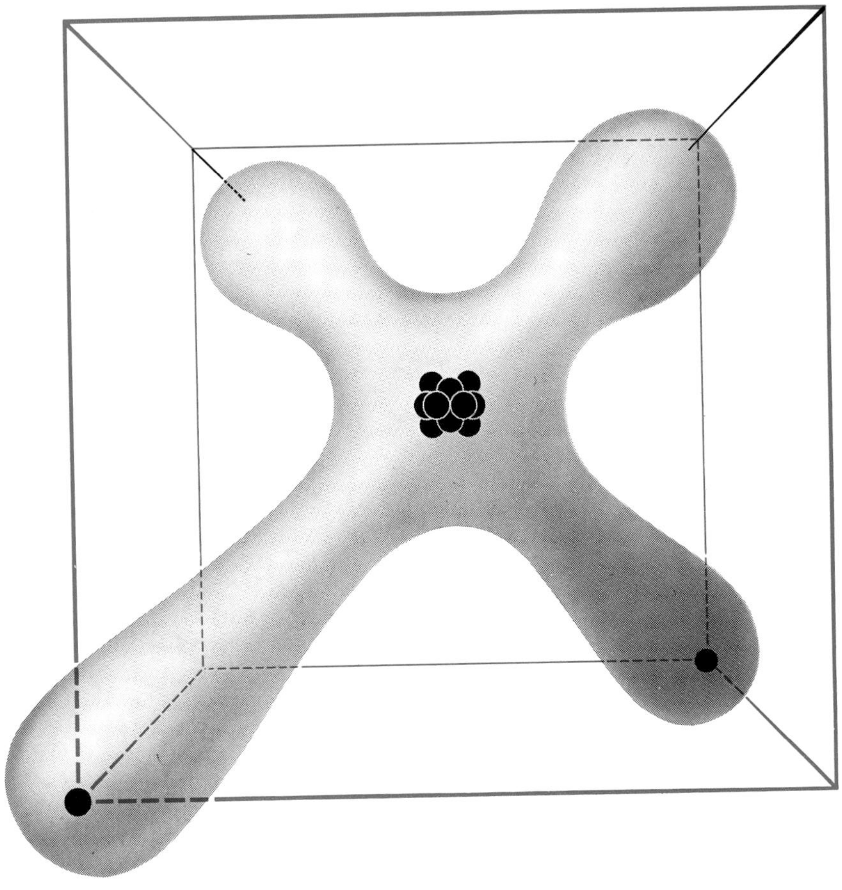

When water molecules are not isolated but packed together, as in the liquid state, these negatively charged arms serve to attach molecules to one another. Each negative arm attracts a hydrogen nucleus in a neighboring water molecule, and the hydrogen atoms thus act to join the molecules with what are called hydrogen bonds. The molecules tend to combine in clusters [see "Water," by Arthur M. Buswell and Worth H. Rodebush; SCIENTIFIC AMERICAN Offprint 262].

In the liquid state the normal thermal motion of the molecules is sufficient to break the bonds as the molecules jostle one another; consequently the clusters are continually being split up and re-formed. When water is cooled to the freezing point, however, the thermal motion is so reduced that the molecules form large, stable clusters: the crystals

MOLECULE OF WATER has four "arms" (actually clouds of electrons) extending from the central nucleus of the oxygen atom. Two of the arms contain hydrogen nuclei (protons) and are positively charged. The other two arms contain no protons and hence are negatively charged. In subsequent illustrations the negative arms of molecules will be omitted.

WATER MOLECULES IN ICE CRYSTAL were shown many years ago by the technique of X-ray diffraction to be arrayed in a hexagonal pattern, viewed here from above (*top*) and from the side (*bottom*). The X-ray analysis revealed that there is a good deal of empty space between the molecules, which explains why ice floats on liquid water. The technique failed to locate the positions of hydrogen atoms, because their power to deflect X rays is too low.

of ice. The first thing that interests us about these crystals is their structure. How are the water molecules arranged in an ice crystal?

The study of ice by the technique of X-ray diffraction showed many years ago that the molecules in the crystal are arrayed in a hexagonal pattern [*see illustration at left*]. This accounts for the six-sided form of snowflakes. More significantly, the analysis revealed that there is a good deal of empty space between the molecules: the crystal has a rather open structure. This finding clearly explained one of the most unusual properties of ice: the fact that, in contrast to almost all other substances, water is less dense in the solid state than in the liquid state, with the result that ice floats on liquid water. In frozen water, it can be seen, the molecules are more loosely packed than in the liquid. When a crystal of ice melts, the breakdown of its structure allows molecules to fill some of the open spaces.

The X-ray studies succeeded in revealing the arrangement of molecules in ice because the oxygen atoms in the molecules serve to deflect the X rays. The technique failed, however, to locate the positions of the hydrogen atoms; their power to deflect X rays is too low. For more than 30 years investigators in Europe and the U.S. have been pursuing the intriguing problem of finding out just where the hydrogen atoms are in the ice crystal, as this information is crucial to understanding the complete structure of ice and some of its most important properties.

The effort took its cue from a suggestion made in 1933 by the British physicists J. D. Bernal and R. H. Fowler. They argued that in all likelihood the orientation of the molecules in the ice crystal is such that each molecule forms good hydrogen bonds with its four nearest neighbors; that is, in each case the molecule's two hydrogen nuclei (protons) are pointed toward its neighbors' negative arms and its own two negative arms are pointed toward its neighbors' protons. This suggestion was accepted as reasonable. Unfortunately, however, it left the problem wide open. It did not specify any particular arrangement of the molecular orientations; indeed, one could picture many different arrangements that would satisfy Bernal and Fowler's rule. Some of these possibilities are shown in the top illustration on the next page, which for the sake of convenience is presented in only two dimensions (instead of the three dimensions of a real

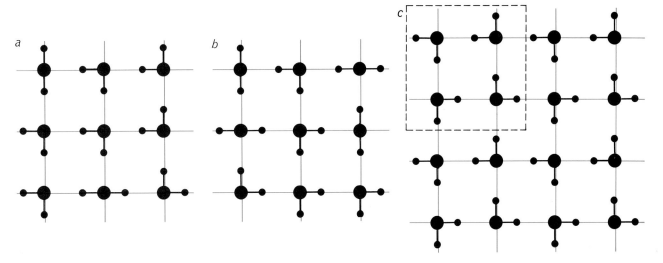

TWO DIFFERENT ORIENTATIONS of water molecules in an ice crystal (a, b) both satisfy the requirement that every pair of neighbors be joined by a hydrogen bond formed when a positive proton faces a negative arm (the hydrogen-bond rule). For convenience the crystals are presented in only two dimensions instead of the three dimensions of a real crystal. A total of 2,604 different arrangements is possible for such an ice crystal containing only nine molecules. In c a hypothetical arrangement is shown with a repeating "unit cell."

crystal) but nevertheless illustrates the essential point.

Does ice have one fixed, invariable structure? In most solids the atoms are arranged in a certain definite order with each type of atom assigned to a particular location in the unit cell, or repeating molecular structure, that makes up the crystal. Experiments on ice soon turned up indirect evidence, however, that its hydrogen atoms are not restricted to a single basic pattern. The evidence had to do with entropy.

The entropy of a system, which fundamentally is defined in terms of its heat capacity (the amount of heat required to raise its temperature), is a measure of the lack of order, or the presence of randomness, within the system. The higher the entropy, the more random the distribution of particles or molecules contained in the system. Now, as a substance is cooled, the reduction of the motions of the molecules within it results in a greater order and a decrease of entropy. At absolute zero the entropy of most substances is zero: they are "frozen" in a certain predictable state of order. Some compounds, however, do not give up all their entropy even when they are cooled to the lowest attainable temperature (almost absolute zero). In one way or another they retain some randomness —a "residual entropy." A good example is carbon monoxide (CO). With a carbon atom at one end and an oxygen atom at the other, this molecule is slightly polarized (the electric charge of one end is slightly more positive than that of the other), but the difference is so small that there is little distinction between the positive and the negative ends of the molecule. Consequently when the substance is cooled to very low tempera-

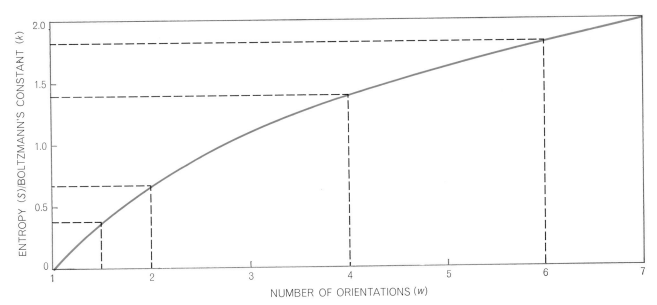

BOLTZMANN RELATIONSHIP, named after the 19th-century Austrian physicist Ludwig Boltzmann, predicts that the entropy (or randomness defined in terms of heat capacity) of any system in which the molecules are independent of one another will be proportional to the natural logarithm of the number of equally likely orientations of the molecules. The values for w of 1.5, 2, 4 and 6 correspond respectively to ice, carbon monoxide, deuterated methane and ice without the hydrogen-bond rule. The experimental value of $S/k = .4$, obtained by measuring the heat capacity of ice to very low temperatures, substantiates hydrogen-bond rule for ice.

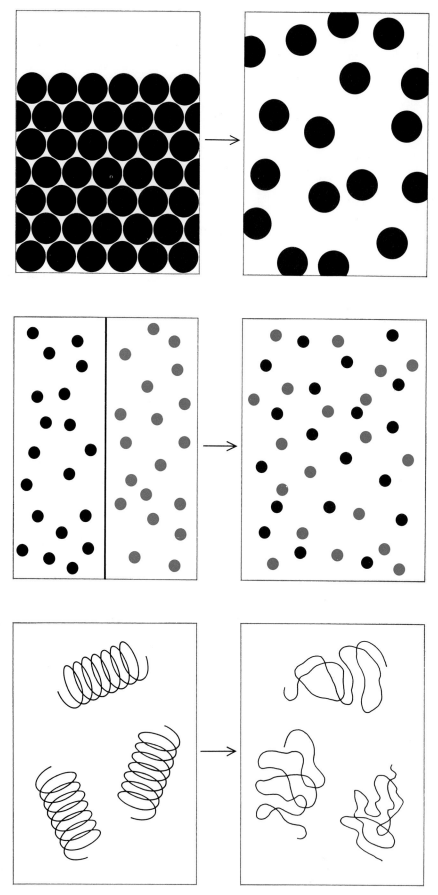

ENTROPY AND DISORDER are closely related, the more random arrangement of a system (*right*) having a higher entropy than the more orderly one (*left*). The three typical entropy-increasing processes shown here are the vaporization of a solid (*top*), the mixing of two gases (*middle*) and the helix-to-random-coil transition of protein molecules (*bottom*).

tures, the orientation of the molecules in the crystal is ruled largely by chance, some molecules pointing in one direction and others in the opposite direction. For this reason carbon monoxide is said to have a residual entropy at absolute zero.

Bernal and Fowler in their discussion of the structure of ice implied that there might be no energy considerations dictating a particular arrangement of the hydrogen atoms. Amplifying this idea, Linus Pauling at the California Institute of Technology argued that ice (like carbon monoxide and a few other compounds) should have a residual entropy at low temperature. Indeed, this had already been observed experimentally by W. F. Giauque and J. W. Stout of the University of California at Berkeley. By measurement of the heat capacity of ice down to very low temperatures they found that the crystals actually did retain a certain amount of entropy.

It became clear, then, that in the crystals of ice the molecules are oriented in more than one way. Could the measured residual entropy shed any light on just how many different arrangements are actually present? A means of tackling this question was available: it was the famous entropy formula of the 19th-century Austrian physicist Ludwig Boltzmann, one of the founders of statistical mechanics. Boltzmann's formula predicts that the entropy (S) of any system in which the molecules are independent of one another will be proportional to the natural logarithm of the number of equally likely orientations of the molecules. In symbolic terms the formula is $S = k \log w$, with w standing for the number of possible orientations and k for the celebrated Boltzmann constant. In the case of carbon monoxide w is 2 (each molecule has two possible orientations), and the formula predicts that the residual entropy divided by the constant (S/k) should therefore be .69. The measured value of carbon monoxide's residual entropy agrees satisfactorily with this figure.

Suppose now we apply the formula to ice. In the ice crystal a molecule has six possible orientations: its hydrogen atoms can be oriented in six different ways. According to the Boltzmann formula, with $w = 6$ the residual entropy (more precisely, S/k) should have the value 1.8 [*see bottom illustration on preceding page*]. Actually in the experimental measurements the value turns out to be only .4. The discrepancy is not surprising; in the Boltzmann formula, as we have noted, w is the number of *independent* alternatives. The orientations of the water molecules in ice are not independent of one another. Once we have selected a particular orientation for one molecule there are only three possible ways a second molecule can orient itself to form a good hydrogen bond with the first, and the molecules that join up with the first have other neighbors that in turn restrict the orientation choices still further. Pauling calculated that the average number of orientation choices for the system (that is, the value of w) is about 1.5 per molecule.

This estimate agreed well with the measurements Giauque and Stout had obtained for the residual entropy of ice; with $w = 1.5$, the value of S/k would be almost exactly .4. Lars Onsager of Yale University suggested in 1939 that it was important to compute as precise a value as possible for the average number of possible orientations. Pauling's estimate of 1.5 was the lower limit for this figure; Onsager noted that the actual value must be somewhat higher. The reason it was important to calculate the number exactly was that, if w turned out to be significantly higher than 1.5, it would indicate that there was some ordering influence within the ice crystals; in other words, that the possible orientations of the molecules were not all equally likely.

For more than 30 years chemists and mathematicians have struggled with this interesting mathematical puzzle, attempting to determine just how many geometric orientations are open to a collection of ice molecules. So far the problem has defied the most sophisticated assaults. Two years ago, however, John Nagle, a student of Onsager's at Yale, pursued some ideas suggested by Edmund A. DiMarzio and Frank H. Stillinger, Jr., of the Bell Telephone Laboratories and arrived at a reliable estimate that narrows the area of uncertainty to very close limits. Nagle found that the exact value must lie between 1.5065 and 1.5068. This result, bringing the theoretical prediction of ice's residual entropy into close agreement with the experimental measurements, shows that if there are any ordering forces in the crystals favoring one arrangement over another, they must be extremely weak.

We have been considering an ideal picture in which the structure of ice is governed strictly by the known rules—in particular the hydrogen-bond rule of Bernal and Fowler, which would lock the molecules in a fixed pattern. Now we must look into certain complications that lead to a surprising new view of

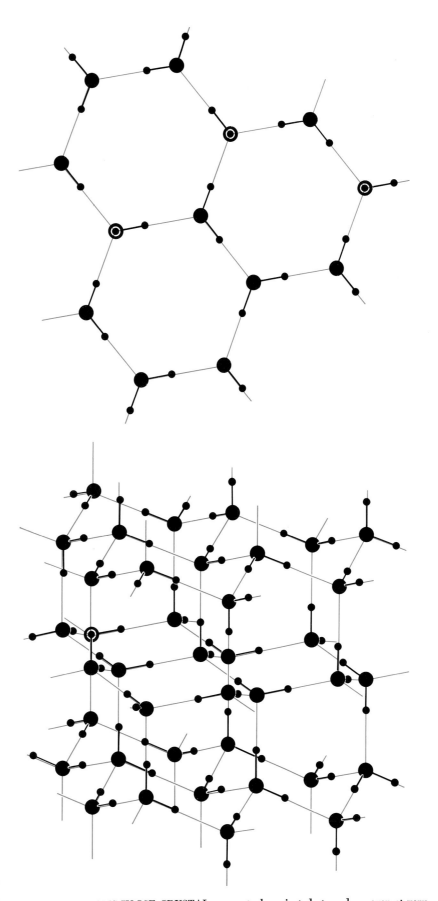

HYDROGEN ATOMS IN ICE CRYSTAL appear to be oriented at random, even at very low temperatures. There is no unit cell in the crystal for hydrogen nuclei. This conclusion is based on measurements of the residual entropy of ice. Crystal is viewed from above (*top*) and from the side (*bottom*). The molecules are somewhat reduced in size for clarity.

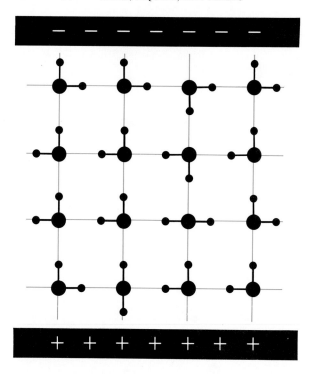

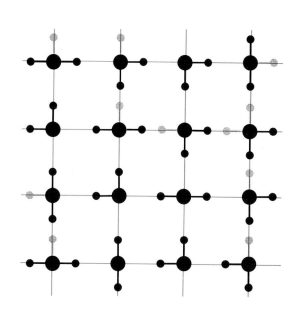

ICE CRYSTALS BECOME POLARIZED when they are placed in an electric field between positively and negatively charged metal plates (*left*). When the field is shut off, the crystals gradually return to their normally "relaxed," or depolarized, state (*right*).

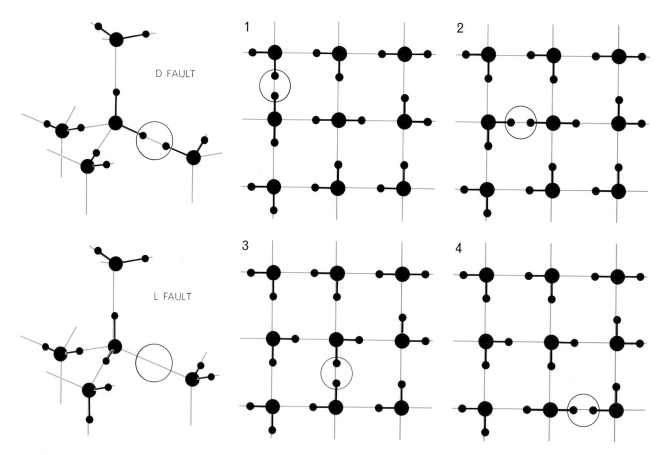

IMPERFECTIONS in the crystal lattice of ice explain how the molecules rotate. The two defects at left, proposed by the Danish chemist Niels Bjerrum in 1951, represent occasional violations of the hydrogen-bond rule. In a *D* fault the positively charged hydrogen arms of two adjacent molecules point toward each other. In an *L* fault the two negatively charged electronic arms do so. At right a *D* fault is shown wandering through a hypothetical two-dimensional crystal, leaving behind a chain of reoriented molecules. The decay of polarization (or dielectric relaxation) illustrated at the top of the page results from the migrations of such Bjerrum faults.

what actually goes on inside the crystals of ice.

The first of these complications was the discovery many years ago that when ice crystals are placed in an electric field between positively and negatively charged metal plates, the crystals become polarized—negative on the side toward the positive plate, positive on the side toward the negative plate. When the field is shut off, the crystals gradually return to their normally depolarized state. Obviously in the polarized condition a majority of the molecules must point a positively charged hydrogen arm toward the negative plate and a negative electronic arm toward the positive plate, and the orientations must become random again when the crystal is depolarized [*see top illustration on opposite page*]. The late Peter J. W. Debye of Cornell University, a pioneer in the chemistry of polar molecules, concluded that the molecules in ice must be able somehow to shift their orientations!

How could they do so? It was difficult to see how the individual molecules could rotate in the interlocked crystal system, where each molecule's orientation is fixed by the hydrogen bonding with four neighbors. In 1951 the Danish chemist Niels Bjerrum found a reasonable explanation. It is well known that many crystals are subject to imperfections, and Bjerrum suggested a form of imperfection in ice crystals that could account for the molecules' ability to change their orientation. In occasional instances, he said, molecules might violate the Bernal-Fowler rule: hydrogen arms of two adjacent molecules might point toward each other, or electronic arms might do so [*see bottom illustration on opposite page*]. He named the first case a *D* fault (from the German word for "doubled," *doppelt*, signifying a doubling of the bond) and the second case an *L* fault (from *leer*, "empty," indicating the absence of a proton). The presence of these faults would enable the molecules involved to pivot and so change their orientation.

There are some who object to Bjerrum's theory, principally on the ground that the electrical repulsion between hydrogen nuclei is too strong to allow a *D* fault to occur. The hypothesis has proved eminently workable, however, and it seems likely to stand up in its basic features. Among other things, it shows how such a fault, passed from one molecule to the next, can cause many molecules in a crystal to shift their orientation.

Experiments suggest that in ice at a temperature just below the melting point

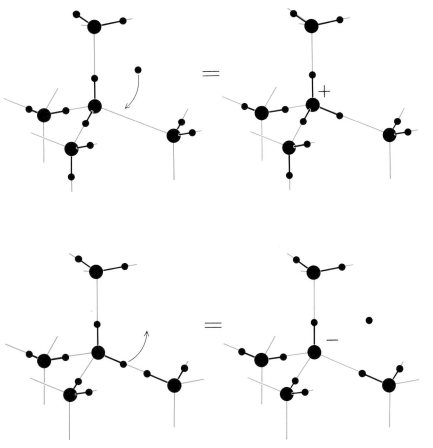

MOLECULAR IONS are additional defects present in an ice crystal. The positive ion at top is called a hydronium ion (H_3O^+). The negative ion at bottom is a hydroxyl ion (OH^-).

one molecule in a million is likely to be involved in a Bjerrum fault. Because of the rapid migration of faults this is enough to effect the polarization of an ice crystal by an applied electric field, and also the "relaxation," or depolarization, of the crystal after the field has been removed. Indeed, the movement of faults and the attendant rotation of molecules in ice crystals have been detected even without the use of electric fields. It has been found that when the molecular orientations in ice are changed by the application of mechanical pressure, which affects some orientations more than others, the orientations return to the normal distribution after the stress is removed. Measurements of the relaxation time indicate that this form of recovery, like electric depolarization, is brought about by the migration of Bjerrum faults.

The most startling finding in the electric-polarization experiments is that in a crystal of ice just below the melting point the "half-life" of the decay of polarization after the shutoff of the electric field is about a hundred-thousandth of a second. We are forced to the conclusion that every molecule in such a crystal normally rotates at the rate of about

100,000 times per second!

There are other indications that ice, far from being a quiescent system, seethes with activity. One is its electrical conductivity. It is true that ice is not a very good conductor, but there are worse; in fact, the conductivity of pure ice is about the same as that of liquid water, although water is much more abundantly supplied with ions as charge carriers. A potential of 200 volts impressed across a one-inch cube of ice will produce a flow of a millionth of an ampere of current through the cube.

In metals and semiconductors electric current is carried by mobile electrons; in ice, as in some other solids and in liquids, it is carried by charged ions. It is convenient to think of the ions in ice as a kind of defect associated with the Bjerrum faults. The ions in ice are the hydronium ion (H_3O^+) and the hydroxyl ion (OH^-); these are also present in liquid water. Onsager and Marc Dupuis have suggested that their role in carrying current in ice resembles the transmission of Bjerrum faults. The hydronium ion, for instance, can transmit current by transferring its extra proton to the next ice molecule, which in turn will pass a proton to the next and so on [*see illus-*

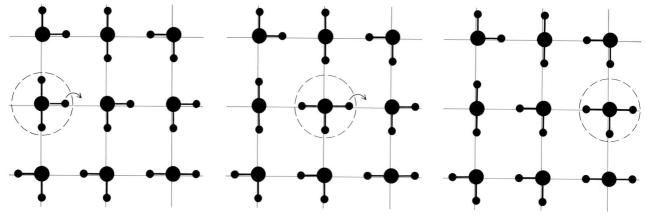

HYDRONIUM ION MOVES at an extremely high speed through the crystal lattice of ice because the entire ion need not jump, just the extra proton (*arrows*). Because of the great mobility of its protons, ice shares many properties with electronic semiconductors.

tration above]. The ions are extremely rare in ice; according to Manfred Eigen and L. De Maeyer of the University of Göttingen, there are only one hydronium ion and one hydroxyl ion for about every million million molecules. The ions move so fast, however, that they convey an appreciable amount of current. Furthermore, in ice a "catcher" is always in position to catch the proton "pitched" by its neighbor, whereas in liquid, amorphous water there may be a wait for a molecule to arrive in a position to receive the pitch. This compensating factor enables ice to conduct electricity about as well as water even though it contains fewer ions.

Ice can be called a semiconductor. Since its current-carriers are protons, whereas in the better-known semiconductors the carriers are electrons, Eigen describes ice as a "protonic" semiconductor. Indeed, Eigen has shown that ice can act as a transistor. The usual practice in making a transistor is to "dope" the semiconducting element with traces of impurities to add extra electrons. Eigen doped ice with small amounts of acids or bases (such as hydrogen fluoride, ammonia or lithium hydroxide) that increased the number of mobile protons,

and he was able to construct a *p-n* junction device that rectified current.

Several years ago two Swiss chemists, W. Kuhn and M. Thürkauf of the University of Basel, reported a curious finding that has led to the discovery of still another kind of movement in the ice crystal. They used labeled molecules of water, tagged with a heavy isotope of hydrogen (deuterium) or an uncommon heavy isotope of oxygen (oxygen 18), as tracers whose travels in the crystals could be followed. Rather surprisingly it turned out that the rate of diffusion through the crystal was the same in both cases: the spread of deuterium was as rapid as that of oxygen 18. This indicated that the tagged atoms moved about in the crystal primarily as members of intact molecules, not as separate hydrogen or hydroxide ions.

How can complete molecules move in the crystal lattice? The most satisfactory explanation seems to be that they are able to make their way through the open spaces, or channels, within the lattice. The normal thermal vibrations of the molecules may occasionally cause a molecule to jump out of its lattice site into an interstitial space between other molecules. From there it may either jump

back or begin to wander about in the interstices of the lattice, perhaps to wind up kicking another molecule out of position and taking its place.

From various experiments, including measurements of the "relaxation" of nuclear magnetic resonance, which is a very sensitive indicator of molecular motion, we can derive an estimate of how often these jumps occur. The calculations lead to the conclusion that in ice at a temperature just below the melting point the average molecule jumps out of its lattice position about once every millionth of a second, and it travels an average distance of about eight molecules before regaining a normal lattice site. Thus the process takes place some 10 times faster than the in-place rotations described earlier!

The placid and symmetrical appearance of a crystal of ice is certainly deceiving. Inside, at the molecular level, the perfection is punctuated by imperfections, the order is disrupted by fascinating varieties of disorder, and an astonishingly busy traffic of molecules and defects is continuously charging about through the lattice. In short, there is a great deal more to ice than meets the eye.

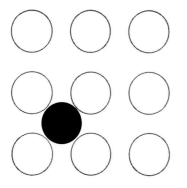
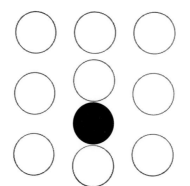
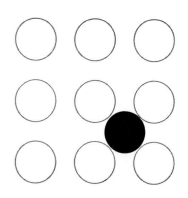

INTERSTITIAL MOLECULE is another type of imperfection in ice crystals. Normal thermal vibrations may occasionally cause a molecule to jump out of its lattice site and begin to wander about through the channels, or open spaces, within the crystal lattice.

V

DYNAMICS OF CHEMICAL SYSTEMS

V

DYNAMICS OF
CHEMICAL SYSTEMS

INTRODUCTION

The first law of thermodynamics states that the energy of the universe is conserved. Why then is there danger of our society running out of energy sources? Why is there an energy crisis? It is the second law of thermodynamics that states that the entropy of the universe is increasing and that all spontaneous processes must be accompanied by an increase in entropy. This means that there are only a limited number of sources of useful low-entropy energy available to society. It also means that the efficiency of many energy transducers, such as the heat engine, is limited to values far less than 100 percent. Thus virtually all energy transducers are heat polluters at the very least and, of course, most present even greater problems of chemical and radioactive pollution.

The energy crisis in the newspapers today is not a crisis of pollution, however (though pollution is a critical problem), but a crisis of availability. M. King Hubbert estimates in article 13, "The Energy Resources of the Earth," that the world's oil resources will be 90 percent gone in 60 years, within the lifetimes of most of our young children. Similarly, he estimates the world's coal resources will last no longer than 700 years.

Major alternatives to the fossil fuels as sources of energy are two in number, nuclear energy and solar energy. Nuclear fission depends upon the naturally available isotope uranium-235, ^{235}U. The amount of energy released by one fission is 200 MeV, 200 million electron volts. This means one gram of ^{235}U is equivalent in energy content to 2.7 metric tons of coal. At present the per capita energy consumption in the United States is equivalent to 11 metric tons of coal/year or about 4 grams ^{235}U/year. Nuclear energy requires far less weight in fuel than does use of fossil fuels. Nevertheless, ^{235}U is a rare isotope, and even though some ^{235}U is easily accessible, its limited availability would provide less than 100 years of fuel at the world's present rate of energy consumption. A part of the world's existing supply of ^{235}U is being used at present in nuclear power plants, which are barely competitive with fossil fuel power plants despite subsidies from the federal government. Although the fuel is comparatively cheap, the capital investment in a nuclear plant is still very high, and waste disposal is a serious problem.

The decreasing supply of ^{235}U can be alleviated by the use of breeder reactors, which will probably be in use commercially within three decades. The breeder reaction exposes natural and relatively abundant uranium-238 (^{238}U), and thorium-232 (^{232}Th), to neutrons, converting them to plutonium-239 (^{239}Pu) and uranium-233 (^{233}U), respectively. These are both fissionable materials; therefore the availability of fission as an energy source will be extended to hundreds of thousands of years. The breeder reaction by which ^{238}U is converted to ^{239}Pu is

$$^{238}_{92}U + {}^{1}_{0}n \longrightarrow {}^{239}_{92}U \searrow \longrightarrow {}^{239}_{93}Np \searrow \longrightarrow {}^{239}_{94}Pu$$
$$\beta^{-} \text{ particle} \qquad \beta^{-} \text{ particle}$$

The breeder reaction for converting ^{232}Th to ^{233}U is

$$^{232}_{90}Th + {}^{1}_{0}n \longrightarrow {}^{233}_{90}Th \searrow \longrightarrow {}^{233}_{91}Pa \searrow \longrightarrow {}^{233}_{92}U$$
$$\beta^{-} \text{ particle} \qquad \beta^{-} \text{ particle}$$

In these notations, the atomic mass number is designated in the upper left superscript (238 at symbol U, for example, is the mass number of that isotope of uranium); the atomic number (equal to the positive charge on the atomic nucleus) is designated in the lower left subscript.

The reactors in which fission takes place are hazardous for two reasons. There is the chance, although perhaps remote, of a failure in the cooling of the reactor, with unpredictable consequences. Second, the radioactive wastes must be disposed of, and it is doubtful if an adequate method has yet been devised. For these reasons, it is hoped that fission reactors are stopgap solutions to the problem of providing energy.

The fusion reactor, on the other hand, by which energy would be provided through the fusion of atoms of deuterium or tritium, would require containment of a plasma at temperatures so high that all matter would be vaporized. The present methods of plasma containment use magnetic fields, but as yet conditions necessary for sustained fusion have not been obtained. The fusion reactions supply energies comparable to those of fission reactions per gram of reacting material, but they present two distinct advantages. There are no radioactive wastes resulting from fusion energy production, and the atoms of deuterium needed for reactions are highly abundant among the hydrogen atoms in seawater. Hubbert discusses these reactions briefly in article 13.

"The Conversion of Energy," article 14, by Summers, contains two major proposals for the capture of solar energy. The recent development of surface coatings which absorb solar energy with a high efficiency and yet block the emission of that energy as infrared radiation provides the basis for one of these proposals. Aden B. Meinel and Marjorie P. Meinel, the proposers of this scheme, believe that 1000-megawatt solar power stations based on their design would be technologically feasible and economical. The energy collection system has been successfully tested on a small scale but the large-scale feasibility of this highly interesting proposal is several years away.

The second proposal involves a five-by-five-mile solar panel in stationary orbit 22,000 miles above the Earth. The resulting electrical energy would then be converted to microwaves which would be captured by a six-by-six-mile antenna on the Earth's surface. The technical feasibility of this idea is somewhat more questionable; it would depend, for example, on the availability of a space shuttle to carry some five million pounds of materials into orbit, a little bit at a time.

The Summers article also contains a description of conventional energy transducers and their efficiencies. It discusses some of the less used sources of energy, such as tidal power and wind power.

Many conversions of energy, such as the combustion of the fossil fuels, require that some work be done through a chemical reaction, and all chemical reactions can be studied in terms of the rates at which the reactions proceed. The rates of chemical reactions cannot be predicted from the energy changes associated with the reaction. If a

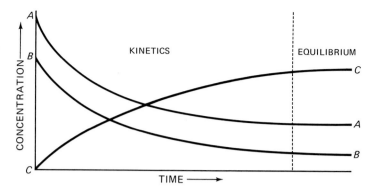

Figure V.1   The rate of a chemical reaction at any time is the slope of curve C at that time. The system of which A, B, and C are components is said to be at equilibrium when concentrations are no longer changing with time.

simple chemical reaction of the form $A + B \rightarrow C$ is considered in a plot of concentrations of each of the components with time, the plot might look like that shown in Figure V.1. After a certain time, the concentrations of A, B, and C reach constant values and the system is said to be at equilibrium. The rate of the chemical reaction is the rate of production of C from A and B. The rate of production of C at any time is just the *slope* of the C curve at that time. So the rate of this reaction starts at some value and gradually drops to zero as the system approaches closer and closer to equilibrium.

If such a simple chemical reaction occurs by the collision of a molecule of A with a molecule of B, then the rate of the forward reaction will be proportional to the instantaneous concentrations of both A and B, and the rate law is: Rate forward $= k_f \bar{A} \bar{B}$, where $\bar{A}$ and $\bar{B}$ signify the instantaneous concentrations of A and B, respectively. If the backward reaction depends only on the rate at which C dissociates, the rate law is: Rate backward $= k_b \bar{C}$, where $\bar{C}$ signifies the instantaneous concentration of C. When these two rates are equal, equilibrium is established. Otherwise a net chemical reaction is occurring and kinetic measurements can be made.

The relaxation measurements discussed in article 15 by Faller are capable of measuring very fast chemical events which occur in 10^{-3} to 10^{-10} seconds. The word relaxation refers to the natural tendency of a chemical reaction to reach a new equilibrium after being perturbed in some way. These methods displace a system from equilibrium by a small amount through a pressure shock, temperature jump, sound absorption, or an electric-field pulse. The simple exponential decay back to a condition of equilibrium is then noted by optical techniques or by monitoring electrical conductivity.

To illustrate what can be studied with fast chemical reactions, Faller presents the equilibrium constant and discusses measurements of the rate of reaction of protons, H^+, and hydroxide ions, OH^-, in water. He also discusses the rate of complex-ion formation. There are therefore numerous examples of the basic principles and reactions of chemistry in this article, making it a valuable learning tool.

As an example of how reaction rates were studied before the development of relaxation methods, Faller discusses the Michaelis-Menten enzyme kinetics, which were proposed in 1913. The test of the validity of this rate law at the molecular level was not possible until 1966, after the development of relaxation methods. This article contains an excellent discussion of enzyme kinetics and of allosteric effects in protein chemistry. Allosteric effects refer to the effects of subtle shape changes upon the biological activity of the protein. Such phenomena are increasingly important in the study of mechanisms of cellular regulation. Both article 1, "Chemistry," by Pauling, and article 31, "The Three-dimensional Structure of an Enzyme Molecule," by Phillips, contain some discussion of the importance of molecular structure to enzyme function.

Reaction rates depend not only on the concentrations of reactants and their temperature; they also depend upon the presence or absence of rate affectors called catalysts. Enzymes are examples of catalysts. Article 16, by Haensel and Burwell, "Catalysis," discusses the importance of catalysis to the petroleum industry as well as to plastics, synthetic fibers, detergents, margarine, and synthetic rubbers.

The basis for catalysis is best described by a figure depicting a hill between the reactants and the products. The hill shown at the center of Figure V.2 is an energy barrier between the two states, which must be overcome for the reaction to occur. A catalyst interacts with one or both reactants, making their subsequent reaction with one another easier. The catalyst lowers the height of the activation energy barrier and makes the reaction occur more rapidly. It in no way affects the energy difference between reactants and products. Nor does it affect the equilibrium constant, because catalysts increase both forward and backward reaction rates by equal factors.

Haensel and Burwell's article discusses many important principles of chemistry. The energy distribution of molecules is discussed (see article 10, "Molecular Motions," by Alder and Wainwright). The temperature dependence of rate constants, activation energies, and many elementary processes are all described. The article is rich in descriptive chemistry as well. Examples are the catalytic cracking of petroleum, the importance of structural isomerization in cracking, the use of catalytic afterburners for automobile exhausts, a definition of an octane number as it applies to gasoline, hydrogenation reactions using a nickel catalyst, the reactions of coordination compounds, and the synthesis of stereospecific polymers using catalysts.

"Why the Sea Is Salt," article 17, by MacIntyre, provides an example of the interaction between chemistry and oceanography. You will recall that in article 4, "The Chemical Elements of Life," there was some discussion of the ocean. The elemental analysis of the ocean MacIntyre presents is far more detailed and accurate than the data in

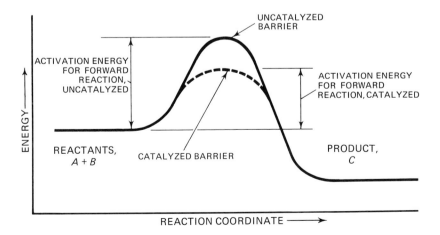

Figure V.2 Barriers that must be overcome before reaction $A + B \rightleftarrows C$ can proceed, with and without the use of a catalyst.

Frieden's article. To understand MacIntyre's data, it is useful to know that a molar solution is a solution containing 1 mole/liter of solution of the component in question. A millimolar solution contains 10^{-3} mole/liter, a micromolar solution contains 10^{-6} mole/liter, and a nanomolar solution contains 10^{-9} mole/liter. Notice that the range of concentration in the table MacIntyre uses to show the composition of seawater is from 55 moles/kg to 2×10^{-9} mole/kg seawater. Notice too that the "thalasso"-chemist's periodic table used by MacIntyre contains the prominent ionic form in which each element is found in the ocean.

A variety of chemical systems are discussed in "Why the Sea Is Salt." The chemical transformations of the major minerals in streams when they "equilibrate" in the ocean provide examples of heterogeneous equilibrium. Specific rate constants are also used to describe the kinetics of these transformations. The Gibbs phase rule is used to explain that there are essentially only two degrees of freedom in seawater, the temperature and the chloride ion concentration.

MacIntyre's article provides some excellent examples of the utility of equilibrium constants. A topic Faller discussed in article 15—the water constant—is presented, and so are the functions of pH and buffers. The Bjerrum diagram for carbonate is a graphic representation of the ionic equilibria of carbonic acid. The first and second dissociation constants for H_2CO_3 are

(1) $$H_2CO_3 \rightarrow H^+ + HCO_3^- \qquad K_1 = \frac{[H^+]\,[HCO_3^-]}{[H_2CO_3]}$$

(2) $$HCO_3^- \rightarrow H^+ + CO_3^{2-} \qquad K_2 = \frac{[H^+]\,[CO_3^{2-}]}{[HCO_3^-]}$$

At the point where $[CO_3^{2-}] = [HCO_3^-]$, $[H^+] = K_2$. Since that point occurs at $pH = 9$, $K_2 = 10^{-9}$. For similar reasons $K_1 = 10^{-6}$. At pH greater than 9, CO_3^{2-} is the dominant form of carbon in seawater, between pH 6 and 9, HCO_3^- is the dominant form, and below pH 6, H_2CO_3 is the dominant form.

Toward the end of his article MacIntyre discusses the importance of the rate at which the carbon dioxide cycle of the atmosphere and ocean equilibrates, and in doing so he emphasizes the effect of burning fossil fuels upon the CO_2 content of the atmosphere. The "greenhouse" effect of added CO_2 in the atmosphere results from a decrease in the radiation of infrared radiation away from the Earth, a decrease caused by the absorption of outgoing infrared radiation by the CO_2. In principle, the average temperature of the Earth would be increased by this mechanism. Finally, MacIntyre turns to a brief discussion of the oxygen and nitrogen cycles of the Earth. The last article in this section (19) will discuss the oxygen cycle in detail. In the last two paragraphs of his article, MacIntyre introduces several ecological problems related to water contamination, and makes a plea for as much scientific concern with sewage chemistry as there is with oceanography.

"Fuel Cells," article 18, by Austin, stresses the poor efficiency of conventional methods of energy conversion, and therefore this article complements article 14, "The Conversion of Energy." These limitations in efficiency are largely the consequence of the second law of thermodynamics. But Austin reports about a recently developed fuel cell that can achieve nearly two times the efficiency of the typical steam power plant. A fuel cell is merely a battery with continually replenished reagents that enable the fuel cell to operate indefinitely. Some of the chemical topics that Austin makes reference to in "Fuel Cells" are oxidation-reduction reactions (in which electrons are transferred from one reacting molecule to another), galvanic cells, electrochemical potentials, limitations due to catalyst poisoning, and the activation energies that limit the rate of electricity production. The fuel cell is a potentially important power source. For example, electric cars would be an easy reality if fuel cells of appropriate power output could convert chemical energy to electrical energy. Unfortunately, as Austin mentions, there are numerous engineering obstacles that must be overcome before such ideas are feasible.

The last article in this section is "The Oxygen Cycle" by Cloud and Gibor. It provides still another example of the interrelationships between geology, biology, and chemistry. The oxygen in the Earth's atmosphere all arose, through photosynthesis, from CO_2, an evolution that implies an equivalent amount of carbon is stored in the Earth as graphite, coal, oil, gas, and other carbon compounds. Apparently, as nearly as it can be measured, this balance does exist. Just as Wald did in article 2, Cloud and Gibor show how oxygen is used in living cells, and they compare the energy efficiencies of respiration and fermentation in living cells. They explain how reduction reactions are necessary for photosynthesis, how the reverse oxidation reaction is necessary for respiration, and the importance of carbon compounds in these processes. The reduction potentials of the cytochromes are important in the process of oxidative metabolism, because they lead to the synthesis of adenosine triphosphate (ATP) and other high-energy phosphate compounds.

The authors discuss the evolution of organisms capable of photosynthesis from organisms dependent on organic food supplies, and again, this presentation is very similar to Wald's, which was written 16 years earlier.

The oxidative metabolism of cells is graphically presented in an effective diagram showing the structures of the principal molecules. That diagram includes glycolysis, respiration, and oxidative phosphorylation—further examples of the oxidation-reduction processes of life.

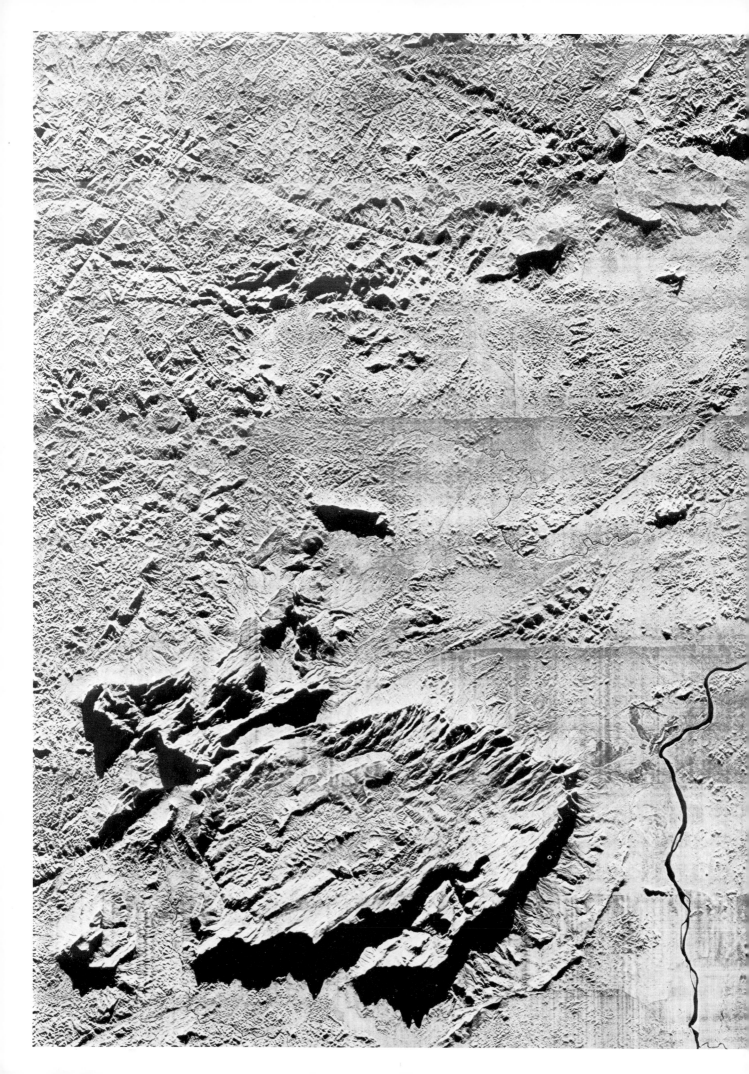

13

THE ENERGY RESOURCES OF THE EARTH

M. KING HUBBERT
September 1971

*They are solar energy (current and stored), the tides, the
earth's heat, fission fuels and possibly fusion fuels.
From the standpoint of human history the epoch of the
fossil fuels will be quite brief*

Energy flows constantly into and out of the earth's surface environment. As a result the material constituents of the earth's surface are in a state of continuous or intermittent circulation. The source of the energy is preponderantly solar radiation, supplemented by small amounts of heat from the earth's interior and of tidal energy from the gravitational system of the earth, the moon and the sun. The materials of the earth's surface consist of the 92 naturally occurring chemical elements, all but a few of which behave in accordance with the principles of the conservation of matter and of nontransmutability as formulated in classical chemistry. A few of the elements or their isotopes, with abundances of only a few parts per million, are an exception to these principles in being radioactive. The exception is crucial in that it is the key to an additional large source of energy.

A small part of the matter at the earth's surface is embodied in living organisms: plants and animals. The leaves of the plants capture a small fraction of the incident solar radiation and store it chemically by the mechanism of photosynthesis. This store becomes the energy supply essential for the existence of the plant and animal kingdoms. Biologically stored energy is released by oxidation at a rate approximately equal to the rate of storage. Over millions of years, however, a minute fraction of the vegetable and animal matter is buried under conditions of incomplete oxidation and decay, thereby giving rise to the fossil fuels that provide most of the energy for industrialized societies.

It is difficult for people living now, who have become accustomed to the steady exponential growth in the consumption of energy from the fossil fuels, to realize how transitory the fossil-fuel epoch will eventually prove to be when it is viewed over a longer span of human history. The situation can better be seen in the perspective of some 10,000 years, half before the present and half afterward. On such a scale the complete cycle of the exploitation of the world's fossil fuels will be seen to encompass perhaps 1,300 years, with the principal segment of the cycle (defined as the period during which all but the first 10 percent and the last 10 percent of the fuels are extracted and burned) covering only about 300 years.

What, then, will provide industrial energy in the future on a scale at least as large as the present one? The answer lies in man's growing ability to exploit other sources of energy, chiefly nuclear at present but perhaps eventually the much larger source of solar energy. With this ability the energy resources now at hand are sufficient to sustain an industrial operation of the present magnitude for another millennium or longer. Moreover, with such resources of energy the limits to the growth of industrial activity are no longer set by a scarcity of energy but rather by the space and material limitations of a finite earth together with the principles of ecology. According to these principles both biological and industrial activities tend to increase exponentially with time, but the resources of the entire earth are not sufficient to sustain such an increase of any single component for more than a few tens of successive doublings.

Let us consider in greater detail the flow of energy through the earth's surface environment [*see illustration on next two pages*]. The inward flow of energy has three main sources: (1) the intercepted solar radiation; (2) thermal energy, which is conveyed to the surface of the earth from the warmer interior by the conduction of heat and by convection in hot springs and volcanoes, and (3) tidal energy, derived from the combined kinetic and potential energy of the earth-moon-sun system. It is possible in various ways to estimate approximately how large the input is from each source.

In the case of solar radiation the influx is expressed in terms of the solar constant, which is defined as the mean rate of flow of solar energy across a unit of area that is perpendicular to the radiation and outside the earth's atmosphere at the mean distance of the earth from the sun. Measurements made on the earth and in spacecraft give a mean value for the solar constant of 1.395 kilowatts per square meter, with a variation of about 2 percent. The total solar radiation intercepted by the earth's diametric plane of 1.275×10^{14} square meters is therefore 1.73×10^{17} watts.

The influx of heat by conduction from the earth's interior has been determined from measurements of the geothermal gradient (the increase of temperature with depth) and the thermal conductivity of the rocks involved. From thousands of such measurements, both on land and on the ocean beds, the average rate of flow of heat from the interior of the earth has been found to be about .063 watt per square meter. For the earth's surface area of 510×10^{12} square meters the total heat flow amounts to

RESOURCE EXPLORATION is beginning to be aided by airborne side-looking radar pictures such as the one on the opposite page made by the Aero Service Corporation and the Goodyear Aerospace Corporation. The technique has advantage of "seeing" through cloud cover and vegetation. This picture, which was made in southern Venezuela, extends 70 miles from left to right.

some 32×10^{12} watts. The rate of heat convection by hot springs and volcanoes is estimated to be only about 1 percent of the rate of conduction, or about $.3 \times 10^{12}$ watts.

The energy from tidal sources has been estimated at 3×10^{12} watts. When all three sources of energy are expressed in the common unit of 10^{12} watts, the total power influx into the earth's surface environment is found to be 173,035 $\times 10^{12}$ watts. Solar radiation accounts for 99.98 percent of it. Another way of stating the sun's contribution to the energy budget of the earth is to note that at $173,000 \times 10^{12}$ watts it amounts to 5,000 times the energy input from all other sources combined.

About 30 percent of the incident solar energy ($52,000 \times 10^{12}$ watts) is directly reflected and scattered back into space as short-wavelength radiation. Another 47 percent ($81,000 \times 10^{12}$ watts) is absorbed by the atmosphere, the land surface and the oceans and converted directly into heat at the ambient surface temperature. Another 23 percent (40,-000×10^{12} watts) is consumed in the evaporation, convection, precipitation and surface runoff of water in the hydrologic cycle. A small fraction, about 370×10^{12} watts, drives the atmospheric and oceanic convections and circulations and the ocean waves and is eventually dissipated into heat by friction. Finally, an even smaller fraction—about

40×10^{12} watts—is captured by the chlorophyll of plant leaves, where it becomes the essential energy supply of the photosynthetic process and eventually of the plant and animal kingdoms.

Photosynthesis fixes carbon in the leaf and stores solar energy in the form of carbohydrate. It also liberates oxygen and, with the decay or consumption of the leaf, dissipates energy. At any given time, averaged over a year or more, the balance between these processes is almost perfect. A minute fraction of the organic matter produced, however, is deposited in peat bogs or other oxygen-deficient environments under conditions that prevent complete decay and loss of energy.

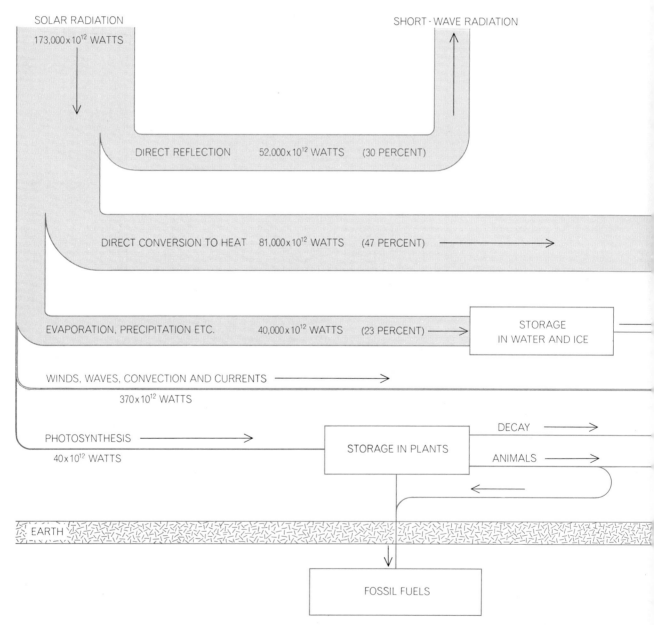

FLOW OF ENERGY to and from the earth is depicted by means of bands and lines that suggest by their width the contribution of each item to the earth's energy budget. The principal inputs are

solar radiation, tidal energy and the energy from nuclear, thermal and gravitational sources. More than 99 percent of the input is solar radiation. The apportionment of incoming solar radiation is

Little of the organic material produced before the Cambrian period, which began about 600 million years ago, has been preserved. During the past 600 million years, however, some of the organic materials that did not immediately decay have been buried under a great thickness of sedimentary sands, muds and limes. These are the fossil fuels: coal, oil shale, petroleum and natural gas, which are rich in energy stored up chemically from the sunshine of the past 600 million years. The process is still continuing, but probably at about the same rate as in the past; the accumulation during the next million years will probably be a six-hundredth of the amount built up thus far.

Industrialization has of course withdrawn the deposits in this energy bank with increasing rapidity. In the case of coal, for example, the world's consumption during the past 110 years has been about 19 times greater than it was during the preceding seven centuries. The increasing magnitude of the rate of withdrawal can also be seen in the fact that the amount of coal produced and consumed since 1940 is approximately equal to the total consumption up to that time. The cumulative production from 1860 through 1970 was about 133 billion metric tons. The amount produced before 1860 was about seven million metric tons.

Petroleum and related products were not extracted in significant amounts before 1880. Since then production has increased at a nearly constant exponential rate. During the 80-year period from 1890 through 1970 the average rate of increase has been 6.94 percent per year, with a doubling period of 10 years. The cumulative production until the end of 1969 amounted to 227 billion (227 × 10^9) barrels, or 9.5 trillion U.S. gallons. Once again the period that encompasses most of the production is notably brief. The 102 years from 1857 to 1959 were required to produce the first half of the cumulative production; only the 10-year period from 1959 to 1969 was required for the second half.

Examining the relative energy contributions of coal and crude oil by comparing the heats of combustion of the respective fuels (in units of 10^{12} kilowatt-hours), one finds that until after 1900 the contribution from oil was barely significant compared with the contribution from coal. Since 1900 the contribution from oil has risen much faster than that from coal. By 1968 oil represented about 60 percent of the total. If the energy from natural gas and natural-gas liquids had been included, the contribution from petroleum would have been about 70 percent. In the U.S. alone 73 percent of the total energy produced from fossil fuels in 1968 was from petroleum and 27 percent from coal.

Broadly speaking, it can be said that the world's consumption of energy for industrial purposes is now doubling approximately once per decade. When confronted with a rate of growth of such magnitude, one can hardly fail to wonder how long it can be kept up. In the case of the fossil fuels a reasonably definite answer can be obtained. Their human exploitation consists of their being withdrawn from an essentially fixed initial supply. During their use as sources of energy they are destroyed. The complete cycle of exploitation of a fossil fuel must therefore have the following characteristics. Beginning at zero, the rate of production tends initially to increase exponentially. Then, as difficulties of discovery and extraction increase, the production rate slows in its growth, passes one maximum or more and, as the resource is progressively depleted, declines eventually to zero.

If known past and prospective future rates of production are combined with a reasonable estimate of the amount of a fuel initially present, one can calculate the probable length of time that the fuel can be exploited. In the case of coal reasonably good estimates of the

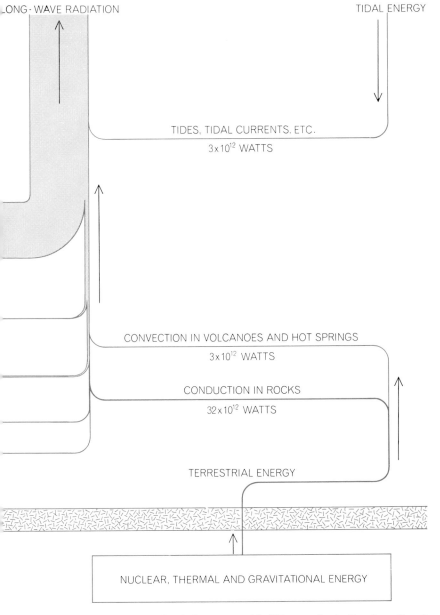

LONG-WAVE RADIATION

TIDAL ENERGY

TIDES, TIDAL CURRENTS, ETC.

3×10^{12} WATTS

CONVECTION IN VOLCANOES AND HOT SPRINGS

3×10^{12} WATTS

CONDUCTION IN ROCKS

32×10^{12} WATTS

TERRESTRIAL ENERGY

NUCLEAR, THERMAL AND GRAVITATIONAL ENERGY

indicated by the horizontal bands beginning with "Direct reflection" and reading downward. The smallest portion goes to photosynthesis. Dead plants and animals buried in the earth give rise to fossil fuels, containing stored solar energy from millions of years past.

amount present in given regions can be made on the basis of geological mapping and a few widely spaced drill holes, inasmuch as coal is found in stratified beds or seams that are continuous over extensive areas. Such studies have been made in all the coal-bearing areas of the world.

The most recent compilation of the present information on the world's initial coal resources was made by Paul Averitt of the U.S. Geological Survey. His figures [*see illustration below*] represent minable coal, which is defined as 50 percent of the coal actually present. Included is coal in beds as thin as 14 inches (36 centimeters) and extending to depths of 4,000 feet (1.2 kilometers) or, in a few cases, 6,000 feet (1.8 kilometers).

Taking Averitt's estimate of an initial supply of 7.6 trillion metric tons and assuming that the present production rate of three billion metric tons per year does not double more than three times, one can expect that the peak in the rate of production will be reached sometime between 2100 and 2150. Disregarding the long time required to produce the first 10 percent and the last 10 percent, the length of time required to produce the middle 80 percent will be roughly

the 300-year period from 2000 to 2300.

Estimating the amount of oil and gas that will ultimately be discovered and produced in a given area is considerably more hazardous than estimating for coal. The reason is that these fluids occur in restricted volumes of space and limited areas in sedimentary basins at all depths from a few hundred meters to more than eight kilometers. Nonetheless, the estimates for a given region improve as exploration and production proceed. In addition it is possible to make rough estimates for relatively undeveloped areas on the basis of geological comparisons between them and well-developed regions.

The most highly developed oil-producing region in the world is the coterminous area of the U.S.: the 48 states exclusive of Alaska and Hawaii. This area has until now led the world in petroleum development, and the U.S. is still the leading producer. For this region a large mass of data has been accumulated and a number of different methods of analysis have been developed that give fairly consistent estimates of the degree of advancement of petroleum exploration and of the amounts of oil and gas that may eventually be produced.

One such method is based on the principle that only a finite number of oil or gas fields existed initially in a given region. As exploration proceeds the shallowest and most evident fields are usually discovered first and the deeper and more obscure ones later. With each discovery the number of undiscovered fields decreases by one. The undiscovered fields are also likely to be deeper, more widely spaced and better concealed. Hence the amount of exploratory activity required to discover a fixed quantity of oil or gas steadily increases or, conversely, the average amount of oil or gas discovered for a fixed amount of exploratory activity steadily decreases.

Most new fields are discovered by what the industry calls "new-field wildcat wells," meaning wells drilled in new territory that is not in the immediate vicinity of known fields. In the U.S. statistics have been kept annually since 1945 on the number of new-field wildcat wells required to make one significant discovery of oil or gas ("significant" being defined as one million barrels of oil or an equivalent amount of gas). The discoveries for a given year are evaluated only after six years of subsequent development. In 1945 it required 26

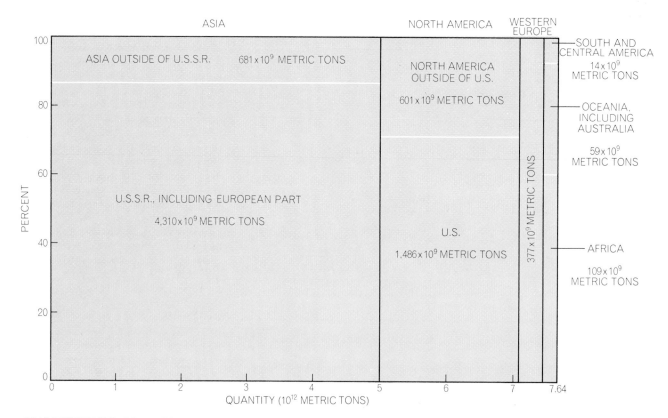

COAL RESOURCES of the world are indicated on the basis of data compiled by Paul Averitt of the U.S. Geological Survey. The figures represent the total initial resources of minable coal, which is defined as 50 percent of the coal actually present. The horizontal scale gives the total supply. Each vertical block shows the apportionment of the supply in a continent. From the first block, for example, one can ascertain that Asia has some 5×10^{12} metric tons of minable coal, of which about 86 percent is in the U.S.S.R.

new-field wildcat wells to make a significant discovery; by 1963 the number had increased to 65.

Another way of illuminating the problem is to consider the amount of oil discovered per foot of exploratory drilling. From 1860 to 1920, when oil was fairly easy to find, the ratio was 194 barrels per foot. From 1920 to 1928 the ratio declined to 167 barrels per foot. Between 1928 and 1938, partly because of the discovery of the large East Texas oil field and partly because of new exploratory techniques, the ratio rose to its maximum of 276 barrels per foot. Since then it has fallen sharply to a nearly constant rate of about 35 barrels per foot. Yet the period of this decline coincided with the time of the most intensive research and development in petroleum exploration and production in the history of the industry.

The cumulative discoveries in the 48 states up to 1965 amounted to 136 billion barrels. From this record of drilling and discovery it can be estimated that the ultimate total discoveries in the coterminous U.S. and the adjacent continental shelves will be about 165 billion barrels. The discoveries up to 1965 therefore represent about 82 percent of the prospective ultimate total. Making

due allowance for the range of uncertainty in estimates of future discovery, it still appears that at least 75 percent of the ultimate amount of oil to be produced in this area will be obtained from fields that had already been discovered by 1965.

For natural gas in the 48 states the present rate of discovery, averaged over a decade, is about 6,500 cubic feet per barrel of oil. Assuming the same ratio for the estimated ultimate amount of 165 billion barrels of crude oil, the ultimate amount of natural gas would be about 1,075 trillion cubic feet. Combining the estimates for oil and gas with the trends of production makes it possible to estimate how long these energy resources will last. In the case of oil the period of peak production appears to be the present. The time span required to produce the middle 80 percent of the ultimate cumulative production is approximately the 65-year period from 1934 to 1999—less than the span of a human lifetime. For natural gas the peak of production will probably be reached between 1975 and 1980.

The discoveries of petroleum in Alaska modify the picture somewhat. In particular the field at Prudhoe Bay appears likely by present estimates to contain

about 10 billion barrels, making it twice as large as the East Texas field, which was the largest in the U.S. previously. Only a rough estimate can be made of the eventual discoveries of petroleum in Alaska. Such a speculative estimate would be from 30 to 50 billion barrels. One must bear in mind, however, that 30 billion barrels is less than a 10-year supply for the U.S. at the present rate of consumption. Hence it appears likely that the principal effect of the oil from Alaska will be to retard the rate of decline of total U.S. production rather than to postpone the date of its peak.

Estimates of ultimate world production of oil range from 1,350 billion barrels to 2,100 billion barrels. For the higher figure the peak in the rate of world production would be reached about the year 2000. The period of consumption of the middle 80 percent will probably be some 58 to 64 years, depending on whether the lower or the higher estimate is used [see bottom illustration on page 147].

A substantial but still finite amount of oil can be extracted from tar sands and oil shales, where production has barely begun. The largest tar-sand deposits are in northern Alberta; they have total recoverable reserves of about 300 billion

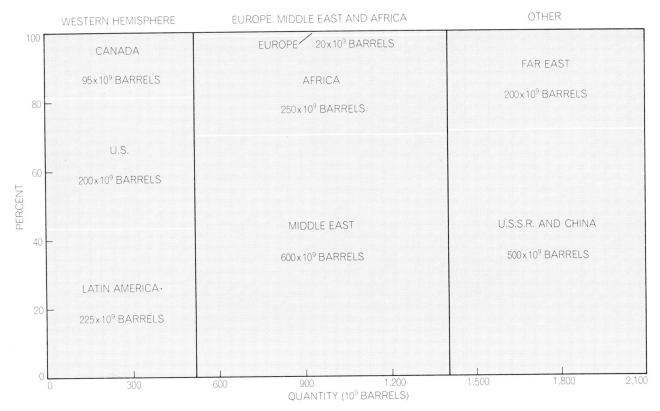

PETROLEUM RESOURCES of the world are depicted in an arrangement that can be read in the same way as the diagram of coal supplies on the opposite page. The figures for petroleum are derived from estimates made in 1967 by W. P. Ryman of the Standard Oil Company of New Jersey. They represent ultimate crude-oil production, including oil from offshore areas, and consist of oil already produced, proved and probable reserves, and future discoveries. Estimates as low as $1,350 \times 10^9$ barrels have also been made.

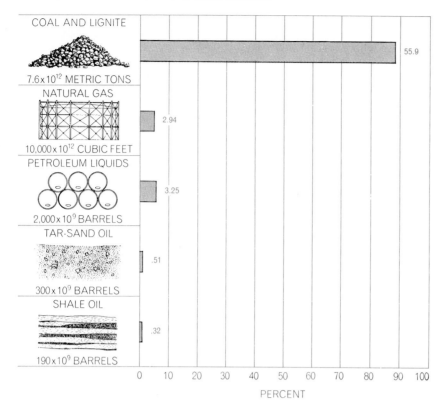

ENERGY CONTENT of the world's initial supply of recoverable fossil fuels is given in units of 10^{15} thermal kilowatt-hours (color). Coal and lignite, for example, contain 55.9 $\times$ 10^{15} thermal kilowatt-hours of energy and represent 88.8 percent of the recoverable energy.

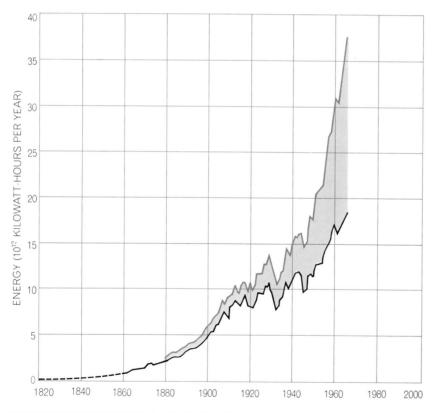

ENERGY CONTRIBUTION of coal (black) and coal plus oil (color) is portrayed in terms of their heat of combustion. Before 1900 the energy contribution from oil was barely significant. Since then the contribution from oil (shaded area) has risen much more rapidly than that from coal. By 1968 oil represented about 60 percent of the total. If the energy from natural gas were included, petroleum would account for about 70 percent of the total.

barrels. A world summary of oil shales by Donald C. Duncan and Vernon E. Swanson of the U.S. Geological Survey indicated a total of about 3,100 billion barrels in shales containing from 10 to 100 gallons per ton, of which 190 billion barrels were considered to be recoverable under 1965 conditions.

Since the fossil fuels will inevitably be exhausted, probably within a few centuries, the question arises of what other sources of energy can be tapped to supply the power requirements of a moderately industrialized world after the fossil fuels are gone. Five forms of energy appear to be possibilities: solar energy used directly, solar energy used indirectly, tidal energy, geothermal energy and nuclear energy.

Until now the direct use of solar power has been on a small scale for such purposes as heating water and generating electricity for spacecraft by means of photovoltaic cells. Much more substantial installations will be needed if solar power is to replace the fossil fuels on an industrial scale. The need would be for solar power plants in units of, say, 1,000 megawatts. Moreover, because solar radiation is intermittent at a fixed location on the earth, provision must also be made for large-scale storage of energy in order to smooth out the daily variation.

The most favorable sites for developing solar power are desert areas not more than 35 degrees north or south of the Equator. Such areas are to be found in the southwestern U.S., the region extending from the Sahara across the Arabian Peninsula to the Persian Gulf, the Atacama Desert in northern Chile and central Australia. These areas receive some 3,000 to 4,000 hours of sunshine per year, and the amount of solar energy incident on a horizontal surface ranges from 300 to 650 calories per square centimeter per day. (Three hundred calories, the winter minimum, amounts when averaged over 24 hours to a mean power density of 145 watts per square meter.)

Three schemes for collecting and converting this energy in a 1,000-megawatt plant can be considered. The first involves the use of flat plates of photovoltaic cells having an efficiency of about 10 percent. A second possibility is a recent proposal by Aden B. Meinel and Marjorie P. Meinel of the University of Arizona for utilizing the hothouse effect by means of selective coatings on pipes carrying a molten mixture of sodium and potassium raised by solar energy to a temperature of 540 degrees Celsius. By

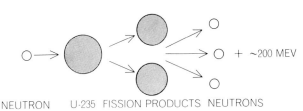

FISSION AND FUSION REACTIONS hold the promise of serving as sources of energy when fossil fuels are depleted. Present nuclear-power plants burn uranium 235 as a fuel. Breeder reactors
now under development will be able to use surplus neutrons from

the fission of uranium 235 (*left*) to create other nuclear fuels: plutonium 239 and uranium 233. Two promising fusion reactions, deuterium-deuterium and deuterium-tritium, are at right. The energy
released by the various reactions is shown in million electron volts.

means of a heat exchanger this heat is
stored at a constant temperature in an
insulated chamber filled with a mixture
of sodium and potassium chlorides that
has enough heat capacity for at least one
day's collection. Heat extracted from this
chamber operates a conventional steam-
electric power plant. The computed efficiency for this proposal is said to be
about 30 percent.

A third system has been proposed by
Alvin F. Hildebrandt and Gregory M.
Haas of the University of Houston. It
entails reflecting the radiation reaching
a square-mile area into a solar furnace
and boiler at the top of a 1,500-foot
tower. Heat from the boiler at a temperature of 2,000 degrees Kelvin would
be converted into electric power by a
magnetohydrodynamic conversion. An
energy-storage system based on the hydrolysis of water is also proposed. An
overall efficiency of about 20 percent is
estimated.

Over the range of efficiencies from 10
to 30 percent the amount of thermal
power that would have to be collected
for a 1,000-megawatt plant would range
from 10,000 to 3,300 thermal megawatts. Accordingly the collecting areas
for the three schemes would be 70, 35
and 23 square kilometers respectively.
With the least of the three efficiencies
the area required for an electric-power
capacity of 350,000 megawatts—the approximate capacity of the U.S. in 1970—
would be 24,500 square kilometers,
which is somewhat less than a tenth of
the area of Arizona.

The physical knowledge and technological resources needed to use solar energy on such a scale are now available.
The technological difficulties of doing
so, however, should not be minimized.

Using solar power indirectly means
relying on the wind, which appears im-
practical on a large scale, or on the
streamflow part of the hydrologic cycle.
At first glance the use of streamflow appears promising, because the world's
total water-power capacity in suitable
sites is about three trillion watts, which

approximates the present use of energy
in industry. Only 8.5 percent of the water power is developed at present, however, and the three regions with the
greatest potential—Africa, South America and Southeast Asia—are the least
developed industrially. Economic problems therefore stand in the way of extensive development of additional water
power.

Tidal power is obtained from the filling and emptying of a bay or an estuary that can be closed by a dam. The
enclosed basin is allowed to fill and
empty only during brief periods at high
and low tides in order to develop as
much power as possible. A number of
promising sites exist; their potential capacities range from two megawatts to
20,000 megawatts each. The total potential tidal power, however, amounts to
about 64 billion watts, which is only 2
percent of the world's potential water
power. Only one full-scale tidal-electric
plant has been built; it is on the Rance
estuary on the Channel Island coast of
France. Its capacity at start-up in 1966
was 240 megawatts; an ultimate capacity of 320 megawatts is planned.

Geothermal power is obtained by extracting heat that is temporarily stored
in the earth by such sources as volcanoes and the hot water filling the sands
of deep sedimentary basins. Only volcanic sources are significantly exploited
at present. A geothermal-power operation has been under way in the Larderel-
lo area of Italy since 1904 and now has
a capacity of 370 megawatts. The two
other main areas of geothermal-power
production are The Geysers in northern
California and Wairakei in New Zea-
land. Production at The Geysers began
in 1960 with a 12.5-megawatt unit. By
1969 the capacity had reached 82 megawatts, and plans are to reach a total installed capacity of 400 megawatts by
1973. The Wairakei plant began operation in 1958 and now has a capacity of
290 megawatts, which is believed to be
about the maximum for the site.

Donald E. White of the U.S. Geological Survey has estimated that the
stored thermal energy in the world's major geothermal areas amounts to about
4×10^{20} joules. With a 25 percent conversion factor the production of electrical energy would be about 10^{20} joules,
or three million megawatt-years. If this
energy, which is depletable, were withdrawn over a period of 50 years, the average annual power production would
be 60,000 megawatts, which is comparable to the potential tidal power.

Nuclear power must be considered under the two headings of fission and
fusion. Fission involves the splitting of
nuclei of heavy elements such as uranium. Fusion involves the combining of
light nuclei such as deuterium. Uranium
235, which is a rare isotope (each 100,-
000 atoms of natural uranium include six
atoms of uranium 234, 711 atoms of uranium 235 and 99,283 atoms of uranium
238), is the only atomic species capable
of fissioning under relatively mild environmental conditions. If nuclear energy depended entirely on uranium 235,
the nuclear-fuel epoch would be brief.
By breeding, however, wherein by absorbing neutrons in a nuclear reactor
uranium 238 is transformed into fissionable plutonium 239 or thorium 232 becomes fissionable uranium 233, it is possible to create more nuclear fuel than
is consumed. With breeding the entire
supply of natural uranium and thorium
would thus become available as fuel for
fission reactors.

Most of the reactors now operating or
planned in the rapidly growing nuclear-
power industry in the U.S. depend essentially on uranium 235. The U.S.
Atomic Energy Commission has estimated that the uranium requirement to
meet the projected growth rate from
1970 to 1980 is 206,000 short tons of
uranium oxide (U_3O_8). A report recently issued by the European Nuclear
Energy Agency and the International
Atomic Energy Agency projects requirements of 430,000 short tons of uranium

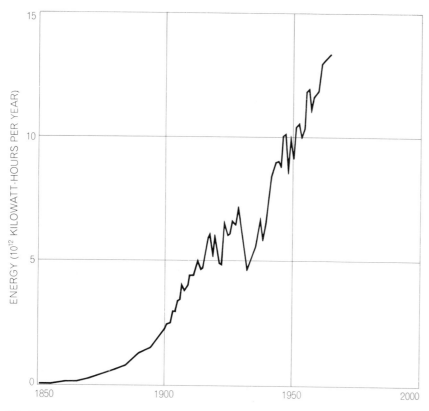

U.S. PRODUCTION OF ENERGY from coal, from petroleum and related sources, from water power and from nuclear reactors is charted for 120 years. The petroleum increment includes natural gas and associated liquids. The dip at center reflects impact of Depression.

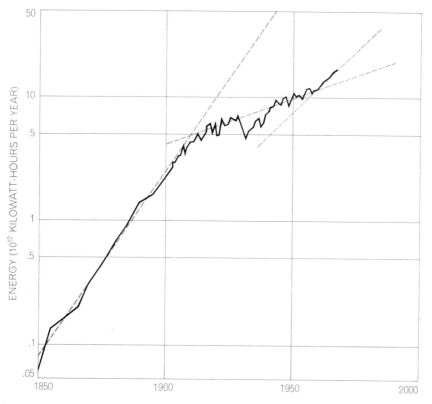

RATE OF GROWTH of U.S. energy production is shown by plotting on a semilogarithmic scale the data represented in the illustration at the top of the page. Broken lines show that the rise had three distinct periods. In the first the growth rate was 6.91 percent per year and the doubling period was 10 years; in the second the rate was 1.77 percent and the doubling period was 39 years; in the third the rate was 4.25 percent with doubling in 16.3 years.

oxide for the non-Communist nations during the same period.

Against these requirements the AEC estimates that the quantity of uranium oxide producible at $8 per pound from present reserves in the U.S. is 243,000 tons, and the world reserves at $10 per pound or less are estimated in the other report at 840,000 tons. The same report estimates that to meet future requirements additional reserves of more than a million short tons will have to be discovered and developed by 1985.

Although new discoveries of uranium will doubtless continue to be made (a large one was recently reported in northeastern Australia), all present evidence indicates that without a transition to breeder reactors an acute shortage of low-cost ores is likely to develop before the end of the century. Hence an intensive effort to develop large-scale breeder reactors for power production is in progress. If it succeeds, the situation with regard to fuel supply will be drastically altered.

This prospect results from the fact that with the breeder reactor the amount of energy obtainable from one gram of uranium 238 amounts to 8.1×10^{10} joules of heat. That is equal to the heat of combustion of 2.7 metric tons of coal or 13.7 barrels (1.9 metric tons) of crude oil. Disregarding the rather limited supplies of high-grade uranium ore that are available, let us consider the much more abundant low-grade ores. One example will indicate the possibilities.

The Chattanooga black shale (of Devonian age) crops out along the western edge of the Appalachian Mountains in eastern Tennessee and underlies at minable depths most of Tennessee, Kentucky, Ohio, Indiana and Illinois. In its outcrop area in eastern Tennessee this shale contains a layer about five meters thick that has a uranium content of about 60 grams per metric ton. That amount of uranium is equivalent to about 162 metric tons of bituminous coal or 822 barrels of crude oil. With the density of the rock some 2.5 metric tons per cubic meter, a vertical column of rock five meters long and one square meter in cross section would contain 12.5 tons of rock and 750 grams of uranium. The energy content of the shale per square meter of surface area would therefore be equivalent to about 2,000 tons of coal or 10,000 barrels of oil. Allowing for a 50 percent loss in mining and extracting the uranium, we are still left with the equivalent of 1,000 tons of coal or 5,000 barrels of oil per square meter.

Taking Averitt's estimate of 1.5 trillion metric tons for the initial minable coal in the U.S. and a round figure of 250 billion barrels for the petroleum liquids, we find that the nuclear energy in an area of about 1,500 square kilometers of Chattanooga shale would equal the energy in the initial minable coal; 50 square kilometers would hold the energy equivalent of the petroleum liquids. Adding natural gas and oil shales, an area of roughly 2,000 square kilometers of Chattanooga shale would be equivalent to the initial supply of all the fossil fuels in the U.S. The area is about 2 percent of the area of Tennessee

and a very small fraction of the total area underlain by the shale. Many other low-grade deposits of comparable magnitude exist. Hence by means of the breeder reactor the energy potentially available from the fissioning of uranium and thorium is at least a few orders of magnitude greater than that from all the fossil fuels combined.

David J. Rose of the AEC, reviewing recently the prospects for controlled fusion, found the deuterium-tritium reaction to be the most promising. Deuterium is abundant (one atom to each

6,700 atoms of hydrogen), and the energy cost of separating it would be almost negligible compared with the amount of energy released by fusion. Tritium, on the other hand, exists only in tiny amounts in nature. Larger amounts must be made from lithium 6 and lithium 7 by nuclear bombardment. The limiting isotope is lithium 6, which has an abundance of only 7.4 percent of natural lithium.

Considering the amount of hydrogen in the oceans, deuterium can be regarded as superabundant. It can also be extracted easily. Lithium is much less

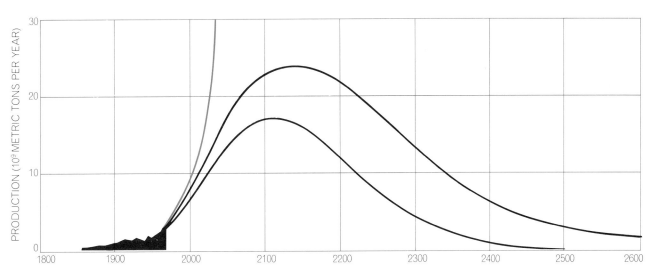

CYCLE OF WORLD COAL PRODUCTION is plotted on the basis of estimated supplies and rates of production. The top curve reflects Averitt's estimate of 7.6×10^{12} metric tons as the initial supply of minable coal; the bottom curve reflects an estimate of 4.3×10^{12} metric tons. The curve that rises to the top of the graph shows the trend if production continued to rise at the present rate of 3.56 percent per year. The amount of coal mined and burned in the century beginning in 1870 is shown by the black area at left.

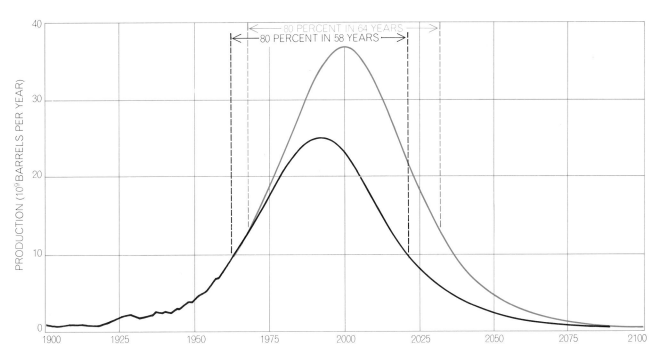

CYCLE OF WORLD OIL PRODUCTION is plotted on the basis of two estimates of the amount of oil that will ultimately be produced. The colored curve reflects Ryman's estimate of $2,100 \times 10^9$ barrels and the black curve represents an estimate of $1,350 \times 10^9$ barrels.

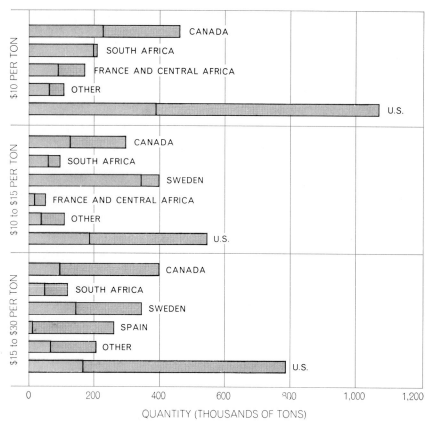

WORLD RESERVES OF URANIUM, which would be the source of nuclear power derived from atomic fission, are given in tons of uranium oxide (U_3O_8). The colored part of each bar represents reasonably assured supplies and the gray part estimated additional supplies.

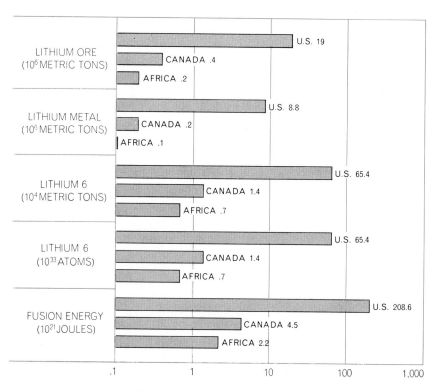

WORLD RESERVES OF LITHIUM, which would be the limiting factor in the deuterium-tritium fusion reaction, are stated in terms of lithium 6 because it is the least abundant isotope. Even with this limitation the energy obtainable from fusion through the deuterium-tritium reaction would almost equal the energy content of the world's fossil-fuel supply.

abundant. It is produced from the geologically rare igneous rocks known as pegmatites and from the salts of saline lakes. The measured, indicated and inferred lithium resources in the U.S., Canada and Africa total 9.1 million tons of elemental lithium, of which the content of lithium 6 would be 7.42 atom percent, or 67,500 metric tons. From this amount of lithium 6 the fusion energy obtainable at 3.19×10^{-12} joule per atom would be 215×10^{21} joules, which is approximately equal to the energy content of the world's fossil fuels.

As long as fusion power is dependent on the deuterium-tritium reaction, which at present appears to be somewhat the easier because it proceeds at a lower temperature, the energy obtainable from this source appears to be of about the same order of magnitude as that from fossil fuels. If fusion can be accomplished with the deuterium-deuterium reaction, the picture will be markedly changed. By this reaction the energy released per deuterium atom consumed is 7.94×10^{-13} joule. One cubic meter of water contains about 10^{25} atoms of deuterium having a mass of 34.4 grams and a potential fusion energy of 7.94×10^{12} joules. This is equivalent to the heat of combustion of 300 metric tons of coal or 1,500 barrels of crude oil. Since a cubic kilometer contains 10^9 cubic meters, the fuel equivalents of one cubic kilometer of seawater are 300 billion tons of coal or 1,500 billion barrels of crude oil. The total volume of the oceans is about 1.5 billion cubic kilometers. If enough deuterium were withdrawn to reduce the initial concentration by 1 percent, the energy released by fusion would amount to about 500,000 times the energy of the world's initial supply of fossil fuels!

Unlimited resources of energy, however, do not imply an unlimited number of power plants. It is as true of power plants or automobiles as it is of biological populations that the earth cannot sustain any physical growth for more than a few tens of successive doublings. Because of this impossibility the exponential rates of industrial and population growth that have prevailed during the past century and a half must soon cease. Although the forthcoming period of stability poses no insuperable physical or biological difficulties, it can hardly fail to force a major revision of those aspects of our current social and economic thinking that stem from the assumption that the growth rates that have characterized this temporary period can somehow be made permanent.

14

THE CONVERSION OF ENERGY

CLAUDE M. SUMMERS
September 1971

The efficiency of home furnaces, steam turbines, automobile engines and light bulbs helps in fixing the demand for energy. A major need is a kind of energy source that does not add to the earth's heat load

A modern industrial society can be viewed as a complex machine for degrading high-quality energy into waste heat while extracting the energy needed for creating an enormous catalogue of goods and services. Last year the U.S. achieved a gross national product of just over $1,000 billion with the help of 69×10^{15} British thermal units of energy, of which 95.9 percent was provided by fossil fuels, 3.8 percent by falling water and .3 percent by the fission of uranium 235. The consumption of 340 million B.t.u. per capita was equivalent to the energy contained in about 13 tons of coal or, to use a commodity now more familiar, 2,700 gallons of gasoline. One can estimate very roughly that between 1900 and 1970 the efficiency with which fuels were consumed for all purposes increased by a factor of four. Without this increase the U.S. economy of 1971 would already be consuming energy at the rate projected for the year 2025 or thereabouts.

Because of steadily increasing efficiency in the conversion of energy to useful heat, light and work, the G.N.P. between 1890 and 1960 was enabled to grow at an average annual rate of 3.25 percent while fuel consumption increased at an annual rate of only 2.7 percent. It now appears, however, that this favorable ratio no longer holds. Since 1967 annual increases in fuel consumption have risen faster than the G.N.P., indicating that gains in fuel economy are becoming hard to achieve and that new goods and services are requiring a larger energy input, dollar for dollar, than those of the past. If one considers only the predicted increase in the use of nuclear fuels for generating electricity, it is apparent that an important fraction of the fuel consumed in the 1980's and 1990's will be converted to a useful form at lower efficiency than fossil fuels are

today. The reason is that present nuclear plants convert only about 30 percent of the energy in the fuel to electricity compared with about 40 percent for the best fossil-fuel plants.

It is understandable that engineers should strive to raise the efficiency with which fuel energy is converted to other and more useful forms. For industry increased efficiency means lower production costs; for the consumer it means lower prices; for everyone it means reduced pollution of air and water. Electric utilities have long known that by lowering the price of energy for bulk users they can encourage consumption. The recent campaign of the utility industry to "save a watt" marks a profound reversal in business philosophy. The difficulty of finding acceptable new sites for power plants underscores the need not only for frugality of use but also for efficiency of use. Having said this, one must emphasize that even large improvements in efficiency can have only a modest effect in extending the life of the earth's supply of fossil and nuclear fuels. I shall develop the point more fully later in this article.

The efficiency with which energy contained in any fuel is converted to

useful form varies widely, depending on the method of conversion and the end use desired. When wood or coal is burned in an open fireplace, less than 20 percent of the energy is radiated into the room; the rest escapes up the chimney. A well-designed home furnace, on the other hand, can capture up to 75 percent of the energy in the fuel and make it available for space heating. The average efficiency of the conversion of fossil fuels for space heating is now probably between 50 and 55 percent, or nearly triple what it was at the turn of the century. In 1900 more than half of all the fuel consumed in the U.S. was used for space heating; today less than a third is so used.

The most dramatic increase in fuel-conversion efficiency in this century has been achieved by the electric-power industry. In 1900 less than 5 percent of the energy in the fuel was converted to electricity. Today the average efficiency is around 33 percent. The increase has been achieved largely by increasing the temperature of the steam entering the turbines that turn the electric generators and by building larger generating units [*see illustration on following page*]. In 1910 the typical inlet temperature was 500 degrees Fahrenheit; today the latest

(overleaf)

STEAM-DRIVEN TURBOGENERATOR at Paradise power plant of the Tennessee Valley Authority near Paradise, Ky., has a capacity of 1,150 megawatts. When placed in operation in February, 1970, it was the largest unit in the world. The turbine, built by the General Electric Company, is a cross-compound design in which steam first enters a high-pressure turbine, below the angled pipe at the left, then flows through the angled pipe to pass in sequence through an intermediate-pressure turbine (*blue casing at rear*) and then through a low-pressure turbine (*blue casing in foreground*). The high-pressure turbine is connected to one generator (*gray housing at left*) and the other two turbine sections to a second generator of the same capacity (*gray housing in foreground*). The entire unit is driven by eight million pounds of steam per hour, which enters the high-pressure turbine at 3,650 pounds per square inch and 1,003 degrees Fahrenheit. The daily coal consumption is 10,572 tons, enough to fill 210 railroad coal cars. The unit has a net thermal efficiency of 39.3 percent. Two smaller turbogenerators, each rated at 704 megawatts, are visible in the background.

units take steam superheated to 1,000 degrees. The method of computing the maximum theoretical efficiency of a steam turbine or other heat engine was enunciated by Nicolas Léonard Sadi Carnot in 1824. The maximum achievable efficiency is expressed by the fraction $(T_1 - T_2)/T_1$, where T_1 is the absolute temperature of the working fluid entering the heat engine and T_2 is the temperature of the fluid leaving the engine. These temperatures are usually expressed in degrees Kelvin, equal to degrees Celsius plus 273, which is the difference between absolute zero and zero degrees C. In a modern steam turbine T_1 is typically 811 degrees K. (1,000 degrees Fahrenheit) and T_2 degrees K. (100 degrees F.). Therefore according to Carnot's equation the maximum theoretical efficiency is about 60 percent. Because the inherent properties of a steam cycle do not allow the heat to be introduced at a constant upper temperature, the maximum theoretical efficiency is not 60 percent but more like 53 percent. Modern steam turbines achieve about 89 percent of that value, or 47 percent net.

To obtain the overall efficiency of a steam power plant this value must be multiplied by the efficiencies of the other energy converters in the chain from fuel to electricity. Modern boilers can convert about 88 percent of the chemical energy in the fuel into heat. Generators can convert up to 99 percent of the mechanical energy produced by the steam turbine into electricity. Thus the overall efficiency is 88 × 47 (for the turbine) × 99, or about 41 percent.

Nuclear power plants operate at lower efficiency because present nuclear reactors cannot be run as hot as boilers burning fossil fuel. The temperature of the steam produced by a boiling-water reactor is around 350 degrees C., which means that the T_1 in the Carnot equation is 623 degrees K. For the complete cycle from fuel to electricity the efficiency of a nuclear power plant drops to about 30 percent. This means that some 70 percent of the energy in the fuel used by a nuclear plant appears as waste heat, which is released either into an adjacent body of water or, if cooling towers are used, into the surrounding air. For a fossil-fuel plant the heat wasted in this way amounts to about 60 percent of the energy in the fuel.

The actual heat load placed on the water or air is much greater, however, than the difference between 60 and 70 percent suggests. For plants with the same kilowatt rating, a nuclear plant produces about 50 percent more waste

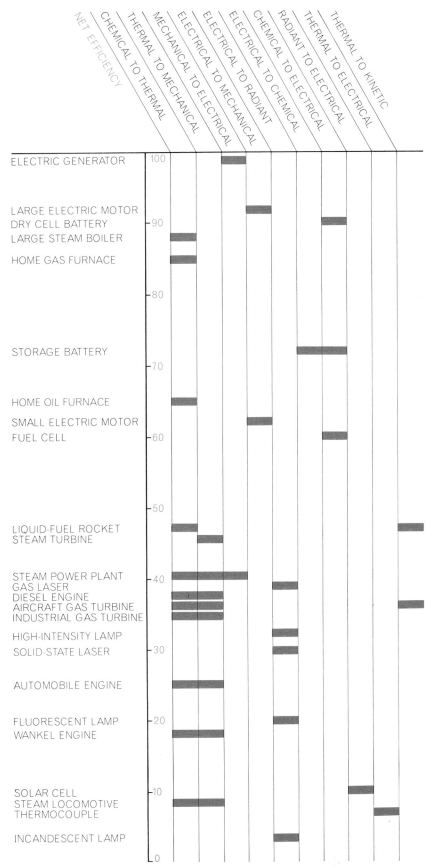

EFFICIENCY OF ENERGY CONVERTERS runs from less than 5 percent for the ordinary incandescent lamp to 99 percent for large electric generators. The efficiencies shown are approximately the best values attainable with present technology. The figure of 47 percent indicated for the liquid-fuel rocket is computed for the liquid-hydrogen engines used in the Saturn moon vehicle. The efficiencies for fluorescent and incandescent lamps assume that the maximum attainable efficiency for an acceptable white light is about 400 lumens per watt rather than the theoretical value of 220 lumens per watt for a perfectly "flat" white light.

heat than a fossil-fuel plant. The reason is that a nuclear plant must "burn" about a third more fuel than a fossil-fuel plant to produce a kilowatt-hour of electricity and then wastes 70 percent of the larger B.t.u. input.

Of course, no law of thermodynamics decrees that the heat released by either a nuclear or a fossil-fuel plant must go to waste. The problem is to find something useful to do with large volumes of low-grade energy. Many uses have been proposed, but all run up against economic limitations. For example, the low-pressure steam discharged from a steam turbine could be used for space heating. This is done in some communities, notably in New York City, where Consolidated Edison is a large steam supplier. Many chemical plants and refineries also use low-pressure steam from turbines as process steam. It has been suggested that the heated water released by power plants might be beneficial in speeding the growth of fish and shellfish in certain localities. Nationwide, however, there seems to be no attractive use for the waste heat from the present fossil-fuel plants or for the heat that will soon be pouring from dozens of new nuclear power plants. The problem will be to limit the harm the heat can do to the environment.

From the foregoing discussion one can see that the use of electricity for home heating (a use that is still vigorously promoted by some utilities) represents an inefficient use of chemical fuel. A good oil- or gas-burning home furnace is at least twice as efficient as the average electric-generating station. In some locations, however, the annual cost of electric space heating is competitive with direct heating with fossil fuels even at the lower efficiency. Several factors account for this anomaly. The electric-power rate decreases with the added load. Electric heat is usually installed in new constructions that are well insulated. The availability of gas is limited in some locations and its cost is higher. The delivery of oil is not always dependable. As fossil fuels become scarcer, their cost will increase, and the production of electrical energy with nuclear fuels will increase. Unfortunately we must expect that a greater percentage of our fuel resources (particularly nuclear fuels) will be consumed in electric space heating in spite of the less efficient use of fuel.

The most ubiquitous of all prime movers is the piston engine. There are two in many American garages, not counting the engines in the power mower, the snowblower or the chain saw. The piston engines in the nation's more than 100 million motor vehicles have a rated capacity in excess of 17 billion horsepower, or more than 95 percent of the capacity of all prime movers (defined as engines for converting fuel to mechanical energy). Although this huge capacity is unemployed most of the time, it accounts for more than 16 percent of the fossil fuel consumed by the U.S. Transportation in all forms—including the propulsion systems of ships, locomotives and aircraft—absorbs about 25 percent of the nation's energy budget.

Automotive engineers estimate that the efficiency of the average automobile engine has risen about 10 percent over the past 50 years, from roughly 22 percent to 25 percent. In terms of miles delivered per gallon of fuel, however, there has actually been a decline. From 1920 until World War II the average automobile traveled about 13.5 miles per gallon of fuel. In the past 25 years the average has fallen gradually to about 12.2 miles per gallon. This decline is due to heavier automobiles with more powerful engines that encourage greater acceleration and higher speed. It takes about eight times more energy to push a vehicle through the air at 60 miles per hour than at 30 miles per hour. The same amount of energy used in accelerating the car's mass to 60 miles per hour must be absorbed as heat, primarily in the brakes, to stop the vehicle. Therefore most of the gain in engine efficiency is lost in the way man uses his machine. Automobile air conditioning has also played a role in reducing the miles per gallon. With the shift in consumer preference to smaller cars the figure may soon begin to climb. The efficiency of the basic piston engine, however, cannot be improved much further.

If all cars in the year 2000 operated on electric batteries charged by electricity generated in central power stations, there would be little change in the nation's overall fuel requirement. Although the initial conversion efficiency in the central station might be 35 percent compared with 25 percent in the piston engine, there would be losses in distributing the electrical energy and in the conversion of electrical energy to chemical energy (in the battery) and back to electrical energy to turn the car wheels. Present storage batteries have an overall efficiency of 70 to 75 percent, so that there is not much room for improvement. Anyone who believes we

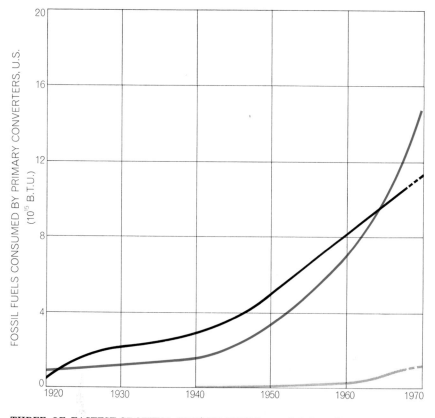

THREE OF FASTEST-GROWING ENERGY USERS are electric utilities (*color*), motor vehicles (*black*) and aircraft (*gray*). Together they now consume about 40 percent of all the energy used in the U.S. As recently as 1940 they accounted for only 18 percent of a much smaller total. The demand for aircraft fuel has more than tripled in 10 years.

would all be better off if cars were electrically powered must consider the problem of increasing the country's electric-generating capacity by about 75 percent, which is what would be required to move 100 million vehicles.

The difficulty of trying to trace savings produced by even large changes in efficiency of energy conversion is vividly demonstrated by what happened when the railroads converted from the steam locomotive (maximum thermal efficiency 10 percent) to diesel-electric locomotives (thermal efficiency about 35 percent). In 1920 the railroads used about 135 million tons of coal, which represented 16 percent of the nation's total energy demand. By 1967, according to estimates made by John Hume, an energy consultant, the railroads were providing 54 percent more transportation than in 1920 (measured by an index of "transportation output") with less than a sixth as many B.t.u. This increase in efficiency, together with the railroads' declining role in the national economy, had reduced the railroads' share of the nation's total fuel budget from 16 percent to about 1 percent. If one looks at a curve of the country's total fuel consumption from 1920 to 1967, however, the impact of this extraordinary change is scarcely visible.

Perhaps the least efficient important use for electricity is providing light. The General Electric Company estimates that lighting consumes about 24 percent of all electrical energy generated, or 6 percent of the nation's total energy budget. It is well known that the glowing filament of an ordinary 100-watt incandescent lamp produces far more heat than light. In fact, more than 95 percent of the electric input emerges as infrared radiation and less than 5 percent as visible light. Nevertheless, this is about five times more light than was provided by a 100-watt lamp in 1900. A modern fluorescent lamp converts about 20 percent of the electricity it consumes into light. These values are based on a practical upper limit of 400 lumens per watt, assuming the goal is an acceptable light of less than perfect whiteness. If white light with a totally flat spectrum is specified, the maximum theoretical output is reduced to 220 lumens per watt. If one were satisfied with light of a single wavelength at the peak sensitivity of the human eye (555 nanometers), one could theoretically get 680 lumens per watt.

General Electric estimates that fluorescent lamps now provide about 70 percent of the country's total illumination and that the balance is divided between incandescent lamps and high-

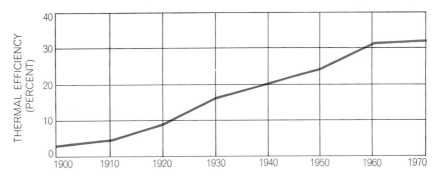

EFFICIENCY OF FUEL-BURNING POWER PLANTS in the U.S. increased nearly tenfold from 3.6 percent in 1900 to 32.5 percent last year. The increase was made possible by raising the operating temperature of steam turbines and increasing the size of generating units.

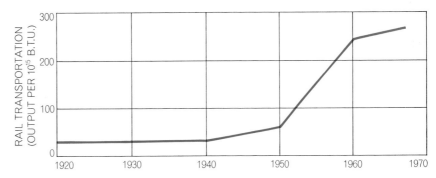

EFFICIENCY OF RAILROAD LOCOMOTIVES can be inferred from the energy needed by U.S. railroads to produce a unit of "transportation output." The big leap in the 1950's reflects the nearly complete replacement of steam locomotives by diesel-electric units.

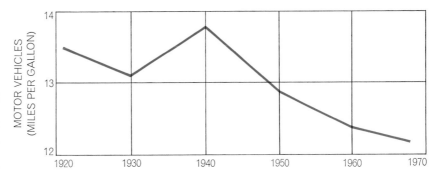

EFFICIENCY OF AUTOMOBILE ENGINES is reflected imperfectly by miles per gallon of fuel because of the increasing weight and speed of motor vehicles. Manufacturers say that the thermal efficiency of the 1920 engine was about 22 percent; today it is about 25 percent.

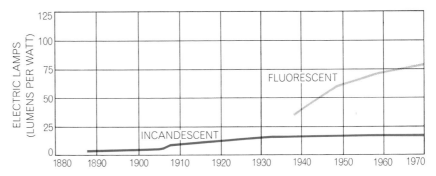

EFFICIENCY OF ELECTRIC LAMPS depends on the quality of light one regards as acceptable. Theoretical efficiency for perfectly flat white light is 220 lumens per watt. By enriching the light slightly in mid-spectrum one could obtain about 400 lumens per watt. Thus present fluorescent lamps can be said to have an efficiency of either 36 percent or 20.

intensity lamps, which have efficiencies comparable to, and in some cases higher than, fluorescent lamps. This division implies that the average efficiency of converting electricity to light is about 13 percent. To obtain an overall efficiency for converting chemical (or nuclear) energy to visible light, one must multiply this percentage times the average efficiency of generating power (33 percent), which yields a net conversion efficiency of roughly 4 percent. Nevertheless, thanks to increased use of fluorescent and high-intensity lamps, the nation was able to triple its "consumption" of lighting between 1960 and 1970 while only doubling the consumption of electricity needed to produce it.

This brief review of changing efficiencies of energy use may provide some perspective when one tries to evaluate

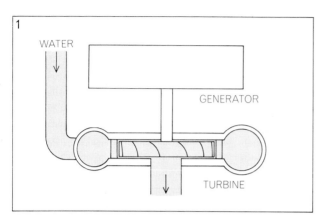

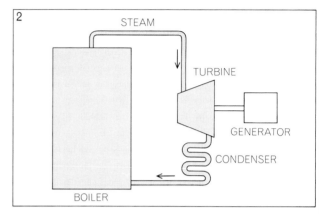

ELECTRIC-POWER GENERATING MACHINERY now in use extracts energy from falling water, fossil fuels or nuclear fuels. The hydroturbine generator (1) converts potential and kinetic energy into electric power. In a fossil-fuel steam power plant (2) a boiler produces steam; the steam turns a turbine; the turbine turns an electric generator. In a nuclear power plant (3) the fission of ura-

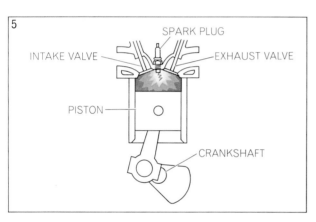

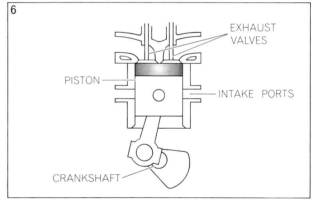

PROPULSION MACHINERY converts the energy in liquid fuels into forms of mechanical or kinetic energy useful for work and transportation. In the piston engine (5) a compressed charge of fuel and air is exploded by a spark; the expanding gases push against the piston, which is connected to a crankshaft. In a diesel engine (6) the compression alone is sufficient to ignite the charge

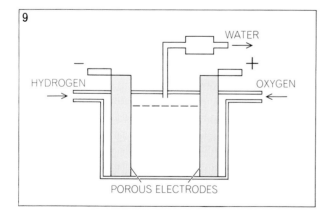

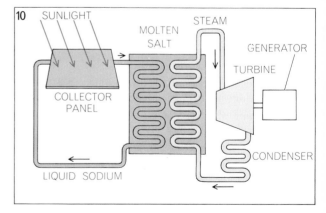

NOVEL ENERGY CONVERTERS are being designed to exploit a variety of energy sources. The fuel cell (9) converts the energy in hydrogen or liquid fuels directly into electricity. The "combustion" of the fuel takes place inside porous electrodes. In a recently proposed solar power plant (10) sunlight falls on specially coated collectors and raises the temperature of a liquid metal to 1,000 degrees F. A heat exchanger transfers the heat so collected to steam, which then turns a turbogenerator as in a conventional power plant. A salt reservoir holds enough heat to keep generating steam during the night and when the sun is hidden by clouds. In a mag-

the probable impact of novel energy-conversion systems now under development. Two devices that have received much notice are the fuel cell and the magnetohydrodynamic (MHD) generator. The former converts chemical energy directly into electricity; the latter is potentially capable of serving as a high-temperature "topping" device to be operated in series with a steam turbine and generator in producing electricity. Fuel cells have been developed that can "burn" hydrogen, hydrocarbons or alcohols with an efficiency of 50 to 60 percent. The hydrogen-oxygen fuel cells used in the Apollo space missions, built by the Pratt & Whitney division of United Aircraft, have an output of 2.3 kilowatts of direct current at 20.5 volts.

A decade ago the magnetohydrodynamic generator was being advanced as

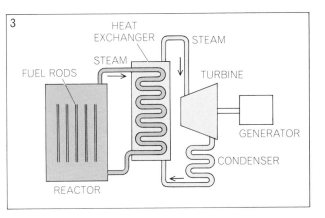

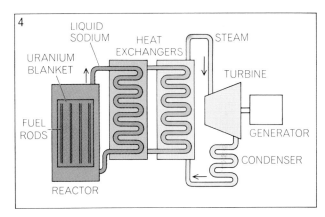

nium 235 releases the energy to make steam, which then goes through the same cycle as in a fossil-fuel power plant. Under development are nuclear breeder reactors (4) in which surplus neutrons are captured by a blanket of nonfissile atoms of uranium 238 or thorium 232, which are transformed into fissile plutonium 239 or U-233. The heat of the reactor is removed by liquid sodium.

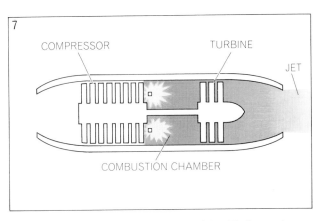

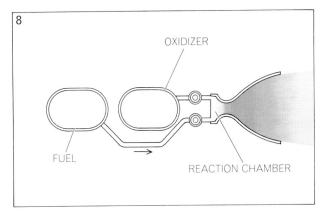

of fuel and air. In an aircraft gas turbine (7) the continuous expansion of hot gas from the combustion chamber passes through a turbine that turns a multistage air compressor. Hot gases leaving the turbine provide the kinetic energy for propulsion. A liquid-fuel rocket (8) carries an oxidizer in addition to fuel so that it is independent of an air supply. Rocket exhaust carries kinetic energy.

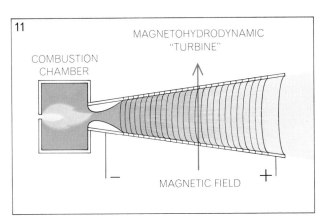

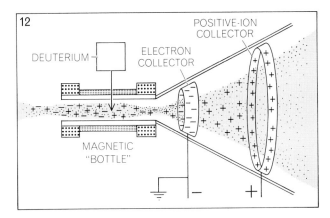

netohydrodynamic "turbine" (11) the energy contained in a hot electrically conducting gas is converted directly into electric power. A small amount of "seed" material, such as potassium carbonate, must be injected into the flame to make the hot gas a good conductor. Electricity is generated when the electrically charged particles of gas cut through the field of an external magnet. A long-range goal is a thermonuclear reactor (12) in which the nuclei of light elements fuse into heavier elements with the release of energy. High-velocity charged particles produced by a thermonuclear reaction might be trapped in such a way as to generate electricity directly.

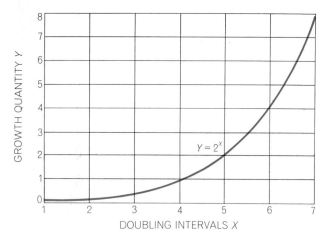

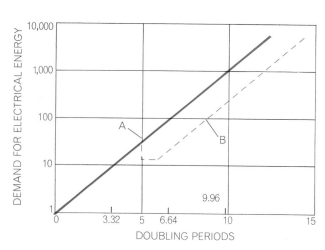

DOUBLING CURVE (*left*) rises exponentially with time. It shows how many doubling intervals are needed to produce a given multiplication of the growth quantity. Thus if electric-power demand continues to double every 10 years, the demand will increase eightfold in three doubling periods, that is, by the year 2001. When ex-

ponential growth curves are plotted on a semilogarithmic scale, the result is a straight line (*right*). If electric-power consumption were cut in half at *A*, held constant for 10 years and allowed to return to the former growth rate, time needed to reach a given demand (*B*) would be extended by only two doubling periods, or 20 years.

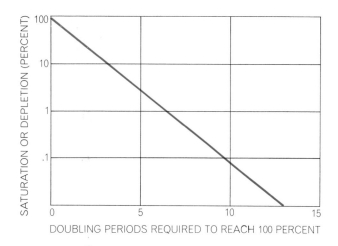

SATURATION OR DEPLETION (PERCENT)	DOUBLING PERIODS TO REACH 100 PERCENT	YEARS FROM NOW
100.0	0.0	0
10.0	3.32	33
1.0	6.64	66
.1	9.96	100
.01	13.28	133
.001	16.60	166
.0001	19.92	199
.00001	23.24	232
.000001	26.56	266
.0000001	29.88	299
.00000001	33.20	332

DEPLETION OF A RESOURCE can be read from the curve at the left. Thus if .1 percent of world's oil has now been extracted, all will be gone in just under 10 doubling periods, or 100 years if the

doubling interval is 10 years. The table at the right shows that the ultimate depletion date is changed very little by large changes in the estimate of amount of resource that has been extracted to date.

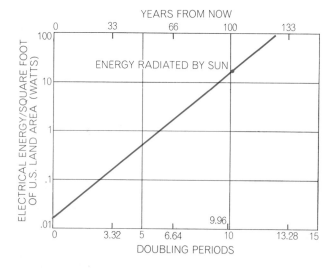

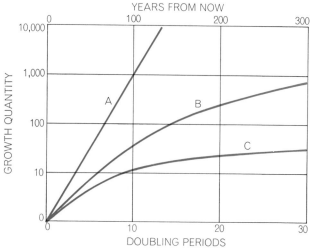

ENERGY RECEIVED FROM SUN on an average square foot of the U.S. will be equaled by production of electrical energy in roughly 100 years if the demand continues to double every 10 years.

THREE GROWTH CURVES are compared. Curve *A* is exponential. In Curve *B* each doubling period is successively increased by 20 percent. In Curve *C* the growth per doubling period is constant.

the energy converter of the future. In such a device the fuel is burned at a high temperature and the gaseous products of combustion are made electrically conducting by the injection of a "seed" material, such as potassium carbonate. The electrically conducting gas travels at high velocity through a magnetic field and in the process creates a flow of direct current [see No. 11 in illustrations on pages 154 and 155]. If the MHD technology can be developed, it should be possible to design fossil-fuel power plants with an efficiency of 45 to 50 percent. Since MHD requires very high temperatures it is not suitable for use with nuclear-fuel reactors, which produce a working fluid much cooler than one can obtain from a combustion chamber fired with fossil fuel.

If ever an energy source can be said to have arrived in the nick of time, it is nuclear energy. Twenty-two nuclear power plants are now operating in the U.S. Another 55 plants are under construction and more than 40 are on order. This year the U.S. will obtain 1.4 percent of its electrical energy from nuclear fission; it is expected that by 1980 the figure will reach 25 percent and that by 2000 it will be 50 percent.

Although a 1,000-megawatt nuclear power plant costs about 10 percent more than a fossil-fuel plant ($280 million as against $250 million), nuclear fuel is already cheaper than coal at the mine mouth. Some projections indicate that coal may double in price between now and 1980. One reason given is that new Federal safety regulations have already reduced the number of tons produced per man-day from the 20 achieved in 1969 to fewer than 15.

The utilities are entering a new period in which they will have to rethink the way in which they meet their base load, their intermediate load (which coincides with the load added roughly between 7:00 A.M. and midnight by the activity of people at home and at work) and peak load (the temperature-sensitive load, which accounts for only a few percent of the total demand). In the past utilities assigned their newest and most efficient units to the base load and called on their older and smaller units to meet the variable daily demand. In the future, however, when still newer capacity is added, the units now carrying the basic load cannot easily be relegated to intermittent duty because they are too large to be easily put on the line and taken off.

There is therefore a need for a new kind of flexible generating unit that may be best satisfied by coupling an industrial gas turbine to an electric generator and using the waste heat from the gas turbine to produce low-pressure steam for a steam turbine–generator set. Combination systems of this kind are now being offered by General Electric and the Westinghouse Electric Company. Although somewhat less efficient than the best large conventional units, the gas-turbine units can be brought up to full load in an hour and can be installed at lower cost per kilowatt. To meet brief peak demands utilities are turning to gas turbines (without waste-heat boilers that can be brought up to full load in minutes) and to pumped hydrostorage systems. In the latter systems off-peak capacity is used to pump water to an elevated reservoir from which it can be released to produce power as needed.

Westinghouse has recently estimated that U.S. utilities must build more than 1,000 gigawatts (GW, or 10^9 watts) of new capacity between 1970 and 1990, or more than three times the present installed capacity of roughly 300 GW. Of the new capacity 500 GW, or half, will be needed to handle the anticipated increase in base load and 75 percent of the 500 GW will be nuclear. More than 400 GW of new capacity will be needed to meet the growing intermediate load, and a sizable fraction of it will be provided by gas turbines. The new peaking capacity, amounting to some 170 GW, will be divided, Westinghouse believes, between gas turbines and pumped storage in the ratio of 10 to seven.

Such projections can be regarded as the conventional wisdom. Does unconventional wisdom have anything to offer that may influence power generation by 2000, if not by 1990? First of all, there are the optimists who believe prototype nuclear-fusion plants will be built in the 1980's and full-scale plants in the 1990's. In a sense, however, this is merely conventional wisdom on an accelerated time scale. Those with a genuinely unconventional approach are asking: Why do we not start developing the technology to harness energy from the sun or the wind or the tides?

Many people still remember the Passamaquoddy project of the 1930's, which is once more being discussed and which would provide 300 megawatts (less than a third the capacity of the turbogenerator shown on page 2) by exploiting tides with an average range of 18 feet in the Bay of Fundy, between Maine and Canada. A working tidal power plant of 240 megawatts has recently been placed in operation by the French government in the estuary of the Rance River, where the tides average 27 feet. How much tidal energy might the U.S. extract if all favorable bays and inlets were developed? All estimates are subject to heavy qualification, but a reasonable guess is something like 100 GW. We have just seen, however, that the utilities will have to add 10 times that much capacity just to meet the needs of 1990. One must conclude that tidal power does not qualify as a major unconventional resource.

What about the wind? A study we conducted at Oklahoma State University a few years ago showed that the average wind energy in the Oklahoma City area is about 18.5 watts per square foot of area perpendicular to the wind direction. This is roughly equivalent to the amount of solar energy that falls on a square foot of land in Oklahoma, averaging the sunlight for 24 hours a day in all seasons and under all weather conditions. A propeller-driven turbine could convert the wind's energy into electricity at an efficiency of somewhere between 60 and 80 percent. Like tidal energy and other forms of hydropower, wind power would have the great advantage of not introducing waste heat into the biosphere.

The difficulty of harnessing the wind's energy comes down to a problem of energy storage. Of all natural energy sources the wind is the most variable. One must extract the energy from the wind as it becomes available and store it if one is to have a power plant with a reasonably steady output. Unfortunately technology has not yet produced a practical storage medium. Electric storage batteries are out of the question.

One scheme that seems to offer promise is to use the variable power output of a wind generator to decompose water into hydrogen and oxygen. These would be stored under pressure and recombined in a fuel cell to generate electricity on a steady basis [see illustration on next page]. Alternatively the hydrogen could be burned in a gas turbine, which would turn a conventional generator. The Rocketdyne Division of North American Rockwell has seriously proposed that an industrial version of the hydrogen-fueled rocket engine it builds for the Saturn moon vehicle could be used to provide the blast of hot gas needed to power a gas turbine coupled to an electric generator. Rocketdyne visualizes that a water-cooled gas turbine could operate at a higher temperature than conventional fuel-burning gas turbines and achieve our overall plant efficiency of 55 percent. If the Rocketdyne concept were successful, it could use hydrogen from any source. A wind-driven hydrogen-rocket gas-turbine power plant should be unconventional enough to please the most exotic taste.

By comparison most proposals for harnessing solar energy seem tame indeed. One fairly straightforward proposal has recently been made to the Arizona Power Authority on behalf of the University of Arizona by Aden B. Meinel and Marjorie P. Meinel of the university's Optical Sciences Center. They suggest that if the sunlight falling on 14 percent of the western desert regions of the U.S. were efficiently collected, it could be converted into 1,000 GW of power, or approximately the amount of additional power needed between now and 1990. The Meinels believe it is within the reach of present technology to design collecting systems capable of storing solar energy as heat at 1,000 degrees F., which could be converted to electricity at an overall efficiency of 30 percent.

The key to the project lies in recently developed surface coatings that have high absorbance for solar radiation and low emittance in the infrared region of the spectrum. To achieve a round-the-clock power output, heat collected during daylight hours would be stored in molten salts at 1,000 degrees F. A heat exchanger would transfer the stored energy to steam at the same temperature. The thermal storage tank for a 1,000-megawatt generating plant would require a capacity of about 300,000 gallons. The Meinels propose that industry and the Government immediately undertake design and construction of a 100-megawatt demonstration plant in the vicinity of Yuma, Ariz. The collectors for such a plant would cover an area of 3.6 million square meters (slightly more than a square mile). The Meinels estimate that after the necessary development has been done a 1,000-megawatt solar power station might be built for about $1.1 billion, or about four times the present cost of a nuclear power plant. As they point out: "Solar power faces the economic problem that energy is purchased via a capital outlay rather than an operating expense." They calculate nevertheless that a plant with an operating lifetime of 40 years should produce power at an average cost of only half a cent per kilowatt hour.

A more exotic solar-power scheme has been advanced by Peter E. Glaser of Arthur D. Little, Inc. The idea is to place a lightweight panel of solar cells in a synchronous orbit 22,300 miles above the earth, where they would be exposed to sunlight 24 hours a day. Solar cells (still to be developed) would collect the radiant energy and convert it to electricity with an efficiency of 15 to 20 percent. The electricity would then be converted electronically in orbit to microwave energy with an efficiency of 85 percent, which is possible today. The microwave radiation would be at a wavelength selected to penetrate clouds with little or no loss and would be collected by a suitable antenna on the earth. Present techniques can convert microwave energy to electric power with an efficiency of about 70 percent, and 80 to 85 percent should be attainable. Glaser calculates that a 10,000-megawatt (10 GW) satellite power station, large enough to meet New York City's present power needs, would require a solar collector panel five miles square.

WIND AS POWER SOURCE is attractive because it does not impose an extra heat burden on the environment, as is the case with energy extracted from fossil and nuclear fuels. Unlike hydropower and tidal power, which also represent the entrapment of solar energy, the wind is available everywhere. Unfortunately it is also capricious. To harness it effectively one must be able to store the energy captured when the wind blows and release it more or less continuously. One scheme would be to use the electricity generated by the wind to decompose water electrolytically. The stored hydrogen and oxygen could then be fed at a constant rate into a fuel cell, which would produce direct current. This would be converted into alternating current and fed into a power line. Off-peak power generated elsewhere could also be used to run the electrolysis cell whenever the wind was deficient.

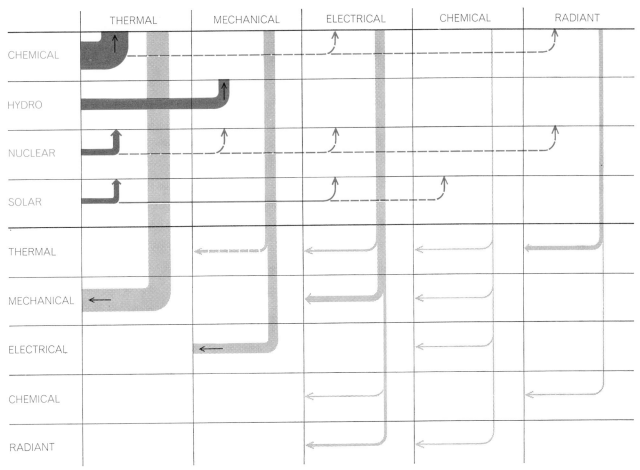

CONVERSION PATHWAYS link many of the familiar forms of energy. The four forms shown in color are either important sources of power today or, in the case of solar energy, potentially important. The broken lines indicate rare, incidental or theoretically useful conversions. The gray lines follow the destiny of intermediate forms of energy. Except for the thermal energy used for space heating, most is converted to mechanical energy. Mechanical energy is used directly for transportation (see illustration below) and for generating electricity. Electrical energy in turn is used for lighting, heating and mechanical work. As a secondary form, chemical energy is found in dry cells and storage batteries. The radiant energy produced by electric lamps ends up chiefly as heat.

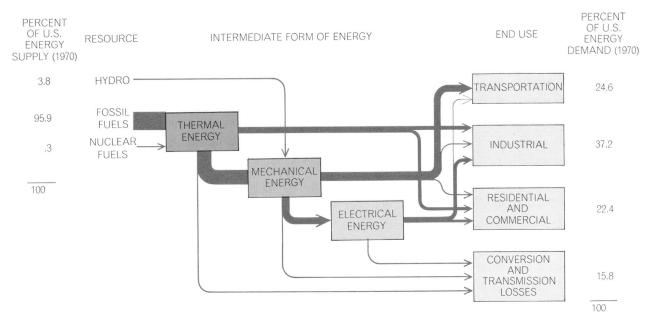

PATHWAYS TO END USES are depicted for the three principal sources of energy. The most direct and most efficient conversion is from falling water to mechanical energy to electrical energy. The energy locked in fossil and nuclear fuels must first be released in the form of thermal energy before it can be converted to mechanical energy and then, if it is desired, to electric power. Conversion and transmission losses include various nonenergy uses of fossil fuels, such as the manufacture of lubricants and the conversion of coal to coke. The biggest loss, however, arises from the generation of electric power at an average efficiency of 32.5 percent.

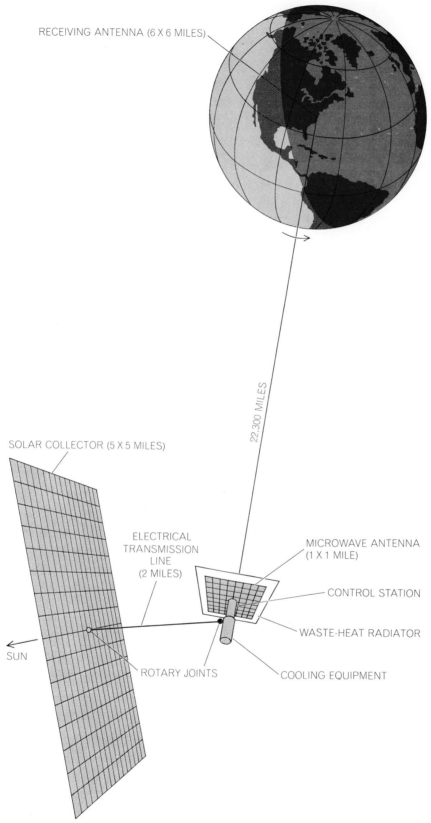

RECEIVING ANTENNA (6 X 6 MILES)

22,300 MILES

SOLAR COLLECTOR (5 X 5 MILES)

ELECTRICAL
TRANSMISSION
LINE
(2 MILES)

MICROWAVE ANTENNA
(1 X 1 MILE)

CONTROL STATION

WASTE-HEAT RADIATOR

SUN

ROTARY JOINTS

COOLING EQUIPMENT

SOLAR COLLECTOR IN STATIONARY ORBIT has been proposed by Peter E. Glaser of Arthur D. Little, Inc. Located 22,300 miles above the Equator, the station would remain fixed with respect to a receiving station on the ground. A five-by-five-mile panel would intercept about 8.5×10^7 kilowatts of radiant solar power. Solar cells operating at an efficiency of about 18 percent would convert this into 1.5×10^7 kilowatts of electric power, which would be converted into microwave radiation and beamed to the earth. There it would be reconverted into 10^7 net kilowatts of electric power, or enough for New York City. The receiving antenna would cover about six times the area needed for a coal-burning power plant of the same capacity and about 20 times the area needed for a nuclear plant.

The receiving antenna on the earth would have to be only slightly larger: six miles square. Since the microwave energy in the beam would be comparable to the intensity of sunlight, it would present no hazard. The system, according to Glaser, would cost about $500 per kilowatt, roughly twice the cost of a nuclear power plant, assuming that space shuttles were available for the construction of the satellite. The entire space station would weigh five million pounds, or slightly less than the Saturn moon rocket at launching.

To meet the total U.S. electric-power demand of 2,500 GW projected for the year 2000 would require 250 satellite stations of this size. Since the demand to 1990 will surely be met in other ways, however, one should perhaps think only of meeting the incremental demand for the decade 1990–2000. This could be done with about 125 power stations of the Glaser type.

The great virtue in power schemes based on using the wind or solar energy collected at the earth's surface, far-fetched as they may sound today, is that they would add no heat load to the earth's biosphere; they can be called invariant energy systems. Solar energy collected in orbit would not strictly qualify as an invariant system, since much of the radiant energy intercepted at an altitude of 22,300 miles is radiation that otherwise would miss the earth. Only the fraction collected when the solar panels were in a line between the sun and the earth's disk would not add to the earth's heat load. On the other hand, solar collectors in space would put a much smaller waste-heat load on the environment than fossil-fuel or nuclear plants. Of the total energy in the microwave beam aimed at the earth all but 20 percent or less would be converted to usable electric power. When the electricity was consumed, of course, it would end up as heat.

To appreciate the long-term importance of developing invariant energy systems one must appreciate what exponential growth of any quantity implies. The doubling process is an awesome phenomenon. In any one doubling period the growth quantity—be it energy use, population or the amount of land covered by highways—increases by an amount equal to its growth during its entire past history. For example, during the next doubling period as much fossil fuel will be extracted from the earth as the total amount that has been extracted to date. During the next 10 years the U.S. will generate as much electricity as it has generated since the beginning of the electrical era.

Such exponential growth curves are usually plotted on a semilogarithmic scale in order to provide an adequate span. By selecting appropriate values for the two axes of a semilogarithmic plot one can also obtain a curve showing the number of doubling periods to reach saturation or depletion from any known or assumed percentage position [*see left half of middle illustration on page 146*]. As an example, let us assume that we have now extracted .1 percent of the earth's total reserves of fossil fuels and that the rate of extraction has been doubling every 10 years. If this rate continues, we shall have extracted all of these fuels in just under 10 doubling periods, or in 100 years. We have no certain knowledge, of course, what fraction of all fossil fuels has been extracted. To be conservative let us assume that we have extracted only .01 percent rather than .1 percent. The curve shows us that in this case we shall have extracted 100 percent in 13.3 doubling periods, or 133 years. In other words, if our estimate of the fuel extracted to this moment is in error by a factor of 10, 1,000 or even 100,000, the date of total exhaustion is not long deferred. Thus if we have now depleted the earth's total supply of fossil fuel by only a millionth of 1 percent (.000001 percent), all of it will be exhausted in only 266 years at a 10-year doubling rate [*see right half of middle illustration on page 146*]. I should point out that the actual extraction rate varies for the different fossil fuels; a 10-year doubling rate was chosen simply for the purpose of illustration.

In estimating how many doubling periods the nation can tolerate if the demand for electricity continues to double every 10 years (the actual doubling rate), the crucial factor is probably not the supply of fuels—which is essentially limitless if fusion proves practical—but the thermal impact on the environment of converting fuel to electricity and electricity ultimately to heat. For the sake of argument let us ignore the burden of waste heat produced by fossil-fuel or nuclear power plants and consider only the heat content of the electricity actually consumed. One can imagine that by the year 2000 most of the power will be generated in huge plants located several miles offshore so that waste heat can be dumped harmlessly (for a while at least) into the surrounding ocean.

In 1970 the U.S. consumed 1,550 billion kilowatt hours of electricity. If this were degraded into heat (which it was) and distributed evenly over the total land area of the U.S. (which it was not), the energy released per square foot would be .017 watt. At the present doubling rate electric-power consumption is being multiplied by a factor of 10 every 33 years. Ninety-nine years from now, after only 10 more doubling periods, the rate of heat release will be 17 watts per square foot, or only slightly less than the 18 or 19 watts per square foot that the U.S. receives from the sun, averaged around the clock. Long before that the present pattern of power consumption must change or we must develop the technology needed for invariant energy systems.

Let us examine the consequences of altering the pattern of energy growth in what may seem to be fairly drastic ways. Consider a growth curve in which each doubling period is successively lengthened by 20 percent [*see bottom illustration at right on page 146*]. On an exponential growth curve it takes 3.32 doubling periods, or 33 years, to increase energy consumption by a factor of 10. On the retarded curve it would take five doubling periods, or 50 years, to reach the same tenfold increase. In other words, the retardation amounts to only 17 years. The retardation achieved for a hundredfold increase in consumption amounts to only 79 years (that is, the difference between 145 years and 66 years).

Another approach might be to cut back sharply on present consumption, hold the lower value for some period with no growth and then let growth resume at the present rate. One can easily show that if consumption of power were immediately cut in half, held at that value for 10 years and then allowed to return to the present pattern, the time required to reach a hundredfold increase in consumption would be stretched by only 20 years: from 66 to 86 years [*see right half of top illustration on page 146*].

For long-term effectiveness something like a constant growth curve is required, that is, a curve in which the growth increases by the same amount for each of the original doubling periods. On such a curve nearly 1,000 years would be required for electric-power generation to reach the level of the radiant energy received from the sun instead of the 100 years predicted by a 10-year doubling rate. One can be reasonably confident that the present doubling rate cannot continue for another 100 years, unless invariant energy systems supply a large part of the demand, but what such systems will look like remains hidden in the future.

15

RELAXATION METHODS IN CHEMISTRY

LARRY FALLER
May 1969

By rapidly upsetting the equilibrium of a chemical reaction one can study the important mechanisms that operate in the interval between a thousandth and a billionth of a second

When a chemist has discovered that two substances react to form a third substance, his job has only begun. His next task is to learn something about the mechanism of the reaction. Did substance *A* dissociate into two subspecies before one of them reacted with *B*? Did *A* and *B* first form a temporary complex that rearranged itself to form *C*? Did the reaction require the help of a catalyst such as an enzyme, and if so, what specific regions of the enzyme were involved in the catalytic mechanism? How fast did the reaction go? Because most chemical reactions take place rapidly, it is not easy to obtain answers to such questions. If the reaction takes place in solution, as a great many reactions do, the reaction time is often very short indeed. Hence much ingenuity has been applied to following the details of reactions on an increasingly short time scale.

Fifteen years ago chemists thought they were doing well to study the rate of reactions in solution that were half-completed in a millisecond. Today half-times as short as a few nanoseconds (billionths of a second) can be directly measured. This dramatic progress has resulted from a new approach to the study of reaction rates to which the cryptic term "relaxation" has been applied. In this approach the equilibrium of a chemical reaction is rapidly upset by changing some important condition, such as temperature or pressure; the change in the concentration of reactants or product is monitored while the reaction reequilibrates (relaxes). This basically simple idea, which could have been implemented technically—had anyone thought of it—at least 30 years ago, was introduced in 1954 by Manfred Eigen of the Max Planck Institute for Physical Chemistry in Göttingen, who received a Nobel prize for his

work in 1967. It has provided a new vision into the elementary steps of chemical reactions and is proving a powerful tool for clarifying biochemical mechanisms.

Traditionally chemical reactions have been studied by mixing separate solutions containing known concentrations of the reactants and timing either the disappearance of one of the reactants or the appearance of the product. As long as the progress of the reaction was followed by extracting samples for quantitative chemical analysis, manual mixing was perfectly adequate. The time required to mix two solutions manually depends ultimately on the dexterity of the experimenter, but it is not less than several seconds.

In 1923 Hamilton Hartridge and F. J. W. Roughton of the University of Cambridge devised a way to study much faster reactions. Two solutions containing the reactants were mechanically forced together in a mixing chamber from which the mixture flowed into a long observation tube. There one could measure changes in color that were related to a change in concentration of either the reactants or the product. The time from the start of the reaction depended on the distance the solution had traveled from the mixing chamber and on the velocity of flow. Hence reaction rates could be determined either by observing the extent of the reaction at different points along the observation tube or by observing the reaction at a single point and varying the flow velocity. In this way it was possible to obtain information on reactions within a few milliseconds of their starting time.

In early applications of this method a galvanometer was used to measure the response of a photocell to changes in the

color of the flowing solution. Because several seconds were needed to make a measurement, a constant flow rate had to be maintained for this period, which meant that fairly large volumes of reactants were required. Therefore the method could not be employed with scarce substances such as biochemicals that were difficult to isolate. The development of cathode ray oscilloscopes that can simultaneously detect, amplify, record and time electronic signals led to "stopped flow" instruments that required much smaller quantities of reactants. In this variation of the continuous-flow method the flow was abruptly stopped after mixing and changes in the color of the stationary reaction solution were recorded spectrophotometrically. By 1940 Britton Chance of the University of Pennsylvania had developed an instrument that required only a tenth of a milliliter of reactants to produce enough readings for plotting a useful curve. Since that time there have been refinements in flow methods but none has reduced the minimum time required to mix two solutions to less than a tenth of a millisecond. Further shortening of the time in studying solution reactions required a radically different approach.

The new approach was foreshadowed in the early 1950's by measurements of sound absorption in solutions containing dissolved salts such as magnesium sulfate. Sound waves of certain frequencies were more strongly absorbed than had been expected. It is now known that the anomalous absorption results from the inability of a reaction involving the magnesium ion (Mg^{++}) to keep pace with the pressure variations in the sound wave. It is of historical interest that Albert Einstein, following a suggestion of the physical chemist Walther Nernst, had predicted in 1920 that a dissociating

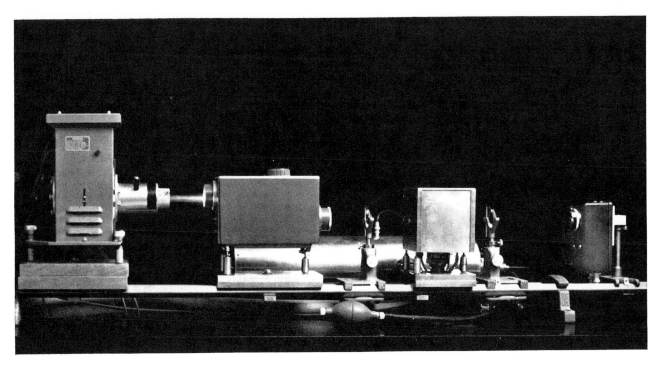

TEMPERATURE-JUMP INSTRUMENT built by the author at Wesleyan University provides information about fast chemical reactions by means of the relaxation technique. "Relaxation" refers to the natural tendency of a chemical reaction to reach a new equilibrium after it has been perturbed in some fashion. In the author's instrument the perturbation is a sharp rise in temperature.

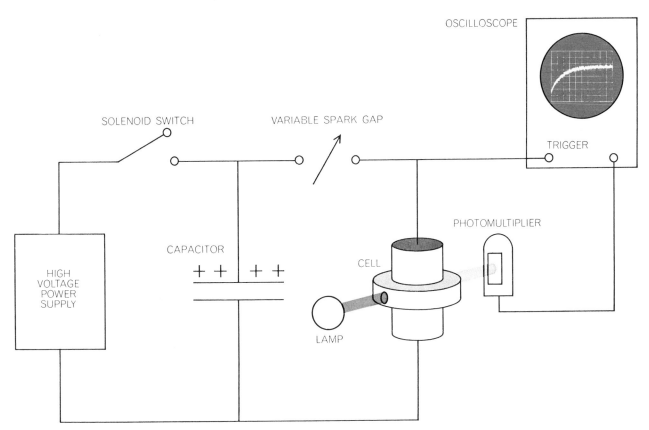

SOURCE OF TEMPERATURE JUMP in the author's instrument is a high-voltage direct-current power supply connected to a capacitor through a solenoid switch. After the capacitor is charged the switch is disconnected. By closing a variable spark gap the energy stored in the capacitor can be discharged through a cell containing the reaction under study. In solution with the reactants are salt ions that carry the discharge current, without entering into the reaction, and rapidly heat the solution. A temperature jump of six to eight degrees Celsius occurs in five microseconds. The temperature jump induces a change in the concentration of reactants and product as the reaction shifts to a new equilibrium. The shift is monitored by a photomultiplier that responds to changes in absorption of one of the reactants. The results are displayed on an oscilloscope, which is triggered by the closing of the spark gap.

TEMPERATURE-JUMP CELL consists of a cylindrical plastic chamber and two electrodes that fit into the ends of the chamber. When a high current is suddenly passed through the cell, the temperature rises sharply and thus alters the equilibrium of a chemical reaction.

gas should absorb sound at high frequencies because the association and dissociation of the gas molecules would be unable to keep pace with the sharp pressure variations in the sound waves. It remained for Eigen to recognize clearly the importance of these observations for the study of fast reactions. He perceived that fast solution reactions could be investigated by rapidly perturbing a reaction at chemical equilibrium and monitoring the rate at which it shifts to a new equilibrium.

Chemical equilibrium is a dynamic state. Let us consider as a prototype the bimolecular reaction between molecules of A and B to form a new chemical species C [see top illustration on opposite page]. At each instant some molecules of A and B are combining to form C and some molecules of C are dissociating into A and B. The reaction is said to be in equilibrium when the forward and backward rates are equal. The rate at which C forms is proportional to the product of the concentration of A and B. The constant of proportionality (that is, the number that balances the equation) is called the forward rate constant. It is a measure of the probability that the reaction will occur. The rate at which C dissociates to re-form A and B depends on the amount of C present and on the backward rate constant. The relative concentrations of A, B and C are fixed by the equilibrium constant K, which is the ratio of the forward to the backward rate constant.

Two important properties of chemical equilibria should be noted. First, they depend on external conditions, the two most familiar being temperature and pressure. For each temperature and pressure there is a particular set of forward and backward rate constants, and therefore a different value of K. Second, if an equilibrium is disturbed by changing the temperature or pressure, the reaction cannot regain equilibrium infinitely fast.

Suppose our prototype reaction shifts to the right when the temperature is raised. Such a shift is reasonable, because as the temperature increases the molecules of A and B move faster. They collide more frequently, increasing the probability of reaction to form a molecule of C. The dissociation of C does not require collision with another molecule, so that the influence of temperature on the forward and backward rate constants is generally different. If the forward rate constant increases faster with temperature than the backward rate constant,

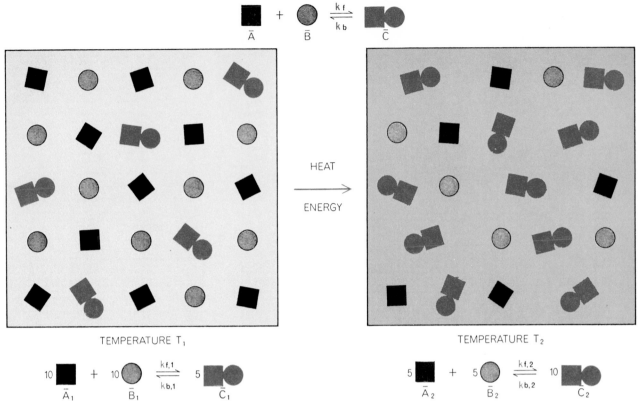

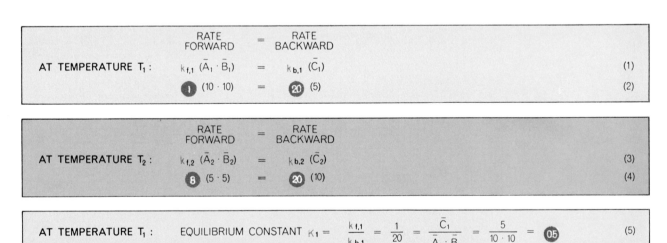

EFFECT OF TEMPERATURE CHANGE on a chemical reaction is usually to shift the concentration of reactants and product that are in equilibrium. Here reactants A and B combine to form C at a certain rate specified by the forward rate constant k_f. The dissociation of C into A and B is governed by the backward rate constant k_b. Letters with bars over them indicate equilibrium concentrations of reactants and product. At temperature T_1 10 molecules of A and 10 molecules of B are in equilibrium with five molecules of C. At temperature T_2 the reaction is driven to the right so that five molecules of A and five of B are now in equilibrium with 10 molecules of C. The values of the rate constants at each temperature can be determined by means of chemical relaxation methods.

RATE AND EQUILIBRIUM CONSTANTS are computed for the reaction depicted at the top of the page. At each temperature, when equilibrium is reached, the rate forward equals the rate backward. The values of $\bar{A}_1$, $\bar{B}_1$ and $\bar{C}_1$ at temperature T_1 are substituted in equation 1. If the forward rate constant $k_{f,1}$ is set equal to 1, it is clear that the backward rate constant must be 20 to make the equation balance (2). A similar substitution is carried out in equation 3 for temperature T_2, giving equation 4. In equation 4 the backward rate constant is kept the same as in equation 2 to indicate that the tendency of molecule C to come apart is not greatly affected by a small change in temperature. The rate at which A and B combine, however, is apt to be significantly accelerated because it depends on the frequency with which they collide. Thus in this hypothetical example $k_{f,2}$ is eight times larger than $k_{f,1}$. With this information for each temperature the equilibrium constants K_1 and K_2 can be computed (equations 5, 6).

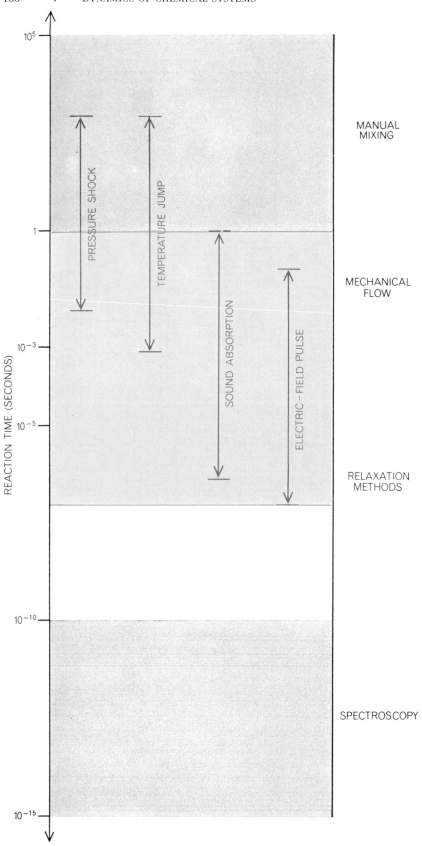

then the relative amount of C in an equilibrium mixture of A, B and C will be greater at higher temperatures.

If our prototype reaction is initially at equilibrium at some temperature T_1, the concentrations of A, B and C will be the appropriate equilibrium concentrations for that temperature [see bottom illustration on preceding page]. Now, suppose the temperature is suddenly increased to a new value T_2. As soon as the temperature is raised the appropriate equilibrium values for the reactants are those corresponding to the higher temperature. In contrast, the net rate at which the actual concentration of C can change is limited by the forward and backward rate constants at the higher temperature and by the instantaneous concentrations of A, B and C. If the temperature is increased rapidly compared with the rate at which the reacting system can respond, the change in the actual concentration of C will lag behind the change in its equilibrium concentration.

For small, stepwise perturbations of the temperature, the reactants are observed to approach their equilibrium values at the higher temperature exponentially, that is, the rate at which C changes is approximately proportional to the difference between its equilibrium concentration and its actual concentration. The reciprocal of the constant of proportionality has units of time and is called the relaxation time (designated by the Greek letter tau). In numerical terms it is the time required for C to approach within approximately a third of its new equilibrium value.

The exact relation between the relaxation time, the forward and backward rate constants and the equilibrium concentrations of the reactants depends on the form of the perturbation and on the reaction mechanism. For a simple bimolecular reaction the reciprocal of the relaxation time is equal to the forward rate constant times the sum of the equilibrium concentrations of A and B plus the backward rate constant. All rate constants and concentrations refer to the higher temperature. The forward and backward rate constants can be evaluated by measuring the relaxation time at different concentrations of A and B. A plot of the reciprocal of the relaxation time against the sum of the equilibrium concentrations of A and B yields a straight line whose slope is the forward rate constant and whose intercept is the backward rate constant [see top illustration on page 169].

If we had chosen as our prototype re-

RELAXATION METHODS fill the gap between more traditional methods for studying the rates of chemical reactions. Reactions with half-times longer than a second can be studied by manually mixing reactants. Mechanical-flow methods extend the range of half-times accessible to study to about a millisecond. At the other extreme, spectroscopy gives information about simple radiation-absorption reactions, which can take place in less than 10^{-10} second (a tenth of a nanosecond). Relaxation techniques fill the gap between 10^{-10} and 10^{-3} second, where many of the elementary steps in chemical reactions take place.

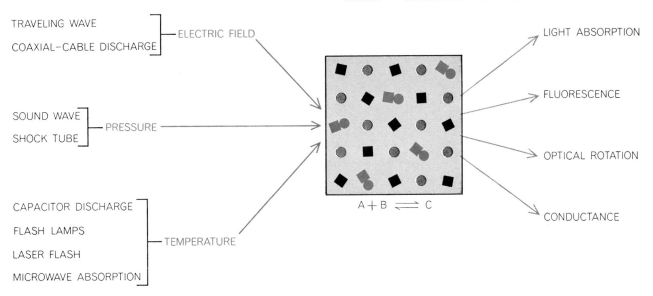

TRAVELING WAVE
COAXIAL-CABLE DISCHARGE ─ ELECTRIC FIELD

SOUND WAVE
SHOCK TUBE ─ PRESSURE

CAPACITOR DISCHARGE
FLASH LAMPS
LASER FLASH
MICROWAVE ABSORPTION ─ TEMPERATURE

A + B ⇌ C

LIGHT ABSORPTION
FLUORESCENCE
OPTICAL ROTATION
CONDUCTANCE

VARIETY OF AGENCIES can be exploited for perturbing the equilibrium of a chemical reaction when employing the relaxation technique. Because chemical equilibria depend on temperature, pressure and strength of the electric field, the rates of fast reactions in solution can be studied by perturbing any one of these variables faster than the chemical system can respond. The illustration indicates some of the ways equilibria can be upset and methods for observing the speed with which equilibrium is reestablished.

action the single-molecule interconversion of two species A and C, the reciprocal of the relaxation time would have equaled the sum of the forward and backward rate constants. Here an independent measurement of the equilibrium constant would be required to evaluate the individual rate constants. More important, the relaxation time would have been independent of the concentrations of reactants [see bottom illustration on page 169]. Relaxation studies not only allow evaluation of rate constants but also discriminate among different possible mechanisms. If we had not known whether C formed directly from A or whether a two-molecule collision with another species B was involved, measurement of the relaxation time at different concentrations of A and B would provide the answer.

In multistep reactions a spectrum of relaxation times may be observed. The number of relaxation times expected depends on the number of independent reactions. If the reactions are sufficiently different, the individual relaxation times can be measured separately by simply changing the time base on an oscilloscope. If not, recourse must be taken to mathematical methods for their separation and interpretation.

The "temperature jump" instrument built by the author at Wesleyan University is illustrated on page 31. Technically it is among the simplest of the relaxation methods. It happens that it is also the most generally useful. In this instru-

ment a capacitor is charged to a high voltage. The energy stored in the capacitor is then discharged through the reaction cell by closing a variable spark gap. The reaction cell contains a neutral salt that does not enter into the reaction but carries the discharge current through the solution, which is rapidly heated by the passage of the current. Changes in the color or transparency of the solution in the cell are monitored by a spectrophotometer and recorded on an oscilloscope. The instrument produces a temperature jump of six to eight degrees Celsius in about five microseconds.

It is not necessary that the perturbation take the form of a simple step. Any perturbation that can be readily expressed mathematically can be used. For example, the pressure in a sound wave varies as a sine wave. Nor is it essential to perturb an equilibrium. Many reactions that go to completion, notably enzyme-catalyzed reactions, pass through a stationary, or quasi-equilibrium, state in which the concentrations of intermediate species are temporarily constant. The rate of formation and dissolution of those intermediates can be studied by rapidly perturbing the stationary state.

In the past decade Eigen, his co-worker Leo de Maeyer and their students have built instruments capable of measuring the fastest possible solution reactions. Instruments built at Göttingen and elsewhere have exploited a variety of agencies for rapidly perturbing chemical reactions [see illustration above]. For example, temperature perturbations can

be produced by the absorption of microwave energy or by the absorption of short bursts of light energy, either from flash lamps or from lasers. Pressure perturbations can be produced in liquid shock tubes as well as by sound waves.

The versatility of these approaches has been extended by coupling them to diverse detection and recording systems. In addition to ultraviolet and visible-light absorption spectroscopy, changes in fluorescence, in optical rotation and in electrical conductivity have been used to follow the progress of chemical reactions. Gordon G. Hammes of Cornell University has successfully coupled stopped-flow and relaxation methods to study the stationary state in an enzyme reaction.

The chemist is not content to know the results of a reaction, the proportions in which reactants combine and the rate of the overall process. He seeks to determine the detailed mechanisms by which chemical transformations occur. It is known that complex reactions involve a series of discrete steps. A variety of techniques allow the identification of reaction intermediates. More formidable is the study of the elementary steps themselves. In general, the transfer of protons (hydrogen ions), the association and dissociation of molecules, the transfer of electrons and the rearrangement of structures—any combination of which comprises the individual steps in complicated reaction mechanisms—are too fast to study by conventional methods.

Consider the association of two mole-

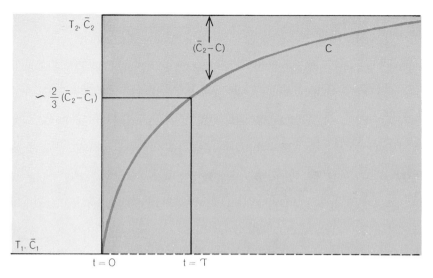

TEMPERATURE-JUMP TECHNIQUE provides data from which rate constants and equilibrium constants can be determined. The apparatus used is the one illustrated on page 31. At the outset, at the initial temperature T_1, product C is present at its equilibrium concentration $\overline{C}_1$. At time zero the temperature is rapidly increased to T_2. The exponential curve (*color*) shows how the instantaneous concentration of C approaches the new equilibrium value $\overline{C}_2$. The relaxation time τ (tau) is approximately equal to the time required for product C to travel two-thirds of the way toward the final equilibrium concentration $\overline{C}_2$.

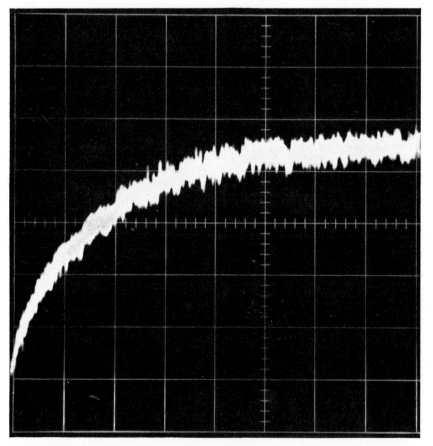

EXAMPLE OF CHEMICAL RELAXATION is shown in this oscilloscope trace made by the author. The reactants are the enzyme alpha-chymotrypsin and proflavin, a strongly colored substance that binds to the enzyme. When subjected to a temperature jump, the equilibrium amount of bound proflavin is abruptly altered, leading to a change in the amount of light at a wavelength of 460 nanometers absorbed by the solution. The trace shows the change in absorption in the 1.6 milliseconds following the temperature perturbation.

cules in water solution as forming some kind of identifiable intermediate complex. How fast can this reaction occur? Of course, since the rate depends on the concentrations of the reactants, it can be made as slow as one wishes by reducing the concentrations. Unfortunately one quickly reaches a point of diminishing returns. If the concentrations are made too small, the reaction is no longer observable. What one would like to estimate is the rate constant for association from which one could calculate the halftime of the reaction in solutions that contain detectable concentrations of the reactants. The theoretical calculation of rate constants is often prohibitively difficult. It is possible, however, to calculate the maximum rate constant for the association of two molecules by assuming that every collision results in the formation of a complex. The collision probability essentially depends on the mobility of the reacting molecules and their size. The faster the molecules move, the more probable a collision is. The bigger the molecules are, the greater the distance at which collision occurs. Both the thermal mobility and the size of the molecules can be measured by the methods of physical chemistry. For small molecules the calculated maximum rate constants range from about 10^9 to 10^{11} liters per mole per second. (A mole is the weight in grams equal to the molecular weight of a substance.) At the concentrations required for optical detection these values correspond to halftimes in the microsecond range. Although such half-times are too short for continuous-flow and stopped-flow techniques, they can be measured by relaxation methods.

The study of proton transfer in water illustrates the use of relaxation methods to investigate a fast elementary process. This was one of the first reactions Eigen and his associates studied by chemical relaxation methods. They found that the rate constant for the bimolecular reaction in which protons (H^+) combine with hydroxyl ions (OH^-) to form water is 1.4×10^{11} liters per mole per second. This value was surprisingly higher than the theoretical maximum calculated by assuming that every collision leads to reaction and by using the measured mobilities of protons and hydroxyl ions. In making such a calculation one also has to assume a certain size for the reacting species. Earlier studies had suggested that a proton in water is associated with a water molecule, forming a hydronium ion (H_3O^+). Because it is much larger than a bare proton, H_3O^+ would collide more frequently with OH^-. Even this assumption, however, led to values smaller than the newly measured rate constant. It turned out that the measured value was in good agreement with the value calculated by assuming that the reacting species were the still larger ions $H_9O_4^+$ and $H_7O_4^-$.

The proton in water solution can be regarded as being hydrated by four water molecules and the hydroxyl ion as being hydrated by three water molecules [*see top illustration on next page*]. Water is a polar molecule, meaning that its electric charges are not evenly distributed. The oxygen has a partial negative charge; the protons (hydrogens) have a partial positive charge. The hydrated proton complex and the hydrated hydroxyl complex are stabilized by the attraction between the charged ions and the oppositely charged ends of the surrounding water molecules. These attractions and the weak interactions between polar water molecules themselves are called hydrogen bonds. The combination of a proton and a hydroxyl ion involves the diffusion together of the hydrated complexes $H_9O_4^+$ and $H_7O_4^-$. Once collision occurs, a hydrogen-bond bridge is formed, and the excess proton is rapidly transferred from $H_9O_4^+$ to $H_7O_4^-$. In ice all the water molecules are bridged to four others by hydrogen bonds. It has been determined from measurements of proton mobility in ice crystals that the mean time an excess proton remains associated with a particular water molecule in a hydrogen-bonded structure is 10^{-13} second. Therefore the slowest step in the combination of protons and hydroxyl ions in water is the random encounter of the hydrated ions. The measured rate constant corresponds to the probability of collision between $H_9O_4^+$ and $H_7O_4^-$.

The unique ability of protons to penetrate spheres of hydration has been confirmed by comparing the rates at which protons combine with negatively charged ions in water solution and the rates at which positively charged metal ions combine with negatively charged ligands (a ligand is anything that associates with metal ions). An example of the former process is the association of acetic acid, in which H^+ reacts with CH_3COO^- to form CH_3COOH. As anticipated, the rate constant for combination (4.5×10^{10} liters per mole per second) equals the calculated collision probability. Once the hydrated hydrogen and acetate ions have collided, pro-

IF $A + B \underset{k_b}{\overset{k_f}{\rightleftharpoons}} C$

THEN, $\dfrac{1}{\tau} = k_{f,2}(\bar{A}_2 + \bar{B}_2) + k_{b,2}$

WHEN PLOTTED:

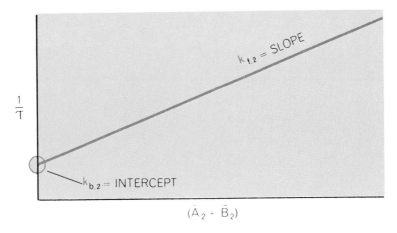

MECHANISM OF REACTION can be clarified by measuring the relaxation time, τ. If the reaction is of the form $A + B \rightleftharpoons C$, $1/\tau$ equals the forward rate constant times the sum of the final equilibrium concentrations of A and B (designated $\bar{A}_2$ and $\bar{B}_2$) plus the backward rate constant. When $1/\tau$ is plotted against $\bar{A}_2 + \bar{B}_2$, the slope of the curve gives the forward rate constant ($k_{f,2}$) and the point of interception gives the backward rate constant ($k_{b,2}$).

IF, HOWEVER, $A \underset{k_b}{\overset{k_f}{\rightleftharpoons}} C$

THEN, $\dfrac{1}{\tau} = k_{f,2} + k_{b,2}$

WHEN PLOTTED:

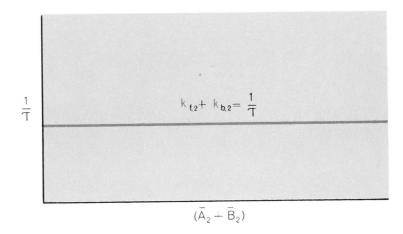

IF REACTION IS UNIMOLECULAR of the form $A \rightleftharpoons C$, it turns out that $1/\tau$ is independent of the concentrations of the reactants, so that the resulting curve is simply a horizontal line. In this case $1/\tau$ equals the sum of the forward and backward rate constants.

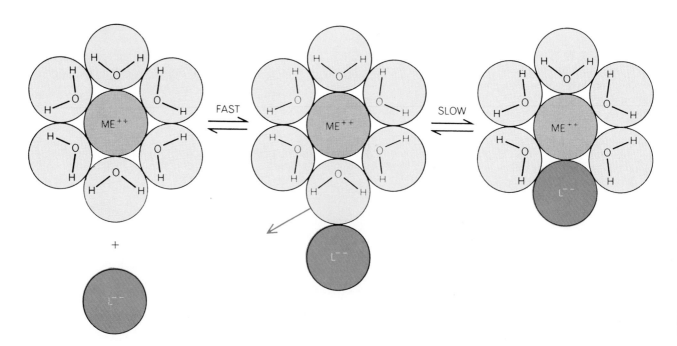

REACTION OF PROTONS AND HYDROXYL IONS in water can be thought of as taking place in two steps, the second much faster than the first. Relaxation experiments and other studies indicate that a proton (H^+) in water is normally hydrated, or surrounded, by four water molecules to form $H_9O_4^+$ and that the hydroxyl ion (OH^-) is hydrated by three water molecules to form $H_7O_4^-$ (*left*). When these complex ions collide, they neutralize each other and form uncharged water molecules. The neutralization actually takes place by a flow of negative charge density, which results in the flipping of weak hydrogen bonds and real bonds (*middle and right*). In the neutral structure that results, each oxygen is strongly bound to two hydrogen atoms and more weakly linked by hydrogen bonds (*broken lines*) to other water molecules. Dots on oxygen atoms represent the potential of forming other hydrogen bonds. This diagram represents in two dimensions structures that actually have three dimensions; thus the circles represent spheres.

FORMATION OF METAL COMPLEX can also be thought of as taking place in two steps, the first being faster than the second. In solution a metal ion (Me^{++}) is often hydrated by six water molecules. The ion with which it combines, called a ligand (L^{--}), is more loosely associated with water molecules (*not illustrated*). The metal ion and ligand come together swiftly (*middle*) but the step in which the ligand displaces a water molecule in the metal's inner hydration sphere takes place much more slowly (*right*).

ton transfer occurs rapidly along a hydrogen-bond bridge.

Metal complexes, for instance the combination of magnesium with sulfate ions, are not formed at a rate controlled by the rate of collision. Typically the rate constants for metal-complex formation are 10,000 to 100,000 times smaller than those for combination with protons. They are generally independent of the ligand but are correlated with the positive-charge density of the metal ion. The greater the ratio of positive charge to the size of the metal ion is, the more slowly the complex is formed. The explanation is that the metal ion and the ligand cannot penetrate each other's hydration spheres. Since the metal ion binds water more tightly, the rate is determined by the charge density of the metal ion. Before a metal-ion complex can form, a water molecule must dissociate out of the metal's inner hydration sphere [see bottom illustration on opposite page].

In addition to proton transfer and metal-complex formation, other elementary reactions have been studied by relaxation methods. The rates of electron transfer, of structural rearrangements and of association reactions have all been successfully measured. The use of relaxation methods to investigate processes involving a series of elementary steps is illustrated by the study of enzyme reactions.

Among the most challenging problems in biochemistry is the detailed explanation of the functioning of enzymes. It has been known for more than a century that enzymes are nature's catalysts. In their absence the myriad chemical reactions in living cells would proceed much too slowly to sustain life. In the past decade it has become clear that some enzymes also function as control units in metabolic pathways, accelerating or decelerating selected regulatory steps, depending on the abundance or deficiency of key metabolites.

Most of the enzymes involved in metabolism have now been identified, isolated and purified. The amino acid sequences of a dozen enzymes are now known. In the past four years the three-dimensional structures of five enzymes have been worked out by X-ray crystallographers. Early this year researchers at the Merck, Sharp & Dohme Research Laboratories and Rockefeller University independently completed the first chemical synthesis of an enzyme (ribonuclease) from its constituent amino acids. The understanding of how an enzyme

functions has proved to be more elusive. Detailed knowledge of the enzyme's structure is an essential first step. In fact, the X-ray analysis of the enzyme lysozyme has given a remarkably clear picture of how the enzyme binds itself to a simple molecule that mimics the structure of the cell wall of a bacterium, and clarifies the mechanism by which lysozyme catalyzes the wall's dissolution. Complete understanding of enzyme function, however, is ultimately a problem in kinetics. The role of enzymes is to regulate the speed of chemical reactions. A complete description of their functioning must therefore include both an identification of the reaction intermediates and an evaluation of the rate constants for the elementary steps in the reaction pathway. So far only three enzymes have been examined in detail by relaxation methods, but it is already apparent that such studies will play an important role in elucidating the detailed mechanisms of enzyme action.

In order to appreciate the power of relaxation methods for studying enzymatic catalysis, it is helpful to understand the limitations inherent in earlier methods. In 1913, 13 years before enzymes were shown to be proteins, Leonor Michaelis and Maud L. Menten explained why the rate of enzyme reactions does not increase indefinitely in linear fashion with substrate concentration but levels off at some maximum rate [see top illustration on next page]. They proposed that an enzyme-substrate intermediate is formed and that it may dissociate either before or after conversion of substrate to product. Since the former is usually much more likely, free enzyme and substrate are virtually in equilibrium with the intermediate complex. The rate of product formation is proportional to the concentration of enzyme-substrate intermediate. As the substrate concentration is increased the concentration of intermediate complex approaches the total enzyme concentration and the rate approaches a maximum value. At substrate concentrations insufficient to saturate the enzyme the rate is the maximum rate times the fraction of the enzyme complexed with substrate.

The difficulty with the Michaelis-Menten formulation of enzyme catalysis is that it postulates three rate constants (k_1, k_{-1} and k_2) but supplies explicit values for only one of them (k_2). The first constant, k_1, determines the rate of formation of the enzyme-intermediate complex. The second, k_{-1}, determines

the rate of dissociation of the complex. The third, k_2, determines the rate at which product is formed from the complex. As shown in the top illustration on the next page, two constants appear in the derived rate expression: the maximum rate, V_m, and the Michaelis constant K_s. After obtaining V_m one can calculate k_2 from a knowledge of the total enzyme concentration (E_0). Called the catalytic rate constant, k_2 is a measure of the number of substrate molecules that an enzyme molecule can convert to product per unit of time. The Michaelis constant (K_s) is a measure of the fraction of enzyme complexed with substrate. It is equal to the substrate concentration required to reach half the maximum rate, or $V_m/2$. The Michaelis constant is a ratio of two rate constants: k_{-1} divided by k_1. Their individual values cannot be extracted in this approach.

The problem is that information is thrown away in the derivation of the rate equation. The enzyme and the substrate are not in equilibrium with the enzyme-substrate intermediate, nor is the intermediate complex in a stationary state throughout the course of the reaction. The derivation of the rate expression without simplifying assumptions requires the solution of a second-order differential equation. The resulting rate expression includes a transient term that corresponds to the buildup of a stationary concentration of the enzyme-substrate intermediate and provides the additional information needed to evaluate the individual rate constants. Generally the transient decays too rapidly to be measured by the older kinetic methods, but even if it could be evaluated, the problem is more complex.

The simple Michaelis-Menten mechanism is chemically unrealistic. It is clear on the basis of many kinds of studies that several functional groups in the enzyme must cooperate in catalyzing the chemical transformation of substrate into product. Any realistic mechanism for enzymatic catalysis must therefore include several identifiable intermediates. It does no good simply to rewrite the Michaelis-Menten equations to provide for an extra intermediate (or more than one), because V_m and K_s then are expressed as ratios of five (or more) different rate constants [see bottom illustration on next page]. The physical meaning of V_m and K_s is thus quite obscure.

The solution to this dilemma is to measure directly the rate of formation and disappearance of the reaction intermediates. Such measurements were seldom possible before the advent of re-

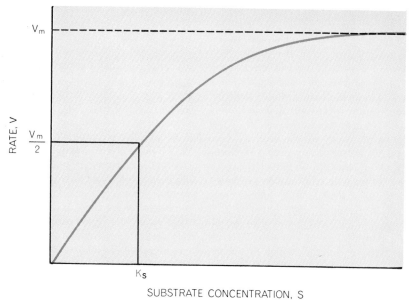

REACTION RATES OF ENZYMES were studied as early as 1913 by Leonor Michaelis and Maud L. Menten. They proposed that an enzyme and its substrate are in virtual equilibrium with an enzyme-substrate complex and that the complex breaks down, in turn, to release the enzyme and the product. The maximum rate of reaction, V_m, depends on the catalytic rate constant (k_2) and the total enzyme concentration (E_0). V_m is reached when the enzyme is saturated with substrate. The equilibrium, or Michaelis, constant (K_s) is equal to the substrate concentration needed to reach half the maximum rate, or $V_m/2$. Constant K_s is a ratio of the backward rate constant (k_{-1}) and forward rate constant (k_1).

ACTUAL ENZYME REACTIONS are more complicated than visualized by Michaelis and Menten. If another intermediate is included in the Michaelis-Menten scheme, V'_m and K'_s can still be determined by experiment, but five different rate constants now appear. V'_m and K'_s provide only ratios of various constants whose individual values remain unknown.

laxation methods. A single enzyme molecule can typically convert 1,000 substrate molecules to product per second. If enough enzyme were used to produce a detectable reaction, the reaction would be over before it could be measured by earlier methods. Relaxation methods now make such high-speed reactions accessible to study. The successful use of relaxation methods to explore the individual steps in an enzyme reaction is nicely illustrated by the study of the enzyme D-glyceraldehyde-3-phosphate dehydrogenase (GAPDH), which catalyzes the phosphorylation of D-glyceraldehyde-3-phosphate to form D-1,3-diphosphoglycerate, one step in the metabolism of sugars.

Kasper Kirschner of the Max Planck Institute in Göttingen has used the temperature-jump method to study how GAPDH is bound to the coenzyme, or cocatalyst, nicotinamide adenine dinucleotide (NAD+). Each enzyme molecule consists of four subunits, and on each subunit is a catalytic site. The binding of the coenzyme to the enzyme is cooperative, that is, the coenzyme is bound more readily at high concentrations than it is at low ones. Since NAD+ binds cooperatively, its availability can regulate the rate at which glyceraldehyde-3-phosphate is phosphorylated. When little NAD+ is available, the phosphorylation goes slowly. When NAD+ reaches a critical concentration, the reaction rate sharply increases. The coenzyme functions rather like an on-off switch.

Jacques Monod, Jeffries Wyman and Jean-Pierre Changeux of the Pasteur Institute in Paris have suggested a mechanism to explain cooperative binding. Because the structure of such a coenzyme, or other effector molecule, that binds cooperatively to the enzyme is distinctly different from that of the substrate, they call their proposal the allosteric model. ("Allo" is the Greek combining form for "other.") In the allosteric model a multiunit enzyme, T, is assumed to be in equilibrium with another form of the enzyme, R, which has a different three-dimensional structure [see illustration on opposite page]. Subunits in the R form are assumed to bind effector molecules more readily than the T form, but the T form is assumed to predominate when no effector molecules are bound. Now if the R form of the fully bound enzyme is favored, binding will be cooperative. At low effector concentrations most of the enzyme is present in the T form. The T form has little affinity for effector, so that most of it remains unbound. As the concentration of effector is increased, more of it is bound, and the equilibrium between R and T shifts toward R. At some point R predominates. Since R binds effector tightly, effector molecules rapidly saturate the enzyme binding sites. The binding process resembles a zipper. Getting it started is difficult, but once started it proceeds rapidly.

The allosteric model predicts as many as nine relaxation times for the binding of an effector to an enzyme of four subunits. Happily Kirschner found that there were only three readily separable relaxation times for the binding of NAD$^+$ to GAPDH. The two faster relaxation times correspond to two-molecule processes. The reciprocal of the fastest relaxation time depends linearly on the sum of the concentration of coenzyme and the concentration of free binding sites on the R form of the enzyme that binds NAD$^+$ more tightly. The other two-molecule process depends on the concentrations of NAD$^+$ and unoccupied T binding sites. Since only two bimolecular relaxation processes were observed, it could be concluded that the binding of effector to either the R or the T form of the enzyme is independent of the number of effector molecules already bound. The third relaxation time does not depend on enzyme concentration. It can therefore be associated with the conformational change between T and R. It does depend on the total concentration of NAD$^+$, reflecting the shift in the equilibrium between T and R with effector concentration, which one would expect if the binding is cooperative. Using the experimentally determined rate constants for the combination of effector with each conformation of the enzyme, and the rate constants for the interconversion of the two forms, Kirschner was able to construct an S-shaped binding curve that was in good agreement with the curve found experimentally.

The fact that the binding of NAD$^+$ to GAPDH can be described by the allosteric model does not mean that all regulatory enzymes conform to this model. It does, however, dramatically illustrate the potential importance of chemical relaxation studies to an understanding of enzyme function. The study of enzyme reactions by relaxation methods is in its infancy. Although the difficulties are formidable, such investigations promise to yield a deeper understanding of enzyme reactions and other complex biochemical processes.

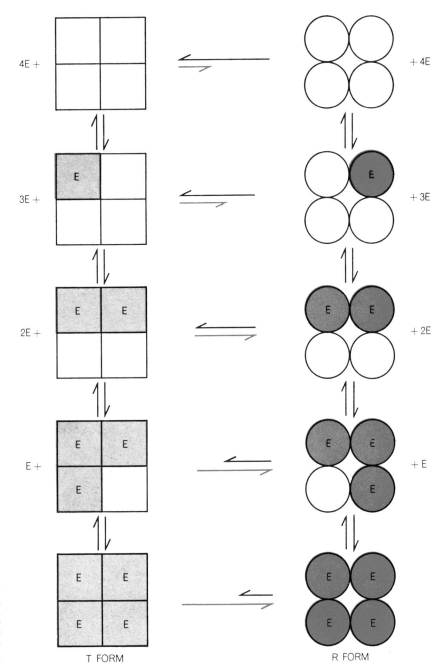

T FORM   R FORM

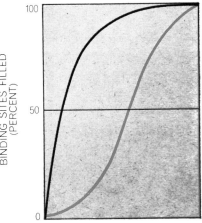

BINDING SITES FILLED (PERCENT)

100

50

0

EFFECTOR CONCENTRATION

ENZYME MODEL depicts an enzyme with four subunits that responds to an effector molecule (E) in addition to its normal substrate. Such an effector is said to be allosteric. The enzyme can exist in two configurations: T form and R form. All four subunits are assumed to change shape together. The R form binds effector molecules more tightly than the T form does. When no effector molecules are bound, the T form is favored. As more and more effector molecules are bound, however, the equilibrium shifts to favor the saturated R form. This type of binding behavior is described as cooperative and is characterized by an S-shaped binding curve (*colored curve at left*). When the binding is noncooperative, the binding curve is hyperbolic (*black curve*).

16

CATALYSIS

VLADIMIR HAENSEL AND ROBERT L. BURWELL, JR.
December 1971

Substances that accelerate chemical reactions without being used up play a major role in producing goods worth more than $100 billion a year. Various techniques help to reveal how a catalyst functions

When a chemist considers a chemical reaction, he generally asks himself three questions: How fast is it? How complete is it? How selective is it? Some reactions are very fast and go to completion to yield a single product. A familiar example is the reaction of sodium and chlorine to form sodium chloride. Other reactions, for instance the reaction of hydrogen and oxygen to form water, go very slowly at room temperature but are extremely fast at higher temperatures. They eventually go to completion to yield a single product. Most reactions are very slow indeed. The chemist has to find ways to speed them up. If he is lucky, he can do that simply by raising the temperature (as in the reaction of hydrogen with oxygen). Unfortunately increasing the temperature frequently produces undesirable side effects.

Two major reasons for seeking alternative means of speeding up chemical reactions can be illustrated by the following examples. One is the reaction of nitrogen and hydrogen to form ammonia (NH_3). The other is a hypothetical reaction between methane (CH_4) and oxygen to form dimethyl ether (CH_3OCH_3) and water. Both reactions are alike in that at room temperature their rate is essentially zero. The laws of thermodynamics tell us, however, that both reactions should go a long way toward completion at room temperature.

Let us see what happens as we raise the temperature with each of these reactions. In the reaction forming ammonia there is a composition (some specific proportion of H_2, N_2 and NH_3) that is in chemical equilibrium at a given temperature and pressure. Other compositions, if they react, must tend toward the equilibrium composition. Suppose one raises the temperature in an attempt to get a reasonably good yield of ammonia in a

reasonable period of time. Before the rate of formation of ammonia becomes fast, the position of equilibrium has shifted to increase the proportion of hydrogen and nitrogen at the expense of ammonia [*see top illustration on page 176*]. For practical purposes, then, one cannot cause nitrogen and hydrogen to combine directly to form ammonia.

In the second reaction other oxidation reactions of methane become fast as the temperature is raised and destroy the methane before detectable amounts of dimethyl ether are formed. In general most possible reactions proceed slowly at room temperature, and the great majority of reactions in this class cannot be practically effected by raising the temperature. For such reactions what one lacks is selectivity.

If nature provided no way to accelerate chemical reactions selectively, our modern technological society could not have arisen, but one must quickly add that its absence would not be noticed because no form of life could exist either. In fact, nature long ago discovered how to effect many reactions of the type that cannot be effected merely by raising the temperature. Man has acquired the knack only recently.

If the mixture of hydrogen and oxygen is exposed to platinum powder at room temperature, a rapid reaction forming water occurs on the surface of the metal particles. A few atoms of platinum can lead to the formation of many molecules of water. This is catalysis, defined as the phenomenon in which a relatively small amount of foreign material, called a catalyst, augments the rate of a chemical reaction without itself being consumed. The chemist now has an alternate means at his disposal for speeding up reactions. How do these catalysts work?

Suppose one knows that two chemical

substances, *A* and *B*, react to form *C* but that the reaction is extremely slow at room temperature. One can demonstrate that no combination of elementary processes involving *A, B* and *C* will result in rapid formation of *C*. Now add a catalyst, designated *Cat*. It provides the possibility of new elementary processes. If processes such as the following are fast, *C* will be formed rapidly:

$$A + Cat \rightarrow ACat$$
$$ACat + B \rightarrow C + Cat$$

The combining tendency of *A* and *Cat* must be adequate to yield the complex *ACat*, yet it must be not so strong as to make *ACat* unreactive. After all, unless *ACat* reacts rapidly with *B* to form *C* and then regenerate *Cat*, we do not have catalysis.

Chemists must by now have discovered the great majority of reactions that proceed without catalysts. Clearly the future of preparative chemistry will heavily involve catalysis. We already have a catalyst (iron) that enables us to manufacture ammonia, but we know of no catalyst that leads to the formation of dimethyl ether from methane and oxygen. There are almost innumerable other reactions that have favorable positions of equilibrium but for which no catalysts are known.

Catalytic reactions can be classified into three principal types. The most common is heterogeneous catalysis, in which the catalyst is a solid and the reactants and products are either gases or liquids. Platinum is a heterogeneous catalyst for the reaction between hydrogen and oxygen. The second type is homogeneous catalysis, in which the reactants, products and catalyst are molecularly dispersed in a single phase, usually the liquid phase. The third type is enzyme catalysis, which is found in living systems. There the catalyst is a complex protein molecule (an enzyme) consisting of scores or hundreds of atoms. Connections between enzyme catalysis and other types are beginning to appear, but at the moment enzyme catalysis forms a separate subject. We shall not take it up here.

The Economic Value of Catalysis

Catalysis has grown at a phenomenal rate because it has made possible the production of new chemicals and the mass production of established chemicals. Today it is an indispensable industrial tool. Catalysis contributes directly and indirectly to products accounting for a sixth of the value of all goods manufactured in the U.S.; therefore it participates in economic activities involving the exchange of more than $100 billion per year. The average consumer has little realization that the synthetics he uses every day have been made possible largely through the use of catalysis. They include plastics and fibers, detergents, hydrogenated fats and synthetic rubber.

The petroleum industry provides an outstanding example of the importance of catalysis. Crude oil contains millions of different hydrocarbon molecules of all shapes and sizes. Those with only one to four carbon atoms in the molecule (for instance methane, ethane, propane and butane) are gases at room temperature. Those with five to ten carbon atoms in the molecule boil in the gasoline range (between 20 and 200 degrees Celsius). This fraction of the crude oil ("virgin naphtha") amounts to less than 20 percent of the total; if it were simply separated by fractionation, it would make a very inferior gasoline. Its properties can be improved by subjecting it to a catalytic process called reforming, which we shall describe below. The bulk of the crude oil (about 80 percent) boils in the range between 200 and 600 degrees C. A part of this material serves as heating fuel; another part, after suitable refining, serves as jet fuel and diesel fuel; a small fraction serves as lubricating oil. By far the major part of the oil boiling in the 200-to-600-degree range is converted into gasoline by one or another type of cracking process.

Cracking can be achieved simply by heating the oil to a temperature of about 500 degrees C. under pressure. A much better gasoline and a higher yield can be obtained by doing the cracking in the presence of a catalyst. Thermal cracking, which involves heating the oil to the decomposition point, can be considered

REFORMING CATALYST converts "virgin naphtha," a naturally occurring fraction of crude oil with a low octane number, into high-octane motor fuel. Known as a dual-function catalyst, it alters the structure of hydrocarbon molecules from less favorable (low octane) to more favorable (high octane) configurations. The sample shown here, magnified about 20 diameters, contains .5 percent platinum and 1 percent chlorine on spheres of alumina.

SURFACE OF DUAL-FUNCTION CATALYST is shown at a magnification of 55,000 diameters in this scanning electron micrograph. One can see the enormous surface area that such a catalyst presents. The catalyst is made by the Universal Oil Products Company.

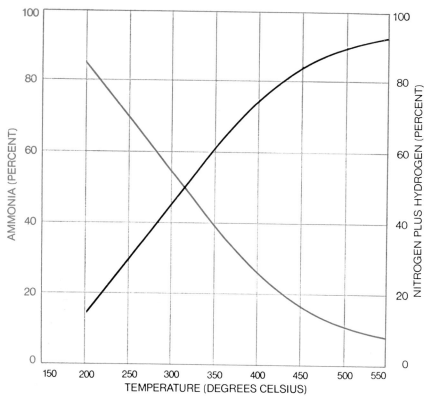

EFFECT OF INCREASING TEMPERATURE on an equilibrium mixture of nitrogen (N$_2$) and hydrogen (H$_2$) is to shift the equilibrium to yield increasingly less ammonia (NH$_3$). Without a catalyst ammonia is formed too slowly even at 550 degrees Celsius to provide a useful process. The equilibrium is shown here for a pressure of 100 atmospheres.

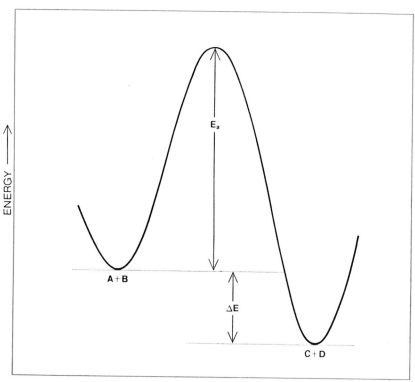

IDEALIZED CHEMICAL REACTION, $A + B = C + D$, requires an activation energy E_a. The potential energy of $A + B$ is at the bottom of the valley at the left, that of $C + D$ is in the valley at the right. A and B can surmount the energy barrier only if they collide with a kinetic energy at least equal to E_a. The barrier is many times greater than the average energy of collision. A catalyst can provide new elementary processes with lower values of E_a.

a sledgehammer approach. Heating in the presence of a catalyst is more sophisticated. Heavy molecules are broken in places dictated by the nature of the catalyst; hence there is a more selective cracking and more pieces boil in the gasoline range. Furthermore, many of the pieces no longer resemble the parts of the original molecule in shape. During catalytic cracking they undergo an isomerization reaction (a molecular rearrangement without change in molecular weight). This change in molecular structure fortunately leads to a better gasoline. If the energy in the fuel is released too quickly, the piston in an automobile engine cannot respond and the energy is expended against the walls of the cylinder, causing a pinging sound ("knock"). Isomerization increases the compactness of gasoline molecules, thereby improving the fuel's antiknock quality [see illustration on page 178].

Catalytic cracking is thus a more selective tool in the fragmentation reaction and the fragments are reassembled into more valuable pieces. Since very few reactions are perfect, a catalyzed reaction is no exception and some pieces are too small to be included in the gasoline. Some of these pieces do not have a full complement of hydrogen, so that two of the adjacent carbon atoms are double-bonded. It has been found that the unsaturated molecule (known as an olefin if it has a single double bond and an open structure) can react with another unsaturated molecule to produce a dimer or a trimer in the presence of a catalyst such as phosphoric acid, so that the resulting larger molecules boil in the gasoline range. It has also been found that the unsaturated molecules can be catalytically condensed with the more compact type of saturated hydrocarbons in the presence of sulfuric acid or hydrofluoric acid, but not with the straight-chain saturated hydrocarbons. On the other hand, the straight-chain saturated hydrocarbons can be catalytically isomerized to more compact structures and then reacted with the unsaturated molecules [see top illustration on page 179]. Thus a good share of the smaller pieces are reconverted into larger molecules and used as gasoline components. In this manner the original large molecules, by a sequence of a number of catalytic steps, are converted into the desired product.

The above discussion has identified only a part of the catalytic participation in the refining of petroleum, but it is clear that if one could tag a few carbon atoms in the original crude oil, they would be seen to go through a complex series of catalytic steps before winding up in a motorist's gas tank. In the U.S. alone some 12 million barrels of crude oil are processed daily, virtually all of it being consumed as an energy source. A small part of this amount, particularly the largest molecular sizes, is used as asphalt and petroleum coke. The smaller fragments from the various cracking steps are gathered, and some, as indicated above, serve to make gasoline. The smallest reactive fragment, ethylene, is used for the production of ethyl alcohol and polyethylene. Much of the polyethylene is made catalytically; it consists of from about 1,500 to 15,000 ethylene molecules strung together in a long chain.

A challenging and timely problem is how to apply catalysis in air-pollution control. At the moment it appears that catalysis will be heavily involved in the treatment of exhaust gas from internal-combustion engines. This will represent the first instance of the direct application by the general public of prepared catalysts, as distinguished from natural catalysts such as yeast.

Early Developments

Having established the substantial contributions of catalysis to the economy, let us briefly trace the history of catalysis. After some 20 years of observing a number of "notable discoveries," Jöns Jakob Berzelius named and defined catalysis in 1836. The contemporary investigator of catalysis is frequently awed by the keen insights of early workers. After defining catalysis Berzelius proposed, among other suggestions, that it involves the development of a force of affinity coming from the catalyst and having an effect on the chemical activity of the reagents.

The "notable discoveries" described by Berzelius included the work of Johann Wolfgang Döbereiner, who in 1823 found that a stream of hydrogen mixed with air would ignite on contact with platinum sponge. In fact, this discovery led to a lamp in which hydrogen was produced from zinc and sulfuric acid, then catalytically ignited to serve as a replacement for the tinderbox for lighting lamps and candles. It is too bad that such an ingenious device had to be replaced with the match. Other evidence of the catalytic properties of platinum were discovered independently by Humphry Davy and Michael Faraday. All these workers recognized that there was something special, an action through touch or contact, that caused a reaction to occur.

CATALYTIC HYDROGENATION was discovered by Paul Sabatier in the 1890's. He found that in the presence of nickel, ethylene, the simplest olefin, reacts rapidly with hydrogen to form ethane (*top*). Benzene reacts with hydrogen to form the saturated ring compound cyclohexane (*bottom*). Many industrial hydrogenation processes use nickel as a catalyst.

OXO PROCESS, developed in Germany in the 1930's, involves the addition of carbon monoxide and hydrogen to olefins (here 1-pentene) to produce long-chain alcohols. The reaction is now used to produce intermediates for synthetic lubricating oils and plasticizers.

Catalysis received little academic attention during the 60 years following the definition by Berzelius. There were, however, a few technological developments. The heterogeneous catalytic oxidation by air of hydrogen chloride to chlorine $(4HCl + O_2 = 2H_2O + 2Cl_2)$ was developed by Henry Deacon in 1868. The direct oxidation of sulfur dioxide with air using a platinum catalyst was observed by Teregrine Phillips in 1831 and restudied by Clemens Winkler in 1875. There was no successful commercial installation of a direct oxidation process, however, until 1901.

As the 19th century ended, the rising importance of the chemical industry stimulated a few chemists to devote their full time to catalysis. Two of the most prominent were Paul Sabatier and Vladimir N. Ipatieff. It is perhaps fair to say that they did more than anyone else to bring catalytic science and technology together.

Sabatier's investigations of catalysis stem from the work of Ludwig Mond, a German-born British chemist, who in 1890 prepared volatile nickel carbonyl: $Ni(CO)_4$. Lord Kelvin remarked that "Mond and his colleagues have giv-

en wings to a heavy metal." Sabatier thought it should be possible to synthesize an analogous compound of nickel and ethylene: $Ni(C_2H_4)_4$. Instead of obtaining the desired compound Sabatier found that in the presence of nickel some of the ethylene (C_2H_4) was converted into its fully hydrogenated analogue, ethane (C_2H_6).

Sabatier quickly recognized that if he deliberately introduced hydrogen into the reaction, his nickel catalyst should convert ethylene into ethane with a high yield. Before long Sabatier demonstrated that nickel acted as a general catalyst for the hydrogenation of a variety of unsaturated (hydrogen-deficient) hydrocarbons [see *upper illustration above*].

Ipatieff, who was only 13 years younger than Sabatier, worked in Russia from 1890 to 1929 and in the U.S. from 1930 to 1952. Ipatieff reported the catalytic dehydrogenation of alcohol in 1901. Subsequently he devised industrial processes for the high-pressure hydrogenation of a wide variety of organic compounds. To design the high-pressure autoclaves required by his process, Ipatieff drew on a knowledge of gun barrels he had acquired as a young officer in the

Russian artillery. Building on the foundations laid by Sabatier and Ipatieff, Friedrich Bergius developed the coal hydrogenation process, which supplied Germany with an important part of its motor fuel during World War II.

A goal of many chemists around 1900 was the synthesis of ammonia from hydrogen and nitrogen. The man who first succeeded was Fritz Haber. He worked out a process in the years between 1905 and 1908, testing thousands of potential catalysts at various temperatures and pressures. The first industrial unit in-

corporating his process was built in 1913. The conversion of ammonia into nitric acid was already commercial in 1906, so that by the start of World War I, Germany was no longer dependent on Chilean saltpeter as a source of nitrates for explosives. Haber's motivation—to make Germany independent of foreign supplies of a crucial industrial material—was a far cry from the innate curiosity that motivated the first workers in the field of catalysis barely 80 years earlier.

The year 1915 marks the entry of physical chemists into the field of ca-

talysis, which had previously been dominated by inorganic and organic chemists. The elegant and precise measurements made by Irving Langmuir of the adsorption strength of various simple molecules on metals gave rise to concepts of chemisorption, and this was further developed by H. S. Taylor, Eric Rideal, P. H. Emmett and a number of other brilliant workers. Their investigations provided further impetus to technological progress.

The 1930's saw a marked spurt in the technological applications of catalysis both in Europe and in the U.S. Significant advances were made by Ipatieff and his co-workers in the use of catalysis for the production of high-octane gasoline. In Europe one of the most interesting new catalytic processes was the Oxo reaction, which involves the addition of carbon monoxide and hydrogen to olefins to produce primary alcohols [see lower illustration on preceding page]. The Oxo catalytic reaction is homogeneous, since a soluble cobalt-carbon monoxide catalyst $[Co_2(CO)_8]$ is employed. It was also the first industrial catalytic process to use complexes of transition metals rather than the metal itself. The transition metals (such as vanadium, cobalt, titanium, manganese, chromium and copper) are unusual in that they can either lend or borrow electrons with equal ease. The Oxo reaction is now used on a substantial scale to produce intermediates for synthetic lubricating oils and plasticizers.

One striking example of the ability of catalysts to perform highly selective molecular alterations is the stereoregular polymerization of olefins pioneered by Karl Ziegler and Giulio Natta, who were awarded the Nobel prize in chemistry in 1963 for their work. Propylene can be converted to three distinct head-to-tail polymers that differ only in the relative placement of their appended methyl groups. By using suitable complexes involving ions of transition metals, one can produce either "isotactic" or "syndiotactic" polypropylene [see top illustration on page 182]. If the starting material is 1,3-butadiene, four different stereotactic polymers can be generated, each requiring a different transition-metal catalyst [see bottom illustration on pages 182 and 183]. It is likely that the future will provide many additional examples of the unique capabilities of such catalysts.

The Mechanism of Catalysis

What makes catalytic reactions go? Like chemical reactions in general, catalytic reactions proceed by a more or less

OCTANE-RATING SCALE assigns to normal heptane a value of zero and to iso-octane a value of 100. Hydrocarbons with side chains burn more slowly than straight-chain hydrocarbons and thus tend to "knock" less in automobile engines. The nine possible isomers (structural variants) of heptane are shown; hydrogen atoms are omitted for clarity.

complicated combination of elementary steps. Each step is elementary in that it takes place without intervening intermediates of significant lifetime. Few chemical reactions involve just a single elementary process. Catalytic reactions, then, proceed by means of a set of one or more intermediates. A simple hypothetical example of this was given at the beginning of the article: the reaction $A + B = C$ by way of the intermediate $ACat$. If a foreign substance is to act as a catalyst, it must provide new elementary processes leading to an overall reaction that is faster than the reaction that would take place if only the pure reactants were present.

A detailed description of a chemical reaction is referred to as its mechanism. The chemist's views concerning the precise nature of the elementary processes that comprise a mechanism are constantly changing as new facts emerge from studies of catalytic systems. It is usual to recognize five stages in a heterogeneous catalytic reaction: (1) diffusion of reactants to the surface of the catalyst, (2) chemisorption of reactants, (3) surface reactions among chemisorbed species, (4) desorption of products, (5) diffusion of products from the surface.

The diffusion steps can limit the overall rate. In laboratory work one usually tries to make steps 1 and 5 faster than the other steps, but it is often impracticable to translate the laboratory techniques into engineering practice in industrial reactors.

The division of adsorption into physisorption and chemisorption follows the usual division of phenomena into physical and chemical. A physisorbed species retains substantially its original structure. Thus its infrared spectrum will be reasonably related to that of the unadsorbed molecule. The heat of adsorption is small, but not much greater than the heat of liquefaction. A chemisorbed species is a surface chemical compound. The heat of adsorption can be very small

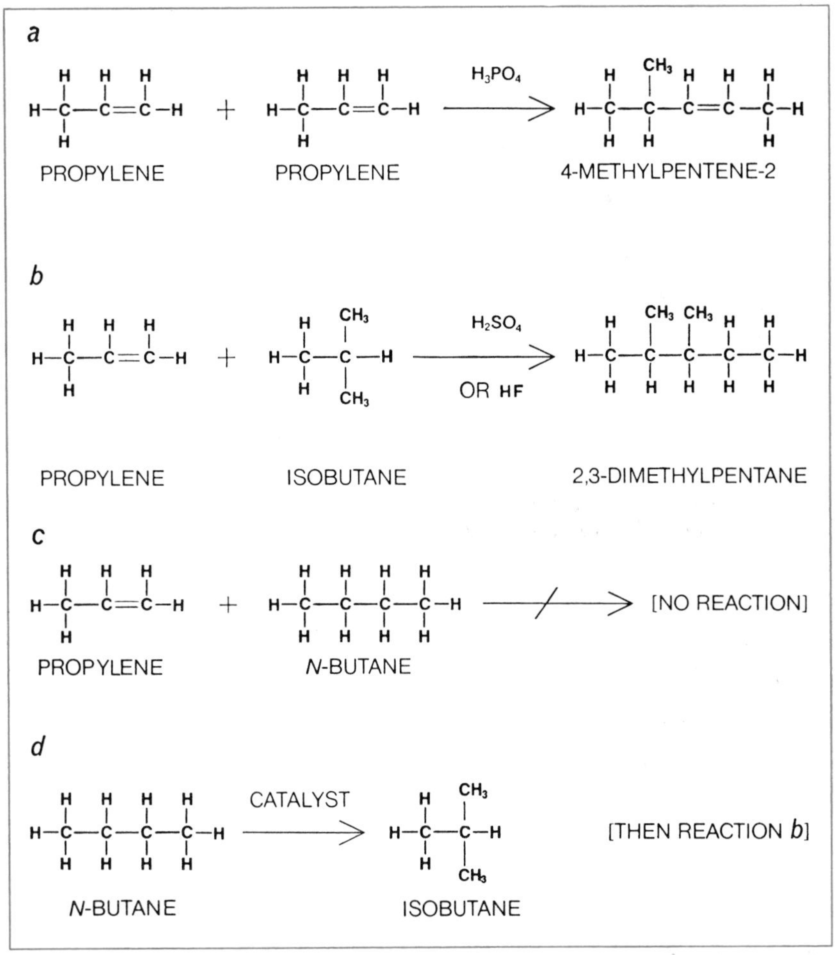

ONE ROUTE TO HIGH-OCTANE FUEL is to assemble small, volatile molecules into larger ones that boil in the gasoline range. Using phosphoric acid as a catalyst (a), light olefins such as propylene can be condensed into branched-chain olefins useful as a motor fuel. Either sulfuric acid, $H_2(SO)_4$, or hydrofluoric acid, HF, will join a light olefin to a saturated branched-chain compound (b), yielding a product with a high octane number. If the saturated molecule has a straight chain (c), no reaction takes place. By isomerization, however, the straight-chain molecule can be rearranged into a molecule that will react (d).

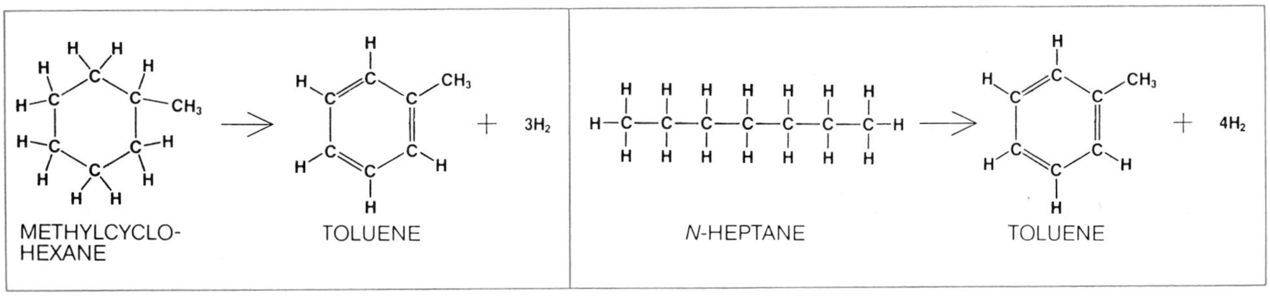

or very large, as with the heat of chemical reactions in general. There may be substantial changes in structure between desorbed and adsorbed species; frequently the adsorbed species is only a fragment of the original molecule. Molecular hydrogen (H_2), for example, of-

ten adsorbs to form a surface hydride [see top illustration on next page]. Chemisorption of this type is called dissociative adsorption. At least one of the reactants must be chemisorbed, but it is difficult to decide when chemisorption is an elementary process and when it

OTHER ROUTES TO HIGH-OCTANE FUEL make use of reforming catalysts that dehydrogenate saturated ring compounds (left) or simultaneously dehydrogenate and cyclize straight-chain hydrocarbons (right). In both examples the product is toluene, which has an octane number of more than 100. The octane number of methylcyclohexane is between 70 and 80. The octane number of n-heptane is zero. Reforming catalysts can also isomerize straight-chain hydrocarbons and crack large molecules into smaller ones.

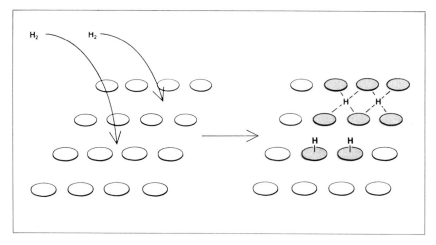

FIRST STEP IN CATALYTIC DEHYDROGENATION is exemplified by the dissociative adsorption of molecular hydrogen on a catalytic surface. Each adsorbed hydrogen atom (H) may be bound to a single atom on the surface or it may be bound to two or more atoms.

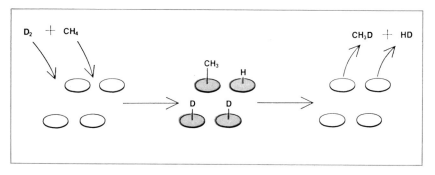

ISOTOPIC EXCHANGE demonstrates that reacting species become dissociated on a catalytic surface. Here deuterium (D) and hydrogen (H) atoms trade places in methane (CH₄).

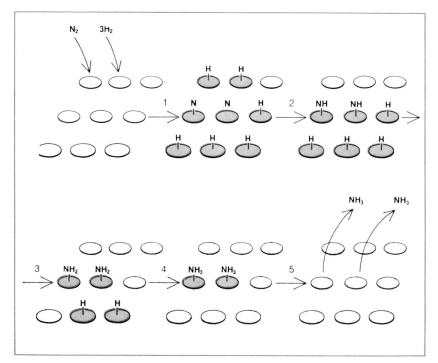

CATALYTIC SYNTHESIS OF AMMONIA is thought to involve dissociation of nitrogen and hydrogen (1) followed by addition of hydrogen to a complex consisting of nitrogen bound to atoms on the surface of the catalyst (2, 3, 4). Finally, ammonia (NH₃) is desorbed.

proceeds by means of the intermediate formation of a physisorbed species.

Step 3 usually involves one or more elementary processes in which chemisorbed species react to form new chemisorbed species. In some cases step 3 may involve reaction between a chemisorbed species and a molecule in the gas or liquid phase to form a new chemisorbed species. In step 4 the chemisorbed species desorbs, often by associative desorption, to give the final product.

Some simple cases may lack a surface-reaction step. Isotopic exchange between deuterium (the hydrogen isotope of mass two) and methane occurs on a number of transition metals (nickel and platinum) and their oxides. The reaction appears to proceed simply by dissociative adsorption followed by associative desorption [see middle illustration at left].

As Haber discovered, pure iron catalyzes the reaction of hydrogen with nitrogen to yield ammonia at about 450 degrees C. Presumably the first step in the process is that the hydrogen molecule (H_2) and the nitrogen molecule (N_2) are dissociatively adsorbed on the surface of the catalyst. Experiments with isotopes of nitrogen show that nitrogen does not dissociate at reasonable rates on iron until the temperature has been raised to about 450 degrees C. Hydrogen, on the other hand, dissociates freely even at the temperature of liquid nitrogen (−196 degrees C.).

It was suggested as long ago as the 1930's that the rate of ammonia synthesis is limited by the low rate of dissociation of nitrogen. It is hypothesized that once the nitrogen dissociates, forming a surface compound that can be designated NCat, hydrogen atoms are added in three rapid steps forming HNCat, H_2NCat and finally H_3NCat, from which NH_3 is readily released [see bottom illustration at left].

Subsequent work has supported and elaborated this mechanism, although many details are unclear and the intermediacy of an adsorbed molecule of nitrogen is favored by some. From the standpoint of industrial importance it has been found that iron plus a few percent of the oxides of potassium and aluminum, which are known as promoters, give a longer-lived catalyst and one with greater resistance to impurities in the feed stream. Catalysts of this type can often be used for several years.

The recently discovered reactions in which coordination complexes of transition elements function as homogeneous catalysts in adding hydrogen to various unsaturated compounds present close similarities to reactions effected by previously known heterogeneous hydrogenation catalysts. When transition elements are incorporated in organometallic compounds, they work as homogeneous catalysts capable of adding hydrogen to various unsaturated compounds. These active compounds are often coordination complexes. The complex must be "coordinatively" unsaturated if it is to activate hydrogen, that is, there must be a vacant or partially vacant position in the "coordination sphere" that surrounds the metal ion to which hydrogen can become bound. For example, cobalt pentacyanide [Co(CN)$_5$] has only five of six potential positions filled and can bind an atom of hydrogen in the empty position. A similar coordination complex can be formed with rhodium, which can hold two atoms of hydrogen in the same state of activation: H$^-$. The rhodium complex is oxidized by hydrogen to give a product in which the hydrogens are formally hydride ions. The reaction is called oxidative addition, and one can readily imagine an analogous oxidative dissociation adsorption. In coordination complexes incorporating ruthenium the hydrogen molecule is split into a positive state (H$^+$) and a negative state (H$^-$); the metal ion is not oxidized [see upper illustration at right].

The last process of activation is essentially the same as the one thought to operate on the surface of certain metal oxides that are used for hydrogenating olefins, for example zinc oxide (ZnO). In the case of ZnO the Zn^{2+} and O^{2-} act as coordinatively unsaturated surface species. (The superscripts refer to the number of electrons removed from or added to the dissociated fragment.) The zinc fragment binds hydrogen as H$^-$ and the oxygen fragment binds hydrogen to form OH$^-$. The olefin is weakly adsorbed on a neighboring site and reacts in turn with the H$^-$ and the hydrogen in the OH$^-$, thereby acquiring two hydrogen atoms [see lower illustration at right]. The process is called heterolytic dissociative adsorption to distinguish it from the homolytic adsorption of hydrogen on metals, in which the hydrogen molecule splits into two equally charged fragments.

Mechanism of Hydrogenation

We have presented several examples of catalytic reactions and some of the current thinking about how they occur. Generally speaking, the mechanism of homogeneous catalysis is easier to elucidate than that of heterogeneous catalysis. In homogeneous catalysis elementary processes involve simple molecules and intermediates. In heterogeneous catalysis we must also consider the effect of surfaces. One cannot isolate a surface intermediate, recrystallize it and determine its structure by X-ray analysis. If surface reactions provide particularly useful possibilities for combinations of elementary processes, they also present special problems in the study of mechanism. We shall give several examples of the techniques that have proved to be helpful in that study.

Many of the extensively investigated

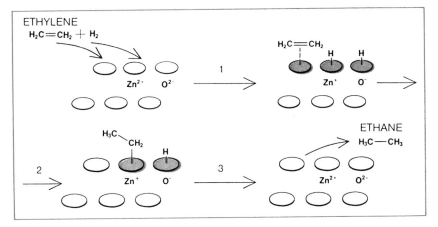

COMPLEXES OF TRANSITION METALS, such as cobalt, rhodium and ruthenium, exhibit unusual catalytic properties. The complexes consist of ions of transition metals coordinated to side groups called ligands (L). Some complexes of this type are soluble in liquids and act as homogeneous catalysts. Pentacyanocobalt (a) is able to activate hydrogen because the "coordination sphere" around the cobalt ion (Co^{2+}) has an empty position that can hold a hydrogen ion that is formally in the negative state (H$^-$). The coordination sphere around the rhodium ion (b) can hold two hydrogen atoms, both H$^-$. In aqueous solution an organometallic compound of ruthenium (c) splits a molecule of hydrogen so that one fragment is negatively charged and one becomes bound to a molecule of water as a proton, forming H$_3$O$^+$. Catalysts of this type can produce the polymers shown on the next two pages.

HYDROGENATION OF ETHYLENE, first achieved by Sabatier on a nickel catalyst, takes place readily on zinc oxide. The key sites appear to be coordinatively unsaturated metal ions (Zn^{2+}) and oxide ions (O^{2-}) on the surface. Ethylene (H$_2$C : CH$_2$) is adsorbed (1) and reacts in sequence (2, 3) with two hydrogen ions. The product is ethane (H$_3$C · CH$_3$).

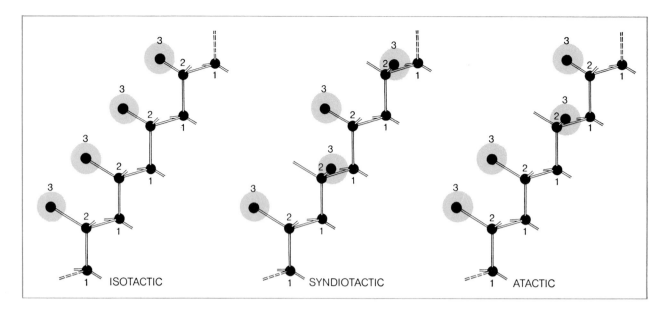

THREE KINDS OF POLYPROPYLENE can be made from propylene (H₂C : CHCH₃) with transition-metal catalysts. At a reaction temperature of 50 degrees C. a catalyst consisting of triethylaluminum and vanadium trichloride yields isotactic polypropylene, in which the methyl (CH₃) groups (*color*) all lie on the same side of the central chain of carbon atoms (*black balls*). Hydrogen atoms have been omitted; their locations are indicated by short bonds projecting from the carbon atoms. At a reaction temperature of −78 degrees C. a catalyst consisting of vanadium tetrachloride and diethyl aluminum chloride yields syndiotactic polypropylene, a molecule in which the methyl groups alternate from side to side. When the methyl groups project at random, the polymer is atactic.

reactions of heterogeneous catalysis are hydrogenations. We have discussed the hydrogenation of nitrogen to ammonia. Hydrogenations of organic compounds such as olefins, acetylenes (hydrocarbons with a triple bond between two adjacent carbon atoms) and ketones (compounds with a terminal oxygen atom doubly bonded to a carbon atom) are usually much easier and occur on a much wider variety of catalysts.

The hydrogenation of dimethylacetylene yields, depending on the catalyst, a number of interesting and useful products [*see top illustration on page 184*]. Dimethylacetylene (CH₃CCCH₃) is a chain of four carbon atoms with the central two carbons joined by a triple bond. If the hydrogenation is carried out at room temperature on nickel or platinum catalysts, the initial products are a mixture of butane (the fully saturated hydrocarbon C₄H₁₀) and two isomers of a four-carbon olefin: *trans*-2-butene and *cis*-2-butene, with *cis*-2-butene predominating. The designation "*trans*" means "located across from each other"; "*cis*" means "located on the same side." Here the terms refer to the location of the two methyl groups with respect to the double bond in the butene molecule.

On a palladium catalyst the initial product is almost exclusively *cis*-2-butene. In studying the mechanism of the reaction, chemists have used deuterium as the hydrogenating agent. By analyzing the product with the mass spectrograph one can separate the product into fractions that differ in the number of deuterium atoms incorporated in each butene molecule. By means of nuclear magnetic-resonance spectroscopy one can then determine the exact location in the molecule of the deuterium atom or atoms.

It turns out that as long as some dimethylacetylene remains unreacted, the initial product consists almost entirely of *cis*-2-butene bearing one deuterium atom on each of its two central carbon atoms. This is what one would expect if dimethylacetylene is chemically adsorbed on the surface of the catalyst by a breaking of one of the bonds between its central carbons [*see second illustration from top on page 184*].

Once all the dimethylacetylene has reacted, *cis*-2-butene begins to be hydrogenated to butane and isomerized to *trans*-2-butene. Actually isomerization is more rapid than hydrogenation. There is considerable evidence that the isomerization involves a mechanism first proposed by Juro Horiuchi and Michael Polanyi. Presumably as the *cis*-2-butene species is held to the catalyst by a single bond one of the two methyl groups rotates to an alternate position, after which a hydrogen replaces the catalytic bond and the molecule—now *trans*-2-butene—leaves the surface [*see third illustration from top on page 184*]. To account for the formation of still another isomer, 1-butene, one can postulate a mechanism that simply shifts the double bond in *cis*-2-butene from its location in the middle of the molecule to a new location between the first and second carbon atoms [*see bottom illustration on page 184*].

Studies with deuterium and a wide variety of hydrocarbons of different

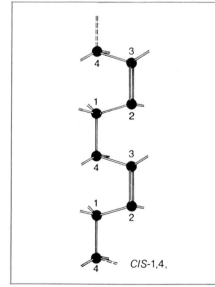

FOUR POLYBUTADIENES, each precisely constructed, can be made from 1,3-butadiene (H₂C : CHCH : CH₂). The designation "1,3" indicates that the first and third carbon atoms in the molecule are joined to the next carbon in sequence by a double bond. If both double bonds open up and take part in

structure have led to a rather good understanding of the nature of the intermediates formed in the reactions of hydrocarbons on metallic catalysts. Relatively little is known, however, about the nature of the chemical bond between metal and carbon, about the exact location of the adsorbed species on the surface of the catalyst and about the origin of the differences in behavior of different metals that act as catalysts.

It will have been noted that all the arguments given above for the surface intermediates were inferential. One might hope to examine the catalyst during reaction by infrared spectroscopy and from the observed infrared spectra deduce what species are present on the surface. Some progress has in fact been made in such a program, but there is a difficulty. Much of the surface of most metals is occupied during hydrogenation by relatively slow-reacting species. That is, most of the surface species are not involved in the hydrogenation but the infrared spectroscope responds to the whole set of adsorbed species. It has not been easy to interpret the results.

More clear-cut results have been obtained with infrared spectroscopy of another type of hydrogenation catalyst, such as the zinc oxide catalyst mentioned above. Chromium oxide also acts as a hydrogenation catalyst. It appears that the key sites for activity on these oxides are coordinatively unsaturated metal ions and oxide ions at the surface.

Heterolytic dissociative adsorption of hydrogen on these sites can be seen by infrared spectroscopy in the case of zinc oxide. The infrared studies indicate the presence of two kinds of adsorbed olefin. Ethylene is rather weakly adsorbed and then reacts with $Zn^{2+}H^-$ [see lower illustration on page 181]. In contrast to the metals, the first reaction has no tendency to reverse. If one starts with ethylene (C_2H_4) and deuterium (D), the product is almost exclusively CH_2DCH_2D. For zinc oxide the deuterium studies provide fairly strong evidence for the illustrated mechanism and the infrared data add additional support. The isotopic distribution over metals is much more complicated.

Evidence from infrared spectroscopy and nuclear magnetic-resonance spectroscopy shows that simple olefins such as ethylene and propylene can be adsorbed on active surface sites in just about all the ways one might imagine. In some cases only a single carbon atom is bonded to the surface; in other cases two carbon atoms are bonded. Sometimes the double bond between the carbons is preserved; in other cases the double bond is opened and provides the link to the surface. It is apparent that quite an organometallic zoo can exist on the surface of metallic catalysts.

The practical aspects of catalysis involve the ability to perform a chemical reaction at a profit. In the simplest commercial installation the reagents and a catalyst are simply placed together in a vessel. The vessel is usually equipped with an indirect steam-heating system and a stirrer. When the reaction is completed, the vessel is emptied and the catalyst is separated from the product and may be reused in the next batch operation. In a continuous operation the reagents are preheated and pumped through a reactor containing the catalyst. The continuous type of operation provides for better control of conditions and ease of handling of product; it also allows operation at higher temperatures and pressures.

Ideally once the catalyst is placed in the reactor it should last indefinitely. In practice catalysts age, they become poisoned, they deteriorate physically and they accumulate reaction products or intermediates that undergo further reaction and finally deactivate the system. The whole idea in continuous operation is to maintain reasonably constant product quality. Usually as the catalyst deactivates, the temperature is raised to maintain a high yield. When these increases reach a limit, the catalyst may either be replaced or subjected to regeneration. For hydrocarbon reactions the regeneration means the removal of carbonaceous deposits. This is normally accomplished by passing a stream of gas with a controlled content of oxygen through the catalyst bed, thus burning off the deposits. While the catalyst is being regenerated or replaced, it is obvi-

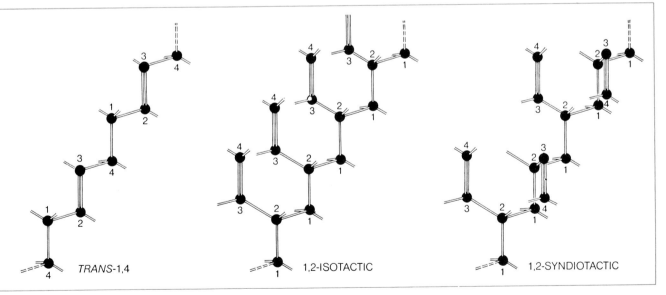

TRANS-1,4   1,2-ISOTACTIC   1,2-SYNDIOTACTIC

the polymerization, the resulting polymer consists of a continuous chain of carbon atoms with a double bond between every other pair of atoms. If the 1–2 and 3–4 single bonds both lie on the same side of the 2–3 double bond, the polymer is termed cis-1,4-polybutadiene. If the single bonds lie on opposite sides, the polymer is trans-1,4-polybutadiene. Here "1,4" indicates that the first and fourth carbon atoms in butadiene link up to form the polymer. If the polymerization involves only the double bond between the first and second carbons in each butadiene molecule, the third and fourth carbon atoms project as a side chain. If the side chains are all on one side of the central chain, the polymer is 1,2-isotactic polybutadiene. If the side chains alternate from side to side, the polymer is 1,2-syndiotactic polybutadiene. Each stereospecific configuration is produced by a distinctive transition-metal catalyst.

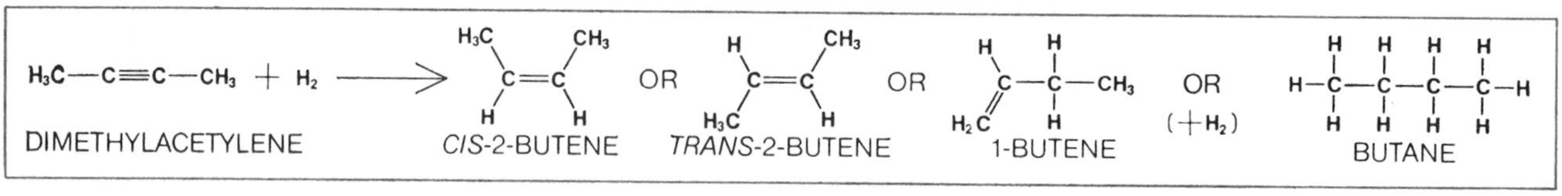

HYDROGENATION OF DIMETHYLACETYLENE can yield three different olefins, cis-2-butene, trans-2-butene and 1-butene, and the saturated hydrocarbon butane. With a palladium catalyst the initial product is chiefly cis-2-butene (see mechanism below).

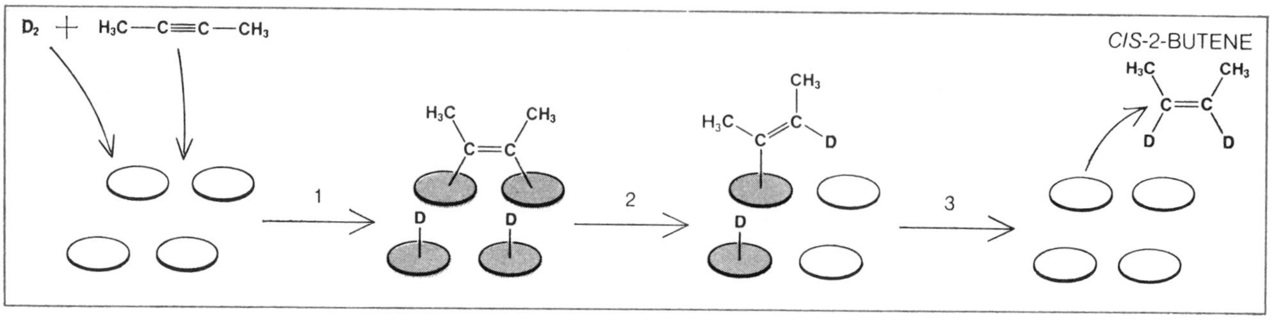

MECHANISM OF HYDROGENATION of dimethylacetylene on palladium has been studied by substituting deuterium (D) for hydrogen. Dimethylacetylene is adsorbed to active sites (color) through its central carbon atoms (1). Two deuterium atoms are then added in sequence (2, 3), yielding cis-2-butene. "Cis" indicates that the two methyl groups are on the same side of the double bond.

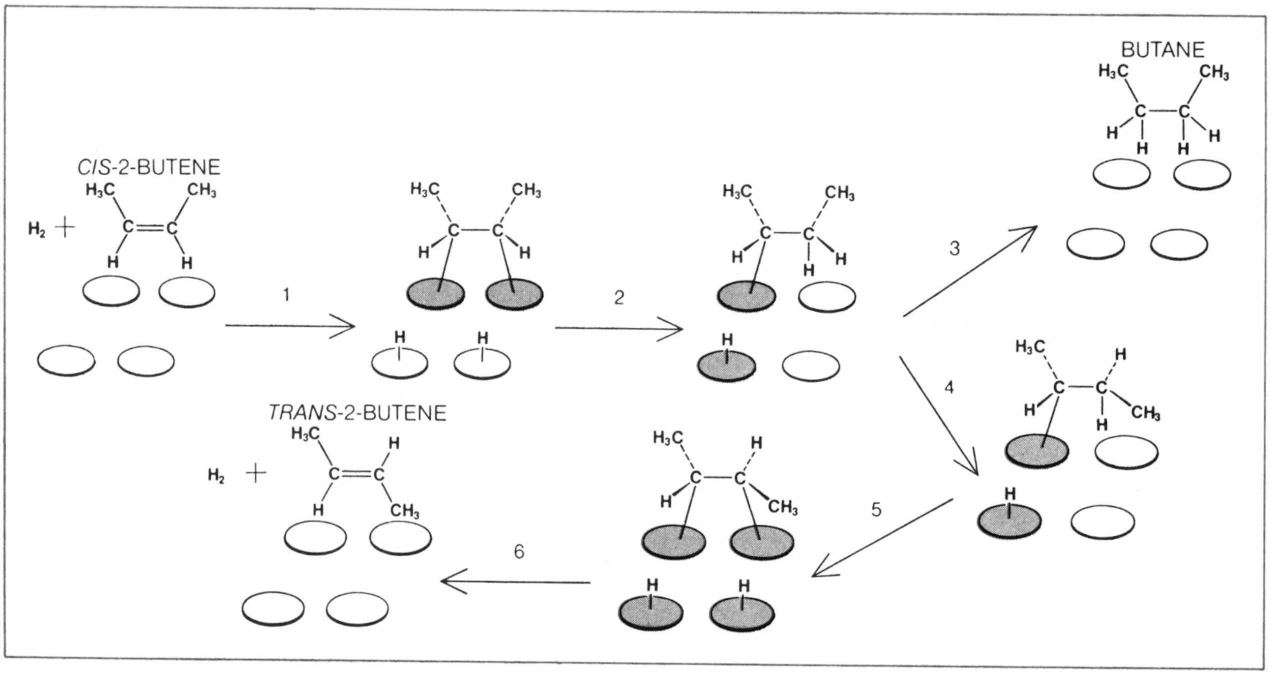

FURTHER HYDROGENATION AND ISOMERIZATION of cis-2-butene can occur once all the dimethylacetylene has reacted. Cis-2-butene is presumably readsorbed and can acquire two more hydro-gens (2, 3) to form butane. Alternatively one methyl group can ro-tate to a new position (4). The hydrocarbon again forms a double bond with the palladium (5) and emerges as trans-2-butene (6).

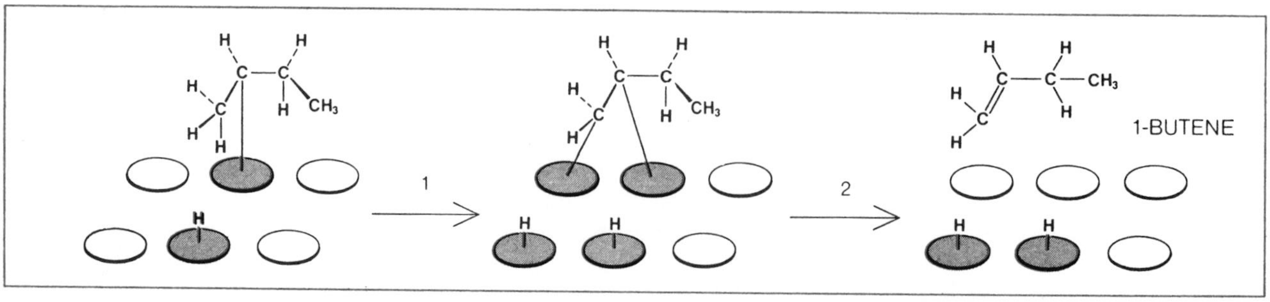

FURTHER ISOMERIZATION converts trans-2-butene to 1-butene. The trans molecule is evidently readsorbed so that its first and sec-ond carbon atoms are linked to active sites (1). On desorption (2) the double bond appears between the first and the second carbon.

ously not making product; ideally, if one has to regenerate, one would prefer to do it continuously and not interfere with the processing of the raw material.

An excellent example of continuous processing and regeneration is the catalytic cracking of oils [*see illustration on next page*]. The catalyst, in finely divided form, is carried into the reactor by the upward flow of vaporized oil, the catalyst-product mixture is withdrawn from the top of the reactor and the catalyst is inertially removed from the product vapors in a cyclone separator. The catalyst is then stripped with steam to remove entrained product and is fed by gravity into a regeneration vessel where it comes in contact with a countercurrent stream of air that burns off the carbonaceous deposits. The clean, hot catalyst is fed through a standpipe and is picked up by the incoming fresh oil. The circulation of catalyst through the reactor and regenerator is very large with respect to the oil processed. Currently more than 6.9 million barrels of oil per day are subjected to catalytic cracking. The catalyst circulation on a daily basis amounts to eight million tons. About .15 pound of catalyst per barrel of oil is added to the unit in order to compensate for loss and withdrawal, thus allowing for the maintenance of equilibrium activity.

Catalysts are prepared in many different ways. A catalyst widely used in catalytic cracking is a composite of silica (SiO_2) and alumina (Al_2O_3), which can be prepared by neutralizing a solution of water glass (Na_2SiO_3) with sulfuric acid to form a gelatinous precipitate that is a polymerized silicic acid [$(H_2SiO_3 \times H_2O)_n$]. This is next treated with aluminum sulfate solution [$Al_2(SO_4)_3$] and ammonia to precipitate aluminum hydroxide [$Al(OH)_3$] on the silicic acid precipitate. The composite is filtered and thoroughly washed to remove all soluble salts. It is then dried in air and calcined at about 500 degrees C. The resulting material contains about 75 percent SiO_2 and 25 percent Al_2O_3 by weight and is believed to have a structure wherein trivalent aluminum atoms are incorporated within a tetrahedral matrix of polymeric silicic acid. Each trivalent aluminum atom thus has to carry a net positive charge to compensate for its enforced tetrahedral coordination.

These positive, or electron-deficient, sites serve as the locus of acidity; in fact, the strength of the acidity thereby created is comparable to the strength of mineral acids such as sulfuric acid. Unlike most mineral acids, the acids created are very stable at high temperatures. The total number of sites is limited and is related to the surface area of the composite. The usual surface area for fresh silica-alumina is about 600 square meters per gram of catalyst. This remarkable surface area is due to a multitude of extremely small channels ranging from five to 100 angstroms in diameter. In service some of the smaller channels collapse, and there is a substantial loss in surface area: down to about 150 square meters per gram. The larger channels survive, however, and the acidity function in these channels is retained.

In general at a given temperature the reaction rate on a heterogeneous catalyst is dependent on two factors: the number of active sites available to the reactant and the turnover rate. The number of active sites is related to the surface area, whereas the turnover rate is a function of the chemistry of the site in relation to the reactant. If the chemistry of the site is such that a reactant stays too long on a site, and during that time is not available to another reactant molecule, the turnover number is low.

Catalytic Reforming

We have pointed out that the gasoline fraction, or virgin naphtha, obtained from crude oil by fractionation needs upgrading to be useful as a motor fuel. Since about 1930 there has been a continuing need to increase the octane number of gasoline. In 1930 the average gasoline had an octane number of about 65; today the average is about 100. To meet this demand new catalytic processes have had to be developed and widely employed.

During the 1950's new platinum-alumina-halogen catalysts were introduced to carry out the catalytic reforming of low-octane oil fractions. The new catalysts performed a dual function by maintaining a proper balance between two catalytic requirements. In general the reactions of catalytic reforming require two types of sites: hydrogenation-dehydrogenation sites (provided by platinum) and acidic sites; catalytic reforming does not occur in the absence of either. The main advantages of the use of this type of catalyst are considerably higher selectivity (higher yield), higher octane number and the ability to run continuously without frequent regeneration.

The chemistry of catalytic reforming using platinum catalysts involves four major reactions. The first is the dehydrogenation of cyclic hydrocarbons, for example converting methylcyclohexane into toluene. The second is dehydrocyclization of paraffins (saturated open-chain hydrocarbons), for example converting normal heptane into toluene [*see bottom illustration on page 179*]. The third is hydrocracking, which means breaking long-chain paraffins into two smaller molecules and adding hydrogen at the sites where the original chain has been broken. The fourth is isomerization, which means converting a straight-chain hydrocarbon into a branched-chain molecule.

All these reactions lead to an octane-number enhancement. For example, the second reaction converts normal heptane, with an octane number of zero, to toluene, which has an octane number in excess of 100. The third reaction can convert decane, for example, with an octane rating of less than zero, into two molecules of isopentane with an octane number of 90. The first and second reactions are strongly endothermic; the third is exothermic; the fourth is essentially thermoneutral. To maintain an overall heat balance one must use multiple reactors with intermediate reheating. Although the overall system produces hydrogen, the process is carried out in the presence of excess hydrogen (recycled from the first and second reactions) at pressures ranging from 150 to 500 pounds per square inch. The excess hydrogen holds down the deposition of carbonaceous material on the catalyst. By this expedient a catalyst can be used continuously for as long as three years. Reaction temperatures are held between 475 and 550 degrees C. Between one volume and three volumes of gasoline are processed per volume of catalyst per hour. The catalyst contains between .3 and .8 percent platinum and up to about 1 percent chlorine or fluorine on a base of highly porous alumina. During the lifetime of the catalyst each platinum atom leads to the reaction of some 20 million molecules of gasoline, a truly catalytic act. The function of the halogen is to induce acid reactions, which convert saturated hydrocarbon rings into benzene (unsaturated) rings.

The formation of benzene-ring, or aromatic, compounds from paraffins (dehydrocyclization) is extremely complex. Some 10 to 15 intermediate steps are postulated. They require the presence of both platinum and acid sites, and adsorbed hydrocarbons evidently migrate from one site to another. In spite of this complexity the efficiency of conversion from the paraffin to an equilibrium mixture of aromatics is of the order of 90 percent. Even higher efficiencies are obtained with the newly developed platinum-rhenium-halogen-alumina catalysts. Since its introduction in 1949, catalytic reforming, using dual-function plati-

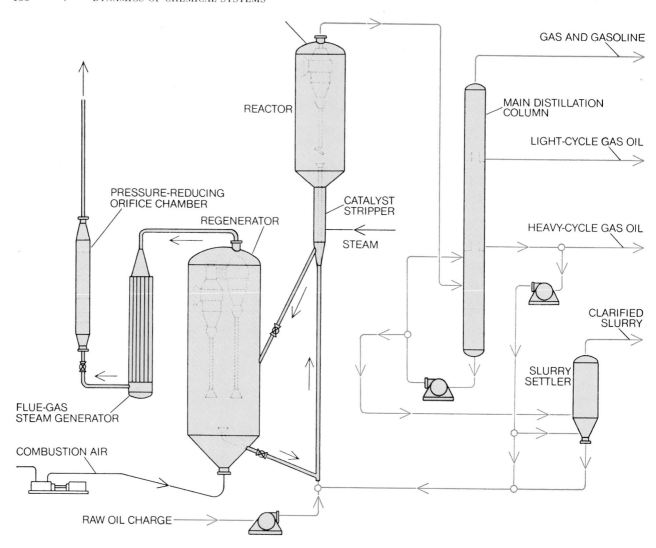

FLUID CATALYTIC CRACKING PROCESS was developed in the late 1930's. The catalyst is in the form of fine grains that behave in the reactor much like a fluid when agitated from below by vaporized oil. The large object depicted in gray inside the reactor is a cyclone separator that removes the catalyst from the reactor product. The reactor product passes to a distillation column where fractions of different volatility are separated. Part of the heaviest fraction is recycled to the reactor. A portion of the catalyst is continuously transferred by gravity from the reactor to the regenerator, where it again behaves like a fluid when it is subjected to a flow of air that burns off carbonaceous deposits. The reactivated catalyst is returned continuously to the reactor. The hot air leaving the regenerator is used to make steam. The catalytic cracker illustrated here was designed by the Universal Oil Products Company.

num-containing catalysts, has grown in importance. Today some 6.8 million barrels per day are processed by this catalytic system in the U.S., western Europe and Japan.

The Future of Catalysis

Although it has attained tremendous practical importance, heterogeneous catalysis has attracted few workers in the chemistry departments of American universities. One may suspect that the need for new areas of research will one day lead organic chemists to extend their interest to heterogeneous catalysis, but only a handful have taken the step so far. In recent years, however, a number of aspects of heterogeneous catalysis have received increased attention in American departments of chemical engineering. There are several other countries where the number of university chemists interested in heterogeneous catalysis is relatively large: Japan, the Netherlands and the U.S.S.R. in particular. It will be interesting to see if this university activity is followed by important technological developments.

One must admit that in spite of the considerable progress in the technological and scientific aspects of catalysis, only experiment can establish whether or not a specific chemical material will make a suitable catalyst for a particular reaction. For many thermodynamically possible reactions no catalysts are known, and many of the catalysts we know were discovered by accident or by a very extensive screening program. The discovery of new promoters has been little helped by theory. The added complexity of the surface makes heterogeneous catalysis an interdisciplinary subject. Both surface chemistry and surface physics must, so to speak, precede heterogeneous catalysis. Therefore a better understanding of many aspects of heterogeneous catalysis must await further advances in the chemistry and physics of surfaces.

New tools are being developed that will enable us to look more closely at the surface of the catalyst and thus possibly gain some additional understanding. The interdisciplinary nature of the study of catalysis, involving as it does both surface chemistry and surface physics, will also require close cooperation with other sciences.

17

WHY THE SEA IS SALT

FERREN MACINTYRE

November 1970

The sea contains more than 70 elements in addition to sodium and chlorine. The global cycles that remove and replenish them involve rainfall, volcanoes and the spreading of the ocean floor

According to an old Norse folktale the sea is salt because somewhere at the bottom of the ocean a magic salt mill is steadily grinding away. The tale is perfectly true. Only the details need to be worked out. The "mill," as it is visualized in current geophysical theory, is the "mid-ocean" rift that meanders for 40,000 miles through all the major ocean basins. Fresh basalt flows up into the rift from the earth's plastic mantle in regions where the sea floor is spreading apart at the rate of several centimeters per year. Accompanying this mantle rock is "juvenile" water—water never before in the liquid phase—containing in solution many of the components of seawater, including chlorine, bromine, iodine, carbon, boron, nitrogen and various trace elements. Additional juvenile water, equally salty but of somewhat different composition, is released by volcanoes that rim certain continental margins, such as those bordering the Pacific, where the sea floor seems to be disappearing into deep trenches [*see illustration on following two pages*].

The elements most abundant in juvenile water are precisely those that cannot be accounted for if the solids dissolved in the sea were simply those provided by the weathering of rocks on the earth's surface. The "missing" elements, such as chlorine, bromine and iodine, were once called "excess volatiles" and were attributed solely to volcanic emanations. It is now recognized that juvenile water may have nearly the same chlorinity as seawater but is much more acid due to the presence of one hydrogen ion (H^+) for every chloride ion (Cl^-). In due course, as I shall explain later, the hydrogen ions are removed and replaced by sodium ions (Na^+), yielding the concentration of ordinary salt (NaCl) that con-

stitutes 90-odd percent of all the "salt" in the sea.

The chemistry of the sea is largely the chemistry of obscure reactions at extreme

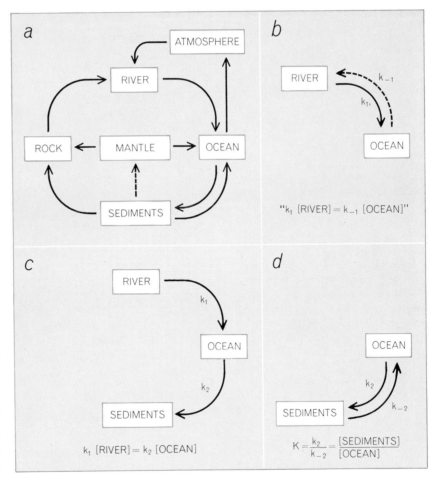

GRAND GEOCHEMICAL CYCLE (*a*) summarizes the global pathways taken sooner or later by the three-score elements that pass through the ocean and maintain its saltiness. The three "thalassochemical" models (*b*, *c*, *d*) abstracted from it are more helpful when trying to understand the rate laws governing the transport of specific elements. The rate constants, k, are expressed as a fraction: one over some number of years. The brackets enclose concentrations of the element being studied, specified according to its environment. The "cyclic" model (*b*) accounts for 90 percent of the chloride in river water. Its rate law is in quotation marks because extra factors, such as the area of the ocean, must be incorporated. The "steady state" model (*c*) works well for reactive trace metals; the reciprocal of k_2 is simply the residence time in the ocean. The "equilibrium" model (*d*) seems the most appropriate for the hydrogen ion (H^+) and the ions of the major metals, such as sodium.

dilution in a strong salt solution, where all the classical chemist's "distilled water" theories and procedures break down. The father of oceanographic chemistry was Robert Boyle, who demonstrated in the 1670's that fresh waters on the way to the sea carry small amounts of salt with them. He also made the first attempt to quantify saltiness by drying seawater and weighing the residue, but his results were erratic because some of the constituents of sea salt are volatile. Boyle found that a better method was simply to measure the specific gravity of seawater and from this estimate the amount of salt present. Since the distribution of density in the sea is important to oceanographers, the same calculation is routinely performed today in reverse: the salinity is deduced by measuring the electrical conductivity of a sample of seawater, and from this and the original temperature of the sample one can compute the density of the seawater at the point the sample was taken.

In 1715 Edmund Halley suggested that the age of the ocean and thus of the world might be estimated from the rate of salt transport by rivers. When this proposal was finally acted on by John Joly in 1899, it gave an age of some 90 million years. The quantity that Joly measured (total amount of x in ocean divided by annual river input of x) is now recognized as the "residence time" of the constituent x, which is an index of an element's relative chemical activity in the ocean. Joly's value is about right for the residence time of sodium; for a more reactive element (in the ocean environment) such as aluminum the residence time is as brief as 100 years.

Not quite 200 years ago Antoine Laurent Lavoisier conducted the first analysis of seawater by evaporating it slowly and obtaining a series of compounds by fractional crystallization. The first compound to settle out is calcium carbonate ($CaCO_3$), followed by gypsum ($CaSO_4 \cdot 2H_2O$), common salt ($NaCl$), Glauber's salt ($Na_2SO_4 \cdot 10H_2O$), Epsom salts ($MgSO_4 \cdot 7H_2O$) and finally the chlorides of calcium ($CaCl_2$) and magnesium ($MgCl_2$). Lavoisier noted that slight changes in experimental conditions gave rise to large shifts in the relative amounts of the various salts crystallized. (In fact, some 54 salts, double salts and hydrated salts can be obtained by evaporating seawater.) To get reproducible results for even the total weight of salt one must remove all organic matter, convert bromides and iodides to chlorides, and carbonates to oxides, before evaporating. The resulting weight, in grams

of salt per kilogram of seawater, is the salinity, $S\%$. (The symbol $\%$ is read "per mil.")

In actual practice the total weight of salt in seawater is nowadays never determined. Instead the amount of chloride ion is carefully measured and a total for all other ions is computed by applying the "constancy of relative proportions." This concept dates back to the middle of the 19th century, when John Murray eliminated confusion about the multiplicity of salts by observing that individual ions are the important thing to talk about when analyzing seawater. Independently A. M. Marcet concluded from many measurements that various ions in the world ocean were present in nearly constant proportions, and that only the absolute amount of salt was variable. This constancy of relative proportions was confirmed by Johann Forchhammer and again more thoroughly by Wilhelm Dittmar's analysis of 77 samples of seawater collected by H.M.S. *Challenger* on the first worldwide oceanographic cruise. These 77 samples are probably the last ever analyzed for all the major constituents. Their average salinity was close to 35%, with a normal variation of only $\pm 2\%$.

In the 86 years since Dittmar reported eight elements, 65 more elements have been detected in seawater. It was recognized more than a century ago that elements present in minute amounts in seawater might be concentrated by sea organisms and thereby raised to the threshold of detectability. Iodine, for example, was discovered in algae 14 years before it was found in seawater. Subsequently barium, cobalt, copper, lead, nickel, silver and zinc were all detected first in sea organisms. More recently the isotope silicon 32, apparently produced by the cosmic ray bombardment of argon, has been discovered in marine sponges.

There are also inorganic processes in the ocean that concentrate trace elements. Manganese nodules (of which more below) are able to concentrate elements such as thallium and platinum to detectable levels. The cosmic ray isotope beryllium 10 was recently discovered in a marine clay that concentrates beryllium. In all, 73 elements (including 13 of the rare-earth group) apart from hydrogen and oxygen have now been detected directly in seawater [see *illustration on page 191*].

It is only in the past 40 years that geochemists have become interested in the chemical processes of the sea for what they can tell us about the history

of the earth. Conversely, only as geophysicists have pieced together a comprehensive picture of the earth's history has it been possible to bring order into marine chemistry.

The earth's present atmosphere and ocean are not primordial but have been liberated from chemical and mechanical entrapment in solid rock. Perhaps four billion years ago, or a little less, there was (according to many geophysicists) a "grand catastrophe" in which the earth's core, mantle, crust, ocean and atmosphere were differentiated from an original homogeneous accumulation of material. Estimates of water released during the catastrophe range from a third to 90 percent of the present volume of the ocean. The catastrophe is not finished even yet, since differentiation of the mantle continues in regions of volcanic activity. Most of the exhalations of volcanoes and hot springs are simply recycled ground water, but if only half of 1 percent of the water released is juvenile, the present production rate is sufficient to have filled the entire ocean in four billion years.

There is evidence that the salinity of the ocean has not changed greatly since

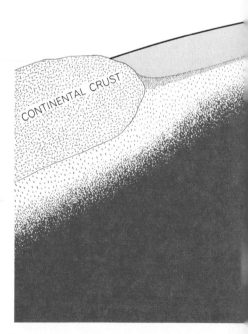

MAGIC SALT MILL at the bottom of the sea, imagined in the old Norse folktale, turns out to be not so fanciful after all. The modern explanation of why the sea is salt invokes the concept of the "mid-ocean" rift and sea-floor spreading, as depicted here in cross section. The rift is a weak point be-

the ocean was formed; in any event the salinity has been nearly constant for the past 200 million years (5 percent of geologic time). The composition of ancient sediments suggests that the ratio of sodium to potassium in seawater has risen from about 1 : 1 to its present value of about 28 : 1. Over the same period the ratio of magnesium to calcium has risen from roughly 1 : 1 to 3 : 1 as organisms removed calcium by building shells of calcium carbonate. It is significant, however, that the total amount of each pair of ions varied much less than the relative amounts.

If we look at rain as it reaches the sea in rivers, we find a distinctly nonmarine mix to its ions. If we catch it even earlier as it tumbles down young mountains, the differences are even more pronounced. This continual input of water of nonmarine composition would eventually overwhelm the original composition of the ocean unless there were corrective reactions at work.

The overall geochemical cycle that keeps the marine ions closely in balance involves a complex interchange of material over decades, centuries and millenniums among the atmosphere, the ocean, the rivers, the crustal rocks, the oceanic sediments and ultimately the mantle [see "a" in illustration on page 187]. Because this overall picture is too general to be of much use, we abstract bits from it and call them thalassochemical models (thalassa is the Greek word for "sea"). One model involves simply the cyclic exchange of sea salt between the rivers and the sea; the cycle includes the transport of salt from the sea surface into the atmosphere, where salt particles act as condensation nuclei on which raindrops grow [see "b" in illustration on page 187]. This process accounts for more than 90 percent of the chloride and about 50 percent of the sodium carried to the sea by rivers.

Another useful abstraction is the "steady state" thalassochemical model. If the ocean composition does not change with time, it must be rigorously true that whatever is added by the rivers must be precipitated in marine sediments [see "c" in illustration on page 187]. Oceanic residence times computed from sedimentation rates, particularly for reactive trace metals, agree well with the input rates from rivers. Unfortunately residence times do not reveal the mechanism by which an element is removed from seawater. For residence times greater than a million years it is often helpful to invoke the "equilibrium" model, which deals only with the rate of exchange between the ocean and its sediments [see "d" in illustration on page 187].

To understand how the earth maintains its geochemical poise over a billion-year time scale we must return to the circle of arrows—the weathering and "unweathering" processes—of the geochemical cycle. This circle starts with primordial igneous rock, squeezed from the mantle. Ignoring relatively minor heavy metals such as iron, we can assume that the rock consists of aluminum, silicon and oxygen combined with the alkali metals: potassium, sodium and calcium. The resulting minerals are feldspars (for example $KAlSi_3O_8$). Rainwater picks up carbon dioxide from the air and falls on the feldspar. The reaction of water, carbon dioxide and feldspar typically yields a solution of alkali ions and bicarbonate ions (HCO_3^-) in which is suspended hydrated silica (SiO_2). The residual detrital aluminosilicate can be approximated by the clay kaolinite:

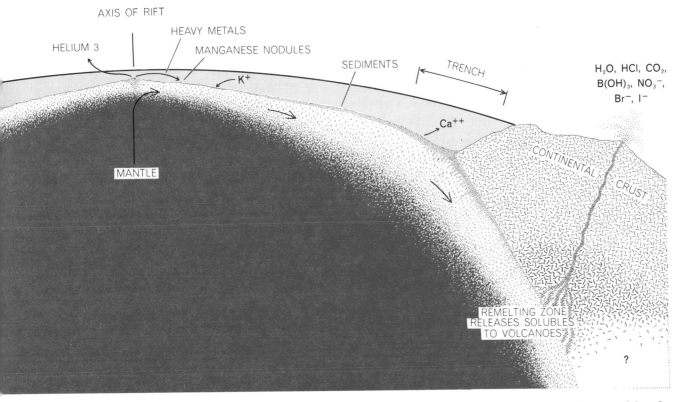

AXIS OF RIFT

HELIUM 3   HEAVY METALS

MANGANESE NODULES

K^+

SEDIMENTS   TRENCH

H_2O, HCl, CO_2,
$B(OH)_3$, NO_3^-,
Br^-, I^-

MANTLE

Ca^{++}

CONTINENTAL CRUST

REMELTING ZONE RELEASES SOLUBLES TO VOLCANOES

?

tween rigid plates, or segments, in the earth's crust. Although the driving mechanism is not yet understood, the plates move apart a few centimeters a year as fresh basalt from the plastic mantle flows up between them. The new basalt releases "juvenile" water (water never before in liquid form) and a variety of elements, including heavy metals that become incorporated in manganese nodules and the rare isotope helium 3, which escapes finally into space. At the continental margin (right) the lithospheric plate is subducted, forming a trench and carrying accumulated sediments with it. (The plate apparently thickens en route as plastic basalt "freezes" to its underside.) As it descends the plate remelts and releases soluble elements and ions that are ejected into the atmosphere by volcanoes. They maintain the saltiness of the sea and together with weathered crustal rock, such as granite, provide the stuff of sediments.

$Al_2Si_2O_5(OH)_4$ [*see Step 1 in illustration on page 194*]. A mountain stream carries off the ions and the silica. The kaolinite fraction lags behind, first as a friable surface on weathering rock, then as soil material and finally as alluvial deposits in river valleys. If the stream evaporates in a closed basin, such as one finds in the western U.S., the result is a "soda lake" containing high concentrations of carbonates and amorphous silica.

In mature river systems the kaolinite fraction reaches the sea as suspended sediment. Encountering an ion-rich environment for the first time, the aluminosilicate must reorganize itself into new minerals. One such mineral, which seems to be forming in the ocean today, is the potassium-containing clay illite [*see Step 2 in illustration on page 194*]. These "clay cation" reactions may take decades or centuries. They are poorly understood because graduate students who study them invariably leave before the reactions are complete. The net effect of such reactions is to tie up and remove some of the potassium and bicarbonate ions, along with aluminum, silicon and oxygen.

A biologically important reaction, usually confined to shallow water, allows marine organisms to build shells of calcium carbonate, which precipitates when calcium (Ca^{++}) and bicarbonate ions react. If dilute hydrochloric acid is present (it is released by volcanoes), it reacts even more rapidly with bicarbonate, forming water and carbon dioxide and leaving free the chloride ion. When marine organisms die and sink to about 4,000 meters, they cross the "lysocline," below which calcium carbonate redissolves because of the high pressure. We have now traced the three metallic ions removed from igneous rock to three separate niches in the ocean. Sodium remains dissolved, potassium precipitates in clays on the deep-sea floor and calcium precipitates in shallow water as biogenic limestone: coral reefs and calcareous oozes.

Ages pass and the geochemical cycle rolls on, converting ocean-bottom clay into hard rock such as granite. When old sea floor finally reaches a region of high pressure and temperature under a continental block, it still contains some free ions that can react with the clay to reconstitute hard rock. A score of reaction schemes are possible. In Step 3 in the illustration on page 194 I have chosen to build a "granite" from equal parts of potassium feldspar, sodium feldspar, potassium mica and quartz. (Notice that calcium is missing because it has dis-

solved from the sediments during their descent into the deep-ocean trenches that carry the sediments under the continental blocks.) The reaction written in Step 3 uses up all the silica formed in Step 1.

The goal of this geochemical exercise has now been reached. First, we have shown that of all the substances that enter the ocean, only sodium and chlorine remain abundantly in solution. Of the other elements, the amount remaining in solution is less than a hundredth of the amount delivered to the ocean and precipitated from it. Second, we have made a start at explaining the observed sodium-potassium ratios: in basalt this ratio is about 1 : 1, in seawater 28 : 1 and in granite 1 : 1.2. If the weight of sodium tied up in granite were about 140 times as great as the weight of sodium dissolved in the sea, the slight excess of potassium over sodium in granite would explain the sea's deficiency in potassium.

We now have working models for thinking about the circulation of the major elements, but we have barely scratched the true complexity and subtlety of seawater. The sources and sinks of the minor elements are now being explored. In many cases we can only guess at what the natural marine form of an element is because our detection techniques either convert all forms to a common form for analysis or miss some forms completely. Moreover, certain ions seem to behave capriciously in the ocean. For example, at the *pH* (hydrogen-ion concentration) of seawater, vanadium should appear as $VO_2(OH)_3^{--}$, an ion with a double negative charge; instead it seems to exist in positively charged form, perhaps as VO_2^+.

Much of what is known about elements in the sea can be summarized in an oceanographer's periodic table [*see illustration on page 195*]. The usefulness of the usual kind of periodic table to the chemist is that it arranges chemically similar elements in vertical columns and presents behavioral trends in horizontal rows. The oceanographer's table shows how these regularities are disrupted in the ocean environment.

First of all, more than a dozen elements have never been detected in seawater, although two of them (palladium and iridium) exist in parts per billion in marine sediments and another (platinum) is present in manganese nodules. The second interesting feature of the oceanographer's table is the tendency for the "upper" and "outer" elements, those in the raised wings, so to speak, to

be the most plentiful in the sea. The "upper" tendency simply reflects the greater cosmic abundance of light elements. (Lithium, beryllium and boron, however, are fairly scarce even cosmically.)

The "outer" trend can be explained in quantum-mechanical terms by the presence or absence of electrons in *d* orbitals, the electron shells principally involved in forming complexes. Elements in the first three columns at the left have no *d* orbitals; those in the last four columns at the right have full *d* orbitals. Both characteristics favor weak chemical bonds, with the result that these two groups of elements tend to ionize readily and remain in solution, either by themselves or in simple combination with oxygen and hydrogen. In contrast, the elements in the center of the table with partially filled *d* orbitals form strong chemical bonds and compounds that precipitate readily; thus they can exist only at low concentration in solution. For silver and the surrounding group of metals the most stable complexes are formed with the most abundant seawater ion: chloride. Most of the other elements that are hungry for *d* electrons form their complexes with oxygen, or oxygen plus some protons (hydrogen nuclei).

Ordinarily the oxidation state of metals

◼️ CURRENTLY RECOVERED FROM SEAWATER

▭ ELEMENTS IN SHORT SUPPLY

▮▮▮▮ RANGE OF BIOLOGICALLY CAUSED CHANGE

▥▥▥ RANGE OF ANALYSES

● METALS CONCENTRATED
● IN MANGANESE NODULES

COMPOSITION OF SEAWATER has been a challenge to chemists since Antoine Laurent Lavoisier made the first analyses. The logarithmic chart on the opposite page shows in moles per kilogram the concentration of 40 of the 73 elements that have been identified in seawater. A mole is equivalent to the element's atomic weight in grams; thus a mole of chlorine is 35 grams, a mole of uranium 238 grams. Only four elements are now recovered from the sea commercially: chlorine, sodium, magnesium and bromine. Recovery of other scarce elements is not promising unless biological concentrating techniques can be developed. Manganese nodules are a potential source of scarce metals but gathering them from the deep-sea floor may not be profitable in this centur

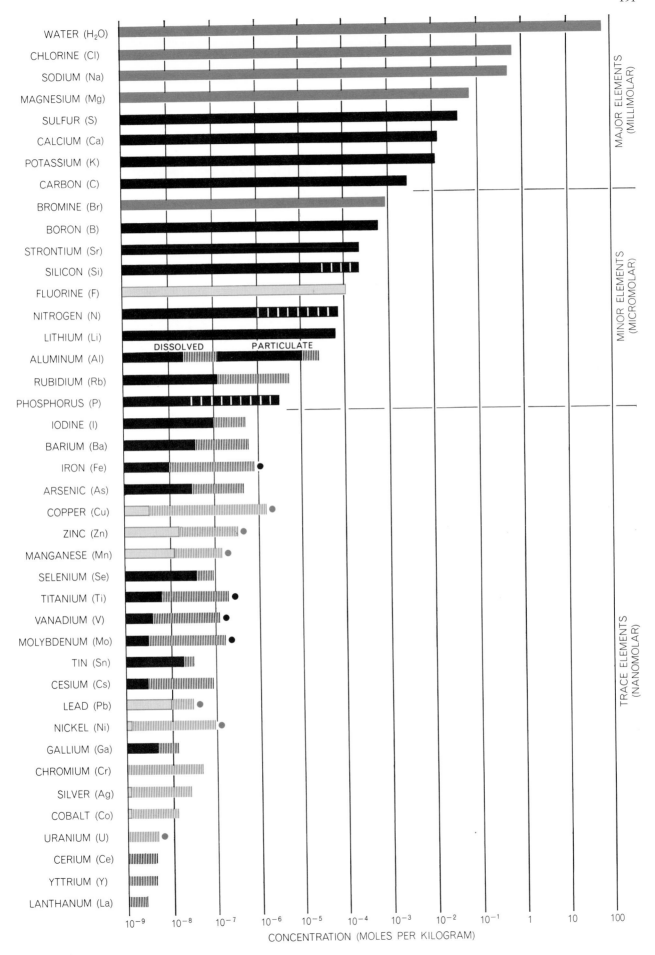

CONCENTRATION (MOLES PER KILOGRAM)

avid for d electrons would be determined by the oxidation potential of seawater, which is a measure of its ability to extract electrons from a substance just as its $p\mathrm{H}$ is a measure of its ability to extract protons. The oxidation potential of seawater has the high value of .75 volt, enabling it to extract the maximum possible number of electrons from nearly all elements except the noble metals (platinum group) and the halogens (fluorine family).

Surprisingly, however, the oxidation potential of seawater does not seem to control the oxidation states of many metals that have partially filled d shells. One reason is that most reactions proceed by a mechanism in which only a

single electron is transferred at a time. Such transfers occur most readily when the reactants are adsorbed on surfaces where atomic geometry and electric-charge distribution are able to expedite the redistribution of electrons (hence the utility of catalysts, which provide such surfaces). But surfaces of any kind are few and far between in the ocean, and (with the exception of manganese nodules) those that do exist are poor catalysts. A second reason for the failure of the sea's oxidation potential to control valence states is that organisms sometimes excrete electron-rich substances, which then remain in that reduced state in spite of seawater's apparent capacity to oxidize them.

Manganese nodules are porous chunks of metallic oxides up to several centimeters in diameter, widely distributed over the ocean floor. They evidently exist because they are autocatalytic for the reaction that produces them. Because of their porous structure, nodules have a surface area of as much as 100 square meters per gram. The autocatalytic property seems to extend to an entire suite of metals that coprecipitate with manganese: iron, cobalt, nickel, copper, zinc, chromium, vanadium, tungsten and lead. Nodules found on the flanks of oceanic ridges contain significant concentrations of metals, such as nickel, that are scarce in seawater itself. This suggests that the nodules are col-

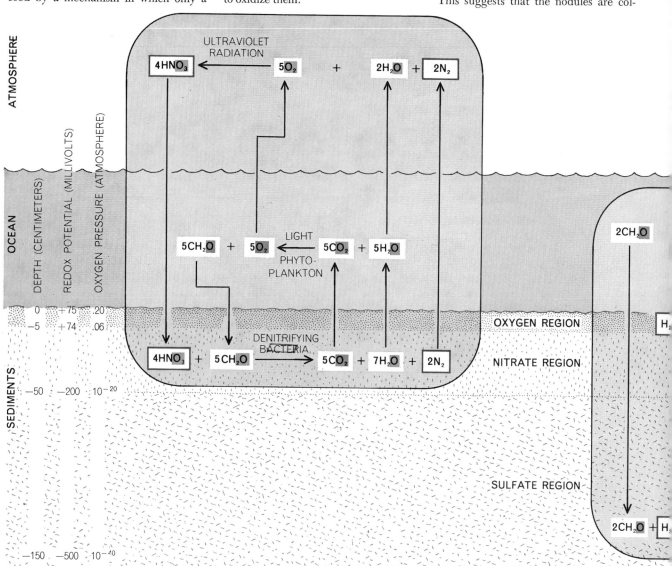

BACTERIA IN MARINE SEDIMENTS, although scarce by terrestrial soil standards, play a major role in replenishing the oxygen of the atmosphere and in limiting the accumulation of organic sediments. The bacteria concerned are buried in fine-grained sediments from several centimeters to several tens of centimeters below the ocean floor, with limited access to free oxygen for respiration.

Thus deprived, they use the oxygen in nitrates and sulfates to oxidize organic compounds, represented by CH_2O. The actual reactions are far more complex than indicated here. The net result, however, is that denitrifying bacteria (*left*) release free nitrogen and convert carbon to a form (carbon dioxide) that can be reutilized by phytoplankton. These organisms, in turn, release free oxy-

lecting juvenile metals as the metals leak from the mantle at the fissure of the ridge. One would like to know why the nodule metals are present in oxide form rather than, as one would expect, in carbonate form.

The level of the discussion so far might best be called thalassopoetry. The discussion can be made more serious in two ways. One approach—the "geochemical balance"—has employed a computer to follow in detail as many as 60 elements as they move through the geochemical cycle, from igneous rock back to metamorphosed sediments. In the second approach the actual chemistry of each element is followed by applying the

thermodynamic methods of Josiah Willard Gibbs to systems regarded as being near equilibrium. This effort was launched by Lars Gunnar Sillén of Sweden and has been pursued by Robert M. Garrels of Northwestern University and by Heinrich D. Holland of Princeton University.

Of course no chemist in his right mind would talk seriously about equilibria in a system of variable temperature, pressure and composition that was poorly stirred, had variable inputs and contained living creatures. On the other hand, the observed uniformity of the ocean and the long periods available for reacting suggest that at least the major components are sufficiently close to equilibrium to make an investigation worthwhile. (We *know* the minor constituents are not in equilibrium.)

The equilibrium approach is based on Gibbs's phase rule, which states that the number of phases (P) possible in a system of C components at equilibrium is given by the equation $P = C + 2 - F$, where F is the number of "degrees of freedom," or quantities that may be independently varied without changing the number of phases or their composition (although F may change their relative proportions). The 2 enters the equation because only two variables, temperature and pressure, are important in most chemical reactions.

One of Sillén's most comprehensive ocean models has nine components: water, hydrochloric acid, silica, three hydroxides (aluminum, sodium and potassium), carbon dioxide and the oxides of magnesium and calcium. Observation of sea-floor sediments, aided by laboratory studies, suggests that a nine-phase ocean will result [see illustrations on page 196]. If C and P both equal nine, the phase rule states that the number of degrees of freedom (F) must equal two. Logically these are temperature (which can vary over the oceanic range from −2 degrees Celsius to 30 degrees) and the chloride ion concentration (which can shift over the normal oceanic range without changing the composition of the stable phases).

A diagrammatic view of how the nine components sort themselves into phases is shown in the bottom illustration on page 196. Note that the liquid phase contains ions not listed either as components or as phases (for example H^+ and OH^-). Thermodynamics need not consider them explicitly because they do not vary independently; their concentrations are fixed by the equilibrium constants that connect the observed phases.

Thus $H_2O = H^+ + OH^-$. Moreover, one knows that the product of H^+ and OH^- is a thermodynamic constant, which equals 10^{-14} mole per liter. Similar relations tie the entire system into a comprehensible whole, so that when all the calculations are performed one has discovered the equilibrium concentrations of five cations (H^+, Na^+, K^+, Mg^{++} and Ca^{++}) and four anions (Cl^-, OH^-, HCO_3^- and CO_3^{--}).

It may seem peculiar to discuss an "atmosphere" containing only water vapor and carbon dioxide. One could easily add oxygen and nitrogen to the list of components. Since they would add no new phases, they would raise the number of degrees of freedom from two to four ($9 = 11 + 2 - 4$). The two new F's would be the total atmospheric pressure and the ratio of oxygen to nitrogen. In the study of the ocean, however, the partial pressure due to carbon dioxide is more significant than the total pressure of the atmosphere. Moreover, the presence of gaseous oxygen and nitrogen has little importance for the inorganic environment of the ocean, so that it is simpler to omit them and just as "real."

Suppose now we perturb the equilibrium of the model ocean by assuming that a submerged volcano has suddenly released enough hydrochloric acid (HCl) to double the amount of chloride ion (Cl^-). The dissociation of hydrochloric acid releases enough H^+ ions to raise the total number of hydrogen ions in the ocean from the former equilibrium value of 10^{-8} mole per liter to $10^{+.3}$. This excess of hydrogen ions almost immediately pushes all the available carbonate ions (CO_3^{--}) to bicarbonate ions (HCO_3^-) and the latter to carbonic acid (H_2CO_3). These shifts, however, only slightly depress the pH, which remains high until the slow circulation of the ocean brings the hydrogen ions in direct contact with the clay sediments on the sea floor.

The structure of clay is such that oxygen atoms at the free corners of polyhedrons carry unsatisfied negative charges, which attract positive ions [see top illustration on page 198]. Because the ocean is so rich in sodium ions (Na^+), they occupy most of the corners of clay polyhedrons. When the excess hydrogen ions come in contact with the clay, they quickly replace the sodium ions and set them adrift. This fast reaction is limited in scope because the surface and interlayer ion-exchange capacity of clay is not very great. Much more capacity is provided when the structure of the clay

LIGHT
PHYTO-PLANKTON
$2H_2O$ + $2CO_2$

+ H_2S
ANIC

ATE
ERIA

H_2S + $2H_2O$ + $2CO_2$

gen. Without the cooperative effort of these two groups of organisms the oxygen in the atmosphere might all be fixed by high-energy processes within some 10 million years. The sulfate bacteria (*right*) play a role in the recycling of sulfur and oxygen.

STEP 1: WEATHERING OF IGNEOUS ROCK

$$\left\{\begin{array}{c} CaAl_2Si_2O_8 \\ \text{ANORTHITE} \\ 2KAlSi_3O_8 \\ \text{POTASSIUM FELDSPAR} \\ 2NaAlSi_3O_8 \\ \text{SODIUM FELDSPAR} \end{array}\right\} + 9H_2O + 6CO_2 \longrightarrow \left\{\begin{array}{c} Ca^{++} \\ 2K^+ \\ 2Na^+ \\ 6HCO_3^- \end{array}\right\} + 8SiO_2(aq) + 3Al_2Si_2O_5(OH)_4 \\ \text{"KAOLINITE"}$$

| IGNEOUS ROCK | + | RAINWATER | $\longrightarrow$ | STREAM WATER | + | DETRITUS |

STEP 2: EQUILIBRATION IN OCEAN

$$3Al_2Si_2O_5(OH)_4 + 2K^+ + 2HCO_3^- \longrightarrow 2K(AlSiO_4)Al_2(OH)_2O_2(Si_2O_4) + 5H_2O + 2CO_2 \uparrow \text{ (DEEP WATER)}$$

| "KAOLINITE" | + | SEAWATER | $\longrightarrow$ | CLAY (ILLITE) |

$$Ca^{++} + 2HCO_3^- \xrightarrow{\text{ORGANISMS}} CaCO_3 \downarrow + H_2O + CO_2 \uparrow \text{ (SHALLOW WATER)}$$

$$2HCl + 2HCO_3^- \xrightarrow{\text{VULCANISM}} 2Cl^- + 2H_2O + 2CO_2 \uparrow$$

STEP 3: METAMORPHOSIS OF SHALE (CLAY)

$$2K(AlSiO_4)Al_2(OH)_2O_2(Si_2O_4) + Na^+ + Cl^- + 8SiO_2 \xrightarrow[\text{PRESSURE}]{\text{HEAT}} \left\{\begin{array}{c} KAlSi_3O_8 \\ \text{POTASSIUM FELDSPAR} \\ NaAlSi_3O_8 \\ \text{SODIUM FELDSPAR} \\ KAl_2(AlSi_3O_{10})(OH)_2 \\ \text{POTASSIUM MICA} \\ SiO_2 \end{array}\right\} + HCl + 2SiO_2 + AlSi_2O_5(OH)$$

| CLAY | + | INTERSTITIAL WATER | $\longrightarrow$ | "GRANITE" + VOLCANIC GAS + QUARTZ + PYROPHYLLITE |

STEP 4: LEFT BEHIND IN OCEAN

$$Na^+ + Cl^-$$

ONLY SALT REMAINS after the ocean "laboratory" has finished processing the complex of chemicals removed from igneous rock by rainwater containing dissolved carbon dioxide. Step 1 yields a solution of alkali ions and bicarbonate (HCO_3^-) ions in which hydrated silica (SiO_2) and aluminosilicate detritus are suspended. In crystalline form the aluminosilicate would be kaolinite. In the ocean (*Step 2*) the "kaolinite" is complexed with potassium ions (K^+) to form illite clay. Marine organisms use the calcium ion (Ca^{++}) to make calcium carbonate shells, which form sediments in shallow water. Hydrochloric acid (HCl), injected by undersea volcanoes, reacts with bicarbonate ions, returning some carbon dioxide to the atmosphere. In Step 3 clay is metamorphosed into "granite." Sodium chloride (*Step 4*) remains. Although some of this sequence is hypothetical, something very similar seems to take place.

is rearranged; for example, the conversion of montmorillonite to kaolinite also consumes hydrogen atoms and releases sodium. Given sufficient time—centuries —such rearrangements inexorably take place, and the *p*H of the ocean slowly drifts back to its equilibrium value. The charge on the excess chloride introduced by the volcano will then be balanced not by H^+ but by Na^+. This slow equilibration mechanism can be regarded as the ocean's "*p*H-stat" (in analogy with "thermostat"). This clay-cation model suggests that the *p*H of the ocean has been constant over the span of geologic time and that hence the carbon dioxide content of the atmosphere has been held within narrow limits.

If only the *p*H-stat were available for leveling surges in *p*H, the ocean might be subjected to violent local fluctuations. For fast response *p*H control is

taken over by a carbonate buffer system [*see bottom illustration on page 198*]. In fact, until recently oceanographers neglected the clay-cation reactions and assumed that the carbonate-buffer system almost completely determined the *p*H of the ocean.

One might think that if the carbon dioxide content of the atmosphere were to decrease, carbon dioxide would flow from the sea into the atmosphere, leading to a general depletion of all carbonate species in the ocean and eventually to the dissolution of some carbonate sediments. In actuality something quite different happens because the carbonate system is its own source of hydrogen ions. Removal of carbon dioxide from water reduces the concentration of carbonic acid (H_2CO_3), the hydrated form of carbon dioxide. Replacement of this acid from bicarbonate ions requires a hydrogen ion, which can only be ob-

tained by converting another bicarbonate ion to carbonate. The overall reaction is $2HCO_3^- \rightarrow H_2CO_3 + CO_3^{--}$. Thus instead of dissolving existing sediments, removing carbon dioxide from the sea may actually precipitate carbonate. This reaction can be seen in the "whitings" of the sea over the Bahama Banks, where cold deep water, rich in dissolved carbon dioxide and calcium, is forced to the surface and warmed. As carbon dioxide escapes into the air, the *p*H drops and aragonite ($CaCO_3$) precipitates, turning large areas of the ocean white with a myriad of small crystals.

The reaction above conserves charge, which means that the "alkalinity"—the traditional name for the concentration of sodium ion ("alkali") needed to balance this negative charge—is also conserved. The "carbonate alkalinity," defined as the bicarbonate concentration plus twice the carbonate concentration,

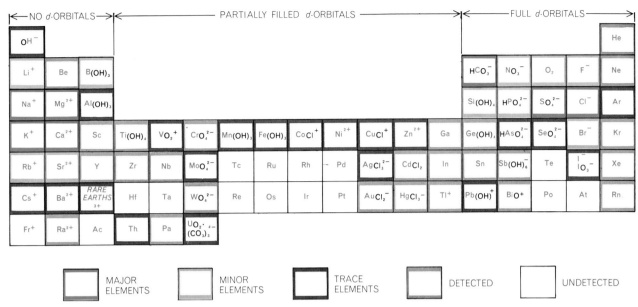

PERIODIC TABLE, as prepared by the "thalassochemist," shows the form in which the detectable elements appear in seawater. In each box the element normally found in that place in the usual periodic table is shown in color; the elements associated with it are in black. Thus carbon appears predominantly as HCO_3^-, arsenic as $HAsO_4^{2-}$ and so on. The superscripts show the number of positive or negative charges carried by each ion. Iodine's two forms, I^- and IO_3^-, are about equally common. Except for the noble gases (*last column at right*), all the elements dissolved in the sea must be present as ions. When an element (other than a noble gas) is shown by itself, without a plus or minus charge, it means that its preferred ionic form in seawater is not yet established.

is useful because it remains fixed even when the relative amounts of the two species vary.

The system can be visualized with the help of the illustration on page 197, which is the "Bjerrum plot" for carbonic acid at constant alkalinity. It takes its name from Niels Bjerrum, who introduced such plots in 1914; it shows the interrelations between the various compounds in the world carbonate system as a function of pH. Although the diagram ignores variations of pressure, temperature and salinity, it displays the essential features of the system.

The Bjerrum plot facilitates a semiquantitative discussion of the relation of atmospheric carbon dioxide to oceanic carbon dioxide. Over the next 20 years we shall burn enough fossil fuel to double the amount of carbon dioxide in the atmosphere from 320 parts per million to 640. On the plot this is indicated by shifting the line A, corresponding to 320 parts per million, to position B, 640 parts per million.

To produce this shift some 2.5×10^{18} grams of carbon dioxide must be added to the atmosphere. If the altered atmosphere were to come to equilibrium with the ocean, the pH of the ocean would drop from its present value of 8.15 to 7.89—still well within the range tolerated by marine organisms. This cannot happen, however, because the total mass of carbon dioxide in the ocean (Σ in the Bjerrum plot) plus the carbon dioxide in the atmosphere would have to increase

from its present value, 128.9×10^{18} grams, to 138.3×10^{18} grams. The difference, 9.4×10^{18} grams, is nearly four times the amount added to the atmosphere.

The long-term equilibration process for such an atmospheric doubling can be broken down into two steps. First the pH-buffer system operates: 2.5×10^{18} grams, or 2 percent of the total mass, is added to the world system at constant alkalinity. The result of this step is the line C in the diagram, corresponding to a total mass of 131.4×10^{18} grams, an atmospheric carbon dioxide content of 390 parts per million and an oceanic pH of 8.08. Next, if the ocean has time to equilibrate with its sediments, the pH-stat will operate, returning the system to pH 8.15 at a constant total mass. The result of this step is that the alkalinity rises by 2 percent, which in terms of the Bjerrum plot means that the system will return to normal except that all the numbers on the concentration axes will be multiplied by 1.02. The long-range effect of a sudden doubling of the atmosphere's carbon dioxide, therefore, is to increase the ultimate value 2 percent, from 320 parts per million to 326, and some of that increase will ultimately find its way into vegetation and humus.

It is obvious that rates are crucial in the global distribution of carbon dioxide. The wind-stirred surface layer of the sea exchanges carbon dioxide rapidly with the atmosphere, requiring less than a decade for equilibration. Because this

layer is only about 100 meters deep it contains only a tiny fraction of the ocean's total volume. Large-scale disposal of atmospheric carbon dioxide therefore requires that the gas be dissolved and transported to deep water.

Such vertical transport takes place almost exclusively in the Weddell Sea off the coast of Antarctica. Every winter, when the Weddell ice shelf freezes, the salt excluded from the newly formed ice increases the salinity and hence the density of the water below. This ice-cold water, capable of containing more dissolved gas than an equal volume of tropical water, cascades gently down the slope of Antarctica to begin a 5,000-year journey northward across the bottom of the ocean. The carbon dioxide in this "antarctic bottom water" has plenty of time to come to equilibrium with clay sediments.

Enough fossil fuel has been burned in the past century to have raised the carbon dioxide content of the atmosphere from about 290 parts per million to 350 parts. Since the actual level is now 320 parts per million, about half of the carbon dioxide put into the air has been removed. Although proof is lacking, a principal removal agent is undoubtedly antarctic bottom water. The process is so slow, however, that the carbon dioxide content of the atmosphere may reach 480 parts per million before the end of the century. By then it should be clear if man's inadvertent global experiment (altering the atmosphere's carbon diox-

COMPONENTS (C)	PHASES (P)	VARIABLES (F)
H_2O	1 GAS	TEMPERATURE
HCl	2 LIQUID	Cl^-
SiO_2	3 QUARTZ (SiO_2)	
$Al(OH)_3$	4 KAOLINITE (t-o CLAY)	
NaOH	5 MONTMORILLONITE (Na-t-o-t CLAY)	
KOH	6 ILLITE (K-t-o-t CLAY)	
MgO	7 CHLORITE (Mg-t-o-t CLAY)	
CO_2	8 CALCITE ($CaCO_3$)	
CaO	9 PHILLIPSITE (Na-K FELDSPAR)	

NINE MAJOR COMPONENTS IN SEA can, to a first approximation, be combined into nine distinctive phases to satisfy the "phase rule" that governs systems in equilibrium. The rule, formulated in the 19th century by Josiah Willard Gibbs, prescribes the number of phases P, components C and degrees of freedom F in such a system: $P = C + 2 - F$. When the number of phases and components are equal, the number of degrees of freedom, F, must be two, which allows both the temperature and the chloride-ion concentration to vary without altering the number of phases. In the clay-containing phases (4, 5, 6, 7) the letter "t" stands for a tetrahedral crystal structure; the letter "o" stands for an octahedral structure.

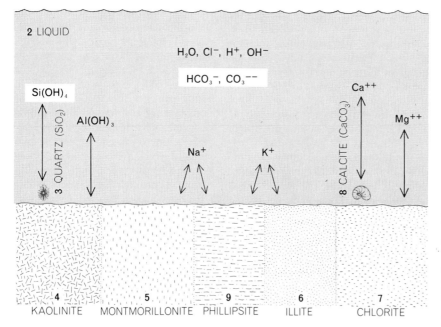

EQUILIBRIUM OCEAN MODEL, consisting of nine phases and nine components, shows how the principal constituents of the ocean distribute themselves among the atmosphere, the ocean and the sediments. Three of the constituents (HCO_3^-, CO_3^{--} and $Si(OH)_4$) are not included among nine listed components but appear as equilibrium products of those that are listed, as do seven ions ($H^+, K^+, Na^+, Ca^{++}, Mg^{++}, Cl^-, OH^-$). Two of the solids are shown as biological "precipitates": "quartz" (3) in the form of silicate structures built by radiolarians and "calcite" (8) in the form of calcium carbonate chambers built by foraminifera. The method of precipitation is unimportant as long as the product is stable. The equilibrium model goes far to explain why the ocean has the composition it does.

ide content) will have the predicted effect of changing the earth's climate. In principle an increase in atmospheric carbon dioxide should reduce the amount of long-wavelength radiation sent back into space by the earth and thus produce a greenhouse effect, slightly raising the average world temperature.

Having described an equilibrium model of the ocean that neglected the atmosphere's content of nitrogen and oxygen, I should not leave the reader with the impression that the continued presence of these two gases in the atmosphere is independent of the ocean. If the ocean were truly in equilibrium with the atmosphere, it would long since have captured all the atmospheric oxygen in the form of nitrates, both in solution and in sediments. This catastrophe has apparently been averted by the intervention of certain marine bacteria that have the happy faculty of releasing nitrogen gas from nitrate compounds and of converting the oxygen to a form that can later be liberated by phytoplankton.

The story is this. A variety of high-energy processes in the atmosphere continuously break the triple chemical bond that holds two nitrogen atoms together in a nitrogen molecule (N_2). The bonds can be broken by ultraviolet photons, by cosmic rays, by lightning and by the explosions in internal-combustion engines. Once dissociated, nitrogen atoms can react with oxygen to form various oxides, which are then carried to the ground by rainfall. In the soil these oxides are useful as fertilizer. Ultimately large amounts of them reach the sea. They do not, however, accumulate there and no one is really sure why.

The best guess is that denitrifying bacteria in oceanic sediments use the oxygen of nitrate to oxidize organic molecules when they run out of free oxygen [*see left half of illustration on pages 192 and 193*]. The nitrogen is released directly as a gas, which goes into solution but is available for return to the atmosphere. The oxygen emerges in molecules of water and carbon dioxide. The carbon dioxide is assimilated by phytoplankton, which build the carbon into organic compounds and release the oxygen as dissolved gas, also available for return to the atmosphere. Without these coupled biological processes the atmospheric fixation of nitrogen would probably exhaust the world's oxygen supply in less than 10 million years. Nevertheless, the amount of nitrogen returned to the atmosphere from the sediments is so small that we may never be able to measure it directly: the yearly return is less

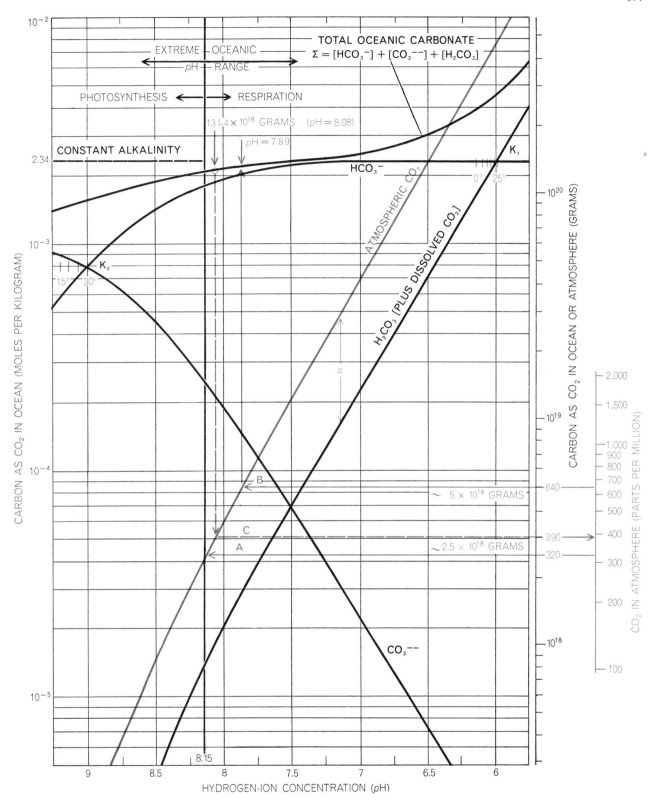

OCEANIC CARBONATE SYSTEM can be represented by a "Bjer-rum diagram" that shows how carbonate in its several forms varies with the ocean's pH, or hydrogen-ion concentration. The diagram is plotted for a constant "carbonate alkalinity" of 2.34×10^{-3} moles of carbonate per kilogram of seawater (*scale at left*). "System point" K_1 shows where the concentrations of bicarbonate ion (HCO_3^-) and carbonic acid (H_2CO_3) are equal. At K_2 the concentrations of bicarbonate and carbonate (CO_3^{--}) are equal. The exact locations of K_1 and K_2 are shown for a range of temperatures (in degrees Celsius) at constant conditions of salinity and pressure. The top curve, Σ, is the sum of oceanic carbonate in all its forms. The normal pH of the ocean is 8.15. The two short arrows at top mark the normal biological limits: at 7.95 the available oxygen has been consumed by respiration; at 8.35 photosynthesis has removed so much carbon dioxide that absorption from the atmosphere rises sharply. The limits of oceanic pH lie between 7.45 and 8.6. The amount of carbon dioxide in the atmosphere (*colored curve and scale at far right*) is related to the amount of carbon dioxide dissolved in the ocean by alpha (α), the average worldwide solubility of carbon dioxide in seawater. The consequences of doubling the carbon dioxide in the atmosphere from 320 parts per million (A) to 640 parts (B) are discussed in the text of the article, as is line C.

than one two-thousandth of the total nitrogen dissolved in the sea.

Another little-known epicycle in the global oxygen cycle probably has the effect of limiting the net accumulation of carbon in the form of oil-bearing shale, tar sands and petroleum. After denitrifying bacteria have consumed the nitrate in young sediments, sulfate bacteria begin oxidizing organic matter with the oxygen contained in sulfates [see right half of illustration on pages 192 and 193]. The product, in addition to water and carbon dioxide, is hydrogen sulfide, the foul-smelling compound that characterizes environments deficient in oxygen. In undisturbed mud the hydrogen sulfide never reaches the surface because it is inorganically reoxidized to sulfate as soon as it comes in contact with free oxygen. It seems likely that the bacterial turnover of oxygen in sulfate is so rapid that half of the world's oxygen passes through this epicycle in about 50,000 years.

The global activities of man have now reached such a scale that they are beginning to have a profound effect on marine chemistry and biology. We are learning that even the ocean is not large enough to absorb all the waste products of industrial society. The experiment involving the release of carbon dioxide is now in progress. DDT, only 25 years on the scene, is now found in the tissues of animals from pole to pole and has pushed several species of birds close to extinction. The concentration of lead in plants, animals and man has increased tenfold since tetraethyl lead was first used as an antiknock agent in motor fuels. And high levels of mercury in fish have forced the abandonment of some commercial fisheries. (Lead and mercury are systemic enzyme poisons.) Of the total petroleum production some .2 percent gets slopped into the sea in half a dozen major accidents each year. (At least six of the rare gray whales died last year after migrating through the oil slick off Santa Barbara caused by the blowout of a well casing belonging to the Union Oil Company.) Conceivably a persistent oil film could change the surface reflectivity of the ocean enough to alter the world's energy balance. The rapid increase in the use of nitrogen fertilizers leaves a nitrate excess that runs into rivers, lakes and ultimately reaches the sea. The sea can probably tolerate the runoff indefinitely but along the way the nitrogen creates algal "blooms" that are hastening the dystrophication of lakes and estuaries.

It is fashionable today to view the ocean as the last global frontier, waiting

only technological "development." Thermodynamically it is easier to extract fresh water from sewage than from seawater. Ecologically it is wiser to keep our concentrated nutrients on land than to dilute them beyond recall in the ocean. Sociologically, and probably economically, it makes more sense to process our junkyards for usable metals than to mine the deep-sea floor. The task is to persuade our engineers and business companies that working with sewage and junk is just as challenging as oceanography and thalassochemistry.

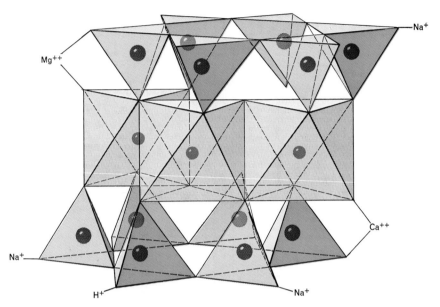

THREE-LAYER CLAY PARTICLE has a layer of octahedrons sandwiched between two layers of tetrahedrons. Each octahedron consists of an atom of aluminum surrounded by six closely packed atoms of oxygen. Each tetrahedron consists of a silicon atom surrounded by four atoms of oxygen. The polyhedrons are tied into layers at shared corners where a single oxygen atom is bonded to a silicon atom on one side and to an aluminum atom on the other. At the free corners the oxygen atoms bear unsatisfied negative charges that attract cations such as sodium $(Na+)$ and potassium $(K+)$. If the hydrogen-ion concentration should rise in the vicinity of clay, free hydrogen ions tend to be exchanged for sodium ions, which are released. In addition, many doubly charged metal ions can replace Si^{4+} at the centers of tetrahedrons and Si^{4+} can replace Al^{3+} in the octahedrons. Whenever this occurs, another cation is bound to the structure to conserve charge. Such reactions apparently exert considerable control over the ocean's composition and hydrogen-ion concentration.

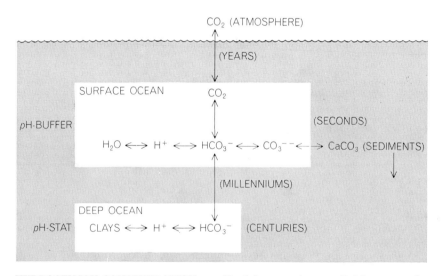

HYDROGEN-ION CONCENTRATION, or pH, of the ocean is controlled by two mechanisms, one that responds swiftly and one that takes centuries. The first, the "pH-buffer," operates near the surface and maintains equilibrium among carbon dioxide, bicarbonate ion (HCO_3^-), carbonate ion (CO_3^{--}) and sediments. The slower mechanism, the "pH-stat," seems to exert ultimate control over pH; it involves the interaction of bicarbonate ions and protons (H^+) with clays. Clay will accept protons in exchange for sodium ions (primarily).

18

FUEL CELLS

LEONARD G. AUSTIN
October 1959

Devices that convert chemical energy directly into electricity, thus circumventing the inefficiency of the heat engines used to drive electric generators, are now under intensive development

Civilization gets most of the energy it consumes from the energy of the chemical bonds in coal, petroleum and natural gas. But in the process of putting that chemical energy to work, it throws most of it away. The energy is first converted, by combustion of

the fuel, into heat. The heat is then converted, by several kinds of heat engine, into mechanical energy, which may in turn be converted into electricity. These transformations yield less than half of the original energy as useful work. But the fault does not lie in the energy-converting machines. Though the most modern central power-stations manufacture electricity at an efficiency of only 35 to 40 per cent, the performance of boilers, turbines and generators has been improved over the years until it now approaches the maximum which can be expected from the heat-steam-electricity cycle. Internal-combustion engines have reached a corresponding peak of efficiency at 25 to 30 per cent, and high-temperature gas turbines are approaching their limit at 40 per cent. The ceiling on efficiency is partly imposed by the second law of thermodynamics, which dictates the downhill flow of energy throughout the cosmos. At

the operating temperatures of heat engines—temperatures set by the strength of materials and the economics of heat transfer—this law decrees that more than half of the original chemical energy must be lost in irrevocably wasted heat. Further energy is lost to the friction that is encountered in any machine.

With conventional energy-converting technology approaching a dead end, power engineers are seeking ways to bypass the heat cycle and to convert the chemical energy of fuels directly into electricity. The notion is not a new one. In 1839 the English investigator Sir William Grove constructed a chemical battery in which the familiar water-forming reaction of hydrogen and oxygen generated an electric current. Fifty years later, also in England, the chemists Ludwig Mond and Carl Langer developed another version of this device which they called a fuel cell. But the dynamo was

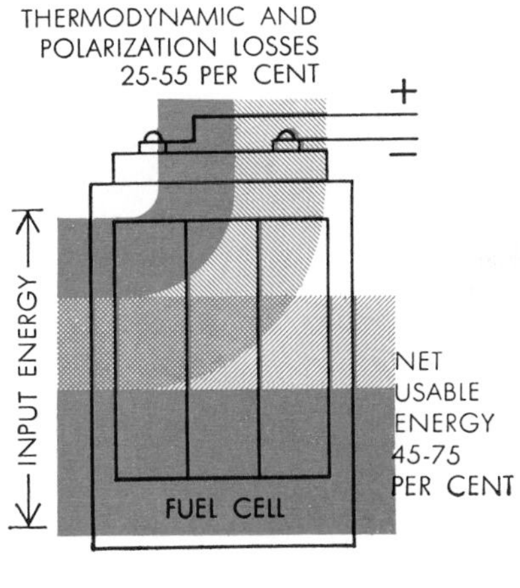

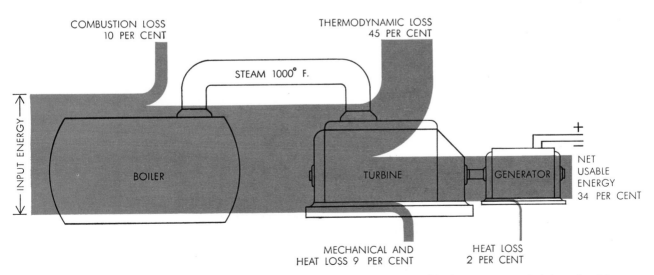

THERMODYNAMIC AND POLARIZATION LOSSES 25-55 PER CENT

INPUT ENERGY

FUEL CELL

NET USABLE ENERGY 45-75 PER CENT

COMBUSTION LOSS 10 PER CENT

THERMODYNAMIC LOSS 45 PER CENT

STEAM 1000° F.

INPUT ENERGY

BOILER

TURBINE

GENERATOR

NET USABLE ENERGY 34 PER CENT

MECHANICAL AND HEAT LOSS 9 PER CENT

HEAT LOSS 2 PER CENT

EFFICIENCY OF FUEL CELL is potentially greater than that of conventional generating equipment. Fuel cell (*top left*) converts 45 to 75 per cent of its input energy (*color*) into electricity compared to 34 per cent for typical steam turbogenerators (*bottom*).

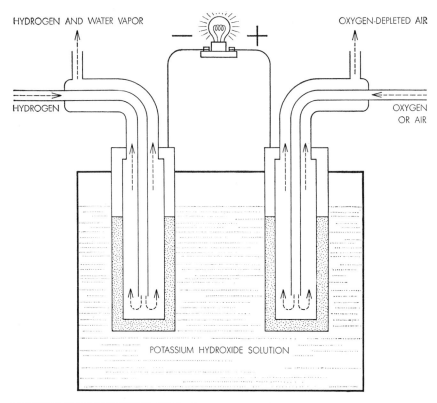

HYDROGEN-OXYGEN FUEL CELL, shown schematically, consists of two porous carbon electrodes (*dotted areas*) separated by an electrolyte such as potassium hydroxide. Hydrogen enters one side of the cell; oxygen, the other. Atoms of both gases diffuse into the electrodes, reacting to form water and to liberate electrons which flow through the circuit.

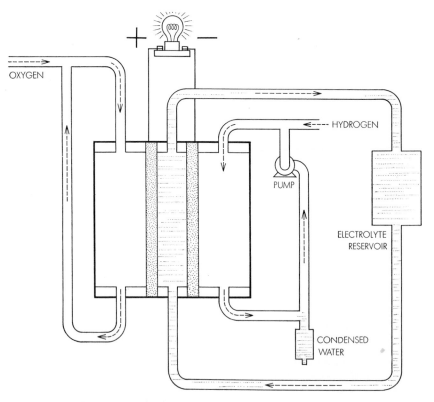

ANOTHER HYDROGEN-OXYGEN CELL was developed recently by Francis T. Bacon of the University of Cambridge. His cell consists basically of an electrolyte solution held between two thin electrodes of porous nickel. Gases under pressure diffuse through the electrodes and react with the electrolyte, which is held in tiny pores in the opposite surface.

then coming into its own, and although research continued spasmodically the difficulties encountered deterred any extensive effort to develop fuel cells. Since 1944, however, the fuel cell has come under active development again, and at least one is now in practical use.

The first voltaic pile and its modern descendant, the dry battery, are fuel cells in a sense: they convert chemical energy directly into electricity. But they use expensive "fuels" such as zinc, lead or mercury that are refined by the expenditure of considerable energy from fossil fuels or hydroelectric power. A true fuel cell uses the basic fuel directly, or almost directly. In theory the fuel cell may approach 100 per cent efficiency in converting the chemical energy of the fuel into electricity; actual efficiencies of 75 per cent—more than twice that of the average steam power-station—are quite feasible.

Fuel cells hold other attractions for contemporary engineering. An artificial satellite, for example, requires a small, light battery that can deliver a high electrical output. The fuel cell can meet these specifications from energy compactly stored in a liquid or gaseous fuel and in oxygen, as opposed to the cumbersome plates of an ordinary battery.

In public transportation the electric motor possesses a number of advantages over the gasoline or Diesel engine, including higher speed, more rapid acceleration, quietness and absence of noxious exhaust gases. However, the high capital cost of the electrical distribution system has caused a decline in electric transport during the past two decades. A few battery-powered delivery trucks still operate in some cities, but they suffer competitively from the low power-to-weight ratio of their lead batteries and from the long periods required for recharging. A fuel cell that could operate efficiently on gasoline or oil and could be "recharged" by the filling of its tank might reverse the present trend toward gasoline and Diesel locomotives, trucks and buses. Ultimately fuel cells might make the quiet, non-air-polluting electric automobile a reality.

The realization of these attractive possibilities will require a great deal of development work. To understand some of the difficulties to be surmounted, let us consider the fuel cell in which hydrogen and oxygen combine to produce an electric current and water.

As everyone knows, hydrogen and oxygen burn to produce water. They do so because separately they possess more energy than water and therefore "prefer" to exist in combination. However, at ordinary temperatures and pressures, additional "activation" energy is needed to raise the molecules to the energy state at which the reaction will ignite; this energy barrier ordinarily prevents the reaction from proceeding at room temperature. Activation energy may be illustrated by the following analogy. In a large number of people there may be one man capable of clearing a seven-foot high-jump bar, several capable of clearing six feet, thousands who can jump five feet, hundreds of thousands who can jump four feet, and so on. Molecules are like that with respect to their individual energy content: only a small fraction of them have high energies at room temperature. If the energy barrier for a reaction is comparable to an eight-foot hurdle, no reaction occurs. Raising the temperature has the effect of increasing the "jumping ability" of the molecules until some can clear the activation-energy barrier. At about 500 degrees centigrade a hydrogen-oxygen mixture will combine explosively, and the chemical energy is converted to heat.

In a hydrogen-oxygen fuel cell essentially the same chemical reaction is made to take place, but the reaction is stepwise at a lower energy of activation for each step. This can be considered as analogous to requiring the molecule to jump several barriers only four feet high, instead of one barrier eight feet high. The reaction thus proceeds quite quickly at room temperature. The cell is also designed so that one of the essential steps in the reaction is the transfer of electrons, from the negative terminal of the cell to the positive terminal, by an electrical connection. The flow of electrons, which is of course an electric current, can be used to drive an electric motor, light a lamp or operate a radio. Instead of the chemical energy of the reaction being immediately converted to heat, a large part of it is carried by the electrons, which can give up the energy as useful electrical work.

The cell consists of two porous electrodes separated by an electrolyte, which in this case is a concentrated solution of sodium hydroxide or potassium hydroxide. On the negative side of the cell, hydrogen gas diffuses through the electrode; hydrogen molecules (H_2), assisted by a catalyst embedded in the electrode surface, are adsorbed on the surface in the form of hydrogen atoms (H). The atoms react with hydroxyl ions (OH^-) in the electrolyte to form water, in the process giving up electrons to the

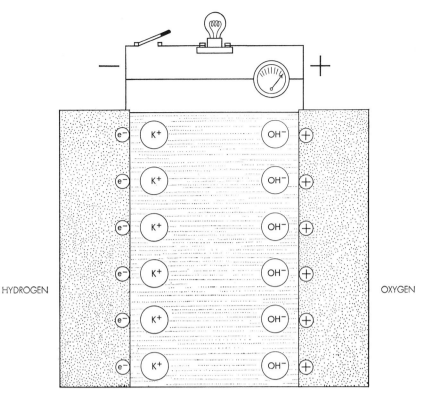

WHEN FUEL-CELL CIRCUIT IS OPEN, the hydrogen electrode accumulates a surface layer of negative charges that attracts positively charged potassium ions in the electrolyte solution. Similarly, the oxygen electrode attracts negative ions to balance its positive charge. These layers prevent further reaction between the gases and the electrolyte.

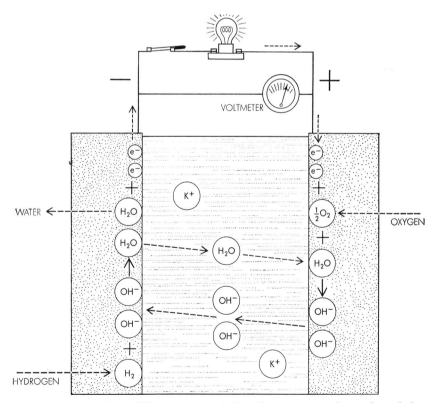

WHEN CIRCUIT IS CLOSED, the gases and electrolyte react to produce a flow of electrons. A catalyst embedded in the electrode dissociates hydrogen gas molecules into individual atoms, which combine with hydroxyl ions in the electrolyte to form water. The process yields electrons to the electrode. The electrons flow through the circuit to the positive electrode, where they combine with oxygen and water to form hydroxyl ions. The ions complete the circuit by migrating through the electrolyte to the negative electrode.

electrode; the water goes into the electrolyte. This reaction is also aided by the catalyst.

The flow of these electrons around the external circuit to the positive electrode constitutes the electric output of the cell and supports the oxygen half of the reaction. On the positive side of the cell oxygen (O_2) diffuses through the electrode and is adsorbed on the electrode surface. In a somewhat indirect reaction the adsorbed oxygen, plus the inflowing electrons, plus water in the electrolyte, form hydroxyl ions. Here again a catalyst helps the reaction to proceed. The hydroxyl ions complete the circle by migrating through the electrolyte to the hydrogen electrode [*see bottom illustration on preceding page*].

If the external circuit is open, the hydrogen electrode accumulates a surface layer of negative charges that attracts a layer of positively charged sodium or potassium ions in the electrolyte; an equivalent process at the oxygen electrode similarly balances its accumulated positive charge. These electrical "double layers" prevent further reaction between the gases and the electrolyte. The presence of the electrical layers provides the potential that forces the electrons through the external circuit when connection is made.

When the circuit is closed and the resistance across the external circuit between the electrodes is high, the reaction proceeds at a moderate rate, and a high percentage of the reaction energy is released as electricity, with only a little lost as heat. Part of the energy is expended at all times, however, in driving the chemical reactions over the barrier of the activation energies of the reactions inside the cell, and this energy appears as heat within the cell. The function of the catalysts in the electrodes is to lower the energy barriers, thus decreasing the amount of useful energy that is converted to heat. As resistance in the external circuit goes down, the current flow increases and a rising proportion of the energy is consumed in overcoming the energy barriers within the cell. With the increase in the reaction rate, heat losses go up rapidly. At zero resistance (short circuit) the reaction proceeds so rapidly that it becomes equivalent to combustion, producing only heat. Thus the reaction energy of the fuel cell resembles the energy of water behind a dam. By allowing the water to escape slowly through the blades of a turbine, we compel it to do useful work. If we open the floodgates, the water gushes out without performing any work.

In addition to the expenditure of energy needed to drive the reaction over activation-energy barriers, the fuel cell must consume some energy to force gas molecules through the electrodes to the reaction area, to transport hydroxyl ions from one electrode to the other and to overcome the electrical resistance of the electrodes themselves. These losses reduce the cell voltage below the theoretical ideal. A common working standard of voltage efficiency for fuel cells, however, is 75 per cent.

In practice, at the present stage of the art, other considerations loom larger than simple efficiency. For instance, a standard criterion is the power output per cubic foot of cell when the cell is converting 75 per cent of the thermodynamically available energy into electricity. Another important factor is the length of time a cell can operate before its performance falls off due to the deterioration of the electrode or the electrolyte.

In a typical hydrogen-oxygen cell the electrodes consist of porous carbon impregnated with catalysts: fine particles of platinum or palladium in the hydrogen electrode and cobalt oxide, platinum or silver in the oxygen electrode. To prevent flooding of the pores by the electrolyte, which would cut down the active surface, the electrodes are waterproofed with a layer of paraffin wax about one molecule thick. This thin film allows ions and individual water molecules to pass through to the internal surfaces of the electrode, but prevents the water from flooding the pores. To bring the electrodes closer together and thus speed ion transport, the electrodes are typically arranged as concentric tubes or adjacent plates. Cells of this type developed by Karl Kordesch of the National Carbon Company have won the distinction of being the first practical fuel cells; the U. S. Army uses them to power its "silent sentry" portable radar sets. Some have been in operation for more than a year with no appreciable decline in performance.

Low-temperature hydrogen-oxygen cells are limited in their applications, although they may find widespread special uses. Hydrogen is a costly fuel and the power-to-volume ratio of the cell (about one kilowatt-hour per cubic foot) makes it too bulky for use in vehicles.

An obvious way to improve the performance of hydrogen-oxygen cells is to

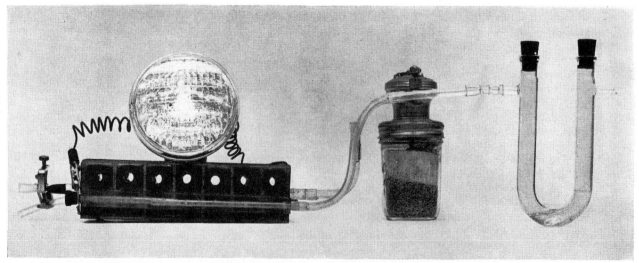

LABORATORY MODEL of a simple fuel cell reacts hydrogen with the oxygen in air. Hydrogen is generated in jar at right by dropping water onto calcium hydride. The gas then flows through carbon tubes in a block of Lucite, where it reacts with an electrolyte. The electrolyte in turn reacts with the oxygen that diffuses into other carbon tubes. The power output of the cell is three watts.

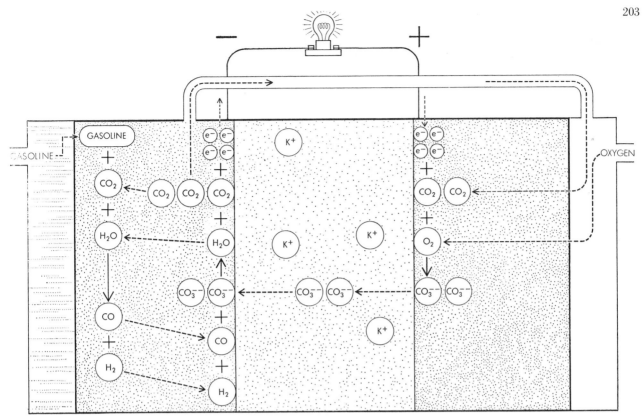

HIGH-TEMPERATURE FUEL CELL operates above 500 degrees centigrade and uses fuels such as gasoline or natural gas. The cell contains two electrodes tightly pressed against a "solid" electrolyte, which is usually a molten salt such as potassium carbonate. The fuel in the cell is usually broken down (by reaction with steam and carbon dioxide) to produce hydrogen and carbon monoxide. These gases then diffuse into the negative electrode, where they react with carbonate ions in the electrolyte, forming carbon dioxide and water and giving up electrons. The electrons flow through the circuit to the positive electrode, where they combine with oxygen and carbon dioxide to form carbonate ions. The carbonate ions complete the cell's electrical circuit by flowing back to the negative electrode.

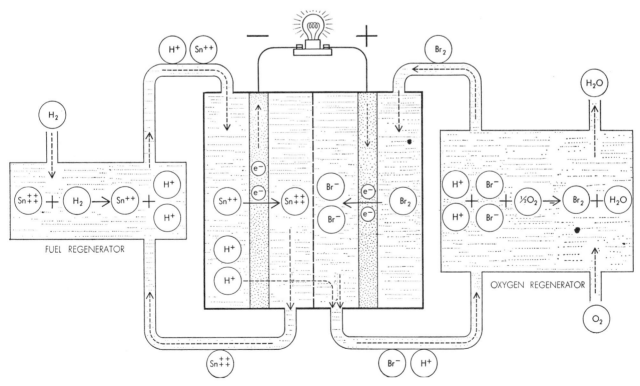

REDOX CELL is so named because in it the fuel and oxygen react with oxidizing and reducing agents in two so-called regenerators. The hydrogen reduces (adds electrons to) tin ions, which then give up electrons to the electrode. The electrons flow to the positive electrode. On the positive side of the cell, oxygen oxidizes (takes electrons from) bromide ions, converting them to bromine. In turn, the electrons flowing into the positive electrode reduce the bromine to bromide ions, which are then returned for regeneration.

operate them at higher pressures (which speed up gas transport through the electrodes) and higher temperatures (which speed up the electrochemical reactions). By appropriate design and insulation the waste heat liberated in the cell can be used to maintain the cell at the proper operating temperature.

The best-known cell of this type has been developed by Francis T. Bacon of the University of Cambridge. It operates at temperatures up to 250 degrees C. with gas pressures up to 800 pounds per square inch. The electrodes are of porous nickel about 1/16-inch thick and are usually in the form of disks or plates. A thin surface layer on the electrode, penetrated by very fine pores, constitutes the reaction area. The electrolyte, a concentrated solution of potassium hydroxide, can enter these pores, but pressure differences within the electrode prevent it from flooding the larger pores in the body of the electrode, through which gas percolates to the reaction area. The Bacon cell produces six times as much power per cubic foot as the low-temperature cell. With this relatively high output, the cell should have bright prospects as a standby source of auxiliary power in airplanes. It can deliver as much as 150 watts per pound, as against 10 watts for the lead-acid storage batteries currently in use.

To produce economical power on a large scale, fuel cells must "burn" cheap fuels such as natural gas, vaporized gasoline or the mixture of gases obtained from the gasification of coal. The extraction of energy from such fuels calls for operating temperatures above 500 degrees C. Since aqueous electrolytes would boil away at these temperatures, the electrolyte consists of some molten salt, usually a carbonate of sodium or potassium mixed with lithium carbonate to lower the melting point. In the most efficient of these cells, the electrolyte is held in a matrix of porous refractory material. The electrodes, made of a variety of metals or metallic oxides, are tightly pressed against the "solid" electrolyte.

In these cells the fuel does not necessarily combine directly with oxygen as hydrogen does in the hydrogen-oxygen cell. Usually the fuel is "cracked" to hydrogen and carbon monoxide by reaction with steam and carbon dioxide, which the fuel cell produces as by-products. This cracking may be conducted outside the cell, or inside the cell on the electrode surface. In the current-generating reaction the hydrogen and carbon mon-

oxide diffuse into the cell at the negative electrode, where they react with carbonate ions in the electrolyte, forming carbon dioxide and water and giving up electrons to the electrode. At the positive electrode, oxygen or air takes up the electrons flowing in from the external circuit and reacts with the carbon dioxide to produce the carbonate ions. The migration of carbonate ions through the electrolyte from the positive to the negative electrode completes the circuit [see top illustration on preceding page].

High-temperature fuel cells, intensively investigated only since World War II, still perform poorly. The best of them produce no more than half a kilowatt per cubic foot—half the yield of the low-temperature hydrogen-oxygen cell and a twelfth the yield of the Bacon cell. However, the progress already made in hydrogen-oxygen cells suggests that further research can improve the performance of high-temperature cells by a factor of 10 or more.

In the "redox" cell—named for reduction and oxidation—the fuel and oxygen do not react directly with each other. Rather, the fuel and oxygen are made to react with other substances in "regenerators" outside the cell to produce chemical intermediates, which in turn generate current in the cell. The over-all reaction is the same as that of combustion, however, because the intermediates are regenerated. A typical cell of this type, developed in England under the leadership of Sir Eric Rideal, utilizes tin salts and bromine as intermediates. The fuel reduces (i.e., adds electrons to) tin ions, which then give up the added electrons to the negative electrode and return to react with more fuel. The oxygen similarly oxidizes (i.e., takes electrons from) bromide ions, converting them to bromine, which then takes up electrons from the positive electrode and returns as bromide ions for regeneration [see bottom illustration on preceding page]. A similar cell, using titanium salts instead of tin, is under development by the General Electric Company.

In principle redox cells should be able to achieve high efficiencies. The intermediates can be chosen so that the electrode reactions are rapid and yield high currents with little energy loss. With suitable catalysts and operating conditions it may be possible to carry out the regeneration reactions at satisfactory efficiencies. However, the problems involved in the regenerators have not yet been solved. Moreover, the two electrolyte systems must be separated from

each other by an impermeable membrane to keep the bromine from mixing and reacting with the tin or titanium ions. All known membranes of this sort have a rather high electrical resistance. It has not yet been demonstrated that the redox cell represents any improvement over simpler types.

Engineers are working on a number of other reaction cycles and combinations of cycles. Each of them presents knotty technical difficulties. But the fundamental processes of electrochemistry are fairly well understood, probably because electrochemical experiments require no expensive apparatus and thus fit well into university budgets. The future development of the fuel cell is thus a question of applied rather than basic research.

Low-temperature and moderate-temperature hydrogen-oxygen cells should come into use during the next few years as low-weight, easily "charged" batteries. The development of strong, lightweight containers, perhaps made of plastic-impregnated glass fibers, would reduce the poundage if not the cubic footage needed to store the reaction gases. Where cost is not too important, the hydrogen could be stored as solid lithium hydride and the oxygen as solid calcium superoxide. Moderate-temperature cells may well be used to power submarines. Such vessels, like nuclear submarines, could cruise for extended periods without surfacing and would be far quieter in operation than nuclear vessels.

Hydrogen-oxygen cells may also furnish a means of capturing the power of the sun. Investigators at the Stanford Research Institute have developed a catalytic process for decomposing water into hydrogen and oxygen by sunlight. Used in conjunction with fuel cells, which would recombine the hydrogen and oxygen into water, a solar photolysis plant covering two square kilometers of desert could provide as much energy as a 100,000-kilowatt power-station in continuous operation. The over-all efficiency of such a plant, estimated at 25 per cent, would be two and a half times that of present solar batteries or solar boilers.

In auxiliary installations at nuclear power-stations, hydrogen-oxygen cells may help to bring down the cost of nuclear power. The high capital cost of nuclear-power plants requires that they be operated at near-peak capacity if they are to yield cheap electricity. Power generated during daily or seasonal periods of low demand might be used to electrolyze water into hydrogen and oxygen, which would then be made to yield the stored energy via fuel cells during peak-demand periods. The large volume of gas generated might be stored in "sausage skins" of plastic film buried underground to eliminate wind damage.

If the performance of high-temperature fuel cells can be substantially improved, large-scale electric power might be generated near sources of cheap natural gas. The power produced would of course be in the form of direct rather than alternating current, and though high-voltage direct current is somewhat easier to transmit than alternating current, fuel cells apparently cannot produce high voltages. Large numbers of cells must be connected in series, and above 700 volts there is electrical leakage through the insulation separating the terminals. Large-scale power from fuel cells should therefore find its first application in electrochemical processes such as the production of aluminum, which utilize large quantities of direct current at low voltage. Electrochemical industries may then congregate near natural-gas sources as they now cluster around hydroelectric installations.

The hydrogen-oxygen cell may make an unorthodox contribution of its own to the chemical industries. With slight modifications a low-temperature cell can employ instead of hydrogen a liquid fuel such as methyl (wood) alcohol, which it oxidizes to formic acid. The power output is very low, but the formic acid is almost free of impurities. Ethyl alcohol can similarly be oxidized to acetic acid, an important raw material in the manufacture of plastics and lacquers. Such processes, amounting to a sort of electrolysis in reverse, may prove useful in the manufacture of other industrial chemicals. Since the energy released would be extracted as electricity rather than heat, unwanted side reactions could be held to a minimum. It would be ironic if fuel cells should find their principal application in the production of chemicals rather than of power.

Attempts to construct cells that would operate directly on coke or coal, the cheapest of fuels, have been disappointing. Far more promising is the mixture of hydrogen, carbon monoxide and hydrocarbons that can be made from coal. With suitable equipment to remove tar and grit, it could be piped directly from the gasification plant and used hot. However, a really low-cost process for generating gas from coal has yet to be devised.

A high-output fuel cell operating on liquid fuel would find immediate application in trucks and locomotives. The technology of electric traction is well developed and is waiting for a compact power-unit utilizing a cheap fuel that can be easily stored and pumped. Designers will have to figure out a simple way to warm up the cell to operating temperature; the high-temperature cell is not a self-starter.

The possibilities of fuel cells are great, but not all these possibilities are going to be realized. Although much small-scale development work remains to be done, some cells have reached the stage where further progress will require large amounts of money and faith. No doubt some of this money will be wasted, and some of the faith will be misplaced. Fuel cell development is not a field for the faint-hearted.

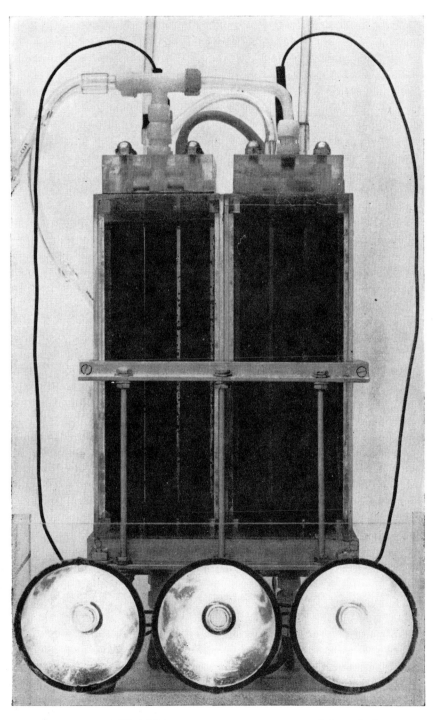

EXPERIMENTAL FUEL CELL operates at room temperature and atmospheric pressure. It produces 20 watts of electrical power, enough to light three bicycle lamps. The two Lucite boxes composing the cell contain an electrolyte solution and nine porous carbon electrodes, four for hydrogen and five for oxygen. This cell, and the one on page 202, were photographed in the laboratories of the National Carbon Company in Parma, Ohio.

19

THE OXYGEN CYCLE

PRESTON CLOUD AND AHARON GIBOR

September 1970

The oxygen in the atmosphere was originally put there by plants. Hence the early plants made possible the evolution of the higher plants and animals that require free oxygen for their metabolism

The history of our planet, as recorded in its rocks and fossils, is reflected in the composition and the biochemical peculiarities of its present biosphere. With a little imagination one can reconstruct from that evidence the appearance and subsequent evolution of gaseous oxygen in the earth's air and water, and the changing pathways of oxygen in the metabolism of living things.

Differentiated multicellular life (consisting of tissues and organs) evolved only after free oxygen appeared in the atmosphere. The cells of animals that are truly multicellular in this sense, the Metazoa, obtain their energy by breaking down fuel (produced originally by photosynthesis) in the presence of oxygen in the process called respiration. The evolution of advanced forms of animal life would probably not have been possible without the high levels of energy release that are characteristic of oxidative metabolism. At the same time free oxygen is potentially destructive to all forms of carbon-based life (and we know no other kind of life). Most organisms have therefore had to "learn" to conduct their oxidations anaerobically, primarily by removing hydrogen from foodstuff rather than by adding oxygen. Indeed, the anaerobic process called fermentation is still the fundamental way of life, underlying other forms of metabolism.

Oxygen in the free state thus plays a role in the evolution and present functioning of the biosphere that is both pervasive and ambivalent. The origin of life

and its subsequent evolution was contingent on the development of systems that shielded it from, or provided chemical defenses against, ordinary molecular oxygen (O_2), ozone (O_3) and atomic oxygen (O). Yet the energy requirements of higher life forms can be met only by oxidative metabolism. The oxidation of the simple sugar glucose, for example, yields 686 kilocalories per mole; the fermentation of glucose yields only 50 kilocalories per mole.

Free oxygen not only supports life; it arises from life. The oxygen now in the atmosphere is probably mainly, if not wholly, of biological origin. Some of it is converted to ozone, causing certain high-energy wavelengths to be filtered out of the radiation that reaches the surface of the earth. Oxygen also combines with a wide range of other elements in the earth's crust. The result of these and other processes is an intimate evolutionary interaction among the biosphere, the atmosphere, the hydrosphere and the lithosphere.

Consider where the oxygen comes from to support the high rates of energy release observed in multicellular organisms and what happens to it and to the carbon dioxide that is respired [*see illustration on page 210*]. The oxygen, of course, comes from the air, of which it constitutes roughly 21 percent. Ultimately, however, it originates with the decomposition of water molecules by light energy in photosynthesis. The 1.5 billion cubic kilometers of water on the earth are split by photosynthesis and reconsti-

tuted by respiration once every two million years or so. Photosynthetically generated oxygen temporarily enters the atmospheric bank, whence it is itself recycled once every 2,000 years or so (at current rates). The carbon dioxide that is respired joins the small amount (.03 percent) already in the atmosphere, which is in balance with the carbon dioxide in the oceans and other parts of the hydrosphere. Through other interactions it may be removed from circulation as a part of the carbonate ion (CO_3^-) in calcium carbonate precipitated from solution. Carbon dioxide thus sequestered may eventually be returned to the atmosphere when limestone, formed by the consolidation of calcium carbonate sediments, emerges from under the sea and is dissolved by some future rainfall.

Thus do sea, air, rock and life interact and exchange components. Before taking up these interactions in somewhat greater detail let us examine the function oxygen serves within individual organisms.

Oxygen plays a fundamental role as a building block of practically all vital molecules, accounting for about a fourth of the atoms in living matter. Practically all organic matter in the present biosphere originates in the process of photosynthesis, whereby plants utilize light energy to react carbon dioxide with water and synthesize organic substances. Since carbohydrates (such as sugar), with the general formula $(CH_2O)_n$, are the common fuels that are stored by plants, the essential reaction of photosynthesis can be written as $CO_2 + H_2O + light energy \rightarrow CH_2O + O_2$. It is not immediately obvious from this formulation which of the reactants serves as the source of oxygen atoms in the carbohydrates and which is the source of free molecular oxygen. In 1941 Samuel Ruben and Mar-

RED BEDS rich in the oxidized (ferric) form of iron mark the advent of oxygen in the atmosphere. The earliest continental red beds are less than two billion years old; the red sandstones and shales of the Nankoweap Formation in the Grand Canyon (*opposite page*) are about 1.3 billion years old. The appearance of oxygen in the atmosphere, the result of photosynthesis, led in time to the evolution of cells that could survive its toxic effects and eventually to cells that could capitalize on the high energy levels of oxidative metabolism.

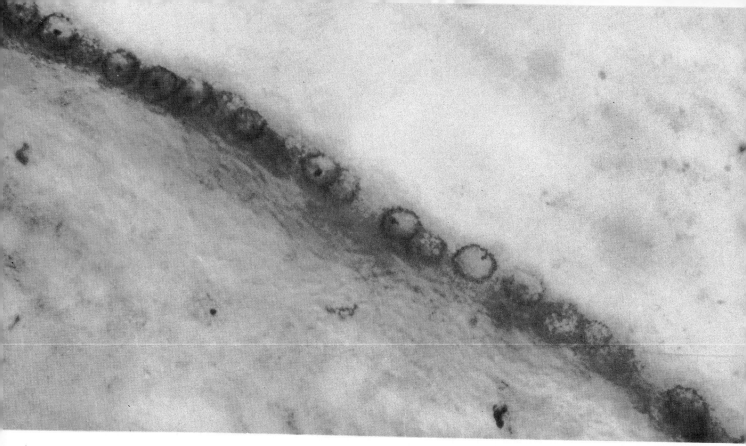

EUCARYOTIC CELLS, which contain a nucleus and divide by mitosis, were, like oxygen, a necessary precondition for the evolution of higher life forms. The oldest eucaryotes known were found in the Beck Spring Dolomite of eastern California by Cloud and his colleagues. The photomicrograph above shows eucaryotic cells with an average diameter of 14 microns, probably green algae. The regular occurrence and position of the dark spots suggest they may be remnants of nuclei or other organelles. Other cell forms, which do not appear in the picture, show branching and large filament diameters that also indicate the eucaryotic level of evolution.

PROCARYOTIC CELLS, which lack a nucleus and divide by simple fission, were a more primitive form of life than the eucaryotes and persist today in the bacteria and blue-green algae. Procaryotes were found in the Beck Spring Dolomite in association with the primitive eucaryotes such as those in the photograph at the top of the page. A mat of threadlike procaryotic blue-green algae, each thread of which is about 3.5 microns in diameter, is seen in the photomicrograph below. It was made, like the one at top of page, by Gerald R. Licari. Cells of this kind, among others, presumably produced photosynthetic oxygen before eucaryotes appeared.

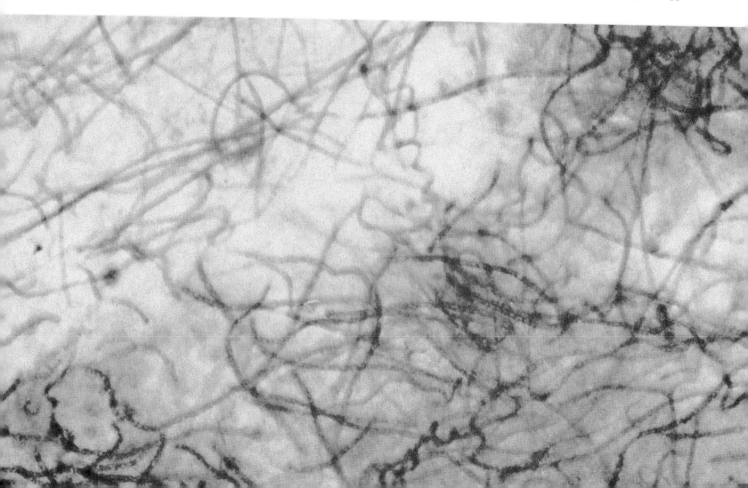

tin D. Kamen of the University of California at Berkeley used the heavy oxygen isotope oxygen 18 as a tracer to demonstrate that the molecular oxygen is derived from the splitting of the water molecule. This observation also suggested that carbon dioxide is the source of the oxygen atoms of the synthesized organic molecules.

The primary products of photosynthesis undergo a vast number of chemical transformations in plant cells and subsequently in the cells of the animals that feed on plants. During these processes changes of course take place in the atomic composition and energy content of the organic molecules. Such transformations can result in carbon compounds that are either more "reduced" or more "oxidized" than carbohydrates. The oxidation-reduction reactions between these compounds are the essence of biological energy supply and demand. A more reduced compound has more hydrogen atoms and fewer oxygen atoms per carbon atom; a more oxidized compound has fewer hydrogen atoms and more oxygen atoms per carbon atom. The combustion of a reduced compound liberates more energy than the combustion of a more oxidized one. An example of a molecule more reduced than a carbohydrate is the familiar alcohol ethanol (C_2H_6O); a more oxidized molecule is pyruvic acid ($C_3H_4O_3$).

Differences in the relative abundance of hydrogen and oxygen atoms in organic molecules result primarily from one of the following reactions: (1) the removal (dehydrogenation) or addition (hydrogenation) of hydrogen atoms, (2) the addition of water (hydration), followed by dehydrogenation; (3) the direct addition of oxygen (oxygenation). The second and third of these processes introduce into organic matter additional oxygen atoms either from water or from molecular oxygen. On decomposition the oxygen atoms of organic molecules are released as carbon dioxide and water. The biological oxidation of molecules such as carbohydrates can be written as the reverse of photosynthesis: $CH_2O + O_2 \rightarrow CO_2 + H_2O +$ energy. The oxygen atom of the organic molecule appears in the carbon dioxide and the molecular oxygen acts as the acceptor for the hydrogen atoms.

The three major nonliving sources of oxygen atoms are therefore carbon dioxide, water and molecular oxygen, and since these molecules exchange oxygen atoms, they can be considered as a common pool. Common mineral oxides such as nitrate ions and sulfate ions are also oxygen sources for living organisms, which reduce them to ammonia (NH_3) and hydrogen sulfide (H_2S). They are subsequently reoxidized, and so as the oxides circulate through the biosphere their oxygen atoms are exchanged with water.

The dynamic role of molecular oxygen is as an electron sink, or hydrogen acceptor, in biological oxidations. The biological oxidation of organic molecules proceeds primarily by dehydrogenation: enzymes remove hydrogen atoms from the substrate molecule and transfer them to specialized molecules that function as hydrogen carriers [see top illustration on pages 212 and 213]. If these carriers become saturated with hydrogen, no further oxidation can take place until some other acceptor becomes available. In the anaerobic process of fermentation organic molecules serve as the hydrogen acceptor. Fermentation therefore results in the oxidation of some organic compounds and the simultaneous reduction of others, as in the fermentation of glucose by yeast: part of the sugar molecule is oxidized to carbon dioxide and other parts are reduced to ethanol.

In aerobic respiration oxygen serves as the hydrogen acceptor and water is produced. The transfer of hydrogen atoms (which is to say of electrons and protons) to oxygen is channeled through an array of catalysts and cofactors. Prominent among the cofactors are the iron-containing pigmented molecules called cytochromes, of which there are several kinds that differ in their affinity for electrons. This affinity is expressed as the oxidation-reduction, or "redox," potential of the molecule; the more positive the potential, the greater the affinity of the oxidized molecule for electrons. For example, the redox potential of cytochrome b is .12 volt, the potential of cytochrome c is .22 volt and the potential of cytochrome a is .29 volt. The redox potential for the reduction of oxygen to water is .8 volt. The passage of electrons from one cytochrome to another down a potential gradient, from cytochrome b to cytochrome c to the cytochrome a complex and on to oxygen, results in the alternate reduction and oxidation of these cofactors. Energy liberated in such oxidation-reduction reactions is coupled to the synthesis of high-energy phosphate compounds such as adenosine triphosphate (ATP). The special copper-containing enzyme cytochrome oxidase mediates the ultimate transfer of electrons from the cytochrome a complex to oxygen. This activation and binding of oxygen is seen as the fundamental step, and possibly the original primitive step, in the evolution of oxidative metabolism.

In cells of higher organisms the oxidative system of enzymes and electron carriers is located in the special organelles called mitochondria. These organelles can be regarded as efficient low-temperature furnaces where organic molecules are burned with oxygen. Most of the released energy is converted into the high-energy bonds of ATP.

Molecular oxygen reacts spontaneously with organic compounds and other reduced substances. This reactivity explains the toxic effects of oxygen above tolerable concentrations. Louis Pasteur discovered that very sensitive organisms such as obligate anaerobes cannot tolerate oxygen concentrations above about 1 percent of the present atmospheric level. Recently the cells of higher organisms have been found to contain organelles called peroxisomes, whose major function is thought to be the protection of cells from oxygen. The peroxisomes contain enzymes that catalyze the direct reduction of oxygen molecules through the oxidation of metabolites such as amino acids and other organic acids. Hydrogen peroxide (H_2O_2) is one of the products of such oxidation. Another of the peroxisome enzymes, catalase, utilizes the hydrogen peroxide as a hydrogen acceptor in the oxidation of substrates such as ethanol or lactic acid. The rate of reduction of oxygen by the peroxisomes increases proportionately with an increase in oxygen concentration, so that an excessive amount of oxygen in the cell increases the rate of its reduction by peroxisomes.

Christian de Duve of Rockefeller University has suggested that the peroxisomes represent a primitive enzyme system that evolved to cope with oxygen when it first appeared in the atmosphere. The peroxisome enzymes enabled the first oxidatively metabolizing cells to use oxygen as a hydrogen acceptor and so reoxidize the reduced products of fermentation. In some respects this process is similar to the oxidative reactions of the mitochondria. Both make further dehydrogenation possible by liberating oxidized hydrogen carriers. The basic difference between the mitochondrial oxidation reactions and those of peroxisomes is that in peroxisomes the steps of oxidation are not coupled to the synthesis of ATP. The energy released in the peroxisomes is thus lost to the cell; the function of the organelle is primarily to protect against the destructive effects of free molecular oxygen.

Oxygen dissolved in water can diffuse

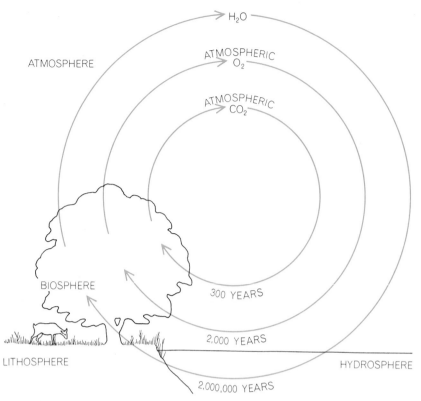

BIOSPHERE EXCHANGES water vapor, oxygen and carbon dioxide with the atmosphere and hydrosphere in a continuing cycle, shown here in simplified form. All the earth's water is split by plant cells and reconstituted by animal and plant cells about every two million years. Oxygen generated in the process enters the atmosphere and is recycled in about 2,000 years. Carbon dioxide respired by animal and plant cells enters the atmosphere and is fixed again by plant cells after an average atmospheric residence time of about 300 years.

across both the inner and the outer membranes of the cell, and the supply of oxygen by diffusion is adequate for single cells and for organisms consisting of small colonies of cells. Differentiated multicellular organisms, however, require more efficient modes of supplying oxygen to tissues and organs. Since all higher organisms depend primarily on mitochondrial aerobic oxidation to generate the energy that maintains their active mode of life, they have evolved elaborate systems to ensure their tissues an adequate supply of oxygen, the gas that once was lethal (and still is, in excess). Two basic devices serve this purpose: special chemical carriers that increase the oxygen capacity of body fluids, and anatomical structures that provide relatively large surfaces for the rapid exchange of gases. The typical properties of an oxygen carrier are exemplified by those of hemoglobin and of myoglobin, or muscle hemoglobin. Hemoglobin in blood readily absorbs oxygen to near-saturation at oxygen pressures such as those found in the lung. When the blood is exposed to lower oxygen pressures as it moves from the lungs to other·tissues, the hemoglobin discharges most of its bound oxygen. Myoglobin, which acts as

a reservoir to meet the sharp demand for oxygen in muscle contraction, gives up its oxygen more rapidly. Such reversible bonding of oxygen in response to changes in oxygen pressure is an essential property of biochemical oxygen carriers.

Lungs and gills are examples of anatomical structures in which large wet areas of thin membranous tissue come in contact with oxygen. Body fluids are pumped over one side of these membranes and air, or water containing oxygen, over the other side. This ensures a rapid gas exchange between large volumes of body fluid and the environment.

How did the relations between organisms and gaseous oxygen happen to evolve in such a curiously complicated manner? The atmosphere under which life arose on the earth was almost certainly devoid of free oxygen. The low concentration of noble gases such as neon and krypton in the terrestrial atmosphere compared with their cosmic abundance, together with other geochemical evidence, indicates that the terrestrial atmosphere had a secondary origin in volcanic outgassing from the earth's interior. Oxygen is not known among the gases so released, nor is it

found as inclusions in igneous rocks. The chemistry of rocks older than about two billion years is also inconsistent with the presence of more than trivial quantities of free atmospheric oxygen before that time. Moreover, it would not have been possible for the essential chemical precursors of life—or life itself—to have originated and persisted in the presence of free oxygen before the evolution of suitable oxygen-mediating enzymes.

On such grounds we conclude that the first living organism must have depended on fermentation for its livelihood. Organic substances that originated in non-vital reactions served as substrates for these primordial fermentations. The first organism, therefore, was not only an anaerobe; it was also a heterotroph, dependent on a preexisting organic food supply and incapable of manufacturing its own food by photosynthesis or other autotrophic processes.

The emergence of an autotroph was an essential step in the onward march of biological evolution. This evolutionary step left its mark in the rocks as well as on all living forms. Some fated eobiont, as we may call these early life forms whose properties we can as yet only imagine, evolved and became an autotroph, an organism capable of manufacturing its own food. Biogeological evidence suggests that this critical event may have occurred more than three billion years ago.

If, as seems inescapable, the first autotrophic eobiont was also anaerobic, it would have encountered difficulty when it first learned to split water and release free oxygen. John M. Olson of the Brookhaven National Laboratory recently suggested biochemical arguments to support the idea that primitive photosynthesis may have obtained electrons from substances other than water. He argues that large-scale splitting of water and release of oxygen may have been delayed until the evolution of appropriate enzymes to detoxify this reactive substance.

We nevertheless find a long record of oxidized marine sediments of a peculiar type that precedes the first evidence of atmospheric oxygen in rocks about 1.8 billion years old; we do not find them in significant amounts in more recent strata. These oxidized marine sediments, known as banded iron formations, are alternately iron-rich and iron-poor chemical sediments that were laid down in open bodies of water. Much of the iron in them is ferric (the oxidized form, Fe^{+++}) rather than ferrous (the reduced form, Fe^{++}), implying that there was a source of oxygen in the column of water above them. Considering the

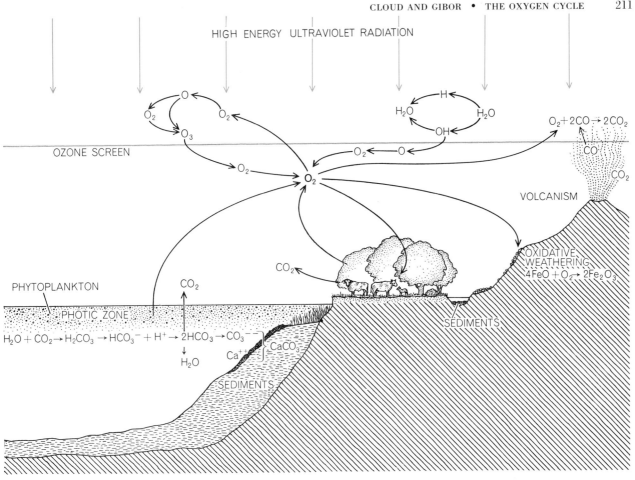

OXYGEN CYCLE is complicated because oxygen appears in so many chemical forms and combinations, primarily as molecular oxygen (O_2), in water and in organic and inorganic compounds. Some global pathways of oxygen are shown here in simplified form.

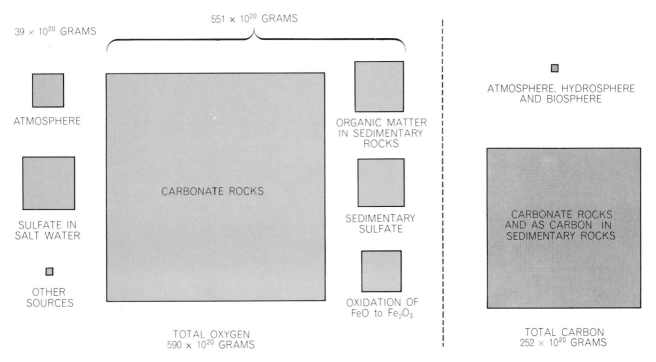

OXYGEN-CARBON BALANCE SHEET suggests that photosynthesis can account not only for all the oxygen in the atmosphere but also for the much larger amount of "fossil" oxygen, mostly in compounds in sediments. The diagram, based on estimates made by William W. Rubey, indicates that the elements are present in about the proportion, 12/32, that would account for their derivation through photosynthesis from carbon dioxide (one atom of carbon, molecular weight 12, to two of oxygen, molecular weight 16).

(top chemical diagrams)

a MYO-INOSITOL →(INOSITOL OXYGENASE, O_2, H_2O)→ D-GLUCURONIC ACID

b SUCCINIC ACID →(SUCCINIC DEHYDROGENASE, FAD, $FADH_2$)→ FUMARIC ACID

OXIDATION involves a decrease in the number of hydrogen atoms in a molecule or an increase in the number of oxygen atoms. It may be accomplished in several ways. In oxygenation (*a*) oxygen is added directly. In dehydrogenation (*b*) hydrogen is re-

---

problems that would face a water-splitting photosynthesizer before the evolution of advanced oxygen-mediating enzymes such as oxidases and catalases, one can visualize how the biological oxygen cycle may have interacted with ions in solution in bodies of water during that time. The first oxygen-releasing photoautotrophs may have used ferrous compounds in solution as oxygen acceptors—oxygen for them being merely a toxic waste product. This would have precipitated iron in the ferric form ($4FeO + O_2 \rightarrow 2Fe_2O_3$) or in the ferro-ferric form (Fe_3O_4). A recurrent imbalance of supply and demand might then account for the cyclic nature and differing types of the banded iron formations.

Once advanced oxygen-mediating enzymes arose, oxygen generated by increasing populations of photoautotrophs containing these enzymes would build up in the oceans and begin to escape into the atmosphere. There the ultraviolet component of the sun's radiation would dissociate some of the molecular oxygen into highly reactive atomic oxygen and also give rise to equally reactive ozone. Atmospheric oxygen and its reactive derivatives (even in small quantities) would lead to the oxidation of iron in sediments produced by the weathering of rocks, to the greatly reduced solubility of iron in surface waters (now oxygenated), to the termination of the banded iron formations as an important sedimentary type and to the extensive formation of continental red beds rich in ferric iron [*see illustration on page 206*]. The record of the rocks supports this succession of events: red beds are essentially restricted to rocks younger than about 1.8 billion years, whereas banded iron formation is found only in older rocks.

So far we have assumed that oxygen accumulated in the atmosphere as a consequence of photosynthesis by green plants. How could this happen if the entire process of photosynthesis and respiration is cyclic, representable by the reversible equation $CO_2 + H_2O + energy$

$\rightleftharpoons CH_2O + O_2$? Except to the extent that carbon or its compounds are somehow sequestered, carbohydrates produced by photosynthesis will be reoxidized back to carbon dioxide and water, and no significant quantity of free oxygen will accumulate. The carbon that is sequestered in the earth as graphite in the oldest rocks and as coal, oil, gas and other carbonaceous compounds in the younger ones, and in the living and dead bodies of plants and animals, is the

equivalent of the oxygen in oxidized sediments and in the earth's atmosphere! In attempting to strike a carbon-oxygen balance we must find enough carbon to account not only for the oxygen in the present atmosphere but also for the "fossil" oxygen that went into the conversion of ferrous oxides to ferric oxides, sulfides to sulfates, carbon monoxide to carbon dioxide and so on.

Interestingly, rough estimates made some years ago by William W. Rubey,

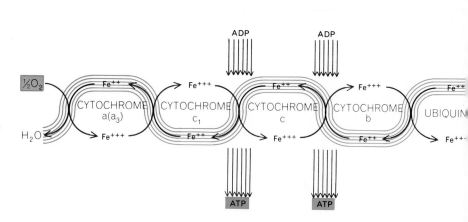

GLUCOSE →(ATP)→ GLUCOSE-6-PHOSPHATE ⇌ FRUCTOSE-6-PHOSPHATE →(ATP)→ FRUCTOSE 1,6-DIPHOSPHATE

● CARBON   ○ OXYGEN   ∘ HYDROGEN   Ⓟ PHOSPHATE

(bottom diagram: ADP, ATP, $\frac{1}{2}O_2$, H_2O, CYTOCHROME a(a_3), CYTOCHROME c_1, CYTOCHROME c, CYTOCHROME b, UBIQUINONE, Fe^{++}, Fe^{+++})

OXIDATIVE METABOLISM provides the energy that powers all higher forms of life. It proceeds in two phases: glycolysis (*top*), an anaerobic phase that does not require oxygen, and aerobic respiration (*bottom*), which requires oxygen. In glycolysis (or fermentation, the anaerobic process by which organisms such as yeast derive their energy) a molecule of the six-carbon sugar glucose is broken down into two molecules of the three-carbon compound pyruvic acid with a net gain of two molecules of adenosine triphosphate, the cellular

C

FUMARIC ACID   L-MALIC ACID   OXALOACETIC ACID

FUMARATE HYDROTASE   MALIC ACID DEHYDROGENASE

moved. In hydration-dehydrogenation (*c*) water is added and hydrogen is removed. Oxygenation does not occur in respiration, in which oxygen serves only as a hydrogen acceptor.

now of the University of California at Los Angeles, do imply an approximate balance between the chemical combining equivalents of carbon and oxygen in sediments, the atmosphere, the hydrosphere and the biosphere [*see bottom illustration on page 211*]. The relatively small excess of carbon in Rubey's estimates could be accounted for by the oxygen used in converting carbon monoxide to carbon dioxide. Or it might be due to an underestimate of the quantities of sul-

fate ion or ferric oxide in sediments. (Rubey's estimates could not include large iron formations recently discovered in western Australia and elsewhere.) The carbon dioxide in carbonate rocks does not need to be accounted for, but the oxygen involved in converting it to carbonate ion does. The recycling of sediments through metamorphism, mountain-building and the movement of ocean-floor plates under the continents is a variable of unknown dimensions, but

it probably does not affect the approximate balance observed in view of the fact that the overwhelmingly large pools to be balanced are all in the lithosphere and that carbon and oxygen losses would be roughly equivalent. The small amounts of oxygen dissolved in water are not included in this balance.

Nonetheless, water does enter the picture. Another possible source of oxygen in our atmosphere is photolysis, the ultraviolet dissociation of water vapor in the outer atmosphere followed by the escape of the hydrogen from the earth's gravitational field. This has usually been regarded as a trivial source, however. Although R. T. Brinkmann of the California Institute of Technology has recently argued that nonbiological photolysis may be a major source of atmospheric oxygen, the carbon-oxygen balance sheet does not support that belief, which also runs into other difficulties.

When free oxygen began to accumulate in the atmosphere some 1.8 billion years ago, life was still restricted to sites

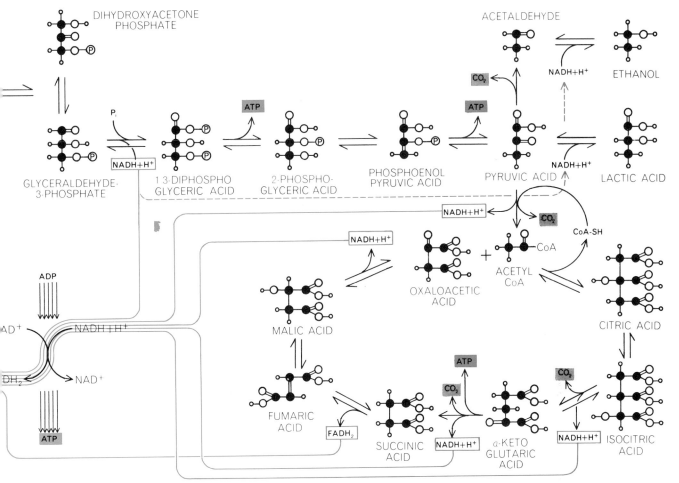

energy carrier. The pyruvic acid is converted into lactic acid in animal cells deprived of oxygen and into some other compound, such as ethanol, in fermentation. In aerobic cells in the presence of oxygen, however, pyruvic acid is completely oxidized to produce carbon dioxide and water. In the process hydrogen ions are removed. The electrons of these hydrogens (and of two removed in

glycolysis) are passed along by two electron carriers, nicotinamide adenine dinucleotide (NAD) and flavin adenine dinucleotide (FAD), to a chain of respiratory enzymes, ubiquinone and the cytochromes, which are alternately reduced and oxidized. Energy released in the reactions is coupled to synthesis of ATP, 38 molecules of which are produced for every molecule of glucose consumed.

shielded from destructive ultraviolet radiation by sufficient depths of water or by screens of sediment. In time enough oxygen built up in the atmosphere for ozone, a strong absorber in the ultraviolet, to form a shield against incoming ultraviolet radiation. The late Lloyd V. Berkner and Lauriston C. Marshall of the Graduate Research Center of the Southwest in Dallas calculated that only 1 percent of the present atmospheric level of oxygen would give rise to a sufficient level of ozone to screen out the most deleterious wavelengths of the ultraviolet radiation. This also happens to be the level of oxygen at which Pasteur found that certain microorganisms switch over from a fermentative type of metabolism to an oxidative one. Berkner and Marshall therefore jumped to the conclusion (reasonably enough on the evidence they considered) that this was the stage at which oxidative metabolism arose. They related this stage to the first appearance of metazoan life somewhat more than 600 million years ago.

The geological record has long made it plain, however, that free molecular oxygen existed in the atmosphere well before that relatively late date in geologic time. Moreover, recent evidence is consistent with the origin of oxidative metabolism at least twice as long ago. Eucaryotic cells—cells with organized nuclei and other organelles—have been identified in rocks in eastern California that are believed to be about 1.3 billion years old [*see top illustration on page 208*]. Since all living eucaryotes depend on oxidative metabolism, it seems likely that these ancestral forms did too. The oxygen level may nonetheless have still been quite low at this stage. Simple diffusion would suffice to move enough oxygen across cell boundaries and within the cell, even at very low concentrations, to supply the early oxidative metabolizers. A higher order of organization and of atmospheric oxygen was required, however, for advanced oxidative metabolism. Perhaps that is why, although the eucaryotic cell existed at least 1.2 billion years ago, we have no unequivocal fossils of metazoan organisms from rocks older than perhaps 640 million years.

In other words, perhaps Berkner and Marshall were mistaken only in trying to make the appearance of the Metazoa coincide with the onset of oxidative metabolism. Once the level of atmospheric oxygen was high enough to generate an effective ozone screen, photosynthetic organisms would have been able to spread throughout the surface waters of the sea, greatly accelerating the rate of oxygen production. The plausible episodes in geological history to correlate with this development are the secondary oxidation of the banded iron formations and the appearance of sedimentary calcium sulfate (gypsum and anhydrite) on a large scale. These events occurred just as or just before the Metazoa first appeared in early Paleozoic time. The attainment of a suitable level of atmospheric oxygen may thus be correlated with the emergence of metazoan root stocks from premetazoan ancestors beginning about 640 million years ago. The fact that oxygen could accumulate no faster than carbon (or hydrogen) was removed argues against the likelihood of a rapid early buildup of oxygen.

That subsequent biospheric and atmospheric evolution were closely interlinked can now be taken for granted. What is not known are the details. Did oxygen levels in the atmosphere increase steadily throughout geologic time, marking regular stages of biological evolution such as the emergence of land plants, of

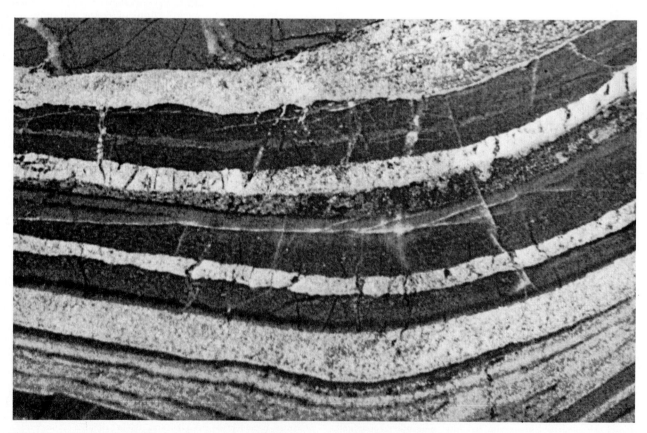

BANDED IRON FORMATION provides the first geological evidence of free oxygen in the hydrosphere. The layers in this polished cross section result from an alternation of iron-rich and iron-poor depositions. This sample from the Soudan Iron Formation in Minnesota is more than 2.7 billion years old. The layers, originally horizontal, were deformed while soft and later metamorphosed.

insects, of the various vertebrate groups and of flowering plants, as Berkner and Marshall suggested? Or were there wide swings in the oxygen level? Did oxygen decrease during great volcanic episodes, as a result of the oxidation of newly emitted carbon monoxide to carbon dioxide, or during times of sedimentary sulfate precipitation? Did oxygen increase when carbon was being sequestered during times of coal and petroleum formation? May there have been fluctuations in both directions as a result of plant and animal evolution, of phytoplankton eruptions and extinctions and of the extent and type of terrestrial plant cover? Such processes and events are now being seriously studied, but the answers are as yet far from clear.

What one can say with confidence is that success in understanding the oxy-

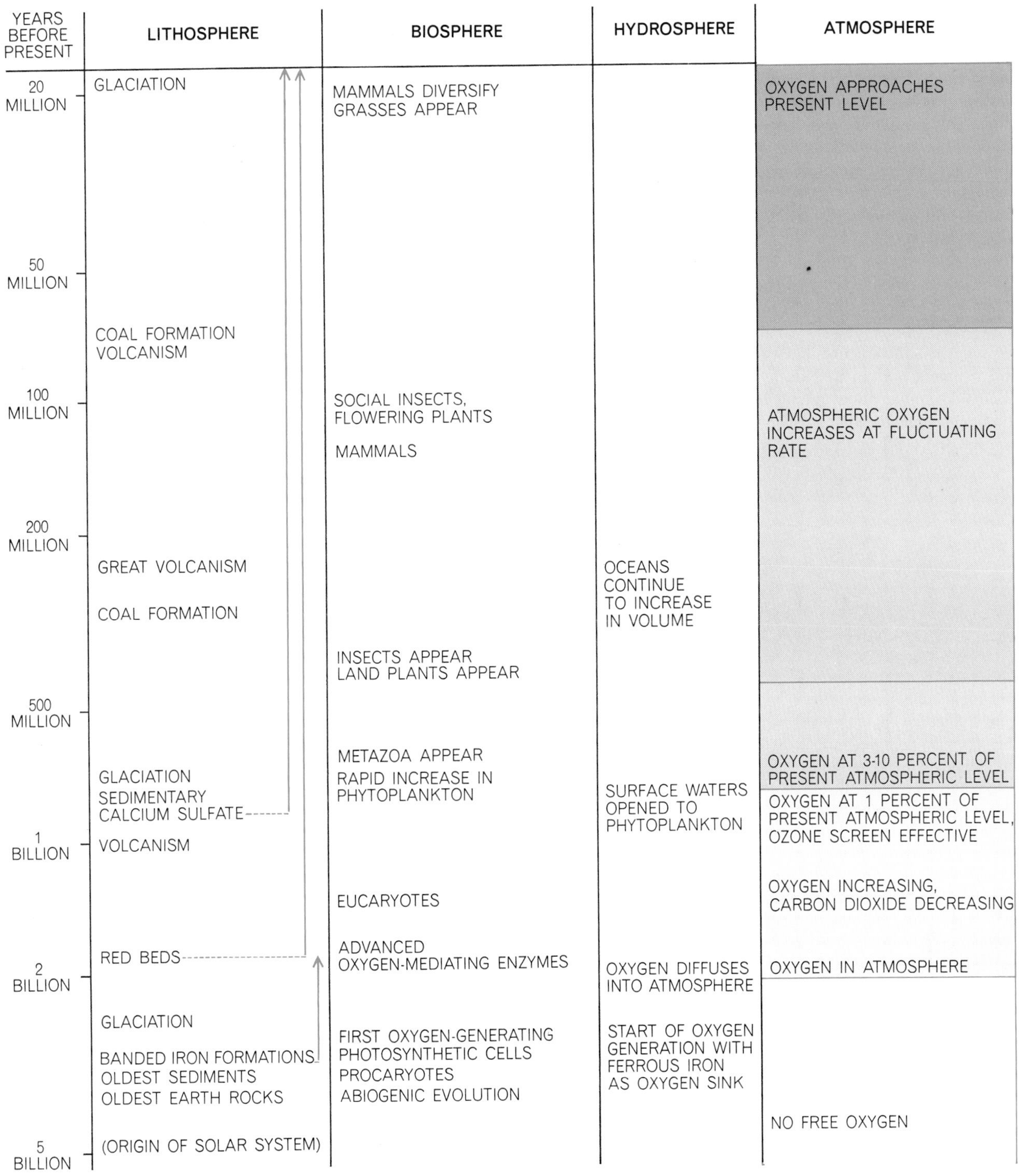

YEARS BEFORE PRESENT	LITHOSPHERE	BIOSPHERE	HYDROSPHERE	ATMOSPHERE
20 MILLION	GLACIATION	MAMMALS DIVERSIFY GRASSES APPEAR		OXYGEN APPROACHES PRESENT LEVEL
50 MILLION				
	COAL FORMATION VOLCANISM			
100 MILLION		SOCIAL INSECTS, FLOWERING PLANTS MAMMALS		ATMOSPHERIC OXYGEN INCREASES AT FLUCTUATING RATE
200 MILLION	GREAT VOLCANISM COAL FORMATION		OCEANS CONTINUE TO INCREASE IN VOLUME	
		INSECTS APPEAR LAND PLANTS APPEAR		
500 MILLION		METAZOA APPEAR RAPID INCREASE IN PHYTOPLANKTON		OXYGEN AT 3-10 PERCENT OF PRESENT ATMOSPHERIC LEVEL
	GLACIATION SEDIMENTARY CALCIUM SULFATE		SURFACE WATERS OPENED TO PHYTOPLANKTON	OXYGEN AT 1 PERCENT OF PRESENT ATMOSPHERIC LEVEL, OZONE SCREEN EFFECTIVE
1 BILLION	VOLCANISM			
		EUCARYOTES		OXYGEN INCREASING, CARBON DIOXIDE DECREASING
2 BILLION	RED BEDS	ADVANCED OXYGEN-MEDIATING ENZYMES	OXYGEN DIFFUSES INTO ATMOSPHERE	OXYGEN IN ATMOSPHERE
	GLACIATION BANDED IRON FORMATIONS OLDEST SEDIMENTS OLDEST EARTH ROCKS	FIRST OXYGEN-GENERATING PHOTOSYNTHETIC CELLS PROCARYOTES ABIOGENIC EVOLUTION	START OF OXYGEN GENERATION WITH FERROUS IRON AS OXYGEN SINK	NO FREE OXYGEN
5 BILLION	(ORIGIN OF SOLAR SYSTEM)			

CHRONOLOGY that interrelates the evolutions of atmosphere and biosphere is gradually being established from evidence in the geological record and in fossils. According to calculations by Lloyd V. Berkner and Lauriston C. Marshall, when oxygen in the atmosphere reached 1 percent of the present atmospheric level, it provided enough ozone to filter out the most damaging high-energy ultraviolet radiation so that phytoplankton could survive everywhere in the upper, sunlit layers of the seas. The result may have been a geometric increase in the amount of photosynthesis in the oceans that, if accompanied by equivalent sequestration of carbon, might have resulted in a rapid buildup of atmospheric oxygen, leading in time to the evolution of differentiated multicelled animals.

gen cycle of the biosphere in truly broad terms will depend on how good we are at weaving together the related strands of biospheric, atmospheric, hydrospheric and lithospheric evolution throughout geologic time. Whatever we may conjecture about any one of these processes must be consistent with what is known about the others. Whereas any one line of evidence may be weak in itself, a number of lines of evidence, taken together and found to be consistent, reinforce one another exponentially. This synergistic effect enhances our confidence in the proposed time scale linking the evolution of oxygen in the atmosphere and the management of the gaseous oxygen budget within the biosphere [see illustration on page 215].

The most recent factor affecting the oxygen cycle of the biosphere and the oxygen budget of the earth is man himself. In addition to inhaling oxygen and exhaling carbon dioxide as a well-behaved animal does, man decreases the oxygen level and increases the carbon dioxide level by burning fossil fuels and paving formerly green land. He is also engaged in a vast but unplanned experiment to see what effects oil spills and an array of pesticides will have on the world's phytoplankton. The increase in the albedo, or reflectivity, of the earth as a result of covering its waters with a molecule-thick film of oil could also affect plant growth by lowering the temperature and in other unforeseen ways. Reductions in the length of growing seasons and in green areas would limit terrestrial plant growth in the middle latitudes. (This might normally be counterbalanced by increased rainfall in the lower latitudes, but a film of oil would also reduce evaporation and therefore rainfall.) Counteracting such effects, man moves the earth's fresh water around to increase plant growth and photosynthesis in arid and semiarid regions. Some of this activity, however, involves the mining of ground water, thereby favoring processes that cause water to be returned to the sea at a faster rate than evaporation brings it to the land.

He who is willing to say what the final effects of such processes will be is wiser or braver than we are. Perhaps the effects will be self-limiting and self-correcting, although experience should warn us not to gamble on that. Oxygen in the atmosphere might be reduced several percent below the present level without adverse effects. A modest increase in the carbon dioxide level might enhance plant growth and lead to a corresponding increase in the amount of oxygen. Will a further increase in carbon dioxide also have (or renew) a "greenhouse effect," leading to an increase in temperature (and thus to a rising sea level)? Or will such effects be counterbalanced or swamped by the cooling effects of particulate matter in the air or by increased albedo due to oil films? It is anyone's guess. (Perhaps we should be more alarmed about a possible decrease of atmospheric carbon dioxide, on which all forms of life ultimately depend, but the sea contains such vast amounts that it can presumably keep carbon dioxide in the atmosphere balanced at about the present level for a long time to come.) The net effect of the burning of fossil fuels may in the long run be nothing more than a slight increase (or decrease?) in the amount of limestone deposited. In any event the recoverable fossil fuels whose combustion releases carbon dioxide are headed for depletion in a few more centuries, and then man will have other problems to contend with.

What we want to stress is the indivisibility and complexity of the environment. For example, the earth's atmosphere is so thoroughly mixed and so rapidly recycled through the biosphere that the next breath you inhale will contain atoms exhaled by Jesus at Gethsemane and by Adolf Hitler at Munich. It will also contain atoms of radioactive strontium 90 and iodine 131 from atomic explosions and gases from the chimneys and exhaust pipes of the world. Present environmental problems stand as a grim monument to the cumulatively adverse effects of actions that in themselves were reasonable enough but that were taken without sufficient thought to their consequences. If we want to ensure that the biosphere continues to exist over the long term and to have an oxygen cycle, each new action must be matched with an effort to foresee its consequences throughout the ecosystem and to determine how they can be managed favorably or avoided. Understanding also is needed, and we are woefully short on that commodity. This means that we must continue to probe all aspects of the indivisible global ecosystem and its past, present and potential interactions. That is called basic research, and basic research at this critical point in history is gravely endangered by new crosscurrents of anti-intellectualism.

THREE ORGANELLES that are involved in oxygen metabolism in the living cell are enlarged 40,000 diameters in an electron micrograph of a tobacco leaf cell made by Sue Ellen Frederick in the laboratory of Eldon H. Newcomb at the University of Wisconsin. A peroxisome (center) is surrounded by three mitochondria and three chloroplasts. Oxygen is produced in the grana (layered objects) in the chloroplasts and is utilized in aerobic respiration in the mitochondria. Peroxisomes contain enzymes involved in oxygen metabolism.

Very few general chemistry texts or laboratory manuals contain a discussion of instrumental analysis. The demands of space and the level of sophistication required to understand these methods generally preclude the inclusion of this material. Unfortunately, this often leaves the beginning student with a somewhat sour taste in his mouth that chemistry consists of calculating solubility products and squinting at burets. The student has no feel for the excitement of modern chemical research. Much of today's research is dependent in part or totally on recent developments in chemical instrumentation. For this reason, we have included this section in the reader to open up for you a few areas of current work.

We have been able to include articles dealing with four instrumental methods. These techniques are X-ray crystallography, infrared spectroscopy, gas chromatography, and mass spectrometry. These methods usually serve one of two purposes. Either they are useful in the determination of the molecular structures of compounds, or they permit the identification of the components of an unknown mixture.

These four techniques of course barely scratch the surface of the plethora of methods now available to the chemist. Although it will not be possible to cover all of the other methods or to treat in depth the ones considered here, you may find it of some interest to browse briefly through some of the other methods currently available.

The widest spectrum of analytical techniques involves the use of electromagnetic radiation. Techniques of ultraviolet (UV) and visible spectroscopy are useful for the study of electronic transitions between the energy states of atoms and molecules. These techniques are also used to determine concentrations of solutes. Usually it is possible to apply Beer's law to spectroscopic data. That law is a relation between the absorbance (A) of the solution, the concentration (c) of solute, the optical path length (l), and (ϵ) the proportionality constant: $A = \epsilon c l$.* The concentration of nucleic acids can be measured by the absorbance at 260 nm and that of proteins at 280 nm. Optical rotatory dispersion (ORD) and circular dichroism (CD) are powerful techniques for the determination of the configuration of biopolymers in solutions. Both methods rely upon the rotation of polarized light as a structural probe. Molecules that display optical rotatory properties are those lacking a center of symmetry, such as molecules with a carbon atom having four different groups attached to it, or molecules containing helices, such as the α-helix of proteins or the Watson-Crick helix of DNA. Nuclear magnetic resonance, NMR, is probably the most important tool available to the organic chemist today for structural determinations. It is not represented in the articles of this section because it is a fairly difficult technique for the beginning student to understand. The nuclei of atoms having odd numbers of nucleons behave like unpaired electrons in that they have a net spin. The proton of the hydrogen nucleus is the best example. If a proton is placed in a magnetic field, **H**, it aligns itself so that its spin is parallel with

---

*A common student version of this law is: "The taller the glass, the thicker the brew, the less the amount of light that gets through."

the field. A radio wave of the appropriate frequency can cause this spin to be inverted. Resonant absorption of the radio wave then occurs and can be recorded. In actual practice, the radio wave frequency is held constant and **H** is varied. The technique is useful because the value of **H** at which resonant absorption occurs depends upon the detailed atomic environment of the H atom. NMR spectrometers of increasing magnetic field strength are being built. As of this writing, the 300-megahertz instrument in Professor Kearn's laboratory at the University of California, Riverside, is the largest instrument in existence. It is yielding detailed information about the nature of the hydrogen bonds in transfer RNA (see Holley's article, 35).

Other techniques involving electromagnetic radiation are electron spin resonance (ESR), light scattering, and Raman, microwave, and atomic absorption spectroscopies.

A number of techniques have been developed primarily for the study of biological macromolecules. Countercurrent distribution is a liquid-liquid partitioning technique. The principles of this method are not unlike those of gas chromatography discussed below. Its use is described by Holley in article 35 for the preparation of pure samples of transfer RNA. Other chromatographic techniques include the use of molecular sieves for hydrocarbons (discussed by Eglinton and Calvin in article 23), ion exchange chromatography (see Keller's article, 22) and diethylaminoethyl cellulose chromatography (used by Holley, article 35). The botanist M. Tswett used an adsorption method to separate plant pigments, circa 1900; it was called chromatography. Paper chromatography was discovered in the 1940's. Electrophoresis requires application of an electrical potential across a supporting medium, such as cellulose acetate or a buffer solution. The technique separates macromolecules on the basis of their electrical charge. The development of the analytical ultracentrifuge by The Svedberg in Uppsala, Sweden in 1923 marked a milestone in the development of protein chemistry. It provided the first unequivocal demonstration of the existence of proteins as discrete molecules. The ultracentrifuges of today can spin 0.7 ml of solution at speeds up to 68,000 rpm and generate a force one-third million times that of gravity. Particles of density greater than the solvent move outward, particles of density less than the solvent move inward, and particles of density equal to the solvent are neutrally buoyant. Such experiments yield information about the shape, molecular weight, and density of a polymer, and provide methods of separation. (See Kornberg's article, 32, and Spiegelman's article, 33, for applications of the ultracentrifuge.) Viscometry is a simple technique that can also be used to learn about the shapes of molecules. The only measurement required is the timed rate of flow through a capillary. Osmotic pressure measurements yield molecular weights. Finally, the electron microscope is a new and powerful addition to the tools available to the biophysical chemist. Article 36, by Miller, displays some of the most elegant micrographs ever taken with the electron microscope. Such instruments use a beam of electrons rather than electromagnetic radiation. The electrons form an image of a molecule that has been deposited on a grid and appropriately stained. The present resolution of 5 Å provides beautiful micrographs of DNA and RNA chains and protein molecules.

An entire range of relaxation techniques has been developed recently. Faller has already described these in some detail in article 15. Systems are rapidly pulsed with a pressure jump, temperature jump, sound absorption, or an electric field. The system is then studied as it "relaxes" back to its original state.

No presentation of instrumental methods, even one as elementary as this one, would be complete without some mention of the computer. The incredibly informative diagrams of Wahl (shown in article 5), the fascinating gyrations of gases, liquids, and solids (described by Alder and Wainwright in article 10), and the seemingly impossible task of providing detailed three-dimensional maps of all of the atoms of protein molecules (such as those shown by Dickerson, in article 30, and Phillips, in article 31), would have been utterly impossible without the capabilities of high-speed computers. Although the computer is little more in concept than an electronic calculator, its enormous memory and the lightning speed of its operation make possible calculations that ten years ago were deemed hopeless.

The above discussion has provided a rapid overview of the salient characteristics of most of the important analytical methods. We now turn to a discussion of the four techniques dealt with in the articles of this section.

One of our criteria for the selection of the articles in this reader was to choose articles written by people who were and are the leaders in their fields. We want you to become acquainted with the men and women who have actually done the work to build the science of chemistry. The author of the first article in this section, "X-ray Crystallography," has one of the longest productive records of any living scientist. Sir Lawrence Bragg submitted his first research paper in 1912. It contained an equation he had derived to describe the new technique of the diffraction of X rays by crystals to determine interionic spacing, d, in crystals. Today we call this equation the Bragg equation: $n\lambda = 2d \sin \theta$. The relation states that an integer, n, multiplied by the wavelength of radiation employed, λ, equals two times d times the sine of the angle of incidence, θ. Sir Lawrence and his father, Sir William Bragg, worked at the Royal Institution and provided much of the experimental developmental work and the theoretical foundations of this technique that has led to such recent spectacular successes as the revelation of the three-dimensional structure of a protein molecule.

In "X-ray Crystallography," Sir Lawrence discusses the conditions for constructive interference of X rays and points out why only X rays have the requisite property, namely a λ that is short enough, to be effective for this method. The structure of NaCl is used to show the several planes of atoms that reflect X rays and why three integers, h, k, and l, must be used to specify these planes. Stressing the importance of a quantity which denotes the intensity of the X-ray beam scattered by each unit of the pattern, Sir Lawrence notes that the repeating units of this pattern can be ions, atoms, or entire molecules, as we also described in the introduction to Section IV. Experimental methods are described for the two basic techniques of crystal photography: in one a single, rotating crystal is photographed; in the other a mass of tiny stationary crystals produce a powder photograph.

An historical development is given to explain why the original method of assuming a particular pattern of atoms has been replaced by the mathematical summation process called Fourier synthesis. This new technique made possible the solution of the 181-atom vitamin B_{12} molecule by Dorothy Crowfoot Hodgkin in 1955. Sir Lawrence concludes his article with a discussion of how the development of another new technique called heavy-atom replacement has led to knowledge of the incredible structures of protein molecules (see Dickerson's article, 30, and Phillips's article, 31).

Infrared spectroscopy has been one of the most important tools of organic chemistry since 1943. One of the pioneers in the development of infrared instruments was Bryce Crawford. His article in this reader

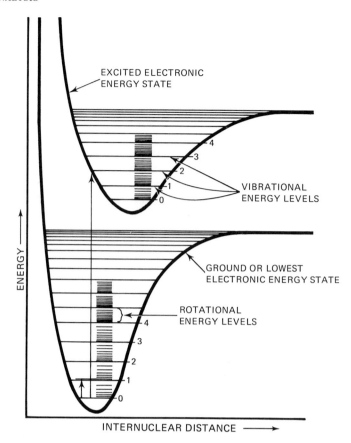

Figure VI.1 Energy levels of a diatomic molecule. The upper curve, and the numbers along it, show vibrational energy levels in the excited electronic energy state. The lower curve, and the numbers along it, show vibrational energy levels in the ground, or lowest, electronic energy state. The long arrow represents an electronic transition from the 0th vibrational level of the ground state to the 2nd vibrational level of the excited electronic state. Ultraviolet light is required for this transition. The short arrow represents a vibrational-rotational transition. Infrared light is required for such a transition, which takes place with a relatively small change in energy. The absorptions of infrared light observed for HCl in Figure VI.2 are caused by such transitions. The short bars above each vibrational energy level represent the rotational energy levels within that vibrational energy level. For example, the bars just above line 4 in the lower curve are the rotational energy levels in the 4th vibrational level of the ground electronic state.

is two decades old, but the material Professor Crawford presents is as relevant and accurate as in 1953.

We have seen in the introduction to Section II that hydrogen atoms have discrete electronic energy levels. The same is true of molecules, and the study of the absorption of radiation by molecules has led to a much deeper understanding of the structure of molecules.

Studying absorption is considerably more complicated with molecules than with atoms. For one thing, the electronic energy levels of a molecule depend on internuclear spacing; for another, the molecule can both rotate and vibrate. Figure VI.1 displays the several types of energy that molecules can possess. The lines for rotation and vibration demonstrate that these energies—as well as the energies of the electrons—are quantized. As in the H atom, absorption of electromagnetic radiation can occur only when the energy, $\epsilon = h\nu$, of the incoming photon exactly matches one of the allowed (some transitions are forbidden) energy differences. If a transition occurs from one electronic state to another, light in the UV or visible region must be absorbed. Transitions between vibrational levels within the same electronic energy state are caused by infrared radiation. Transitions between rotational levels within the same vibrational level are brought about by microwave and far-IR radiation. Thus it is well to keep in mind that Crawford's article deals only with vibrational changes induced by infrared radiation.

Molecules of more than two or three atoms can undergo quite complex vibrations. Two atoms undergo simple harmonic oscillation. Three-atom molecules can experience a variety of vibrations. All of these vibrations can be reduced to just three fundamental modes of vibration. Crawford provides good pictorial representations of the three vibrational modes of CO_2. It is important to note that only those modes which alter the electron distribution in the molecule lead to absorption of infrared radiation.

A number of infrared spectra are presented. These spectra are plots of the absorbance of IR radiation versus either the wavelength or the wave number of the radiation. You may recall that the frequency, ν, and wavelength, λ, of electromagnetic radiation are related through their product to the speed of light, c: $c = \lambda\nu$. Light comes in packets called photons which have energy $\epsilon = h\nu$, where h is Planck's constant. These two equations can be combined to give: $\epsilon = hc/\lambda$. Because energy is inversely proportional to wavelength and not directly proportional, a new unit called the wave number, $\tilde{\nu}$, has been defined as the reciprocal of the wavelength, $1/\lambda$. A check of the units shows that $\epsilon = hc\tilde{\nu}$. Older and less sophisticated instruments are calibrated in wavelengths that range from about 2 to 25 microns (1 micron = 10^{-6} meter). Newer and more sophisticated instruments are calibrated in wave numbers and have ranges from 200 to 4000 cm^{-1}.

Historically, the principal use of IR spectroscopy has been the identification of molecules. Each specific group of atoms in a molecule absorbs radiation at a particular frequency. The figure at the bottom of page 243 gives some of these characteristic bands. Far more extensive tables are available; these give the wave-number ranges of absorption of a large number of groups of atoms. Such tables enable the chemist to interpret the spectra shown on page 242. Each molecule has its own characteristic fingerprint region; frequently this can lead to a fairly definite identification of the molecule.

Recently, high-resolution IR spectrometers have been developed which resolve the broad bands mentioned above into their component peaks, revealing detailed information about the structure of small molecules. Gaseous hydrogen chloride gives a rather broad band around 2200 cm^{-1} when examined with a fairly low-resolution spectrometer. The spectrum shown in Figure VI.2 was obtained with a Beckman IR-12 spectrometer, which has rather good resolution. The multiple peaks represent transitions from one rotational level of the lowest vibrational state to another rotational level of the next highest state. A detailed, quantum mechanical analysis of this spectrum yields the internuclear distance between the H and Cl atoms, the fundamental frequency at which the two atoms vibrate, and how anharmonic the vibrations of the HCl molecule are.

The technique of gas chromatography, which Keller describes in article 22, is not designed to reveal the structures of molecules whose properties have never been studied. Instead, it is an exceedingly useful tool to determine the composition of unknown mixtures provided that the components of the mixture are molecules whose properties in the chromatographic system are known.

The chromatographic process begins with the injection of a tiny amount (as small as 5×10^{-6} ml) of a liquid into a flowing stream of gas. The carrier gas carries the vapor through a long column, or tube. The tube is generally filled with small pellets that are coated with a non-

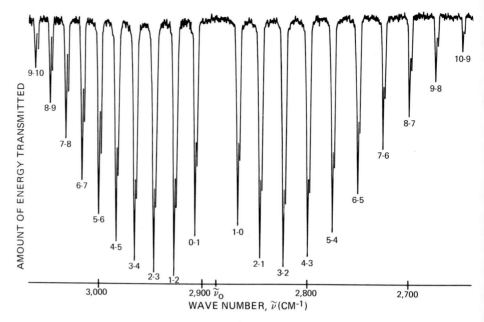

Figure VI.2 Spectrum produced by gaseous hydrogen chloride, as measured with a Beckman IR-12 spectrometer. All the absorptions in this spectrum arise from transitions of the HCl molecule from the 0th vibrational state to the 1st vibrational state. The numbers below each absorption signify the change in the rotational state during the transition. For example, 1-2 shows absorption by a molecule starting in the 1st rotational state in the 0th vibrational state and ending in the 2nd rotational state of the 1st vibrational state. The vibrational quantum number, v, and the rotational quantum number, J, are used to identify such transitions: $v = 0, J = 1 \rightarrow v = 1, J = 2$. In a low-resolution infrared spectrometer all of these absorptions smear into one broad band centered at the fundamental frequency, $\tilde{\nu}_0$. The absorptions at frequencies higher than $\tilde{\nu}_0$ (to the left of $\tilde{\nu}_0$) are $\Delta J = +1$ transitions. The absorptions at frequencies lower than $\tilde{\nu}_0$ are $\Delta J = -1$ transitions. Transitions such that $\Delta J = 0$ or $\Delta J = +2, +3, \ldots$, cannot occur. Each absorption is a doublet because 76 percent of natural chlorine is ^{35}Cl while 24 percent is the heavier isotope ^{37}Cl. The smaller peaks associated with the $^{1}H\ ^{37}Cl$ molecules are at slightly lower frequencies than the corresponding peaks associated with the $^{1}H\ ^{35}Cl$ because of the larger mass of the ^{37}Cl isotope.

volatile liquid. Even molecules of very similar structure have different solubilities in the stationary liquid phase. Thus different compounds move through the tube at different rates. A detector placed at the end of the tube records each component as it emerges. Each peak is identified by measuring the elution time of known components under identical conditions of temperature, column packing, and gas flow. The areas under the several peaks yield the concentrations of each component within a few percent.

The separations achievable by gas chromatography are astonishing. Keller cites one case of the resolution of 14 components in a mixture, which was completed in two seconds, and which used a liquid sample of 5×10^{-6} ml volume. This technique is very sensitive in measuring trace components and is probably the most satisfactory technique available for the determination of extremely low molecular concentrations. Plots are given in this article of the gas-chromatographic analysis of samples of bourbon, scotch, and gin, which contain volatile components that influence taste and aroma. The data indicate that bourbon contains the most of these components in addition to the expected content of water and ethanol.

In the decade since this article was written, one prediction in the article has not been borne out. Keller discusses a variety of other chromatographic techniques such as liquid-solid and gas-solid chromatography, and predicts that gas-liquid chromatography might prove superior to liquid-solid or ion-exchange resin chromatography in the analysis of amino acid mixtures, but this has not been realized.

The last article, by Eglinton and Calvin, introduces another technique and provides an exciting example of how new instrumental advances have opened up entirely new fields of research. The new technique is mass spectrometry and the new field is paleochemotaxonomy.

Paleochemotaxonomy is the classification of ancient organisms by means of chemical analysis of the remains of those organisms. Classical paleontology relies on finding the actual remains of the organism. Generally only the skeleton can be found, except in rare instances when preserved bodies have been found—as happened in the peat bogs of Jutland, Denmark. More often, though, all visible traces of the tissue portions of the organism have vanished. But that does not mean that *all* traces have disappeared; namely, some molecules may still exist. Proteins, nucleic acids, and polysaccharides don't weather particularly well over a period of a million years. They are all formed by dehydration reactions and they hydrolyze fairly readily to give back their constituent monomers, and these give no clues as to the structure of the parent organism. One class of molecules weathers extremely well. These are hydrocarbons—molecules consisting of only carbon and hydrogen. These compounds arise from a class of biopolymers called lipids, which were discussed previously in article 8, by Fox. The relation of a simple straight-chain lipid to its constituent parts is shown in Figure VI.3. The reaction shown here is the same as that employed by pioneer women when they boiled fat with ashes, which contain lye (NaOH), to obtain soap. Weathering of lipids over several millennia leads to the same reaction, plus the loss of one carbon atom from the chain by decarboxylation. Because the fatty acids were originally synthesized from two-carbon fragments, the resulting hydrocarbons contain an odd number of carbon atoms. Knowing this, Eglinton and Calvin have looked for such straight-chain hydrocarbons as well as other branched and cyclic hydrocarbons in ancient rocks, as a delicate probe into the existence of earlier life forms. The types and concentrations of such molecules are characteristic of a particular organism.

A problem encountered with this approach is the identification of trace amounts of a mixture of quite similar molecules. We would expect the technique of gas chromatography we have just encountered to be an integral part of such an analysis, and indeed it is.

The material to be analyzed is pulverized and the lipids are extracted. This fraction is passed through a column to obtain the alkane fraction. The resultant alkane mixture is further fractionated with molecular sieves. These sieves are synthetically produced aluminosilicate solids, which are ideally suited for the separation of the several classes of hydrocarbons. They consist of networks of channels of rather uniform size. Sieves having a pore size of 5 Å are used. This permits the straight-chain alkanes of diameter 4.5 Å to enter the sieve partially. They are recovered from the sieve when the sieve is dissolved in hydrogen fluoride. The branched and cyclic hydrocarbons do not enter the sieve but stay in the solution. The linear alkanes are then separated by gas chromatography.

The final problem is that it is difficult to identify the linear alkanes positively from their position on the chromatogram because they are so large. Eglinton and Calvin solved this by feeding each compound emerging from the GC one at a time into a mass spectrometer. This device is in common use to identify all kinds of organic molecules. It directs a beam of ionizing electrons at the gaseous sample. This literally knocks the molecule to pieces. The result is a series of small positively charged fragments. These charged fragments pass through a magnetic field that separates the particles on the basis of their mass-to-charge ratios. These mass spectra give the amounts of each fragment present. Each molecule has a unique mass-distribution pattern and can thus be identified.

Chemistry has come a long way from the days of a double pan balance, some litmus paper, and a couple of test tubes.

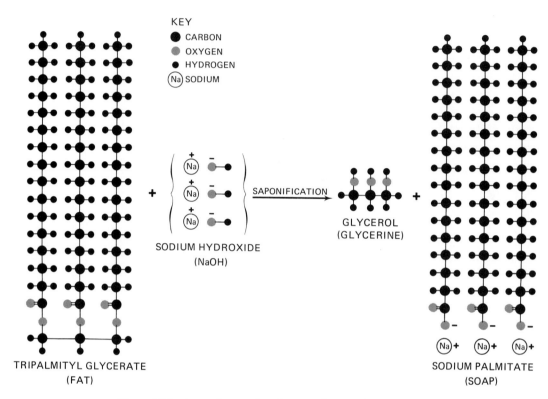

Figure VI.3   A reaction used in the manufacture of soap.

20

X-RAY CRYSTALLOGRAPHY

SIR LAWRENCE BRAGG
July 1968

The new knowledge of the atomic structure of matter uncovered over the past half-century by the X-ray-diffraction technique has led to a fundamental revision of ideas in many sciences

Fifty-six years ago a new branch of science was born with the discovery by Max von Laue of Germany that a beam of X rays could be diffracted, or scattered, in an orderly way by the orderly array of atoms in a crystal. At first the main interest in von Laue's discovery was focused on its bearing on the controversy about the nature of X rays; it proved that they were waves and not particles. It soon became clear to some of us, however, that this effect opened up a new way of studying matter, that in fact man had been presented with a new form of microscope, several thousand times more powerful than any light microscope, that could in principle resolve the structure of matter right down to the atomic scale. The development of X-ray crystallography since 1912 has more than fulfilled our early expectations. It not only has revealed the way atoms are arranged in many diverse forms of matter but also has cast a flood of light on the nature of the forces between the atoms and on the large-scale properties of matter. In many cases this new knowledge has led to a fundamental revision of ideas in other branches of science. A culmination of sorts has been reached in the past few years with the successful structural analysis of several of the basic molecules of living matter—the proteins—each of which consists of thousands of atoms held together by an incredibly intricate network of chemical bonds.

The purpose of this article is to go back to the beginning and broadly summarize the course of X-ray crystallography over the past half-century or so. In so doing I shall try to answer two key questions: Why X rays? Why crystals?

X-ray crystallography is a strange branch of science. The result of an investigation lasting many years can be summed up in a "model." I have often been asked: "Why are you always showing and talking about models? Other kinds of scientists do not do this." The answer is that what the investigator has been seeking all along is simply a structural plan, a map if you will, that shows all the atoms in their relative positions in space. No other branch of science is so completely geographical; a list of spatial coordinates is all that is needed to tell the world what has been discovered.

The atomic structure of a crystal is deduced from the way it diffracts a beam of X rays in different directions. A crystal is built of countless small structural units, each consisting of the same arrangement of atoms; the units are repeated regularly like the pattern of a wallpaper, except that in a crystal the pattern extends in three dimensions in space. The directions of the diffracted beams depend on the repeat distances of the pattern. The strengths of the diffracted beams, on the other hand, depend on the arrangement of atoms in each unit. The wavelets scattered by the atoms interfere to give a strong resultant in some directions and a weak resultant in others. The goal of X-ray analysis is to find the atomic arrangement that accounts for the observed strengths of the many diffracted beams.

This brings us to the question of why X rays, of all the available forms of electromagnetic radiation, are indispensable for this method of investigation. In order for the interference of the diffracted beams to produce marked changes in the amount of scattering in different directions, the differences in the paths taken by reflected beams must be on the order of a wavelength. Only X rays have wavelengths short enough to satisfy this condition. For example, the distance between neighboring sodium and chlorine atoms in a crystal of sodium chloride (ordinary table salt) is 2.81 angstrom units (an angstrom is 10^{-10} meter), whereas the most commonly used wavelength in X-ray analysis is 1.54 angstroms.

Actually crystals came into the picture only because they are a convenient means to an end. The resultant scattering of X rays would be hopelessly confused and impossible to interpret if the scattering units were randomly distributed in all orientations. In a crystal the units are all similarly oriented and hence scatter the X rays in the same way; as a result a total scattering measurement made with a whole crystal leads directly to a determination of the amount scattered by an individual unit.

The Condition for Diffraction

The easiest way to approach the optical problem of X-ray diffraction is to consider the X-ray waves as being reflected by sheets of atoms in the crystal. When a beam of monochromatic (uniform wavelength) X rays strikes a crystal, the wavelets scattered by the atoms in each sheet combine to form a reflected wave. If the path difference for waves reflected by successive sheets is a whole number of wavelengths, the wave trains will combine to produce a strong reflected beam. In more formal geometric terms, if the spacing between the reflecting planes is d and the glancing angle of the incident X-ray beam is θ, the path difference for waves reflected by successive planes is $2d \sin \theta$ [*see upper illustration on page 228*]. Hence the condition for diffraction is $n\lambda = 2d \sin \theta$, where n is an integer and λ is the wavelength.

I first stated the diffraction condition in this form in my initial adventure into research in a paper presented to the Cambridge Philosophical Society in

1912, and it has come to be known as Bragg's law. It is, I have always felt, a cheaply earned honor, because the principle had been well known for some time in the optics of visible light.

The atoms of a given crystal can be arranged in sheets in a number of differ-ent ways; three possible arrangements of the sheets in a crystal of sodium chlo-ride are indicated in the illustration on the opposite page. The equation for re-flection can be satisfied for any set of planes whose spacing is greater than half the wavelength of the X rays used; this condition sets a limit on how many or-ders of diffracted waves can be obtained from a given crystal using an X-ray beam of a given wavelength.

In the case of an optical diffraction grating with an interlinear spacing a, the orders of the diffracted waves are de-fined by a single integer n in the equa-tion $n\lambda = a \sin \theta$; the diffracted waves are referred to as first-order waves, sec-ond-order waves and so forth [*see lower illustration at left*]. In the case of a crys-tal, on the other hand, the pattern re-peats in three dimensions, and so the order of the diffracted waves must be defined by three integers, which are rep-resented generally by the letters h, k and l.

In the structural diagrams of sodium chloride on the opposite page the axes of the structure are denoted by the letters OA, OB and OC, these being the inter-vals at which the pattern repeats. In the diagram at the left the first reflection to appear from the planes perpendicular to OA will be one for which there is a path difference of the two wavelengths between O and A, since there are two sheets of atoms in this distance. With re-spect to the spacing OA, then, this ini-tial reflection is a second-order reflec-tion; with respect to the spacings OB and OC, however, the same reflection is a zero-order reflection, since the reflect-ing planes are parallel to both OB and OC. Therefore this type of reflection, or diffraction, is assigned the order (200), indicating $h = 2$, $k = 0$ and $l = 0$. Sim-ilarly, the initial reflection to appear from the planes in the diagram at the center is (220), whereas for the diagram at the right it is (111). Higher orders of reflec-tion would of course have higher integer values of h, k and l.

An Example of X-Ray Analysis

The structure of sodium chloride is a simple arrangement of cubic symmetry in which sodium and chlorine atoms oc-cur alternately in three directions at right angles, like a chessboard in three dimen-sions. How was this structure derived?

The analysis will be gone into in some detail here, because it is generally repre-sentative, even though this particular case is so very simple. A glance at the structural diagrams of sodium chloride shows that the planes represented there are of two kinds. The reflections, or or-ders, designated (200), (400), (600) and so on, and those designated (220), (440), (660) and so on, arise from sheets of at-oms that are identical, each containing equal numbers of sodium and chlorine

BRAGG'S LAW, first formulated by the author in 1912, states the condition for diffraction of an incident beam of monochromatic X rays by the successive sheets of atoms in a crystal. In general terms the law states that if the path difference for waves reflected by successive sheets of atoms is a whole number of wavelengths, the wave trains will combine to produce a strong reflected beam. In more formal geometric terms, if the spacing between the reflect-ing planes of atoms is d and the glancing angle of the incident X-ray beam is θ, the path dif-ference for waves reflected by successive planes is $2d \sin \theta$. In this diagram the extra path followed by the lower ray (*heavy colored line at bottom*) is four wavelengths long, which is exactly equal to the path difference of $2d \sin \theta$ between the two diffracted rays (*upper right*).

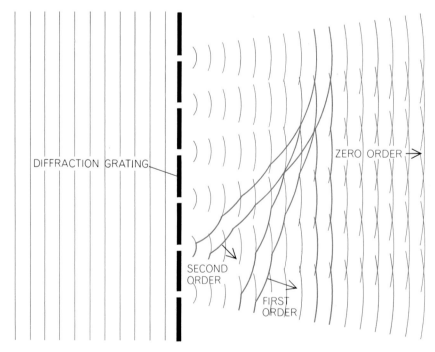

DIFFRACTION ORDERS are illustrated here for the comparatively simple case of a ruled optical diffraction grating. In this case the diffracted waves are defined by a single integer n in the equation $n\lambda = a \sin \theta$, where λ is the wavelength of the incident radiation and a is the spacing between the lines of the grating. In the case of a crystal, on the other hand, the pattern repeats in three dimensions, and so the order of the diffracted X-ray waves must be defined by three integers, which are represented generally by the letters h, k and l.

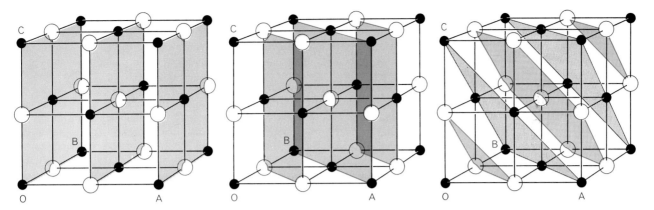

THREE POSSIBLE ARRANGEMENTS of the reflecting sheets of atoms in a sodium chloride crystal are indicated by the colored planes. The axes of this simple cubic crystal are denoted by the letters OA, OB and OC. In the diagram at left the first reflection to appear from the planes perpendicular to OA is assigned the order (200), since there is a path difference of two wavelengths between O and A while the reflecting planes are parallel to OB and OC. (In this case $h = 2$, $k = 0$ and $l = 0$.) Similarly, the initial reflection to appear from the planes in the diagram at center is designated (220), whereas for the diagram at right it is (111) with respect to chlorine planes. In general, orders with even indices arise from sheets that are identical and hence result in strong, in-phase reflections, whereas orders with odd indices arise from alternately occupied sheets and hence result in weak, out-of-phase reflections.

atoms. One would expect the sequence of successive orders to fall off regularly in intensity. As the diagram at the right on this page shows, however, the reflection (111) comes from a more complex set of planes, in that the sheets are alternately occupied by sodium and chlorine atoms. Since for (111) there is a path difference of one wavelength for the strongly reflecting chlorine planes, the waves reflected from the weaker sodium planes halfway between them will be opposite in phase. The order (111) will be weak, since the sodium contribution partially offsets the chlorine contribution. On the other hand, for (222) the contributions will be in phase and the order will be strong.

In this type of space lattice, as it is called, there are identical points at the face centers as well as at the corners of the cube; this implies that the indices must be either all odd or all even. These observations can be generalized by stating that orders with even indices, such as (200), (220) and (222), should form a strong sequence, whereas those with odd indices, such as (111), (113) and (333), should be comparatively weak.

This is the effect that is actually observed. The illustration on page 7 shows a very early set of measurements of sodium chloride and potassium chloride made with the ionization spectrometer, a device invented by my father, W. H. Bragg, in 1913. The abscissas measure the glancing angle, the ordinates the strength of the reflection. The two peaks seen on each order are the $K\alpha$ and $K\beta$ "lines" in the spectrum of the palladium anticathode, the $K\alpha$ line being the stronger of the two. The orders are re-

flected from crystal faces with crystallographic indices (100), (110) and (111). As the curves show, the order (111) for sodium chloride is anomalously small, whereas (222) fits into the same sequence with (200) and (220). For potassium chloride, on the other hand, the scattering powers of the potassium atoms and the chlorine atoms are so nearly the same that the order (111) is too weak to be observed. It was on the basis of such evidence that the structural arrangement of both of these alkaline halides was confirmed.

Although the preceding analysis is somewhat simplified, it is a typical example of the method used in the early determinations of crystal structure. A number of orders of diffracted waves were measured, either with the ionization spectrometer or on a photographic plate, and an attempt was then made to find an atomic arrangement that accounted for the relative intensities of the various orders.

The Significance of $F(hkl)$

The quantity $F(hkl)$ is the cornerstone of X-ray analysis, and its determination is the final aim of all experimental methods. This quantity is a measure, for each order (hkl), of the intensity of the beam scattered by the whole unit of a pattern expressed in terms of the amount scattered by a single electron as a unit. For instance, the quantity $F(000)$ is scattered in the forward direction through zero angle, so that there are no path differences to cause interference; $F(000)$ is therefore the total number of electrons in the unit of pattern. For higher orders

there is a reduction in intensity owing to interference.

It is important to note that $F(hkl)$ is a dimensionless ratio, characteristic only of the crystal structure. It is independent of the wavelength of the X rays. If a smaller wavelength is used, the orders appear at lower angles and path differences are reduced, but phase differences remain the same. Thus $F(hkl)$ depends only on the distribution of scattering matter in the unit cell, which it is the object of X-ray analysis to determine.

The theoretical basis for measuring values of $F(hkl)$ was laid down by C. G. Darwin in two brilliant papers soon after the discovery of X-ray diffraction. In those early days the experimental observations were too approximate for a test of his theory, and a number of years elapsed before it could be applied.

Darwin's first calculation assumed the crystal to be "ideally perfect." Rough tests showed, however, that the efficiency of reflection was many times stronger than his theory indicated. Darwin correctly reasoned that the cause of the discrepancy was the departure of the crystal from perfection. It is a curious paradox that imperfect crystals reflect more efficiently than perfect crystals. In the latter case the reflection, which is almost complete over a few seconds of arc, comes from a thin superficial layer only; planes at greater depths cannot contribute because the uppermost layers have robbed the radiation, so to speak, of the component the lower layers would otherwise have reflected. Actual crystals, however, are in general far from perfect. They are like a three-dimensional crazy quilt of small blocks that differ slightly in orientation; as a result the crystal reflects over an appreciable angular range. Within this range rays penetrate into the crystal until they encounter a block at the correct angle for reflection, and the contributions from all such blocks add to the total reflection.

Darwin's second formula, therefore, applies to what is called an "ideally imperfect" crystal, and it is the formula always used. The intensity of the incident beam, or the amount of radiation per unit of time, is compared with the total amount of radiation received by the recorder as the crystal is swept through the reflecting range at a constant angular rate; this enables all elements of the mosaic to make their contribution to the reflection.

When calculating a value of $F(hkl)$ for a postulated atomic arrangement, it is necessary to know the contributions from individual atoms, which depend on the characteristic distribution of electrons in each atom. These distributions were calculated by Douglas R. Hartree in 1925 and are expressed as "F curves" typical of each atom. Intensity measurements were the subject of an extensive study by the University of Manchester

school, culminating in the paper by Reginald James, Ivar Waller and Hartree on the zero-point energy of the rock-salt lattice. Amplitudes of thermal vibration can be measured by their effects in reducing F values; by extrapolating to absolute zero it was found that the atoms still had a vibration corresponding to a half-quantum, as theoretical studies had indicated.

Experimental Measurements

When a diffracted X-ray beam is recorded by an ionization chamber, a Geiger counter or a proportional counter, the orders are recorded one by one, by setting the crystal and the chamber at suitable angles. Alternatively the beams can be recorded as spots on a photographic plate or film by turning the crystal during the exposure so that a number of planes can reflect. In the early crystal determinations the ionization spectrometer measured orders individually. As more complex crystals were attempted and more orders had to be measured, the photographic method was favored because a single exposure registered a large number of orders. Recently auto-

mation has obviated the tedium of making numerous individual measurements, and the most advanced analyses are now performed with counters as recorders.

The original X-ray spectrometer designed by W. H. Bragg is a typical example of the first method [*see illustration below*]. A collimated beam from the X-ray tube fell on the face of the crystal and was reflected through slits into the recording ionization chamber, which was filled with a heavy gas (methyl bromide) to increase ionization. The outer case was at a potential of several hundred volts, and the ionization was measured by driving the charge onto a coaxial wire connected to a tilted gold-leaf electroscope. It was with this instrument that my father made his pioneer investigations on the X-ray spectra from anticathodes of a number of different metals, a project that formed the basis for H. G. J. Moseley's subsequent work on atomic number; the early determinations of crystal structure, for which I was mainly responsible, were also made with this instrument.

Considering the crudeness of the apparatus by modern standards, it gave surprisingly accurate results. A main trouble arose from the vagaries of the X-ray

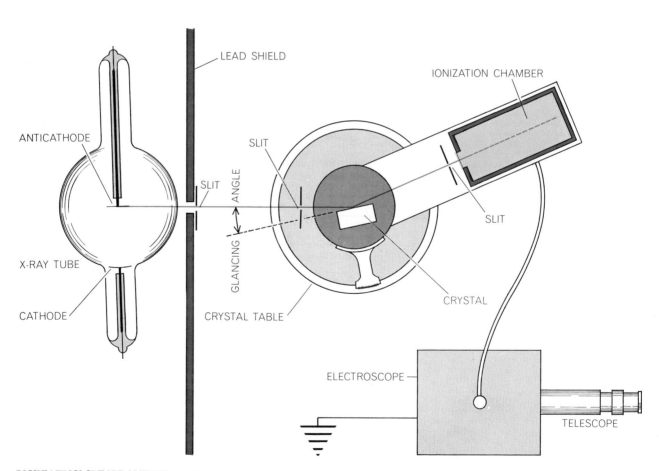

IONIZATION SPECTROMETER was the instrument used by the author's father, W. H. Bragg, to conduct the first investigations of

the X-ray spectra from various metallic anticathodes and later by the author to make the early determinations of crystal structure.

tubes in those days. The tube was energized by a Rumkorff coil, first with a hammer switch and later with a mercury switch, which gave a steadier discharge. If the X rays from the tube got too hard, one held a match under a fine palladium tube attached to the main tube, which allowed some gas to diffuse in; if they got too soft, one sparked to a bunch of mica sheets inside, which absorbed some gas. The gold-leaf electroscope was also a tricky instrument for accurate work. I think that one of the main reasons why X-ray analysis developed in my father's laboratory at the University of Leeds, even though the fundamental discovery had been made in Germany, was that my father had so much experience in making accurate ionization measurements with the primitive apparatus then available.

When the diffracted beams are measured one by one, the indexing presents no difficulty because the crystal orientation that produces each beam is known. When many beams are recorded at the same time on a photographic plate, however, each of them must be identified. A number of ingenious methods have been devised for this purpose.

In general two types of X-ray photograph have been widely used. One type is called a "rotation" photograph [see top illustration on next page]. In this method the X rays fall on a small crystal, which is rotated around an axis that coincides with one of its principal crystal axes, and the diffracted beams are recorded on a cylindrical film. The images of the diffracted beams all lie on "layer lines"; for instance, if the crystal axis is along OC, the layer lines correspond to $l = 0, l = 1, l = 2$, and the spots have all values of h and k. If the spots are very numerous, it may be too difficult to sort them out and a Weissenberg camera is used. In this technique one layer line is singled out by a slit, and the film is translated as the crystal turns. If the film has been translated horizontally, the displacement of a spot along the horizontal axis tells the angle of setting of the crystal when it was recorded and so defines the other indices [see illustration on page 234].

Another elegant device in this category is the precession camera. Crystal, photographic plate and screen perform a sinuous dance in such a way that those spots, for instance, with a definite value of l and all values of h and k are recorded as a rectilinear net [see illustration on page 238]. This method is particularly suitable for crystals with large unit cells and consequently numerous values of the indices.

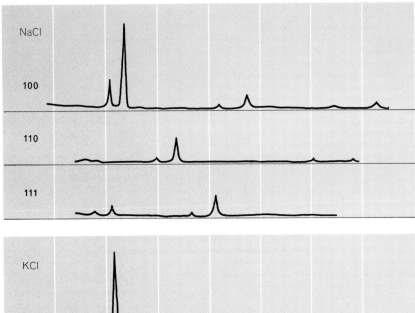
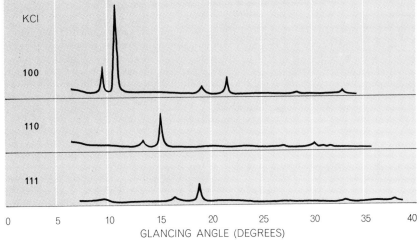

EARLY MEASUREMENTS of the intensity of the reflected X rays from sodium chloride (*top*) and potassium chloride (*bottom*) were made with the ionization spectrometer. The orders were reflected from crystal faces with crystallographic indices (100), (110) and (111). The two peaks seen on each order are the $K\alpha$ and $K\beta$ "lines" in the spectrum of the palladium anticathode, the $K\alpha$ line being the stronger of the two. For sodium chloride the order (111) is anomalously small, because the weak sodium contribution partially offsets the strong chlorine contribution, whereas for potassium chloride the order (111) is too weak to be observed, because the scattering powers of the potassium atoms and the chlorine atoms are so nearly similar. This comparison confirmed the structures assigned to the crystals.

The second general method of X-ray photography is the powder method, developed independently in 1916 by Peter J. W. Debye and Paul Scherrer in Switzerland and by Albert W. Hull in the U.S. [see bottom illustration on next page]. The powder method is used when the material is available only in microcrystalline form. The X rays fall on a mass of tiny crystals in all orientations, and the beams of each order (hkl) form a cone. Arcs of the cones are intercepted by a film surrounding the specimen. In the powder photographs of sodium chloride and potassium chloride on page 9 one can see that in sodium chloride there is a weak series for odd (hkl) values, whereas in potassium chloride these disappear because the scattering powers of the potassium atoms and the chlorine

atoms are so similar. In addition potassium chloride has a larger spacing than sodium chloride, hence the displacement of the arcs to smaller angles. The powder method has found its main use in the study of alloys.

Symmetry

The pattern of every crystal has certain symmetry elements that form a three-dimensional scaffolding on which the atoms are arranged, and these elements can be uniquely determined by the X-ray-diffraction method. In the early days of X-ray analysis, when the ionic compounds that were being studied often had a high symmetry, this fact was of great assistance in arriving at a solution. The possible schemes of symmetry

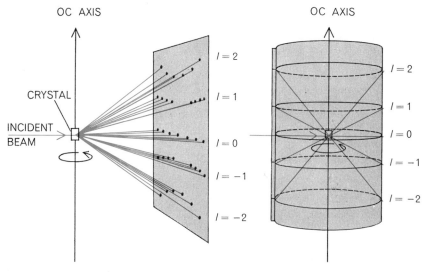

ROTATION PHOTOGRAPHS ARE MADE by aiming the X rays at a small crystal, which is rotated around an axis that coincides with one of its principal symmetry axes; the diffracted beams are recorded on either a flat plate (*left*) or a cylindrical film (*right*). The images of the diffracted beams all lie on "layer lines"; in this case the crystal axis is along OC, the layer lines correspond to $l = 0, l = 1, l = 2$ and the spots have all values of h and k.

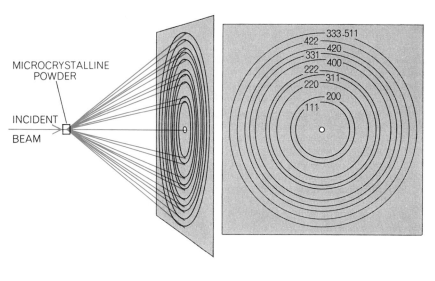

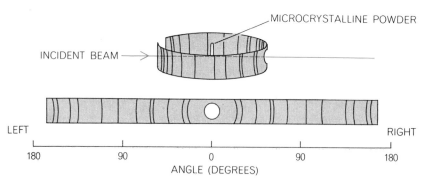

POWDER PHOTOGRAPHS ARE MADE by aiming the X rays at a mass of tiny crystals in all orientations. The diffracted beams of each order (hkl) will then form a cone. If recorded on a plate perpendicular to the incident beam, each diffraction order will appear as a ring surrounding the central spot (*top*); the positions of the rings shown are typical of a face-centered-cubic crystal lattice. It is usually more convenient to employ a cylindrical photographic film whose axis is perpendicular to the incident radiation (*bottom*). Arcs of the cones are intercepted at all angles up to nearly 180 degrees; the film is then unrolled.

are limited by geometry, just as the possible number of regular solid figures are limited, although in the case of crystal symmetries the number, 230, is quite large. Symmetry axes and symmetry planes can be identified by noting regular absences of diffracted-wave orders; the presence or absence of symmetry centers can be determined by a statistical survey of intensities, as was first shown by A. J. Wilson; crystals with symmetry centers characteristically have many more weak reflections than crystals with no symmetry centers.

Finally, X rays can tell "which way around" a structure is. Optically active molecules can have two forms, one of which is the reflection of the other (the dextro and levo forms of the chemist). In general when the waves scattered by the atoms have phases as if coming from atomic centers, these two forms give identical X-ray diffraction, that is, the reflection from the right-hand side has the same amplitude as that from the left, although the phase is reversed. When the wavelength of the X rays is close to an absorption edge of an atom, however, there is an appreciable phase change. The atom scatters as if at one location for the one side and at another location for the other side, so that the resultant amplitudes are different. This enables dextro and levo to be distinguished; for instance, in the classic case of a tetrahedron with four different corners one could tell for each orientation whether one was looking at an apex or a base. J. M. Bijvoet was the first to distinguish between dextro and levo forms of the tartrate ion. There was a 50 : 50 chance that the traditional chemical convention for representing dextro and levo was correct; luckily it turned out to be right!

Inorganic Compounds

The first crystals to be analyzed by means of X rays were simple types. An approximate measure of the complexity of a crystal is the number of parameters that must be determined in order to define the positions of the atoms. In the case of an atom at a symmetry center, for instance, no parameters are needed; it must be exactly at the center. If the atom lies on an axis, its position along the axis is fixed by one parameter; if on a reflection plane, by two; if in a position of no symmetry, by three.

The early determinations were limited to one or two parameters; in fact, it was doubted whether more complicated crystals would ever be analyzed. The break-

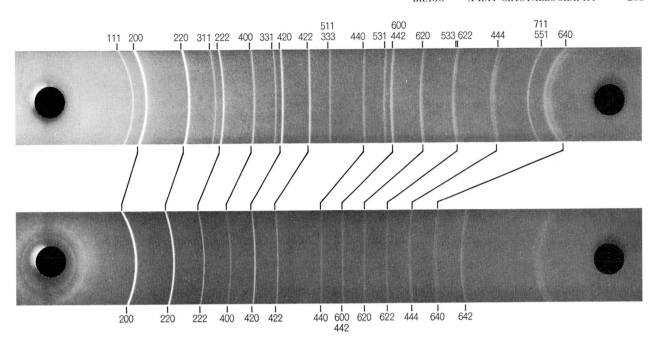

POWDER PHOTOGRAPHS of the diffracted X rays from sodium chloride (*top*) and potassium chloride (*bottom*) confirm the earlier findings made with the ionization spectrometer: In sodium chloride there is a weak series for odd (*hkl*) values, whereas in potassium chloride these orders disappear. Potassium chloride has a larger spacing; hence the arcs are displaced to smaller angles.

through into much more complex structures was made in the early 1920's by the Manchester school, where analysis was extended to cases of 10 or 20 parameters, a great advance at that time. It was made possible by quantitative measurements and increasing experience in the nature of inorganic compounds.

One of the first successes of X-ray analysis was to show that these compounds are not built of molecules. They are ionic in character, with a regular alternation of positive and negative ions held together by electrical attraction. For instance, in the sodium chloride structure there are no sodium chloride groups but rather a chessboard pattern of positive sodium ions and negative chlorine ions. It was difficult in the early days to reconcile the new view of ionic compounds with classical chemical ideas, but once accepted the ionic view afforded a much fuller understanding of the construction of such compounds.

In an ionic compound the ions pack together as if they had characteristic sizes. Their dimensions are not completely fixed, but they vary only over a small range. On the whole the negative ions are by far the largest, because their electrons are more loosely held.

The packing of ions of characteristic size is a very useful concept when postulating various atomic arrangements, particularly when combined with a knowledge of the symmetry elements. The hexagonal symmetry of the crystal beryl ($Be_3Al_2Si_6O_{18}$), for instance, is of a high order, with sixfold, threefold and twofold axes, symmetry planes and symmetry centers. An atom cannot overlap itself, so that it must either lie exactly on one of these symmetry elements or be just off it. This restriction is so demanding that the structure of beryl could be immediately deduced once the symmetry was determined. The top illustration on page 235 shows the only possible way of packing the atoms of the beryl formula into the network of symmetry elements.

The laws governing the structure of inorganic compounds, established by Linus Pauling in 1929, afford another guide in seeking a solution. They also explain why some compounds are stable, whereas others that seem equally plausible from a chemical point of view do not actually exist. Pauling's rule is based on the requirement that for stability the energy of the compound must be as low as possible. Each small positive ion lies inside a cluster of larger negative ions; for instance, very small positive ions such as beryllium or boron are each surrounded by three oxygen atoms; silicon is surrounded by four oxygens, magnesium and iron by six oxygens and still larger ions by eight or more oxygens. If we picture electric fields in terms of lines of force, suppose the number of lines representing the charge on the positive ion is divided equally between the negative ions coordinated around it. Pauling's rule states that the total number of lines coming to the negative ion from all its positive neighbors just balances its charge. This might seem at first sight a simple rule, but it is powerful in excluding improbable structures. Its significance is that when Pauling's rule is obeyed, the lines of force between positive and negative ions stretch only over the very short distances between nearest neighbors, so that the energy of the electric field is at a minimum and the structure is stable.

The study of inorganic compounds culminated in the determination of all the common mineral forms, in particular the silicates. These compounds obey Pauling's rule rigorously, because they must be very stable in order to exist as minerals. The explanation of their composition proved to be very interesting. Their nature is determined by the silicon-to-oxygen ratio, ranging from SiO_4 in the basic rocks to SiO_2 in quartz. Although the ratio varies widely, the silicon ion is always surrounded by four oxygen atoms; the silicon-oxygen tetrahedrons may, however, share no corners, one corner, two corners, three corners or four corners by having an oxygen atom in common. In the SiO_3 silicates, for example, there are long strings of SiO_4 groups that run endlessly through the structure, representing infinite linear negative ions bound laterally by the positive ions. The silicate groups occur as sheet ions in the micas and clays, and as three-dimensional-network ions with metals in their interstices in the feldspars. This unexpected feature of minerals explains their composition in a simple and elegant way; it is one of the important new conceptions introduced by X-ray analysis.

Alloys

After the nature of inorganic compounds had been clarified, the next achievement of X-ray analysis was to explain the nature of metallic alloys, which had hitherto been so mysterious. The pioneer investigations into alloy structure were made by Arne F. Westgren in Sweden. They were developed by Albert J. Bradley and his pupils at Manchester, for the most part between 1925 and 1935. The powder method was used perforce, because the material was in microcrystalline form, and in Bradley's hands it reached a peak of perfection that has hardly been equaled since. The ground covered was so extensive that it is only possible to give the briefest summary.

In the first place, the determination of alloy structure provided the foundation on which a theory of alloy chemistry could be developed. When two metals unite in varying composition, they form a series of phases. These compounds are non-Daltonian, that is, they are not composed of some simple ratio of elements; on the contrary, each exists over a range of composition. William Hume-Rothery first pointed out that in different binary systems phases with very similar physical properties tend to have the same ratio of free electrons to atoms in their composition. Structure determinations showed that such phases have a closely similar atomic arrangement but with curious characteristics. The essential similarity lies in the positions occupied by atoms, not in the way the kinds of atoms are distributed among the places. Apparently the relation of atom to atom is relatively immaterial; it is the position of the atoms that is all-important. This in turn was explained by theoretical physicists in terms of Brillouin zones. Treating the free electrons as standing waves, the system has a low energy if the electrons of shortest wavelength are just too long to be reflected by the most marked reflecting planes in the phase structure. In the stable phase the atoms take up arrangements that create the strongly reflecting planes required for low energy. To put it broadly, an alloy is not a compound between one metal and another but rather a compound between all the metal atoms on the one hand and all the free electrons on the other. It is perhaps not too much to say that the X-ray determination of alloy structures led for the first time to a rational theory of metal chemistry.

Equilibrium systems, or the phases produced by variations of composition, had previously been deduced from studies of polished and etched specimens, but they can be mapped out far more directly by X-ray analysis. Phases can be recognized by their powder photographs and the composition of the phase in any given case can, after some preliminary trials, be fixed by noting which spacings of the unit cell vary over the range. Ternary and even quaternary systems, which would hardly be amenable to metallographic methods, can be tackled by X-ray methods.

An interesting application in this area was found in the "order-disorder" systems. An example of such a system is the copper-gold alloy Cu_3Au, first studied by Gudmund Borelius of Sweden. At high temperatures all the points of the face-centered-cubic lattice of this alloy are occupied at random by gold or copper atoms; at low temperatures after slow annealing the gold atoms segregate into the cube corners, leaving the face centers to the copper atoms. The progress of the segregation can be followed by the appearance of new lines, corresponding to greater spacings, in the powder photographs. The variation of segregation with temperature presents interesting thermodynamic problems of second-order phase change, and the X-ray work did much to stimulate the study of corresponding changes in other systems.

Another phenomenon studied intensively by Bradley was the splitting of a phase into regions with slightly different composition, which were nonetheless united in having a continuous crystal lattice. Such segregation sets up intense internal strains. It is characteristic of alloys used as strong permanent magnets, because the strains give the material a high magnetic retentivity.

In general, X-ray analysis has provided a powerful new tool for examining the properties of alloy systems, an achieve-

WEISSENBERG PHOTOGRAPH is a type of rotation photograph that is used when the spots are too numerous to sort out by the conventional method. In this technique one layer line is singled out by a slit, and the film is translated as the crystal turns. The displacement of a spot along the translation axis (in this case the horizontal axis) tells the angle of setting of the crystal when it was recorded and so defines the other reflection indices of the crystal.

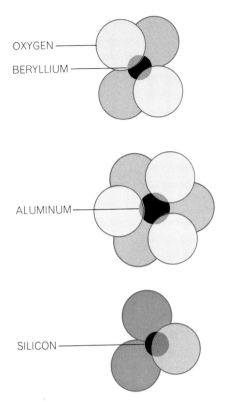

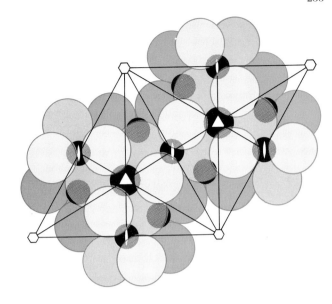

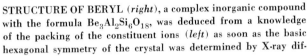

VERTICAL-ROTATION AXES

⬭ TWO-FOLD AXIS

△ THREE-FOLD AXIS

⬡ SIX-FOLD AXIS

STRUCTURE OF BERYL (*right*), a complex inorganic compound with the formula $Be_3Al_2Si_6O_{18}$, was deduced from a knowledge of the packing of the constituent ions (*left*) as soon as the basic hexagonal symmetry of the crystal was determined by X-ray diffraction. Since an atom must lie exactly on one of the symmetry elements or be just off it, there could be only one way of packing the atoms of the beryl formula into the network of symmetry elements. A key to the symmetry axes of the crystal is given at bottom.

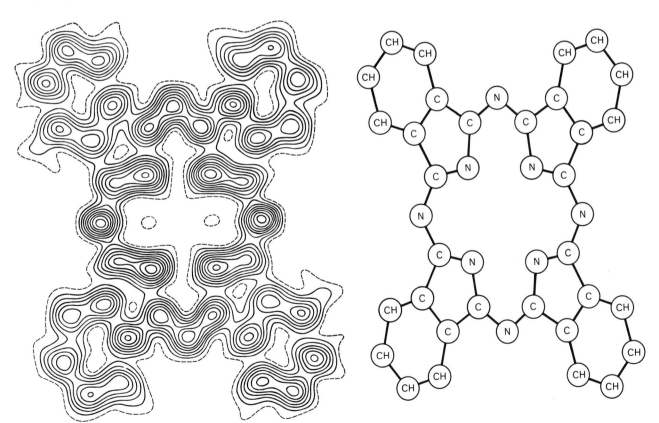

FOURIER REPRESENTATION of the electron-density distribution in a molecule of phthalocyanine (*left*) was used to construct the atomic model of the molecule (*right*). The Fourier "density map" is arrived at by treating a molecular structure not as a cluster of individual atoms but as a continuous electron distribution capable of scattering X rays. The density distribution of the molecule as a whole is obtained by adding together the terms of a Fourier series, a mathematical expression that can be used to represent the periodic variation of sets of electron-density sheets in all directions. The Fourier method is ideal for analyzing organic molecules.

ment that has great technical importance as well as scientific interest.

The Fourier Method

Another method of X-ray analysis attacks the solution of the crystal structure from a quite different angle. So far the crystal has been regarded as a pattern of atoms, each of which scatters X rays as if from its center with an efficiency determined by Hartree's F curves. The resultant of the waves scattered by these atoms is then compared with the observed amplitude of reflection, the position of the atoms being adjusted to give the best fit.

This method was successful as long as the number of atoms in the unit cell was small. As increasingly complex crystals were studied, however, it became more and more difficult to try adjustments of so many parameters simultaneously,

even when the structure was approximately known. The refining of the structure to get the best fit became extremely laborious.

The Fourier method is in a sense a complete reversal of this process. A structure is treated not as a cluster of atoms but as a continuous electron distribution capable of scattering X rays. The investigator seeks to map this continuous distribution, and, if he is successful, he can then recognize the positions of the atoms by noting where the electron density rises to peak values. There is no juggling with the positions of atoms one by one; the density map shows the best position for all of them, however large their number.

The density distributions are mapped by adding together the terms of a "Fourier series," a mathematical expression that can be used to represent any quan-

tity that varies periodically. Since a crystal is a periodic pattern in three dimensions, the electron density can be represented by a three-dimensional Fourier series. Each element of the series is a set of electron sheets, or strata, that vary periodically in density, and if the amplitudes and phases of these sheets (which crisscross in all directions) are known, they can be added and the result is a plot of the density distribution.

My father first pointed out, in his Bakerian Lecture to the Royal Society of London in 1915, that each of these periodic components reflects one corresponding order and only that order; moreover, the amplitude of the reflected waves is proportional to the amplitude of the Fourier component. Since changes in the phase of the reflected waves can still give the same X-ray effects, however, the only way of choosing the right phase

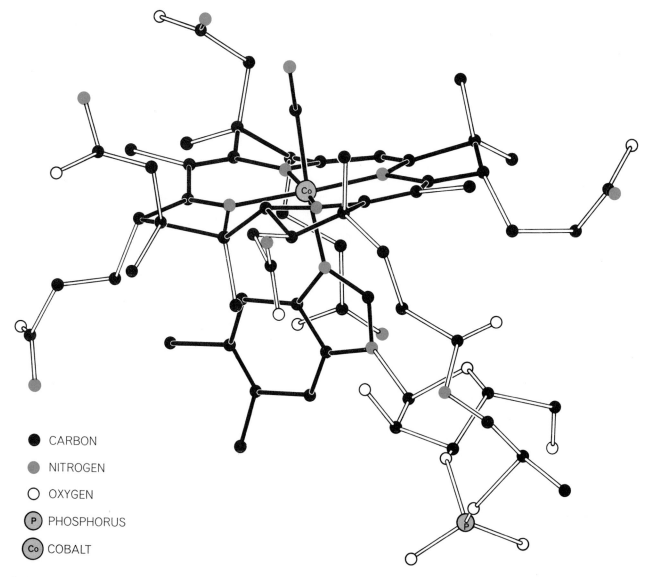

● CARBON

● NITROGEN

○ OXYGEN

(P) PHOSPHORUS

(Co) COBALT

STRUCTURE OF VITAMIN B-12, solved by Dorothy Crowfoot Hodgkin in 1955, represented one of the outstanding achieve-   ments of what might be called the classical methods of X-ray analysis. The formula of vitamin B-12 molecule is $C_{63}H_{84}N_{14}O_{14}PCo$.

is to introduce some criterion of reality leading to a picture with the right number of atoms of the right kind in the unit cell. The Fourier method therefore centers around a "phase hunt." Once the phases are known the structure is "in the bag."

Organic Molecules

From 1930 onward the Fourier method was recognized as being ideal for analyzing organic molecules, and many were determined in this way. The first organic structures, naphthalene and anthracene, had been outlined by W. H. Bragg in 1922 at the same time that the study of inorganic structures was pursued at Manchester, and his laboratory at the Royal Institution concentrated on the organic field.

Initially most of the studies dealt with crystals that had symmetry centers. This is a great simplification of the problem, because the phases of the Fourier components with respect to such a center must by symmetry be either 0 or π; in other words, $F(hkl)$ must be either plus or minus. The advantage here is that calculations on the basis of a quite rough approximation of the structure generally make it clear which is the right sign, particularly in the case of the strong and therefore important orders. A Fourier series can then be calculated, the position of the atoms can be improved, further signs that were formerly doubtful can be fixed and a new Fourier can be summed. This process of refinement is rapidly convergent, and after a few stages the structure is accurately determined.

A three-dimensional Fourier series is a formidable affair, and in these attacks on organic structures the less ambitious task of using two-dimensional series and getting a projection of the structure on a plane was more usually undertaken. Projections on the three principal planes define the positions of the atoms in space. A pretty example of this was the analysis of phthalocyanine with 60 parameters, where the signs of the F's were found without any guesswork as to the nature of the structure [see bottom illustration on page 235].

It is possible to substitute a heavy atom at the center of the phthalocyanine molecule without any alteration in the crystal lattice. If a spot becomes stronger when the heavy atom is introduced, its original F value with respect to the center of the molecule must have been positive; if the spot becomes weaker, the F value must have been negative. All the signs were determined in this way and

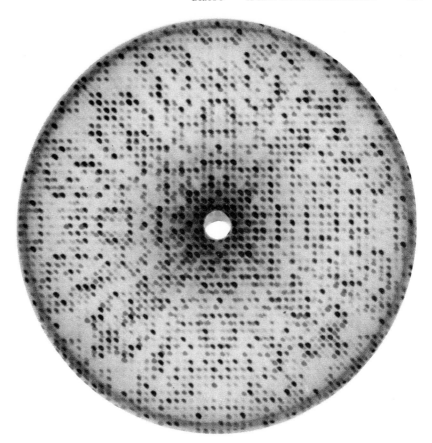

ADDITION OF A HEAVY ATOM at a specific place among the molecules in a protein crystal provided the key that has led to the recent solution of several protein structures. In this illustration of the method two precession photographs of the same crystal of lysozyme are superposed with a slight relative displacement. The set of spots due to the native protein is printed in black. The set of spots due to the protein with the heavy atom is printed in color. Numerous changes in the intensities of corresponding spots can be detected.

used in a two-dimensional Fourier series. The result is shown as a contour plot of the electron densities.

In most cases the results of X-ray analysis confirmed the topography assigned to the molecules by organic chemistry, but the X-ray findings determined the bond distances and bond angles with great accuracy and so cast much light on the nature of the bonds.

The next stage in the study of organic molecules was to tackle far more complex structures, some of which had defeated the efforts of organic chemists to elucidate their stereochemistry. A forerunner of this stage was the solution of strychnine by Biyvoet in 1948, published independently at almost the same time that Sir Robert Robinson's researchers at the University of Oxford arrived at an identical structure by purely chemical reasoning. The outstanding examples are the solution of penicillin and vitamin B-12; in each case Dorothy Crowfoot Hodgkin of Oxford was the leader of the research. The latter investigation was a saga of X-ray analysis that took eight years. Not only was much chemical information about the molecule lacking

but also conclusions arrived at on chemical grounds were actually misleading. The molecule, of formula $C_{63}H_{84}N_{14}O_{14}PCo$, is shown in the illustration on the opposite page.

The solution illustrates the curious and unique character of X-ray analysis, which is reminiscent of the solution of a code, or of an ancient form of writing such as Egyptian hieroglyphics or Minoan Linear B. It was first assumed that the phases of the F's were those of the cobalt atom (Co) at the center of the molecule and a Fourier series was formed on this hypothesis. Although it turned out that this is far from true, the phases are, as it were, weighted in this direction because the cobalt atom is so heavy compared with the other atoms. Fourier series have a surprisingly obliging way of trying to tell the investigator something with the most sketchy basis of information, and in this case the series outlined in a shadowy way the molecular structure immediately surrounding the cobalt atom. The information was used to adjust the phases, and a further series was formed and so the structure gradually began to emerge from the cobalt

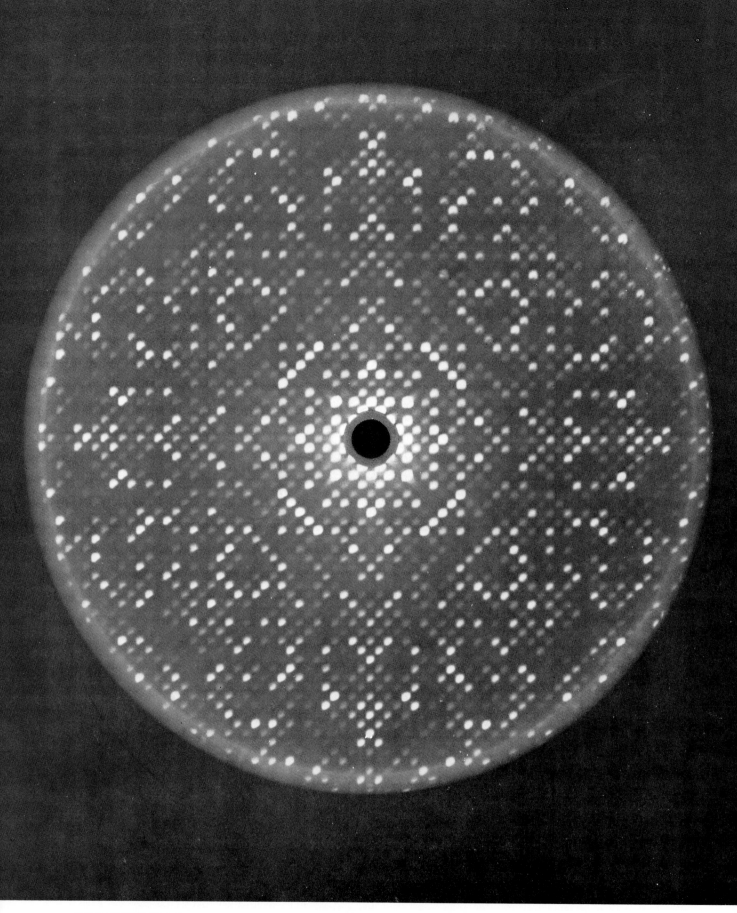

X-RAY PHOTOGRAPH of lysozyme, the second protein and first enzyme to have its molecular structure determined by X-ray analysis, symbolizes the recent achievements of the X-ray technique. The bright spots correspond to various orders of diffracted waves produced by irradiating the lysozyme crystal with a beam of monochromatic X rays. This particular type of X-ray photograph is called a precession photograph; it is produced by manipulating the crystal, the photographic plate and an intervening screen in such a way as to hold one of the three indices of the diffracted beams constant while recording the values of the other two indices in the form of a rectilinear pattern. The lysozyme molecule contains 1,950 atoms and measures approximately 40 angstrom units in its largest dimension.

atom outward. The calculations would have been impossibly onerous had it not been for the availability of electronic computing of structure factors and three-dimensional Fourier series. The solution of vitamin B-12 represented the highest flight of what might be called the classical methods of X-ray analysis, and the Nobel prize awarded to Mrs. Hodgkin in 1964 was a well-deserved acknowledgment of her achievement.

Biochemical Molecules

We now come to a most dramatic turning point in the history of X-ray analysis. When vitamin B-12 was analyzed, with 181 atoms in the molecule, it seemed hard to imagine that much more complex structures could ever be tackled; it had taken eight years to complete and the difficulties increase as a high power of the number of atoms. And then, as the result of an investigation that had lasted for some 20 years, a way was finally found to solve the structure of the immensely more complicated molecules of living matter, the proteins. The first of these to be analyzed, by John C. Kendrew in 1955, was myoglobin, which has 2,500 atoms in its molecule.

By a curious paradox, the very size of such a molecule has opened up a new line of attack that is not possible in the case of simpler types, leading to a direct determination of phases without any element of guesswork or trial and error. The principle is the same as in the case of phthalocyanine described above, where the substitution of a heavy atom in the molecule enabled the signs of the F values to be found. A discovery made by M. F. Perutz in 1953 made it possible to generalize this method for the proteins.

He found that heavy atoms such as mercury and gold, or complexes containing such atoms, can be incorporated at specific places in the framework of the protein crystal without affecting the arrangement of the molecules, which are so large and loosely packed that the added groups find places in the interstices. Further, Perutz showed that the added heavy atom produces changes in the F values large enough to be accurately measurable. It might at first sight appear strange that the addition of one heavy atom of mercury to a protein molecule with 2,500 atoms of carbon, oxygen and hydrogen should make an appreciable difference; it does so because the scattering comes from one center, whereas the random contribution from n atoms ranging over all phases is proportional to $\sqrt{n}$, not to n. The changes in intensity caused by the addition of a heavy atom can be seen when two precession photographs made with the same crystal are superposed with a slight relative displacement [see illustration on page 237]. The black set of spots is due to the native protein, the colored set to the protein with a heavy atom, and a close examination will show numerous changes in the intensities of corresponding spots.

It is necessary to find some three or four heavy atoms that can be attached to definite sites on the molecule. Although the structure is initially unknown, direct methods are available for finding the relative positions of these "staining" atoms. The phase difference between the F value H due to the heavy atom and the F value P due to the protein can be found by comparing $F(P)$ with $F(P + H)$, because $F(P + H)$ must be the vectorial resultant of $F(P)$ and $F(H)$; the knowledge of the phase difference for several heavy atoms pins down the position of the Fourier component.

Because the solution is direct it can be found by giving instructions to a computer. The computer is essential because the complexity is so great. Some 100,000 or 200,000 intensities must be measured accurately by means of an automatic machine that sets the crystal and the recorder at the right positions for one order after another and lists the results. A corresponding number of equations must be solved to find the phases, and the Fourier series of many thousands of terms must be formed. This long series must be summed at perhaps a quarter of a million places in the unit cell to give the density at each point. The information is then automatically turned into contours, which are plotted on stacked transparent sheets, and the investigator has then to translate the density distribution into atomic arrangement.

The final result is impressive. Some half-dozen protein structures have so far been analyzed and they are already beginning to yield valuable information on such vital biochemical processes as the operation of enzymes. The second protein molecule to be analyzed successfully (after myoglobin) was the enzyme lysozyme (by David C. Phillips). The most recent success has been hemoglobin (by Perutz); the model of this protein contains 10,000 atoms. I confess that when I contemplate one of these models, I can still hardly believe that it has been possible to work out all its details by the optical principles of X-ray analysis, which half a century ago claimed sodium chloride as its first success.

21

CHEMICAL ANALYSIS BY INFRARED

BRYCE CRAWFORD, JR.
October 1953

Electromagnetic waves make the atoms in a compound vibrate. The frequencies to which a substance responds can tell much about its structure and about the nature of chemical bonds

Infrared spectroscopy has grown like a mushroom in the past 10 years. Before the war it was employed by only a few chemists and physicists, using home-built or custom-built infrared spectrometers. Now the instrument is a standard commercial item supplied by a competitive industry to chemical and medical researchers all over the country.

More than 1,300 commercial infrared spectrometers, each representing an investment of two to six Cadillacs, are earning their way in scientific laboratories and industrial plants.

To an old-timer in the field—"old-timer" by virtue of having studied infrared as long as, say, 15 years—this burgeoning use of infrared by his colleagues is most gratifying, but hardly surprising. The power of infrared as a tool for chemical characterization and analysis has been known for some 20 years. The infrared spectrum of a substance is related to its chemical structure in a uniquely convenient way, and organic chemists have been heard to say that the infrared spectrum of an organic compound is its most important physical property.

Infrared radiation itself was discovered more than 150 years ago, long before scientists had any clear understanding of radiation. Sir William Herschel, who started life as an organist at Bath and became an astronomer, made the discovery. In 1800 he reported to the Royal Society certain experiments in heat radiation. He resolved sunlight into its spectrum with a glass prism and placed a thermometer at successive positions in the spectrum. He found heat radiation not only in the visible spectrum but in the longer wavelengths beyond the red. Herschel even crudely measured the absorption of this radiation by various substances, including tap water, distilled water, sea water, gin and brandy. He could not know, could not even suspect, how revealing the absorption could be about chemical structure. Before anyone could appreciate the significance of infrared absorption, light radiation itself had to be understood. A century passed before the theory of the nature of light was worked out and the necessary techniques and instruments for infrared analysis were developed.

Sir William had found that the most intense heat radiations were outside the visible part of the spectrum. Conse-

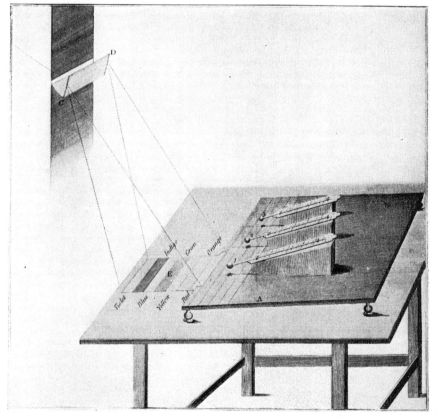

FIRST INFRARED SPECTROMETER, improvised by Sir William Herschel in 1800, was the means by which he discovered infrared radiation. Thermometers placed in successive bands of solar spectrum registered more heat outside and beyond red region than within.

quently for many years it was believed that heat and light were two quite different radiations. But some 35 years later Sir William's son, Sir John Herschel, and other investigators advanced the idea that heat and light might merely be different manifestations of the same radiation—that is, light waves of different wavelengths. Before the end of the century James Clerk Maxwell had shown theoretically and Heinrich Hertz had proved experimentally the essential identity of light, heat and other electromagnetic radiation.

The infrared region, as spectroscopists usually define it, lies between the visible and the radio portions of the spectrum—that is, the wavelengths between one thousandth of a millimeter and one millimeter. (Infrared of course is not synonymous with the common meaning of the word "heat," for, as we have seen, there are heat radiations in the visible part of the spectrum.) The region most useful to the chemist is the range of wavelengths from 2 to 20 microns (thousandths of a millimeter). This is sometimes called the vibrational region.

We shall confine ourselves to this region and to considering how absorption spectra in it give information about matter. The other infrared regions, and the physics of the radiation itself, are just as interesting: it was in the near infrared that Max Planck, by instinct and gentle nature one of the most classical of physicists, was led to the disturbing idea of the quantum, which upset the whole beautiful pattern of 19th-century physics. But we cannot cover the whole subject of infrared in one article.

The chemically useful study of infrared absorption spectra was really begun in 1903 at Cornell University. A graduate student named William W. Coblentz, under the physics professor Edward Nichols, had undertaken research on infrared absorption. After improving the available experimental techniques, he received an appointment as a research associate of the Carnegie Institution of Washington, which enabled him to set about measuring absorption spectra of pure substances. He mapped spectra for two years and in 1905 published a collection of accurate infrared absorption spectra for 131 substances. Even today, after 48 years of improvements in infrared technique (to which Coblentz himself contributed much until his retirement from the National Bureau of Standards in 1945) the monumental work that he did at Cornell still stands worthy of study. Subsequent studies

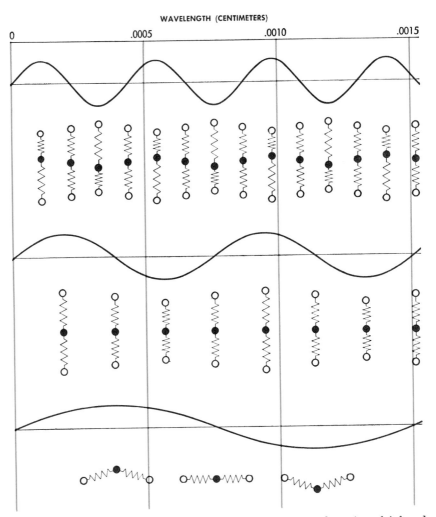

WAVELENGTH (CENTIMETERS)

VIBRATION OF MOLECULAR BONDS accounts for selective absorption of infrared energy by various compounds. Each of three modes of vibration of carbon dioxide bonds shown here is resonant to infrared at wavelength indicated. Energy is absorbed, however, only by modes at top and bottom, which disturb geometric and electric symmetry of atom.

have confirmed more than they have corrected Coblentz's observations.

When sunlight falls upon a green leaf of a tree, the leaf absorbs the red wavelengths of the light and reflects the green wavelengths: this is what gives the leaf its color. The absorption tells something about the leaf's molecular composition. Essentially the same principle is involved in studying infrared absorption, except that the radiation is not visible; if our eyes were attuned to the infrared, we could recognize a compound by its characteristic color. Instead we measure its absorption of heat. Infrared radiation is passed through a solution of a compound, and the compound characterizes itself by the wavelengths it absorbs and those it transmits. Each compound has its own absorption spectrum.

The absorption of infrared is due to some disturbance within the molecule; this Coblentz established by observing that he got different spectra from isomers—molecules which are composed of

the same atoms but in different arrangements. Coblentz found further that certain subgroupings of atoms within molecules identified themselves by absorbing characteristic wavelengths: for example, phenyl compounds, containing the benzene ring, absorbed at 3.25 and 6.75 microns, while mustard oils, containing the thiocyanate group, absorbed at 4.78 microns. These absorptions were additive: in phenyl mustard oil Coblentz found "the characteristic vibration of the mustard oils superposed upon the vibration of the benzene nucleus." He concluded that "there is a something—call it 'particle,' 'group of atoms,' 'ion' or 'nucleus'—in common, with many of the compounds studied, which causes absorption bands that are characteristic of the great groups of organic compounds, but we do not know what that 'something' is."

With half a century's progress since Coblentz, we now have a clear idea of the "something." It lies in the bonds between the atoms in a molecule. These bonds are written in chemical formulas as single, double and triple lines connecting the atoms. As the Danish chemist Niels Bjerrum once said, they "summarize, in a very compact form, chemistry's knowledge of the creation and destruction of compounds. Nowhere in science has a shorthand notation been developed which summarizes such an abundance of exact knowledge in so small a space." In 1914 Bjerrum showed that, if we think of the atoms as small masses and the bonds as springs holding the atoms together, we can account correctly for the vibrational behavior of molecules, as observed in their infrared spectra and in their heat capacities.

The bonds hold the atoms in position fairly tightly, but not rigidly. The response of the atoms to a light wave is much like the response of cork balls, floating on a lake, to waves on the lake. As waves move past the ball, they push

the ball alternately up and down. So a light wave moving past an atom sweeps an oscillating electric field over it, and if the atom carries an electrical charge it will be pushed first one way and then the other. Atoms in general do carry a charge, greater or lesser according to the molecule; thus in the hydrogen chloride molecule the hydrogen atom carries a small positive charge, the chlorine atom a corresponding negative one. Because of these opposite charges, the electric field of the light wave will push the two atoms in opposite directions, and will tend to set them into vibration, stretching and compressing the H-Cl bond alternately.

The bond has a natural frequency of vibration, determined by the masses of the two atoms and the restoring force of the bonds. A light wave with this frequency of oscillation will have most effect on the bond: its energy will greatly increase the natural vibrations of the

atoms. The molecule will absorb part of the energy of the light at this resonant frequency, and an absorption detector will show an absorption peak for that wavelength.

To obtain the infrared spectrum of a sample, we illuminate the sample with infrared radiation of successive wavelengths from 2.5 to 25 microns and measure with the spectrometer the amount of light transmitted by the sample at each wavelength. A modern spectrometer automatically computes the percentage of the light transmitted at each wavelength and in 10 or 15 minutes produces a curve of transmittance against wavelength, or, more commonly nowadays, transmittance against frequency. Frequency, meaning the number of waves that sweep past per second, is easily computed from the wavelength and the speed of light. The measure of frequency is called the "wave number"—it is actually the number of waves in one centimeter of light beam. In the range of infrared that we are considering the wave numbers run from 400 to 4,000 per centimeter.

Modern atomic theory at once shows the chemical importance of the frequency of absorption. Atoms and molecules absorb light in quanta, the energy in each quantum being proportional to the frequency of the light. The absorbed energy is in fact Planck's constant h times the frequency, or hc times the wave number. Einstein enunciated the principle that, when molecules absorb light, each quantum is wholly absorbed by one molecule.

Now a molecule can safely absorb only so much energy: chemical bonds are not unbreakable—and a good thing, or there'd be no chemistry! The energy required to break the bonds is known fairly accurately; it is comparable to the energy of a quantum of ultraviolet or visible light. When a molecule absorbs such light, the bonds are either broken or profoundly altered, and the "excited" molecule is a very different entity from the original. The deduction of atomic structure from ultraviolet spectra has been compared to deducing the structure of a piano from the sounds emitted as it falls down a flight of stairs.

The absorption of a quantum in the infrared, on the other hand, is not so rough a process. At these wavelengths a quantum of energy is only about one twentieth that in the ultraviolet. Hence the radiation merely sets the bond into vibration. In infrared spectroscopy we don't push the piano down the stairs,

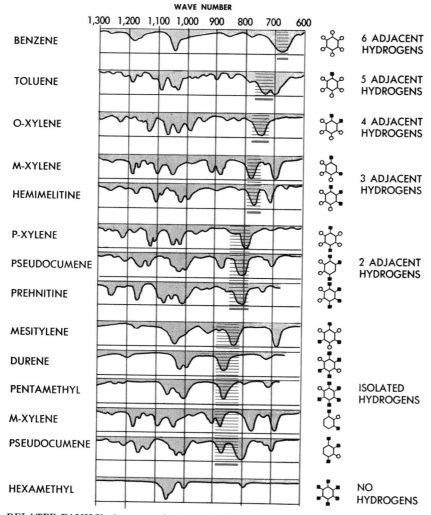

WAVE NUMBER

RELATED FAMILY of compounds, the methyl benzenes, exhibit both an over-all similarity and distinctive differences in their infrared spectra. The absorption dips which are marked in red are due to hydrogen wagging (see middle diagram on the facing page). The hydrogen atoms are vibrating in and out of, in a direction perpendicular to, the page surface.

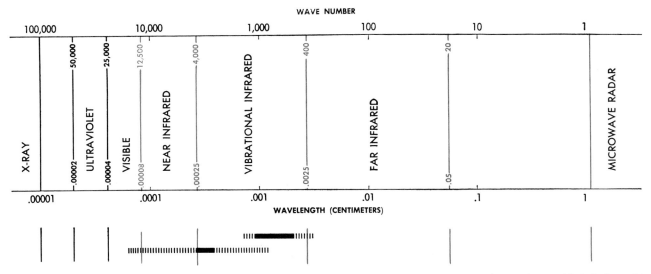

ELECTROMAGNETIC SPECTRUM is diagrammed above. Wave number means the number of wavelengths per centimeter and is proportional to frequency. Wave numbers from 400 to 4,000 have the right frequency to excite molecular vibrations. Bars below locate concentrations of heat radiation from a black body at 80 degrees F. (*upper bar*) and 1.800 degrees (*lower*). In each bar 50 per cent of the total heat energy is in the black sections, 40 per cent in the broken section, and the rest tapers off from the ends.

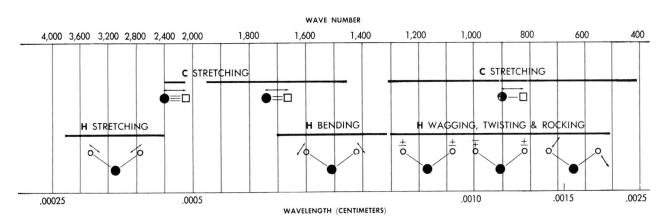

TYPES OF VIBRATION that cause most common absorption bands are here located in spectrum. Carbon atoms are represented by solid black circles, hydrogen atoms by small open circles, other atoms by squares. Carbon bonds stretch at the indicated frequencies whether running between two carbon atoms or between carbon and a different atom. Hydrogen wagging, twisting and rocking cover the 500 to 1,300 region jointly. Plus and minus signs for wagging and twisting indicate motion in and out of plane of page.

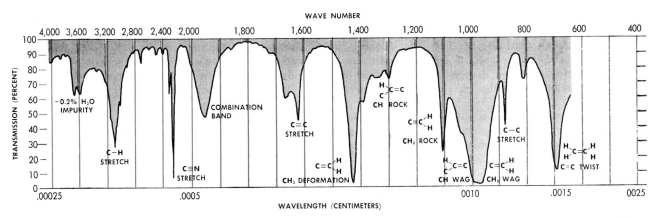

ACRYLONITRILE, whose molecular formula appears at left, gives the infrared spectrum shown above. The curve represents the amount of energy transmitted by a sample of the material as the wavelength increases (*from left to right*). Dips indicate frequencies at which energy is strongly absorbed, showing that a natural frequency of molecular vibration has been reached. Vibrations causing dips are indicated for comparison with chart above.

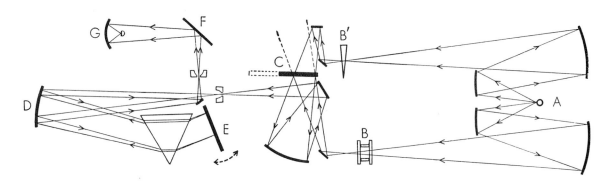

INFRARED SPECTROPHOTOMETER made by the Perkin-Elmer Corporation is shown on opposite page. Above are a photograph and diagram of its optical system. Infrared rays from a heated carbide rod (A) are split into two beams, one passing through the sample to be analyzed (B) and the other through a wedge whose transparency can be adjusted. A rotating mirror and diaphragm (C) alternately passes the two beams, by way of a pair of mirrors, to the curved mirror (D). This reflects the beams through a prism

we just plunk the keyboard a bit, and the sounds are a bit easier to relate to the piano.

We can carry this image further. The infrared spectrum of a two-atom molecule such as HCl shows absorption not only at the single natural frequency of the bond, but also at the overtones of that frequency. A molecule containing several atoms has several bonds and several natural frequencies. Here we cannot match up the frequencies with the individual bonds. We may think of such a molecule as a mechanical system of several masses connected by springs, and the resonant frequencies will be those of such a set of coupled oscillators. When one bond is set into vibration, the rest of the molecule also is involved through the interconnecting bonds. Hence the resonant frequencies are characteristic of the *whole molecule*. They are determined by (1) the masses, which means the specific atoms involved, (2)

the spring forces, which means the bonds in the molecule, and (3) the way in which these are coupled, which means the specific geometrical arrangement of the bonds. If any of these changes, the set of resonant frequencies will change.

The infrared spectrum is simply a display of the resonant frequencies of the sample. Actually not all resonant frequencies give rise to absorption. Only those which, like the simple H-Cl vibration, cause some net change in the separation between positive and negative charges will interact with the oscillating electric field. Thus a molecule like Cl_2, which can have no charge separation no matter how the bond length changes (one Cl is like another!), does not absorb anywhere in the infrared. One of the resonant frequencies of CO_2, in which the two oxygen atoms move symmetrically, causes no charge displacement and hence does not appear in the infrared spectrum. But in general there

are enough active resonant frequencies in a molecule so that the infrared spectrum is characteristic of the atoms, the bonds and the geometrical arrangement.

And here we see one reason for the great value of infrared: its *specificity*. The infrared spectrum is the most nearly unique property of a substance. Even geometrical isomers, which have the same atoms and the same bonds but differ in arrangement, can be distinguished by infrared. Such isomers are very hard to distinguish by ordinary chemical methods, yet their "slight" structural differences can give rise to profound differences in biological activity. Many, perhaps most, biological phenomena are highly sensitive to such differences in compounds. Consequently medical investigators need a sensitive analytical method. The infrared spectrometer was a key tool for the late cancer researcher Konrad Dobriner at Memorial Hospital in New York City in his

which separates their wavelengths. A rocking mirror (E) sweeps the dispersed beams across the detector, by way of (D), the small plane mirror, (F), and (G). The strengths of the alternately received beams are compared at successive wavelengths. When, the detector senses a difference between the sample and the reference beams, an alternating current is generated which adjusts the movable wedge to equalize the signals. The motion of the wedge in turn is transmitted to pen that traces the absorption curve on drum.

work of unraveling the metabolism of steroids in the body.

Inanimate objects—if internal-combustion engines can be so classified—also are sometimes sensitive to isomeric differences. The difference between "knock" and "anti-knock" gasoline components is a problem nicely suited to the infrared spectrometer. Indeed, the need in the petroleum industry for a fast, reliable and convenient method of distinguishing isomeric hydrocarbons and analyzing mixtures of them has played a large part in the recent upsurge of infrared. The war gave rise to pressing demands for high-test gasoline and for synthetic-rubber intermediates; both of these involved analysis of isomeric hydrocarbons. Infrared had been used in a few industrial laboratories in the late 1930s. Under the wartime challenge it was soon shown that infrared spectra offered the rapid and accurate analytical method so badly needed. And to meet the need for instruments the first commercial infrared spectrometer appeared in 1943.

The infrared also has unusual sensitivity to atomic mass; it can distinguish not only isomers but also isotopes. Practically the only methods for analyzing the stable isotopes used in tracer work are infrared and mass spectroscopy. Which is the better method in a given case will depend on many factors; in cases calling for a fast, nondestructive analytical method applicable *in situ*, the specificity of infrared can be very useful.

Yet all this is not the whole story. The greatest advantage of infrared as a research tool is that an infrared spectrum can be interpreted in terms of the same concepts chemists use in studying chemical properties—bonds and bond groupings. In classifying compounds and thinking about them and working with them, chemists have long spoken of "functional groups"—of olefins, of acid chlorides and so on. The functional groups provide a broad chemical description and a clue to the specific chemical properties of a compound—and they can be related to its infrared spectrum.

Let us consider for a moment the mechanical model. When two or more high-frequency springs are coupled together tightly, the vibrations of the resulting system form a new pattern in general quite distinct from that of the uncoupled springs; but when two high-frequency springs are coupled by means of a low-frequency spring, the resulting vibration pattern will include the frequencies of the high-frequency springs with only small shifts. Now chemical bonds fall approximately into two classes: (1) "high-frequency," which includes all multiple bonds and the single bonds involving hydrogen, and (2) "low-frequency," which includes all other single bonds, such as C-C, C-O, C-N. In

DETECTOR for spectrophotometer is shown in front view (*top*) and side view (*bottom*) about 16 times actual size. Beam of radiation is focused on thermocouple, the fine gold ribbon mounted on the contacts.

studying infrared spectra we may therefore try out the idea of a "vibrational functional group"—any set of high-frequency bonds directly connected. According to this, all molecules with the group $-N=C=S$, for example, should have some frequencies in common, but they should differ from those with the $-N=C-$ or $C=S$ groups. Here indeed we find the "something" for which Coblentz was groping!

Now we find a piece of good fortune—one of the rare presents Nature bestows on investigators. Chemical functional groups and vibrational functional groups run parallel. Therefore certain characteristic absorption bands in an infrared spectrum give direct and strong evidence on the chemical nature of the sample. It is not really conclusive evidence, for the functional group idea is an approximation, and the parallelism of chemical and vibrational functional groups is not too strict. The infrared spectrum is not a magic crystal ball in which one reads the structural formula of an unknown sample. But its clues, when wisely used, can shorten by weeks the time required to complete the job. And the direct applicability of the chemical-bond concept and the functional-group concept means that the chemist can understand infrared spectra without having to learn a new language. He *does* have to learn a new dialect of his chemical language—and incautious chemists who have overlooked this have made some serious blunders. But the dialect can be picked up relatively quickly.

The knowledge of molecular structure required to make use of infrared spectroscopy had been achieved by about 1930 or 1935. Why, then, did the blossoming of chemical infrared start only in 1943? Some pioneer applications of infrared *had* been made in the chemical industry as early as 1936. But the technique is difficult. The cost of the modern commercial instruments and their need of maintenance still remind the user that the infrared spectrometer is doing a basically harder job than its ultraviolet counterpart. It has been well said that "no other spectrometric region of the electromagnetic range is so beset with experimental difficulties." Basically these arise from the very fact that infrared radiation is resonant with atomic vibrations. We have seen the advantages stemming from this; let us glance at the price we must pay.

As a source of infrared, we must use vibrating atoms, and the only practical way to set atoms vibrating is through heat. To get a reasonable intensity of

infrared, we must heat our source to a high temperature; a common source is an electrically heated rod of carborundum at 2,200 degrees Fahrenheit. Refractory materials able to withstand such temperatures are brittle, which can be a nuisance! Moreover, the laws of physics worked out by Planck show that most of the energy goes into the near-infrared and even the visible region, wasting our power and filling our instruments with a lot of unwanted wavelengths which must be filtered out or shunted off.

The same basic trouble arises in the measurement of infrared intensity—detection, as it is appropriately called. The infrared quanta do not disturb molecules enough even to affect a photographic plate! All we can measure is a slight rise in the temperature of the absorber, and we measure it today as Herschel did—by the effect on a blackened thermometer. The "thermometer," however, is a very sensitive thermocouple.

We must find prisms to disperse the infrared spectrum and optics to focus it, windows to let the radiation through the cells holding our samples and solvents in which to dissolve materials we want to study. Since all atoms can vibrate, it is not easy to find substances transparent to the infrared. Glass and water are completely opaque to it; ordinary organic solvents have so many absorption bands themselves that they obscure the spectrum of the substance dissolved in them. So we use mirrors instead of lenses, make our prisms and cell windows of rock salt, and think hard about the choice of solvents.

For infrared, the wartime need was a blessing in disguise—a very perfect disguise, to quote Mark Twain—because urgent necessity stimulated great improvements in instrumentation. The availability of better instruments today is a great blessing not only to analytical chemists and workers in the chemical industry but also to infrared spectroscopists interested in fundamental chemistry. Infrared spectroscopy has helped us win our present understanding of molecular structure, notably the geometry and dynamics summarized in the ball-and-spring model. Nowadays more people than ever before are at work on this fundamental use of infrared, studying the nature of those springs and the distribution of electronic charges in the bonds. Without depreciating the more widespread use of infrared for the analysis of compounds, we may well feel that in the long run the fundamental use will be more exciting. For the nature of the chemical bond is the problem at the heart of all chemistry.

22

GAS CHROMATOGRAPHY

ROY A. KELLER
October 1961

*A simple analytical method sharply separates
components of complex mixtures. When used to
characterize perfumes and flavors, its sensitivity rivals
that of the human nose*

The most pervasive problem in chemistry and biochemistry is that of determining the composition of complex mixtures of matter. The research chemist would like to know at all times exactly what his test tubes and flasks contain; the industrial chemist has an equally urgent need to know the composition of materials flowing through reaction vessels and distillation columns and into product tanks. Within the past 20 to 30 years, in response to this need for fast and accurate analyses, a number of powerful analytical tools have been developed. They include instruments that measure how compounds absorb ultraviolet and infrared radiation, instruments that determine atomic or molecular mass and instruments that determine how the magnetic properties of atomic nuclei are influenced by different molecular configurations.

One of the newest and most versatile analytical techniques is gas-liquid chromatography, usually referred to simply as gas chromatography. The name is somewhat misleading; there is nothing chromatic about the method or its results. The name comes from the original method of liquid-solid chromatography described in 1906 by Michael Tswett, a Russian botanist. Tswett found that when a solution of chlorophyll was allowed to filter through a column firmly packed with pulverized calcium carbonate, the various fractions of the chlorophyll mixture separated into distinctively colored bands. He called the result a chromatogram [see "Chromatography," by William H. Stein and Stanford Moore; Scientific American Offprint 81]. A useful variation of Tswett's original concept is paper chromatography, in which compounds in solution migrate at different speeds across a sheet of porous paper.

In gas chromatography the compounds in a mixture migrate at differing speeds when carried along by an inert gas through a tube that has been packed or treated in a special way. The method was first suggested in 1941 by the British chemists A. T. James and A. J. P. Martin. The method underwent development in many laboratories beginning around 1950, and the first commercial instrument came on the market in 1955. Today some 40 models of gas chromatographs are being built by 20 U.S. and a dozen European manufacturers. The instruments are widely used for analyzing complex organic mixtures such as those commonly found in petroleum products, essential oils, perfumes, flavors and other substances of biological origin. Samples containing as many as 76 dif-

SAMPLE IS INJECTED into a gas chromatograph by means of a syringe that can deliver liquid samples as small as one hundred-thousandth of a cubic centimeter. Gas samples require a special valve arrangement to meter the sample through a chamber of known volume.

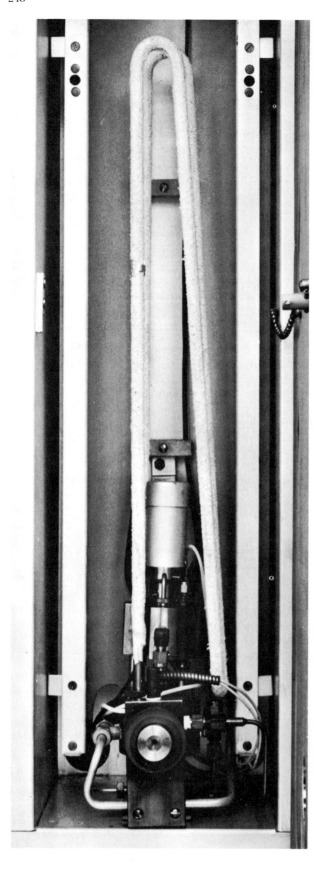

TWO TYPES OF GAS CHROMATOGRAPH differ in the design of the column where fractionation takes place. The instrument at left uses the original type of column, a quarter-inch in diameter and one to several meters long, packed with a pulverized inert substance. The column illustrated is four meters long and is folded twice. Instrument at right uses a capillary column, 150 to 300 feet long, without packing, proposed by Marcel J. E. Golay, a consultant to the Perkin-Elmer Corporation. A nonvolatile liquid carried on the packing, or on the inside wall of the capillary tube, acts as a partitioning agent and promotes fractionation of the sample.

ferent substances have been successfully analyzed in one pass. Analysis time is typically a few minutes and sometimes only a few seconds. Samples usually range in size from a few hundredths to a few thousandths of a gram. Some instruments can handle samples weighing not much more than a millionth of a gram, and in such samples they can detect the presence of substances that weigh no more than a trillionth of a gram—about the weight of a single bacterium.

The fractionating column of a typical gas chromatograph consists of a copper or stainless steel tube about a quarter of an inch in diameter and from one to four meters long, though occasionally much longer tubes are used. The tube is packed with an inert material such as firebrick or diatomaceous earth that has been pulverized and coated with a nonvolatile liquid called a partitioner [see illustration at right]. After the tube has been packed it is usually bent into a series of U turns or wound into a helix so that it can be fitted easily into an insulated box, the temperature of which is thermostatically controlled.

The liquid selected as the partitioner largely determines the performance of the chromatograph. The partitioner must not react with the sample being analyzed and it must not be volatilized by the stream of carrier gas that propels the sample through the column. Above all, the partitioner must show different affinities (to use an old-fashioned term) for each of the substances likely to be found in the sample mixture.

Partitioners that work well with one type of sample may be completely useless for another type. As the sample is moved through the column by the carrier gas, the partitioner must interfere in a selective fashion with the progress of each compound present, slowing up the progress of some and letting others travel through the column more swiftly. At the outlet of the column a detecting device signals the emergence of each different compound by activating a recording pen on a strip chart.

In the search for good partitioners, builders of gas chromatographs have experimented with virtually every viscous fluid, grease and low-melting-point solid in the laboratory, including such substances as silicone rubber, stopcock grease and hydrogenated shark oil. This eclecticism has sometimes had unhappy results when excellent separations were achieved with substances that could never be duplicated, even when re-

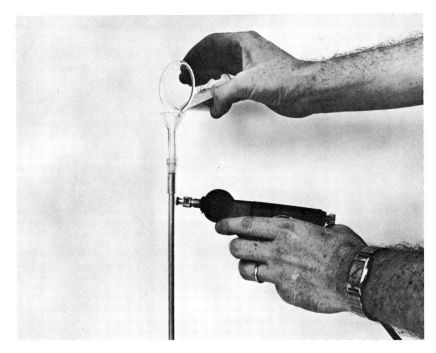

PREPARATION OF CHROMATOGRAPHIC COLUMN begins (top) with pulverized diatomaceous earth, to which is added a viscous partitioning liquid dissolved in a volatile carrier. The solvent is evaporated by heat while the coated powder is gently agitated (middle). The dried coated powder is packed into a tube with the aid of a vibrator (bottom).

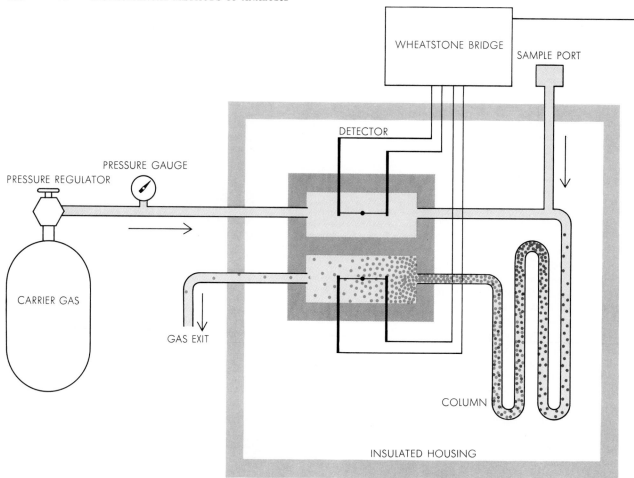

SIMPLICITY OF GAS CHROMATOGRAPH is among the virtues of the instrument. The sample is swept by a carrier gas through a specially treated column in which various components of the sample migrate at different speeds. A detector measures the electrical conductivity of the gas leaving the column as well as that of the carrier gas entering the column. The difference, as determined by

ordered by the same lot or batch number. In order to deposit the viscous partitioner on the pulverized support, the partitioner is usually dissolved in a volatile solvent that can be evaporated, leaving the partitioner behind. Normally the partitioner weighs from about a quarter to a third as much as the pulverized packing in the column. The coated particles should flow freely, and there is some advantage in holding the amount of partitioner to a minimum.

Let us now look more closely at what happens to a collection of molecules injected into the carrier gas moving through the column of a gas chromatograph. Some of the molecules rapidly dissolve in the liquid partitioner, and a dynamic equilibrium is soon established as they pass back and forth between the liquid and vapor filling the interstices of the column packing. At equilibrium the concentration of molecules of each type is a constant ratio in the two phases. Molecules of compound A, for example, may partition themselves equally between the liquid and vapor phase; molecules of compound B, on the other hand,

may be highly soluble in the liquid phase and therefore relatively few of them will be found in the vapor phase once equilibrium is reached. In this case the moving gas will tend to sweep molecules of compound A down the tube, leaving those of compound B behind in the liquid. Once the molecules of A have been carried to a region containing fresh liquid, however, some of them will redissolve until a new equilibrium is reached. By the same token, when fresh gas passes over the liquid containing molecules of B in solution, some of the B molecules will enter the gas phase in order to establish equilibrium. If we regard the sample as a plug of molecules moving through the column, we can visualize a sort of molecular leapfrog. The volatile molecules are continuously being swept to the head of the plug, where they redissolve; the less volatile molecules fall to the tail of the plug, but they too are continuously being picked up and inched forward by the gas stream pressing from behind. Eventually, if conditions are right, all the molecules of the more volatile components will be carried clear ahead of those of the less

volatile and a clean separation will be achieved.

A major problem in the early days of gas chromatography was to find a detector that would respond quickly as the separated compounds emerged in rapid succession from the end of the fractionating column. The job of the detector is not to identify the emerging compound but merely to signal when the output gas is carrying foreign molecules and when it is not. Once this is known it is easy enough to calibrate the output readings by feeding samples of known composition into the instrument.

In their original gas chromatography of fatty acids or amines (which are organic bases) James and Martin passed the output of their column through a solvent that extracted the acids or amines from the effluent gas. The collecting was done in a series of small batches so that the time of emergence from the column could be recorded. Using a color indicator, James and Martin could tell when an acid or an amine emerged from the column. They then titrated each sample to determine how much acid or base was present. The job not only was

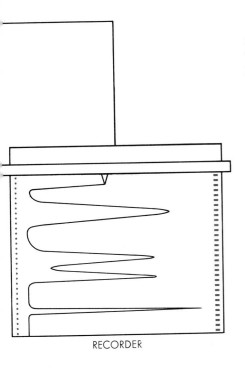

RECORDER

a Wheatstone bridge, is recorded on a strip chart. The instrument can be calibrated by analyzing samples of known composition.

laborious but also it lent itself to titratable samples only.

A much simpler and more universal solution to the detection problem was finally found in the thermal-conductivity cell. The cell utilizes the fact that the electrical resistance of a heated wire varies with its temperature. If a gas of constant composition and flow rate is allowed to pass over a heated wire, the wire will be cooled a constant amount and so register a constant resistance. If a gas of different thermal conductivity appears in the stream striking the wire, the wire will change in temperature and in electrical resistance, and this change can be recorded in ink on a strip chart.

At least two other detection devices have been developed for gas chromatography. In both devices a change in gas composition is signaled by a change in the ionization—and hence the electrical conductivity—of the gas stream. In one device the gas stream is ionized (broken up into electrically charged fragments) by being passed through a hydrogen flame; in the other device the stream is ionized by bombardment with radiation from a bit of radioactive material.

To obtain sharply defined gas-chromatograph records, called fractograms, the instrument designer can vary the pressure and flow rate of the carrier gas, the operating temperature of the column, the structure and particle size of the column packing and, of course, the nature of the liquid partitioner. A considerable body of theory and empirical art has grown up around the solute-solvent interactions that underlie effective partitioning.

For example, if one wishes to separate a hydrocarbon and an alcohol having nearly the same boiling point (e.g., 3-methylpentane and methyl alcohol, both of which boil at about 64 degrees centigrade), it is desirable to use a partitioner that resembles the alcohol in containing hydroxyl (OH) groups in its structure. The hydrogen atom of the hydroxyl group in the alcohol will tend to form a bond with the oxygen atom of the hydroxyl group in the partitioner. The hydrocarbon will not form such a bond and will therefore move through the column faster. A relatively nonvolatile liquid containing hydroxyl groups that can be used for this separation is polyethylene glycol.

Offhand one might think that a partitioner, to be effective, should show a differential solvent action on each component in a sample mixture. One might then conclude that the solutes would be retained by a partitioner in order of their solubility. In practice, however, this is not always true. The reason is that solubility as conventionally measured with solvents in bulk—say in a laboratory beaker—is quite different from the solubility shown when a solvent is thinly distributed over an enormous surface area, as it is in a chromatographic column. In the latter case a new factor appears: the effect of surface adsorption. A solute is said to be adsorbed if its concentration in the immediate region of the surface exceeds that in the bulk liquid. Adsorption can arise either at the interface between the solid support and the partitioning liquid, or at the interface between liquid and gas, or it can occur in both regions. Sometimes adsorption enhances the desired separation; at other times it interferes with it. For example, when alcohols are chromatographed on a hydrocarbon partitioner, they tend to displace the hydrocarbon and fasten themselves to the support.

Considerable work has been done on gas chromatographs that achieve separations strictly by differential adsorption on the solid packing, without help from a liquid partitioner. This is called gas-solid chromatography, and for certain

sample mixtures in which the molecules have much the same architecture it produces even sharper separations than gas-liquid chromatography.

Until recently gas-liquid chromatography was successful only with sample mixtures whose components boiled within about 50 degrees C. of one another. When the boiling range was greater than that, it was usually impossible to choose a column operating temperature that would sharply resolve both the least volatile and most volatile components. If the temperature was held low, the more volatile substances would be resolved, but the less volatile would lag behind and become spread out by diffusion. By raising the temperature the resolution of the less volatile could be sharpened, but the more volatile would then rush through the column and emerge in a poorly resolved bunch. The

IONIZATION DETECTOR, one of three principal types of detector used in gas chromatography, employs a hydrogen flame to break up chemical compounds into electrically charged fragments (ions). By measuring the electrical conductivity of the ionized gas at the column exit the detector signals the passage of various fractions of the sample. In this photograph the cover of the detector has been removed and the normally colorless flame has been made visible.

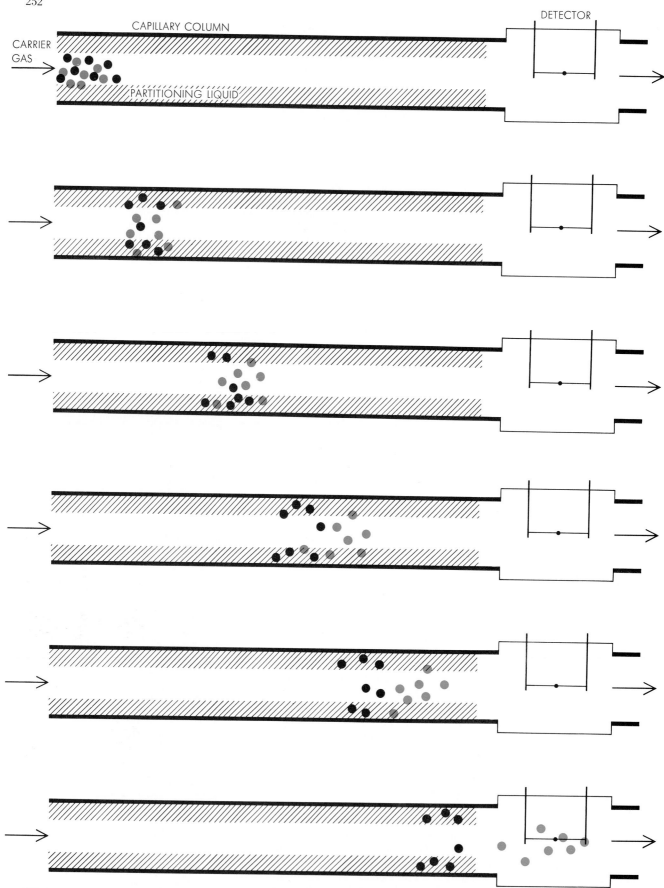

CHROMATOGRAPHIC SEPARATION takes place when a sample mixture (*black and colored balls*) is driven by an inert gas through a capillary tube coated with an immobile liquid called a partitioner. In the original form of gas-liquid chromatography the partitioner was deposited on a pulverized packing. The role of the partitioner is to dissolve (and adsorb) various components of the sample in differential fashion. After fractionation the separated components pass through a detector, whose output is recorded on a chart.

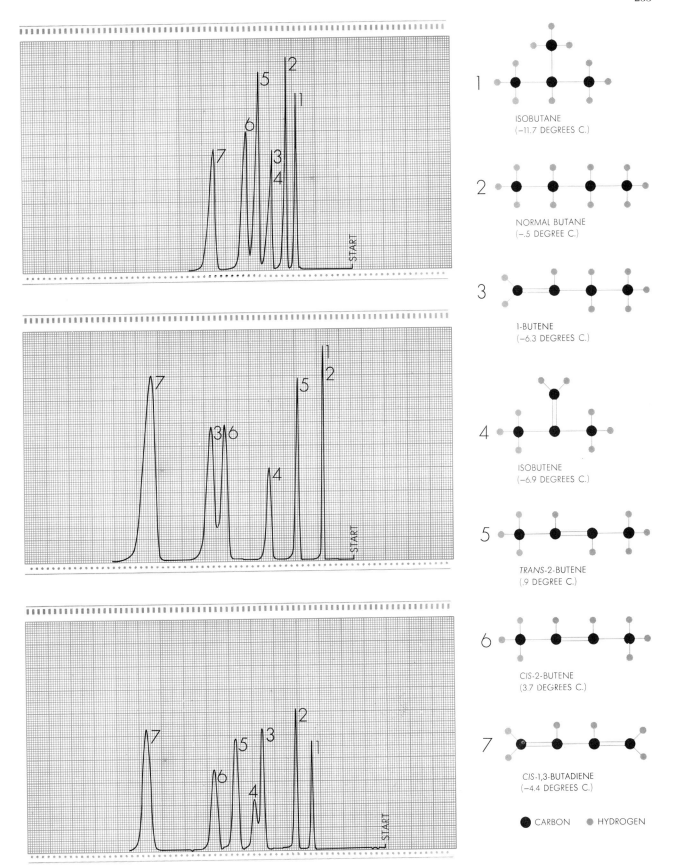

1 ISOBUTANE
(−11.7 DEGREES C.)

2 NORMAL BUTANE
(−.5 DEGREE C.)

3 1-BUTENE
(−6.3 DEGREES C.)

4 ISOBUTENE
(−6.9 DEGREES C.)

5 TRANS-2-BUTENE
(.9 DEGREE C.)

6 CIS-2-BUTENE
(3.7 DEGREES C.)

7 CIS-1,3-BUTADIENE
(−4.4 DEGREES C.)

● CARBON   ● HYDROGEN

SEPARATION OF FOUR-CARBON HYDROCARBONS, which boil within a narrow range, is a familiar problem in the oil and synthetic rubber industries. The molecular structure and boiling points of the principal four-carbon hydrocarbons appear at right. The fractograms (*left*) show how these compounds are fraction- ated by different partitioning liquids: dimethylsulfolane (*top*), silver nitrate in diethylene glycol (*middle*) and hexanedione (*bottom*). The upper two fractograms were made on columns four meters long operated at 25 degrees centigrade. Bottom fractogram was made on a column two meters long operated at zero degrees.

LEMON OIL

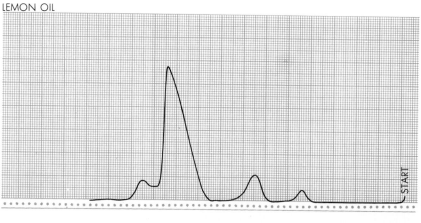

LIME OIL

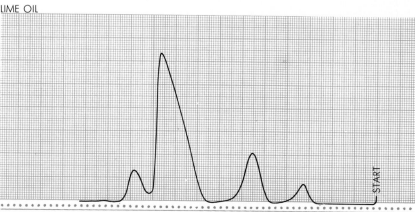

LEMON OIL

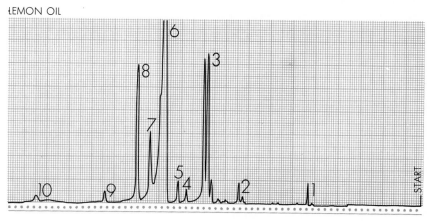

LIME OIL

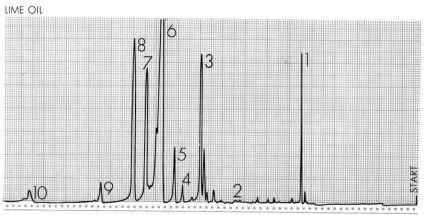

PERFORMANCE OF OLD AND NEW INSTRUMENTS is demonstrated by fractograms of two similar essential oils: the terpene fractions of lemon oil and lime oil. When analyzed on a standard two-meter packed column, the two oils produced the upper pair of fractograms. When analyzed on a 175-foot capillary (Golay) column, they produced the detailed lower pair. The peaks labeled "6" are made by limonene ($C_{10}H_{16}$), which has a lemon-like odor.

answer to this problem has been found in "temperature programing," which simply means starting the separation at a low temperature and raising it in regular steps until the job is done. By this procedure one can fractionate mixtures whose components have a boiling range as broad as 200 degrees C.

Actually it is not necessary for the components of a sample to be anywhere near their boiling points for gas chromatography to work. At the University of Arizona we have analyzed samples of volatile inorganic halides (for example, niobium chloride) at temperatures near their melting point, or in the vicinity of 250 degrees C. Other investigators have operated machines at 500 degrees C. that will chromatograph compounds boiling as high as 625 degrees C.

An important advance in the design of gas chromatographs was made about 1956 by Marcel J. E. Golay, a consultant to the Perkin-Elmer Corporation. Golay, a student of information theory, conceived the improvement after making a theoretical study of the migration of a solute through a packed column. In the fashion characteristic of theorists he sought a simplified model to substitute for the actual highly complex situation prevailing in a tube packed with porous particles of random shapes and sizes. He selected for his model a bundle of capillary tubes, equal in diameter to the granule size of the packing, which were evenly coated with the partitioning medium. Upon analyzing his calculations Golay concluded that a single, very long coated capillary tube would achieve separations equivalent, if not superior, to those produced by packed columns and do so in a much shorter time and under less severe conditions of temperature and driving pressure. Golay met his skeptics by preparing some of the first columns himself. They amply fulfilled his predictions.

Commercial columns of his design usually employ a capillary tube having an inside diameter of .01 inch. The tubes, ranging from 150 to 1,000 feet in length, are coiled into a compact helix [*see illustration at right on page 241*]. If these fine capillary columns are not to be flooded, the sample size must be extremely small, usually about five-millionths of a cubic centimeter in liquid volume, or about five-thousandths of a cubic centimeter after vaporization. To obtain such minute volumes a sample some 20 times larger is vaporized and shot through a stream splitter that ad-

1 ACETALDEHYDE
2 FORMALDEHYDE
3 ETHYL FORMATE
4 ETHYL ACETATE
5 METHANOL
6 PROPANOL
7 ISOAMYL ALCOHOL

mits about 5 per cent to the column and discards the remainder. Ordinarily a few minutes to half an hour are required for such samples to migrate through the column. As an extreme example of what can be achieved with a capillary column and a highly sensitive detector, it has been possible to record 14 peaks—each representing a separate compound in a sample of closely related hydrocarbons —in less than two seconds.

The areas under different peaks in a fractogram are roughly proportional to the fractional amounts of each substance in the original sample. With care the method is accurate to about 2 per cent. In general the investigator has two methods for discovering exactly what substance is represented by a particular peak. The commonest method is to use samples of known composition to calibrate the machine. Alternatively he may isolate the column effluent that produced a given peak and characterize it by some suitable analytical technique, for example by using an infrared spectrophotometer or a mass spectrometer. The gas chromatograph will separate not only compounds with closely similar properties but also various forms of individual compounds. In organic chemistry almost all but the simplest compounds can exist in two or more forms known as isomers. These are molecules containing the same number and kind of atoms fitted together in different geometrical arrangements [see illustration on page 253].

In some cases gas chromatography provides direct clues to compound identification. The volume of gas, called the retention volume, that precedes a particular solute through the column depends on the nature of the solute, the choice of partitioner and the temperature. Within limits, retention volume is not overly sensitive to length of column, driving pressure, flow rate or the amount of partitioner employed. As a result one can determine retention volumes for various compounds of interest and use these volumes for identifying unknown samples [see illustration on following page]. A considerable effort is now being made to utilize gas chromatography for determining the structure of molecules by

BOURBON

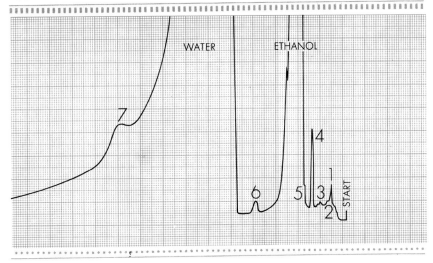

SCOTCH

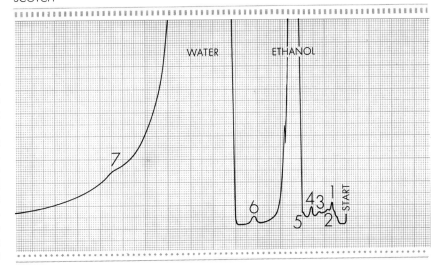

GIN

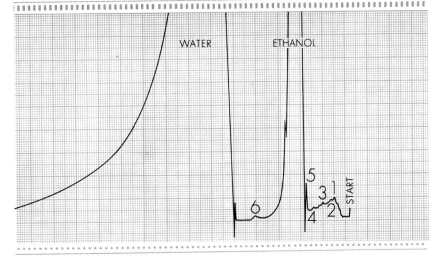

FRACTOGRAMS OF POTABLE SPIRITS provide a sensitive measure of volatile components that influence taste and aroma. These analyses confirm that Scotch and gin contain fewer such components than bourbon. (Isoamyl alcohol is fusel oil.) Samples were run on a two-meter packed column. Peaks were identified by adding various known compounds to a whiskey that had previously been analyzed. Analyses were made by Robert B. Carroll, a chemical consultant, and Lawrence C. O'Brien, then at the Perkin-Elmer Corporation.

relating retention volume to particular molecular configurations.

Much of the explosive growth of gas chromatography can be attributed to the petroleum industry, which has to deal with materials of extraordinary complexity. Crude oils commonly contain more than 150 different hydrocarbons, many of them isomers of each other. For separating them gas chromatography has proved a powerful tool. Oil firms have now begun to use gas chromatographs for continuous monitoring of process streams in the refinery. Chromatographic analyses will be transmitted to a computer, which will automatically calculate the optimum operating conditions for catalytic cracking towers. Gas chromatography is almost the only method available for detecting certain catalytic poisons that impair the polymerization process when present in concentrations of only 50 parts per million.

Automotive engineers use gas chromatography for analyzing the exhaust of engines under a variety of operating conditions with various fuels. The results are used to improve both engines and fuels and also contribute to a reduction in the air pollution created by engine exhausts.

Elsewhere, fractograms are rapidly replacing the opinions of "sniff and taste" panels as a method of assaying the uniformity of instant coffees, blends of whiskey and many other products whose commercial acceptance depends on subtle flavors and aromas. In such products gas chromatography can often detect trace components present in only one part per billion. It is, in fact, the first analytical instrument to rival the human nose in sensitivity.

Before long one can reasonably expect that gas chromatography will be used not only as an analytical tool but also as a method of preparing ultrapure materials. Columns are now operating that can refine and separate the components in batches about one ounce in size. If the process can be made continuous, it should be able to provide laboratory chemicals and even commercial compounds having a wholly new order of purity.

Recently Donald E. Johnson, Sara Jo Scott and Alton Meister of Tufts University announced success in a long-sought goal: separation by gas chromatography of derivatives of the amino acids. The method may prove superior to the liquid-solid chromatographic method, employing a column packed with ion-exchange resins, that has been widely used for the analysis of the amino acids in proteins [see "The Chemical Structure of Proteins," by William H. Stein and Stanford Moore; SCIENTIFIC AMERICAN Offprint 80].

A gas chromatograph of remarkable capabilities is now being developed by the Aerojet-General Corporation and the Jet Propulsion Laboratory of the California Institute of Technology; it will be placed aboard the *Surveyor* moon probe scheduled for launching in 1963. Upon reaching the moon the probe will pick up samples of lunar crust, grind them and feed them into a heater, where they will be pyrolized. The gas chromatograph will include two types of detector to analyze samples pyrolized at 150, 325, 500 and 1,000 degrees C. The heater and chromatograph must consume no more than 10 watts of current. The whole apparatus must weigh no more than 11.5 pounds and must fit inside a box measuring 8 by 8 by 10 inches. Finally, to withstand the shock of landing on the moon, it must tolerate a deceleration force equivalent to 100 times the force of gravity on the earth. An important objective of the lunar chromatograph is to learn whether or not the moon's crust contains complex organic compounds of the type associated with living matter. This is an impressive assignment for an analytical tool that has come into general use only within the past five or six years.

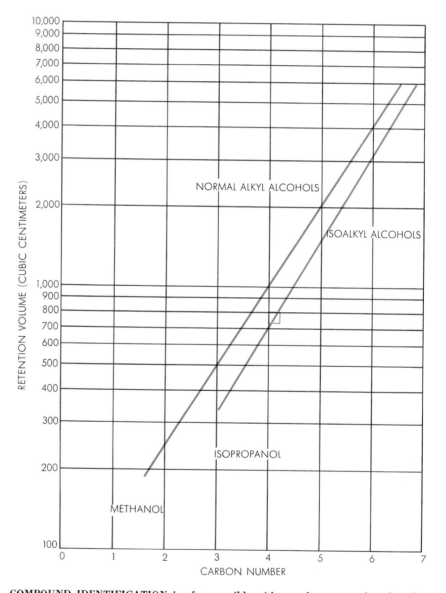

COMPOUND IDENTIFICATION is often possible with gas chromatography when the sample is known to contain one of a limited number of compounds. For example, normal alkyl alcohols, whose carbon atoms form a straight chain, require a greater volume of carrier gas to drive them through a given column than isoalkyl alcohols having the same number of carbon atoms in a branched chain. Thus if an unknown four-carbon alcohol has a retention volume of 750 cubic centimeters of carrier gas (*square*), one can be sure it is isobutanol and not normal butanol, which would have a retention volume of about 1,000 c.c.

23

CHEMICAL FOSSILS

GEOFFREY EGLINTON AND MELVIN CALVIN
January 1967

Certain rocks as much as three billion years old have been found to contain organic compounds. What these compounds are and how they may have originated in living matter is under active study

If you ask a child to draw a dinosaur, the chances are that he will produce a recognizable picture of such a creature. His familiarity with an animal that lived 150 million years ago can of course be traced to the intensive studies of paleontologists, who have been able to reconstruct the skeletons of extinct animals from fossilized bones preserved in ancient sediments. Recent chemical research now shows that minute quantities of organic compounds—remnants of the original carbon-containing chemical constituents of the soft parts of the animal—are still present in some fossils and in ancient sediments of all ages, including some measured in billions of years. As a result of this finding organic chemists and geologists have joined in a search for "chemical fossils": organic molecules that have survived unchanged or little altered from their original structure, when they were part of organisms long since vanished.

This kind of search does not require the presence of the usual kind of fossil—a shape or an actual hard form in the rock. The fossil molecules can be extracted and identified even when the organism has completely disintegrated and the organic molecules have diffused into the surrounding material. In fact, the term "biological marker" is now being applied to organic substances that show pronounced resistance to chemical change and whose molecular structure gives a strong indication that they could have been created in significant amounts only by biological processes.

One might liken such resistant compounds to the hard parts of organisms that ordinarily persist after the soft parts have decayed. For example, hydrocarbons, the compounds consisting only of carbon and hydrogen, are comparatively resistant to chemical and biological attack. Unfortunately many other biologi-

cally important molecules such as nucleic acids, proteins and polysaccharides contain many bonds that hydrolyze, or cleave, readily; hence these molecules rapidly decompose after an organism dies. Nevertheless, several groups of workers have reported finding constituents of proteins (amino acids and peptide chains) and even proteins themselves in special well-protected sites, such as between the thin sheets of crystal in fossil shells and bones [see "Paleobiochemistry," by Philip H. Abelson; SCIENTIFIC AMERICAN, July, 1956].

Where complete destruction of the organism has taken place one cannot, of course, visualize its original shape from the nature of the chemical fossils it has left behind. One may, however, be able to infer the biological class, or perhaps even the species, of organism that gave rise to them. At present such deductions must be extremely tentative because they involve considerable uncertainty. Although the chemistry of living organisms is known in broad outline, biochemists even today have identified the principal constituents of only a few small groups of living things. Studies in comparative biochemistry or chemotaxonomy are thus an essential parallel to organic geochemistry. A second uncertainty involves the question of whether or not the biochemistry of ancient organisms was generally the same as the biochemistry of present-day organisms. Finally, little is known of the chemical changes wrought in organic substances when they are entombed for long periods of time in rock or a fossil matrix.

In our work at the University of California at Berkeley and at the University of Glasgow we have gone on the assumption that the best approach to the study of chemical fossils is to analyze geological materials that have had a relatively simple biological and geological history.

The search for suitable sediments requires a close collaboration between the geologist and the chemist. The results obtained so far augur well for the future.

Organic chemistry made its first major impact on the earth sciences in 1936, when the German chemist Alfred Treibs isolated metal-containing porphyrins from numerous crude oils and shales. Certain porphyrins are important biological pigments; two of the best-known are chlorophyll, the green pigment of plants, and heme, the red pigment of the blood. Treibs deduced that the oils were biological in origin and could not have been subjected to high temperatures, since that would have decomposed some of the porphyrins in them. It is only during the past decade, however, that techniques have been available for the rapid isolation and identification of organic substances present in small amounts in oils and ancient sediments. Further refinements and new methods will be required for detailed study of the tiny amounts of organic substances found in some rocks. The effort should be worthwhile, because such techniques for the detection and definition of the specific architecture of organic molecules should not only tell us much more about the origin of life on the earth but also help us to establish whether or not life has developed on other planets. Furthermore, chemical fossils present the organic chemist with a new range of organic compounds to study and may offer the geologist a new tool for determining the environment of the earth in various geological epochs and the conditions subsequently experienced by the sediments laid down in those epochs.

If one could obtain the fossil molecules from a single species of organism, one would be able to make a direct correlation between present-day biochemistry

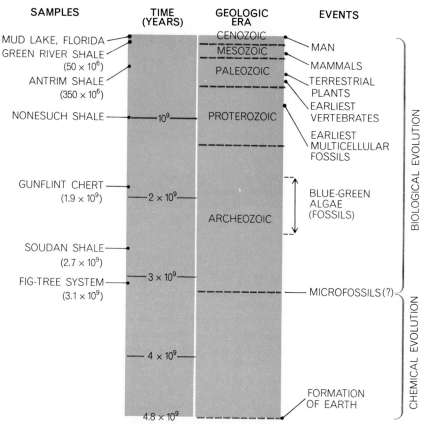

GEOLOGICAL TIME SCALE shows the age of some intensively studied sedimentary rocks (*left*) and the sequence of major steps in the evolution of life (*right*). The stage for biological evolution was set by chemical evolution, but the period of transition is not known.

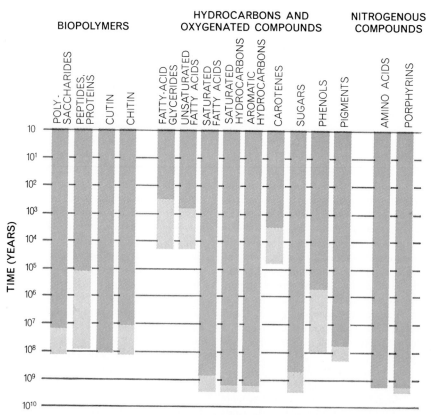

ORGANIC COMPOUNDS originally synthesized by living organisms and more or less modified have now been found in many ancient rocks that began as sediments. The dark bars indicate reasonably reliable identification; the light bars, unconfirmed reports. Cutin and chitin are substances present respectively in the outer structures of plants and of insects.

and organic geochemistry. For example, one could directly compare the lipids, or fatty compounds, isolated from a living organism with the lipids of its fossil ancestor. Unfortunately the fossil lipids and other fossil compounds found in sediments almost always represent the chemical debris from many organisms.

The deposition of a compressible fine-grained sediment containing mineral particles and disseminated organic matter takes place in an aquatic environment in which the organic content can be partially preserved; an example would be the bottom of a lake or a delta. The organic matter makes up something less than 1 percent of many ancient sediments. The small portion of this carbon-containing material that is soluble in organic solvents represents a part of the original lipid content, more or less modified, of the organisms that lived and died while the sediment was being deposited.

The organic content presumably consists of varying proportions of the components of organisms—terrestrial as well as aquatic—that have undergone chemical transformation while the sediment was being laid down and compressed. Typical transformations are reduction, which has the effect of removing oxygen from molecules and adding hydrogen, and decarboxylation, which removes the carboxyl radical (COOH). In addition, it appears that a variety of reactive unsaturated compounds (compounds having available chemical bonds) combine to form an insoluble amorphous material known as kerogen. Other chemical changes that occur with the passage of time are related to the temperature to which the rock is heated by geologic processes. Thus many petroleum chemists and geologists believe petroleum is created by progressive degradation, brought about by heat, of the organic matter that is finely disseminated throughout the original sediment. The organic matter that comes closest in structure to the chains and rings of carbon atoms found in the hydrocarbons of petroleum is the matter present in the lipid fraction of organisms. Another potential source of petroleum hydrocarbons is kerogen itself, presumably formed from a wide variety of organic molecules; it gives off a range of straight-chain, branched-chain and ring-containing hydrocarbons when it is strongly heated in the laboratory. One would also like to know more about the role of bacteria in the early steps of sediment formation. In the upper layers of most newly formed sediments there is strong bacterial activity, which must surely result in extensive alteration of the initially deposited organic matter.

In this article we shall concentrate on the isolation of fossil hydrocarbons. The methods must be capable of dealing with the tiny quantity of material available in most rocks. Our general procedure is as follows.

After cutting off the outer surface of a rock specimen to remove gross contaminants, we clean the remaining block with solvents and pulverize it. We then place the powder in solvents such as benzene and methanol to extract the organic material. Before this step we sometimes dissolve the silicate and carbonate minerals of the rock with hydrofluoric and hydrochloric acids. We separate the organic extract so obtained into acidic, basic and neutral fractions. The compounds in these fractions are converted, when necessary, into derivatives that make them suitable for separation by the technique of chromatography. For the initial separations we use column chromatography, in which a sample in solution is passed through a column packed with alumina or silica. Depending on their nature, compounds in the sample pass through the column at different speeds and can be collected in fractions as they emerge.

In subsequent stages of the analysis finer fractionations are achieved by means of gas-liquid chromatography. In this variation of the technique, the sample is vaporized into a stream of light gas, usually helium, and brought in contact with a liquid that tends to trap the compounds in the sample in varying degree. The liquid can be supported on an inorganic powder, such as diatomaceous earth, or coated on the inside of a capillary tube. Since the compounds are alternately trapped in the liquid medium and released by the passing stream of gas they progress through the column at varying speeds, with the result that they are separated into distinct fractions as they emerge from the tube. The temperature of the column is raised steadily as the separation proceeds, in order to drive off the more strongly trapped compounds.

The initial chromatographic separation is adjusted to produce fractions that consist of a single class of compound, for example the class of saturated hydrocarbons known as alkanes. Alkane molecules may consist either of straight chains of carbon atoms or of chains that include branches and rings. These subclasses can be separated with the help of molecular sieves: inorganic substances, commonly alumino-silicates, that have a fine honeycomb structure. We use a sieve whose

mesh is about five angstrom units, or about a thousandth of the wavelength of green light. Straight-chain alkanes, which resemble smooth flexible rods about 4.5 angstroms in diameter, can enter the sieve and are trapped. Chains with branches and rings are too big to enter and so are held back. The straight-chain alkanes can be liberated from the sieve for further analysis by dissolving the sieve in hydrofluoric acid. Other families of molecules can be trapped in special crystalline forms of urea and thiourea, whose crystal lattices provide cavities with diameters of five angstroms and seven angstroms respectively.

The families of molecules isolated in this way are again passed through gas

chromatographic columns that separate the molecular species within the family. For example, a typical chromatogram of straight-chain alkanes will show that molecules of increasing chain length emerge from the column in a regularly spaced sequence that parallels their increasing boiling points, thus producing a series of evenly spaced peaks. Although the species of molecule in a particular peak can often be identified tentatively on the basis of the peak's position, a more precise identification is usually desired. To obtain it one must collect the tiny amount of substance that produced the peak—often measured in micrograms—and examine it by one or more analytical methods such as ultraviolet and infrared

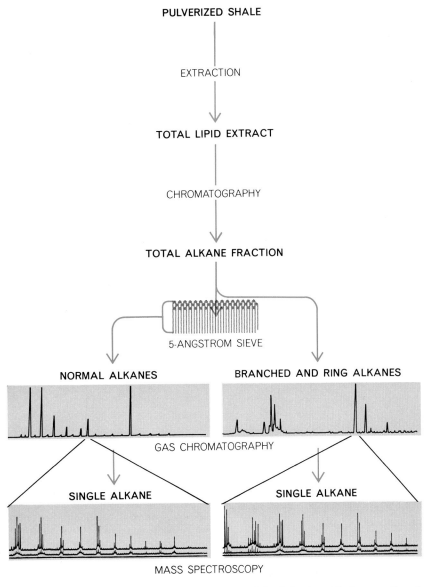

ANALYTICAL PROCEDURE for identifying chemical fossils begins with the extraction of alkane hydrocarbons from a sample of pulverized shale. In normal alkanes the carbon atoms are arranged in a straight chain. (Typical alkanes are illustrated on page 261.) Molecular sieves are used to separate straight-chain alkanes from alkanes with branched chains and rings. The two broad classes are then further fractionated. Compounds responsible for individual peaks in the chromatogram are identified by mass spectrometry and other methods.

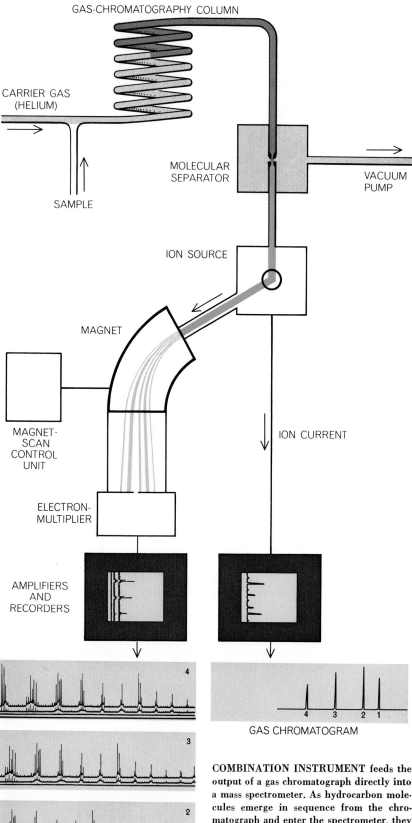

GAS-CHROMATOGRAPHY COLUMN

CARRIER GAS
(HELIUM)

SAMPLE

MOLECULAR
SEPARATOR

VACUUM
PUMP

ION SOURCE

MAGNET

MAGNET-
SCAN
CONTROL
UNIT

ION CURRENT

ELECTRON-
MULTIPLIER

AMPLIFIERS
AND
RECORDERS

GAS CHROMATOGRAM

MASS SPECTRA

COMBINATION INSTRUMENT feeds the output of a gas chromatograph directly into a mass spectrometer. As hydrocarbon molecules emerge in sequence from the chromatograph and enter the spectrometer, they are ionized, or broken into charged fragments. The size of the ionization current is proportional to the amount of material present at each instant and can be converted into a chromatogram. In the spectrometer the charged fragments are directed through a magnetic field, which separates them according to mass. Each species of molecule produces a unique mass-distribution pattern.

spectroscopy, mass spectrometry or nuclear magnetic resonance. In one case X-ray crystallography is being used to arrive at the structure of a fossil molecule.

A new and useful apparatus is one that combines gas chromatography and mass spectrometry [*see illustration at left*]. The separated components emerge from the chromatograph and pass directly into the ionizing chamber of the mass spectrometer, where they are broken into submolecular fragments whose abundance distribution with respect to mass is unique for each component. These various analytical procedures enable us to establish a precise structure and relative concentration for each organic compound that can be extracted from a sample of rock.

How is it that such comparatively simple substances as the alkanes should be worthy of geochemical study? There are several good reasons. Alkanes are generally prominent components of the soluble lipid fraction of sediments. They survive geologic time and geologic conditions well because the carbon-hydrogen and carbon-carbon bonds are strong and resist reaction with water. In addition, alkane molecules can provide more information than the simplicity of their constitution might suggest; even a relatively small number of carbon and hydrogen atoms can be joined in a large number of ways. For example, a saturated hydrocarbon consisting of 19 carbon atoms and 40 hydrogen atoms could exist in some 100,000 different structural forms that are not readily interconvertible. Analysis of ancient sediments has already shown that in some cases they contain alkanes clearly related to the long-chain carbon compounds of the lipids in present-day organisms [*see illustration on next page*]. Generally one finds a series of compounds of similar structure, such as the normal, or straight-chain, alkanes (called *n*-alkanes); the compounds extracted from sediments usually contain up to 35 carbon atoms. Alkanes isolated from sediments may have been buried as such or formed by the reduction of substances containing oxygen.

The more complicated the structure of the molecule, the more valuable it is likely to be for geochemical purposes: its information content is greater. Good examples are the alkanes with branches and rings, such as phytane and cholestane. It is unlikely that these complex alkanes could be built up from small subunits by processes other than biological ones, at least in the proportions found. Hence we are encouraged to look for biological precursors with appropriate preexisting carbon skeletons.

In conducting this kind of search one makes the assumption, at least at the outset, that the overall biochemistry of past organisms was similar to that of present-day organisms. When lipid fractions are isolated directly from modern biological sources, they are generally found to contain a range of hydrocarbons, fatty acids, alcohols, esters and so on. The mixture is diverse but by no means random. The molecules present in such fractions have structures that reflect the chemical reaction pathways systematically followed in biological organisms. There are only a few types of biological molecule wherein long chains of carbon atoms are linked together; two examples are the straight-chain lipids, the end groups of which may include oxygen atoms, and the lipids known as isoprenoids.

The straight-chain lipids are produced by what is called the polyacetate pathway [see illustration on next page]. This pathway leads to a series of fatty acids with an even number of carbon atoms; the odd-numbered molecules are missing. One also finds in nature straight-chain alcohols (n-alkanols) that likewise have an even number of carbon atoms, which is to be expected if they are formed by simple reduction of the corresponding fatty acids. In contrast, the straight-chain hydrocarbons (n-alkanes) contain an odd number of carbon atoms. Such a series would be produced by the decarboxylation of the fatty acids.

The second type of lipid, the isoprenoids, have branched chains consisting of five-carbon units assembled in a regular order [see illustration on page 263]. Because these units are assembled in head-to-tail fashion the side-chain methyl groups (CH$_3$) are attached to every fifth carbon atom. (Tail-to-tail addition occurs less frequently but accounts for several important natural compounds, for example beta-carotene.) When the isoprenoid skeleton is found in a naturally occurring molecule, it is reasonable to assume that the compound has been formed by this particular biological pathway.

Chlorophyll is possibly the most widely distributed molecule with an isoprenoid chain; therefore it must make some contribution to the organic matter in sediments. Its fate under conditions of geological sedimentation is not known, but it may decompose into only two or three large fragments [see illustration on page 266]. The molecule of chlorophyll a consists of a system of intercon-

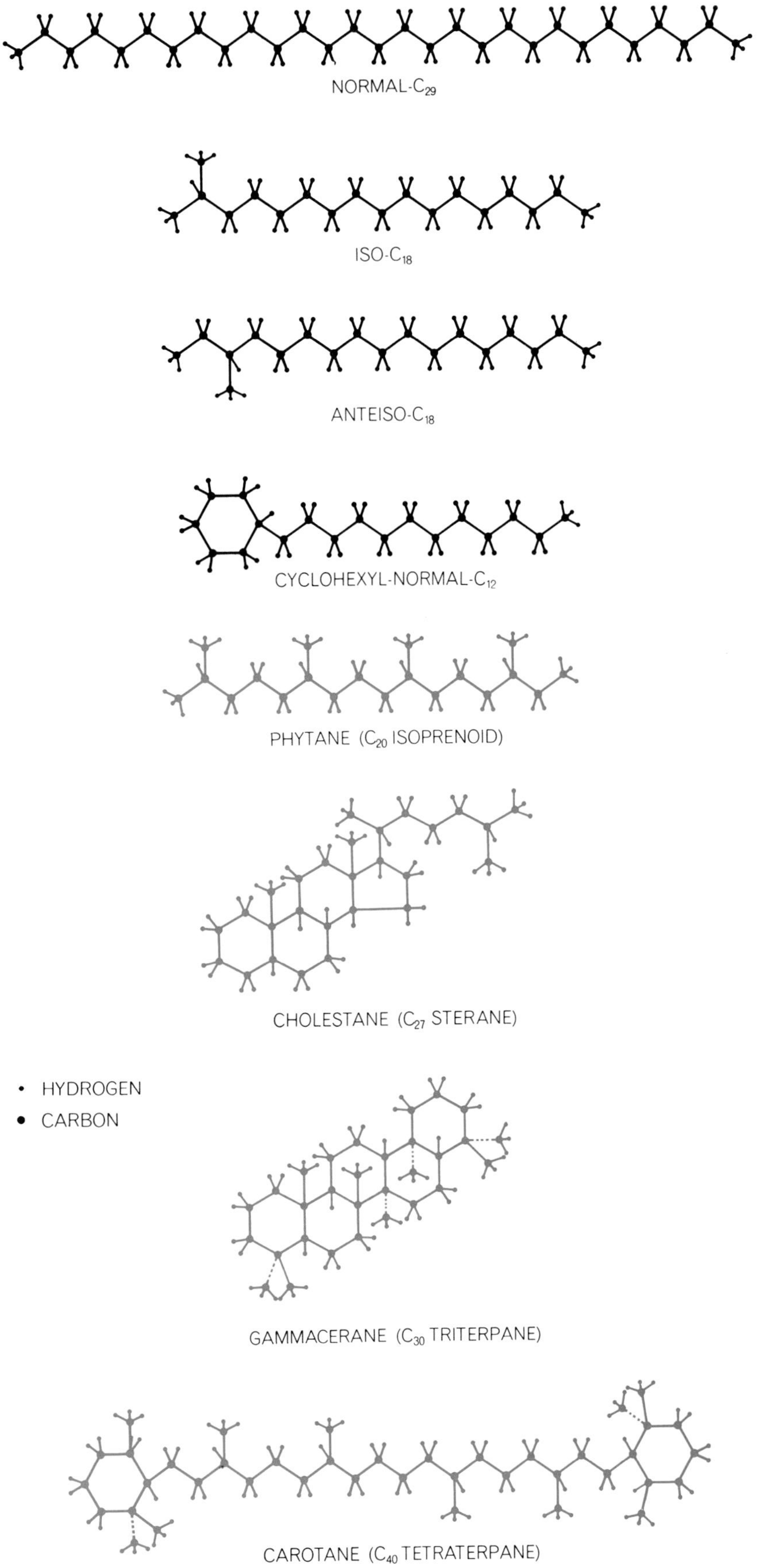

NORMAL-C$_{29}$

ISO-C$_{18}$

ANTEISO-C$_{18}$

CYCLOHEXYL-NORMAL-C$_{12}$

PHYTANE (C$_{20}$ ISOPRENOID)

CHOLESTANE (C$_{27}$ STERANE)

• HYDROGEN
• CARBON

GAMMACERANE (C$_{30}$ TRITERPANE)

CAROTANE (C$_{40}$ TETRATERPANE)

ALKANE HYDROCARBON MOLECULES can take various forms: straight chains (which are actually zigzag chains), branched chains and ring structures. Those depicted here have been found in crude oils and shales. The molecules shown in color are so closely related to well-known biological molecules that they are particularly useful in bespeaking the existence of ancient life. The broken lines indicate side chains that are directed into the page.

nected rings and a phytyl side chain, which is an isoprenoid. When chlorophyll is decomposed, it seems likely that the phytyl chain is split off and converted to phytane (which has the same number of carbon atoms) and pristane (which is shorter by one carbon atom). When both of these branched alkanes are found in a sediment, one has reasonable presumptive evidence that chlorophyll was once present. The chlorophyll ring system very likely gives rise to the metal-containing porphyrins that are found in many crude oils and sediments.

Phytane and pristane may actually enter the sediments directly. Max Blumer of the Woods Hole Oceanographic Institution showed in 1965 that certain species of animal plankton that eat the plant plankton containing chlorophyll store quite large quantities of pristane and related hydrocarbons. The animal plankton act in turn as a food supply for bigger marine animals, thereby accounting for the large quantities of pristane in the liver of the shark and other fishes.

An indirect source for the isoprenoid alkanes could be the lipids found in the outer membrane of certain bacteria that live only in strong salt solutions, an environment that might be found where ancient seas were evaporating. Morris Kates of the National Research Council of Canada has shown that a phytyl-containing lipid (diphytyl phospholipid) is common to bacteria with the highest salt requirement but not to the other bacteria examined so far.

This last example brings out the point that in spite of the overall oneness of present-day biochemistry, organisms do differ in the compounds they make. They also synthesize the same compounds in different proportions. These differences are making it possible to classify living species on a chemotaxonomic, or chemical, basis rather than on a morphological, or shape, basis. Eventually it may be possible to extend chemical classification to ancient organisms, creating a discipline that could be called paleochemotaxonomy.

Our study of chemical fossils began in 1961, when we decided to probe the sedimentary rocks of the Precambrian period in a search for the earliest signs of life. This vast period of time, some four billion years, encompasses the beginnings of life on this planet and its early development to the stage of organisms consisting of more than one cell [see illustrations on page 258]. We hoped that our study would complement the efforts being made by a number of work-

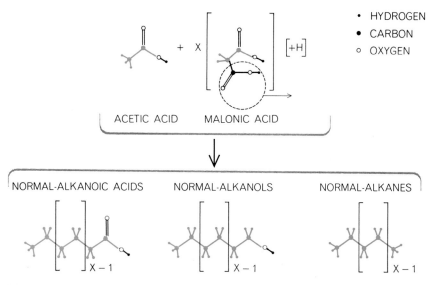

• HYDROGEN
• CARBON
○ OXYGEN

ACETIC ACID   MALONIC ACID

NORMAL-ALKANOIC ACIDS   NORMAL-ALKANOLS   NORMAL-ALKANES

STRAIGHT-CHAIN LIPIDS are created in living organisms from simple two-carbon and three-carbon compounds: acetate and malonate, shown here as their acids. The complex biological process, which involves coenzyme A, is depicted schematically. The fatty acids (n-alkanoic acids) and fatty alcohols (n-alkanols) produced in this way have an even number of carbon atoms. The removal of carbon dioxide from the fatty acids, the net effect of decarboxylation, would give rise to a series of n-alkanes with an odd number of carbon atoms.

ers, including one of us (Calvin), to imitate in the laboratory the chemical evolution that must have preceded the appearance of life on earth. We also saw the possibility that our work could be adapted to the study of meteorites and of rocks obtained from the moon or nearby planets. Thus it even includes the possibility of uncovering exotic and alien biochemistries. The exploration of the ancient rocks of the earth provides a testing ground for the method and the concepts involved.

We chose the alkanes because one might expect them to resist fairly high temperatures and chemical attack for long periods of time. Moreover, J. G. Bendoraitis of the Socony Oil Company, Warren G. Meinschein of the Esso Research Laboratory and others had already identified individual long-chain alkanes, including a range of isoprenoid types, in certain crude oils. Even more encouraging, J. J. Cummins and W. E. Robinson of the U.S. Bureau of Mines had just made a preliminary announcement of their isolation of phytane, pristane and other isoprenoids from a relatively young sedimentary rock: the Green River shale of Colorado, Utah and Wyoming. Thus the alkanes seemed to offer the biological markers we were seeking. Robinson generously provided our laboratory with samples of the Green River shale, which was deposited some 50 million years ago and constitutes the major oil-shale reserve of the U.S.

The Green River shale, which is the

remains of large Eocene lakes in a rather stable environment, contains a considerable fraction (.6 percent) of alkanes. Using the molecular-sieve technique, we split the total alkane fraction into alkanes with straight chains and those with branched chains and rings and ran the resulting fractions through the gas chromatograph [see illustration at top left on page 264]. The straight-chain alkanes exhibit a marked dominance of molecules containing an odd number of carbon atoms, which is to be expected for straight-chain hydrocarbons from a biological source. The other fraction shows a series of prominent sharp peaks; we conclusively identified them as isoprenoids, confirming the results of Cummins and Robinson. The large proportion of phytane, the hydrocarbon corresponding to the entire side chain of chlorophyll, is particularly noteworthy. The oxygenated counterparts of the steranes and triterpanes (27 to 30 carbon atoms) and the high-molecular-weight n-alkanes (29 to 31 carbons) are typical constituents of the waxy covering of the leaves and pollen of land plants, leading to the inference that such plants made major contributions to the organic matter deposited in the Green River sediments.

Although the gross chemical structure (number of rings and side chains) of the steranes and triterpanes was established in this work, it was only recently that the precise structure of one of these hydrocarbons was conclusively established. E. V. Whitehead and his associates in the British Petroleum Company and Robinson and his collaborators in the Bureau of Mines have shown that one of the triterpanes extracted from the Green River shale is identical in all respects with gammacerane [see illustration on page 261]. Conceivably it is produced by the reduction of a compound known as gammaceran-3-beta-ol, which was recently isolated from the common protozoon *Tetrahymena pyriformis*. Other derivatives of gammacerane are rather widely distributed in the plant kingdom.

At our laboratory in Glasgow, Sister Mary T. J. Murphy and Andrew McCormick recently identified several steranes and triterpanes and also the tetraterpane called perhydro-beta-carotene, or carotane [see top illustration on page 268]. Presumably carotane is derived by reduction from beta-carotene, an important red pigment of plants. A similar reduction process could convert the familiar biological compound cholesterol into cholestane, one of the steranes found in the Green River shale [see same illustration on page 268]. The mechanism and sedimentary site of such geochemical reduction processes is an important problem awaiting attack.

W. H. Bradley of the U.S. Geological Survey has sought a contemporary counterpart of the richly organic ooze that presumably gave rise to the Green River shale. So far he has located only four lakes, two in the U.S. and two in Africa, that seem to be reasonable candidates. One of them, Mud Lake in Flor-

ida, is now being studied closely. A dense belt of vegetation surrounding the lake filters out all the sand and silt that might otherwise be washed into it from the land. As a result the main source of sedimentary material is the prolific growth of microscopic algae. The lake bottom uniformly consists of a grayish-green ooze about three feet deep. The bottom of the ooze was deposited about 2,300 years ago, according to dating by the carbon-14 technique.

Microscopic examination of the ooze shows that it consists mainly of minute fecal pellets, made up almost exclusively of the cell walls of blue-green algae. Some pollen grains are also present. Decay is surprisingly slow in spite of the ooze's high content of bound oxygen and the temperatures characteristic of Florida. Chemical analyses in several laboratories, reported this past November at a meeting of the Geological Society of America, indicate that there is indeed considerable correspondence between the lipids of the Mud Lake ooze and those of the Green River shale. Eugene McCarthy of the University of California at Berkeley has also found beta-carotene in samples of Mud Lake ooze that are about 1,100 years old. The high oxygen content of the Mud Lake ooze seems inconsistent, however, with the dominance of oxygen-poor compounds in the Green River shale. The long-term geological mechanisms that account for the loss of oxygen may have to be sought in sediments older than those in Mud Lake.

Sediments much older than the Green River shale have now been examined by our groups in Berkeley and Glasgow, and by workers in other universities and in oil-industry laboratories. We find that the hydrocarbon fractions in these more ancient samples are usually more complex than those of the Green River shale; the gas chromatograms of the older samples tend to show a number of partially resolved peaks centered around a single maximum. One of the older sediments we have studied is the Antrim shale of Michigan. A black shale probably 350 million years old, it resembles other shales of the Chattanooga type that underlie many thousands of square miles of the eastern U.S. Unlike the Green River shale, the straight-chain alkane fraction of the Antrim shale shows little or no predominance of an odd number of carbon atoms over an even number [see middle illustration of three at top of pages 264 and 265]. The alkanes with branched chains and rings, however, continue to be rich in isoprenoids.

The fact that alkanes with an odd number of carbon atoms are not predominant in the Antrim shale and sediments of comparable antiquity may be owing to the slow cracking by heat of carbon chains both in the alkane component and in the kerogen component. The effect can be partially reproduced in the laboratory by heating a sample of the Green River shale for many hours above 300 degrees centigrade. After such treatment the straight-chain alkanes show a reduced dominance of odd-carbon molecules and the branched-chain-and-ring fraction is more complex.

The billion-year-old shale from the Nonesuch formation at White Pine, Mich., exemplifies how geological, geochemical and micropaleontological techniques can be brought to bear on the problem of detecting ancient life. With the aid of the electron microscope Elso S. Barghoorn and J. William Schopf of Harvard University have detected in the Nonesuch shale "disaggregated particles of condensed spheroidal organic matter." In collaboration with Meinschein the Harvard workers have also found evidence that the Nonesuch shale contains isoprenoid alkanes, steranes and porphyrins. Independently we have analyzed the Nonesuch shale and found that it contains pristane and phytane, in addition to iso-alkanes, anteiso-alkanes and cyclohexyl alkanes.

Barghoorn and S. A. Tyler have also detected microfossils in the Gunflint chert of Ontario, which is 1.9 billion years old, almost twice the age of the

ACETIC ACID   MEVALONIC ACID   ISOPENTENYL PYROPHOSPHATE
(3 UNITS)

DIMER, $(C_5)_2$

TRIMER, $(C_5)_3$

TETRAMER, $(C_5)_4$

POLYMER $(C_5)_n$

BRANCHED-CHAIN LIPIDS are produced in living organisms by an enzymatically controlled process, also depicted schematically. In this process three acetate units link up to form a six-carbon compound (mevalonic acid), which subsequently loses a carbon atom and is combined with a high-energy phosphate. "Head to tail" assembly of the five-carbon subunits produces branched-chain molecules that are referred to as isoprenoid structures.

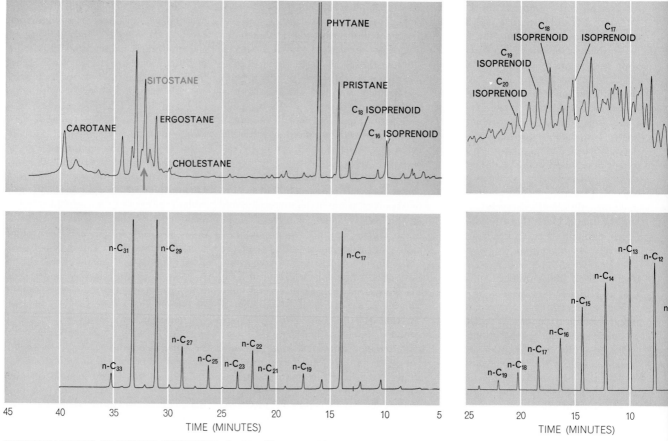

HYDROCARBONS IN YOUNG SEDIMENT, the 50-million-year-old Green River shale, produced these chromatograms. Alkanes with branched chains and rings appear in the top curve, normal alkanes in the bottom curve. The alkanes in individual peaks were identified by mass spectrometry and other methods. Such alkanes as phytane and pristane and the predominance of normal alkanes with an odd number of carbon atoms affirm that the hydrocarbons are biological in origin. The bimodal distribution of the curves is also significant.

OLDER SEDIMENTS are represented by the Antrim shale (*left*), which is 350 million years old, and by the Soudan shale (*right*), which is 2.7 billion years old. Alkanes with branched chains and rings again are shown in the top curves, normal alkanes

Nonesuch shale. They have reported that the morphology of the Gunflint microfossils "is similar to that of the existing primitive filamentous blue-green algae."

One of the oldest Precambrian sediments yet analyzed is the Soudan shale of Minnesota, which was formed about 2.7 billion years ago. Although its total hydrocarbon content is only .05 percent, we have found that it contains a mixture of straight-chain alkanes and branched-chain-and-ring alkanes not unlike those present in the much younger Antrim shale [*see third illustration of three at top of these two pages*]. In the branched-chain-and-ring fraction we have identi-

fied pristane and phytane. Steranes and triterpanes also seem to be present, but we have not yet established their precise three-dimensional structure. Preston E. Cloud of the University of California at Los Angeles has reported that the Soudan shale contains microstructures resembling bacteria or blue-green algae,

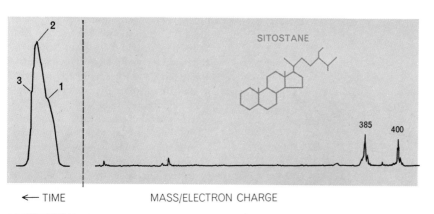

IDENTIFICATION OF SITOSTANE in the Green River shale was accomplished by "trapping" the alkanes that produced a major peak in the chromatogram (*colored arrow at top left on this page*) and passing them through the chromatograph-mass spectrometer. As the chromatograph drew the curve at the left, three scans were made with the spectrometer. Scan 1 (*partially shown at right*) is identical with the scan produced by pure sitostane.

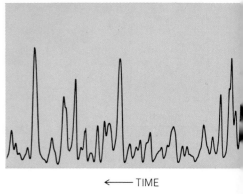

NINETEEN-CARBON ISOPRENOID was identified in the Antrim shale by using a co-injection technique together with a high-resolution gas chromatograph. These two high-resolution curves, each taken from a

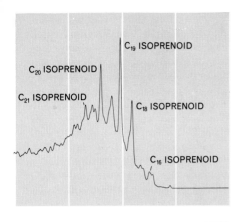

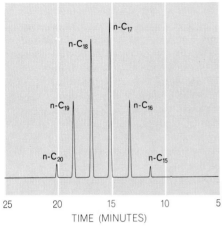

in the bottom curves. These chromatograms lack a pronounced bimodal distribution and the normal alkanes do not show a predominance of molecules with an odd number of carbon atoms. Nevertheless, the prevalence of isoprenoids argues for a biological origin.

but he is not satisfied that the evidence is conclusive.

A few reports are now available on the most ancient rocks yet examined: sediments from the Fig Tree system of Swaziland in Africa, some 3.1 billion years old. An appreciable fraction of the alkane component of these rocks consists

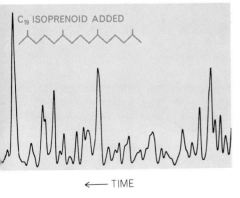

much longer trace, show the change in height of a specific peak when a small amount of pure 19-carbon isoprenoid was coinjected with the sample. Other peaks can be similarly identified by coinjecting known alkanes.

of isoprenoid molecules. If one assumes that isoprenoids are chemical vestiges of chlorophyll, one is obliged to conclude that living organisms appeared on the earth only about 1.7 billion years after the earth was formed (an estimated 4.8 billion years ago).

Before reaching this conclusion, however, one would like to be sure that the isoprenoids found in ancient sediments have the precise carbon skeleton of the biological molecules from which they are presumed to be derived. So far no sample of pristane or phytane—the isoprenoids that may be derived from the phytyl side chain of chlorophyll—has been shown to duplicate the precise three-dimensional structure of a pure reference sample. Vigorous efforts are being made to clinch the identification.

Assuming that one can firmly establish the presence of biologically structured isoprenoid alkanes in a sediment, further questions remain. The most serious one is: Were the hydrocarbons or their precursors deposited when the sediment was formed or did they seep in later? This question is not easily answered. A sample can be contaminated at any point up to—and after—the time it reaches the laboratory bench. Fossil fuels, lubricants and waxes are omnipresent, and laboratory solvents contain tiny amounts of pristane and phytane unless they are specially purified.

One way to determine whether or not rock hydrocarbons are indigenous is to measure the ratio of the isotopes carbon 13 and carbon 12 in the sample. (The ratio is expressed as the excess of carbon 13 in parts per thousand compared with the isotope ratio in a standard: a sample of a fossil animal known as a belemnite.) The principle behind the test is that photosynthetic organisms discriminate against carbon 13 in preference to carbon 12. Although we have few clues to the abundance of the two isotopes throughout the earth's history, we can at least test various hydrocarbon fractions in a given sample to see if they have the same isotope ratio. As a simple assumption, one would expect to find the same ratio in the soluble organic fraction as in the insoluble kerogen fraction, which could not have seeped into the rock as kerogen.

Philip H. Abelson and Thomas C. Hoering of the Carnegie Institution of Washington have made such measurements on sediments of various geological ages and have found that the isotope ratios for soluble and insoluble fractions in most samples agree reasonably well. In some of the oldest samples, however,

there are inconsistencies. In the Soudan shale, for example, the soluble hydrocarbons have an isotope ratio expressed as −25 parts per thousand compared with −34 parts per thousand for the kerogen. (In younger sediments and in present-day marine organisms the ratio is about midway between these two values: −29 parts per thousand.) The isotope divergence shown by hydrocarbons in the Soudan shale may indicate that the soluble hydrocarbons and the kerogen originated at different times. But since nothing is known of the mechanism of kerogen formation or of the alterations that take place in organic matter generally, the divergence cannot be regarded as unequivocal evidence of separate origin.

On the other hand, there is some reason to suspect that the isoprenoids did indeed seep into the Soudan shale sometime after the sediments had been laid down. The Soudan formation shows evidence of having been subjected to temperatures as high as 400 degrees C. The isoprenoid hydrocarbons pristane and phytane would not survive such conditions for very long. But since the exact date, extent and duration of the heating of the Soudan shale are not known, one can only speculate about whether the isoprenoids were indigenous and survived or seeped in later. In any event, they could not have seeped in much later because the sediment became compacted and relatively impervious within a few tens of millions of years.

A still more fundamental issue is whether or not isoprenoid molecules and others whose architecture follows that of known biological substances could have been formed by nonbiological processes. We and others are studying the kinds and concentrations of hydrocarbons produced by both biological and nonbiological sources. Isoprene itself, the hydrocarbon whose polymer constitutes natural rubber, is easily prepared in the laboratory, but no one has been able to demonstrate that isoprenoids can be formed nonbiologically under geologically plausible conditions. Using a computer approach, Margaret O. Dayhoff of the National Biomedical Research Foundation and Edward Anders of the University of Chicago and their colleagues have concluded that under certain restricted conditions isoprene should be one of the products of their hypothetical reactions. But this remains to be demonstrated in the laboratory.

It is well known, of course, that complex mixtures of straight-chain, branched-chain and even ring hydrocarbons can readily be synthesized in the

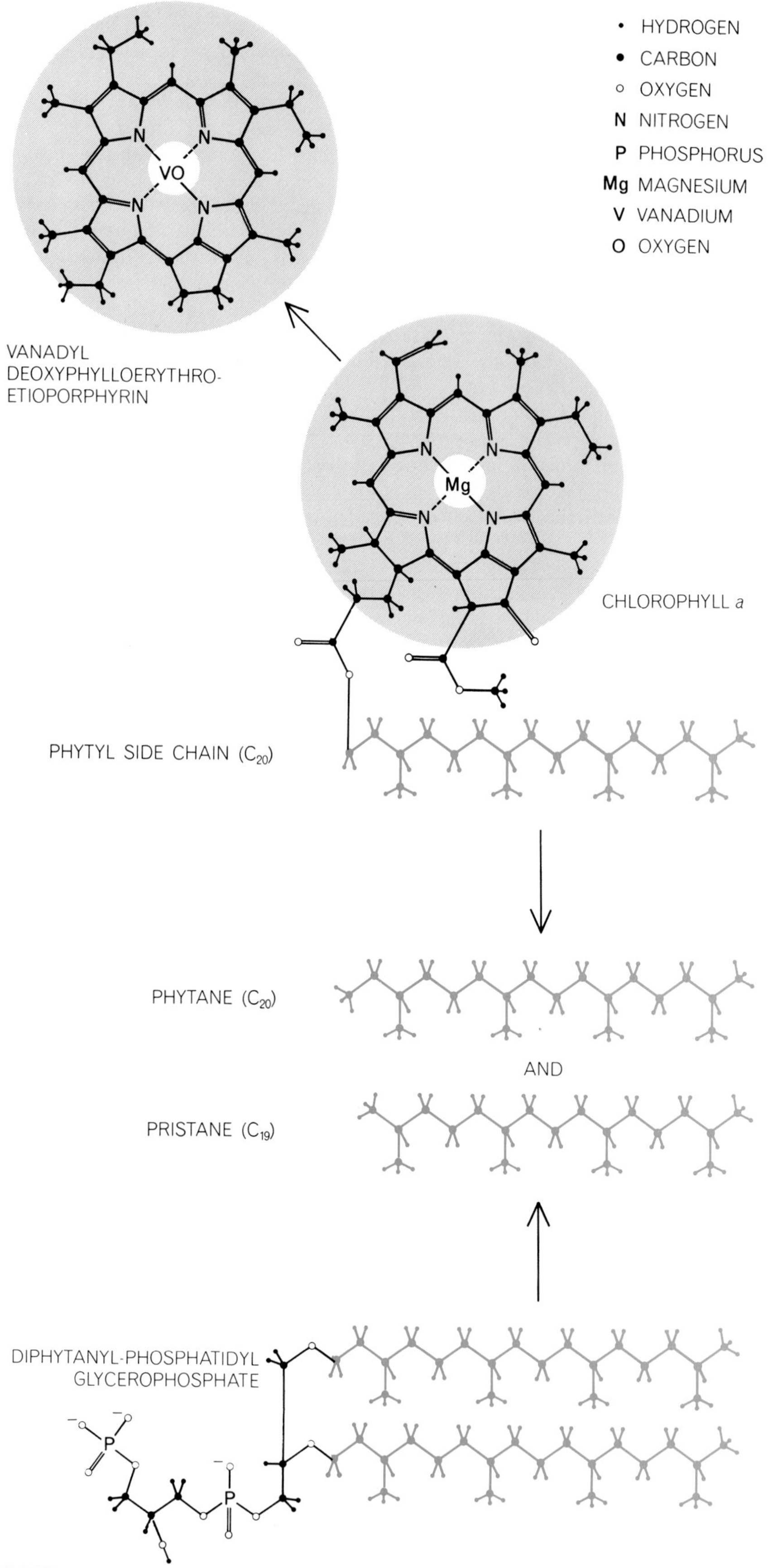

- • HYDROGEN
- • CARBON
- ○ OXYGEN
- N NITROGEN
- P PHOSPHORUS
- Mg MAGNESIUM
- V VANADIUM
- O OXYGEN

VANADYL
DEOXYPHYLLOERYTHRO-
ETIOPORPHYRIN

CHLOROPHYLL a

PHYTYL SIDE CHAIN (C$_{20}$)

PHYTANE (C$_{20}$)

AND

PRISTANE (C$_{19}$)

DIPHYTANYL-PHOSPHATIDYL
GLYCEROPHOSPHATE

DEGRADATION OF CHLOROPHYLL *A*, the green pigment in plants, may give rise to two kinds of isoprenoid molecules, phytane and pristane, that have been identified in many ancient sediments. It also seems likely that phytane and pristane can be derived from the isoprenoid side chains of a phosphate-containing lipid (*bottom structure*) that is a major constituent of salt-loving bacteria. The porphyrin ring of chlorophyll *a* is the probable source of vanadyl porphyrin (*upper left*) that is widely found in crude oils and shales.

laboratory from simple starting materials. For example, the Fischer-Tropsch process, used by the Germans as a source of synthetic fuel in World War II, produces a mixture of saturated hydrocarbons from carbon monoxide and water. The reaction requires a catalyst (usually nickel, cobalt or iron), a pressure of about 100 atmospheres and a temperature of from 200 to 350 degrees C. The hydrocarbons formed by this process, and several others that have been studied, generally show a smooth distribution of saturated hydrocarbons. Many of them have straight chains but lack the special characteristics (such as the predominance of chains with an odd number of carbons) found in the similar hydrocarbons present in many sediments. Isoprenoid alkanes, if they are formed at all, cannot be detected.

Paul C. Marx of the Aerospace Corporation has made the ingenious suggestion that isoprenoids may be produced by the hydrogenation of graphite. In the layered structure of graphite the carbon atoms are held in hexagonal arrays by carbon-carbon bonds. Marx has pointed out that if the bonds were broken in certain ways during hydrogenation, an isoprenoid structure might result. Again a laboratory demonstration is needed to support the hypothesis. What seems certain, however, is that nonbiological syntheses are extremely unlikely to produce those specific isoprenoid patterns found in the products of living cells.

Another dimension is added to this discussion by the proposal, made from time to time by geologists, that certain hydrocarbon deposits are nonbiological in origin. Two alleged examples of such a deposit are a mineral oil found enclosed in a quartz mineral at the Abbott mercury mine in California and a bitumen-like material called thucolite found in an ancient nonsedimentary rock in Ontario. Samples of both materials have been analyzed in our laboratory at Berkeley. The Abbott oil contains a significant isoprenoid fraction and probably constitutes an oil extracted and brought up from somewhat older sediments of normal biological origin. The thucolite consists chiefly of carbon from which only a tiny hydrocarbon fraction can be extracted. Our analysis shows, however, that the fraction contains trace amounts of pristane and phytane. Recognizing the hazards of contamination, we are repeating the analysis, but on the basis of our preliminary findings we suspect that the thucolite sample represents an oil of biological origin that has been almost completely carbonized. We are aware, of course, that one runs the risk of invoking circular arguments in such discussions. Do isoprenoids demonstrate biological origin (as we and others are suggesting) or does the presence of isoprenoids in such unlikely substances indicate that they were formed nonbiologically? The debate may not be quickly settled.

There is little doubt, in any case, that organic compounds of considerable variety and complexity must have accumulated on the primitive earth during the prolonged period of nonbiological chemical development—the period of chemical evolution. With the appearance of the first living organisms biological evolution took command and presumably the "food stock" of nonbiological compounds was rapidly altered. If the changeover was abrupt on a geological time scale, one would expect to find evidence of it in the chemical composition of sediments whose age happens to bracket the period of transition. Such a discontinuity would make an intensely exciting find for organic geochemistry. The transition from chemical to biological evolution must have occurred earlier than three billion years ago. As yet, however, no criteria have been established for distinguishing between the two types of evolutionary process.

We suggest that an important distinction should exist between the kinds of molecules formed by the two processes. In the period of chemical evolution autocatalysis must have been one of the dominant mechanisms for creating large molecules. An autocatalytic system is one in which a particular substance promotes the formation of more of itself. In biological evolution, on the other hand, two different molecular systems are involved: an information-bearing system based on nucleic acids and a catalytic system based on proteins. The former directs the synthesis of the latter. A major problem, subject to laboratory experiment, is visualizing how the two systems originated and were linked.

The role of lipids in the transition may have been important. Today lipids form an important part of the membranes of all living cells. A. I. Oparin, the Russian investigator who was among the first to discuss in detail the chemical origin of life, has suggested that an essential step in the transition from chemical to biological evolution may have been the formation of membranes around droplets, which could then serve as "reaction vessels." Such self-assembling membranes might well have required lipid constituents for their function, which would be to allow some compounds to enter and leave the "cell"

more readily than others. These membranes might have been formed nonbiologically by the polymerization of simple two-carbon and three-carbon units. According to this line of reasoning, the compounds that are now prominent constituents of living things are prominent precisely because they were prominent products of chemical evolution. We scarcely need add that this is a controversial and therefore stimulating hypothesis.

What one can say with some confidence is that autocatalysis alone seems unlikely to have been capable of producing the distribution pattern of hydrocarbons observed in ancient Precambrian rocks, even when some allowance is made for subsequent reactions over the course of geologic time. That it could have produced compounds of the observed type is undoubtedly possible, but

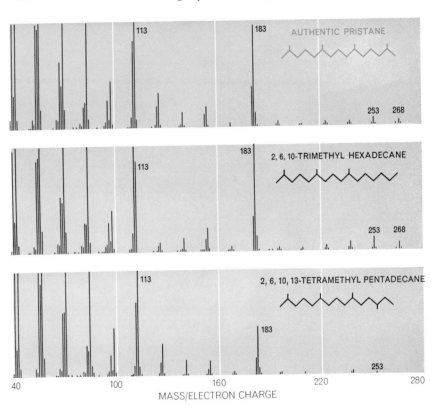

SIMILARITY OF MASS SPECTRA makes it difficult to distinguish the 19-carbon isoprenoid pristane from two of its many isomers (molecules with the same number of carbon and hydrogen atoms). The three records shown here are replotted from the actual tracings produced by pure compounds. When the sample contains impurities, as is normally the case, the difficulty of identifying authentic pristane by mass spectrometry is even greater.

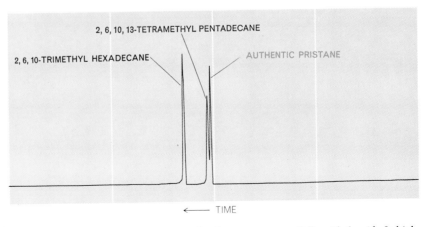

IDENTIFICATION OF PRISTANE can be done more successfully with the aid of a high-resolution gas chromatograph. When pure pristane and the isomers shown in the illustration above are fed into such an instrument, they produce three distinct peaks. This curve and the mass spectra were made by Eugene McCarthy of the University of California at Berkeley. He also made the isoprenoid study shown at the bottom of pages 264 and 265.

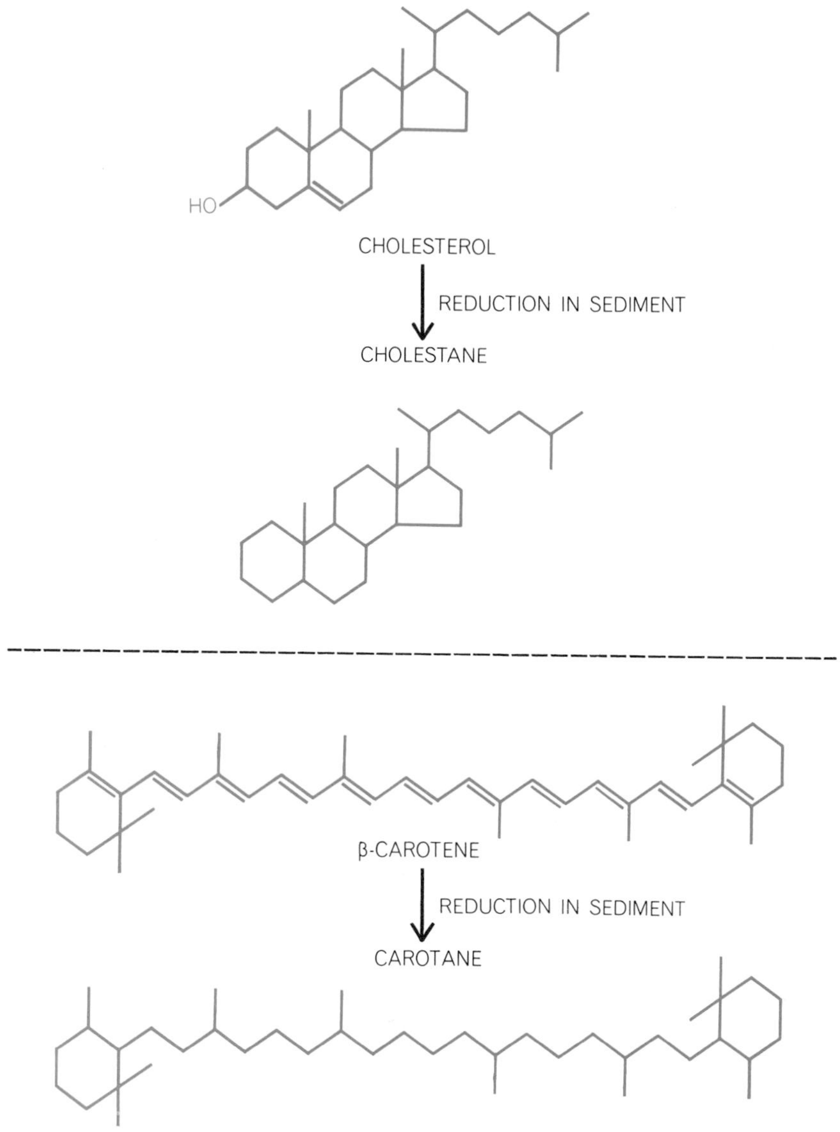

CHOLESTEROL

↓ REDUCTION IN SEDIMENT

CHOLESTANE

β-CAROTENE

↓ REDUCTION IN SEDIMENT

CAROTANE

TWO ALKANES IN GREEN RIVER SHALE, cholestane and carotane, probably have been derived from two well-known biological substances: cholesterol and beta-carotene. The former is closely related to the steroid hormones; the latter is a red pigment widely distributed in plants. These two natural substances can be converted to their alkane form by reduction: a process that adds hydrogen at the site of double bonds and removes oxygen.

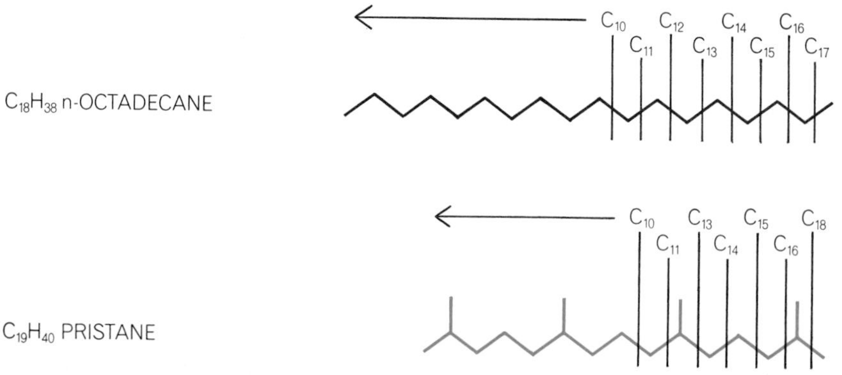

$C_{18}H_{38}$ n-OCTADECANE

$C_{19}H_{40}$ PRISTANE

EFFECT OF HEATING ALKANES is to produce a smoothly descending series of products (normal alkenes) if the starting material is a straight-chain molecule such as n-octadecane. (The term "alkene" denotes a hydrocarbon with one carbon-carbon double bond.) If, however, the starting material is an isoprenoid such as pristane, heating it to 600 degrees centigrade for .6 second produces an irregular series of alkenes because of the branched chain. Such degradation processes may take place in deeply buried sediments. These findings were made by R. T. Holman and his co-workers at the Hormel Institute in Minneapolis, Minn.

it seems to us that the observed pattern could not have arisen without the operation of those molecular systems we now recognize as the basis of living things. Eventually it should be possible to find in the geological record certain molecular fossils that will mark the boundary between chemical and biological evolution.

Another and more immediate goal for the organic geochemist is to attempt to trace on the molecular level the direction of biological evolution. For such a study one would like to have access to the actual nucleic acids and proteins synthesized by ancient organisms, but these are as yet unavailable (except perhaps in rare instances). We must therefore turn to the geochemically stable compounds, such as the hydrocarbons and oxygenated compounds that must have derived from the operation of the more perishable molecular systems. These "secondary metabolites," as we have referred to them, can be regarded as the signatures of the molecular systems that synthesized them or their close relatives.

It follows that the carbon skeletons found in the secondary metabolites of present-day organisms are the outcome of evolutionary selection. Thus it should be possible for the organic geochemist to arrange in a rough order of evolutionary sequence the carbon skeletons found in various sediments. There are some indications that this may be feasible. G. A. D. Haslewood of Guy's Hospital Medical School in London has proposed that the bile alcohols and bile acids found in present-day vertebrates can be arranged in an evolutionary sequence: the bile acids of the most primitive organisms contain molecules nearest chemically to cholesterol, their supposed biosynthetic precursor.

Within a few years the organic geochemist will be presented with a piece of the moon and asked to describe its organic contents. The results of this analysis will be awaited with immense curiosity. Will we find that the moon is a barren rock or will we discover traces of organic compounds—some perhaps resembling the complex carbon skeletons we had thought could be produced only by living systems? During the 1970's and 1980's we can expect to receive reports from robot sampling and analytical instruments landed on Mars, Venus and perhaps Jupiter. Whatever the results and their possible indications of alien forms of life, we shall be very eager to learn what carbon compounds are present elsewhere in the solar system.

VII

ORGANIC CHEMISTRY

ORGANIC CHEMISTRY

Throughout its history, chemistry has been divided into a number of separate branches. These classical branches have been organic, inorganic, physical, and analytical. Fortunately these boundaries are blurring. The boundaries have in large measure been artificial and have actually impeded progress in developing the science of chemistry. The present authors consider themselves to be biophysical chemists. To work in our respective research areas, it is necessary to transcend not only the boundaries within chemistry but to have some familiarity with the disciplines of biology and physics as well. The following article from the journal *Geotimes* (July-August, 1962) makes this point in a rather humorous way:

GEOCHEMISTRY OFFERS RANGE OF SPECIALTIES

A hydromicrobiogeochemist is one who studies small underwater flora and their relationship to underlying rock strata by using chemical methods.

A microhydrobiogeochemist is one who studies flora in very small bodies of water and their relationship to underlying rock strata by using chemical methods.

A microbiohydrogeochemist is one who studies small flora and their relationship to underlying rock strata by using chemical methods and SCUBA equipment.

A biohydromicrogeochemist is a very small geochemist who studies the effects of plant life in hydrology.

A hydrobiomicrogeochemist is a very small geochemist who studies wet plants.

A biomicrohydrogeochemist is a very small, wet geochemist who likes lettuce.

Despite this blurring of the separate disciplines within one science and between several sciences, many chemists would answer the question "What kind of scientist are you?" with the response "I'm an organic chemist," or "I'm a physical chemist," and so on. Most college chemistry curricula include a course entitled "organic chemistry" as a separate offering. One of the tasks expected of most general chemistry texts and assigned to teachers of general chemistry courses is to provide an introduction to organic chemistry. Thus we have felt that it is appropriate to include a section on organic chemistry in this reader. It is an important discipline in that much of the chemical industry relies on the knowledge and techniques of this discipline and it provides much of the necessary groundwork for the last section of this reader.

Organic chemistry is that branch which deals with the chemistry of compounds that contain carbon. The word organic was derived in the past century from the notion that only organisms could produce compounds containing carbon atoms and hence that these were "organic."* This notion was placed to rest decisively by Wöhler in 1828 when he synthesized urea, a carbon-containing compound, from ammonia and carbon dioxide.

*Some of the same fuzzy thinking is present today with respect to "organic" gardening. The compounds added to the soil, such as manure and compost, are indeed "organic." However, these plant foods are of no value as nutrients until they are broken down into simpler compounds, which are distinctly "nonorganic" or inorganic.

There are 105 known elements. It may seem surprising that one branch of chemistry is named after the compounds of only one element. Why not silicon chemistry? The answer lies in the fact that no other elements, including silicon, form very strong covalent bonds with each other. Silicon-silicon bonds are readily cleaved. Silicon-oxygen-silicon bonds are quite strong but do not yield the incredible diversity available with carbon compounds. Carbon atoms can bond to one, two, three, or four other carbon atoms. This capability, plus the possibility of single, double, and triple bonds, and the ability to form very strong covalent bonds with a wide variety of other atoms, such as H, O, S, N, Cl, Br, I, etc., yield a staggering number of compounds. The current edition of the *Handbook of Chemistry and Physics* by the Chemical Rubber Company lists 13,620 compounds. Undoubtedly this is a small fraction of the presently known compounds and a very small fraction of the possible organic compounds.

Organic chemistry was the first branch of chemistry to develop into an extensive science. It had a rough beginning, as described in the quote of Wöhler given at the beginning of Roberts's article, "Organic Chemical Reactions." Organic chemistry was "like a primeval forest . . . into which one may dread to venture." By the end of the nineteenth century, thousands of compounds had been synthesized and their properties measured. The development of the octet rule by Lewis provided some theoretical basis for what compounds could exist and what reactions could occur. Subsequent theoretical developments, especially molecular orbital theory, and subsequent instrumental advances, especially IR and NMR, have made the thicket far more penetrable. In fact, hundreds of thousands of freshman chemistry students routinely spend a few weeks of their general chemistry course pondering over the outlines of this forest of compounds and reactions. Their study generally embraces three facets—nomenclature and structure, types of reactions, and an introduction to reaction mechanisms. The four articles in this section have been selected to give you a brief exposure to each of these three areas of study.

Before you plunge into these articles, it may prove useful for you to review (or perhaps see for the first time) some of the classes of organic compounds. In most reactions involving organic compounds, only a small portion of the molecule—the part called a functional group—undergoes chemical reaction. Organic molecules are grouped and named according to these functional groups. At the beginning of each of the first three articles of this section, you will find some examples of the structure and nomenclature of organic compounds. Table VII.1 may be of some use to you in providing an overall perspective of compounds, functional groups, and names.

"Organic Chemical Reactions" by Roberts is placed first in this section not because reaction mechanisms are the simplest to understand but because the introductory diagrams should give you a good feel for the geometry of simple carbon compounds. It is important for you to visualize the tetrahedral nature of the four bonds about the central carbon atom. It may be of help for you to refer back to article 12, "Ice," by Runnels. The structure of water shown on the first page of that article displays the four electron clouds around the oxygen atom. If four atoms were bonded to the clouds rather than the two shown, this would be a good representation of the tetrahedrally bonded carbon atom. The four representations at the bottom of the first page of Roberts's article show four ways to display mono-chloro-substituted methane.

The principal thrust of Roberts's article is a description of the mechanisms of organic reactions. Most of the discussion centers on S_N1 and S_N2 reactions. The S stands for substitution reactions. One group is replaced by another. The N stands for nucleophilic. This implies that the entering group has an unshared electron pair and is attracted to an atom, such as the carbon atom of methyl chloride, which has a partial positive charge. The numbers one and two refer to whether the reaction is first order or second order: whether the rate of the reaction depends on one or two terms specifying concentration (the importance of concentrations is discussed in the introduction to Section V).

Of these two mechanisms, the S_N2 mechanism is discussed in the greatest detail. The reaction studied is the substitution of OH^- for Cl^- on CH_3Cl. The rationale is presented for why the hydroxide ion attacks the side of the methyl chloride away from the chlorine atom. The principal evidence for this mechanism comes from studies with optical isomers — mirror-image molecular structures. Optical isomerism is not dealt with in Roberts's article in great detail. However, you will encounter the topic again in article 28, by Doty, and an introduction to it was given in the first article in the reader, by Pauling. Where the hydroxide ion attacks determines whether the resulting molecule will assume one form or its mirror image.

Other evidence includes steric effects. Roberts uses three-dimen-

Table VII.1 Classes of organic compounds

Class	Functional group		Example	
	Structure	Name	Structure	Name
Alcohols	—OH	Hydroxyl	CH_3—CH—CH_3 with OH below CH	2-Propanol
Aldehydes	—C(=O)—H	–	CH_3—CH_2—C(=O)—H	Propanal
Amides	—C(=O)—NH_2	Amide	CH_3—C(=O)—NH_2	Acetamide
Amines	—NH_2	Amino	CH_3—CH_2—NH_2	Ethyl amine
Carboxylic acids	—C(=O)—OH	Carboxyl	CH_3—C(=O)—OH	Acetic acid
Esters	—C(=O)—O—	–	CH_3—C(=O)—O—CH_3	Methyl acetate
Ethers	—O—	–	CH_3—CH_2—O—CH_2—CH_3	Diethyl ether
Hydrocarbons	–	–	CH_3—CH_2—CH—C—CH_3 (with CH_3 above and CH_3, CH_3 below)	2,2,3-Trimethylpentane
Ketones	—C(=O)—	Carbonyl	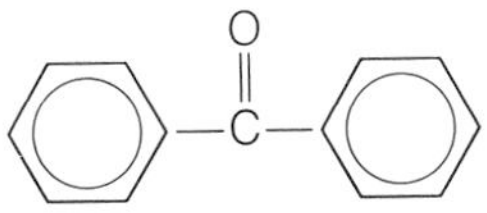 (phenyl)—C(=O)—(phenyl)	Diphenyl ketone
Mercaptans	—SH	Sulfhydryl	CH_3—SH	Methanethiol

sional models effectively to demonstrate that atoms have finite sizes represented by their covalent radii. Models such as those shown on 279 and 281 in Roberts's article are useful in answering numerous questions about molecular structure. In concluding the details of S_N2 reaction mechanisms, Roberts provides the opportunity for studying electrical effects, which are important in many reactions, by mentioning their role in S_N2 reactions.

The S_N1 mechanism is discussed in somewhat less detail. The point Roberts emphasizes is that these reactions often result in the rearrangement of the carbon skeleton of the reactant molecule. We have already seen numerous examples of this in "Catalysis," the Haensel and Burwell article (16).

A number of ions have been described at several points in this reader. In Roberts's article, a new type of ion called a carbonium ion is introduced. The name applies to an organic molecule in which one of the carbon atoms forms only three bonds and bears a positive charge. The most stable carbonium ions are those with three carbon atoms bonded to one carbon atom, called a tertiary carbon. Such ions are important in many other organic reactions beside the S_N1 reaction discussed here.

"Organic Chemical Reactions" concludes with a number of examples of other, more complex mechanisms, which are required to explain the unexpected products of certain reactions.

The second article, by Breslow, deals with another aspect of organic chemistry. This branch of chemistry deals with carbon-containing compounds that can be broadly classified as aromatic and nonaromatic. We have seen a number of nonaromatic compounds in the articles by Roberts (24), Haensel and Burwell (16), Eglinton and Calvin (23), and Fox (8). These compounds include straight-chain, branched, cyclic, and unsaturated compounds. That encompasses a lot of molecules but leaves out the parent compound of all aromatic molecules, benzene.

When benzene was first isolated, it was realized that it and its derivatives had distinctive odors or aromas. Thus, the name aromatic was applied to these compounds. Although it is of very little scientific use, the term has been retained.

The structure of benzene is introduced by means of Lewis diagrams. The concept displayed by these diagrams was developed in 1916 by G. N. Lewis. Although newer and more powerful theories have been developed, the basic idea that atoms form covalent bonds by sharing two electrons remains essentially correct. Lewis also postulated the octet rule. Each atom strives to acquire or to share eight electrons. Because this rule has wide applicability among the elements of the second row of the periodic table, carbon compounds can be represented in general by four bonds to each carbon atom. We have seen an exception in the carbonium ions of Roberts's article and will encounter another in the carbanions discussed below. Just as Roberts did in his article, Breslow makes use of a number of representations of organic structures. Those shown on page 290 are, from top to bottom, Lewis diagrams, Bohr diagrams, and ball-and-stick models. The Bohr diagrams (those with dots on circles) are helpful in counting electrons, but of course they bear no resemblance to the atoms of the present quantum mechanical structural view.

Breslow depicts the structure of benzene with two resonance Lewis diagrams. These represent Kekulé's idea that the site of unsaturation in benzene cannot be localized. Benzene consists of six carbon atoms arranged in a planar ring with a hydrogen atom attached to each carbon in the plane of the hexagon. Using two forms and the term

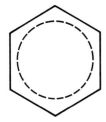

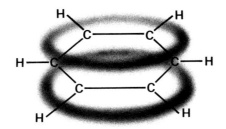

a MODIFIED LEWIS DIAGRAM   *b*   MOLECULAR ORBITAL DIAGRAM

Figure VII.1 Two representations of the structure of benzene. *a*, Modified Lewis diagram; *b*, Molecular orbital diagram.

resonance in describing benzene is not to imply that the structure jumps back and forth from one form to the other. The diagrams instead imply that benzene actually has a structure intermediate between these two forms, which cannot be represented adequately with simplicity. Two ways around this difficulty are shown in Figure VII.1. The diagram on the left displays the three alternating double and single bonds as a circle. The diagram on the right is a molecular orbital representation, as discussed later in Breslow's article.

The best clue to the special nature of the aromatic system is a comparison of the chemical reactivity of benzene with that of unsaturated compounds. We saw earlier in Haensel and Burwell's article (16) that hydrogen adds readily to a double bond, as in ethylene, on a ZnO surface. Because benzene is portrayed with three double bonds, it would seem that it should undergo hydrogenation readily also. However, it does not.

This and other clues point to the fact that the electrons in benzene are delocalized. They are spread out over the entire molecule as in the conjugated compounds discussed earlier by Hubbard and Kropf in article 9. Molecular orbital theory provides the best description of such systems. This theory provided the explanation for a baffling observation. If the six-carbon system of benzene was particularly stable, why were the four-carbon system of cyclobutadiene, and other three-, four-, and five-carbon molecular and ionic ring systems so reactive? Erich Hückel provided the answer in 1931 by using molecular orbital theory. He found that systems having $4n + 2$ delocalized electrons have a special stability. Thus any ring system with 2, 6, 10, 14, etc., delocalized electrons would be particularly stable and therefore unreactive.

Breslow's article concludes with a brief discussion of a powerful rule formulated by Woodward and Hoffmann. They extended the use of the $4n + 2$ rule to explain why some reactions occur easily and others do not. Those reactions which pass through an intermediate aromatic state having $4n + 2$ delocalized electrons are favored over those that use $4n$ delocalized electrons.

The next article, by Lambert, is one of the most difficult articles included in the reader. It deals with a particular kind of molecular isomer. We have already seen D- and L- isomers in articles 2 and 24. These are optical isomers. They are mirror images of each other and

$$COOH \qquad\qquad HOOC$$

$$\overset{|}{C}$$

$$H_2N \qquad\qquad H$$

$$CH_3$$

$$\overset{|}{C}$$

$$H \qquad\qquad NH_2$$

$$H_3C$$

L-Alanine   D-Alanine

rotate the plane of polarized light equally but in opposite directions. These compounds have identical chemical and physical properties with this one exception. The conversion of one form to the other requires the exchange of any two groups around the central carbon atom.

Basically, pairs of compounds that are isomeric have the same molecular formulae, but the arrangement of the parts within each compound is different. *Cis-trans* isomers are compounds having two substituents on either side of a double bond. Unlike optical isomers, these isomers have somewhat different chemical and physical properties. Interconversion of *cis-trans* isomers requires conversion of the

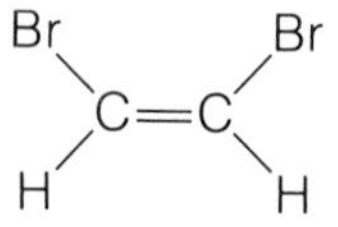

Br   Br   Br   H
  \   /   \   /
   C = C   C = C
  /   \   /   \
 H   H   H   Br

cis-Dibromoethylene 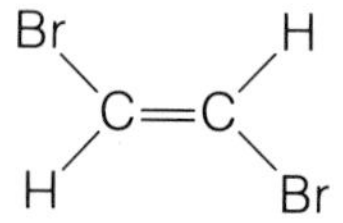   *trans*-Dibromoethylene

double bond to a single bond, rotation about the carbon-carbon single bond, and re-formation of the double bond.

Positional isomers also differ somewhat in their properties: a func-

$CH_3—CH_2—CH_2—OH$ $CH_3—CH—CH_3$
                                        |
                                        OH

1-Propanol   2-Propanol

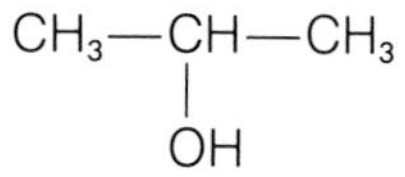

tional group can occupy several sites on the carbon skeleton. Finally, structural isomers are completely different molecules, but are de-

$CH_3—O—CH_3$ $CH_3—CH_2—OH$
Dimethyl ether   Ethanol

scribed with the same formula. Interconversions of positional and structural isomers again require breaking covalent bonds.

The above four classes of isomers hold three things in common. These pairs of compounds all have the same empirical formulae, or the same atoms in equal numbers. They differ in physical or chemical properties (or both), and they require that enough energy be available to break covalent bonds if interconversion between the two compounds of the pair is to occur.

Lambert, in "The Shapes of Organic Molecules," deals with a more subtle kind of isomer—conformational isomers. In his introductory discussion Lambert explains the properties of single bonds, and he describes these bonds as allowing atoms to have free rotation. The ball-and-stick models in Lambert's article convey this concept. We know, however, that rotation about single bonds is actually somewhat restricted. The configuration in which the groups attached to the two adjacent carbons are as far apart as possible or staggered (when viewed along the carbon-carbon bond) has the lowest energy. When the groups are only slightly unaligned (or eclipsed), the configuration has the highest energy. These compounds with slightly different configurations are called conformers to distinguish them from the isomers mentioned earlier, in which bond breaking must occur to interconvert the two forms. The outstanding Corey-Pauling-Koltun models (which chemists call CPK models) reflect this idea very nicely. The face of the single-bonded carbon atom and the configuration of the connector are shown in Figure VII.2. Because the two locking pins on the connector are 180° out of phase with each other and because the holes into which the pins fit on the atomic

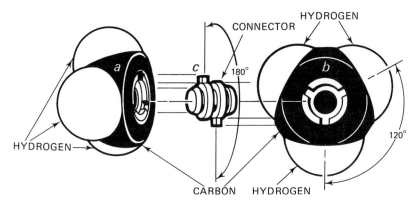

Figure VII.2 A Corey-Pauling-Koltun model representing the breakage of the carbon-carbon single bond in ethane. *a* and *b*, Single-bond methyl groups (CH₃); *c*, Connector, or bond, allowing restricted rotation of two carbon atoms.

face are only 120° apart, the pins fit in the holes only when the carbon atoms are rotated 60° relative to each other. This corresponds to the low-energy conformer shown in the top figure of page 300, in Lambert's article. Intermediate positions are higher in energy, and can be represented with the CPK models by compressing the rubber pin.

Much of Lambert's discussion centers on the conformers of closed six-membered rings. Be careful to note that these are not aromatic compounds. They contain no double bonds and therefore are saturated compounds. The substituents on the carbon atoms lie approximately in the plane of the ring and are called equatorial atoms, or they are above or below the ring and are called axial atoms. Nuclear magnetic resonance (NMR) spectroscopy (discussed in the introduction to Section VI) has been used to demonstrate the existence of the two kinds of hydrogen atoms in cyclohexane.

Several types of heterocyclic compounds are also discussed. These compounds bear some similarities to the purines and pyrimidines extensively discussed in the latter half of Section VIII.

As the number of rings increases, the complexity of the conformers rises rapidly. It reaches a peak with the structures of proteins. As we will see in the next section, protein structure and function depend critically on conformation. A very slight change in structure can destroy the activity of an enzyme. Lambert points out that for lysozyme to operate on a polysaccharide as described by Dickerson in article 30, the protein must somewhat flatten out the chair configuration of the sugar ring. Again a knowledge of conformers was needed to understand the details of this enzyme's functioning.

The final article, with its pictures of beetles, flies, fleas, lice, bugs, and worms, may seem somewhat out of place in this section. We believe quite the contrary. The first three articles present some illustrations of the science of organic chemistry. We feel it is important to provide an illustration of how organic chemistry interacts with the problems of society.

One such problem, about which we all have become acutely aware, is the one caused by the application of DDT to control insect populations. It has also caused serious declines in the populations of several species of birds and has become dispersed throughout the globe, reaching all the way to the penguins of Antarctica. Just as life can be measured in generations, so can the compounds that destroy life. DDT and its derivatives represent one level of sophistication higher than the arsenates that comprised the first generation of pesticides. Both generations, however, were far too persistent in the environment and far too indiscriminate in their lethal effects. A third generation of pesticides is needed and is presently under development. This development is the subject of Williams's article.

These new pesticides will probably be hormones. All insects require a particular hormone called a juvenile hormone for the normal development of the insect through its stages: larva $\rightarrow$ pupa $\rightarrow$ adult. The juvenile hormone must be present for the larva to develop. However, for the larva to pass through the pupal stage to adulthood, this hormone must be absent. The method of using juvenile hormones as pesticides is to spray a few grams of these substances per acre of infested crop. The hormone would readily enter insects in all stages of development and would either stop the eggs from hatching or would limit the growth to the larval form. Williams observes that the insects could not develop a resistance to their own hormone without committing suicide.

Each insect probably has a unique juvenile hormone. Thus insect control could be limited precisely to selected species. The juvenile hormone of the Cecropia moth has been isolated and its structure identified. It consists of a thirteen-carbon chain having two double bonds. At one end is a methyl ester (see Table VII.1 at the beginning of this introduction). Near the other end is a somewhat unusual structure $-\overset{\displaystyle O}{\overset{\displaystyle /\backslash}{C-C}}-$. It would of course require tons and tons of Cecropia moths to supply the amounts of hormones necessary for the control of that species. And this brings us back to where we began this introduction. Namely, once the structure of a natural product has been determined, it is the job of the organic chemist to find a way to synthesize the desired compound in a satisfactory yield of high purity from readily available starting materials. There is hope that organic chemists will be able to synthesize juvenile hormones to control those insects which destroy crops and are the vectors of disease, in ways that do not, at the same time, endanger man and other creatures which inhabit the Earth.

24

ORGANIC CHEMICAL REACTIONS

JOHN D. ROBERTS
November 1957

*Chemists are re-examining familiar reactions of
simpler organic molecules, even tracing the courses of
individual atoms. These studies help in understanding
the behavior of complex compounds*

A century and a quarter ago Friedrich Wöhler of Germany, the leading pioneer of organic chemistry, wrote to his Swedish friend and co-pioneer Jöns J. Berzelius: "Organic chemistry is now enough to drive one mad. It appears like a primeval forest of the tropics, full of the most remarkable things, a monstrous and boundless thicket, with no escape, into which one may dread to venture."

It cannot be said that chemists today have come to know every tree of that forest, or even penetrated all of its thickets. But the wilderness now looks somewhat less formidable. We know what organic substances are made of, what kinds of structure they have and how to make some of them. And we are beginning to find paths leading toward an understanding of why organic compounds behave as they do and what mechanisms govern their reactions. These reaction mechanisms are the topic of my article.

The names of organic compounds are a fearsome thicket for nonchemists, but fortunately most compounds can be portrayed simply with ball-and-stick models like a child's take-apart toy [*see drawing below*]. The sticks represent the electronic bonds linking the atoms together. In the model of the carbon atom they are set at an angle of 109.5 degrees to one another, because this is the normal angle between the atoms attached to carbon, but in some organic compounds the connections between the atoms are bent, and the bonds in such molecules act like bent steel springs rather than rigid sticks.

Let us try to picture a common type of reaction, well illustrated by the one between methyl chloride (CH_3Cl) and sodium hydroxide ($NaOH$). When the two compounds are mixed in a hot solution, the hydroxyl group (OH) from sodium hydroxide replaces the chlorine of the methyl chloride and we get methyl alcohol (CH_3OH). This is known as a substitution reaction. Our question is: How does the substitution take place?

In solution sodium hydroxide dissociates into sodium ions and hydroxide ions. We can disregard the sodium ions; in fact, the reaction goes equally well if we use lithium hydroxide or potassium hydroxide instead of sodium hydroxide. The rate of the reaction, as measured by the formation of the products, is directly proportional to the concentration of hydroxide ions and of methyl chloride molecules. Now the rate of collisions between hydroxide ions and methyl chloride molecules in the mixture also depends directly on these concentrations. Therefore we can take it as highly likely that a hydroxide ion displaces the chlorine atom by colliding with the methyl chloride molecule.

Not every collision is effective: in fact, the chlorine atom yields its place to hydroxide in only about one of every 10 million collisions. This suggests that the hydroxide ion has to approach the methyl chloride molecule in a particular

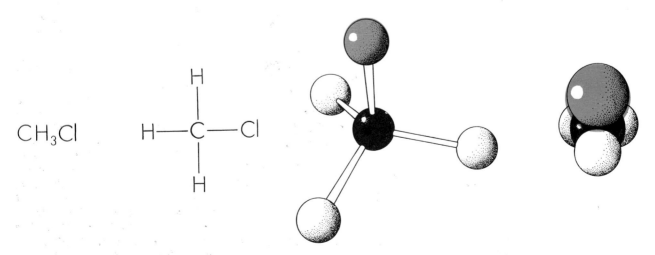

METHYL CHLORIDE molecule may be visualized in several ways. At left is condensed formula; next, the structural formula; next, ball and stick model; at right, a model showing relative sizes of atoms. Carbon atoms are black; hydrogen, white; chlorine, color.

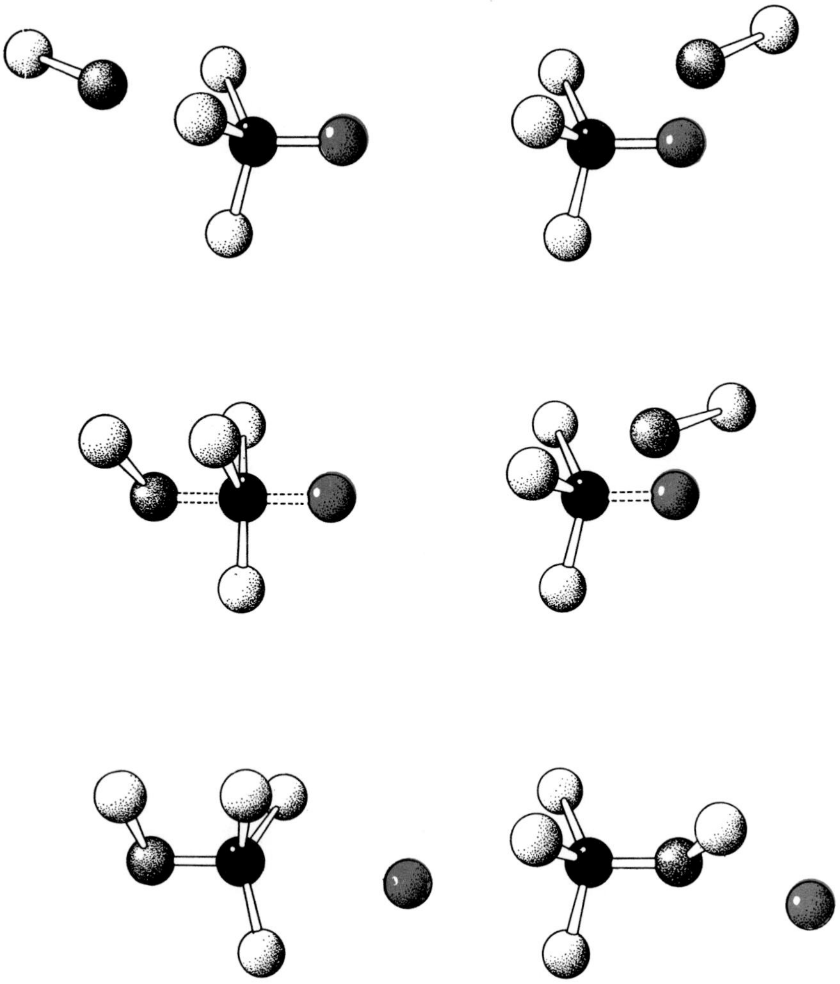

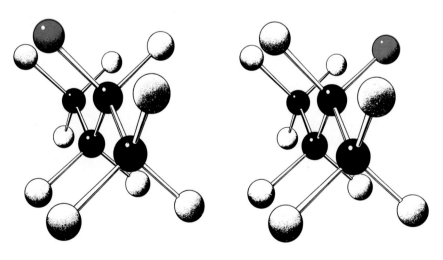

SUBSTITUTION REACTION, in which an OH group (*dumbbell*) displaces chlorine atom (*color*) on a methyl chloride molecule, is shown as a "back-side" reaction at left and a "front-side" reaction at right. Gray ball shows oxygen atom. Bonds shown in broken lines represent halfway point at which OH group attaches itself and pushes off chlorine atom.

OPTICAL ISOMERS are molecules which are identical except for their opposite "handedness." These are models of s-butyl chloride isomers. The molecular structure at left is a mirror image of the one at the right, just as a left hand is a mirror image of a right hand.

way to effect the substitution. We can think at once of two possibilities. The hydroxide ion might attack the molecule on the side where the chlorine atom is attached (the "front side") or on the opposite side (the "back side"). In the first case it would dislodge the chlorine atom directly and seize its position; in the second, it would attack the back side of the carbon atom and in so doing cause the molecule to expel the chlorine atom [*see diagrams at left*].

We have no way of testing which of these attacks actually works on the methyl chloride molecule itself, because, as the diagrams show, either process would result in exactly the same structure for the transformed molecule (methyl alcohol). But we can test the issue on a somewhat more complicated chloride also containing methyl groups— the compound called s-butyl chloride. This compound can exist in two forms, each the mirror image of the other [*see diagrams below*]. We can consider one "right-handed," the other "left-handed." Now if a hydroxide ion displaces the chlorine atom in the right-handed version from the front side, it will produce a right-handed alcohol molecule; if it attacks from the back side, the product will be a left-handed molecule. Experiments leave no doubt about what actually occurs. The direct reaction between right-handed s-butyl chloride and hydroxide ions always produces left-handed s-butyl alcohol, which means that the ion must attack the back side.

This type of mechanism has been found to cover a whole class of organic reactions. It is called the S_N2 mechanism: S for substitution, N for "nucleophilic" (meaning that the substituting ion is attracted to the nuclear carbon atom itself, via the back side) and 2 for "bimolecular" (meaning that the reaction is effected by a collision between two molecules).

Let us look further into the mechanism. We can assume that hydroxide ions attack methyl chloride, as they do s-butyl chloride, from the back side. Now if we substitute methyl groups for the hydrogen atoms attached to the carbon atom of methyl chloride, the rate of the S_N2 reaction falls drastically. Why? The most reasonable answer is that the methyl groups, being much bulkier than hydrogen atoms, obstruct the access of hydroxide ions to the back side of the carbon atom. Strong support for this inference has been furnished by the finding that the measured reaction rates agree with theoretical calculations of the difficulty of access. To reach the carbon atom the incoming ion must push aside the methyl groups. This means it must compress them and bend their bonds. The stiffness of chemical bonds can be measured by means of spectroscopy, and the compressibility of molecules or atoms can be estimated from the extent to which they scatter a very fast beam of atoms.

Frank H. Westheimer of Harvard University worked out a problem of this sort in meticulous detail. He studied a reaction which converts a biphenyl compound from the right-handed to the left-handed form by twisting the bond that links the two phenyl rings [*see diagrams at right*]. This involves considerable bond-stretching and distortion of the molecule, because as one ring turns around the other, their projecting atoms must push past one another. Westheimer computed how much bond-stretching, bending and compression of atoms would be required, and found that the over-all resistance corresponded closely to the rate of the reaction.

Besides the bulk effect, an electrical effect—that is, the effect of electric charges in a molecule—must play a large role in substitution reactions. Indeed, in some cases the electrical influence is stronger than the bulk effect.

What is the electrical effect? We can best approach it by considering an organic acid, such as acetic acid (CH_3COOH). This molecule contains a methyl group and a carboxyl group ($COOH$), the latter making it an acid. In solution the hydrogen atom splits off as a positive ion, leaving the negatively charged acetate ion. The proportion of the molecules that shed hydrogen atoms is a measure of the strength of the acid: if a high proportion drop the carboxyl hydrogen atom, the acid is strong; if a small proportion, it is weak.

Acetic acid is a comparatively weak acid. But if we replace one of the hydrogen atoms of the methyl group with a fluorine atom, the acid becomes much stronger—200 times as strong. That is to say, the carboxyl hydrogen splits off much more readily [*see upper diagrams on page 286*]. Why so? Experiments with different compounds show that this result cannot be a bulk effect. We are forced to the conclusion that it is an electrical effect. The fluorine atom strongly attracts electrons. As a result electrons are pulled away from the hydroxyl group on the other side of the carbon atom to which it is attached.

The hydroxyl group therefore becomes more positive; it then holds its hydrogen atom less firmly, and so the atom departs more readily as a positive ion.

Such experiments on acids show two things: (1) that fluorine and other strong attractors of electrons probably exert their influence on S_N2 reactions by the electrical effect, and (2) that methyl groups exert mainly a bulk effect, for they have little electrical effect.

How does the electrical effect work in S_N2 reactions? In a general sense, by much the same process as in acetic acid, but in the opposite direction. If a fluorine atom is substituted for a hydrogen atom on the carbon atom of methyl chloride, it pulls electrons away from the carbon atom, so that the carbon atom becomes more positive. The carbon therefore holds the negative chloride ion more firmly, and as a result replacement of the chlorine by hydroxide ions is slowed down.

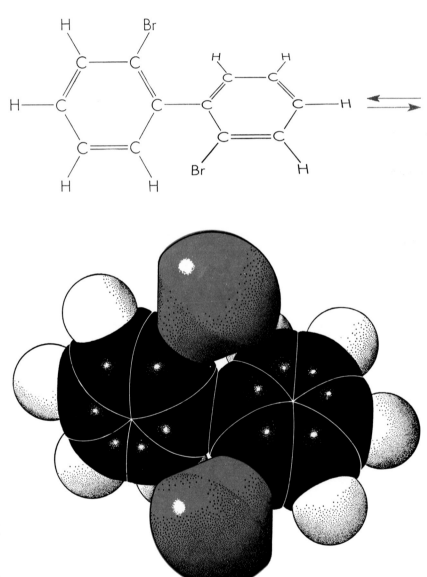

BULK EFFECT of atoms in interfering with a reaction is illustrated in these drawings. If it were not for the large bromine atoms (*in color*), the two parts of this double-ring compound (2,2'-dibromobiphenyl) would rotate freely with respect to one another, changing from right-handed form shown here to left-handed form in which bromine atoms are behind planes of rings rather than in front. Actually the bromines cannot slide past the opposite hydrogen atoms into this position unless some energy is added to distort the molecule.

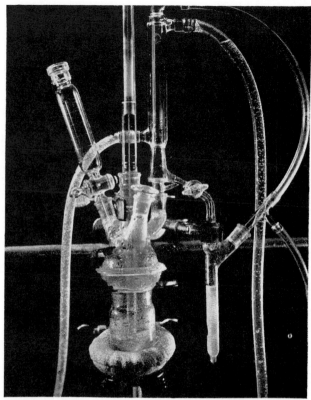

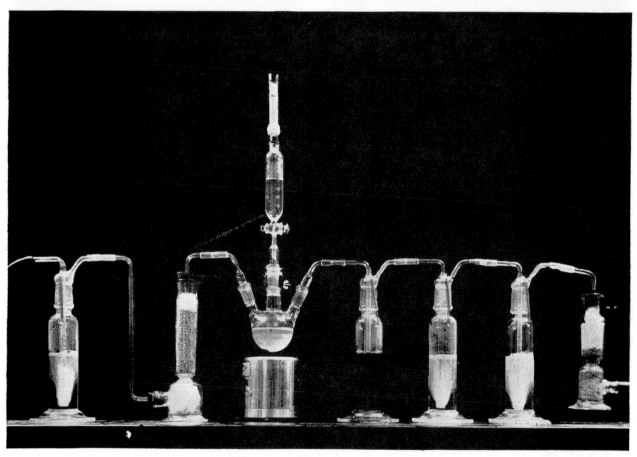

RADIOACTIVE-TRACER TECHNIQUE is illustrated in the photographs on this and the following page. The experiment shown is the one described by the formulas on page 284. A three-carbon ring is tagged with a carbon-14 tail by allowing it to react with radio-active carbon dioxide in the apparatus at top left. It is then converted to a four-carbon ring in the flask at top right. Two carbons are split off and converted to carbon dioxide in the central flask of the array at bottom. The gas is collected in some of flasks to right.

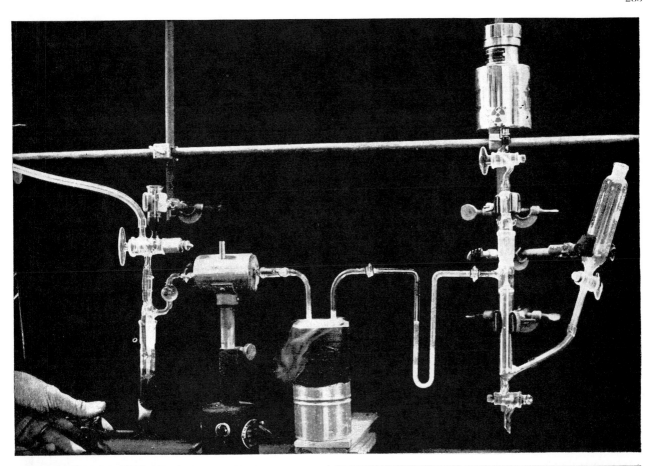

COMPLETE CONVERSION TO CARBON DIOXIDE of the four-carbon ring compound is carried out in the heated tube at the left of the top photograph. The gas is then assayed for radioactive carbon in the ion chamber at bottom. The chamber contains a central electrode insulated from the walls. When a radioactive carbon atom disintegrates it ionizes the gas in the chamber and a small pulse of current passes between electrode and walls. The total current, which depends on the amount of carbon 14 present, is recorded on chart.

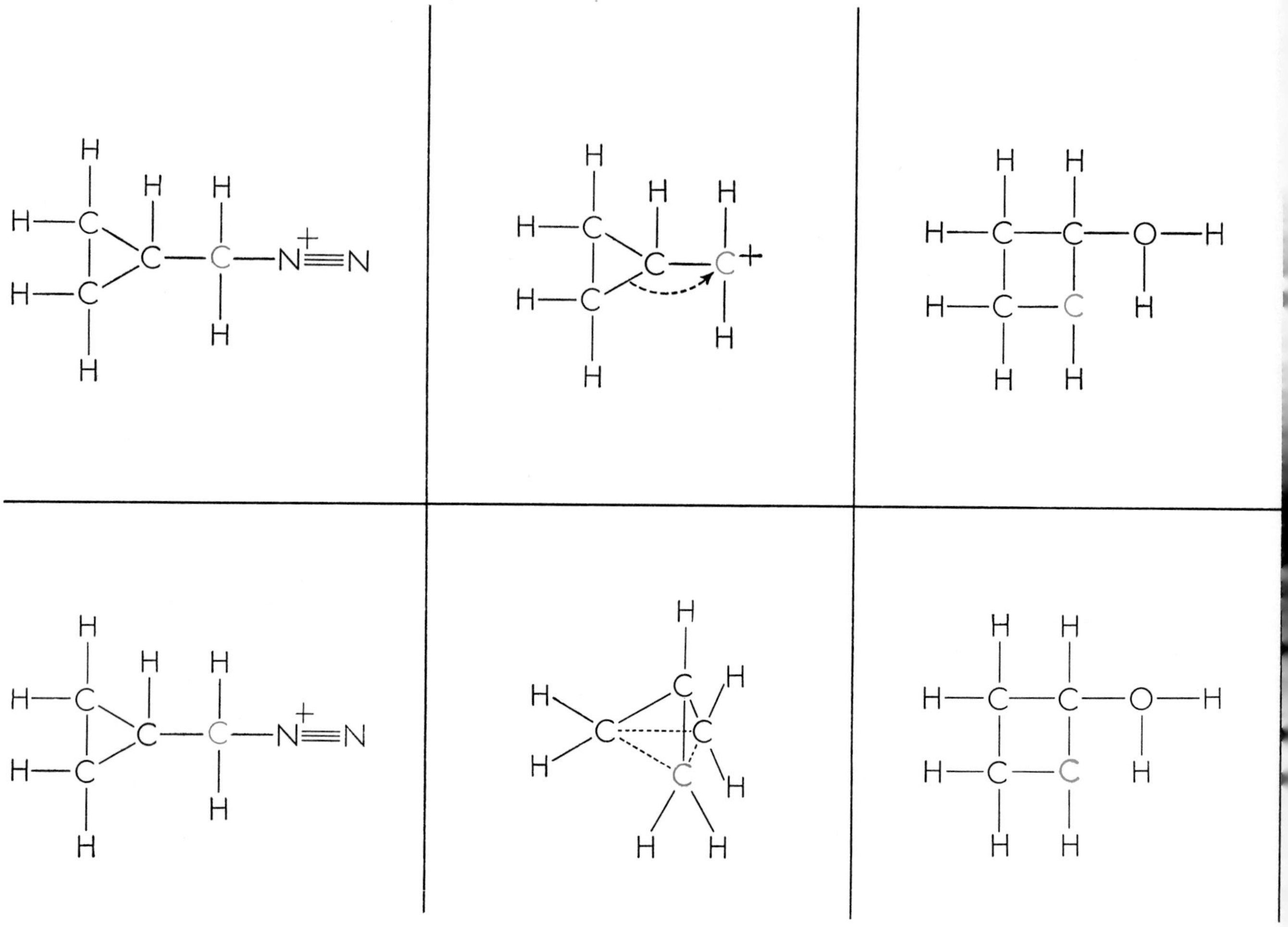

COMPLEX CARBON SHUFFLING is illustrated in these diagrams. Top row shows possible steps in which a three-carbon ring (*top left*) is converted to a four-carbon ring (*third from left*), apparently by a simple rearrangement of a bond (*second from left*). If this were the mechanism, then carbon-14 atom (*color*) should always be next to the carbon which lost the ring bond and gained an OH group. Actually one third of the product has the labeled atom diagonally across from this carbon, as in the diagram at right. This is now explained

Let us turn to another type of substitution reaction, which plainly proceeds differently from S_N2 because its rate is totally independent of the concentration of the hydroxide ion. It is called S_N1, the 1 representing the fact that its rate depends only on the concentration of the molecule that is to be converted. This mechanism, like S_N2, converts organic chlorides to alcohols, but in a different way: water now supplies the hydroxide ion.

In the simplest form it is a two-step process. First the molecule ionizes slowly in water, releasing the chlorine as a negative ion. The molecule then reacts rapidly with water (HOH), taking on the OH and leaving the H as a positive ion, which promptly joins up with free OH to re-form water.

An important feature of the S_N1

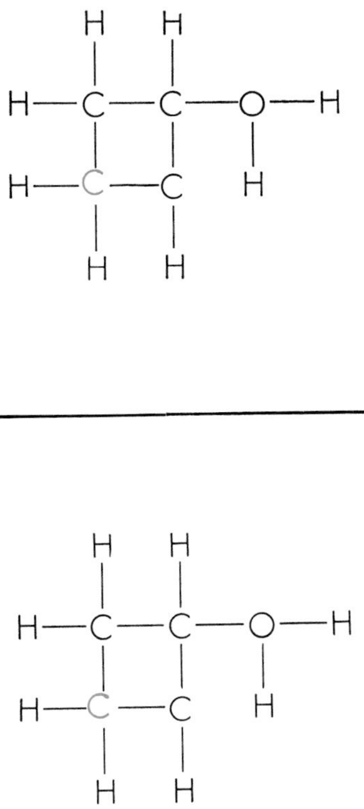

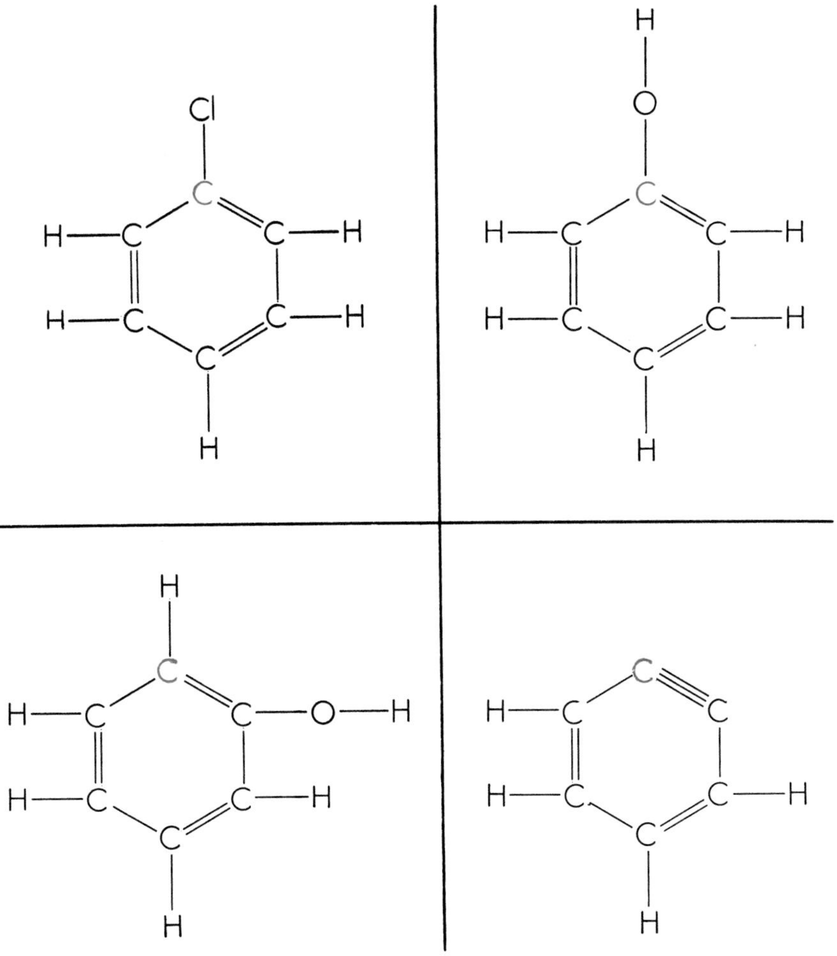

REARRANGEMENT REACTION, in which phenyl chloride (top left) is converted to phenol by reaction with sodium hydroxide, is diagrammed. The OH group would be expected to take position of the chlorine (top right). Sometimes, however, it appears one atom away (bottom left). Explanation is formation of an intermediate compound with triple bond (bottom right).

by the sequence in the bottom row, in which the original ring forms an intermediate pyramid-like ion (second from left) from which it falls into either of the final configurations.

process is that it often rearranges the structure of the reacting molecule. An example is the action of water on a five-carbon compound called neopentyl chloride. After the molecule has shed the chlorine atom, one of its methyl groups swings around from the side to the end of the molecule, and the hydroxide ion takes its place [see lower diagrams on next page]. Thus the reaction of neopentyl chloride with water yields not neopentyl alcohol but an alcohol with a different structure, called t-amyl alcohol.

The structure-changing type of reaction is important in the petroleum industry. It is used to convert "straight-chain" gasolines with poor antiknock properties into "branched-chain," high-octane gasolines.

Sometimes the carbon atoms of a

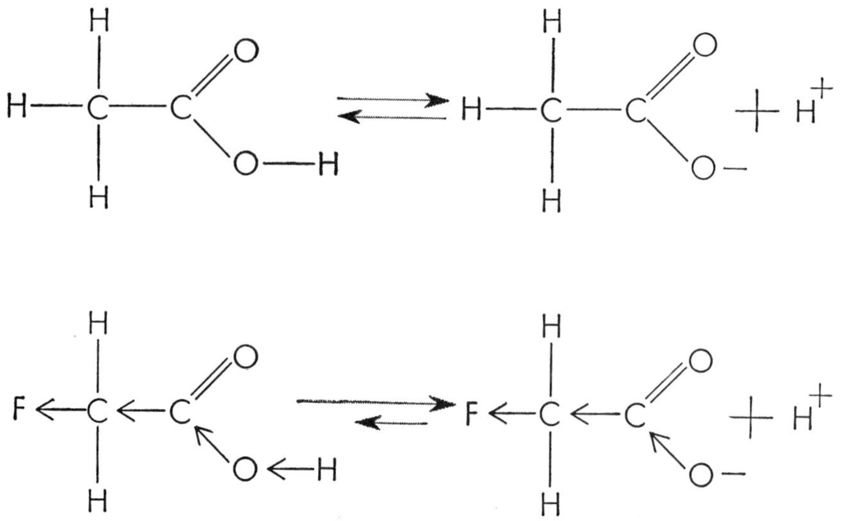

ELECTRICAL EFFECT of fluorine is shown diagrammatically. Acetic acid (*top left*) ionizes weakly. When fluorine is substituted for hydrogen at left end (*bottom*) it attracts electrons from other atoms, as suggested by arrows, so that H ion can break away more easily.

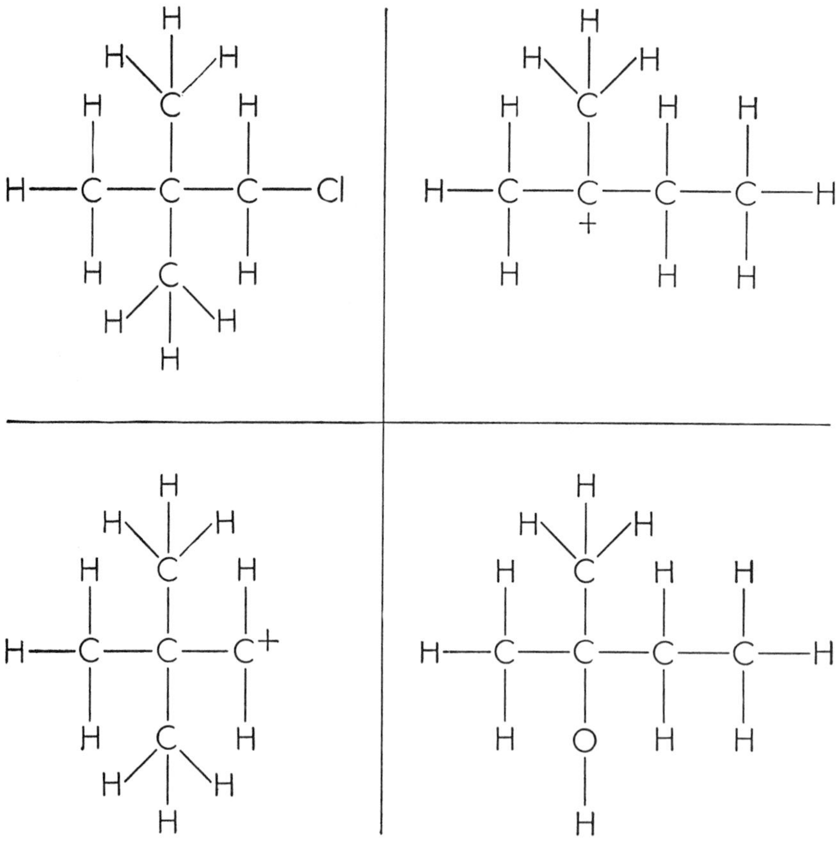

REARRANGEMENT OF CARBON SKELETON is demonstrated when neopentyl chloride (*top left*) is converted to alcohol. When Cl breaks off, producing an ion (*bottom left*), a neighboring carbon moves to the ionized site (*top right*), leaving a charge on the central carbon. This atom now picks up the OH group to form the alcohol shown at bottom right.

molecule are thoroughly scrambled in an S_N1 reaction. Although the final product may not suggest this, investigations with radiocarbon 14 as a marker bring the reorganization to light. It becomes an intriguing challenge to try to reconstruct the sequence of events. An example is illustrated by the accompanying diagrams [*see page 284*]. Here a three-carbon ring with a tail is converted into a four-carbon ring. It would appear that the ring had simply opened and picked up the tail. But we discover that in about one third of the product the tagged carbon atom turns up in an unexpected place. Attempting to picture what has happened, we deduce that the molecule first twisted itself into a pyramid. Any of the three bonds running to the peak of the pyramid may break to take on the hydroxide group; since all three events are equally likely, two thirds of the product molecules should have the tagged carbon next to the OH and one third opposite it. This is almost exactly the proportion we find experimentally.

As a final example of an intricate reaction, consider the conversion of chlorobenzene to phenol [*see diagrams; 284–285*]. In this reaction there appears to be no structural rearrangement, and the reaction rate is proportional to the concentrations of both the chloride molecule and the hydroxyl ion. It would seem to be an ordinary S_N2 reaction. Yet experiments on molecules labeled with carbon 14 show that the OH group often turns up not on the carbon atom vacated by the chlorine but on the one next to it. We now know that when the chlorine atom breaks off, the adjacent carbon simultaneously loses a hydrogen atom, and a triple bond is fleetingly established. To begin and end on a pair of common atoms these bonds must be distorted from the normal 109.5-degree bond angle. The ring shape of the molecule distorts them much further. Chemically, this means that the bonds break easily. When they do, the opened bonds take up H and OH from the water. Since the triple bond between the two carbon atoms may open in either direction by chance, the OH is equally likely to appear on the carbon atom vacated by the chlorine or on the one adjacent.

The "boundless forest" of organic chemistry has grown extraordinary foliage of which Wöhler never even dreamed. Much of the forest has been well mapped, but we still have far to go before we shall be able to understand the reaction mechanisms by which living cells make organic compounds.

25

THE NATURE OF AROMATIC MOLECULES

RONALD BRESLOW

August 1972

*"Aromaticity" refers to the exceptional stability of
certain ring-shaped organic molecules. Whether a given
system is aromatic, nonaromatic or antiaromatic
depends on how many of its electrons are "delocalized"*

In common with the other natural sciences chemistry is fundamentally concerned with understanding the laws of nature. Part of the effort is directed at understanding general laws that govern the common behavior of all matter at the molecular level. Most basic chemical research, however, is directed at understanding the properties of individual chemical compounds and how molecular structure determines these properties. In trying to correlate structure with properties the chemist is not limited to substances presented to him by nature. Perhaps more than investigators in any other discipline, the chemist is concerned with the properties of novel substances and substances prepared artificially. Chemistry remains a "natural" science, but it is concerned not only with structures and phenomena preexisting in nature but also with innumerable other structures and phenomena that are possible through the operation of nature's chemical laws.

For this reason synthesis has always been an important activity in chemical investigation. The modern chemical industry abundantly illustrates the practical results that can flow from the synthesis of new chemical substances with unusual properties. The large majority of these substances owe their existence to the work of investigators who were interested only in exploring the natural laws of chemistry and who perceived that the synthesis and study of new structures were more rewarding than simple examination of those nature has already provided. The rewards of this approach are well illustrated by the history of the development of the concepts of aromaticity and antiaromaticity in organic chemistry.

Chemistry is cursed with unsuitable ancient words and names that are understood by practitioners and confuse the uninitiated. "Organic" chemistry is the name for any chemistry concerned with carbon compounds. Before the 1820's it was believed the carbon compounds that can be isolated from living organisms could only be made "organically." No one has doubted for a long time that chemists can synthesize all such compounds in the laboratory, but the name lingers on.

By the same token it was recognized early that the compound benzene and substances related to it had special properties. One of the most obvious was that such compounds had an odor different from that of other carbon compounds; hence the group of substances including benzene was called "aromatic." The aroma of these substances is the least interesting special thing about them, but unfortunately chemists then used the term "aromaticity" to describe the much more interesting chemical properties of benzene and its relatives. The term is so deeply embedded in the chemical literature that it would be fruitless to try to change it now, but clearly what the chemist understands by aromaticity has nothing to do with the layman's interpretation of the word.

Benzene is a colorless liquid composed of molecules with the formula C_6H_6. The six carbon atoms of benzene are arranged in a regular hexagon; each carbon atom has one hydrogen atom attached to it by a single covalent bond consisting of two electrons, one supplied by the carbon atom and the other by the hydrogen atom [*see the illustration on page 290*]. In order to appreciate why the chemical properties of benzene were surprising, one must consider what is known about other hydrocarbons (compounds consisting only of carbon and hydrogen). The simplest hydrocarbon is methane (CH_4) [*see top illustration on*

page 288]. In this molecule the four hydrogen atoms are attached to the central carbon atom by four covalent bonds, each consisting of two shared electrons between the carbon atom and each hydrogen atom. Therefore each hydrogen atom has a filled electronic valence shell containing two electrons: the two it is sharing with a carbon atom. The carbon atom also has a filled valence shell of eight electrons, consisting of four of its own electrons together with the four provided by the four hydrogen atoms. (The carbon atom also has two electrons bound close to its nucleus in an inner shell, but these electrons do not participate in chemical reactions.)

Stable compounds of carbon almost always have eight electrons surrounding carbon in its valence shell; when the eight electrons participate in four single bonds to another atom, so that each single bond incorporates two shared electrons, one says that the carbon is saturated. The molecule ethane (C_2H_6) is also a saturated hydrocarbon, since it has eight electrons around each carbon in the form of four two-electron single bonds. In ethane, however, one of the single bonds joins the two carbon atoms.

In contrast, the molecule ethylene (C_2H_4) is the simplest member of the class of unsaturated hydrocarbons [*see bottom illustration on page 288*]. In this molecule each carbon atom is attached to two hydrogen atoms by two-electron single bonds, but the two central carbon atoms are held together by a double bond, which means that two electron pairs are shared by these carbons. The result again is that each hydrogen shares one electron pair and has a filled valence shell, whereas each carbon shares four electron pairs and also has a filled valence shell. The difference is simply that in ethylene two of the electron pairs are shared by the same two atoms.

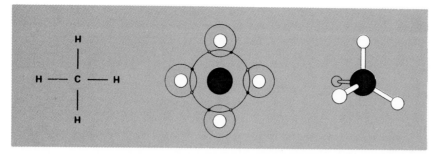

METHANE (CH$_4$), the simplest hydrocarbon, is composed of four hydrogen atoms attached to the central carbon atom by four covalent bonds, each consisting of two shared electrons. As the electron-shell diagram shows, each hydrogen atom has a filled valence shell containing two electrons (its one electron together with one provided by the carbon atom), whereas each carbon atom has a filled valence shell of eight electrons (four of its own electrons together with the four provided by the four hydrogen atoms). Stable hydrocarbon compounds of this type, in which the eight valence electrons of carbon participate in four single bonds to another atom, so that each single bond incorporates two shared electrons, are said to be saturated. As the ball-and-stick diagram shows, the four hydrogens of the methane molecule lie at the corners of a regular tetrahedron, with the carbon at the center.

Long experience has made it clear that such double bonds are more easily broken than single bonds. Hence it is fairly easy to react unsaturated molecules with a variety of chemical reagents. For example, in the presence of a catalyst such as platinum or nickel one molecule of ethylene will readily react with one molecule of hydrogen (H$_2$), a process that breaks and eliminates one of the two carbon-carbon bonds. Ethylene is then chemically transformed into ethane. This process, catalytic hydrogenation, is widely used to convert relatively reactive unsaturated compounds into saturated ones by the addition of hydrogen atoms at the site of carbon-carbon double bonds. In the manufacture of margarine or of peanut butter it is common

practice to hydrogenate some of the unsaturated fats, thereby removing at least some of the carbon-carbon double bonds.

Unsaturated compounds participate in other chemical reactions with great ease. For instance, ethane is essentially inert when it is mixed with bromine molecules (Br$_2$), whereas ethylene reacts rapidly to add the two bromine atoms across the carbon-carbon double bond and again produce a compound containing only saturated carbon, in this case 1,2-dibromoethane. (By numbering the carbon atoms in an organic compound the chemist can tell where various substituent atoms are attached; thus "1,2-dibromoethane" tells him that two bromine atoms are attached to two separate carbon atoms.)

Returning now to the benzene molecule, we can first of all describe the kind of structure we expect it to have and then see how its properties differ from those one might expect by simple analogy with more familiar compounds. We know that each carbon atom of the benzene molecule can form only a single bond to a neighboring hydrogen atom. If each carbon formed only single bonds to each of its two neighboring carbons, it would have only three single bonds, accounting for only six electrons. We conclude that benzene needs one more bond at each of its carbons, so that each will have a filled valence shell of eight electrons. This requirement is easily satisfied if we simply put three double bonds into the molecule: a double bond between carbons 1 and 2, another between carbons 3 and 4 and a third between carbons 5 and 6.

Early chemists drew this unsaturated structure but were troubled because benzene did not behave like a typical unsaturated molecule [see illustration on opposite page]. Although benzene can be hydrogenated to cyclohexane (C$_6$H$_{12}$) by the addition of three molecules of hydrogen, the reaction requires significantly higher temperatures and pressures than the hydrogenation of other unsaturated molecules. Similarly, the reaction of benzene with bromine is slow compared with what one expects from an unsaturated hydrocarbon. More significantly, the bromine does not add to one of the carbon-carbon double bonds but ends up replacing a hydrogen atom, yielding bromobenzene (C$_6$H$_5$Br).

These examples indicate the kind of special properties that chemists came to associate with benzene and that led them to believe this unsaturated hydrocarbon and others in the same family have unusual chemical properties. That set of properties came to be known as "aromaticity." Perhaps the best way to describe the term is to say that it means that the benzene ring, with its three double bonds, is a unit of surprising stability.

Because of this stability benzene and various derivatives containing the benzene ring are slow to react with hydrogen or other reagents that ordinarily react with unsaturated molecules. The stability also means that when benzene and its derivatives do react, as in the slow reaction of benzene with bromine, there is a tendency for the unsaturated ring structure, with its three double bonds, to be preserved.

These properties were known more than 100 years ago. The structural ex-

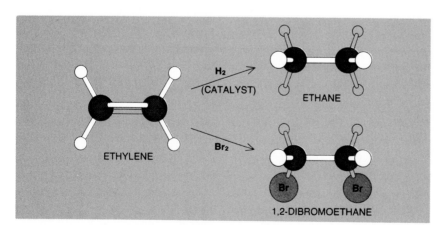

ETHYLENE (C$_2$H$_4$), the simplest member of the class of unsaturated hydrocarbons, contains two central carbon atoms held together by an easily broken double bond. Such unsaturated molecules react quite readily with a variety of chemical reagents. Two typical reactions are shown here. In the process called catalytic hydrogenation one molecule of ethylene (left) reacts with one molecule of hydrogen (H$_2$) in the presence of a catalyst; the reaction eliminates one of the two carbon-carbon bonds, chemically transforming the ethylene in ethane (top right). In the presence of bromine molecules (Br$_2$) ethylene reacts rapidly to add the two bromine atoms across the carbon-carbon double bond, again producing a compound containing only saturated carbon, in this case 1,2-dibromoethane (bottom right).

planation of them first began to emerge in the 1860's, primarily through the insight of the German chemist August Kekulé. He realized that the structure of benzene I have described is arbitrary, since the decision was made to put double bonds between carbon atoms 1 and 2, 3 and 4, and 5 and 6. An alternative would have been to leave single bonds in those positions and to put the double bonds between carbons 2 and 3, 4 and 5, and 6 and 1. There is thus an ambiguity in the structure, and Kekulé reasoned that this might be related to the special properties of benzene. By his own account he awoke from a dream in which snakes were chasing each other around a circle and realized that the double bonds in benzene could move around the benzene ring so as to spend part of their time between alternate pairs of carbon atoms.

At the time of Kekulé's dream, of course, electrons were unknown. Today we would say that the electrons that form the covalent bonds are delocalized: spread out over the entire benzene ring rather than localized between two particular carbons. Another description is to say that benzene is a "resonance hybrid." That is, it does not have a structure in which double bonds are localized between any particular neighboring pair of carbons but instead has a structure that is a hybrid of the two possible configurations. In this new average structure each carbon has one and a half bonds to each of its neighbor carbons rather than a double bond on one side and a single bond on the other. The same type of structure can be assigned to many other molecules with unsaturated six-member rings, such as naphthalene, pyridine and their derivatives, thus explaining their stability, or aromaticity. When Kekulé proposed that the double bonds in benzene chase themselves around the ring, it was not at all understood how this could stabilize the molecule, but modern quantum mechanics can handle such phenomena quantitatively and readily accounts for the increased stability.

One may ask if it is possible to remove a proton (the nucleus of a hydrogen atom) from a hydrocarbon and leave the hydrogen atom's electron behind in the valence shell of the neighboring carbon atom. It turns out that removing a proton from most hydrocarbons is quite difficult. When it is accomplished, the negatively charged compound that results is called a carbanion.

It has been known for some time, however, that a proton can be removed rather easily from a particular unsaturated hydrocarbon with a five-member

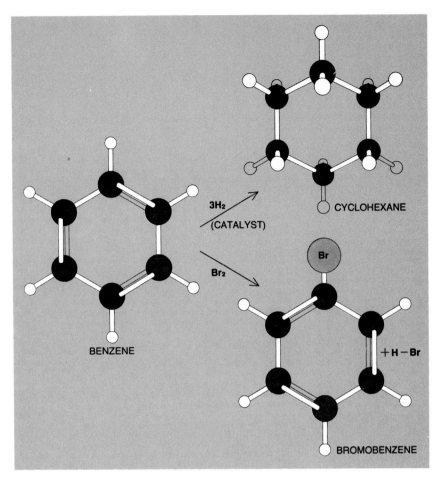

CHEMICAL REACTIONS OF BENZENE are not like those of a typical unsaturated hydrocarbon. For example, benzene can be hydrogenated to form cyclohexane (C_6H_{12}) by the addition of three molecules of hydrogen (*top right*), but the reaction requires significantly higher temperatures and pressures than the hydrogenation of other unsaturated molecules. Similarly, the reaction of benzene with bromine (*bottom right*) is slow compared with what one expects from an unsaturated hydrocarbon; moreover, the bromine does not add to one of the carbon-carbon bonds but ends up replacing a hydrogen atom, yielding bromobenzene (C_6H_5Br). The unusual chemical properties exhibited by benzene (and similar hydrocarbons) in such reactions are what is described by the term aromaticity.

ring: 1,3-cyclopentadiene. The resulting carbanion (the cyclopentadienyl anion) resembles a benzene ring from which one carbon atom has been removed, so that the ring contains only two double bonds instead of three. In addition, one of the carbon atoms is drawn with an unshared pair of electrons [see illustration on page 291].

It is evident that each carbon atom has eight valence electrons, thus possessing a filled outer shell. The reader will also see what chemists realized: It is possible to draw five structures for the cyclopentadienyl anion in which the unshared pair of electrons is successively placed on each of the five carbon atoms. As in benzene itself, the electrons are delocalized, and again it was quickly apprehended that such delocalization is undoubtedly the reason for the anion's unusual stability. As a result the cyclopentadienyl anion can be prepared un-

der conditions far milder than those required for the preparation of other anions.

More recently a positively charged hydrocarbon, a carbonium ion, has been prepared that is also strikingly stable. The compound, the cycloheptatrienylium cation ($C_7H_7^+$), which has seven carbons in a ring and a net positive charge, is remarkably stable for a positively charged molecule. The compound resembles a benzene molecule into which an extra carbon and hydrogen have been inserted [see top illustration on page 292].

The seven-member ring contains three double bonds, so that six of the seven carbon atoms have eight valence electrons. The seventh carbon atom is left with only six electrons in its valence shell and hence carries a net positive charge. Again the choice of how to draw the double bonds and where to put the

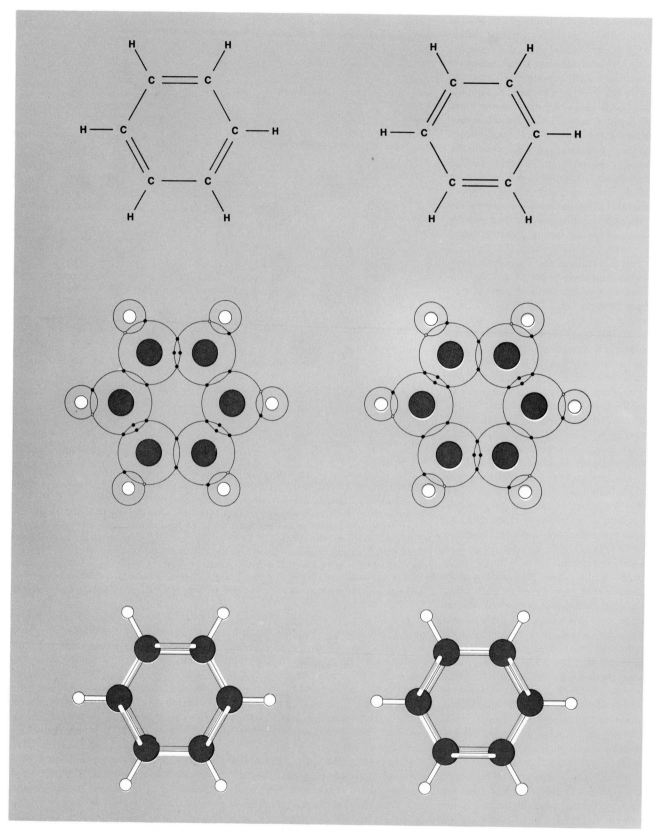

BENZENE, a colorless liquid composed of molecules with the formula C_6H_6, is the historical prototype of the aromatic compounds. The six carbon atoms of benzene all lie in the same plane, situated at the corners of a regular hexagon. Each carbon atom has one hydrogen atom attached to it by a single covalent bond consisting of two electrons, one supplied by the carbon atom and the other by the hydrogen atom. The carbon atoms are joined to form the closed benzene ring by a combination of three single bonds and three double bonds. The positions of the double bonds can be as-signed arbitrarily in two different ways (*left*, *right*), and the true structure is regarded as a hybrid of these two resonance forms. Each form is in turn represented in this illustration in three differ-ent ways, beginning with the familiar chemical diagrams (*top*). In the electron-shell diagrams (*middle*) electrons contributed by the hydrogen atoms are indicated by white dots; electrons contributed by carbon atoms are indicated by black dots. (Only the six valence, or outer-shell, electrons of each carbon are shown.) Ball-and-stick convention (*bottom*) is used in most of the following illustrations.

positive charge is arbitrary; the reader will see, in fact, that seven structures can be written for the cycloheptatrienylium cation. The true structure is a hybrid of the seven, in which each carbon atom has one-seventh of a positive charge and the electrons of the double bonds are spread uniformly around the ring.

Emboldened by examples of this type, chemists set out to prepare new aromatic systems other than those found in nature or already known. One of the earliest surprises came with the synthesis of cyclooctatetraene, which has eight carbons in a ring and four double bonds [see bottom illustration on next page]. It was expected that this would also be an exceptionally stable, or "aromatic," molecule with the electrons delocalized around the ring.

The actual properties of cyclooctatetraene, when it was finally synthesized by Richard Willstätter in 1913, were not at all like those of benzene. Instead cyclooctatetraene was as reactive as any ordinary unsaturated molecule and showed no sign of any special stability. This finding was so contrary to what was expected that for a long time most chemists believed that the synthesis had been unsuccessful and that a different molecule had actually been prepared.

Further surprises awaited those who tried to prepare cyclobutadiene (C_4H_4), a compound with four carbons in a ring containing two double bonds, which could be located in either of two arbitrary positions. Once more it was expected that the real compound would be a hybrid of the two structures and that the molecule would be stabilized, just as benzene is, by electron delocalization. In actuality all attempts to prepare cy-

clobutadiene by conventional chemical methods led to other materials; the compound proved impossible to prepare. It quickly became apparent that this was because the molecule, far from being strongly stabilized, is extraordinarily reactive. Quite recently it has been possible to generate cyclobutadiene for a fleeting instant in the gaseous state and to observe its reactions with other compounds.

To summarize up to this point, we have seen that an unsaturated six-member ring (benzene) is unusually stable, as is a negatively charged five-member ring (the cyclopentadienyl anion) and a positively charged seven-member ring (the cycloheptatrienylium cation). On the other hand, unsaturated neutral hydrocarbon rings with four members (cyclobutadiene) and eight members (cyclooctatetraene) are notably unstable. Various studies have also shown that positively charged five-member rings and negatively charged seven-member rings are unstable.

Observations of this kind led to the hypothesis that there is something special about the "aromatic sextet." Ignoring the electrons that participate in the single bonds and considering only the electron pairs that get delocalized in the various stable and unstable structures, it is apparent that there are six such electrons in benzene, corresponding to the three electron pairs of the double bonds. (We count only one electron pair in each of the double bonds: the electron pair that can be put elsewhere in an alternative structure.) Therefore in benzene six electrons are delocalized and the others stay essentially fixed between the atoms involved in single bonding. Similarly, in the cyclopentadienyl anion six electrons

are delocalized: two electrons for each of the two double bonds and an electron pair that is unshared and leads to the negative charge on one of the carbons.

In the cycloheptatrienylium cation the six electrons of the three double bonds are delocalized. (The positive charge, which is spread around the ring, represents an absence of electron pairs, so that it does not contribute anything more to the total number of electrons delocalized.) In the cyclopentadienyl cation only four electrons are delocalized, and in the cycloheptatrienylium anion eight electrons are delocalized; both configurations are unstable.

What is so magical about the number six? The answer came out of some studies in the 1930's by Erich Hückel on a new way to apply quantum mechanics to molecules, which led to the development of what is called molecular-orbital theory. In this theory the delocalized electrons I have been describing are thought about in a new way. Rather than first assigning electrons to particular locations between pairs of carbon atoms and later deciding that they can be spread out over the entire ring, molecular-orbital theory starts off with the idea that in a system such as a benzene ring there are orbitals, or places where electrons can be put, that spread out over the entire six carbon atoms. Therefore the set of six carbons and six hydrogens of a benzene ring is regarded as a kind of superatom, and electrons are added to it, just as in describing the structure of an atom we consider adding electrons to the atomic nucleus. When this form of molecular quantum mechanics was applied to molecules exhibiting (or lacking) aromaticity, it quickly emerged that

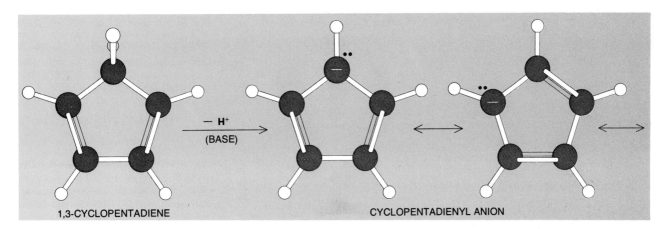

1,3-CYCLOPENTADIENE   CYCLOPENTADIENYL ANION

CONVERSION of 1,3-cyclopentadiene, an unsaturated hydrocarbon with a five-member ring, to the aromatic cyclopentadienyl anion can be readily accomplished by the removal of a hydrogen nucleus, or proton (H^+), leaving the hydrogen atom's electron behind in the valence shell of the neighboring carbon atom. The resulting negatively charged compound, called a carbanion, resembles a benzene ring from which one carbon atom has been removed. Five resonance structures are possible for the cyclopentadienyl anion, two of which are shown here. In each case one of the carbon atoms is drawn with an unshared pair of electrons.

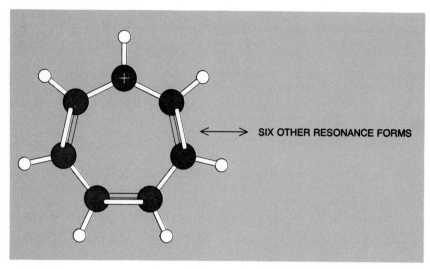

SIX OTHER RESONANCE FORMS

CARBONIUM ION ($C_7H_7{}^+$), a positively charged aromatic hydrocarbon with remarkable stability, has recently been prepared. The compound, which has seven carbons in a ring and hence is called a cycloheptatrienylium cation, resembles a benzene molecule into which an extra carbon and hydrogen have been inserted. The seven resonance forms differ according to how the double bonds are drawn and where the positive charge is put.

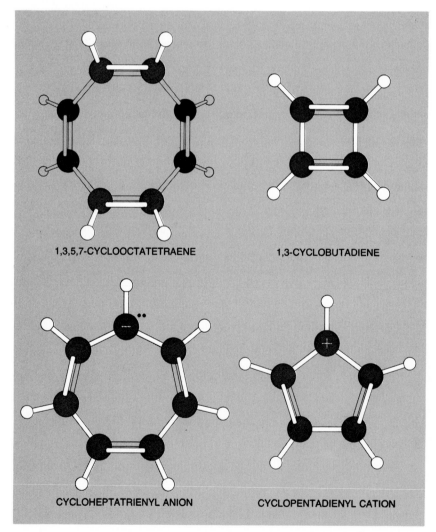

1,3,5,7-CYCLOOCTATETRAENE

1,3-CYCLOBUTADIENE

CYCLOHEPTATRIENYL ANION

CYCLOPENTADIENYL CATION

FOUR NONAROMATIC COMPOUNDS, all molecules with the electrons potentially delocalized around a ring, were originally sought in the expectation that they would also be exceptionally stable, or "aromatic." In fact their properties were found to be not at all like those of benzene; all four are at least as reactive as any ordinary unsaturated molecule.

"magic numbers" of delocalized electrons are needed to fill up stable shells, just as in ordinary atoms two electrons are needed to fill a valence shell for hydrogen and eight electrons are needed to fill one for carbon.

According to Hückel's theory, there should be stable delocalized ring systems where the number of electrons needed to fill molecular-orbital shells have the magic values 2, 6, 10, 14 and so on. This has become known as the $4n + 2$ rule, in which n is zero or any integer. The number 6 is the second member of the magic series, and all the aromatic systems discussed up to this point have had six delocalized electrons in a ring.

In recent years there has been an intensive effort in many laboratories to explore the implications of this theoretical work. One approach has involved looking for the predicted new aromatic molecules that fit the requirements of the Hückel $4n + 2$ rule. A second important activity has been to try to get further information about the properties of those molecules, such as cyclobutadiene, that do not fit the Hückel rule. Of course, molecular quantum mechanics has progressed since the 1930's, and these experimental observations are guided and interpreted by various new and more sophisticated versions of the original Hückel molecular-orbital theory.

My own work in this area began in 1956 when I arrived as a new instructor in chemistry at Columbia University. I was intrigued by the prediction that there was an aromatic system still to be discovered that would have only two delocalized electrons and so be the simplest system of all. The way to get two delocalized electrons into a ring structure is to construct a three-member ring with one double bond and one positive charge. It would be expected that the electron pair of the double bond would be delocalized over the three carbon-carbon single bonds and that the positive charge would also be spread out over the three carbons. This structure, the cyclopropenyl cation ($C_3H_3{}^+$), was our synthetic goal [see top illustration on next page].

For a variety of technical reasons my colleagues and I concluded that it would be easiest to start off by trying to make a derivative of this system in which the three carbon atoms carried groups other than hydrogen atoms. In 1959 we succeeded in preparing a derivative of the system, incorporating a central cyclopropenyl cation ring, in which each of the three carbon atoms carried what are called phenyl groups (C_6H_5) rather than hydrogen atoms. The properties of this molecule, including its remarkable stability, made it quite clear that the central cyclopropenyl cation ring was indeed a new aromatic system. In work extending over the next 10 years our laboratory prepared a variety of derivatives of this simplest aromatic system, whose properties make it fully clear that the cyclopropenyl cation is strongly stabilized by electron delocalization. In 1969,

10 years after we prepared the first derivative of this system, we were finally able to prepare the parent compound $C_3H_3^+$ itself. Its properties were fully consistent with those we had deduced from a study of the other derivatives of the cation and confirmed the prediction that it is indeed the simplest of the aromatic systems.

Other chemists have been concerned with extending Hückel's predictions in the other direction. For example, Thom-

as J. Katz of Columbia has prepared a dianion by adding two more electrons to cyclooctatetraene [see middle illustration below]. Whereas the neutral hydrocarbon, with only eight delocalized electrons, was not aromatic, Katz found that the new dianion was in fact an aromatic system with remarkable stability. The two additional electrons fill up the next shell predicted in the molecular-orbital theory, making a system with 10 delocalized electrons in all.

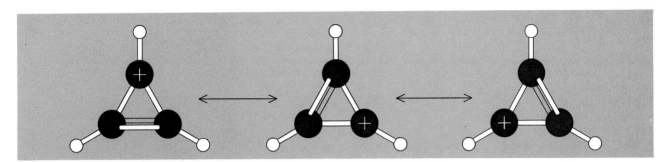

SIMPLEST AROMATIC SYSTEM, the cyclopropenyl cation ($C_3H_3^+$), was first synthesized in 1969 by the author and his colleagues at Columbia University. The two delocalized electrons are spread over a three-member carbon ring that has one double bond and one positive charge. The three possible resonance forms of the molecule are shown. The cyclopropenyl cation and its derivatives (in which the three carbon atoms carry groups other than hydrogen atoms) are all strongly stabilized by electron delocalization.

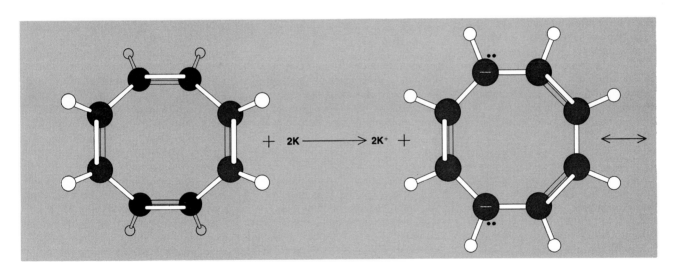

AROMATIC DIANION ($C_8H_8^{--}$) (*right*) was prepared by Thomas J. Katz of Columbia by adding a pair of extra electrons (provided by two potassium atoms) to the nonaromatic compound cyclooctatetraene, a neutral eight-membered hydrocarbon ring (*left*).

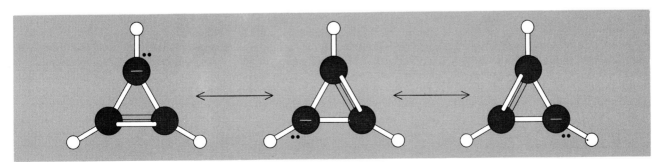

SIMPLEST ANTIAROMATIC SYSTEM, the cyclopropenyl anion ($C_3H_3^-$), is so unstable that it has never been possible to observe its properties directly. Derivatives of the anion have been prepared, however, and from their properties it has been possible to show that the antiaromaticity, or instability, of the system is caused by the delocalization of the "wrong" number of electrons.

Franz Sondheimer of University College London has been working with much larger systems that fit the Hückel predictions. For example, he has prepared a compound with 18 carbons in a ring, each carbon bearing one hydrogen atom [*see illustration at right*]. The nine double bonds in this neutral hydrocarbon can be formally written in two different schemes, indicating that 18 electrons in all are delocalized. The number 18, of course, fits the $4n + 2$ rule.

The properties of this hydrocarbon, which Sondheimer calls [18]annulene, are consistent with its being an aromatic stabilized system. Although the primary interest in these systems was at first simply the attempt to discover a new part of the chemical world predicted by theory, a number of chemists in industry are now exploring the possible applications of the new aromatic systems to the solution of practical problems.

The Hückel rule makes qualitative predictions about 25 percent of the possible delocalized systems one could create but says essentially nothing about what to expect for the other 75 percent, the compounds with $4n$, $4n + 1$ or $4n + 3$ delocalized electrons. Molecular-orbital theory, on the other hand, being a quantitative theory, makes specific predictions about the energies and reactivities of all the delocalized systems. Even the newest forms of molecular-orbital theory, however, contain a number of approximations, and it is not clear how well they do predict the properties of unknown molecules in detail. For this reason there has been considerable interest in exploring the chemistry of the molecules with $4n$ delocalized electrons experimentally to see what kind of properties they have.

I have already referred to the chemistry of cyclooctatetraene, which appeared to be similar to the chemistry of ordinary molecules with carbon-carbon double bonds. On that basis one might conclude that a molecule of this kind, with eight delocalizable electrons, has normal chemical properties rather than the unusual stability of the aromatic compounds. It is now known, however, that cyclooctatetraene is not a flat eight-membered ring but is strongly puckered. Theory indicates that under these circumstances the electrons cannot be easily delocalized, so that cyclooctatetraene is not a good example of a molecule with $4n$ delocalized electrons.

Perhaps the molecule cyclobutadiene, which I have also mentioned, would be more relevant. From other parts of chemical theory it is quite clear that this

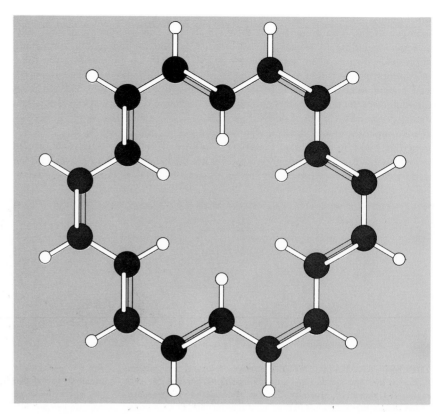

LARGE AROMATIC RING, designated [18]annulene, was prepared by Franz Sondheimer of University College London. The nine double bonds in this neutral hydrocarbon can be drawn in two different ways, indicating that 18 electrons in all can be delocalized.

molecule should be flat and that the four electrons in the ring should be delocalizable. The observations over the years that cyclobutadiene is an extremely unstable molecule suggested that perhaps the systems with $4n$ delocalized electrons do not have normal properties but are for some reason unstable. One could not exclude, however, a number of other explanations for the high reactivity of cyclobutadiene.

We became interested in this question and decided to investigate the chemistry of the cyclopropenyl anion ($C_3H_3^-$), the negatively charged analogue of the cyclopropenyl cation ($C_3H_3^+$). Like the cation, the anion has one double bond in the three-member ring [*see bottom illustration on preceding page*]. The placement of the double bond, of course, is arbitrary, as is the placement of the negative charge. In the actual delocalized molecule one would expect to find one-third of a negative charge on each of the three carbons, with the electron pair of the double bond similarly spread out among the three carbons.

It is evident that the cyclopropenyl anion, with four delocalized electrons, does not satisfy Hückel's $4n + 2$ rule and thus does not meet the requirements for aromaticity. We hoped to learn whether a system with four delocalized

electrons has normal stability or whether the molecule is destabilized by the four delocalized electrons. The prediction that it might be destabilized did not come out of the simple version of molecular-orbital theory originally described by Hückel but was suggested by more recent methods.

It turns out that the cyclopropenyl anion is so extremely unstable that it has never been possible to prepare it and observe its properties directly. In a series of studies extending over the past 10 years we have detected derivatives of the cyclopropenyl anion and have been able to show from their properties that the system is indeed strongly destabilized. Alternative explanations of the instability have been ruled out, so that it seems quite clear that the cyclopropenyl anion is actually destabilized by the delocalization of the four electrons. We have termed this phenomenon antiaromaticity; it reflects unusual instability because of the presence of a delocalized electron system, just as aromaticity reflects unusual stability because of the delocalization of a different number of electrons.

We and others have also studied derivatives of cyclobutadiene and have good evidence that here too the obvious instability of the system is partly due to antiaromaticity, reflecting the energetic effect of delocalizing the "wrong" number of electrons. Studies in our laboratory and elsewhere have also found such destabilizing effects for the cyclopentadienyl cation and the cycloheptatrienylium anion, as well as for some other systems with $4n$, rather than $4n + 2$, delocalized electrons.

The fact that in molecules there are magic numbers of electrons that can be delocalized with a gain in stability of the molecule, and other numbers whose delocalization results in instability, has extensive implications. One of the most striking advances involving such concepts is due to Robert B. Woodward of Harvard University and Roald Hoffmann of Cornell University, who recently formulated a very general rule about which chemical reactions can proceed easily and which can proceed only with difficulty.

As an example, it is known that two molecules of ethylene (C_2H_4) do not react readily to form the new molecule cyclobutane (C_4H_8) even though the product would be more stable than the starting materials [see upper illustration below]. The difficulty is that part of the way through the reaction the two electron pairs of the two double bonds in the two ethylene molecules must be delocalized in order to make the new carbon-carbon single bonds that are required to form cyclobutane. Since four electrons need to be delocalized, the situation is extremely unfavorable.

In contrast, it is relatively easy for ethylene, a molecule with one double bond, to combine with 1,3-butadiene, a molecule with two double bonds [see lower illustration below]. Here three double bonds are available to supply six electrons for delocalization as the reaction proceeds, which satisfies the $4n + 2$

rule for stability. Since the work of Woodward and Hoffmann is actually much more explicit and inclusive than these simple examples indicate, it has led chemists to predict reactions that had not previously been observed.

Thus we see that the exploration of aromaticity, and more recently of antiaromaticity, has stimulated important intellectual developments in the theory of molecular structure. At the same time these studies have resulted in the preparation of a number of new types of molecule that are under active exploration as building blocks for future useful materials. Finally, the theoretical advances accompanying this exploration have greatly increased the ability of chemists to predict precisely what chemical reactions will occur and how they will occur, thereby strengthening the entire field of synthetic organic chemistry.

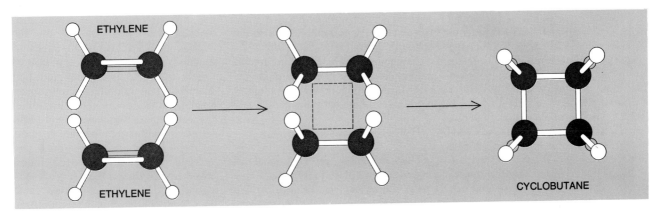

TWO MOLECULES OF ETHYLENE react only with great difficulty to form the new molecule cyclobutane (C_4H_8), even though the product would be more stable than the starting materials, because the reaction must pass through an antiaromatic intermediate state with four delocalized electrons (broken lines). In general, ring systems with $4n$ delocalized electrons are strongly destabilized.

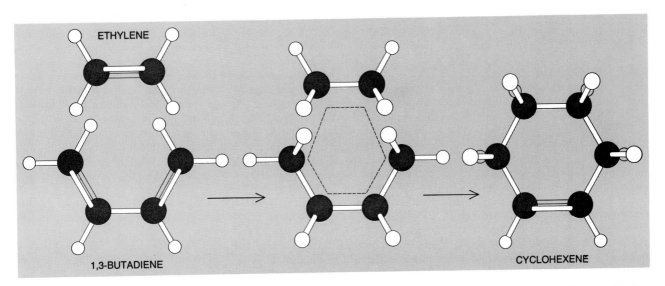

ETHYLENE AND 1,3-BUTADIENE, a molecule with two double bonds, can in contrast be combined rather easily to form the new molecule cyclohexene, because the reaction passes through an aromatic intermediate state with six delocalized electrons (broken lines). Ring systems whose delocalized electrons satisfy the "magic" $4n + 2$ rule are in general extremely stable and hence are aromatic.

26

THE SHAPES OF ORGANIC MOLECULES

JOSEPH B. LAMBERT
January 1970

*The 1969 Nobel prize in chemistry was given to men
who showed that certain molecules can assume
different shapes simply by rotations around single
bonds. Such differences influence chemical reactivity*

Chemists had quite remarkable insights about the shapes of molecules many years before the electronic theory of matter was fully developed by physicists. In 1865 Friedrich August Kekulé had the inspired idea that the six carbon atoms in the benzene molecule (C_6H_6) are joined in a ring. In 1890 Hermann Sachse proposed that the ring of six carbon atoms in cyclohexane, a molecule with six more hydrogen atoms than benzene, would be free of strain if the carbon atoms were located alternately above and below the plane of the ring instead of being located in the plane itself. Such a structure has the appearance of a reclining chair. Because the benzene ring has double bonds between alternate pairs of carbon atoms, the atoms are forced to remain in a plane. Sachse's chair hypothesis, and its implications for other organic molecules, was rejected for more than 30 years until physical experiments by Walter Hückel, Jacob Boeseken and Odd Hassel of Norway provided corroboration. Sachse's model was not universally accepted, however, until after 1950, when persuasive chemical evidence in its favor was supplied by Derek H. R. Barton of the Imperial College of Science and Technology. Last month Hassel and Barton received the Nobel prize in chemistry "for their work to develop and apply the concept of conformation in chemistry."

The concept of conformation, or conformational analysis, is concerned with the different three-dimensional forms that can be assumed by certain molecules whose atoms are free to rotate around one or more bonds. Such interconvertible forms are known as conformational isomers, or conformers, to distinguish them from other isomers in which two or more molecules of the same formula have different three-dimensional structures that are interconvertible only if bonds are broken and re-formed. It was Barton more than anyone else who showed that different conformational isomers can exhibit distinctly different chemical reactivities. In the past 20 years conformational analysis has been invaluable in the synthesis of pharmacological agents, notably steroids and antibiotics of the tetracycline and penicillin families. In biochemistry conformational analysis is helping to clarify the mechanisms by which enzymes promote chemical reactions in the living cell.

Some Simple Hydrocarbons

The principles of conformational analysis can be illustrated by describing the structures of several simple carbon-containing compounds. In methane (CH_4) each hydrogen atom is singly bonded to a central carbon atom and the four hydrogens form the corners of a tetrahedron. The angle formed by the C–H bonds between any two adjacent hydrogen atoms is 109.5 degrees [*see illustration on page 299*]. Since this tetrahedral shape is inflexible, methane has no conformational isomers. In ethylene (C_2H_4) the two carbon atoms are connected by a double bond, and two hydrogen atoms are bonded to each carbon in such a way that the structure is rigid and completely planar, or flat. Again no conformers are possible.

Now let us replace one hydrogen atom on each carbon atom of ethylene by a methyl group (CH_3), thus creating the hydrocarbon called 2-butene (CH_3–CH= CH–CH_3). Two different forms of 2-butene can exist: in one the methyl groups are both on the same side of the molecule (the *cis* form) and in the other the methyls are on opposite sides (the *trans* form). Rotation is not possible around the rigid double bond that connects the central carbon atoms. Inter-conversion of the *cis* and *trans* isomers of 2-butene requires breaking the double bond, rotating one of the methyl groups 180 degrees around the residual single bond and re-forming the double bond. Since conformers must be interconvertible without breaking bonds, *cis*- and *trans*-2-butene do not qualify. Isomers that require bond breakage for interconversion are termed geometrical isomers.

Ethane (C_2H_6) is the simplest hydrocarbon that exhibits conformational isomerism. The two carbon atoms are connected by a single bond and each carbon is bonded tetrahedrally to three hydrogen atoms. In other words, the molecule consists of two methyl groups linked by a carbon-carbon single bond, around which rotation is allowed and does occur. The methyl groups thus resemble propellers rotating in a well-understood and definable fashion. A view along the carbon-carbon bond, known as the Newman projection, clearly reveals the conformational isomers [*see illustration on page 299*]. At one instant the hydrogen atoms on the front carbon eclipse those on the back. At another instant each front hydrogen falls midway between two rear hydrogens. These "eclipsed" and "staggered" arrangements in ethane are conformational isomers. Nearly all the principles of conformational analysis can be illustrated by ethane and its derivatives.

Molecules that Rotate

Until about 1935 it was thought that all rotational forms of ethane—eclipsed, staggered and intermediate—occurred with equal probability, implying that all possessed equal energy. From the work of Edward Teller, Kenneth S. Pitzer and others it became evident, however, that the staggered conformation has the least potential energy and therefore is the

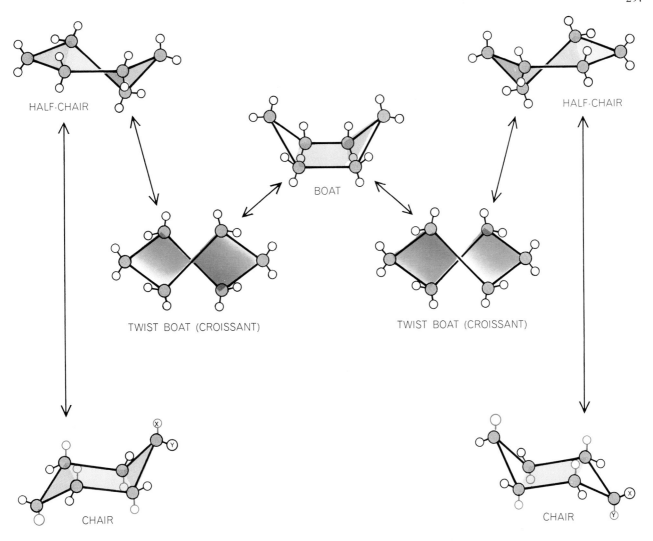

SIX-MEMBER RING OF CYCLOHEXANE, C_6H_{12}, can exist in various shapes known as conformational isomers, or conformers. The most stable conformers are the chairs; the least stable are the half-chairs (*see energy diagram below*). The hydrogen atoms in the molecule can assume two distinguishable positions: "axial," when they are more or less perpendicular to the average plane of the ring; "equatorial," when they lie close to the plane. When the chair flips from one conformation to its mirror image, axial hydrogens become equatorial and vice versa. This reversal is shown for two atoms, X and Y. In the chairs axial atoms are shown in color.

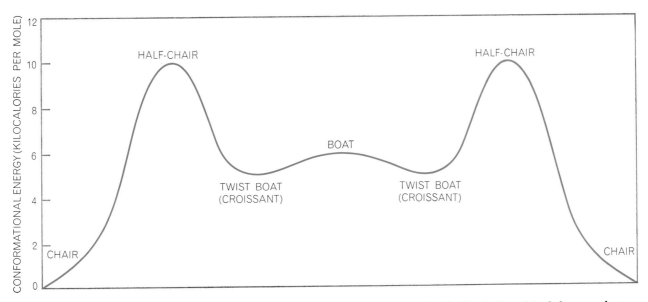

INVERSION OF CYCLOHEXANE takes place through a sequence of ring distortions involving a spectrum of energy states. At room temperature the molecule flips back and forth between the two stable chair conformations. The boat conformers are quasi-stable.

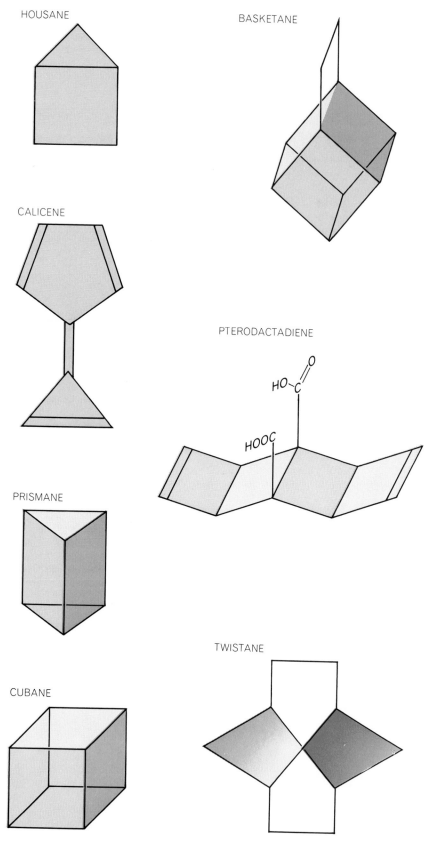

HOUSANE

BASKETANE

CALICENE

PTERODACTADIENE

PRISMANE

TWISTANE

CUBANE

DESCRIPTIVE AND HUMOROUS NAMES have been assigned to the structures of some recently synthesized organic compounds. The straight lines represent bonds between carbon atoms, which form the corners and intersections of the figures. Each carbon atom is assumed to have enough hydrogen atoms attached to it to satisfy any of its four valence bonds that are not linked to another carbon atom; thus housane would have eight hydrogens. Calicene looks like a chalice (*calix* in Latin). In pterodactadiene carboxyl groups form the "head" and the "tail" of a structure that resembles a prehistoric winged reptile, the pterodactyl.

most stable. As the hydrogen atoms move closer to the eclipsed conformation there is an increase in the repulsion between atoms and in other energy factors. The eclipsed form thus represents an unstable energy maximum [*see top illustration on page 300*]. Since a given hydrogen atom must pass three opposing hydrogen atoms in making a full 360-degree revolution, the energy barrier is said to be threefold. It follows that there are three distinct but energetically identical conformations, since any hydrogen atom can be staggered between three different pairs of opposing hydrogens. The increase in energy of the eclipsed forms is referred to as Pitzer strain.

A molecule with a similar but slightly more complex set of conformations is *n*-butane (CH_3–CH_2–CH_2–CH_3), which can be formed by adding two atoms of hydrogen to replace the double bond of either *cis*- or *trans*-2-butene. Rotation around the resulting carbon-carbon single bond (–CH_2–CH_2–) produces different conformational isomers [*see bottom illustration on page 300*]. The conformer of highest energy results when the two methyl groups eclipse each other. When the methyls are staggered 60 degrees apart, the energy drops to a certain minimum (not the lowest one), producing a stable conformer called the *gauche* isomer. When the methyl groups eclipse hydrogen atoms, the energy rises and the conformer is again unstable. The minimum of lowest energy and the most stable arrangement result when the methyl groups are 180 degrees apart, an arrangement called the *anti* isomer. Thus in a 360-degree circuit stable conformations are produced when the angles between one methyl group (imagined as being fixed at 0 degrees) and the second methyl group (imagined as being in rotation) are 60, 180 and 300 degrees. At room temperature about 80 percent of *n*-butane is in the 180-degree, or *anti*, conformation. The remaining 20 percent is divided between the two equivalent *gauche* conformers in which the methyls are 60 or 300 degrees apart.

What no one fully appreciated before the work of Barton and others is that there is a powerful relation between conformational arrangement and chemical reactivity. For example, if one of the hydrogen atoms is removed from the second carbon in *n*-butane and replaced by bromine (Br), the resulting compound is 2-bromobutane (CH_3–$CHBr$–CH_2–CH_3), which yields the same variety of conformational isomers as *n*-butane. One way to prepare *cis*- and *trans*-2-butene is to remove hydrogen bromide (HBr) from the central carbon atoms of 2-bromobutane. HBr is most easily removed when the hydrogen and bromine atoms are 180 degrees apart [see illustration on page 301]. When the methyl group rather than a hydrogen atom is anti to the bromine, the molecule is "sterile": HBr cannot be removed. When the methyl groups are anti to each other, removal of HBr yields trans-2-butene; when the methyls are 60 degrees apart (gauche),

the reaction yields cis-2-butene. Thus each stable conformational isomer has its own distinctive reaction path: one leads to the trans product, one to the cis product and one to no reaction at all.

Carbon Atoms in Rings

The chemistry of carbon compounds, historically referred to as organic chem-

istry, owes its complexity to the remarkable stability of the carbon-carbon bond. Carbon chains, both straight and branched, can be constructed in a seemingly infinite variety of lengths and patterns. Regardless of length or pattern, however, the conformational properties of a particular bond will be determined by the principles described above.

Let us now see what these principles

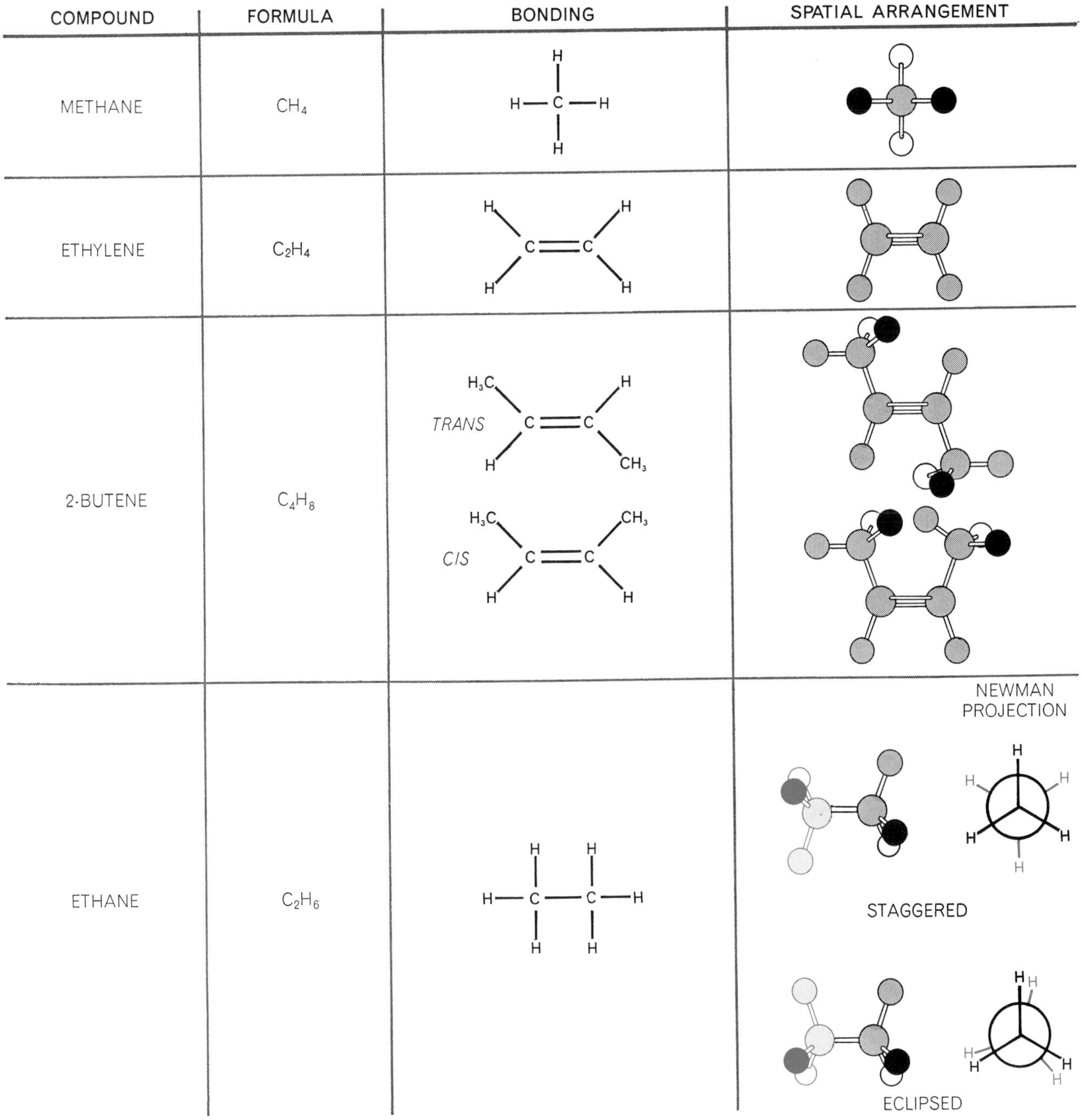

COMPOUND	FORMULA	BONDING	SPATIAL ARRANGEMENT
METHANE	CH_4		
ETHYLENE	C_2H_4		
2-BUTENE	C_4H_8	TRANS / CIS	
ETHANE	C_2H_6		NEWMAN PROJECTION / STAGGERED / ECLIPSED

RIGID AND FLEXIBLE MOLECULES are represented by four simple hydrocarbon compounds. In methane the atoms form a rigid tetrahedron; the angle between any two hydrogen atoms, with carbon at the vertex, is 109.5 degrees. In more complex hydrocarbons, particularly ring structures, this angle often cannot be conserved, producing what is called Baeyer strain. Ethylene also has a rigid structure, but all its atoms lie in a plane. In 2-butene there are two rigid arrangements: the cis form, in which both methyl

(CH_3) groups lie on the same side of the double bond, and the trans form, in which they lie on opposite sides. To interconvert these two forms the double bond must be broken and rejoined, hence they are defined as geometrical isomers. In contrast, ethane has a continuous set of conformational isomers, produced by simple bond rotation. At one extreme the hydrogen atoms are staggered 60 degrees; at the other extreme they are "eclipsed," or in line, as represented in the "Newman projections" at the lower right.

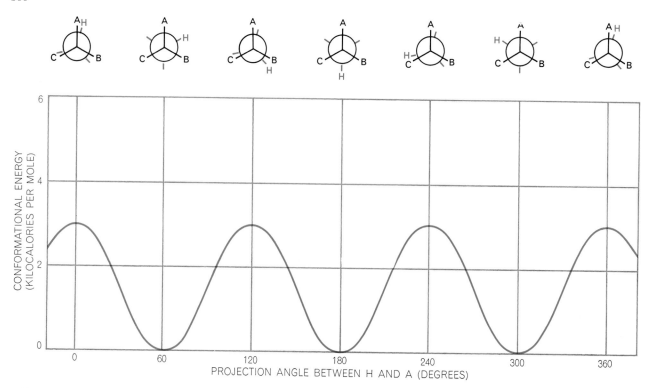

ENERGY OF ETHANE CONFORMERS traces out a simple sine curve when the front carbon atom and its three hydrogen atoms (A, B, C) are assumed to be stationary while the rear carbon atom and its hydrogens execute a 360-degree clockwise rotation. During this journey the molecule alternates between eclipsed states of high energy and staggered states of low energy. A given rear hydrogen (H) passes through three stable conformations at 60, 180 and 300 degrees that are spatially different but energetically equivalent.

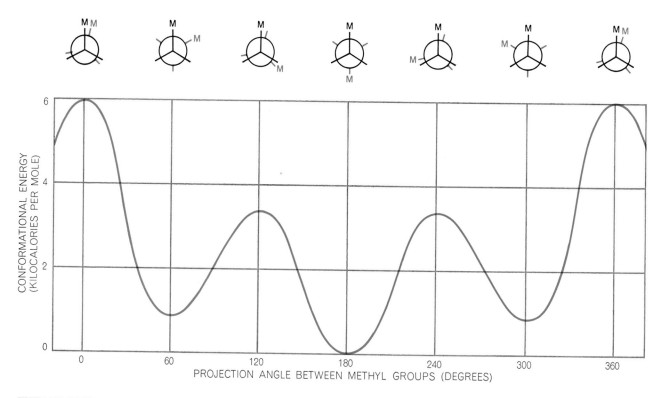

ENERGY OF N-BUTANE CONFORMERS traces out a curve similar to that for ethane. The molecule of n, or normal, butane (C_4H_{10}) can be thought of as a molecule of ethane in which one hydrogen atom on each carbon has been replaced by a methyl (CH_3) group. These groups are represented by the two M's in the Newman projections. The stable conformation at 180 degrees is called the anti arrangement because the two methyls are antipodal. The slightly more energetic, and thus less stable, conformations at 60 and 300 degrees are termed gauche. In the gauche arrangements the methyl groups are closer together and consequently interfere with each other more than in the stable anti arrangement. Maximum steric interference occurs when the two methyl groups are fully eclipsed.

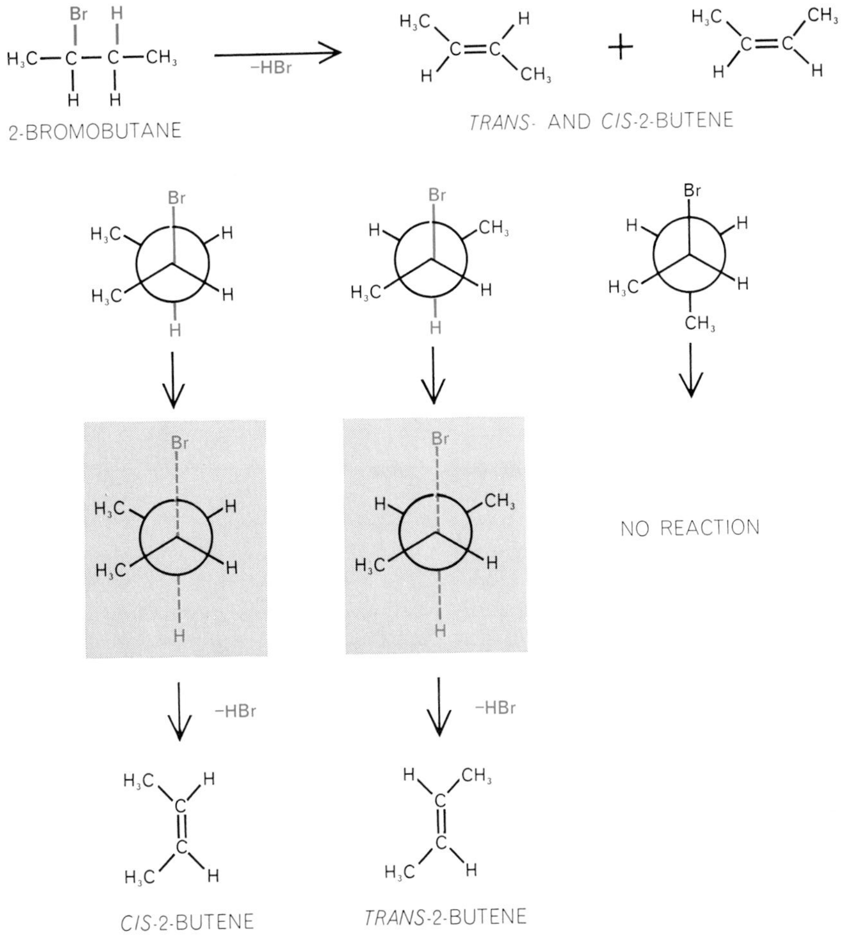

imply when a carbon chain is formed into a ring. The smallest and simplest ring is cyclopropane (C_3H_6), in which three carbon atoms form the corners of an equilateral triangle [see top diagram in illustration on next page]. The 60-degree angle between adjacent carbon atoms is far from the preferred angle of 109.5 degrees. For this reason the molecule, as Adolf von Baeyer was the first to perceive, is under considerable strain. Furthermore, when the hydrogen atoms are viewed down any carbon-carbon bond, they eclipse one another all around the ring. In spite of the strain in the cyclopropane ring, it is so rigid that no conformational isomers can exist.

The four-member analogue of this system, cyclobutane (C_4H_8), has a lesser amount of "Baeyer strain" between adjacent carbon atoms because they form 90-degree angles. Furthermore, the Pitzer, or eclipsing, strain can be relieved by deformations in the ring [see middle diagram in illustration on next page]. In the most stable arrangement opposite pairs of carbon atoms lie either above or below a plane containing the other pair. Between these stable extremes there is a continuum of unstable conformational isomers.

In the five-member system, cyclopentane (C_5H_{10}), the angle at each corner of the pentagon would be 108 degrees if the molecule were planar. Although almost no Baeyer strain would exist in the planar state, there would be strong eclipsing interactions among the subtended hydrogen atoms. If a single carbon atom, with its two hydrogens, is lifted out of the plane, only two pairs of hydrogen atoms remain eclipsed [see bottom diagram in illustration on next page]. This distortion travels around the ring like a wave, so that each carbon atom is out of the plane a fifth of the time, thereby equalizing the strain throughout the ring.

We come now to the important six-member system, cyclohexane (C_6H_{12}), which, as Sachse suspected long ago, is distinctly nonplanar. If the carbon atoms in cyclohexane were forced to lie in a plane, they would form a simple hexagon with bond angles of 120 degrees; the Baeyer strain would be considerable. There would also be strong eclipsing interactions among hydrogens on the ring. Sachse pointed out that strainless bond angles of 109.5 degrees would result if the carbon atoms were in alternate positions above and below the plane of the ring. Moreover, the hydrogen atoms would be staggered and thus would not eclipse one another [see top illustration

ACTIVE AND INERT CONFORMERS are illustrated by 2-bromobutane, a molecule of n-butane in which a bromine atom replaces a hydrogen atom on one of the interior carbon atoms. When 2-bromobutane reacts with a base such as potassium hydroxide, a molecule of hydrogen bromide (HBr) is removed, forming cis- and trans-2-butene. It turns out that for the reaction to take place the departing bromine and hydrogen must be opposite each other. One of three stable conformers, the one at the far right in the Newman projections, does not meet this requirement, so that it is unreactive. When the two methyls are gauche (diagram at left), the reaction yields cis-2-butene. When the methyls are anti (middle), the product is trans-2-butene. The structures in the colored panels represent transition states in which the bonds to the bromine atom and the opposite hydrogen have weakened, prior to breaking.

on page 297]. Such a chair conformation should be free of both Baeyer (angle) strain and Pitzer (eclipsing) strain.

Why did Sachse's plausible model meet with so much resistance from chemists? In the presumed planar form all the hydrogens in cyclohexane are chemically and physically indistinguishable. In Sachse's chair arrangement, however, there should be two different types of hydrogens. Half of the 12 hydrogen atoms should extend directly up and down from the average plane of the chair; the other half should lie more or less in the plane. The former hydrogens are termed axial, the latter equatorial. Each carbon atom should therefore be bonded to one axial hydrogen atom and one equatorial hydrogen atom. If another atom, such as chlorine or bromine, replaced a hydrogen atom, two isomers

should be possible: one with the replacement-atom axial, the other with it equatorial. Chemists were unable to find the two isomers.

The reason is now understood. Sachse himself described a second conformation of cyclohexane, the "boat conformation," that is free of angle strain but not of eclipsing strain. It is obtained conceptually by forcing the carbon atom that forms either the head or the foot of the chair to flip to the other side of the plane of the molecule. The chair can be re-created by performing the same operation on the carbon atom directly opposite to the first. When the original chair is inverted in this manner, all the hydrogen atoms that were·axial in the first conformation become equatorial, and all the equatorial hydrogen atoms become axial. At ordinary temperatures cyclohexane chairs invert so rapidly that one cannot replace a hydrogen atom with, say, chlorine and hope to isolate two different isomers. Nonetheless, a few years ago, by working at very low temperatures, Frederick R. Jensen and his co-workers at the University of California at Berkeley succeeded in separating isomers of monochlorocyclohexane in which the chlorine atoms were either axial or equatorial.

Making Isomers Visible

A striking confirmation of the existence and interconversion of the axial and equatorial positions in cyclohexane has been provided by nuclear magnetic resonance (NMR) spectroscopy. During the past decade NMR has emerged as the organic chemist's most valuable tool for conformational analysis. Theoretically each chemically distinct hydrogen atom resonates at a different natural frequency. The frequency of the signal varies with the structural environment of the hydrogen atom; the strength of the signal varies with the total number .

of hydrogen atoms of a given chemical type in the molecule. As the spectrum is usually recorded, peaks correspond to resonances. Thus methane produces an NMR spectrum with only one peak because the four hydrogen atoms are indistinguishable. Ethane and ethylene also produce spectra with one peak, but the spectrum of *cis*-2-butene has two peaks: one large peak for the six hydrogen atoms that belong to methyl groups and a peak one-third as large for the two hydrogens attached to the carbon atoms joined by a double bond.

One would expect the NMR spectrum of cyclohexane to exhibit two peaks of equal size, one for the six axial hydrogens and one for the six equatorial hydrogens. At room temperature, however, the spectrum shows only one large peak. This observation implies that the molecular chair is inverting so rapidly that the spectrometer can record only the average signal from the two kinds of hydrogen atom. As the temperature is lowered, however, the single peak broadens and finally splits into two distinct peaks, each half the height of the original peak

[*see illustration on opposite page*]. The two peaks correspond to the axial and equatorial hydrogen atoms in a ring that no longer appears to be inverting. When the NMR technique is used to look for the conformational isomers of cyclopentane and cyclobutane, one finds that the warping motions are so rapid that both compounds give only one sharp peak even at extremely low temperatures.

Although NMR spectroscopy clearly shows that at room temperature and above cyclohexane exists in a rapidly flipping chair conformation, neither NMR nor any other technique has demonstrated the presence of the boat conformation. Evidence for its being an intermediate during ring inversion is mainly presumptive. Because of eclipsing strain the boat conformers have more energy and therefore are less stable than the chair [*see bottom illustration on page 297*]. Some strain can be released if the eclipsed hydrogens are twisted away from one another. A slightly more stable form of cyclohexane known as the twist boat (or, in French terminology, the *croissant*) is thereby produced. The molecule twistane, for example, was synthesized specifically because it incorporates the structure of the twist boat, held in position by carbon-carbon bridges [*see illustration on page 298*]. This flexing motion that relieves eclipsing strain can be carried around the ring so that each carbon atom in turn serves as the bow or the stern of a boat. Thus when discussing cyclohexane conformers, it is more appropriate to speak of the flexible boat family—even of the conformational fleet—than of a single boat-shaped conformer, since so many nearly equivalent forms are interconverting.

When one or more hydrogen atoms of cyclohexane are replaced by a bulky substituent such as a methyl group, one can see more clearly how the occupation of an axial or equatorial position determines the preferred conformation of the molecule. If the methyl group is in the axial position, its hydrogen atoms are so close to nearby axial hydrogens that the hydrogens are unnaturally crowded [*see illustration on page 303*]. In the equatorial position the methyl group has plenty of room. As a result the axial isomer of methylcyclohexane is less stable than the equatorial isomer. At room temperature only about 5 percent of methylcyclohexane is present as the axial isomer. As one can imagine, the size of the substituent determines how much of the compound will be in the axial form or the equatorial one.

Let us now consider what happens when cyclohexane is modified so that

CYCLOPROPANE
C$_3$H$_6$

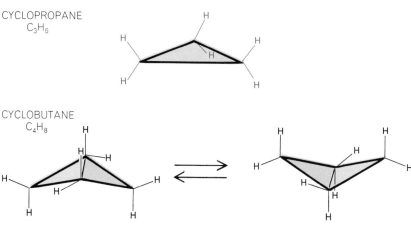

CYCLOBUTANE
C$_4$H$_8$

CYCLOPENTANE
C$_5$H$_{10}$

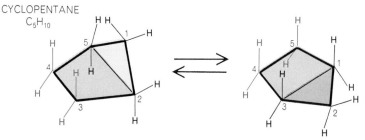

CYCLIC COMPOUNDS form a general class in which conformational isomerism is important. Cyclopropane, held rigid by geometrical constraints, is an exception. Its carbon-carbon bonds are under high Baeyer strain and the hydrogen atoms eclipse one another all around the ring. Eclipsed hydrogens are shown in color. In cyclobutane Baeyer strain is relieved by bending of the ring, which also serves to stagger all the hydrogen atoms. The "creases" in bent rings are shown by thin black lines; they are not bonds. The cyclopentane ring could be planar if Baeyer strain were the only consideration since the angles of a regular pentagon are 108 degrees, which is close to the preferred value of 109.5 degrees. But to relieve the eclipsing strain of hydrogen atoms one corner of the ring is pushed out of the plane so that only four hydrogens eclipse one another. If carbon atom No. 1 points upward at one instant, carbon atom No. 2 points downward the next and so on around the ring.

two hydrogen atoms on adjacent carbon atoms are replaced by two methyl groups; such a compound is 1,2-dimethylcyclohexane. Three distinct arrangements are possible: both methyls can be axial (*ax-ax*), both can be equatorial (*eq-eq*) or one can be in each position (*ax-eq*) [*see illustration on next page*]. Inasmuch as ring inversion exchanges the axial and equatorial positions, the *ax-ax* and *eq-eq* isomers must be interconvertible; similarly, the *ax-eq* isomer must interconvert with an equivalent *eq-ax* isomer.

The *eq-eq* and *ax-ax* isomers are commonly referred to as *trans* isomers because in each case one methyl group points down from the plane of the ring and the other points up. The *ax-eq* and *eq-ax* isomers are *cis* because both methyls point to the same side of the plane. The *ax-ax* form of the *trans* isomer must be almost nonexistent because two methyls in the axial position would be extremely crowded. The *cis* and *trans* forms cannot be converted into each other without breaking and re-forming a carbon-carbon bond; therefore they must be geometrical isomers.

The Multiplicity of Rings

Instead of adding two methyl groups to a cyclohexane ring one can add a complete second ring so that the two rings share a common side. The compound decalin, which is used as a paint and lacquer solvent, is one of the most important examples of a molecule with two rings "fused" in this fashion. From the point of view of each ring the other appears to be a 1,2-disubstituent [*see illustration on page 304*]. If the rings are fused in a *cis* fashion, one bond from each ring is axial and one is equatorial. Of two conceivable *trans* arrangements, *eq-eq* and *ax-ax*, only the former is structurally possible. Because axial-axial bonds point 180 degrees away from each other it is physically impossible to bridge them with only four carbon atoms (the four that complete the second ring). The rings of *cis*-decalin can invert simul-

EVIDENCE FOR CHAIR conformation in cyclohexane is supplied by nuclear magnetic resonance (NMR) spectroscopy, which supplies a separate signal for each chemically distinguishable hydrogen atom. The series of curves shows how the inversion rate of the cyclohexane ring slows down as the temperature of the molecule is lowered. Below about —60 degrees Celsius (*bottom curve*) the distinct axial and equatorial hydrogen atoms produce two very sharp peaks.

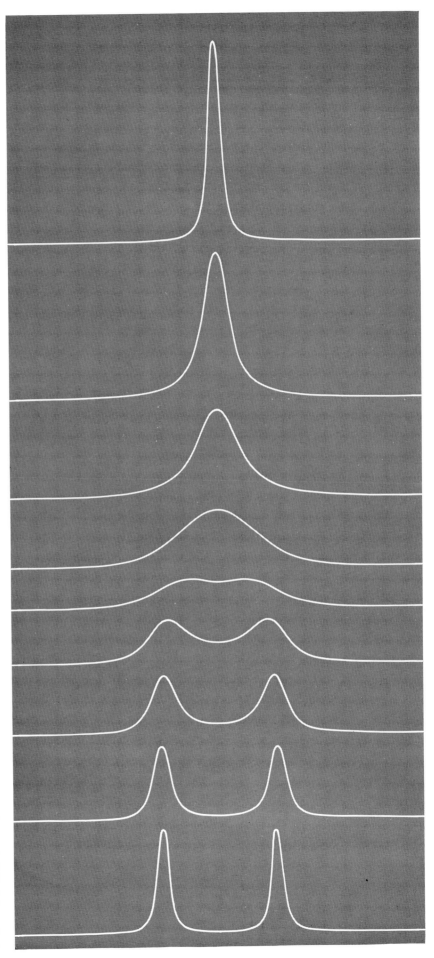

taneously to give an identical conformer in which the axial-equatorial roles of the substituent positions are reversed.

In principle it is possible to keep fusing ring on ring indefinitely. Anthracene, a well-known coal-tar derivative, consists of a sequence of three rings in a row. Because the rings in anthracene possess alternating double and single bonds, as in the benzene ring, the three rings are forced to remain in a plane. If, however, all the double bonds are hydrogenated, producing what is called a saturated molecule, the resulting three-ring product is perhydroanthracene, a molecule that has five principal conformational isomers. The most stable is the one in which all the fusion bonds are equatorial so that the structure is *trans-trans* [see illustration on page 305].

If the third ring is offset from the axis of the first two rings, in which case the

points of fusion are adjacent to the carbons that join the first and second ring, the skeleton of perhydrophenanthrene is produced. Again the all-equatorial, or *trans-trans*, arrangement of bonds is the most stable. If a cyclopentane ring is added to the offset ring of the perhydrophenanthrene skeleton, one obtains perhydrocyclopentanophenanthrene, the skeleton of a family of important biological substances: the steroids. The four rings are labeled *A, B, C* and *D*, with *A* representing the six-member ring most distant from the five-member ring, which is *D*. Although many geometries of ring fusion are possible in steroids, the linkage is invariably *trans* between the *B* and *C* rings and usually *trans* between the *C* and *D* rings. The steroids are such an important class of compounds that an enormous amount of study has been devoted to them in the past 50 years. They

include such natural products as cholesterol, the sex hormones, cortisone and the related adrenocortical hormones, and recently an entire family of synthetic birth-control substances.

Although rings of more than six members are common in organic chemistry, not much is known about their conformational characteristics. It is known, however, that the saturated seven-member rings resemble cyclopentane and the boat family of cyclohexane in their flexibility. The properties of larger rings will undoubtedly be closely examined during the next few years.

Some Modified Chairs

The cyclohexane chair is frequently modified when one of the carbon atoms in the ring is replaced by the atom of another element or when various substituents are added to the ring. In general one can expect the altered molecule to assume a spatial arrangement that will minimize steric interactions, that is, to adopt the "most comfortable" shape. A series of modified cyclohexane structures is illustrated on page 306.

When one of the carbon atoms is replaced by oxygen, the resulting structure, tetrahydropyran, is the basis for a large class of sugars. In the most common hexose sugar, β-D-glucose, four of the carbon atoms in the ring carry a hydroxyl (OH) group in place of one hydrogen and the fifth carbon carries a hydroxymethyl (CH_2OH) group. The chair is very little distorted, and all the substituents are equatorial. The abundance of β-D-glucose is undoubtedly due to its conformational stability. The shapes of sugar molecules were among the earliest subjects of conformational analysis; in fact, Norman Haworth first defined "conformation" as it is used today in his 1929 book on sugars.

Another "heteroatom" that creates little distortion in the cyclohexane ring is nitrogen. When a CH_2 unit is replaced by NH, the resulting structure is the piperidine ring, found in most alkaloids, a large class of polycyclic compounds that includes morphine, lysergic acid, strychnine and codeine. The piperidine chair inverts so rapidly that the H of the NH group flips continuously from equatorial to axial. At room temperature the axial and equatorial conformers are present in almost equal amounts.

When a sulfur atom rather than an oxygen or nitrogen atom is inserted in the cyclohexane ring, producing thiane, the chair is significantly distorted because the carbon-sulfur bond is appre-

METHYLCYCLOHEXANE

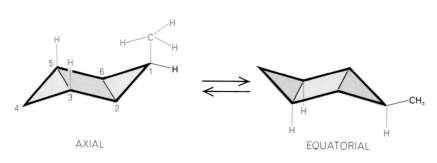

AXIAL   EQUATORIAL

1,2-DIMETHYLCYCLOHEXANE

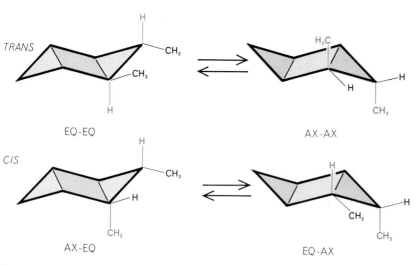

TRANS   *CIS*

EQ-EQ   AX-AX   AX-EQ   EQ-AX

INTERCONVERTING ISOMERS are produced when one or more methyl groups are substituted for hydrogen atoms in cyclohexane. In methylcyclohexane an axial methyl group on carbon No. 1 is crowded by axial hydrogens on carbon No. 3 and No. 5. Thus the conformer with the methyl in the equatorial position is favored. When there are two methyl groups on adjacent carbon atoms (1,2-dimethylcyclohexane), three arrangements are possible. The *trans* geometrical isomer consists of two conformational isomers, one with both methyls equatorial (*eq-eq*) and one with both axial (*ax-ax*). Only one arrangement is possible for the *cis* isomer: *ax-eq*. Ring inversion yields an identical substitution pattern (*eq-ax*).

ciably longer than the carbon-carbon or carbon-nitrogen bond. Moreover, the normal C-S-C angle is close to 90 degrees, rather than 109.5 degrees. As a result the thiane ring is puckered, so that the molecule resembles a beach chair that has been raised from a reclining position to a more upright one.

The reverse distortion, a flattening, takes place when one of the carbon atoms in the cyclohexane ring is doubly bonded to an oxygen atom, forming cyclohexanone. If another oxygen atom is added to the carbon atom at the opposite side of the ring, producing 1,4-cyclohexanedione, one might expect still further flattening to occur. Instead the

molecule flips into a twist-boat structure, probably to relieve eclipsing strain that would be present if the molecule were extremely flattened. The half-chair conformation, which corresponds to the state of highest energy in the inversion of cyclohexane, can be achieved as a preferred molecular conformation by removing two hydrogen atoms from adjacent carbon atoms in cyclohexane and joining the carbons by a double bond; the resulting molecule is cyclohexene.

The Conformation of Proteins

Although steroids, alkaloids and sugars are not small molecules, the proteins

that, in the form of enzymes, catalyze the most important processes in the living cell are many times larger. Proteins are long chains of the 20-odd different amino acid molecules. The conformation of protein chains depends on complex interactions among the amino acids and on their specific sequence. In many cases the chain first develops a helical structure, and the helix as a whole folds into a conformation that probably minimizes the steric and electrostatic interactions among all the various groups. The final conformation is so important to the molecule's biological function that any alteration in shape may reduce its activity or destroy it altogether. The complete

DECALIN

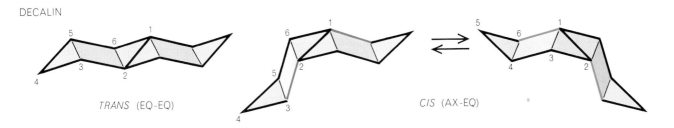

TRANS (EQ-EQ)   *CIS* (AX-EQ)

PERHYDROANTHRACENE

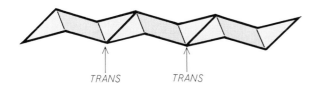

TRANS   *TRANS*

PERHYDROPHENANTHRENE

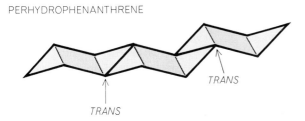

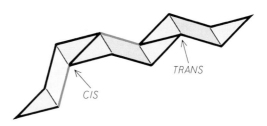

TRANS   *TRANS*   *TRANS*   *CIS*

PERHYDROCYCLOPENTANOPHENANTHRENE (STEROID SKELETON)

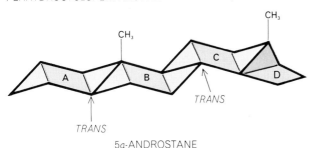

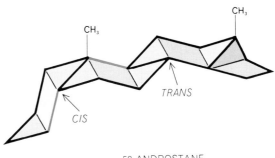

5α-ANDROSTANE   5β-ANDROSTANE

FUSION OF CYCLOHEXANE RINGS produces complex structures in which the geometry of ring fusion can be either *trans* (*eq-eq*) or *cis* (*ax-eq*). An *ax-ax trans* conformer cannot be built with six-member rings. Axial bonds are shown in color. Perhydroanthracene has four conformers, including one boat, besides the one shown. Perhydrophenanthrene also has four other conformers.

DISTORTION	COMPOUND	SHAPE
1 NONE	TETRAHYDROPYRAN	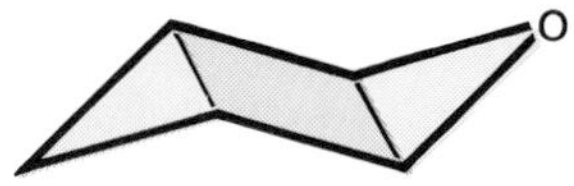
2 NONE	PIPERIDINE	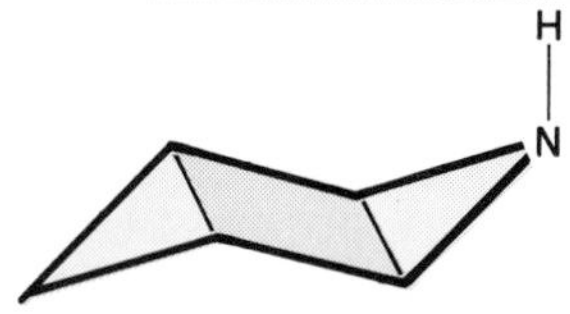
3 PUCKERING	THIANE	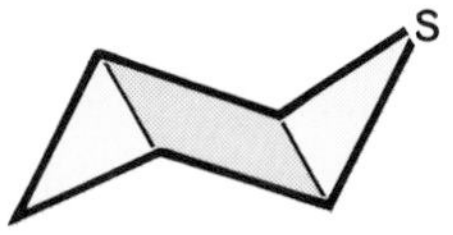
4 FLATTENING	CYCLOHEXANONE	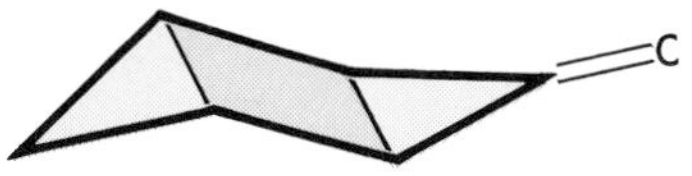
5 TWIST BOAT	1,4-CYCLOHEXANEDIONE	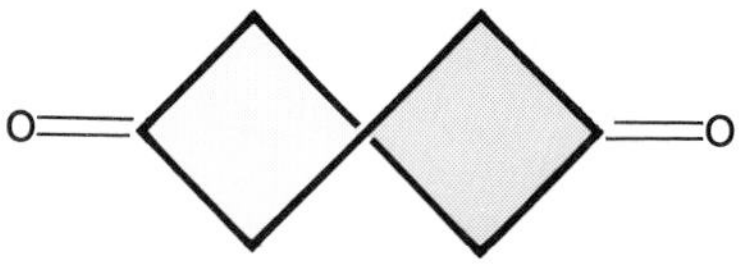
6 HALF-CHAIR	CYCLOHEXENE	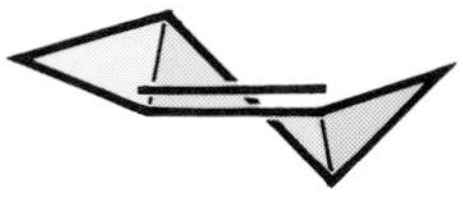

DISTORTED CHAIRS are often produced when the basic structure of cyclohexane is modified by substituent atoms. The chair may be puckered (3) or flattened (4). Addition of two external oxygen atoms produces a twist boat (5). Removal of two hydrogens creates a double bond, producing a half-chair (6). There is good evidence for all these conformations.

three-dimensional structures of about a dozen proteins have now been worked out by the X-ray-diffraction analysis of proteins in crystalline form. One of the proteins whose structure is known—the enzyme lysozyme—destroys the cell wall of bacteria by cleaving long polysaccharide chains, which consist of many sugar units tied together. For the enzyme to operate on the sugar linkages, one of its tetrahydropyran rings must first flip from its normal chair conformation to a much flattened structure [see "The Three-dimensional Structure of an Enzyme Molecule," by David C. Phillips, beginning on page 360].

Flavor chemists of the Monsanto Company have recently proposed a conformational dependence between a particular protein and substances that taste sweet. When the protein combines with "sweet" molecules, its conformation changes; conceivably the change is relayed to the brain as a signal of sweetness. The intensity of the signal may be related to the amount of change.

Presumably DNA, the double-strand helical molecule that embodies the genetic message in living cells, has a structure susceptible to conformational change. One of the greatest mysteries in genetic chemistry is how the DNA helix unwinds so that each strand can act as a template for the construction of a complementary strand of DNA (or for the construction of a strand of messenger RNA, the molecule that directs the synthesis of protein molecules). Rotation around carbon-carbon, carbon-nitrogen and carbon-oxygen bonds must be involved in this elaborate and precise process. A present goal of conformational analysis is a bond-by-bond description of the rotational processes that accompany conformational changes in biochemical systems such as proteins and nucleic acids. Much of the current work of this type is based on X-ray-diffraction studies of crystalline solids. One of the most pressing needs today is for new methods for obtaining structural information about molecules in liquids, the state of matter in which most chemical and biochemical reactions take place.

27

THIRD-GENERATION PESTICIDES

CARROLL M. WILLIAMS

July 1967

*The first generation is exemplified by arsenate of lead;
the second, by DDT. Now insect hormones promise to
provide insecticides that are not only more specific but
also proof against the evolution of resistance*

Man's efforts to control harmful insects with pesticides have encountered two intractable difficulties. The first is that the pesticides developed up to now have been too broad in their effect. They have been toxic not only to the pests at which they were aimed but also to other insects. Moreover, by persisting in the environment—and sometimes even increasing in concentration as they are passed along the food chain—they have presented a hazard to other organisms, including man. The second difficulty is that insects have shown a remarkable ability to develop resistance to pesticides.

Plainly the ideal approach would be to find agents that are highly specific in their effect, attacking only insects that are regarded as pests, and that remain effective because the insects cannot acquire resistance to them. Recent findings indicate that the possibility of achieving success along these lines is much more likely than it seemed a few years ago. The central idea embodied in these findings is that a harmful species of insect can be attacked with its own hormones.

Insects, according to the latest estimates, comprise about three million species—far more than all other animal and plant species combined. The number of individual insects alive at any one time is thought to be about a billion billion (10^{18}). Of this vast multitude 99.9 percent are from the human point of view either innocuous or downright helpful. A few are indispensable; one need think only of the role of bees in pollination.

The troublemakers are the other .1 percent, amounting to about 3,000 species. They are the agricultural pests and the vectors of human and animal disease. Those that transmit human disease are the most troublesome; they have joined with the bacteria, viruses and protozoa in what has sometimes seemed like a grand conspiracy to exterminate man, or at least to keep him in a state of perpetual ill health.

The fact that the human species is still here is an abiding mystery. Presumably the answer lies in changes in the genetic makeup of man. The example of sickle-cell anemia is instructive. The presence of sickle-shaped red blood cells in a person's blood can give rise to a serious form of anemia, but it also confers resistance to malaria. The sickle-cell trait (which does not necessarily lead to sickle-cell anemia) is appreciably more common in Negroes than in members of other populations. Investigations have suggested that the sickle cell is a genetic mutation that occurred long ago in malarial regions of Africa. Apparently attrition by malaria-carrying mosquitoes provoked countermeasures deep within the genes of primitive men.

The evolution of a genetic defense, however, takes many generations and entails many deaths. It was only in comparatively recent times that man found an alternative answer by learning to combat the insects with chemistry. He did so by inventing what can be called the first-generation pesticides: kerosene to coat the ponds, arsenate of lead to poison the pests that chew, nicotine and rotenone for the pests that suck.

Only 25 years ago did man devise the far more potent weapon that was the first of the second-generation pesticides. The weapon was dichlorodiphenyltrichloroethane, or DDT. It descended on the noxious insects like an avenging angel. On contact with it mosquitoes, flies, beetles—almost all the insects—were stricken with what might be called the "DDT's." They went into a tailspin, buzzed around upside down for an hour or so and then dropped dead.

The age-old battle with the insects appeared to have been won. We had the stuff to do them in—or so we thought. A few wise men warned that we were living in a fool's paradise and that the insects would soon become resistant to DDT, just as the bacteria had managed to develop a resistance to the challenge of sulfanilamide. That is just what happened. Within a few years the mosquitoes, lice, houseflies and other noxious insects were taking DDT in their stride. Soon they were metabolizing it, then they became addicted to it and were therefore in a position to try harder.

Fortunately the breach was plugged by the chemical industry, which had come to realize that killing insects was —in more ways than one—a formula for

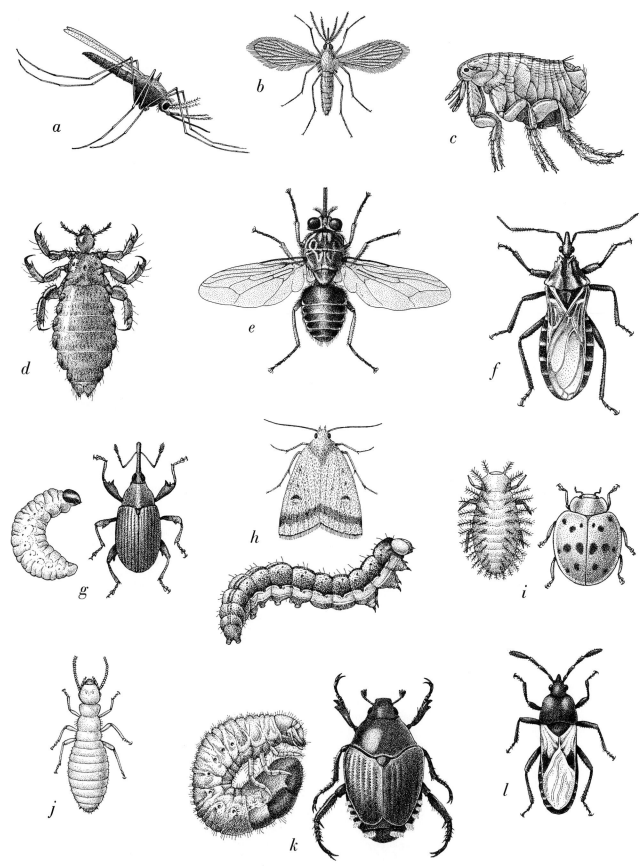

INSECT PESTS that might be controlled by third-generation pesticides include some 3,000 species, of which 12 important examples are shown here. Six (a–f) transmit diseases to human beings; the other six are agricultural pests. The disease-carriers, together with the major disease each transmits, are (a) the *Anopheles* mosquito, malaria; (b) the sand fly, leishmaniasis; (c) the rat flea, plague; (d) the body louse, typhus; (e) the tsetse fly, sleeping sickness, and (f) the kissing bug, Chagas' disease. The agricultural pests, four of which are depicted in both larval and adult form, are (g) the boll weevil; (h) the corn earworm; (i) the Mexican bean beetle; (j) the termite; (k) the Japanese beetle, and (l) the chinch bug. The species in the illustration are not drawn to the same scale.

getting along in the world. Organic chemists began a race with the insects. In most cases it was not a very long race, because the insects soon evolved an insensitivity to whatever the chemists had produced. The chemists, redoubling their efforts, synthesized a steady stream of second-generation pesticides. By 1966 the sales of such pesticides had risen to a level of $500 million a year in the U.S. alone.

Coincident with the steady rise in the output of pesticides has come a growing realization that their blunderbuss toxicity can be dangerous. The problem has attracted widespread public attention since the late Rachel Carson fervently described in *The Silent Spring* some actual and potential consequences of this toxicity. Although the attention thus aroused has resulted in a few attempts to exercise care in the application of pesticides, the problem cannot really be solved with the substances now in use.

The rapid evolution of resistance to pesticides is perhaps more critical. For example, the world's most serious disease in terms of the number of people afflicted continues to be malaria, which is transmitted by the *Anopheles* mosquito—an insect that has become completely resistant to DDT. (Meanwhile the protozoon that actually causes the disease is itself evolving strains resistant to antimalaria drugs.)

A second instance has been presented recently in Vietnam by an outbreak of plague, the dreaded disease that is conveyed from rat to man by fleas. In this case the fleas have become resistant to pesticides. Other resistant insects that are agricultural pests continue to take a heavy toll of the world's dwindling food supply from the moment the seed is planted until long after the crop is harvested. Here again we are confronted by an emergency situation that the old technology can scarcely handle.

The new approach that promises a way out of these difficulties has emerged during the past decade from basic studies of insect physiology. The prime candidate for developing third-generation pesticides is the juvenile hormone that all insects secrete at certain stages in their lives. It is one of the three internal secretions used by insects to regulate growth and metamorphosis from larva to pupa to adult. In the living insect the juvenile hormone is synthesized by the corpora allata, two tiny glands in the head. The corpora allata are also responsible for regulating the flow of the hormone into the blood.

At certain stages the hormone must be

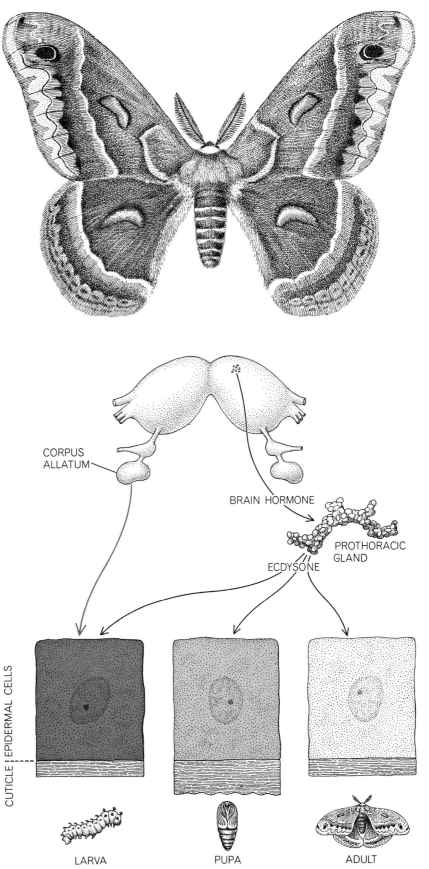

HORMONAL ACTIVITY in a Cecropia moth is outlined. Juvenile hormone (*color*) comes from the corpora allata, two small glands in the head; a second substance, brain hormone, stimulates the prothoracic glands to secrete ecdysone, which initiates the molts through which a larva passes. Juvenile hormone controls the larval forms and at later stages must be in low concentration or absent; if applied then, it deranges insect's normal development. The illustration is partly based on one by Howard A. Schneiderman and Lawrence I. Gilbert.

CHEMICAL STRUCTURES of the Cecropia juvenile hormone (*left*), isolated this year by Herbert Röller and his colleagues at the University of Wisconsin, and of a synthetic analogue (*right*) made in 1965 by W. S. Bowers and others in the U.S. Department of Agriculture show close similarity. Carbon atoms, joined to one or two hydrogen atoms, occupy each angle in the backbone of the molecules; letters show the structure at terminals and branches.

JUVENILE HORMONE ACTIVITY has been found in various substances not secreted by insects. One (*left*) is a material synthesized by M. Romanuk and his associates in Czechoslovakia. The other (*right*), isolated and identified by Bowers and his colleagues, is the "paper factor" found in the balsam fir. The paper factor has a strong juvenile hormone effect on only one family of insects, exemplified by the European bug *Pyrrhocoris apterus*.

secreted; at certain other stages it must be absent or the insect will develop abnormally [*see illustration on preceding page*]. For example, an immature larva has an absolute requirement for juvenile hormone if it is to progress through the usual larval stages. Then, in order for a mature larva to metamorphose into a sexually mature adult, the flow of hormone must stop. Still later, after the adult is fully formed, juvenile hormone must again be secreted.

The role of juvenile hormone in larval development has been established for several years. Recent studies at Harvard University by Lynn M. Riddiford and the Czechoslovakian biologist Karel Sláma have resulted in a surprising additional finding. It is that juvenile hormone must be absent from insect eggs for the eggs to undergo normal embryonic development.

The periods when the hormone must be absent are the Achilles' heel of insects. If the eggs or the insects come into contact with the hormone at these times, the hormone readily enters them and provokes a lethal derangement of further development. The result is that the eggs fail to hatch or the immature insects die without reproducing.

Juvenile hormone is an insect invention that, according to present knowledge, has no effect on other forms of life. Therefore the promise is that third-generation pesticides can zero in on in-

sects to the exclusion of other plants and animals. (Even for the insects juvenile hormone is not a toxic material in the usual sense of the word. Instead of killing, it derails the normal mechanisms of development and causes the insects to kill themselves.) A further advantage is self-evident: insects will not find it easy to evolve a resistance or an insensitivity to their own hormone without automatically committing suicide.

The potentialities of juvenile hormone as an insecticide were recognized 12 years ago in experiments performed on the first active preparation of the hormone: a golden oil extracted with ether from male Cecropia moths. Strange to say, the male Cecropia and the male of its close relative the Cynthia moth remain to this day the only insects from which one can extract the hormone. Therefore tens of thousands of the moths have been required for the experimental work with juvenile hormone; the need has been met by a small but thriving industry that rears the silkworms.

No one expected Cecropia moths to supply the tons of hormone that would be required for use as an insecticide. Obviously the hormone would have to be synthesized. That could not be done, however, until the hormone had been isolated from the golden oil and identified.

Within the past few months the difficult goals of isolating and identifying the hormone have at last been attained by a team of workers headed by Herbert Röller of the University of Wisconsin. The juvenile hormone has the empirical formula $C_{18}H_{36}O_2$, corresponding to a molecular weight of 284. It proves to be the methyl ester of the epoxide of a previously unknown fatty-acid derivative [*see upper illustration on this page*]. The apparent simplicity of the molecule is deceptive. It has two double bonds and an oxirane ring (the small triangle at lower left in the molecular diagram), and it can exist in 13 different molecular configurations. Only one of these can be the authentic hormone. With two ethyl groups ($CH_2 \cdot CH_3$) attached to carbons No. 7 and 11, the synthesis of the hormone from any known terpenoid is impossible.

The pure hormone is extraordinarily active. Tests the Wisconsin investigators have carried out with mealworms suggest that one gram of the hormone would result in the death of about a billion of these insects.

A few years before Röller and his colleagues worked out the structure of the authentic hormone, investigators at several laboratories had synthesized a number of substances with impressive juvenile hormone activity. The most potent of the materials appears to be a crude mixture that John H. Law, now at the University of Chicago, prepared by a simple one-step process in which hydrogen chloride gas was bubbled through an alcoholic solution of farnesenic acid. Without any purification this mixture was 1,000 times more active than crude Cecropia oil and fully effective in killing all kinds of insects.

One of the six active components of Law's mixture has recently been identified and synthesized by a group of workers headed by M. Romaňuk of the Czechoslovak Academy of Sciences. Romaňuk and his associates estimate that from 10 to 100 grams of the material would clear all the insects from 2½ acres. Law's original mixture is of course even more potent, and so there is much interest in its other five components.

Another interesting development that preceded the isolation and identification of true juvenile hormone involved a team of investigators under W. S. Bowers of the U.S. Department of Agriculture's laboratory at Beltsville, Md. Bowers and his colleagues prepared an analogue of juvenile hormone that, as can be seen in the accompanying illustration [*top of opposite page*], differed by only two carbon atoms from the authentic Cecropia hormone (whose structure was then, of course, unknown). In terms of the dosage required it appears that the Beltsville compound is about 2 percent as active as Law's mixture and about .02 percent as active as the pure Cecropia hormone.

All the materials I have mentioned are selective in the sense of killing only insects. They leave unsolved, however, the problem of discriminating between the .1 percent of insects that qualify as pests and the 99.9 percent that are helpful or innocuous. Therefore any reckless use of the materials on a large scale could constitute an ecological disaster of the first rank.

The real need is for third-generation pesticides that are tailor-made to attack only certain predetermined pests. Can such pesticides be devised? Recent work that Sláma and I have carried out at Harvard suggests that this objective is by no means unattainable. The possibility arose rather fortuitously after Sláma arrived from Czechoslovakia, bringing with him some specimens of the European bug *Pyrrhocoris apterus*—a species that had been reared in his laboratory in Prague for 10 years.

To our considerable mystification the bugs invariably died without reaching sexual maturity when we attempted to rear them at Harvard. Instead of metamorphosing into normal adults they continued to grow as larvae or molted into adult-like forms retaining many larval characteristics. It was evident that the bugs had access to some unknown source of juvenile hormone.

Eventually we traced the source to the paper toweling that had been placed in the rearing jars. Then we discovered that almost any paper of American origin—including the paper on which *Scientific American* is printed—had the same effect. Paper of European or Japanese manufacture had no effect on the bugs. On further investigation we found that the juvenile hormone activity originated in the balsam fir, which is the principal source of pulp for paper in Canada and the northern U.S. The tree synthesizes what we named the "paper factor," and this substance accompanies the pulp all the way to the printed page.

Thanks again to Bowers and his associates at Beltsville, the active material of the paper factor has been isolated and characterized [*see lower illustration on opposite page*]. It proves to be the methyl ester of a certain unsaturated fatty-acid derivative. The factor's kinship with the other juvenile hormone analogues is evident from the illustrations.

Here, then, is an extractable juvenile hormone analogue with selective action against only one kind of insect. As it happens, the family Pyrrhocoridae includes some of the most destructive pests of the cotton plant. Why the balsam fir should have evolved a substance against only one family of insects is unexplained. The most intriguing possibility is that the paper factor is a biochemical memento of the juvenile hormone of a former natural enemy of the tree—a pyrrhocorid predator that, for obvious reasons, is either extinct or has learned to avoid the balsam fir.

In any event, the fact that the tree synthesizes the substance argues strongly that the juvenile hormone of other species of insects can be mimicked, and perhaps has been by trees or plants on which the insects preyed. Evidently during the 250 million years of insect evolution the detailed chemistry of juvenile hormone has evolved and diversified. The process would of necessity have gone hand in hand with a retuning of the hormonal receptor mechanisms in the cells and tissues of the insect, so that the use as pesticides of any analogues that are discovered seems certain to be effective.

The evergreen trees are an ancient lot. They were here before the insects; they are pollinated by the wind and thus, unlike many other plants, do not depend on the insects for anything. The paper factor is only one of thousands of terpenoid materials these trees synthesize for no apparent reason. What about the rest?

It seems altogether likely that many of these materials will also turn out to be analogues of the juvenile hormones of specific insect pests. Obviously this is the place to look for a whole battery of third-generation pesticides. Then man may be able to emulate the evergreen trees in their incredibly sophisticated self-defense against the insects.

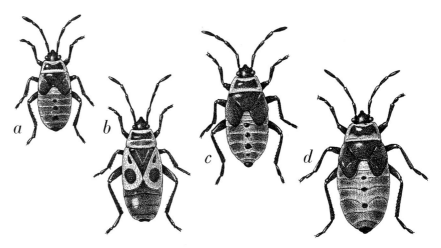

EFFECT OF PAPER FACTOR on *Pyrrhocoris apterus* is depicted. A larva of the fifth and normally final stage (*a*) turns into a winged adult (*b*). Contact with the paper factor causes the insect to turn into a sixth-stage larva (*c*) and sometimes into a giant seventh-stage larva (*d*). The abnormal larvae usually cannot shed their skin and die before reaching maturity.

VIII

THE CHEMISTRY OF LIFE

This reader began with discussions by Pauling and Wald of the historical perspectives of chemistry and biology and predictions about the future of their respective disciplines. We have treated most of the principles of general chemistry in Sections II through VII. It is therefore appropriate for us to close this book with a section on the applications of chemistry in biology. This field is known by a variety of terms. Molecular biology and biophysical chemistry are terms frequently applied to the discipline discussed in this section.

Wald introduced the four classes of biopolymers. These are large compounds which were originally derived from biological sources. Although scientists have been able to synthesize most of these compounds, we still use the term *bio*polymers because living cells continue to manufacture the bulk of these compounds and do so with comparative ease.

The polysaccharides were mentioned by Wald, who illustrated one at the bottom of page 12. In addition, some of the properties of polysaccharides will be explored in Phillips's article, 31. These are large molecules composed of repeating units of sugars, monosaccharides. Glucose, one of these units, designated as monomers, is the most abundant organic carbon compound on Earth. Two important polymers are comprised exclusively of glucose. Cellulose, the structural element of most plants, and starches, a major energy source for animals, are polyglucose compounds. They differ in a subtle way as to how the glucose rings are joined together. This subtle difference has profound consequences for a person lost in the woods without food, who must improvise a diet by foraging. Because the polysaccharides are not as interesting structurally as the proteins and nucleic acids, they have received relatively little study up until now. However, recent findings indicate that carbohydrates are important in the stimulation of an antibody response and that some carbohydrates do contain structural information which is recognized by other cellular components and which provides specificity to cellular biochemistry. It appears that more attention will be directed to the carbohydrates in the near future than has been in the past.

The second class of biopolymers mentioned by Wald are the fats. Fats belong to the class of compounds known as lipids. We have had an extensive introduction to the lipids in article 8, by Fox. We saw the great excitement that the technique of freeze-etching in the production of electron micrographs has brought to the study of cell membranes. These membranes, composed of lipids, phosphates, and proteins, are important in the transmission of nerve impulses.

Neither of these classes of compounds is receiving as much attention from scientists at present as the two classes Wald showed at the bottom of pages 14 and 15. The proteins and nucleic acids possess a great deal of structural variability and information content. We will devote this entire section to a discussion of just these two classes of biopolymers. This would appear to be justified in terms not only of the current research and of our knowledge of their interrelations, but also the incredible variety of functions which each class of compounds performs that are necessary for the sustenance of life.

PROTEINS

Life as we know it today on Earth would be impossible without proteins. Enzymes, which make metabolic processes proceed in a reasonable time span, are proteins. Large organisms such as horses and humans could not have evolved without the oxygen transfer and storage system of the two proteins hemoglobin and myoglobin. The highly efficient energy-yielding process of respiration depends on the cytochromes, which are proteins, for electron transfer. The buffering of blood depends in part on proteins such as the albumins that blood contains. As Fox discussed in article 8, proteins are involved in as yet unknown ways in the functioning of membranes. Collagen is a protein which comprises the bulk of the structural and motile material of muscle. Hair and nail are made of the protein keratin. The all-important immune defense system depends upon the proteins called immunoglobulins.

The first article, by Doty, is one of the best introductions to protein chemistry available to a student. It assumes no prior knowledge in the field and the level of the discussion seldom becomes too sophisticated for the beginning chemistry student.

Proteins are made up of amino acids. This name implies that they contain an amine group, $-NH_2$, and an acid group, $-COOH$. Indeed all amino acids share the common structure:

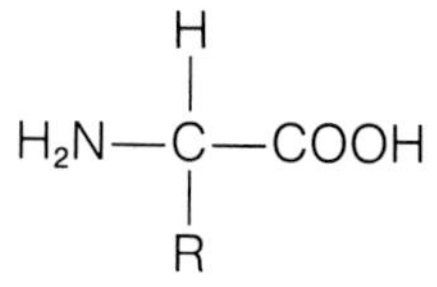

$$H_2N-\underset{\displaystyle R}{\overset{\displaystyle H}{C}}-COOH$$

There are twenty different amino acids commonly found in proteins. They differ only in the nature of the R group attached to the central or α-carbon atom. Doty gives the side-chain structures for twenty amino acids. The structures of two common amino acids not on this list are given below:

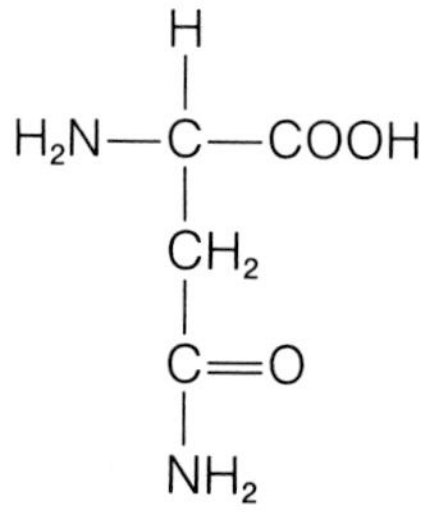

$$H_2N-\overset{\displaystyle H}{\underset{\displaystyle CH_2}{C}}-COOH$$
$$\underset{\displaystyle NH_2}{\overset{\displaystyle CH_2}{C=O}}$$

Asparagine

$$H_2N-\overset{\displaystyle H}{\underset{\displaystyle CH_2}{C}}-COOH$$
$$CH_2$$
$$\underset{\displaystyle NH_2}{C=O}$$

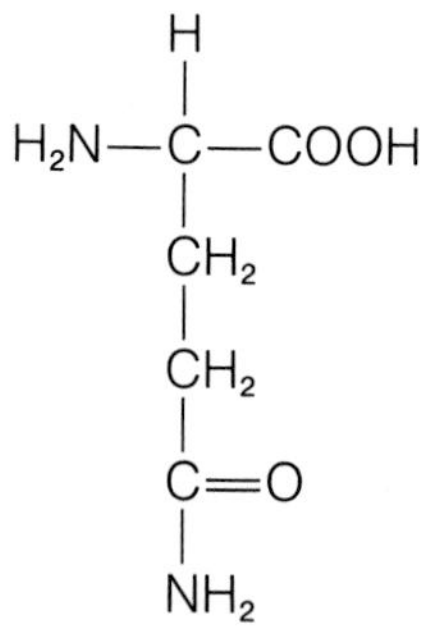

Glutamine

These compounds are clearly derivatives of aspartic acid and glutamic acid.

It is an important and perhaps surprising fact that evolution has determined that all organisms contain only L-amino acids in their proteins. As discussed by Pauling in article 1 and in the introduction to Section VII, the absolute configuration that is at the core of the amino acids in proteins corresponds to the structure:

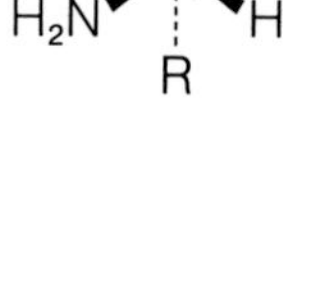

$$\underset{\displaystyle R}{H_2N \diagdown \overset{\displaystyle COOH}{\underset{}{C}} \diagup H}$$

In this depiction, as in the depictions of the optical isomers in the introduction to Section VII, wedge shapes show bonds directed toward the viewer; dashed bonds project away.

The amino acids are linked together to form larger molecules called peptides by means of the most common reaction found in the synthesis of biopolymers: dehydration condensation. This means that when two molecules are condensed together, a molecule of water is lost. An example of a dehydration condensation is

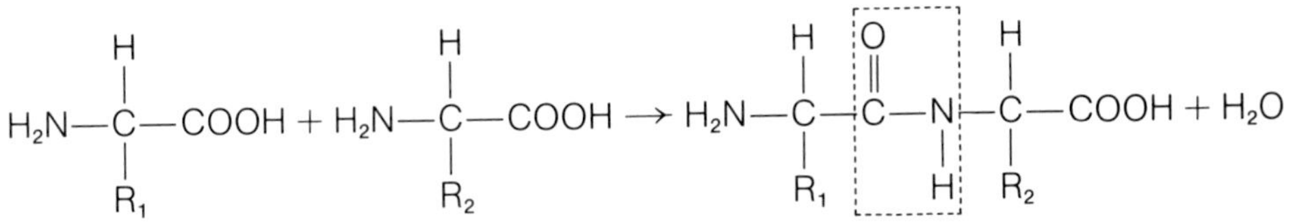

$$
\underset{R_1}{\overset{H}{H_2N-C-COOH}} + \underset{R_2}{\overset{H}{H_2N-C-COOH}} \rightarrow \underset{R_1}{\overset{H}{H_2N-C}}\underset{H}{\overset{O}{-C-N-}}\underset{R_2}{\overset{H}{C-COOH}} + H_2O
$$

The $-\overset{O}{\underset{H}{C-N}}-$ bond is called the peptide bond. It is a planar group of atoms. This fact is of importance in the secondary structure of proteins. The interatomic distances in the peptide group are known to $\pm 0.01\text{Å}$ and the bond angles to $\pm 1°$. The $-S-S-$ bond formed between two cysteine residues is the only type of covalent bond found between amino acids in proteins other than the peptide bond. The structure

$$
\underset{R}{\overset{H \quad H \quad O}{-N-C-C-}}
$$

corresponds to the amino acid minus a water molecule and hence is called a residue.

The molecule created in the above reaction is called a dipeptide. It is important to observe that the molecule has an $-NH_2$ group at one end and a $-COOH$ at the other. Thus the structure can be enlarged by the addition of more amino acids. A large number of amino acids joined by peptide bonds is called a polypeptide. One or more chains of polypeptides are called a protein. The sequence of R groups along the $-N-C-C-N-C-C-$ backbone is called the primary structure. At the time Doty wrote his article, the primary structure of only one protein, the pancreatic hormone insulin, was known. Today hundreds of polypeptide sequences are known.

Several higher structural levels are used by protein chemists to describe the arrangement of atoms in a protein. The secondary structure describes the regular, repeated arrangement of the backbone in space. The two principal configurations displayed by proteins were correctly described in 1951 through the brilliant work of Pauling and Corey at the California Institute of Technology. Their work was based upon very careful X-ray crystallographic studies of the properties of oligopeptides. These structures are shown on pages 328 and 329 in Doty's article. The α-helix is a right-handed helix held together by hydrogen bonds parallel to the axis of the helix. They form between the $-\overset{H}{N}-$ group in one peptide bond and the $-\overset{O}{C}-$ of the third peptide bond further along the helix. There are 3.6 residues/turn with a rise/turn of 5.4 Å. Note that the R groups project outward from the core of the helix. A space-filling model of this helix shows that the core of the helix is solid. There is not even space for a water molecule, as might be implied by Doty's ball-and-stick diagram.

The diagram below the α-helix is the β-pleated sheet. In this structure, polypeptide chains are held together side-by-side by $\diagup C{=}O \cdots H{-}N\diagdown$ hydrogen bonds which are perpendicular to the length of the chain. A sheet of chains results. It has a zig-zag or crinkled appearance ($\diagdown\diagup\diagdown\diagup\diagdown\diagup$) when viewed from the end and hence is called the pleated sheet configuration. The R groups alternately project above and below the plane of the sheet. The two chains shown by Doty are parallel. This means that the direction of the two sequences in the chains is the same:

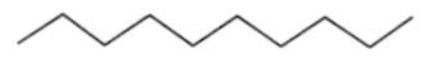

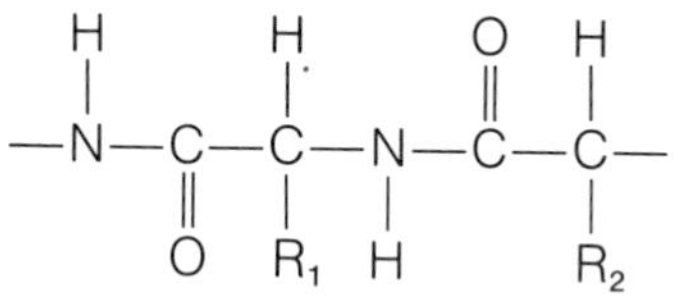

$$
\begin{array}{ccccccc}
 & \text{H} & \text{H} & & \text{O} & \text{H} & \\
 & | & | & & \| & | & \\
-\text{N}-\text{C}-\text{C}-\text{N}-\text{C}-\text{C}- \\
 & \| & | & | & & | & \\
 & \text{O} & \text{R}_1 & \text{H} & & \text{R}_2 &
\end{array}
$$

Most proteins containing β-pleated sheet structures, such as silk, have the antiparallel structure. Adjacent chains run in opposite directions.

Most globular proteins contain some sections of α-helix and β-pleated sheet. The color drawings in article 31, by Phillips, display short segments of these two configurations. You will need to understand Doty's figures thoroughly in order to find the α and β segments in the rather complicated lysozyme structure. Doty used optical rotatory dispersion to study the percentage of helical configuration of proteins in solution. The X-ray structures obtained in recent years provide more accurate and direct data.

The triple helical structure of the collagen that makes up structural tissue is also discussed by Doty. An excellent extension of this discussion can be found in the 1969 *Scientific American* article by Fraser entitled "Keratins" (Offprint 1155).

Protein chemistry entered a very exciting phase following the predictions of the secondary structures of proteins by Corey and Pauling in 1951. This work progressed along several lines. One of these was to find ways in which proteins could be synthesized. Apart from the intrinsic interest to the scientist in accomplishing this task, there was the possibility that synthetic proteins, such as insulin, could be made to serve a medical need, just as we saw earlier that synthetic pheromones could meet an agricultural need.

Between 1963 and 1965, three research groups announced the synthesis of insulin. This protein consists of two chains joined by two disulfide bridges. The A chain contains 21 amino acid residues and the B chain contains 30 residues. As mentioned above, both the composition and sequence of the chains had been known for some time. Because the chains are not particularly long, classical methods of chemical synthesis and purification were employed successfully. Small segments of the chains were synthesized and then linked in the proper order. The problem with this approach is the immensity of the labor that is required for the work and, consequently, the difficulty of obtaining a sufficient amount of product of high purity.

Merrifield was the leader in the development of a new technique to overcome many of the problems of the classical methods of protein synthesis. This technique is called solid phase peptide synthesis, because the synthesis occurs on the surface and in the interior of beads of polystyrene. These beads are about 50 microns (50×10^{-6} meter) in diameter. Reactive groups, $-CH_2Cl$, are attached to the benzene rings of the polystyrene beads to provide the sites where chain growth occurs. Next, the amino group of the first amino acid (at the $-COOH$ end of the peptide chain) is blocked with the compound *tert*-butyloxycarbonyl (this compound is known also simply as Boc),

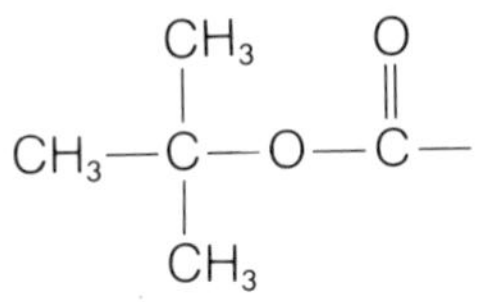

$$\begin{array}{c} \quad\ \ CH_3 \quad\ \ O \\ \quad\ \ | \qquad\ \ || \\ CH_3-C-O-C- \\ \quad\ \ | \\ \quad\ \ CH_3 \end{array}$$

Note that this is a tertiary group, as discussed in the introduction to the preceding section. This Boc-protected amino acid is then bound to the resin at the —CH_2Cl site. The Boc group is then removed. A second blocked amino acid is added and a peptide bond formed between the two amino acids in the presence of a condensing or dehydrating agent. And so the chain is built, one unit at a time, from the C-terminal end. It should be noted that the incoming —COOH group must be activated in order for it to react with the —NH_2 group of the preceding amino acid, and that reactive groups in side chains, such as the —COOH of aspartic and glutamic acids and the —NH_2 groups of arginine and lysine, must be protected before the coupling reaction occurs. A final requirement, which was met, was that the coupling reaction had to be nearly 100 percent effective.

Except for the first step, the above description is not qualitatively different from the classical, wet chemical methods of synthesis. Why has this method attracted so much attention? The power of this method is that it solves the separation and purification problems inherent in all organic reactions. The growing peptide chains, about a trillion of them per bead, are attached to the solid beads. Thus, when a reaction is completed, the excess reagent can be quantitatively removed by a simple filtration and washing process. The next reagent then can be added and the process continued. When the polypeptide chains are complete, they are cleaved from the bead and the blocking groups are removed.

Another tremendous advantage of this method is that all of these steps can be done mechanically. Valves selectively add the appropriate reagent or the activated, protected amino acid at the proper time. Precision pumps can supply the correct amounts of reagents. The reaction vessel is made of a sealed glass tube with a fritted glass filter disc in the middle, and can be shaken mechanically. Most important, all of these operations can be placed under the automatic control of a programming unit. Since the successful synthesis of insulin in 1967 by Merrifield, he and other workers have synthesized β-lactoglobulin and ribonuclease.

The lengthy article by Dickerson provides a fascinating account of the application of the study of the primary sequences of proteins to a new field called paleochemistry. This effort is similar to that of Eglinton and Calvin described in article 23, where the amount and type of hydrocarbon molecules provided clues as to the type of organisms present in rock formations from earlier eras. In that article and now in Dickerson's, we see how the creative use of chemistry has led to significant advances in a totally separate discipline.

Dickerson deals with two aspects of the chemistry of a small but very important protein, cytochrome c. This protein, which contains 104 residues, is found in all organisms using respiration, or consumption of oxygen, as their source of energy. It, together with a number of other enzymes, such as cytochromes b, a_1, a, and a_3, provide a pathway for the transfer of electrons and hydrogen atoms from food molecules such as glucose.

The first aspect treated is the comparison of the primary structures of cytochrome c from 38 species of plants and animals ranging from a pumpkin to a California gray whale. No other protein from so many different organisms has been analyzed this fully. There is a very high degree of correlation between these sequences. Thirty-five of the amino acid sites are invariant among all of these species. The similarities are even more pronounced upon examination of the color coding.

The color coding employed here is a very useful tool to help us understand homologies in the cytochrome c structures. However, care must be exercised, because two different codes are employed. The first diagram (all of these diagrams were done by the noted scientific illustrator Irving Geis) uses the rate of evolutionary variation as the basis of the color scheme. If very few changes have occurred in evolutionary time, the amino acids are indicated with the "hot" colors of red and orange. The "cool" colors, yellow-green, blue-green, blue, and purple represent sites where more variation has occurred.

The color code of the two-page table of sequences is based upon the chemical properties of the amino acid side chains. Hydrophobic (water-hating) or "oily" amino acids are depicted with red and orange, hydrophilic (water-loving) residues are displayed in blue and violet, and intermediate amino acids are located by the intermediate colors of green and yellow. This color scheme is useful when comparing amino acids at a particular site in the polypeptide chains. A mutation between two species that resulted in the change from one hydrophilic residue to another (say aspartic acid to glutamic acid) would have little impact upon the functioning of the molecule. The color scheme would show this similarity by printing them both in violet.

These sequences have been combined with the geological record to provide a graph of the number of amino acid changes against time for several proteins. The graphs are remarkably linear. There are significant differences in the slopes of the lines, however. Dickerson discusses this in terms of the functions that these four proteins serve in the living organism.

The second aspect of the structure of cytochrome c that is discussed is the actual three-dimensional geometry of this protein as it is derived from the heart of the horse. The technique of X-ray crystallography was used by Dickerson and his colleagues to determine the position and orientation in space of all of the residues of this protein. As has been found with the several other proteins whose three-dimensional structures have been determined, an analogy between these and an "oil drop" describes the model rather well. The hydrophobic or oily residues are found in the interior of the molecule and the hydrophilic amino acids are on the surface. This arrangement is effective in removing the nonpolar residues from the aqueous environment. The clustering of hydrophobic groups within proteins has been termed the hydrophobic bond by Walter Kauzmann of Princeton.

The location of a number of the invariant residues in the structure of cytochrome c is discussed in detail. The dominant feature of the molecule is the heme group, with a central iron atom that is oxidized and reduced between the +3 and +2 states as the molecule shuttles electrons to other molecules. The heme group is held in a crevice, and in Dickerson's article it is colored primarily in red and orange. This indicates that selection exerts a strong pressure to keep the geometry about the crevice constant so that the heme group can be held in the correct orientation to function. Dickerson discusses the roles of a large number of amino acids in detail and gives special attention to the role of phenylalanine (residue 82), which may be a part of the electron-transfer mechanism. No β-pleated sheet structures were found; only two short stretches of α-helix were found.

One of the most interesting aspects of the next article, by Phillips, in which he presents the three-dimensional structure of lysozyme, is that this protein has three α-helical regions and a short stretch of β-pleated sheet. The excellent drawing, again by Irving Geis, showing the full molecule, permits us to locate these ordered regions. The β-pleated sheet is the easiest to find. It consists of a loop in the backbone, which is depicted in green, in the lower left area of the drawing. Hydrogen bonds are depicted as gray rods. You should be able to find the four nearly vertical gray rods holding the loop together. Reference back to Doty's detailed atomic drawing of the β-pleated sheet may be helpful at this point.

The three stretches of α-helix are harder to locate because of their positions in the molecule. Again, reference to Doty's article indicates that you must look for a series of gray rods all pointing in the same direction along a cylinder. One stretch of helix lies across the very center of the figure and tilts slightly to the right. The other two sections form a cross in the lower right area of the drawing.

As was shown with cytochrome c, the oil drop model again accurately describes lysozyme. The polar and charged residues are shown with red and blue sticks and are found mainly on the surface. The hydrophobic residues are depicted in yellow and primarily occupy the center of the molecule.

The real excitement in this article, however, lies not in the fact that another (and this is the third) protein structure has been determined, but in the details it reveals about how an enzyme works. This is the first detailed accounting that has ever been available to show us how an enzyme operates on its substrate.

From 1913 until 1966, the scheme for enzyme catalysis proposed by Leonor Michaelis and Maude Menten has been the central dogma for generations of enzymologists. You will recall that Faller described tests of the validity of their scheme in his article in Section V. The symbol E represents the enzyme in the Michaelis-Menten description of enzyme catalysis, S is the substrate or molecule which is changed, and P is the product:

$$E + S \rightleftarrows ES \longrightarrow E + P$$

But the central questions of what the ES complex looks like and how the enzyme alters the substrate were basically unknown until Phillips's work. He shows the substrate neatly fitted into a crevice on the surface of the molecule, in the diagrams on pages 364 and 366. The article closes with a detailed look at molecular mechanics, showing how the amino acid side chains lining the cleft probably affect the cleavage of the hexasaccharide chain.

NUCLEIC ACIDS

The last five articles in this volume describe what has become the central dogma of molecular biology, that is, that DNA specifies certain codes for replication of DNA, DNA also specifies codes for RNA, and RNA specifies codes for proteins. These articles also provide an introduction to many of the experimental techniques of molecular biology.

Clearly DNA is near the very beginning of all life. In "The Synthesis of DNA" by Kornberg, article 32, the Watson-Crick structure of the DNA helix is presented, as well as the chemical structures of its component parts. Kornberg used an electron microscope to provide a "picture" of DNA which demonstrated that these giant polymers are closed, circular molecules. The enzymatic synthesis of DNA is described, with reference to the monomers—nucleotide-5'-triphosphates—and an enzyme—now called DNA polymerase.

Kornberg describes the use of density-gradient sedimentation in the separation of the natural DNA strand from the newly synthesized

DNA strand. By substituting 5′-bromouracil, a denser replacement for thymine, the synthetic product can be separated from the natural DNA. The sensitivity of such techniques makes it possible to separate linear molecules from circular molecules, as well as density-labelled molecules from unlabelled molecules. The end product of Kornberg's experiments is a totally synthetic double-strand DNA molecule capable of biological activity.

In closing his article Kornberg discusses genetic engineering. The introduction of an appropriate synthetic DNA to a cell incapable of performing a certain function because it lacks a necessary DNA could cure disease. The actual impact upon evolution and natural selection, as these apply to man—when man can introduce new genetic information in this way—is uncertain. But the *potential* impact is great. It is conceivable that man is on the threshold of controlling his evolution at least partially by application of his knowledge.

The details of the molecular mechanism of natural, *in vivo* DNA *replication* are still not worked out. The ways in which many enzymes are involved are still not known, for example. Although the DNA of bacteria and viruses is replicated from one end to the other, or around one closed circle, the DNA of the higher organism is so long it replicates in sections called replicating units.

The next step in the flow of genetic information is discussed by Spiegelman in article 33. *Transcription*, the production of RNA from DNA, results in the formation of all cellular RNA, transfer RNA (tRNA), ribosomal RNA (rRNA), and messenger RNA (mRNA). Spiegelman uses the terms dictionary RNA for tRNA, and translatable RNA for mRNA, though neither of these terms is presently in use. The final step in what we spoke of earlier as the central dogma of molecular biology, *translation*, which is the synthesis of protein from messenger RNA, is also diagrammed, and in describing this, Spiegelman introduces the functions of the ribosome, of tRNA, and of mRNA.

Single-strand DNA can form a double helix with RNA of the correct complementary sequence. Hybridization experiments which exploit this fact are described by Spiegelman. These prove that all three major classes of RNA originate from a DNA sequence. Hybridization has shown, furthermore, that only one of the two strands of DNA is copied. Hybridization remains the most important method for determining the presence or location of specific sequences of DNA or RNA. Density-gradient sedimentation is again shown to be useful in these determinations. The genetic information that is embodied in DNA is transcribed in triplets of chemical compounds, called bases. The bases are adenine (A), thymine (T), guanine (G), and cytosine (C) in DNA, and adenine, uracil (U), guanine, and cytosine in RNA. The coding triplets are called codons. The three-letter codons specify the order of amino acids which are present in proteins. The mechanism by which this process proceeds is described in some detail by Crick in "The Genetic Code: III."

Crick diagrams two binding sites in the ribosomes where charged transfer RNA's form temporary bonds. The ribosome moves along the messenger RNA reading triplets (codons), attaching appropriate amino acids at binding site *A*, and then adding the growing protein chain that is anchored to the tRNA at the *B* site to the new amino acid on the tRNA at the *A* site. Crick details the experiments leading to the discovery of the code. It is a very redundant code; there are many three-letter codons for the same amino acid. Part of this redundancy results from imperfect reading of the messenger by the tRNA (for which Crick has devised a wobble hypothesis), but mostly it results from the existence of more than one tRNA for a given amino acid. Crick discusses the effects of various changes in the sequence of DNA bases and the resulting changes in the protein coded by that part of the DNA. Such changes in DNA base sequence are called mutations.

"The Nucleotide Sequence of a Nucleic Acid" by Holley, article 35, presents the base sequence of the molecule of transfer RNA by which alanine is brought into the binding sites of the ribosome. The method of sequencing involves cleaving the molecule at a number of different points with different enzymes to get small pieces which can be separated and sequenced. Then by overlap the entire sequence is pieced together again. Holley's method of separation involved diethylaminoethyl cellulose chromatography, though more recent experiments in this kind of separation use paper chromatography and electrophoresis.

The isolation of one gram of pure alanine tRNA required that Holley start with 300 pounds of yeast. Over a period of three years even this small amount of transfer RNA was sufficient to show how tRNA forms a specific covalent bond to alanine and then again incorporates that amino acid specifically into a growing protein chain. Just recently the crystal structure of yeast phenylalanine tRNA was determined by Alex Rich. This structure shows the tRNA contains many helical regions, as shown in the figure in Holley's article entitled "hypothetical models." The anticodon loop (IGC) is a common feature of all transfer RNA's, though the anticodon is different for each different tRNA. The CCA end of the molecule is the site of amino acid attachment.

The last article, "The Visualization of Genes in Action," by Miller, provides electron microscope pictures of the entire process of transcription and translation. These remarkable electron micrographs provide a genuine visualization of these most remarkable chemical processes.

28

PROTEINS

PAUL DOTY

September 1957

· The principal substance of living cells, these giant molecules have identical backbones. Each is adapted to its specific task by a unique combination of side groups, size, folding and shape

Thousands of different proteins go into the make-up of a living cell. They perform thousands of different acts in the exact sequence that causes the cell to live. How the proteins manage this exquisitely subtle and enormously involved process will defy our understanding for a long time to come. But in recent years we have begun to make a closer acquaintance with proteins themselves. We know they are giant molecules of great size, complexity and diversity. Each appears to be designed with high specificity for its particular task. We are encouraged by all that we are learning to seek the explanation of the function of proteins in a clearer picture of their structure. For much of this new understanding we are indebted to our experience with the considerably simpler giant molecules synthesized by man. High-polymer chemistry is now coming forward with answers to some of the pressing questions of biology.

Proteins, like synthetic high polymers, are chains of repeating units. The units are peptide groups, made up of the monomers called amino acids [*see diagram below*]. There are more than 20 different amino acids. Each has a distinguishing cluster of atoms as a side group [*see next two pages*], but all amino acids have a certain identical group. The link-

ing of these groups forms the repeating peptide units in a "polypeptide" chain. Proteins are polypeptides of elaborate and very specific construction. Each kind of protein has a unique number and sequence of side groups which give it a particular size and chemical identity. Proteins seem to have a further distinction that sets them apart from other high polymers. The long chain of each protein is apparently folded in a unique configuration which it seems to maintain so long as it evidences biological activity.

We do not yet have a complete picture of the structure of any single protein. The entire sequence of amino acids has been worked out for insulin [see "The Insulin Molecule," by E. O. P. Thompson, SCIENTIFIC AMERICAN Offprint 42]; the determination of several more is nearing completion. But to locate each group and each atom in the configuration set up by the folded chain is intrinsically a more difficult task; it has resisted the Herculean labors of a generation of X-ray crystallographers and their collaborators. In the early 1930s W. T. Astbury of the University of Leeds succeeded in demonstrating that two X-ray diffraction patterns, which he called alpha and beta, were consistently associated with certain fibers, and he identified a third with collagen, the pro-

tein of skin, tendons and other structural tissues of the body. The beta pattern, found in the fibroin of silk, was soon shown to arise from bundles of nearly straight polypeptide chains held tightly to one another by hydrogen bonds. Nylon and some other synthetic fibers give a similar diffraction pattern. The alpha pattern resisted decoding until 1951, when Linus Pauling and R. B. Corey of the California Institute of Technology advanced the notion, since confirmed by further X-ray diffraction studies, that it is created by the twisting of the chain into a helix. Because it is set up so naturally by the hydrogen bonds available in the backbone of a polypeptide chain [*see top diagram on page 328*], the alpha helix was deduced to be a major structural element in the configuration of most proteins. More recently, in 1954, the Indian X-ray crystallographer G. N. Ramachandran showed that the collagen pattern comes from three polypeptide helixes twisted around one another. The resolution of these master plans was theoretically and esthetically gratifying, especially since the nucleic acids, the substance of genetic chemistry, were concurrently shown to have the structure of a double helix. For all their apparent general validity, however, the master plans did not give us the complete configuration in three dimensions

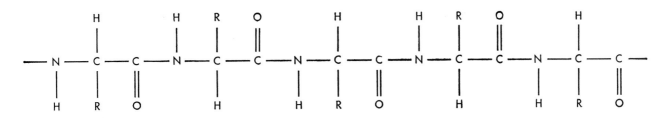

POLYPEPTIDE CHAIN is a repeating structure made up of identical peptide groups (CCONHC). The chain is formed by amino acids, each of which contributes an identical group to the backbone plus a distinguishing radical (R) as a side group.

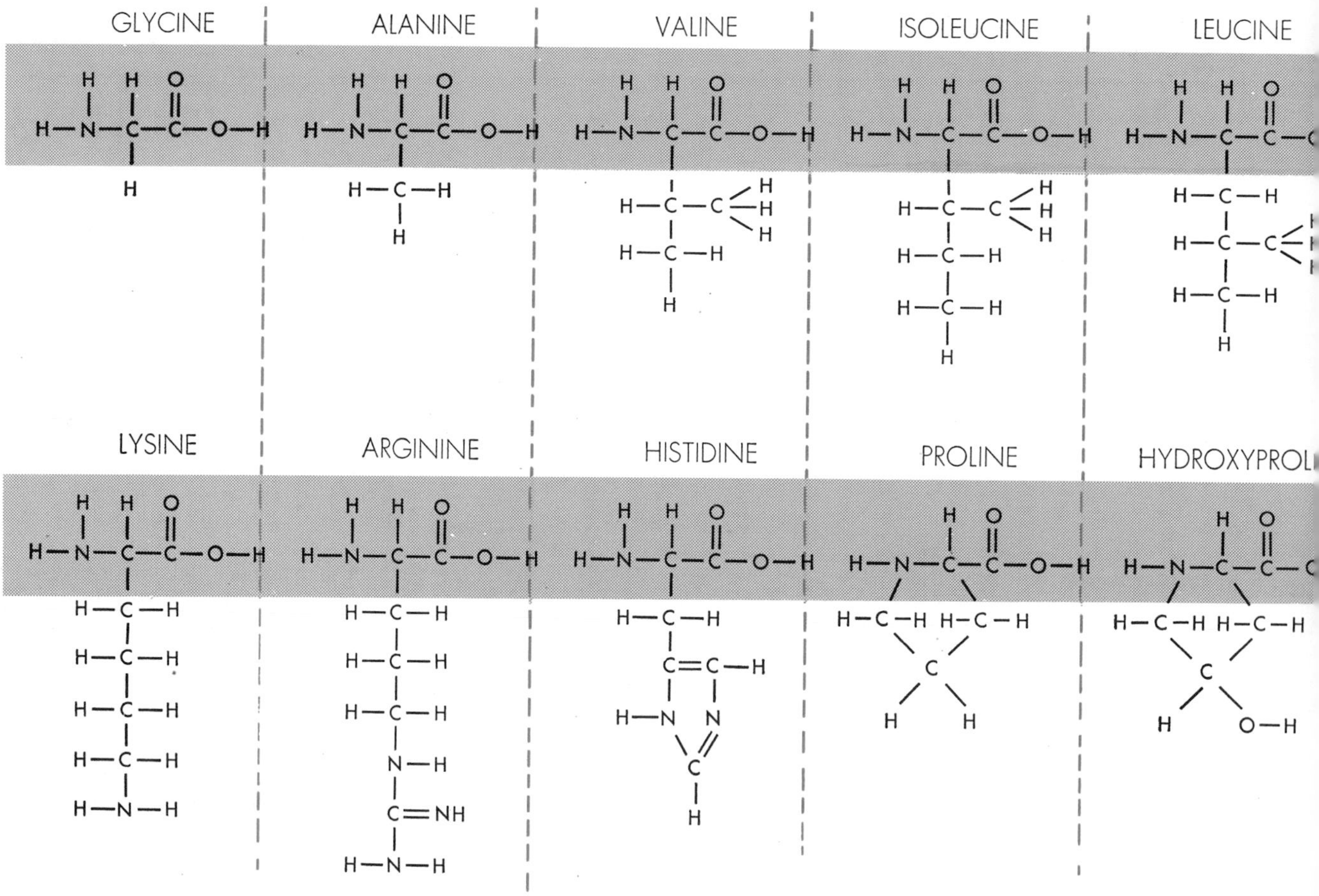

GLYCINE   ALANINE   VALINE   ISOLEUCINE   LEUCINE

LYSINE   ARGININE   HISTIDINE   PROLINE   HYDROXYPROL[INE]

AMINO ACIDS, the 20 commonest of which are shown in this chart, have identical atomic groups (*in colored bands*) which react to form polypeptide chains. They are distinguished by their unique side groups. In forming a chain, the amino group (NH_2) of one

of any single protein.

The X-ray diffraction work left a number of other questions up in the air. Since the alpha helix had been observed only in a few fibers, there was no solid experimental evidence for its existence elsewhere. There was even a suspicion that it could occur only in fibers, where it provides an economical way to pack polypeptides together in crystalline structures. Many proteins, especially chemically active ones such as the enzymes and antibodies, are globular, not linear like those involved in fibers and structural tissues. In the watery solutions which are the natural habitat of most proteins, it could be argued, the affinity of water molecules for hydrogen bonds would disrupt the alpha helix and reduce the chain to a random coil. These doubts and suppositions have prompted investigations by our group at Harvard University in collaboration with E. R. Blout of the Children's Cancer Research Foundation in Boston.

In these investigations we have em-

ployed synthetic polypeptides as laboratory models for the more complex and sensitive proteins. When Blout and coworkers had learned to polymerize them to sufficient length—100 to 1,000 amino acid units—we proceeded to observe their behavior in solution.

Almost at once we made the gratifying discovery that our synthetic polypeptides could keep their helical coils wound up in solutions. Moreover, we found that we could unwind the helix of some polypeptides by adjusting the acidity of our solutions. Finally, to complete the picture, we discovered that we could reverse the process and make the polypeptides wind up again from random coils into helixes.

The transition from the helix to the random coil occurs within a narrow range as the acidity is reduced; the hydrogen bonds, being equivalent, tend to let go all at once. It is not unlike the melting of an ice crystal, which takes place in a narrow .temperature range. The reason is the same, for the ice crys-

tal is held together by hydrogen bonds. To complete the analogy, the transition from the helix to the random coil can also be induced by heat. This is a true melting process, for the helix is a one-dimensional crystal which freezes the otherwise flexible chain into a rodlet.

From these experiments we conclude that polypeptides in solution have two natural configurations and make a reversible transition from one to the other, depending upon conditions. Polypeptides in the solid state appear to prefer the alpha helix, though this is subject to the presence of solvents, especially water. When the helix breaks down here, the transition is to the beta configuration, the hydrogen bonds now linking adjacent chains. Recently Blout and Henri Lenormant have found that fibers of polylysine can be made to undergo the alpha-beta transition reversibly by mere alteration of humidity. It is tempting to speculate that a reversible alpha-beta transition may underlie the process of muscle contraction and other types of

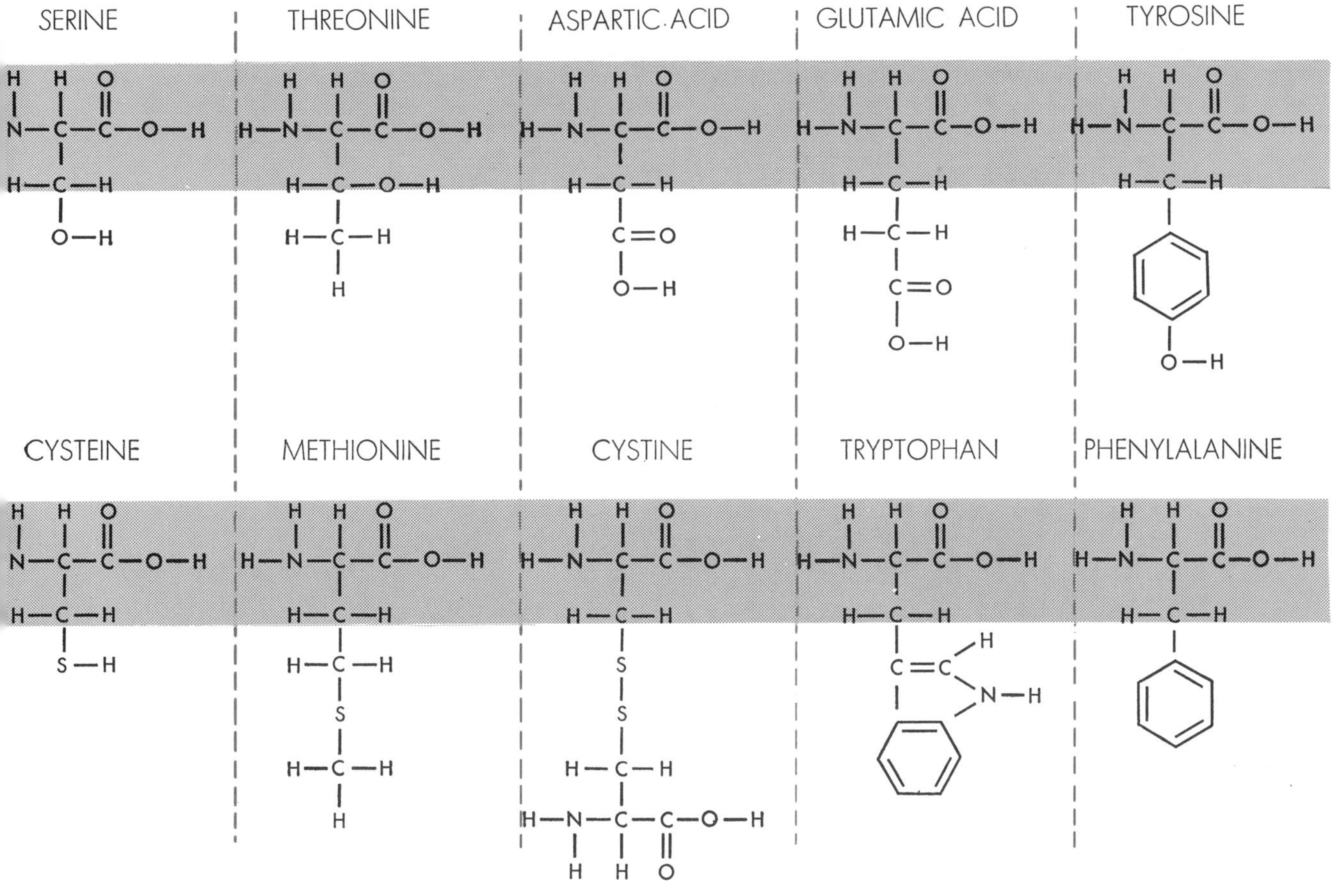

SERINE | THREONINE | ASPARTIC ACID | GLUTAMIC ACID | TYROSINE

CYSTEINE | METHIONINE | CYSTINE | TRYPTOPHAN | PHENYLALANINE

molecule reacts with the hydroxyl group (OH) of another. This reaction splits one of the amino hydrogens off with the hydroxyl group to form a molecule of water. The nitrogen of the first group then forms the peptide bond with the carbon of the second.

movement in living things.

Having learned to handle the polypeptides in solution we turned our attention to proteins. Two questions had to be answered first: Could we find the alpha helix in proteins in solution, and could we induce it to make the reversible transition to the random coil and back again? If the answer was yes in each case, then we could go on to a third and more interesting question: Could we show experimentally that biological activity depends upon configuration? On this question, our biologically neutral synthetic polypeptides could give no hint.

For the detection of the alpha helix in proteins the techniques which had worked so well on polypeptides were impotent. The polypeptides were either all helix or all random coil and the rodlets of the first could easily be distinguished from the globular forms of the second by use of the light-scattering technique. But we did not expect to find that any of the proteins we were going to investigate were 100 per cent helical in configura-tion. The helix is invariably disrupted by the presence of one of two types of amino acid units. Proline lacks the hydrogen atom that forms the crucial hydrogen bond; the side groups form a distorting linkage to the chain instead. Cystine is really a double unit, and forms more or less distorting cross-links between chains. These units play an important part in the intricate coiling and folding of the polypeptide chains in globular proteins. But even in globular proteins, we thought, some lengths of the chains might prove to be helical. There was nothing, however, in the over-all shape of a globular protein to tell us whether it had more or less helix in its structure or none at all. We had to find a way to look inside the protein.

One possible way to do this was suggested by the fact that intact, biologically active proteins and denatured proteins give different readings when observed for an effect called optical rotation. In general, the molecules that exhibit this effect are asymmetrical in atomic structure. The side groups give rise to such asymmetry in amino acids and polypeptide chains; they may be attached in either a "left-handed" or a "right-handed" manner. Optical rotation provides a way to distinguish one from the other. When a solution of amino acids is interposed in a beam of polarized light, it will rotate the plane of polarization either to the right or to the left [see diagrams at top of page 330]. Though amino acids may exist in both forms, only left-handed units, thanks to some accident in the chemical phase of evolution, are found in proteins. We used only the left-handed forms, of course, in the synthesis of our polypeptide chains.

Now what about the change in optical rotation that occurs when a protein is denatured? We knew that native protein rotates the plane of the light 30 to 60 degrees to the left, denatured protein 100 degrees or more to the left. If there was some helical structure in the protein, we surmised, this shift in rotation might be induced by the disappearance of the helical structure in the denaturation process. There was reason to believe that the helix, which has to be either left-handed or right-handed, would have optical activity. Further, although it appeared possible for the helix to be wound either way, there were grounds for assuming that nature had chosen to make all of its helixes one way or the other. If it had not, the left-handed and right-handed helixes would mutually cancel out their respective optical rotations. The change in the optical rotation of proteins with denaturation would then have some other explanation entirely, and we would have to invent another way to look for helixes.

To test our surmise we measured the optical rotation of the synthetic poly-peptides. In the random coil state the polypeptides made an excellent fit with the denatured proteins, rotating the light 100 degrees to the left. The rotations in both cases clearly arose from the same cause: the asymmetry of the amino acid units. In the alpha helix configuration the polypeptides showed almost no rotation or none at all. It was evident that the presence of the alpha helix caused a

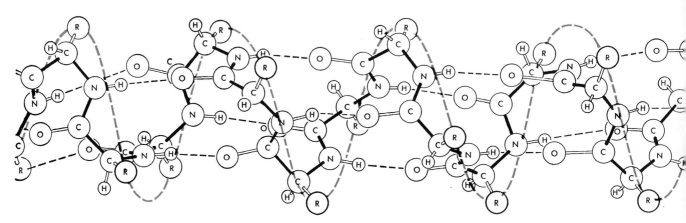

ALPHA HELIX gives a polypeptide chain a linear structure shown here in three-dimensional perspective. The atoms in the repeating unit (CCONHC) lie in a plane; the change in angle between one unit and the next occurs at the carbon to which the side group

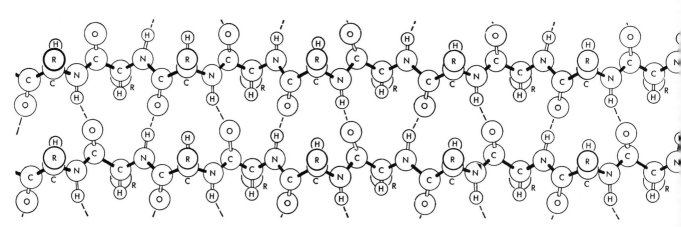

BETA CONFIGURATION ties two or more polypeptide chains to one another in crystalline structures. Here the hydrogen bonds do not contribute to the internal organization of the chain, as in the alpha helix, but link the hydrogen atoms of one chain to the oxygen

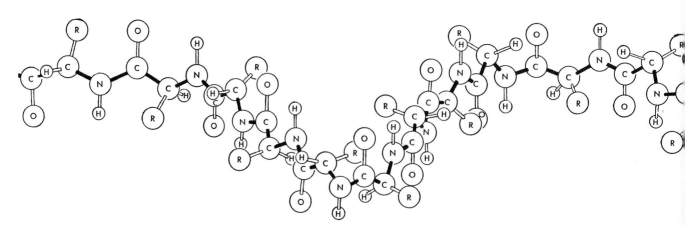

RANDOM CHAIN is the configuration assumed by the polypeptide molecule in solution, when hydrogen bonds are not formed. The flat configuration of the repeating unit remains, but the chain rotates about the carbon atoms to which the side groups are at-

counter-rotation to the right which nearly canceled out the leftward rotation of the amino acid units. The native proteins also had shown evidence of such counter-rotation to the right. The alpha configuration did not completely cancel the leftward rotation of the amino acid units, but this was consistent with the expectation that the protein structures would be helical only in part. The experiment

thus strongly indicated the presence of the alpha helix in the structure of globular proteins in solution. It also, incidentally, seemed to settle the question of nature's choice of symmetry in the alpha helix: it must be right-handed.

When so much hangs on the findings of one set of experiments, it is well to double check them by observa-

tions of another kind. We are indebted to William Moffitt, a theoretical chemist at Harvard, for conceiving of the experiment that provided the necessary confirmation. It is based upon another aspect of the optical rotation effect. For a given substance, rotation varies with the wavelength of the light; the rotations of most substances vary in the same way. Moffitt predicted that the presence of

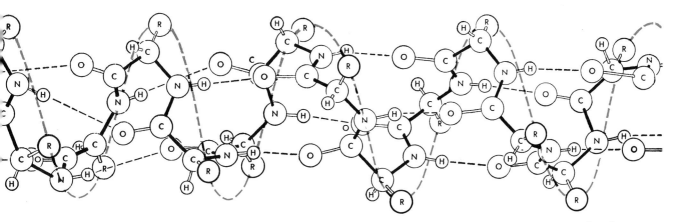

(R) is attached. The helix is held rigid by the hydrogen bond (*broken black lines*) between the hydrogen attached to the nitro-

gen in one group and the oxygen attached to a carbon three groups along the chain. The colored line traces the turns of the helix.

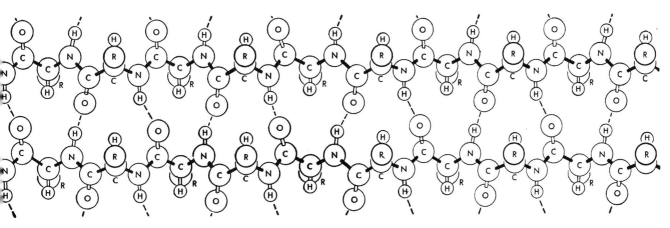

atoms in the adjoining chain. The beta configuration is found in silk and a few other fibers. It is also thought that polypeptide

chains in muscle and other contractile fibers may make reversible transitions from alpha helix to beta configuration when in action.

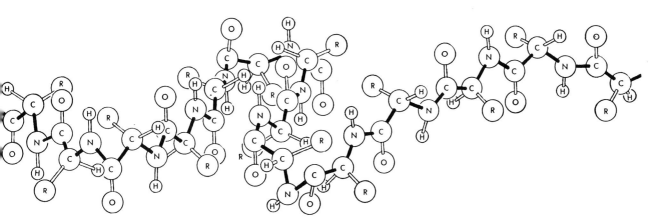

tached. The random chain may be formed from an alpha helix when hydrogen bonds are disrupted in solution. A polypeptide

chain may make a reversible transition from alpha helix to random chain, depending upon the acid-base balance of the solution.

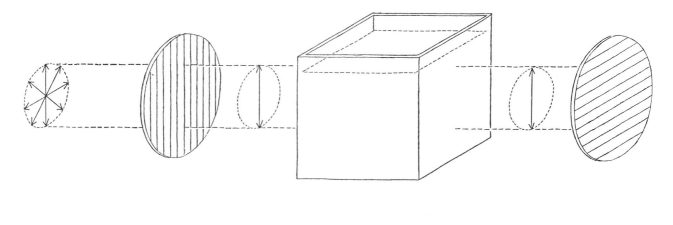

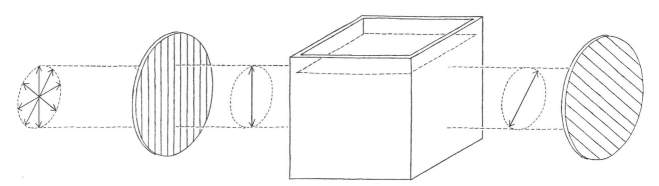

OPTICAL ROTATION is induced in a beam of polarized light by molecules having certain types of structural asymmetry. At top a beam of light is polarized in the vertical plane and transmitted unchanged through a neutral solution. At bottom asymmetrical molecules in the solution cause the beam to rotate from the vertical plane. The degree of rotation may be determined by turning the second polarizing filter (*right*) to the point at which it cuts off the beam. The alpha helix in a molecule causes such rotation.

the alpha helix in a substance would cause its rotation to vary in a different way. His prediction was sustained by observation: randomly coiled polypeptides showed a normal variation while the helical showed abnormal. Denatured and native proteins showed the same contrast. With the two sets of experiments in such good agreement, we could conclude with confidence that the alpha helix has a significant place in the structure of globular proteins. Those amino acid units that are not involved in helical configurations are weakly bonded to each other or to water molecules, probably in a unique but not regular or periodic fashion. Like synthetic high-polymers, proteins are partly crystalline and partly amorphous in structure.

The optical rotation experiments also provided a scale for estimating the helical content of protein. The measurements indicate that, in neutral solutions, the helical structure applies to 15 per cent of the amino acid units in ribonuclease, 50 per cent of the units in serum albumin and 85 per cent in tropomyosin. With the addition of denaturing agents to the solution, the helical content in each case can be reduced to zero. In

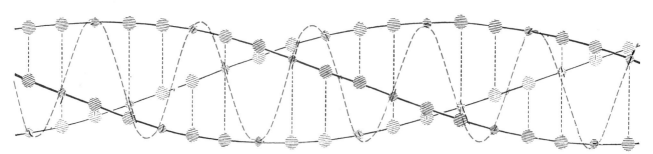

COLLAGEN MOLECULE is a triple helix. The colored broken line indicates hydrogen bonds between glycine units. The black broken lines indicate hydrogen bonds which link hydroxyproline units and give greater stability to collagens in which they are found.

some proteins the transition is abrupt, as it is in the synthetic polypeptides. On the other hand, by the use of certain solvents we have been able to increase the helical content of some proteins—in the case of ribonuclease from 15 to 70 per cent. As in the polypeptides, the transition from helix to random coil is reversible. The percentage of helical structure in proteins is thus clearly a variable. In their natural environment, it appears, the percentage at any given time represents the equilibrium between the inherent stability of the helix and the tendency of water to break it down.

In a number of enzymes we have been able to show that biological activity falls off and increases with helical content. Denaturation is now clearly identified with breakdown of configuration, certainly insofar as it involves the integrity of the alpha helix. This is not surprising. It is known that catalysts in general must have rigid geometrical configurations. The catalytic activity of an enzyme may well require that its structure meet similar specifications. If this is so, the rigidity that the alpha helix imposes on the otherwise flexible polypeptide chain must play a decisive part in establishing the biological activity of an enzyme. It seems also that adjustability of the stiffness of structure in larger or smaller regions of the polypeptide chain may modify the activity of proteins in response to their environment. Among other things, it could account for the versatility of the gamma globulins; without any apparent change in their amino acid make-up, they are able somehow to adapt themselves as antibodies to a succession of different infectious agents.

The next step toward a complete anatomy of the protein molecule is to determine which amino acid units are in the helical and which in the nonhelical regions. Beyond that we shall want to know which units are near one another as the result of folding and cross-linking, and a myriad of other details which will supply the hues and colorings appropriate to a portrait of an entity as intricate as protein. Many such details will undoubtedly be supplied by experiments that relate change in structure to change in function, like those described here.

In the course of our experiments with proteins in solution we have also looked into the triple-strand structure of collagen. That structure had not yet been resolved when we began our work, so we did not know how well it was designed for the function it serves in structural tissues. Collagen makes up one third of the proteins in the body and 5 per cent of its total weight; it occurs as tiny fibers or fibrils with bonds that repeat at intervals of about 700 Angstroms. It had been known for a long time that these fibrils could be dissolved in mild solvents such as acetic acid and then reconstituted, by simple precipitation, into their original form with their bandings restored. This remarkable capacity naturally suggested that the behavior of collagen in solution was a subject worth exploring.

Starting from the groundwork of other investigators, Helga Boedtker and I were able to demonstrate that the collagen molecule is an extremely long and thin rodlet, the most asymmetric molecule yet isolated. A lead pencil of comparable proportions would be a yard long. When a solution of collagen is just slightly warmed, these rodlets are irreversibly broken down. The solution will gel, but the product is gelatin, as is well known to French chefs and commercial producers of gelatin. The reason the dissolution cannot be reversed was made clear when we found that the molecules in the warmed-up solution had a weight about one third that of collagen. It appeared that the big molecule of collagen had broken down into three polypeptide chains.

At about the same time Ramachandran proposed the three-strand helix as the collagen structure. Not long afterward F. H. C. Crick and Alexander Rich at the University of Cambridge and Pauline M. Cowan and her collaborators at King's College, London, worked out the structure in detail, It consists of three polypeptide chains, each incorporating three different amino acid units—proline, hydroxyproline and glycine. The key to the design is the occurrence of glycine, the smallest amino acid unit, at every third position on each chain. This makes it possible for the bulky proline or hydroxyproline groups to fit into the links of the triple strand, two of these nesting in each link with the smaller glycine unit [*see diagram on page 330*].

One question, however, was left open in the original model. Hydroxyproline has surplus hydrogen bonds, which, the model showed, might be employed to reinforce the molecule itself or to tie it more firmly to neighboring molecules in a fibril. Independent evidence seemed to favor the second possibility. Collagen in the skin is irreversibly broken down in a first degree burn, for example, at a temperature of about 145 degrees Fahrenheit. This is about 60 degrees higher than the dissolution temperature of the collagen molecule in solution. The obvious inference was that hydroxyproline lends its additional bonding power to the tissue structure. Moreover, tissues with a high hydroxyproline content withstand higher temperatures than those with lower; the skin of codfish, with a low hydroxyproline content, shrivels up at about 100 degrees. Tomio Nishihara in

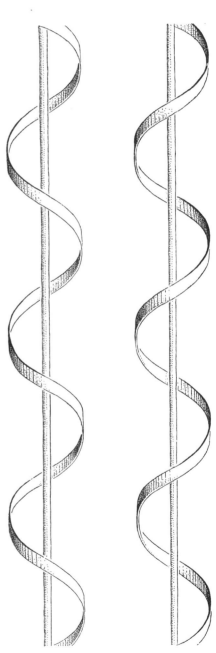

ASYMMETRY of a helix is either left-handed (*left*) or right-handed. Helix in proteins appears to be exclusively right-handed.

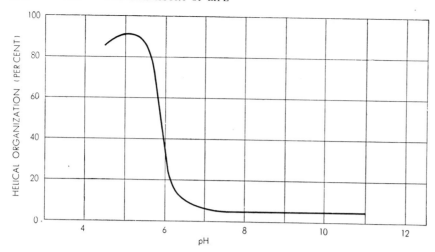

ALPHA HELIX BREAKDOWN is induced in solutions of some polypeptides when the pH (acidity or alkalinity) reaches a critical value at which hydrogen bonds are disrupted.

our laboratory has compared the breakdown temperatures of collagen molecules and tissues from various species and found that the tissue temperature is uniformly about 60 degrees higher. Thus we must conclude that the extra stability conferred by hydroxyproline goes directly to the molecule and not to the fibril.

The structure of collagen demonstrates three levels in the adaptation of polypeptide chains to fit the requirements of function. First there are the chains as found in gelatin, with their three amino acids lined up in just the right sequence. These randomly coiled and quite soluble molecules are transformed into relatively insoluble, girderlike building units when united into sets of three by hydrogen bonds. The subtly fashioned collagen molecules are still too fragile to withstand body temperatures. When arranged side by side, however, they form a crystalline structure which resists comparatively high temperatures and has fiber-like qualities with the vast range of strengths and textures required in the different types of tissues that are made of collagen.

The story of collagen, like that of other proteins, is still far from complete. But it now seems that it will rank among the first proteins whose molecular structure has been clearly discerned and related in detail to the functions it serves.

29

THE AUTOMATIC SYNTHESIS OF PROTEINS

R. B. MERRIFIELD
March 1968

*By anchoring an amino acid to a plastic bead one can
add other amino acids one by one in automatically
controlled steps. This method has already been used
to make the small protein insulin*

The synthesis of proteins is one of the primary functions of the living cell, and the intricate series of operations by which the cell accomplishes the task has recently become known in considerable detail. Long before the biosynthesis of proteins was at all understood, at the turn of the century, the great chemist Emil Fischer believed proteins could be synthesized in the laboratory. It took a long time to accumulate the knowledge and techniques required to put together one of these enormously complex substances. Beginning in 1963, as the result of concerted work by many individuals, three groups of chemists—one in the U.S., one in Germany and one in China—succeeded in making in the laboratory a comparatively simple protein: the pancreatic hormone insulin.

The conventional process for the chemical synthesis of proteins, or of the smaller chains of amino acids called peptides, is a slow and painstaking affair. In our laboratory at Rockefeller University we set out some years ago to look for a simpler and more efficient process—one that might lend itself to automatic operation. The result was a new technique, "solid phase" peptide synthesis, in which the peptide chains are assembled not in solution but on small, solid beads of polystyrene. In 1965 the solid phase method was successfully applied to the synthesis of insulin.

The availability through controlled chemical synthesis of insulin with all the properties of the natural hormone is of great value because the road is now open for making related molecules that differ from the parent compound in precisely known ways. Such analogues should help to clarify the manner in which the natural hormone functions and may in time lead to insulin derivatives that exhibit greater or more prolonged activity for the treatment of diabetes.

Even more exciting than the synthesis of insulin is the possibility of synthesizing an enzyme—one of the proteins that catalyze metabolic processes. That goal is now in sight and will surely be attained. It will add significantly to our understanding of this important class of proteins and of the mechanisms by which the living cell carries out its essential functions.

Peptide Synthesis

To understand how the chemical synthesis of a protein can be approached it is best to look first at how peptides are synthesized. Peptides are simpler models of proteins; they consist of the same structural elements (but fewer of them) linked together in the same way. The linkage between the amino acid subunits is known as the peptide bond; a series of such bonds constitutes the primary "backbone" structure of peptides and proteins. The formation of these bonds is the principal problem in peptide synthesis and is also the first step in protein synthesis. In the case of proteins, however, there are also secondary and tertiary bonds that control the cross-linking and folding of the molecule and are responsible for its three-dimensional shape. Because peptides are shorter and lack these complicating additional bonds, they are simpler compounds to study. They have been used to develop the chemistry needed to begin the synthetic work on proteins.

Amino acids are compounds containing several reactive groups: one amino group (NH_2), one carboxyl group ($COOH$) and in many instances another reactive group located on a side chain [*see top illustration on page 336*]. In general all but one of these groups must be protected against undesired combinations during a chemical reaction if a specific, pure product of known structure is to be obtained. In order to prepare even the simplest chain of two amino acid units (a dipeptide) the basic amino group of one unit and the acidic carboxyl group of the other must be blocked. It is then possible to activate the carboxyl group of the first amino acid—that is, to increase its energy level—so that it will couple with the free amino group of the second one to form the peptide bond.

Now let us consider extending the chain to form a longer peptide. There are two general approaches. In the "fragment" method short peptide chains are built up and are combined to form the larger final molecule; in the "stepwise" method single amino acid units are added one at a time until the final molecule is completed. (In both methods the successive additions can in principle be made at either end of the molecule, although in practice there are certain limitations.) The fragment technique is the older and until recently was the more common approach. Its advantages are that more of the intermediate peptides are of small size and that there are greater differences between the properties of the reactants and of the products than there are in the stepwise procedure. On the other hand, the coupling yields are generally lower in the fragment method and there is a greater chance of unwanted side reactions.

Before any of the chain-lengthening processes can be carried out it is necessary to remove one of the blocking groups from the initial dipeptide. To "deprotect" selectively in the presence of the other protecting group (or of several such groups) and without damage to the peptide chain requires careful planning. The choice of the protecting groups and the activating, or coupling, agent for each amino acid has been a major concern of peptide chemists.

At each step of the synthesis it is usually necessary to isolate, purify and characterize the products of the reaction, and it is at this point that the greatest difficulties are often encountered. Crystallization is the classical procedure for purifying peptides, as it is for most other

organic compounds. It depends on the formation of an orderly array of molecules that grows in size until it precipitates from solution. Ideally only molecules of one kind will be in the crystalline precipitate and all undesired substances will remain in solution and be washed away. Sometimes one can obtain from a reaction mixture quite pure peptides that do crystallize readily. Particularly when one is working with long peptide chains, however, the yield may often be amorphous material or crude crystalline precipitates contaminated with various by-

products. One must then resort to special purification procedures that may require many days each. Consider for a moment the time and effort involved in the synthesis of a 100-unit peptide if such coupling and purification steps must be performed 99 times!

The Solid Phase Approach

To synthesize molecules of the size and complexity of proteins, it seemed clear, methods of the greatest efficiency and simplicity would have to be devel-

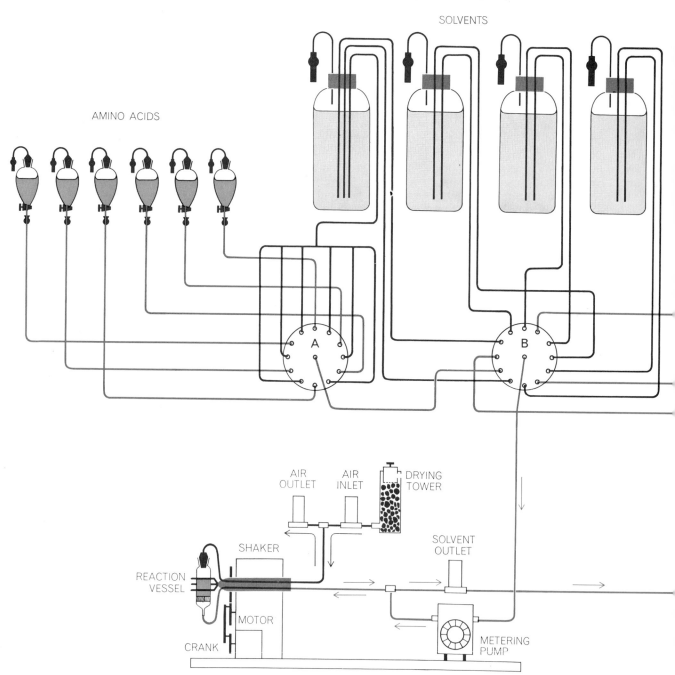

SCHEMATIC DIAGRAM of the automatic apparatus shows the "plumbing" circuits. The proper amino acid, other reagent or solvent is pumped from its reservoir through a selector valve (*A or B*) into the reaction vessel while air is displaced at the top of the vessel. A mechanical shaker rocks the vessel to mix the reactants. Solvents and by-products are removed by vacuum through the filter in the bot-

oped. In 1959, with these requirements in mind, a new approach to peptide synthesis was conceived. The new idea was to synthesize the long chains one unit at a time, but without stopping to isolate the individual intermediate peptides; to make this feasible the plan was to anchor the chain to an insoluble solid support.

The first amino acid would be chemically bonded to a solid particle and the rest of the amino acid units would be added to it stepwise in the proper order. Since the solid support would be com-

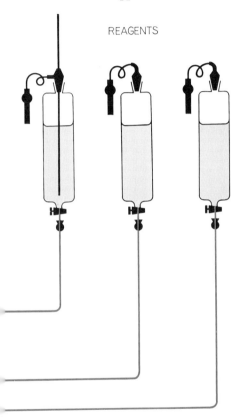

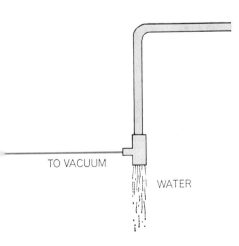

pletely insoluble in the various solvents, all the intermediate peptide products would also be held in an insoluble state; they could therefore be purified simply by dissolving the unwanted by-products and reagents and washing them away. This would involve only an elementary filtration step, but it would accomplish essentially the same kind of purification as the classical recrystallization: the growing peptide would be an insoluble precipitate, whereas the undesired reagents would be in solution. Filtration is much easier and faster than crystallization. Most important, it can be done in the same way at each step, whereas crystallization is necessarily an individualized procedure that is different for each new intermediate peptide.

The general scheme for solid phase peptide synthesis is straightforward. A suitable solid support is selected and a reactive site is produced on it. The first amino acid—actually the terminal amino acid of the proposed peptide chain—is attached by its carboxyl group to the reactive site. Now the second amino acid, with all but one of its reactive groups protected, is activated and coupled to the first amino acid, leaving a protected dipeptide firmly bound to the support. The solid can be filtered and washed thoroughly to remove all the excess reagents and any by-products without the slightest danger of losing the desired peptide.

Next the protecting group on the amino end is removed and the whole process is repeated exactly as before but with a new amino acid. After the required sequence of amino acids has been assembled in this manner the peptide chain is finally removed from the support by selectively breaking the bond that has been holding the two together throughout the synthesis. Now for the first time the peptide chain is free and can be dissolved and separated from the solid support. Once it is in solution conventional purification procedures can be carried out.

Before this scheme could be developed into a workable procedure a number of rather severe requirements had to be met, having to do with the nature of the solid support, the type of bond linking the peptide to the support and the choice of protecting groups and coupling reagents. The solid support had to be completely insoluble in all the solvents that might be used in the synthetic reactions or in the washing steps; it had to be physically stable and in a convenient form to permit filtration and other manipulations; it had to have a reactive site at which the peptide chains could be at-

tached but should otherwise be chemically inert and stable; finally, in order to allow the synthesis of a sufficient quantity of peptide, it should either have a very large surface-to-volume ratio or be readily permeable to the soluble reagents.

After considerable exploration a substance that met these requirements was found. It is a polystyrene resin, a linear polymer of styrene in which the styrene chains are loosely linked together with divinylbenzene, and it is in the form of beads about 50 microns (.002 inch) in diameter. The amount of cross-linking agent was selected to give a resin of high molecular weight that would be completely insoluble but at the same time free to swell in organic solvents. This makes the beads permeable to reagents dissolved in the solvent, so that reactions can occur not only on the surface but also within the interstices of the gel-like matrix. Although we are talking about beads that are barely visible as separate particles to the unaided eye, they are actually enormous compared with the dimensions of amino acids or even of proteins: each bead can support some 10^{12} (one trillion) peptide chains!

The Anchor Bond

Polystyrene itself has no convenient reactive site for anchoring the peptide chain, but it can be readily modified in many ways to make such a site. The choice of the modification was dictated by the kind of bond needed to hold the peptide to the resin. The bond must be easy to form, it must be completely stable during the dozens of reactions involved in assembling the peptide chain and it must be readily cleaved under relatively mild conditions at the end of the synthesis. The anchor we chose was prepared by attaching chloromethyl groups ($ClCH_2$) to the six-carbon rings of the polystyrene and then reacting them with the first amino acid to form what is known as a benzyl ester [*see illustration on page 337*]. The benzyl group also served to protect many of the reactive side chains.

The choice of the benzyl group in turn influenced the choice of the protecting group for the amino ends of the successive amino acids. That is, it had to be possible to remove the amine protection selectively at every cycle of the synthesis without detaching the peptide from the resin or deprotecting the side chains. The protecting agent we chose was tertiary butyloxycarbonyl ("Boc"), a group that had been developed a few

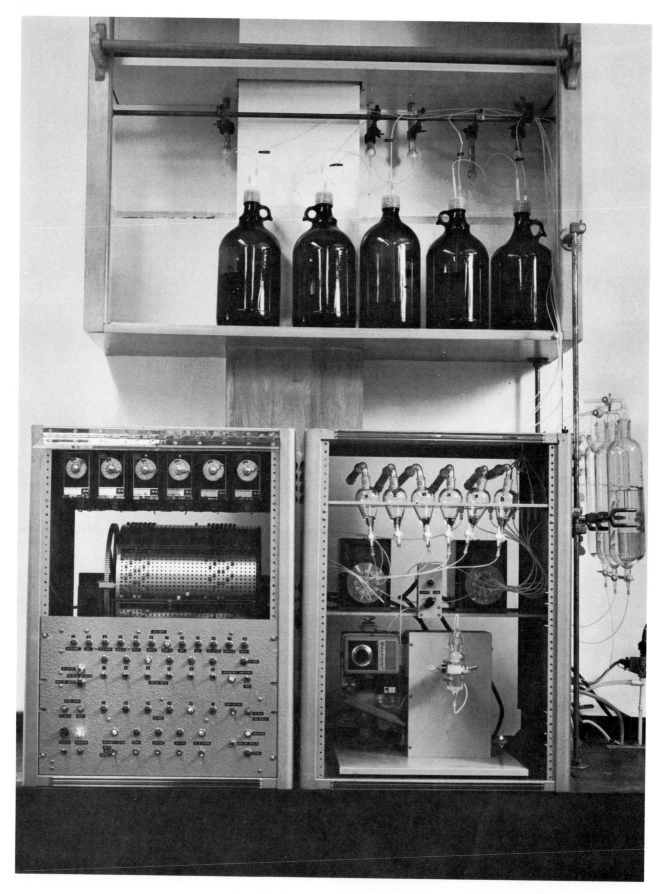

APPARATUS for the automatic synthesis of peptide chains is seen in the author's laboratory. It includes the small glass reaction vessel (*lower right*) with its attendant "plumbing" and a programming unit (*left*). The rectangular pins on the rotating drum operate switches that control the pump, valves, timers and shaker that fill and empty the vessel and mix the reagents. Amino acids are supplied from the six glass vessels (*middle right*). Solvents and other reagents are supplied from the larger containers above and at right.

years earlier for use in conventional syntheses. It was sensitive to certain anhydrous acids that would not affect the ester and it could therefore be removed without disturbing the anchor.

The principal remaining problem was the peptide-forming reaction itself, which had to be rapid and must not give rise to side reactions. Most important of all, it had to go to completion; it was absolutely crucial to the success of the process that this step give essentially 100 percent yields. Suppose, for example, the second amino acid were to couple to the extent of only 90 percent, which is considered a most acceptable yield in organic chemistry. What will happen when the third amino acid is coupled? It will react with the amino end of the dipeptide to form a tripeptide, but it will also couple with the unreacted 10 percent of single amino acids to produce dipeptides that lack the second amino acid. In an ordinary synthesis the intermediate products are isolated at each stage; in the solid phase method they are simply washed free of soluble impurities, and the abnormal chain will be carried through the entire synthesis, giving at the end a mixture consisting of 90 percent of the correct peptide and 10 percent of a peptide with a missing link. In this simple case the two products could probably be separated and purified, but if incomplete reactions were to occur several times during the synthesis of a long peptide, a complex mixture would result that might not be so easy to separate [see upper illustration on page 338].

We therefore put great stress on finding reagents and conditions that would lead to complete coupling reactions, and we have fortunately been able to achieve that goal in practice. Many ways to activate amino acids have been developed for use in peptide synthesis. For the solid phase method the most successful procedures have been to activate with the reagent dicyclohexylcarbodiimide or to use the nitrophenyl esters of the amino acids. These activated forms are highly effective, but only if they can reach the proper sites. To ensure their rapid, unimpaired penetration into the resin, where most of the peptide chains are located, a solvent of high swelling capacity, such as methylene chloride, is necessary.

This solvent causes the beads to swell to approximately twice their original diameter, which means that the polymer chains are then distributed in eight times the initial volume. The polymer molecules occupy only about 12 percent of the total space of each swollen bead, and the remaining 88 percent is filled with solvent containing the activated amino acid molecules. The diffusion of these small molecules is relatively little inhibited by the polymer, so that the reactions take place almost as fast as they would in solution.

The efficiency of coupling also depends on the concentrations of the reactants: the peptide chain and the amino acid being added to it. If one begins with equal amounts of the reactants, their concentrations will decrease to very low levels as the reaction nears completion

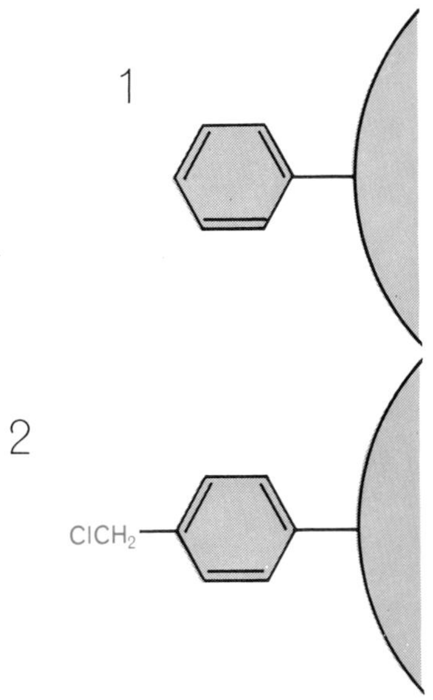

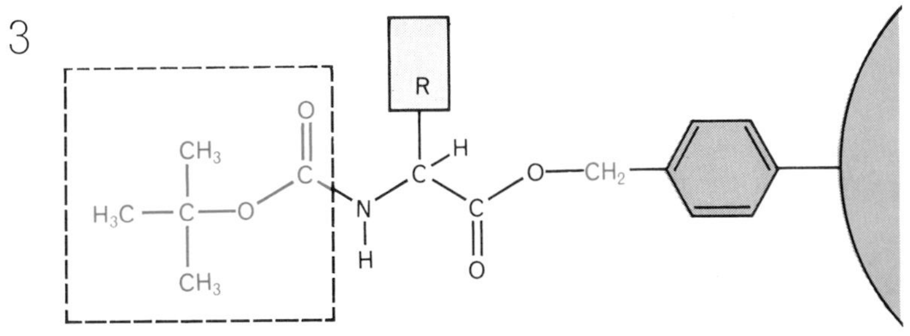

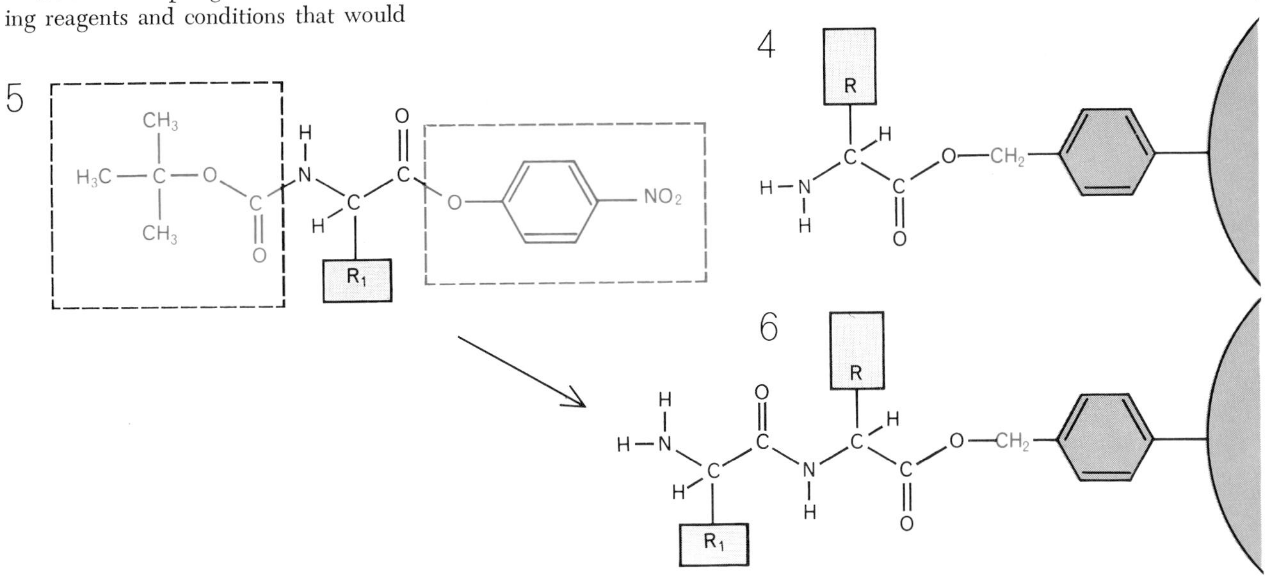

SOLID PHASE METHOD is carried out stepwise from the carboxyl end toward the amino end of the peptide. An aromatic ring of the polystyrene (1) is activated by attaching a chloromethyl group (2). The first amino acid (black), protected by a butyloxycarbonyl (Boc) group (black box), is coupled to the site (3) by a benzyl ester bond and is then deprotected (4). Subsequent amino acid units are supplied in one of two activated forms; a second unit is shown in one of these forms, the nitrophenyl ester of the amino acid (5). The ester (colored box) is eliminated as the second unit couples to the first. Then the second unit is deprotected, leaving a dipeptide (6).

and the reaction rate, which is proportional to the product of the concentrations, will gradually approach zero. The practical consequence is that some of the amino acids never do link up, and the reaction never quite goes to completion. This is the usual situation in conventional syntheses.

If, on the other hand, there is a rather large excess of one of the reactants, a significant rate can be maintained until essentially all the limiting reactant (the peptide chain) has entered into bond formation. To illustrate, suppose we begin with 100 parts of each reactant and

call the initial reaction rate 10,000 (100 × 100). After the reaction is 99 percent completed, one part of each reactant would remain, and the relative rate would be only 1, or 1/10,000 as fast as at the beginning. Even with a very fast initial rate it would take a long time to complete the reaction.

Suppose instead we begin with 100 parts of the peptide chain as before but with 400 parts of the activated amino acid. Now after 99 percent of the chain is used up the relative rate would still be 301 (1 × 301). In the presence of the fourfold excess of the amino acid, one

can calculate, the reaction can go to 99.99 percent of completion in the same time it would take to go to only 75 percent if equal amounts of the reactants had been used. One of the important advantages of solid phase peptide synthesis is that such an excess of the amino acid derivative can be used without complicating the subsequent purification procedure, since at the end of each reaction the excess is simply removed by filtration and washing. Thus we can force the reaction to completion and leave essentially no free, unreacted peptide chains.

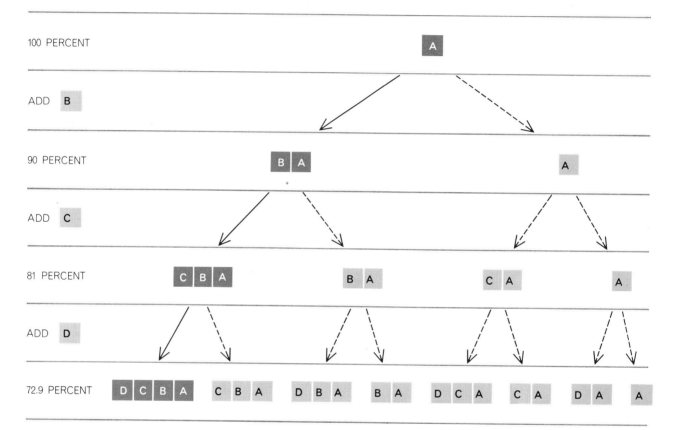

HIGH YIELD from each coupling reaction is important in the solid phase method because products of incomplete reactions persist through the filtering steps. If each amino acid were to couple with even a relatively high efficiency, say 90 percent, the yield of pure peptide (*dark color*) would be down to 72.9 percent by the time the fourth unit was added. More important, the seven different peptide fragments that lack one amino acid unit or more would have to be separated chemically. In practice, yields are close to 100 percent.

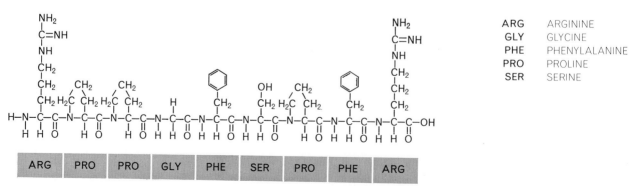

BRADYKININ, a small peptide hormone, was one of the first peptides synthesized in the author's laboratory by the solid phase method. Its nine amino acid subunits (five different amino acids) were assembled stepwise from the carboxyl (*right*) end of the chain.

The deprotection and coupling reactions just described can be repeated alternately until the desired peptide chain has been assembled on the resin beads. The final step is the cleavage of the benzyl ester bond that has been holding the chain to the resin throughout the synthesis. As indicated earlier, this bond was chosen because it is stable during the synthesis but can be selectively split at the right time without damaging the peptide chain. The resin is suspended in anhydrous trifluoroacetic acid, and dry hydrogen bromide gas is bubbled through to effect the splitting. (More recently anhydrous liquid hydrogen fluoride has been successfully employed for this step.) These reagents also remove the protecting groups on side chains. The peptide, now in a free and soluble state, is separated from its resin support by filtration and is purified. It is then ready for analysis and, where possible, for biological assay.

Those who are familiar with the mechanism of protein synthesis in the living cell will see some superficial resemblance between it and the system just described. Both depend on a particulate support (in the cell the support is the ribosome), both involve activation of the amino acid (in the cell the amino acid is activated by the energy-rich molecule adenosine triphosphate, or ATP) and both are stepwise processes. It even has been learned recently that in bacteria the synthesis starts with one end of the chain protected and that an enzyme later carries out a deprotection step just as we do in the laboratory [see "How Proteins Start," by Brian F. C. Clark and Kjeld A. Marcker; SCIENTIFIC AMERICAN, Offprint 1092].

The analogy should not be pushed too far, however. The natural synthesis is much more elegant and efficient than the laboratory one, and no one presumes to duplicate or even approach the complexities of the cell's scheme at this point. As a matter of fact, the chemical synthesis was not even patterned after nature; it was only in retrospect that the similarities became evident. Nevertheless, it could well be that in the future organic chemists may benefit from a more complete understanding of the way in which living systems perform their tasks.

An Automatic System

It was clear to us from the outset that an automatic mechanized process was needed for making large peptides, and that the solid phase approach would be well suited to such a process. This is so

BEADS OF POLYSTYRENE, on which amino acid subunits are assembled in the "solid phase" method to make a peptide chain, are enlarged 300 diameters. They average 50 microns (.002 inch) in diameter but become about twice as large when swollen in a solvent to make them more reactive. About a trillion peptide chains can be "grown" on a single bead.

because the intermediate products in the synthesis need not be isolated but are purified by simple filtration and washing reactions that can be carried out in a single vessel; the manipulations required to transfer products from one container to another have been eliminated. Once the resin beads with an amino acid attached are placed in the vessel it is only necessary to introduce the appropriate liquid solvent or reagent, allow it to react, remove the excess reagent and by-products by filtration and then repeat the process with the next reagent.

It is easy to visualize how all these steps can be accomplished automatically by a rather simple device. We constructed a machine that consists essentially of two parts: a reaction vessel with the plumbing necessary to introduce and remove the solvents in the right order at the right times and a programmer that controls these operations. The solvents and the reagents, contained in a series of reservoirs, are selected one at a time by a specially designed rotary valve. The solvents are introduced into the bottom of the reaction vessel by a metering pump while air is displaced at the top. Valves to the vessel are then closed and a mechanical shaker mixes the reactants for a predetermined time. Next a vacuum withdraws solvent through a porous glass filter disk in the bottom of the ves-

sel while dry air enters at the top; the beads, with the peptide attached, remain in the vessel. One cycle of the synthesis (the lengthening of the peptide chain by one amino acid) requires 12 different reagents, one of which is a protected amino acid. The next cycle calls for the same series of reagents except for a different amino acid, which is selected by a second rotary valve.

All the steps just described are controlled by a "stepping-drum" programmer. It is like an old-fashioned music box. Once the pins have been positioned on the drum to play the proper tune the machine takes over and directs the chemical synthesis. The pins activate switches that turn the pump and shaker on and off and open and close the valves at the proper times and in the proper sequence. One cycle of the synthesis requires 100 steps of the drum and takes about four hours, so that it is now possible to carry out automatically all the operations required for the assembly of a peptide chain at the rate of six amino acids a day.

After the details of the synthetic scheme had been worked out by the synthesis of small peptides the procedure was given a more demanding test: the preparation of the hormone bradykinin, a nine-amino-acid peptide with several physiological activities that served as sensitive criteria to demonstrate the identity and purity of the final product. The synthetic bradykinin was identical in all respects, both chemically and biologically, with the natural hormone. During the past four years the solid phase technique has been applied to the preparation of bradykinin analogues (nearly 100 of which have been made by J. M. Stewart at Rockefeller University) and other small peptide hormones such as angiotensin and oxytocin. It has also been used in our laboratory and others for the synthesis of the antibiotics gramicidin-S and tyrocidin, and for synthetic studies of the immunological determinants of hemoglobin and tobacco mosaic virus. It was clear that peptides containing 10 or 20 amino acids could be made by the solid phase method as well as by classical procedures. The important

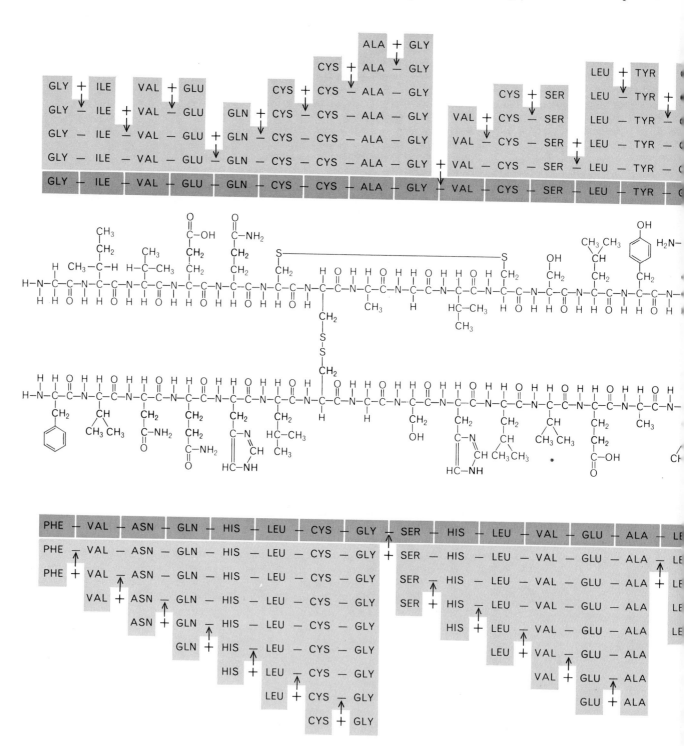

FIRST SYNTHESIS of insulin chains was accomplished by the fragment method, as illustrated here. The primary structure of the two chains is diagrammed at the center, together with the names of the amino acid units (*see key at top right*). Note the disulfide (S—S) bonds between cysteine units, two of which link the *A* (*top*) and *B* (*bottom*) chains. The general pattern of the synthesis of the

question then became whether or not molecules as large and complex as proteins could be synthesized by this method.

Synthesis of Insulin

The smallest molecule that qualifies as a true protein is insulin. It naturally became the object of intensive synthetic work by several groups of chemists when, in the late 1950's, it seemed likely that the synthesis of a protein was a feasible goal. Insulin was chosen not only for its size but also for several other important reasons. The availability of synthetic hormone would help to answer many questions about its mechanism of action. Most important, the composition and complete primary structure (the sequence of amino acid units) had become known a few years before through the work of the group led by Frederick Sanger at the University of Cambridge [see "The Insulin Molecule," by E. O. P. Thompson; SCIENTIFIC AMERICAN, Offprint 42].

The insulin molecule is much more complex than a simple peptide such as bradykinin [see illustration on this page]. It not only has nearly six times as many amino acids but also has a greater variety of them: 17 rather than five. This introduces many new problems of side-chain protection. Particularly complicating is the presence of three disulfide-bond (S—S) cross-links between cysteine units. Insulin consists of two linear peptide chains: an A chain with 21 amino acids and a B chain with 30. They are held together by two interchain disulfide bridges, and in addition one of the chains has an intrachain disulfide loop. Furthermore, the molecule has a definite three-dimensional conformation. Although the X-ray structure has not yet

ALA	ALANINE
ARG	ARGININE
ASN	ASPARAGINE
CYS	CYSTEINE
GLN	GLUTAMINE
GLU	GLUTAMIC ACID
GLY	GLYCINE
HIS	HISTIDINE
ILE	ISOLEUCINE
LEU	LEUCINE
LYS	LYSINE
PHE	PHENYLALANINE
PRO	PROLINE
SER	SERINE
THR	THREONINE
TYR	TYROSINE
VAL	VALINE

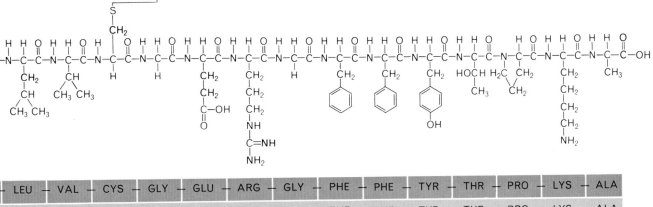

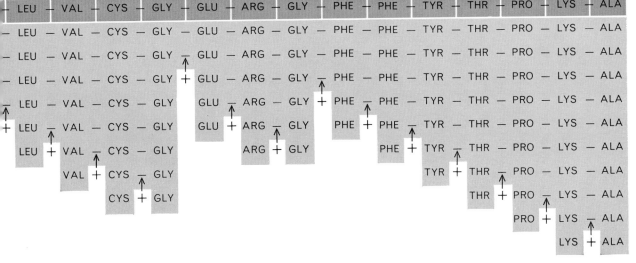

A chain by P. G. Katsoyannis' group at the University of Pittsburgh is shown above the insulin formula. The 21 amino acid units were assembled stepwise into intermediate fragments of from two to five units and the fragments were then coupled in stages to assemble the complete chain. The similar synthesis of the B chain by Helmut Zahn's group in Germany is shown below the insulin formula.

been worked out in detail, it is clear from the fact that insulin forms characteristic crystals that it is composed of molecules with a precise structure.

How can one hope to build up the long peptide chains, to form the three disulfide bonds between the correct cysteine units and then to fold the entire assembly into its proper shape? This is asking a lot, because there are many possible ways for the S—S bonds to form, and the possible variations in the conformation of the molecule are enormous. It is only possible at this time because nature comes to the chemist's aid. If we simply make the two chains with the six cysteine units all in the reduced (SH) form (in which a hydrogen atom is attached to each sulfur atom) and mix them under the proper oxidizing conditions, they will preferentially form the correct S—S bridges and fold into the characteristic insulin conformation all by themselves!

The discovery in 1960, by G. H. Dixon and A. C. Wardlaw of the University of Toronto, that this would happen provided the key peptide chemists needed to undertake the synthesis of insulin. All four laboratories that have made insulin have depended on this fact and have made the two chains separately. From the point of view of the chemist this is really peptide synthesis rather than protein synthesis, but when the two chains are combined, the final product meets all the usual criteria for a protein and justifies the conclusion that a real protein has been synthesized in the laboratory.

The first published synthesis of an individual insulin chain was made by P. G. Katsoyannis and his colleagues at the University of Pittsburgh School of Medicine in 1963. They made the 21-residue A chain of sheep insulin by the fragment method. When the synthetic A chain was linked with the natural B chain, the combination gave rise to a small but definite amount of insulin activity. Later that same year a large group of chemists at the Technische Hochschule at Aachen in Germany, under the direction of Helmut Zahn, reported the synthesis of both the A and the B chain and the successful combination of the two for the first total synthesis of insulin. The overall yields (2.9 percent for the A chain and 7 percent for the B) and the extent of the combination (.2 to 1 percent) were still low, but true insulin activity was obtained. These syntheses, which also followed the fragment approach, required 89 reaction steps for the A chain, 132 steps for the B chain and three more steps for the combination of the two. Each step, of course,

required numerous operations.

During the same period a third group was working on insulin at the Academy of Science in Shanghai and the University of Peking. Their first important contribution to the problem was the development of improved methods for the separation and recombination of natural insulin chains. Their yield was eventually increased to about 50 percent, which meant that the combination was far from a random process. Their major contribution was the preparation in 1965 of the first crystalline, all-synthetic insulin. The crystals were obtained in low yield, but they had the same form as the native molecule and, most important, the full biological activity (more than 20 units per milligram). This was a crucial element of the proof that insulin had in fact been synthesized.

Automatic Synthesis of Insulin

There remained a very real problem. Large numbers of chemists had to work for several years to produce tiny quantities of the peptides. In order to produce useful amounts of insulin and to be able to make modifications in the structure in a more efficient way, we undertook in 1965 to apply the solid phase method to the task. The results have been very encouraging. Although more than 5,000

separate operations were required to assemble the 51 amino acids into the two chains of bovine insulin, most of these were performed automatically under the control of the drum programmer, so that it was possible for one man to carry out the synthesis of both chains in only a few days.

Beginning with three grams of resin, Arnold Marglin of Rockefeller University was able to prepare approximately two grams of protected A chain. A total of eight grams of the protected B chain was made on eight grams of the resin. The reaction that detached the peptide chains from their polystyrene support also removed most of the side-chain protecting groups, leaving only the benzyl groups on the cysteine and histidine side

SOLID PHASE SYNTHESIS of the *A* chain (*left*) and the *B* chain (*right*) is diagrammed. An amino acid protected by a Boc group (*vertical bar*) is coupled to a polystyrene bead (*top*), then deprotected. Activated amino acids, protected at the amino end and if necessary

chains. These could be removed by reduction with metallic sodium dissolved in liquid ammonia, a reaction that was discovered many years ago by Vincent du Vigneaud of the Cornell University Medical College and was the key to his historic syntheses of the pituitary hormones oxytocin and vasopressin. Applied to insulin, however, the sodium treatment at first broke some of the bonds between the amino acids threonine and proline in the B chain. Once we recognized what was happening it was possible to keep the chains from splitting by careful modification of the conditions of the reaction.

This deprotecting step left the cysteine groups in the reduced (SH) form. Although it was just this SH form that

we would later want for the final oxidation step to link the two chains, the SH groups were too unstable to undergo the purification procedures that were now necessary; they were stabilized by conversion to S-sulfonates (SSO_3^-). Then the two peptide chains could be purified by three methods: filtration, which depends on molecular size; countercurrent distribution, which depends on differential solubility, and free-flow electrophoresis, which depends on electric charge. The resulting products could be shown to be homogeneous by other electrophoretic and chromatographic criteria. Amino acid analyses showed that the chains had the compositions characteristic of the A and B chains of insulin.

The final step in the synthesis of in-

sulin was the combination of the two purified chains. First the cysteine sulfonates were converted back to the SH form. Then the chains were combined by the method developed by the Chinese, which involves the slow oxidation by air of the SH forms of both chains

ALA- ●

LYS-ALA- ●

PRO-LYS-ALA- ●

THR-PRO-LYS-ALA- ●

TYR-THR-PRO-LYS-ALA- ●

PHE-TYR-THR-PRO-LYS-ALA- ●

PHE-PHE-TYR-THR-PRO-LYS-ALA- ●

GLY-PHE-PHE-TYR-THR-PRO-LYS-ALA- ●

ARG-GLY-PHE-PHE-TYR-THR-PRO-LYS-ALA- ●

GLU-ARG-GLY-PHE-PHE-TYR-THR-PRO-LYS-ALA- ●

GLY-GLU-ARG-GLY-PHE-PHE-TYR-THR-PRO-LYS-ALA- ●

CYS-GLY-GLU-ARG-GLY-PHE-PHE-TYR-THR-PRO-LYS-ALA- ●

VAL-CYS-GLY-GLU-ARG-GLY-PHE-PHE-TYR-THR-PRO-LYS-ALA- ●

LEU-VAL-CYS-GLY-GLU-ARG-GLY-PHE-PHE-TYR-THR-PRO-LYS-ALA- ●

TYR-LEU-VAL-CYS-GLY-GLU-ARG-GLY-PHE-PHE-TYR-THR-PRO-LYS-ALA- ●

LEU-TYR-LEU-VAL-CYS-GLY-GLU-ARG-GLY-PHE-PHE-TYR-THR-PRO-LYS-ALA- ●

ALA-LEU-TYR-LEU-VAL-CYS-GLY-GLU-ARG-GLY-PHE-PHE-TYR-THR-PRO-LYS-ALA- ●

GLU-ALA-LEU-TYR-LEU-VAL-CYS-GLY-GLU-ARG-GLY-PHE-PHE-TYR-THR-PRO-LYS-ALA- ●

VAL-GLU-ALA-LEU-TYR-LEU-VAL-CYS-GLY-GLU-ARG-GLY-PHE-PHE-TYR-THR-PRO-LYS-ALA- ●

LEU-VAL-GLU-ALA-LEU-TYR-LEU-VAL-CYS-GLY-GLU-ARG-GLY-PHE-PHE-TYR-THR-PRO-LYS-ALA- ●

HIS-LEU-VAL-GLU-ALA-LEU-TYR-LEU-VAL-CYS-GLY-GLU-ARG-GLY-PHE-PHE-TYR-THR-PRO-LYS-ALA- ●

SER-HIS-LEU-VAL-GLU-ALA-LEU-TYR-LEU-VAL-CYS-GLY-GLU-ARG-GLY-PHE-PHE-TYR-THR-PRO-LYS-ALA- ●

GLY-SER-HIS-LEU-VAL-GLU-ALA-LEU-TYR-LEU-VAL-CYS-GLY-GLU-ARG-GLY-PHE-PHE-TYR-THR-PRO-LYS-ALA- ●

CYS-GLY-SER-HIS-LEU-VAL-GLU-ALA-LEU-TYR-LEU-VAL-CYS-GLY-GLU-ARG-GLY-PHE-PHE-TYR-THR-PRO-LYS-ALA- ●

LEU-CYS-GLY-SER-HIS-LEU-VAL-GLU-ALA-LEU-TYR-LEU-VAL-CYS-GLY-GLU-ARG-GLY-PHE-PHE-TYR-THR-PRO-LYS-ALA- ●

HIS-LEU-CYS-GLY-SER-HIS-LEU-VAL-GLU-ALA-LEU-TYR-LEU-VAL-CYS-GLY-GLU-ARG-GLY-PHE-PHE-TYR-THR-PRO-LYS-ALA- ●

GLN-HIS-LEU-CYS-GLY-SER-HIS-LEU-VAL-GLU-ALA-LEU-TYR-LEU-VAL-CYS-GLY-GLU-ARG-GLY-PHE-PHE-TYR-THR-PRO-LYS-ALA- ●

ASN-GLN-HIS-LEU-CYS-GLY-SER-HIS-LEU-VAL-GLU-ALA-LEU-TYR-LEU-VAL-CYS-GLY-GLU-ARG-GLY-PHE-PHE-TYR-THR-PRO-LYS-ALA- ●

VAL-ASN-GLN-HIS-LEU-CYS-GLY-SER-HIS-LEU-VAL-GLU-ALA-LEU-TYR-LEU-VAL-CYS-GLY-GLU-ARG-GLY-PHE-PHE-TYR-THR-PRO-LYS-ALA- ●

PHE-VAL-ASN-GLN-HIS-LEU-CYS-GLY-SER-HIS-LEU-VAL-GLU-ALA-LEU-TYR-LEU-VAL-CYS-GLY-GLU-ARG-GLY-PHE-PHE-TYR-THR-PRO-LYS-ALA- ●

HBr in NH3
PHE-VAL-ASN-GLN-HIS-LEU-CYS-GLY-SER-HIS-LEU-VAL-GLU-ALA-LEU-TYR-LEU-VAL-CYS-GLY-GLU-ARG-GLY-PHE-PHE-TYR-THR-PRO-LYS-ALA

Na in NH3
PHE-VAL-ASN-GLN-HIS-LEU-CYS-GLY-SER-HIS-LEU-VAL-GLU-ALA-LEU-TYR-LEU-VAL-CYS-GLY-GLU-ARG-GLY-PHE-PHE-TYR-THR-PRO-LYS-ALA

Na2SO3, Na2S4O6 SO_3^- SO_3^-
PHE-VAL-ASN-GLN-HIS-LEU-CYS-GLY-SER-HIS-LEU-VAL-GLU-ALA-LEU-TYR-LEU-VAL-CYS-GLY-GLU-ARG-GLY-PHE-PHE-TYR-THR-PRO-LYS-ALA

at the side chain (*horizontal bars*), are then coupled stepwise. When the chain is complete, it is cleaved from the bead, and most protecting groups are removed, by hydrogen bromide treatment. Sodium treatment removes groups protecting the cysteine and histidine side chains. Then the cysteines are changed from the sulfhydryl form to the stable S-sulfonate form in preparation for purification.

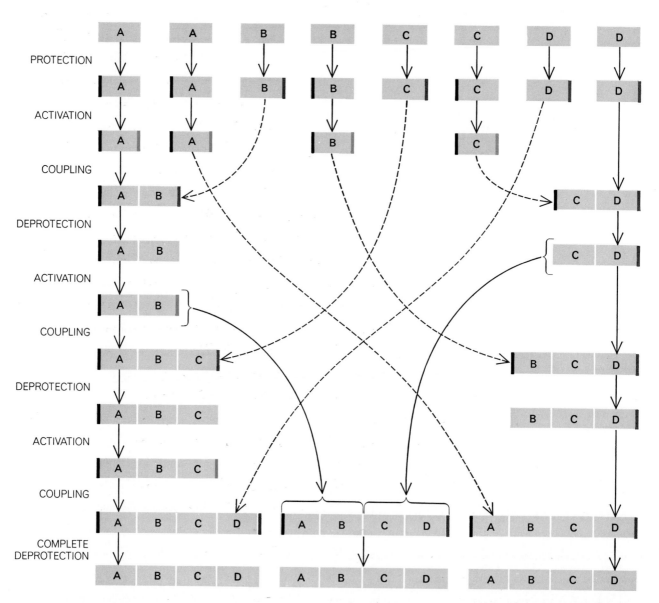

a *b* *c*

AMINO ACID (*a*) has an amino (NH$_2$) and a carboxyl (COOH) group separated by a carbon atom that carries a side chain (*R*). The peptide bond (*b*) forms between the carboxyl and the amino ends of two units with the elimination of a molecule of water. A series of amino acid subunits held together by such bonds (*color*) constitutes the primary ("backbone") structure of peptides and proteins (*c*).

TWO GENERAL APPROACHES to peptide synthesis are diagrammed. In the "stepwise" method amino acid units are added one at a time starting at the amino end of the peptide (*left*) or the carboxyl end (*right*) until the final peptide (of four amino acids, *A, B, C* and *D*, in this case) is assembled (*bottom left and right*). In the "fragment" method small peptides are prepared stepwise and are then combined (*long solid arrows*) to form the final peptide (*bottom center*). In each case it is necessary to protect the amino ends (*black bars*) and carboxyl ends (*gray bars*) against unwanted reactions, to activate the carboxyl ends (*dark-color bars*) for coupling, to "deprotect" selectively in preparation for the next step and finally to deprotect completely. The peptide can then be purified.

[*see illustration below*]. We were able to show that either of our synthetic chains could be combined with the complementary natural chain to produce biologically active, semisynthetic insulin. Then the two synthetic chains were combined to form all-synthetic insulin. The synthetic hormone was active in the standard biological assay, which is based on the amount that must be injected to lower the blood sugar enough to cause

convulsions in 50 percent of a group of experimental mice. The response to the synthetic preparations was shown to be due to low blood sugar rather than to some nonspecific toxic effect because the animals recovered rapidly after they were given glucose. In addition, the synthetic material behaved like insulin in various physical and chemical tests. For example, its mobility in paper electrophoresis was the same as that of the

natural hormone.

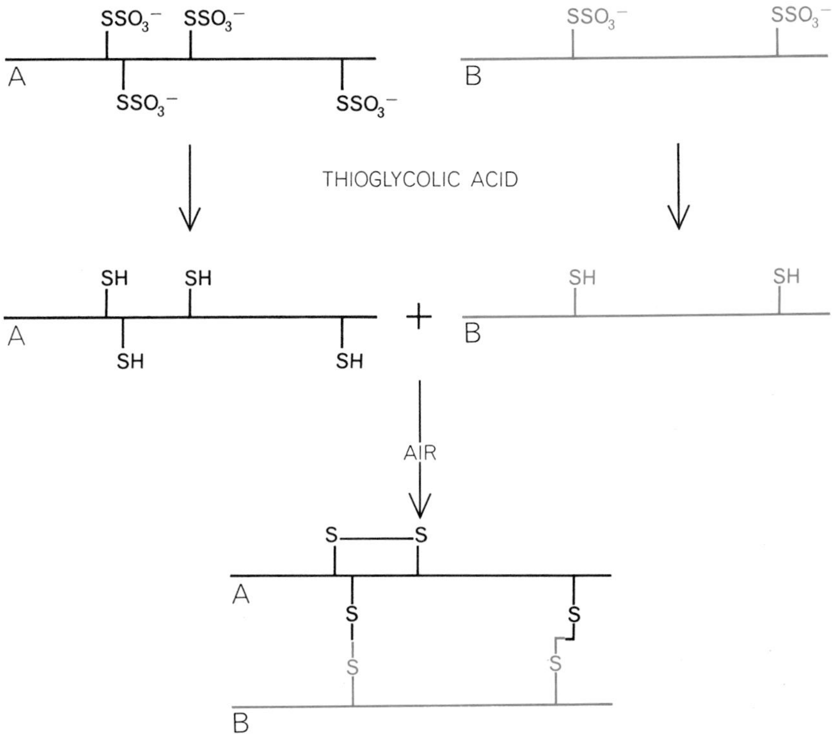

The yield and the purity of the chains themselves, which constitute the peptide synthesis portion of the work, are quite good, but our combination yields are still poor. Important progress in that regard has been made recently by Katsoyannis and his co-workers. They have modified one of the earlier Chinese methods (using the SH form of the *A* chain and the SSO_3^- form of the *B* chain) and can now obtain much improved yields in the final step of the synthesis. It is fair to conclude that the synthesis of one protein has been accomplished and that another major hurdle has been cleared by chemists in their continuing struggle to duplicate nature.

Current developments in peptide chemistry should bring within reach other small protein molecules that are of great biological interest. The structures of myoglobin, cytochrome *c*, ferredoxin and growth hormone are known, for example, and we can expect some of them to be synthesized. A major step toward the synthesis of living systems will come with the synthesis of virus-coat protein; that of the tobacco mosaic virus may be the first.

It is the enzymes that are probably of greatest current interest. It is important to learn how these complex protein molecules function in their control of biochemical reactions. Why are they such enormously active catalysts and why are they so specific in their action? What factors are responsible for the "active centers" of enzymes and for their specific binding sites? How does the primary structure of the protein control its three-dimensional structure and its function; how will changes in the amino acid sequence influence biological activity? Automatic solid phase syntheses should help to answer some of these questions.

COMBINATION of the *A* and *B* chains, now purified and in the S-sulfonate form, was carried out by the method developed in China. The chains were mixed and reduced to the sulfhydryl form with thioglycolic acid. On exposure to air in an alkaline solution the sulfhydryl groups oxidized slowly and the three disulfide bonds characteristic of natural insulin were formed. By-products formed by incorrect cross-linking could be removed by extraction.

30

THE STRUCTURE AND HISTORY OF AN ANCIENT PROTEIN

RICHARD E. DICKERSON
April 1972

*To oxidize food molecules all organisms from yeasts
to man require a variant of cytochrome c. Differences
in this protein from species to species provide a
1.2-billion-year record of molecular evolution*

Between 1.5 and two billion years ago a profound change took place in some of the single-celled organisms then populating our planet, a change that in time would contribute to the rise of many-celled organisms. The machinery evolved for extracting far more energy from foods than before by combining food molecules with oxygen. One of the central components of the new metabolic machinery was cytochrome *c*, a protein whose descendants can be found today in every living cell that has a nucleus. By studying the cytochrome *c* extracted from various organisms it has been possible to determine how fast the protein has evolved since plants and animals diverged into two distinct kingdoms and in fact to provide an approximate date of 1.2 billion years ago for the event. For example, the cytochrome *c* molecules in men and chimpanzees are exactly the same: in the cells of both the molecule consists of 104 amino acid units strung together in exactly the same order and folded into the same three-dimensional structure. On the other hand, the cytochrome *c* in man has diverged from the cytochrome *c* in the red bread mold *Neurospora crassa* in 44 out of 104 places, yet the three-dimensional structures of the two cytochrome *c* molecules are essentially alike. We think we can now explain how it is that so many of the 104 amino acid units in cytochrome *c* are interchangeable and also why certain units cannot be changed at all without destroying the protein's activity.

Let us try to visualize the earth before cytochrome *c* first appeared. The first living organisms on the planet were little more than scavengers, extracting energy-rich organic compounds (includ-

ing their neighbors) from the water around them and releasing low-energy breakdown products. We still have the "fossils" of this life-style in the universal process of anaerobic (oxygenless) fermentation, as when a yeast extracts energy from sugar and releases ethyl alcohol, or when an athlete who exercises too rapidly converts glucose to lactic acid and gets muscle cramps. Anaerobic fermentation is part of the common biochemical heritage of all living things.

The upper limit on how much life the planet could support with only fermentation as an energy source was determined by the rate at which high-energy compounds were synthesized by nonbiological agencies: ultraviolet radiation, lightning discharges, radioactivity or heat. When some organisms developed the ability to tap sunlight for energy, photosynthesis was born and the life-carrying capacity of the earth increased enormously. This was the age of the bac-

SKELETON OF CYTOCHROME *c* MOLECULE is depicted in the illustration by Irving Geis on the opposite page. A variant of this protein molecule is found in the cells of every living organism that utilizes oxygen for respiration. The illustration shows in simplified form how 104 amino acid units are linked in a continuous chain that grips and surrounds a heme group, a complex rosette with an atom of iron (*Fe*) at its center. The picture is color-coded to indicate how much variation has been tolerated by evolution at each of the 104 amino acid sites in the molecule. Some species lack the 104th amino acid, and all species except vertebrates have as many as eight extra amino acids at the beginning of the chain (*see table on pages 4 and 5*). The amino acids that are most invariant throughout evolution, and presumably the most important, are shown in red and orange; the more variable sites appear in yellow-green, blue-green, blue and purple. The indispensable heme group is crimson. Each amino acid is represented only by its "alpha" carbon atom: the atom that carries a side chain unique for each of the 20 amino acids. The upper drawing at left below shows how two amino acids link up through an amide group (*colored panel*); the side chains connected to the alpha carbons (*color*) are represented by the balls labeled R. The lower drawing at left below shows the scheme used in the cytochrome *c* skeleton on the opposite page; all amide linkages (—CO—NH—) are omitted and the only side groups shown are those that are attached to the heme. The amino acids at the 35 invariant sites of the cytochrome *c* molecule (*red*) are designated in abbreviated form (*see key at right below*).

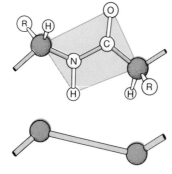

Ala	Alanine	Leu	Leucine
Asp	Aspartic acid	Lys	Lysine
Asn	Asparagine	Met	Methionine
Arg	Arginine	Phe	Phenylalanine
Cys	Cysteine	Pro	Proline
Gly	Glycine	Ser	Serine
Glu	Glutamic acid	Thr	Threonine
Gln	Glutamine	Trp	Tryptophan
His	Histidine	Tyr	Tyrosine
Ile	Isoleucine	Val	Valine

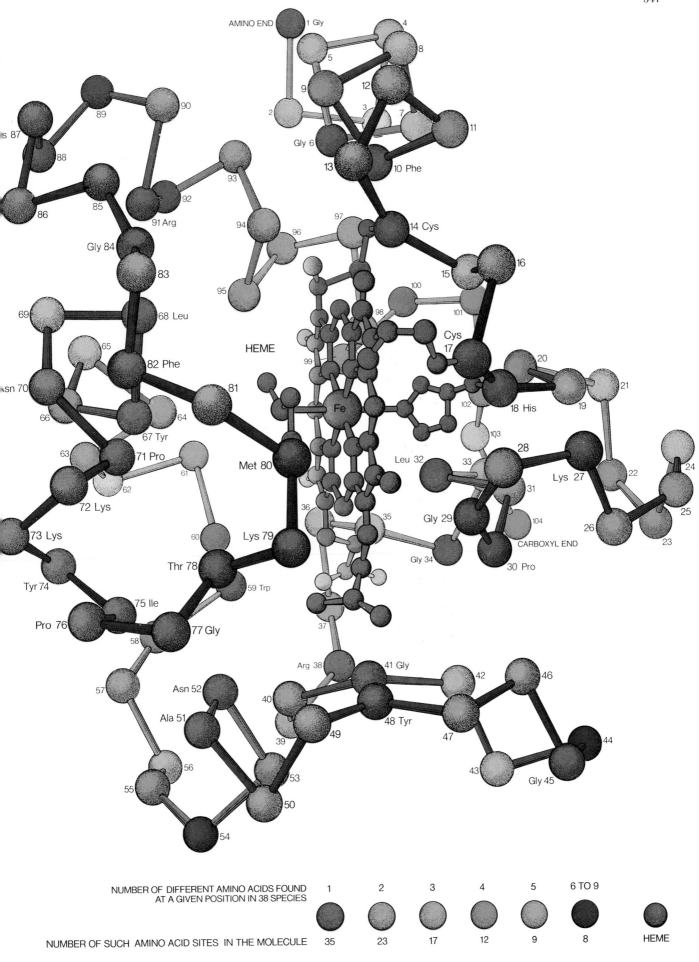

AMINO END 1 Gly
4
8
5
9 12
2
Gly 6 3 7 11
13 10 Phe
14 Cys
97
15 16
100
98 101
Cys
17
99 20
18 His 21
102 19
HEME
Leu 32 103
69 28
68 Leu 33 31 22
82 Phe 81 104 24
Gly 29
Asn 70 65 Lys 27
66 67 Tyr 64 CARBOXYL END 25
63 71 Pro 61 Met 80 36 35 30 Pro 26
62 23
72 Lys Gly 34
73 Lys 60 Lys 79
Thr 78 59 Trp
Tyr 74 75 Ile 77 Gly
Pro 76 58 37
57 Arg 38 41 Gly 42 46
Asn 52 40 48 Tyr 47 44
Ala 51 39 49 43 Gly 45
56
55 53
50
54

89 90
s 87 88
86 85 92
91 Arg 93
94 96
95
83
Gly 84
Fe

NUMBER OF DIFFERENT AMINO ACIDS FOUND
AT A GIVEN POSITION IN 38 SPECIES

1	2	3	4	5	6 TO 9	

NUMBER OF SUCH AMINO ACID SITES IN THE MOLECULE   35   23   17   12   9   8   HEME

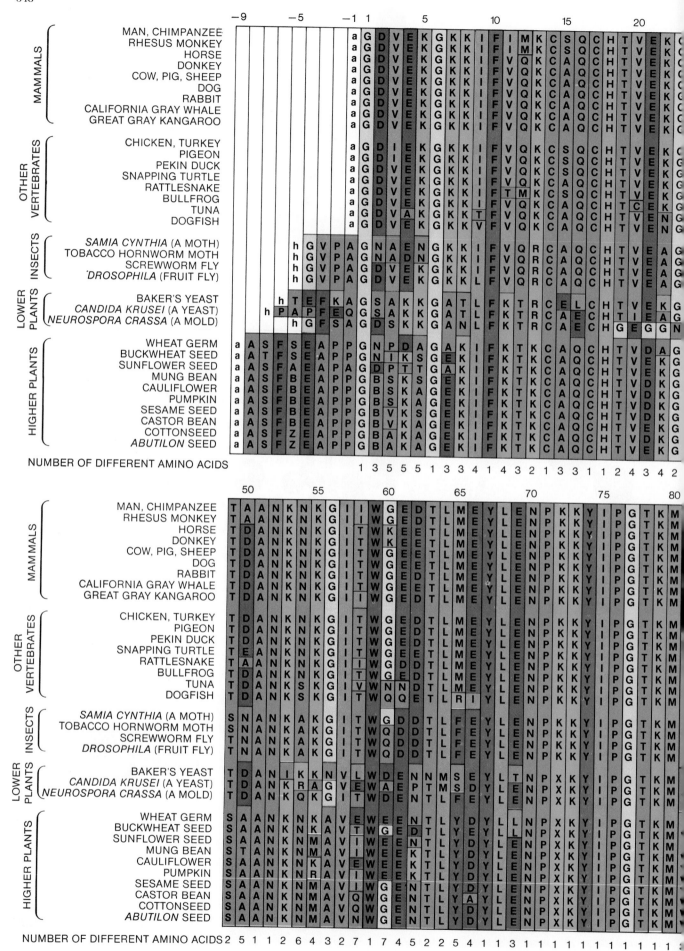

NUMBER OF DIFFERENT AMINO ACIDS

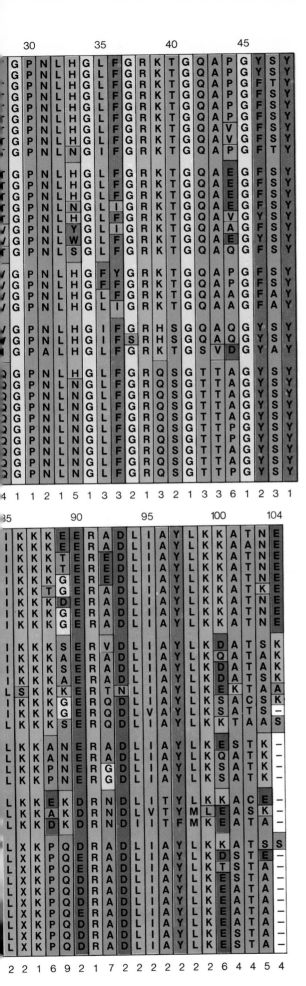

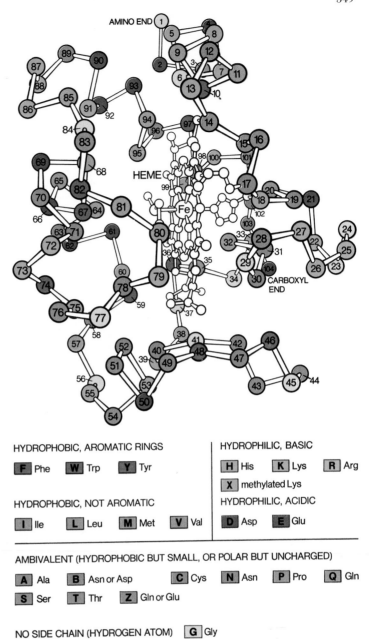

HYDROPHOBIC, AROMATIC RINGS

| F | Phe | W | Trp | Y | Tyr |

HYDROPHOBIC, NOT AROMATIC

| I | Ile | L | Leu | M | Met | V | Val |

HYDROPHILIC, BASIC

| H | His | K | Lys | R | Arg |
| X | methylated Lys |

HYDROPHILIC, ACIDIC

| D | Asp | E | Glu |

AMBIVALENT (HYDROPHOBIC BUT SMALL, OR POLAR BUT UNCHARGED)

| A | Ala | B | Asn or Asp | C | Cys | N | Asn | P | Pro | Q | Gln |
| S | Ser | T | Thr | Z | Gln or Glu |

NO SIDE CHAIN (HYDROGEN ATOM) | G | Gly

COMPOSITION OF CYTOCHROME *c* IN 38 SPECIES is presented in the table at left. No other protein has been so fully analyzed for so many different organisms. The color code used here differs from the one used in the molecular skeleton on page 347. On these two pages color is employed to classify amino acids according to their chemical properties (*see key directly above*). Thus the three "oily" (hydrophobic) amino acids with aromatic benzene rings in their side chains (phenylalanine, tryptophan and tyrosine) are shown in red. Four other amino acids that are hydrophobic but nonaromatic are shown in orange. At the other extreme, amino acids that are hydrophilic, or water-loving, are shown in blue or violet. Amino acids that can be found in either aqueous or nonaqueous environments (and hence are ambivalent) are green or yellow. Polar amino acids can have asymmetric distributions of positive and negative charge. The detailed structure of the side chains of the amino acids can be found on page 351. It is easy to pick out from the table at left the amino acid sites where evolution has allowed no change or has allowed substitution only by chemically similar amino acids; these sites are identified by vertical bands of a single color. A letter *a* at the beginning of the chain indicates that a methyl group (CH₃) is attached to the amino end of the molecular chain. A letter *h* indicates that the methyl group is absent.

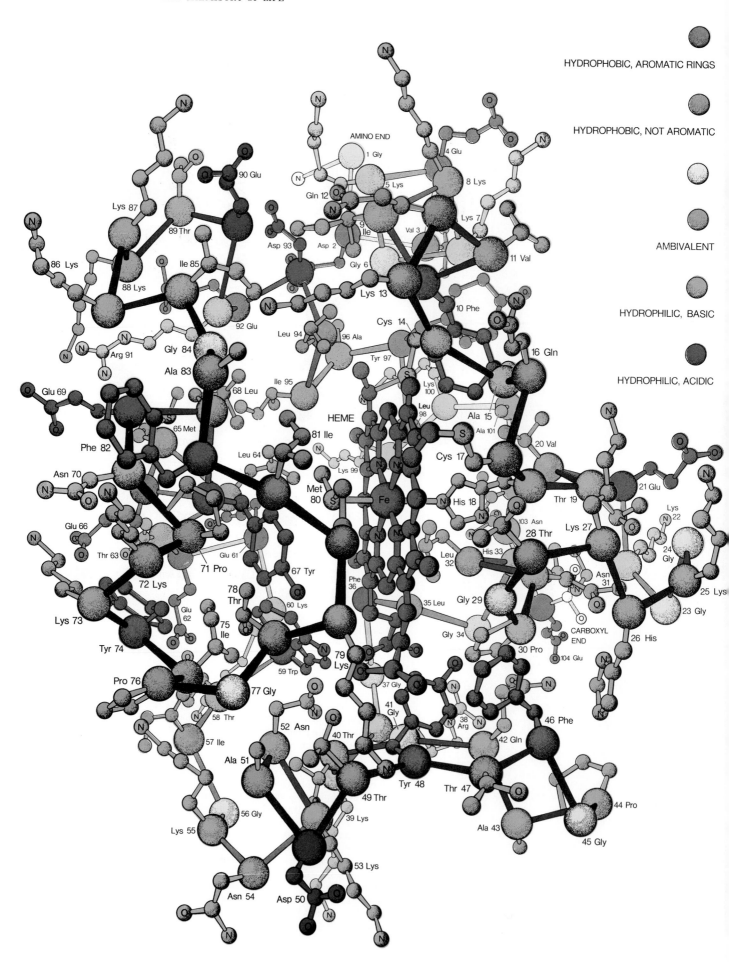

HYDROPHOBIC, AROMATIC RINGS

HYDROPHOBIC, NOT AROMATIC

AMBIVALENT

HYDROPHILIC, BASIC

HYDROPHILIC, ACIDIC

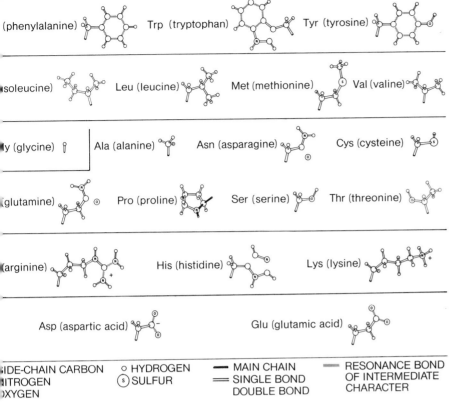

(phenylalanine)   Trp (tryptophan)   Tyr (tyrosine)

soleucine)   Leu (leucine)   Met (methionine)   Val (valine)

y (glycine)   Ala (alanine)   Asn (asparagine)   Cys (cysteine)

glutamine)   Pro (proline)   Ser (serine)   Thr (threonine)

arginine)   His (histidine)   Lys (lysine)

Asp (aspartic acid)   Glu (glutamic acid)

IDE-CHAIN CARBON	○ HYDROGEN	— MAIN CHAIN	░░ RESONANCE BOND
IITROGEN	Ⓢ SULFUR	= SINGLE BOND	OF INTERMEDIATE
OXYGEN		DOUBLE BOND	CHARACTER

CYTOCHROME *c* MOLECULE WITH SIDE CHAINS appears in the illustration on the opposite page. The picture shows the structure of horse-heart cytochrome *c* in the oxidized state as determined through X-ray crystallography by the author and his colleagues at the California Institute of Technology. Alpha-carbon atoms are numbered and the amide groups (−CO−NH−) connecting the alpha carbons are represented only by a solid bond, as in the preceding drawings. For clarity three side chains at the "back" of the molecule have been left out: leucine 35, phenylalanine 36 and leucine 98. The color coding follows the coding in the illustration on pages 348 and 349. One can see from the three-dimensional structure that side chains in the interior of the molecule, around the heme group (*crimson*), tend to be hydrophobic (*red and orange*), whereas amino acids with hydrophilic side chains (*blue and violet*) are found on the outside, where they are ordinarily in contact with water. A major exception to this rule is the hydrophobic side chain of phenylalanine 82, which sits on the surface of the molecule at the left of the heme. The region between the hydrophilic chains, and above isoleucine 81, is a cavity that is apparently open to solvent molecules. Lysine 13, above this cavity, is known to interact with a large oxidase complex when cytochrome *c* is oxidized. The structures of the side chains of 20 amino acids appear above.

within an organized nucleus, and their respiratory and photosynthetic machinery (if it is present) is similarly dispersed. Green algae and all the higher plants and animals are eukaryotes (cells with "good" nuclei); their DNA is organized within a nucleus, and their respiration is carried out in the organelles called mitochondria. In eukaryote plants photosynthesis is conducted in still other organelles called chloroplasts. Mitochondria are the powerhouse of all eukaryote cells. Their role is to break down the energy-rich molecules obtained from foods, combine them with oxygen and store the energy produced by harnessing it to synthesize molecules of adenosine triphosphate (ATP). The mitochondria of all eukaryotes are alike in their chemistry, as if once the optimum chemical mechanism had been arrived at it was never changed.

Biological oxidation involves at least a score of special enzymes that act first as acceptors and then as donors of the electrons or hydrogen atoms removed from food molecules. In the last part of the process one finds a series of cytochrome molecules (identified by various subscript letters), all of which incorporate a heme group containing iron, the same heme group found in hemoglobin. Electrons are passed down a chain of cytochrome molecules: from cytochrome b to cytochrome c_1, from cytochrome c_1 to cytochrome c, to cytochromes a and a_3 and finally to oxygen atoms, where they are combined with hydrogen ions to produce water. This is a stepwise process designed to release energy in small parcels rather than all at once. In the transfer of electrons from cytochrome b to cytochrome c_1 and again in the transfer between the cytochromes a, a_3 and oxygen, energy is channeled off to synthesize ATP, which acts as a general-purpose energy source for cell metabolism.

Most of the cytochromes are bound tightly to the mitochondrial membrane, but one of them, cytochrome c, can easily be solubilized in aqueous mediums and can be isolated in pure form. The other components can be isolated as multienzyme complexes: b and c_1 as a cytochrome reductase complex, and a and a_3 as a cytochrome oxidase. The reductase donates electrons to cytochrome c; the oxidase accepts them again. To illustrate how similar all eukaryotes really are to one another, it has been found that cytochrome c from any species of plant, animal or eukaryotic microorganism can react in the test tube with the cytochrome oxidase from any other species. Worm or primate, whale

teria and the blue-green algae [see "The Oldest Fossils," by Elso S. Barghoorn; SCIENTIFIC AMERICAN, Offprint 895].

The more advanced forms of photosynthesis released a corrosive and poisonous gas into the atmosphere: oxygen. Some bacteria responded by retreating to oxygen-free corners of the planet, where their descendants are found today. Other bacteria and blue-green algae developed ways to neutralize gaseous oxygen by combining it with their own waste products. The next step was to harness the energy released by oxidation of these waste compounds. (If you are going to burn your garbage, you might as well keep warm by the fire.) This was the beginning of oxidation, or respiration, the second big breakthrough in increasing the supply of energy available to life on the earth.

When a yeast cell oxidizes sugars all the way to carbon dioxide and water instead of stopping short at ethyl alcohol, it gets 19 times as much energy per gram of fuel. When oxygen combines with lactic acid in the athlete's muscles and the cramps are dissipated, he receives a correspondingly greater energy return from his glucose. Any improvement in metabolism that multiplies the supply of energy available by such a large factor would be expected to have a revolutionary effect on the development of life. We now believe the specialization of cells and the appearance of multicelled plants and animals could only have come about in the presence of such a large new supply of energy.

Bacteria and blue-green algae are prokaryotes (prenuclear cells); their genetic material, DNA, is not confined

or wheat are all alike under the mito-chondrial membrane.

The Evolution of Cytochrome *c*

Since cytochrome *c* is so ancient and at the same time so small and easily purified, it has received much attention from protein chemists interested in the evolutionary process. The complete amino acid sequence of cytochrome *c* has been determined for more than 40 species of eukaryotic life. Thirty-eight of these sequences are compared in the illustration on pages 348 and 349. We have more information on the evolution of this molecule than on the evolution of any other protein.

Emanuel Margoliash of Northwestern University and Emil Smith of the University of California at Los Angeles were among the first to notice that the amino acid sequences from various species are different and that the degree of difference corresponds quite well with the distance that separates the two species on the evolutionary tree. Detailed computer analyses of these differences by Margoliash, by Walter Fitch of the University of Wisconsin and by others have led to the construction of elaborate family trees of living organisms entirely without recourse to the traditional anatomical data. The family trees agree remarkably well with those obtained from classical morphology; it is obvious that

comparison of amino acid sequence is a powerful tool for studying the process of evolution.

Another result may at first be surprising. Cytochrome *c* is still evolving slowly and is doing so at a rate that is approximately constant for all species, when the rate is averaged over geological time periods. This kind of analysis of molecular evolution was first carried out on hemoglobin a decade ago by Linus Pauling and Emile Zuckerkandl at the California Institute of Technology. If we compare hemoglobin and cytochrome *c*, we find that cytochrome *c* is changing much more slowly. Why should this be? The protein chains are synthesized from instructions that are embodied in DNA, and it is in the DNA that mutations take place. Do mutations occur more often in the DNA that makes hemoglobin than in the DNA that makes cytochrome *c*? There is no reason to think so. The explanation therefore must lie in the natural-selection, or screening, process that tests whether or not mutant molecules can do their job.

Before discussing the various "formulas" that have passed the test of making a successful cytochrome *c*, I shall describe briefly the structure of proteins. All protein molecules are built up by linking amino acids end to end. Each of the 20 different amino acids has a carboxyl group (–COOH) at one end and an amino group (–NH$_2$) at the other. To link the carboxyl group of one amino acid with the amino group of another amino acid a molecule of water must be removed, producing an amide linkage (–CO–NH–). Because only a part (although, to be sure, the distinctive part) of an amino acid enters a protein chain, the chemist refers to it as a "residue." Thus he speaks of a glycine residue or a phenylalanine residue at such-and-such a position in a protein chain.

The carbon adjacent to the amide linkage is called the alpha carbon. It is important because each amino acid has a distinctive side chain at this position. The side chain may be nothing more than a single atom of hydrogen (as it is in the case of the amino acid glycine) or it may consist of a number of atoms, including a six-carbon "aromatic" ring (as it does in the case of phenylalanine, tryptophan and tyrosine).

The 20 amino acids can be grouped into three broad classes, depending on the character of their side chains [*see illustration on preceding page*]. Five are hydrophilic, or water-loving, and tend to acquire either a positive or a negative charge when placed in aqueous solution; three of the five are basic in character

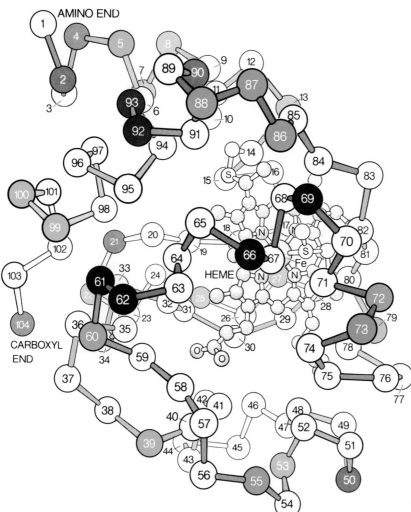

PLOT OF DISTRIBUTION OF ELECTRIC CHARGES on the back of horse-heart cytochrome *c* reveals that most of the 19 hydrophilic lysines (*color*), which carry positive charges (and hence are basic), are distributed on the two flanks of the molecule. Nine of 12 negatively charged (acidic) side chains (*gray*) are clustered in one zone in the upper center of the molecule. The electrically negative character of this zone has been maintained throughout evolution, although the specific locations of the acidic side groups vary. No organism, from wheat germ to man, has a cytochrome *c* with fewer than six acidic amino acids in this zone and no organism has more than five acidic amino acids everywhere else on the molecule. Furthermore, these extreme values are not found in the same species. It is highly likely that these charged zones participate in binding cytochrome *c* to other large molecules.

(arginine, histidine and lysine) and the other two are acidic (aspartic acid and glutamic acid). Seven are not readily soluble in water and hence are termed hydrophobic; they include the three amino acids mentioned above that have rings in their side chains plus leucine, isoleucine, methionine and valine. The remaining eight amino acids react ambivalently to water: alanine, asparagine, cysteine, glutamine, glycine, proline, serine and threonine.

Now let us see how much the successful formulas for cytochrome c differ from species to species. The cytochrome c molecules of men and horses differ by 12 out of 104 amino acids. The cytochrome c's of the higher vertebrates—mammals, birds and reptiles—differ from the cytochrome c's of fishes by an average of 19 amino acids. The cytochrome c's of vertebrates and insects differ by an average of 27 amino acids; moreover, the cytochrome c molecules of insects and plants have a few more amino acid residues at the beginning of the chain than the equivalent molecules of vertebrates. The greatest disparity between two cytochrome c's is the one between man and the bread mold *Neurospora;* they differ at more than 40 percent of their amino acid positions. How can two molecules with such large differences in amino acid composition perform identical chemical functions?

We begin to see an answer when we look at where these changes are. Some parts of the amino acid sequence, as indicated on pages 4 and 5, never vary. Thirty-five of the 104 amino acid positions in cytochrome c are completely invariant in all known species, including a long sequence from residue 70 through residue 80. The 35 invariant sites are occupied by 15 different amino acids; they are shown in red in the structural drawing on page 3. Another 23 sites are occupied by only one of two different but closely similar amino acids. There are 18 different sets of interchangeable pairs at these 23 sites; they are shown in orange in the illustration. At 17 sites natural selection has evidently accepted only sets of three different amino acids; these 17 interchangeable triplets are colored yellow-green.

It was already known from sequence studies, before the X-ray structural analysis, that where such substitutions are allowed the interchangeable amino acids almost always have the same chemical character. In general all must be either hydrophilic or hydrophobic or else neutral with respect to water. Such interchanges are called conservative substitutions because they conserve the

overall chemical nature of that part of the protein molecule.

In only a few places along the chain can radical changes be tolerated. Residue 89, for example, can be acidic (aspartic acid or glutamic acid), basic (lysine), polar but uncharged (serine, threonine, asparagine and glutamine), weakly hydrophobic (alanine) or devoid of a side chain (glycine). Almost the only type of side chain that appears to be forbidden at this point in the molecule is a large hydrophobic one. Such "indifferent" regions are rare, however, and cytochrome c overall is an evolutionarily conservative molecule.

We have no reason to think the gene for cytochrome c mutates more slowly than the gene for hemoglobin, or that the invariant, conservative and radical regions of the sequence reflect any difference in mutational rate within the cytochrome c gene. The mutations are presumably random, and what we see in these species comparisons are the molecules that are left after the rigid test of survivability has been applied. Invariant regions evidently are invariant because any mutational changes there are lethal and are weeded out. Conservative changes can be tolerated elsewhere as long as they preserve the essential chemical properties of the molecule at that point. Radical changes presumably indicate portions of the molecule that do not matter for the operation of the protein.

This is as far as we can go from sequence comparisons alone. The explanation of variability in terms of the essential or nonessential character of different parts of the molecule is plausible, yet science has always been plagued by plausible but incorrect hypotheses. To progress any further we need to know how the amino acid sequence is folded to make an operating molecule. In short, we need the three-dimensional structure of the protein.

The Molecule in Three Dimensions

With the active collaboration of Margoliash, who was then working at the Abbott Laboratories in North Chicago, I began the X-ray-crystallographic analysis of horse-heart cytochrome c at Cal Tech in 1963, with the sponsorship of the National Science Foundation and the National Institutes of Health. As cytochrome c transfers electrons in the mitochondrion, it oscillates between an oxidized form (ferricytochrome) and a reduced form (ferrocytochrome); the iron atom in the heme group is alternately in the +3 and +2 oxidation state.

We decided to begin our analysis with the oxidized form, a decision that was largely tactical since both oxidation states would ultimately be needed if we were to try to decipher the electron-transfer process.

In X-ray crystallography one directs a beam of X rays at a purified crystal of the substance under study and records the diffraction pattern produced as the beam strikes the sample from different angles. X rays entering the sample are themselves deflected at various angles by the distribution of electron charges within the crystal. Highly sophisticated computer programs have been devised for deducing from tens of thousands of items of X-ray-diffraction data the three-dimensional distribution of electronic charge. From this distribution one can infer, in turn, the distribution of the amino acid side chains in the protein molecule.

We obtained our first low-resolution map of the oxidized form of horse-heart cytochrome c five years ago and the first high-resolution map three years ago. These maps have been used to construct detailed three-dimensional models of the protein. One can also feed the three-dimensional coordinates into a computer and obtain simple ball-and-stick drawings that can be viewed stereoptically, enabling one to visualize the folded chain of the protein in three dimensions [see illustrations on next two pages]. Just a year ago we calculated the first high-resolution map for the reduced form of cytochrome c. We are now improving this model and comparing the two oxidation states.

Several striking features of the amino acid sequences of cytochrome c were in the back of our minds as we worked out the first high-resolution structure. We knew that the most strongly conserved sites throughout evolution were those occupied by three distinctive types of residue: the positively charged (basic) residues of lysine; the three hydrophobic and aromatic residues of phenylalanine, tryptophan and tyrosine, and the four hydrophobic but nonaromatic residues of leucine, isoleucine, methionine and valine. These sites can now be located with the help of the illustration on page 350, whose color coding differs from the coding of the illustration on page 347. Here hydrophobic residues are shown in warm colors (red and orange) whereas neutral residues and hydrophilic residues, both basic and acidic, are shown in cool colors (green, yellow, blue and violet).

It had been known from the chemical analysis of the molecule's amino acid se-

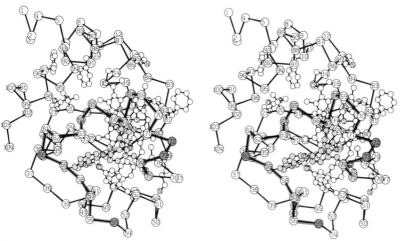

STEREOSCOPIC PAIR OF LEFT SIDE of oxidized cytochrome *c* molecule, drawn by computer, shows only a few key side chains for clarity. The main chain (*color*) from sites 55 to 75 defines a loop, the "left channel," which is filled with strongly hydrophobic side chains; their alpha carbons are in light color. Three of these side chains include aromatic rings: tryptophan 59, tyrosine 67 and tyrosine 74, also shown in light color. Alpha carbons with hydrophilic, positively charged side chains around the left channel are shown in dark color. This pair and one on opposite page can be viewed with standard stereoscopic viewer.

quence that the basic and hydrophobic groups tend to appear in clusters along the chain. For example, basic residues are found in the regions of sites 22 through 27, sites 38 and 39, sites 53 through 55 and sites 86 through 91. Hydrophobic residues are found in regions 9 through 11, 32 through 37, 80 through 85 and 94 through 98. The residues at sites 14 and 17 (cysteine) and site 18 (histidine) are invariant, which is understandable since they form bonds to the heme group. Less understandably, the long stretch from site 70 to site 80 is equally invariant. Before the structural evidence was available it had been suspected that methionine, at site 80, might be bonded to the iron atom on the other side of the heme from the histidine at site 18, but it was impossible to be sure from chemical evidence alone.

It was also known from chemical analysis that horse cytochrome *c* incorporates 12 glycines (the residues with only hydrogen as a side chain) and that these glycines were either invariant or else conserved in the great majority of species. It was known too that of the eight phenylalanines or tyrosines (with aromatic rings in their side chains) seven are either invariant in all species or replaceable only by one another. In the case of residue 36, phenylalanine or tyrosine is replaced in three species by isoleucine, whose side chain, although it is nonaromatic, is at least as large and hydrophobic as the side chains it replaces.

All these similarities and conservatisms were known before the X-ray analysis, but none could be explained in terms of structure. It was assumed that every residue had been placed where it was by natural selection and that it contained potentially important information about the working parts of the cytochrome molecule. Natural selection, however, does not act on an amino acid sequence but rather on the folded and operating molecule in its association with other biological molecules. Having a sequence without the folding instructions is like having a list of parts without a blueprint of the entire machine.

Cytochrome *c* and Evolution

Now that the blueprint for cytochrome *c* is revealed, let us look more closely at its representation on page 3. To keep the illustration simple no side chains have been included except for those that are bonded to the heme group. Moreover, along the main chain the illustration depicts only the alpha-carbon atoms from which side chains would, if they were shown, branch off. The amide groups (–CO–NH–) that connect alpha carbons are represented simply by straight lines. The picture is therefore a simplified folding diagram of the cytochrome molecule.

We see that the flat heme group, a symmetrical rosette of carbon and nitrogen atoms with an atom of iron at its center, sits in a crevice with only one edge exposed to the outside world. If the heme participates directly in shuttling electrons in and out of the molecule, the transfer probably takes place along this edge. Cysteines 14 and 17 and histidine 18 hold the heme in place

from the right as depicted, and the other heme-binding group on the left is indeed methionine 80, as had been suspected.

It was known from earlier X-ray studies of proteins that sequences of amino acids frequently fold themselves into the helical configuration known as the alpha helix; in other cases the amino acids tend to assume a rippled or corrugated configuration called a beta sheet. Cytochrome *c* has no beta sheets and only two stretches of alpha helix, formed by residues 1 through 11 and 89 through 101. For the most part the protein chain is wrapped tightly around the heme group, leaving little room for the alpha and beta configurations that are prominent in other proteins.

Just as one can use cytochrome *c* to learn about evolution, one can also use evolution to learn about cytochrome *c*. As I have noted, the illustration on page 3 is color-coded to indicate the amount of variability in the kind of amino acid tolerated at each site. The structure is "hot" (red and orange) in the functionally important places in the molecule when differences among species are absent or rare, and it is "cool" (green, blue and violet) in regions that vary widely from one species to another and thus are presumably less important to a viable molecule of cytochrome *c*.

The heme crevice is hot, indicating that strong selection pressures tend to keep the environment of the heme group constant throughout evolution. The invariant residues 70 through 80 are also hot, and we now see that they are folded to make the left side of the molecule and the pocket in which the heme sits. The right side of the molecule is warm, consisting of sites where only one, two or three different amino acids are tolerated. The back of the molecule is its cool side; residues 58 and 60 and four more residues on the back of the alpha helix are each occupied by six or more different amino acids in various species. These are powerful clues to the important parts of the molecule, whether for electron transfer or for interaction with two large molecular complexes, the reductase and the oxidase.

How the Molecule Folds Itself

If we now turn to the illustration on page 350, which shows all the side chains of horse-heart cytochrome *c*, many of the evolutionary conservatisms become understandable. (As before, amide groups are still shown only as straight lines; their atomic positions are known but are not particularly relevant to this article.) In this illustration the colors are selected to classify the various sites according to the character of the amino acid tolerated (hydrophilic, hydrophobic or ambivalent); the same color coding applies in the illustration on pages 348 and 349 showing the amino acid sequences in the cytochrome *c*'s of 38 different species.

Nonpolar, hydrophobic groups are found predominantly on the inside of the molecule, away from the external aqueous world, whereas charged groups, acidic or basic, are always on the outside. This arrangement is a good example of the "oil drop" model of a folded protein. According to this model, when an amino acid chain is synthesized inside a cell, it is helped to fold in the proper way by the natural tendency of hydrophobic, or "oily," side chains to retreat as far as possible from the aqueous environment and cluster in the center of the molecule. An even stronger statement can be made: If it is necessary for the successful operation of a protein molecule that certain portions of the polypeptide chain be folded into the interior, then natural selection will favor the retention of hydrophobic side chains at that point so that the proper folding is achieved. A charged, or hydrophilic, side chain can be pushed into the interior of a protein molecule, but a considerable price must be paid in terms of energy. Thus in most cases the presence of a charged group at a given site helps to ensure that the chain at that point will be on the outside of the folded molecule. (Charged groups inside a protein are known only in one or two cases where they play a role in the catalytic mechanism of the protein.)

We can now see the reason for the evolutionary conservatism of hydrophobic side chains, and one of the reasons for the conservatism of the hydrophilic residue lysine: they help to make the molecule fold properly. Radical changes of side chain that prevent proper folding are lethal. No folding, no cytochrome; no cytochrome, no respiration; no respiration, no life. It is seldom that cause and effect in evolution are quite so clear-cut.

There is still more to the lysine story. The lysines are not only on the outside; they are clustered in two positively charged regions of the molecular surface, separated by another zone of negative charge. This segregation of charge has not been found in any other protein structure and, as we shall see, probably occurs because cytochrome *c* interacts with two molecular complexes (the reductase and the oxidase) rather than with small substrate molecules as an en-

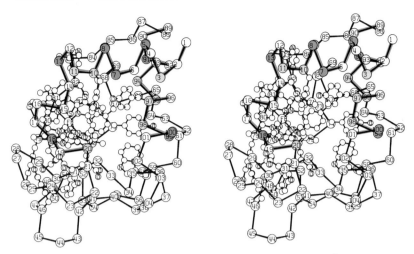

STEREOSCOPIC PAIR OF RIGHT SIDE of oxidized cytochrome *c* shows two sequences forming alpha helixes: the sequence from 1 through 11 and the sequence from 89 through 101. The two alpha helixes and the chain from 12 through 20 outline the right channel. Like the left channel it is lined with hydrophobic side groups, but it apparently contains a slot large enough to receive a hydrophobic side chain from another molecule. As in the stereoscopic drawing on the opposite page, alpha carbons with positively charged side chains around this channel are indicated in dark color; alpha carbons with strongly hydrophobic side chains are indicated in light color. The computer program for preparing the stereoscopic pictures was written by Carroll Johnson of the Oak Ridge National Laboratory.

zyme does. The charge arrangements are believed to be part of the process by which large molecules recognize each other.

Most of the 19 lysines are found on the left and right sides of the molecule, viewed from the back on page   . The left and right sides of the molecule can be examined separately in the two stereoscopic pairs on these two pages. On the left side eight lysines surround a loop of chain from sites 55 through 75 that is tightly packed with hydrophobic groups, including the invariant tyrosine 74, tryptophan 59 and tyrosine 67 farther inside. Although we do not yet know the electron-transfer mechanism, it has been suggested that the aromatic rings of the three invariant residues could provide an inward path for the electron when cytochrome *c* is reduced. Another eight lysines are found on the right side, on the periphery of what appears to be a true channel large enough to hold a hydrophobic side chain of another large molecule. This right-side channel is bounded by the two alpha helixes and by the continuation from the first alpha helix through residues 12 to 20. Within this channel are found two large aromatic side chains: phenylalanine 10 (which cannot change) and tyrosine 97 (which can also be phenylalanine but nothing else). In summary, on the right side is a channel lined with hydrophobic groups (including two aromatic rings) and surrounded by an outer circle of positive charges. As someone in our laboratory remarked on looking at the

model, it resembles a docking ring for a spaceship.

This remark may not be entirely frivolous. It is known from chemical work that the attraction between cytochrome *c* and the cytochrome oxidase complex is largely electrostatic, involving negatively charged groups on the oxidase and positively charged basic groups on cytochrome *c*. Either the left or the right cluster of lysines must be involved in this binding. Moreover, Kazuo Okunuki of the University of Osaka has shown that if just one positive charge, lysine 13, is blocked with a bulky aromatic chemical group, the reactivity of cytochrome with its oxidase is cut in half. Chemically blocking lysine 13 means physically blocking the upper part of the heme crevice. Lysine 13 is closer to the right cluster of positive charges than to the left; thus it would appear more likely that the heme crevice and the right channel together are the portions of the molecular surface that "see" the oxidase complex.

What, then, are the roles of the positive zone on the left side and the negative patch at the rear? The positive zone, with its three aromatic rings, may be the binding site to the reductase; we know virtually nothing about the chemical nature of this binding. The negative patch may be a "trash dump," an unimportant part of the molecule's surface where there are enough negative charges to prevent an excessively positive overall charge. The fact that the six most variable amino acid sites are in this part

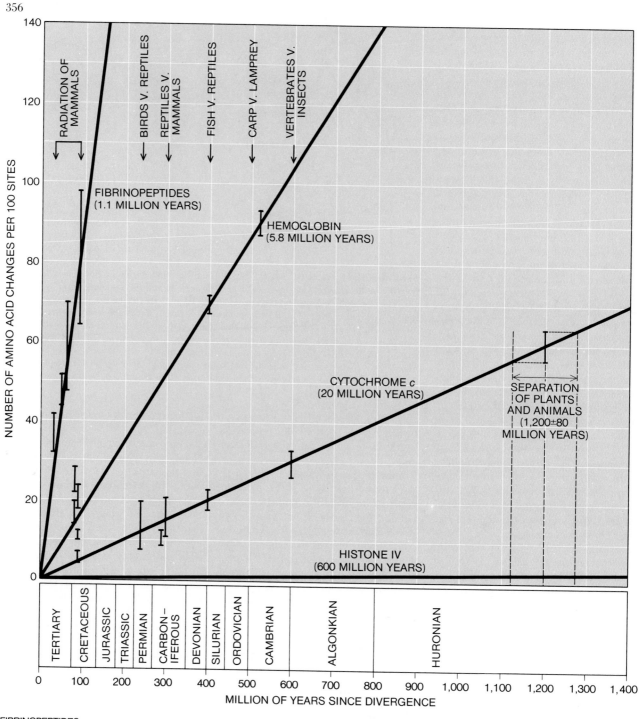

NUMBER OF AMINO ACID CHANGES PER 100 SITES

RADIATION OF MAMMALS

BIRDS V. REPTILES

REPTILES V. MAMMALS

FISH V. REPTILES

CARP V. LAMPREY

VERTEBRATES V. INSECTS

FIBRINOPEPTIDES
(1.1 MILLION YEARS)

HEMOGLOBIN
(5.8 MILLION YEARS)

CYTOCHROME c
(20 MILLION YEARS)

SEPARATION
OF PLANTS
AND ANIMALS
(1,200±80
MILLION YEARS)

HISTONE IV
(600 MILLION YEARS)

TERTIARY | CRETACEOUS | JURASSIC | TRIASSIC | PERMIAN | CARBON-IFEROUS | DEVONIAN | SILURIAN | ORDOVICIAN | CAMBRIAN | ALGONKIAN | HURONIAN

MILLION OF YEARS SINCE DIVERGENCE

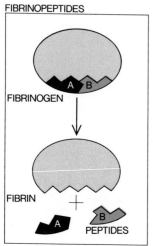

FIBRINOPEPTIDES

FIBRINOGEN

A B

FIBRIN

A

B
PEPTIDES

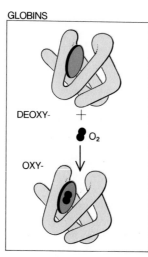

GLOBINS

DEOXY- +

O₂

OXY-

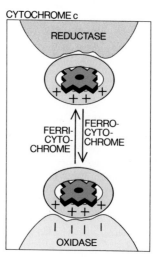

CYTOCHROME c

REDUCTASE

FERRI-
CYTO-
CHROME

FERRO-
CYTO-
CHROME

OXIDASE

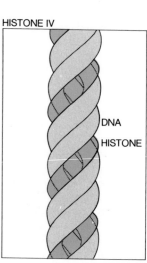

HISTONE IV

DNA

HISTONE

of the molecule would support such an idea.

On the other hand, it is equally possible that this collection of negative charges has a function. Acidic amino acids are actually conserved throughout the various species, although in a subtle way that was overlooked in the earlier sequence comparisons. Selection pressures have kept this zone of the molecular surface negative, even though the individual residues that carry the negative charges differ from one species to another. Because several sections of the protein chain bend into and out of this acidic region, the conservation of negative charge is not immediately obvious if one looks only at the stretched-out sequence. This is a good illustration of the principle of molecular evolution that natural selection acts on the folded, functioning protein and not on its amino acid sequence alone.

If we look carefully at where the glycine residues are, we can appreciate why such a large number are evolutionarily invariant. The heme group is so large that 104 amino acids are barely enough to wrap around it. There are many places where a chain comes too close to the heme or to another chain for a side chain to fit in. It is just at these points that we find the glycines with their single hydrogen atom as a side chain.

The last type of conservatism, the conservatism of the aromatic side chains, is more difficult to explain. Tyrosines and phenylalanines tend to occur in nearby pairs in the folded cytochrome c molecule: residues 10 and 97 in the right channel, 46 and 48 below the heme crevice, 67 and 74 along with tryptophan 59 in the hydrophobic left channel. Only residue 36, which can be tyrosine, phenylalanine or isoleucine, seems to have merely a space-filling role on the back of the molecule; it is an "oily brick."

The three aromatic rings in the left channel may be involved in electron transfer during reduction. The two rings in the right channel could also be employed in electron transfer, or might only help to define the hydrophobic slot in the middle of that channel. Tyrosine 48 at the bottom of the molecule helps to hold the heme in place by making a hydrogen bond to one of the heme's propionic acid side chains. In cytochrome c from the tuna and the bonito, where residue 46 is tyrosine, electron-density maps have shown that this residue also holds the heme by a hydrogen bond to its other propionic acid group. These two tyrosines, along with cysteines 14 and 17, help to lock the heme in place in a way not seen in hemoglobin or myoglobin.

Phenylalanine 82 is the enigma. Never tyrosine or anything else, it extends its oily side chain out into the aqueous world on the left side of the heme crevice, where it has no visible role. A price must be paid in energy for its being there. Why should such a large hydrophobic group be on the outside of the molecule, and why should it be absolutely unchanging through the entire course of evolution? Viewing the oxidized molecule alone, it is impossible to say, but when at the end of this article we look briefly at the recently revealed structure of reduced cytochrome, we shall see the answer fall into place at once.

The structural reasons for the evolutionary conservatism in cytochrome c throughout the history of eukaryotic life can now largely be explained. Cytochrome c is unique among the structurally analyzed proteins in that it has segregated regions of charge on its surface. The roles assigned to these regions in the foregoing discussions have been speculative and may be quite wrong. What we can be sure of is that these regions do have roles in the operation of the molecule. Chance alone, or even common ancestry, could not maintain

these positive and negative regions, along with paired and exposed aromatic groups, in all species through more than a billion years of molecular evolution. The conservative sequences are shouting to us, "Look!" Now we have to be clever enough to know what to look for.

Rates of Protein Evolution

With this background we are equipped to return to a question raised earlier: What determines the rates of evolution of different proteins? We begin by making a graph where the vertical axis represents the average difference in amino acid sequence between two species of organism on two sides of an evolutionary branch point, for example the branch point between fish and reptiles or between reptiles and mammals. The horizontal axis represents the time elapsed since the divergence of the two lines as determined by the geological record. If such a graph is plotted for cytochrome c, one finds that all the branch points fall close to a straight line, indicating a constant average rate of evolutionary change [see illustration on opposite page].

How can this be? How can cytochrome c change at so nearly a constant rate during the long period in which the external morphology of the organism was diversifying toward the present-day cotton plant, bread mold, fruit fly, rattlesnake and chimpanzee? This is an illustration of a fundamental advantage of proteins as tools in studying evolution. Natural selection ultimately operates on populations of whole living organisms, the only criterion of success being the ability of the population to survive, reproduce and leave behind a new generation. The farther down toward the molecular level one goes in examining living organisms, the more similar they become and the less important the morphological differences are that separate a clam from a horse. One kind of chemical machinery can serve many diverse organisms. Conversely, one external change in an organism that can be acted on by natural selection is usually the effect not of a single enzyme molecule but of an entire set of metabolic pathways.

The observed uniform rate of change in cytochrome c simply means that the biochemistry of the respiratory package, the mitochondrion, is so well adjusted, and the mitochondrion is so well insulated from natural selection, that the selection pressures become smoothed out at the molecular level over time spans of millions of years. A factory can convert

RATES OF EVOLUTION OF PROTEINS (opposite page) can be inferred by plotting average differences in amino acid sequences between species on two sides of an evolutionary branch point that can be dated, for example the branch point between fish and reptiles or between reptiles and mammals. The average differences (vertical axis) have been corrected to allow for the occurrence of more than one mutation at a given amino acid site. The length of the vertical data bars indicates the experimental scatter. Times since the divergence of two lines of organisms from a common branch point (horizontal axis) have been obtained from the geological record. The drawings below the graph show schematically the function (described in the text) of the molecules whose evolutionary rate of change is plotted. The rate of change is proportional to the steepness of the curve. It can be represented by a number called the unit evolutionary period, which is the time required for the amino acid sequence of a protein to change by 1 percent after two evolutionary lines have diverged. For fibrinopeptides this period is about 1.1 million years, whereas for histone IV it is 600 million years. The probable reasons for these differences are discussed in the text.

from making military tanks to making sports cars and keep the same machine tools and power source. Similarly, a primitive eukaryote cell line can lead to such diverse organisms as sunflowers and mammals and still retain a common metabolic chemistry, including the respiratory package that comprises cytochrome c. One of the advantages of proteins in studying the process of evolution is just this relative insulation from the immediate effects of external selection. Protein structure is farther removed from selection pressures and closer to the sources of genetic variation in DNA than gross anatomical features or inherited behavior patterns are.

The only other proteins for which enough sequence information is available to allow this kind of analysis are hemoglobin and the fibrinopeptides: the short amino acid chains left over when fibrinogen is converted to fibrin in the process of blood clotting. One hemoglobin chain consists of approximately 140 amino acids. Fibrinopeptides A and B, on the other hand, consist of only about 20 amino acids, which are cut out of fibrinogen and discarded during the clotting process. The hemoglobins and the fibrinopeptides also appear to be evolving individually at a uniform average rate, but their rates are quite different. Whereas 20 million years are required to produce a change of 1 percent in the amino acid sequence of two diverging lines of cytochrome c, the same amount of change takes a little less than six million years in hemoglobins and just over one million years in the fibrinopeptides, as indicated on page 356. The approximate time required for a 1 percent change in sequence to appear between diverging lines of the same protein is defined as the unit evolutionary period. That period has been roughly estimated for a number of proteins for which only two or three sequences from different species are known. Most simple enzymes evolve approximately as fast as hemoglobin and much more rapidly than cytochrome c. Does all of this mean that the genes for these proteins are mutating at different rates? Are we looking at differences in variation or in selection?

Since there is no evidence to suggest variable rates of mutation, one asks what case can be made for differences in selection pressure among the different proteins. The case appears to be quite convincing [see inset figures in illustration on page 12]. The fibrinopeptides are "spacers" that prevent fibrinogen from adopting the fibrin configuration before the clotting mechanism is trig-

gered. As long as they can be cut out by an enzyme when the time comes for the blood to clot, they would seem to have few other requirements. Thus one would expect a fibrinogen molecule to tolerate many random changes in the fibrinopeptide spacers. If the unit evolutionary period measures not the rate of appearance of mutations but the rate of appearance of harmless mutations, then it is not surprising that a 1 percent change can occur in the sequence of fibrinopeptides in just over a million years.

A successful hemoglobin molecule has more constraints. Each hemoglobin molecule embodies four heme groups that not only bind oxygen but also cooperate in such a way that the oxygen is released more rapidly into the cell when the local acidity, created by the presence of carbon dioxide, builds up. The structural basis for this "breathing" mechanism has only recently been explained with the help of X-ray crystallography by M. F. Perutz and his co-workers at the Medical Research Council Laboratory of Molecular Biology in England. If a random mutation is five times as likely to be harmful in hemoglobin as in the fibrinopeptides, one can account for hemoglobin's having a unit evolutionary period that is five times as long.

The chances of randomly damaging cytochrome c are evidently three to four times greater than they are for hemoglobin. Why should this be, and why should the unit evolutionary period for cytochrome c be greater than the period for enzymes of comparable size? The X-ray structure has given us a clue to the answer. Cytochrome c is a small protein that interacts over a large portion of its surface with molecular complexes that are larger than itself. It is virtually a "substrate" for the reductase and oxidase complexes. A large fraction of its surface is subject to strong conservative selection pressures because of the requirement that it mate properly with other large molecules, each with its own genetic blueprint. This evidently explains why the patches of positive and negative charge are preserved so faithfully throughout the history of eukaryotic life. Hemoglobin and most enzymes, in contrast, interact principally with smaller molecules: with oxygen in the case of hemoglobin or with small substrate molecules at the active sites of enzymes. As long as these restricted regions of the molecule are preserved the rest of the molecular surface is relatively free to change. Mutations are weeded out less rigorously and sequences diverge faster.

A satisfying confirmation of these

ideas comes from the amino acid sequences of histone IV, one of the basic proteins that binds to DNA in the chromosomes and that may play a role in expressing or suppressing genetic information. When molecules of histone IV from pea seedlings and calf thymus are compared, one finds that they differ in only two of their 102 amino acids. If we adopt an approximate date of 1.2 billion years ago from the cytochrome study for the divergence of plants and animals, we find that histone IV has a unit evolutionary period of 600 million years. Clearly the conservative selection pressure on histone IV must be intense. Since histone IV participates in the control processes that are at the heart of the genetic mechanism, its sensitivity to random changes is hardly surprising.

The date of 1.2 billion years ago for the divergence of plants and animals is based on the cytochrome-sequence comparisons, assuming that the observed linear rate of evolution of cytochrome c in more recent times can be extrapolated back to that remote epoch. Is this a fair extrapolation? It probably is for cytochrome c because the biochemistry of the mitochondrion evolved still earlier; the great similarity in respiratory reactions among all eukaryotes argues that there has not been much innovation in cytochrome systems since. The respiratory chain had probably "settled down" by 1.2 billion years ago. It is reassuring that the cytochrome figure of 1.2 billion years is in harmony with the relatively scarce fossil record of Precambrian life.

If one accepts the provocative suggestion that eukaryotes developed from a symbiotic association of several prokaryotes, one of which was a respiring bacterium that became the ancestor of present-day mitochondria, one is obliged to conclude that the respiratory machinery had stabilized in essentially its present form before or during this symbiosis. The same thing cannot be said for hemoglobin and its probable ancestor, myoglobin. They can provide no clue to the date when animals and plants diverged, since the globins were evolving to play several different roles during and after this period, as multicelled organisms arose. In no sense had the globins settled down 1.2 million years ago. Nevertheless, if the right proteins are selected, and if the data are not overextended, it should be possible to use the rates of protein evolution to assign times to events in the evolution of life that have left only faint traces in the geological record.

So far we have mentioned the electron-transfer mechanism of cytochrome *c* only in passing, virtually ignoring the structure of the reduced molecule. The mechanism is another story in itself, and one that cannot yet be written. One hopes that the clues supplied by X-ray analysis will suggest the best chemical experiments to try next, in order to learn the mechanism of the oxidation-reduction process. The reduced cytochrome structure has been obtained so recently that it would be premature to base many deductions on it.

A Glimpse of Molecular Dynamics

One obvious structural feature, which undoubtedly has great physiological significance, is that in the reduced molecule the top of the heme crevice is closed. The chain from residues 80 to 83 swings to the right (as the molecule is depicted on page 6), the exposed phenylalanine 82 slips into the heme crevice to the left of the heme and nearly parallel to it, and the heme becomes less accessible to the outside world. The absolute preservation of this phenylalanine side chain throughout evolution, in an environment that is energetically unfavorable in the oxidized molecule, argues that closing of the heme crevice in the reduced molecule is important for its biological activity.

Several explanations might be offered. The aromatic ring of phenylalanine 82 may be part of the electron-transfer mechanism, or its removal from the heme crevice may be necessary to permit an electron-transferring group to enter beside the heme or just to approach the edge of the heme. At a minimum the refolding of the chain from residues 80 to 83 may be a "convulsive" motion that pushes the oxidase complex away from the protein after electron transfer is achieved by some other pathway.

This article has been speculative enough without making a choice between these or other alternatives. At this stage, as both oxidized and reduced cytochrome analyses are being extended to higher resolution, it is enough to say that we can see more refolding of the protein chain in passing between the two states than has been observed in any other protein. Phenylalanine 82 swings to an entirely new position and several other aromatic rings change orientation, including the three in the left channel. As the molecule is reduced, the right channel apparently is partly blocked by residues 20 and 21. We now have pictures of both strokes of a very ancient two-stroke molecular engine. We hope in time to be able to figure out how it operates.

31

THE THREE-DIMENSIONAL STRUCTURE OF AN ENZYME MOLECULE

DAVID C. PHILLIPS
November 1966

The arrangement of atoms in an enzyme molecule has been worked out for the first time. The enzyme is lysozyme, which breaks open cells of bacteria. The study has also shown how lysozyme performs its task

One day in 1922 Alexander Fleming was suffering from a cold. This is not unusual in London, but Fleming was a most unusual man and he took advantage of the cold in a characteristic way. He allowed a few drops of his nasal mucus to fall on a culture of bacteria he was working with and then put the plate to one side to see what would happen. Imagine his excitement when he discovered some time later that the bacteria near the mucus had dissolved away. For a while he thought his ambition of finding a universal antibiotic had been realized. In a burst of activity he quickly established that the antibacterial action of the mucus was due to the presence in it of an enzyme; he called this substance lysozyme because of its capacity to lyse, or dissolve, the bacterial cells. Lysozyme was soon discovered in many tissues and secretions of the human body, in plants and most plentifully of all in the white of egg. Unfortunately Fleming found that it is not effective against the most harmful bacteria. He had to wait seven years before a strangely similar experiment revealed the existence of a genuinely effective antibiotic: penicillin.

Nevertheless, Fleming's lysozyme has proved a more valuable discovery than he can have expected when its properties were first established. With it, for example, bacterial anatomists have been able to study many details of bacterial structure [see "Fleming's Lysozyme," by Robert F. Acker and S. E. Hartsell; SCIENTIFIC AMERICAN, June, 1960]. It has now turned out that lysozyme is the first enzyme whose three-dimensional structure has been

determined and whose properties are understood in atomic detail. Among these properties is the way in which the enzyme combines with the substance on which it acts—a complex sugar in the wall of the bacterial cell.

Like all enzymes, lysozyme is a protein. Its chemical makeup has been established by Pierre Jollès and his colleagues at the University of Paris and by Robert E. Canfield of the Columbia University College of Physicians and Surgeons. They have found that each molecule of lysozyme obtained from egg white consists of a single polypeptide chain of 129 amino acid subunits of 20 different kinds. A peptide bond is formed when two amino acids are joined following the removal of a molecule of water. It is customary to call the portion of the amino acid in-

corporated into a polypeptide chain a residue, and each residue has its own characteristic side chain. The 129-residue lysozyme molecule is cross-linked in four places by disulfide bridges formed by the combination of sulfur-containing side chains in different parts of the molecule [see illustration on opposite page].

The properties of the molecule cannot be understood from its chemical constitution alone; they depend most critically on what parts of the molecule are brought close together in the folded three-dimensional structure. Some form of microscope is needed to examine the structure of the molecule. Fortunately one is effectively provided by the techniques of X-ray crystal-structure analysis pioneered by Sir Lawrence Bragg and his father Sir William Bragg.

ALA	ALANINE	GLY	GLYCINE	PRO	PROLINE
ARG	ARGININE	HIS	HISTIDINE	SER	SERINE
ASN	ASPARAGINE	ILEU	ISOLEUCINE	THR	THREONINE
ASP	ASPARTIC ACID	LEU	LEUCINE	TRY	TRYPTOPHAN
CYS	CYSTEINE	LYS	LYSINE	TYR	TYROSINE
GLN	GLUTAMINE	MET	METHIONINE	VAL	VALINE
GLU	GLUTAMIC ACID	PHE	PHENYLALANINE		

TWO-DIMENSIONAL MODEL of the lysozyme molecule is shown on the opposite page. Lysozyme is a protein containing 129 amino acid subunits, commonly called residues (*see key to abbreviations above*). These residues form a polypeptide chain that is cross-linked at four places by disulfide (–S–S–) bonds. The amino acid sequence of lysozyme was determined independently by Pierre Jollès and his co-workers at the University of Paris and by Robert E. Canfield of the Columbia University College of Physicians and Surgeons. The three-dimensional structure of the lysozyme molecule has now been established with the help of X-ray crystallography by the author and his colleagues at the Royal Institution in London. A painting of the molecule's three-dimensional structure appears on pages 4 and 5. The function of lysozyme is to split a particular long-chain molecule, a complex sugar, found in the outer membrane of many living cells. Molecules that are acted on by enzymes are known as substrates. The substrate of lysozyme fits into a cleft, or pocket, formed by the three-dimensional structure of the lysozyme molecule. In the two-dimensional model on the opposite page the amino acid residues that line the pocket are shown in dark green.

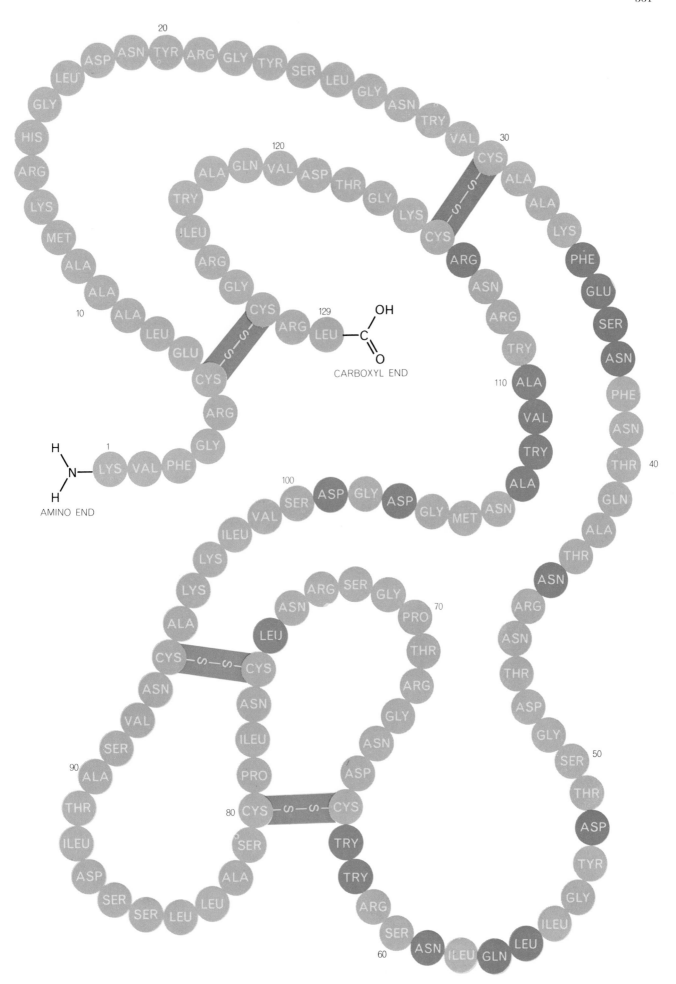

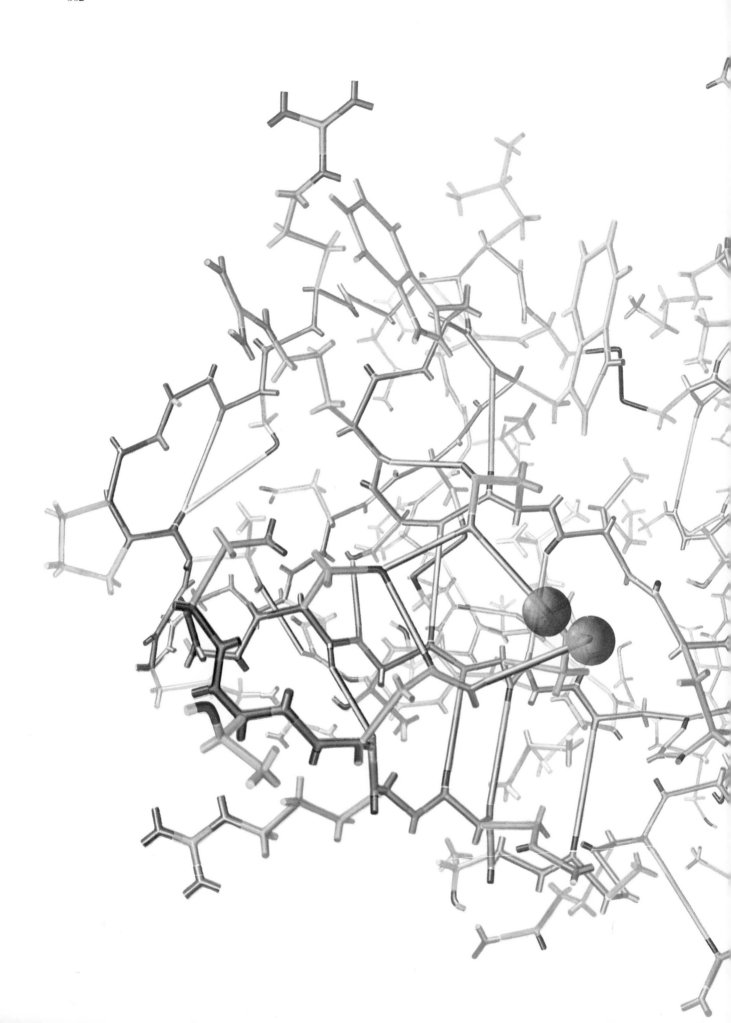

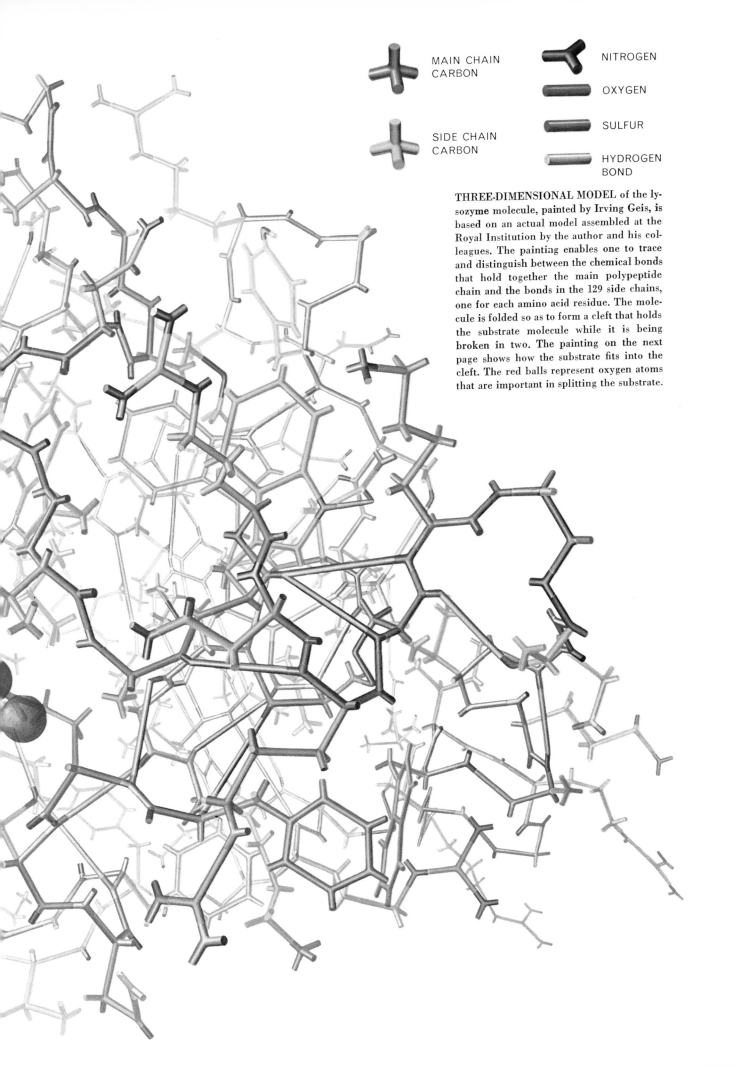

MAIN CHAIN CARBON

SIDE CHAIN CARBON

NITROGEN

OXYGEN

SULFUR

HYDROGEN BOND

THREE-DIMENSIONAL MODEL of the lysozyme molecule, painted by Irving Geis, is based on an actual model assembled at the Royal Institution by the author and his colleagues. The painting enables one to trace and distinguish between the chemical bonds that hold together the main polypeptide chain and the bonds in the 129 side chains, one for each amino acid residue. The molecule is folded so as to form a cleft that holds the substrate molecule while it is being broken in two. The painting on the next page shows how the substrate fits into the cleft. The red balls represent oxygen atoms that are important in splitting the substrate.

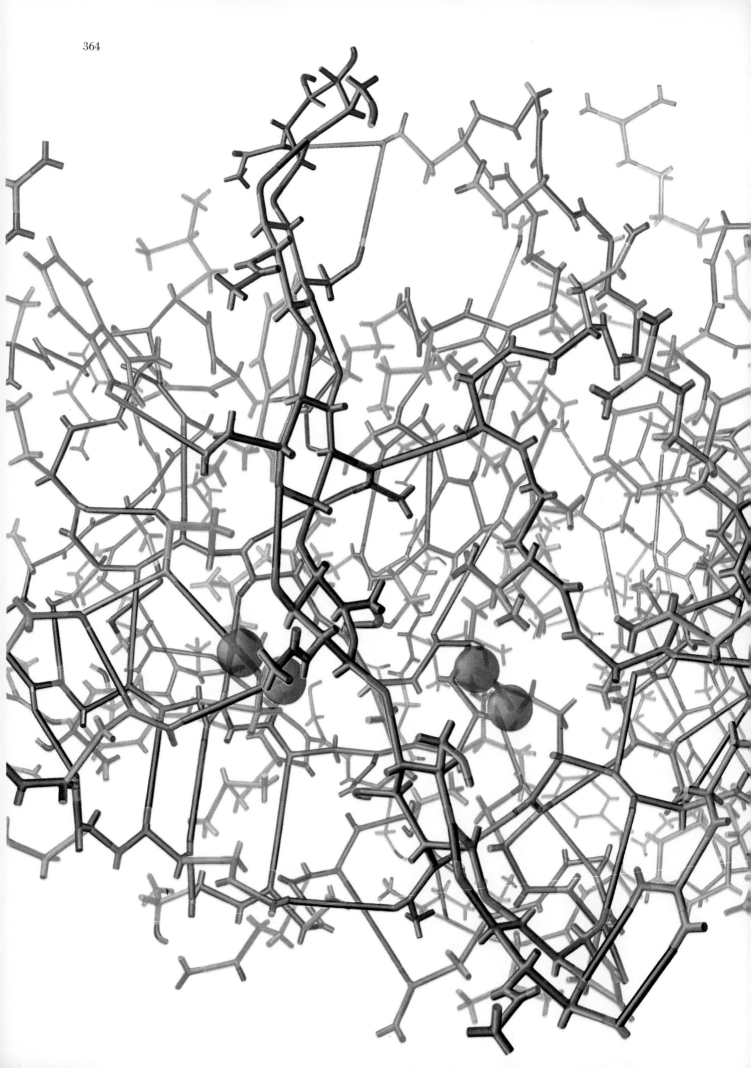

The difficulties of examining molecules in atomic detail arise, of course, from the fact that molecules are very small. Within a molecule each atom is usually separated from its neighbor by about 1.5 angstrom units (1.5×10^{-8} centimeter). The lysozyme molecule, which contains some 1,950 atoms, is about 40 angstroms in its largest dimension. The first problem is to find a microscope in which the atoms can be resolved from one another, or seen separately.

The resolving power of a microscope depends fundamentally on the wavelength of the radiation it employs. In general no two objects can be seen separately if they are closer together than about half this wavelength. The shortest wavelength transmitted by optical microscopes (those working in the ultraviolet end of the spectrum) is about 2,000 times longer than the distance between atoms. In order to "see" atoms one must use radiation with a much shorter wavelength: X rays, which have a wavelength closely comparable to interatomic distances. The employment of X rays, however, creates other difficulties: no satisfactory way has yet been found to make lenses or mirrors that will focus them into an image. The problem, then, is the apparently impossible one of designing an X-ray microscope without lenses or mirrors.

Consideration of the diffraction theory of microscope optics, as developed by Ernst Abbe in the latter part of the 19th century, shows that the problem can be solved. Abbe taught us that the formation of an image in the microscope can be regarded as a two-stage process. First, the object under examination scatters the light or other radiation falling on it in all directions, forming a diffraction pattern. This pattern arises because the light waves scattered from different parts of the object combine so as to produce a wave of large or small amplitude in any direction

according to whether the waves are in or out of phase—in or out of step—with one another. (This effect is seen most easily in light waves scattered by a regularly repeating structure, such as a diffraction grating made of lines scribed at regular intervals on a glass plate.) In the second stage of image formation, according to Abbe, the objective lens of the microscope collects the diffracted waves and recombines them to form an image of the object. Most important, the nature of the image depends critically on how much of the diffraction pattern is used in its formation.

X-Ray Structure Analysis

In essence X-ray structure analysis makes use of a microscope in which the two stages of image formation have been separated. Since the X rays cannot be focused to form an image directly, the diffraction pattern is recorded and the image is obtained from it by calculation. Historically the method was not developed on the basis of this reasoning, but this way of regarding it (which was first suggested by Lawrence Bragg) brings out its essential features and also introduces the main difficulty of applying it. In recording the intensities of the diffracted waves, instead of focusing them to form an image, some crucial information is lost, namely the phase relations among the various diffracted waves. Without this information the image cannot be formed, and some means of recovering it has to be found. This is the well-known phase problem of X-ray crystallography. It is on the solution of the problem that the utility of the method depends.

The term "X-ray crystallography" reminds us that in practice the method was developed (and is still applied) in the study of single crystals. Crystals suitable for study may contain some

10^{15} identical molecules in a regular array; in effect the molecules in such a crystal diffract the X radiation as though they were a single giant molecule. The crystal acts as a three-dimensional diffraction grating, so that the waves scattered by them are confined to a number of discrete directions. In order to obtain a three-dimensional image of the structure the intensity of the X rays scattered in these different directions must be measured, the phase problem must be solved somehow and the measurements must be combined by a computer.

The recent successes of this method in the study of protein structures have depended a great deal on the development of electronic computers capable of performing the calculations. They are due most of all, however, to the discovery in 1953, by M. F. Perutz of the Medical Research Council Laboratory of Molecular Biology in Cambridge, that the method of "isomorphous replacement" can be used to solve the phase problem in the study of protein crystals. The method depends on the preparation and study of a series of protein crystals into which additional heavy atoms, such as atoms of uranium, have been introduced without otherwise affecting the crystal structure. The first successes of this method were in the study of sperm-whale myoglobin by John C. Kendrew of the Medical Research Council Laboratory and in Perutz' own study of horse hemoglobin. For their work the two men received the Nobel prize for chemistry in 1962 [see "The Three-dimensional Structure of a Protein Molecule," by John C. Kendrew, SCIENTIFIC AMERICAN Offprint 121, and "The Hemoglobin Molecule," by M. F. Perutz, SCIENTIFIC AMERICAN Offprint 196].

Because the X rays are scattered by the electrons within the molecules, the image calculated from the diffraction pattern reveals the distribution of electrons within the crystal. The electron density is usually calculated at a regular array of points, and the image is made visible by drawing contour lines through points of equal electron density. If these contour maps are drawn on clear plastic sheets, one can obtain a three-dimensional image by assembling the maps one above the other in a stack. The amount of detail that can be seen in such an image depends on the resolving power of the effective microscope, that is, on its "aperture," or the extent of the diffraction pattern that has been included in the formation of the image. If the waves diffracted through sufficiently high angles are included

MODEL OF SUBSTRATE shows how it fits into the cleft in the lysozyme molecule. All the carbon atoms in the substrate are shown in purple. The portion of the substrate in intimate contact with the underlying enzyme is a polysaccharide chain consisting of six ringlike structures, each a residue of an amino-sugar molecule. The substrate in the model is made up of six identical residues of the amino sugar called N-acetylglucosamine (NAG). In the actual substrate every other residue is an amino sugar known as N-acetylmuramic acid (NAM). The illustration is based on X-ray studies of the way the enzyme is bound to a trisaccharide made of three NAG units, which fills the top of the cleft; the arrangement of NAG units in the bottom of the cleft was worked out with the aid of three-dimensional models. The substrate is held to the enzyme by a complex network of hydrogen bonds. In this style of model-making each straight section of chain represents a bond between atoms. The atoms themselves lie at the intersections and elbows of the structure. Except for the four red balls representing oxygen atoms that are active in splitting the polysaccharide substrate, no attempt is made to represent the electron shells of atoms because they would merge into a solid mass.

(corresponding to a large aperture), the atoms appear as individual peaks in the image map. At lower resolution groups of unresolved atoms appear with characteristic shapes by which they can be recognized.

The three-dimensional structure of lysozyme crystallized from the white of hen's egg has been determined in atomic detail with the X-ray method by our group at the Royal Institution in Lon-

don. This is the laboratory in which Humphry Davy and Michael Faraday made their fundamental discoveries during the 19th century, and in which the X-ray method of structure analysis was developed between the two world wars by the brilliant group of workers led by William Bragg, including J. D. Bernal, Kathleen Lonsdale, W. T. Astbury, J. M. Robertson and many others. Our work on lysozyme was begun in 1960

when Roberto J. Poljak, a visiting worker from Argentina, demonstrated that suitable crystals containing heavy atoms could be prepared. Since then C. C. F. Blake, A. C. T. North, V. R. Sarma, Ruth Fenn, D. F. Koenig, Louise N. Johnson and G. A. Mair have played important roles in the work.

In 1962 a low-resolution image of the structure was obtained that revealed the general shape of the molecule and

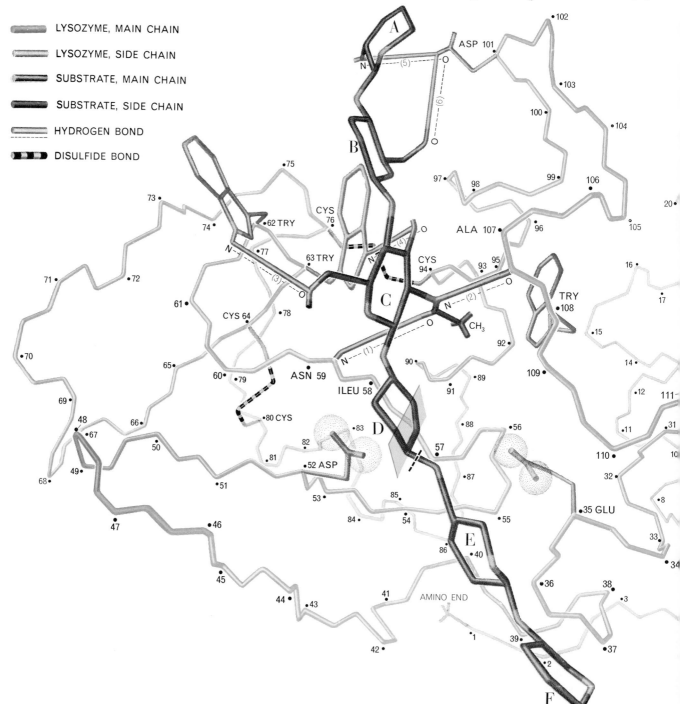

LYSOZYME, MAIN CHAIN
LYSOZYME, SIDE CHAIN
SUBSTRATE, MAIN CHAIN
SUBSTRATE, SIDE CHAIN
HYDROGEN BOND
DISULFIDE BOND

MAP OF LYSOZYME AND SUBSTRATE depicts in color the central chain of each molecule. Side chains have been omitted except for those that produce the four disulfide bonds clipping the lysozyme molecule together and those that supply the terminal connections for hydrogen bonds holding the substrate to the lysozyme. The top three rings of the substrate (A, B, C) are held to the underlying enzyme by six principal hydrogen bonds, which are identified by number to key with the description in the text. The lyso-

showed that the arrangement of the polypeptide chain is even more complex than it is in myoglobin. This low-resolution image was calculated from the amplitudes of about 400 diffraction maxima measured from native protein crystals and from crystals containing each of three different heavy atoms. In 1965, after the development of more efficient methods of measurement and computation, an image was calculated

on the basis of nearly 10,000 diffraction maxima, which resolved features separated by two angstroms. Apart from showing a few well-separated chloride ions, which are present because the lysozyme is crystallized from a solution containing sodium chloride, the two-angstrom image still does not show individual atoms as separate maxima in the electron-density map. The level of resolution is high enough, however, for many of the groups of atoms to be clearly recognizable.

The Lysozyme Molecule

The main polypeptide chain appears as a continuous ribbon of electron density running through the image with regularly spaced promontories on it that are characteristic of the carbonyl groups (CO) that mark each peptide bond. In some regions the chain is folded in ways that are familiar from theoretical studies of polypeptide configurations and from the structure analyses of myoglobin and fibrous proteins such as the keratin of hair. The amino acid residues in lysozyme have now been designated by number; the residues numbered 5 through 15, 24 through 34 and 88 through 96 form three lengths of "alpha helix," the conformation that was proposed by Linus Pauling and Robert B. Corey in 1951 and that was found by Kendrew and his colleagues to be the most common arrangement of the chain in myoglobin. The helixes in lysozyme, however, appear to be somewhat distorted from the "classical" form, in which four atoms (carbon, oxygen, nitrogen and hydrogen) of each peptide group lie in a plane that is parallel to the axis of the alpha helix. In the lysozyme molecule the peptide groups in the helical sections tend to be rotated slightly in such a way that their CO groups point outward from the helix axes and their imino groups (NH) inward.

The amount of rotation varies, being slight in the helix formed by residues 5 through 15 and considerable in the one formed by residues 24 through 34. The effect of the rotation is that each NH group does not point directly at the CO group four residues back along the chain but points instead between the CO groups of the residues three and four back. When the NH group points directly at the CO group four residues back, as it does in the classical alpha helix, it forms with the CO group a hydrogen bond (the weak chemical bond in which a hydrogen atom acts as a

bridge). In the lysozyme helixes the hydrogen bond is formed somewhere between two CO groups, giving rise to a structure intermediate between that of an alpha helix and that of a more symmetrical helix with a three-fold symmetry axis that was discussed by Lawrence Bragg, Kendrew and Perutz in 1950. There is a further short length of helix (residues 80 through 85) in which the hydrogen-bonding arrangement is quite close to that in the three-fold helix, and also an isolated turn (residues 119 through 122) of three-fold helix. Furthermore, the peptide at the far end of helix 5 through 15 is in the conformation of the three-fold helix, and the hydrogen bond from its NH group is made to the CO three residues back rather than four.

Partly because of these irregularities in the structure of lysozyme, the proportion of its polypeptide chain in the alpha-helix conformation is difficult to calculate in a meaningful way for comparison with the estimates obtained by other methods, but it is clearly less than half the proportion observed in myoglobin, in which helical regions make up about 75 percent of the chain. The lysozyme molecule does include, however, an example of another regular conformation predicted by Pauling and Corey. This is the "antiparallel pleated sheet," which is believed to be the basic structure of the fibrous protein silk and in which, as the name suggests, two lengths of polypeptide chain run parallel to each other in opposite directions. This structure again is stabilized by hydrogen bonds between the NH and CO groups of the main chain. Residues 41 through 45 and 50 through 54 in the lysozyme molecule form such a structure, with the connecting residues 46 through 49 folded into a hairpin bend between the two lengths of comparatively extended chain. The remainder of the polypeptide chain is folded in irregular ways that have no simple short description.

Even though the level of resolution achieved in our present image was not enough to resolve individual atoms, many of the side chains characteristic of the amino acid residues were readily identifiable from their general shape. The four disulfide bridges, for example, are marked by short rods of high electron density corresponding to the two relatively dense sulfur atoms within them. The six tryptophan residues also were easily recognized by the extended electron density produced by the large double-ring structures in their

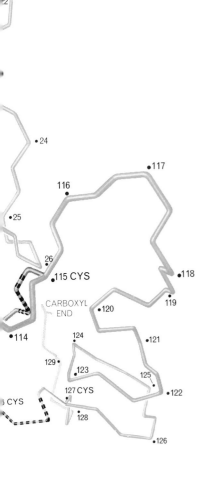

zyme molecule fulfills its function when it cleaves the substrate between the *D* and the *E* ring. Note the distortion of the *D* ring, which pushes four of its atoms into a plane.

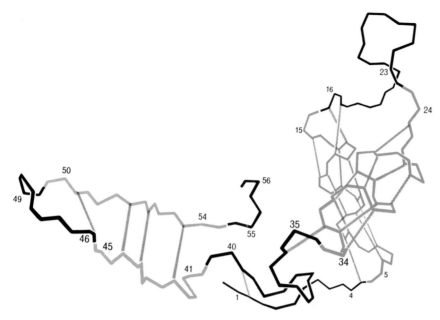

FIRST 56 RESIDUES in lysozyme molecule contain a higher proportion of symmetrically organized regions than does all the rest of the molecule. Residues 5 through 15 and 24 through 34 (*right*) form two regions in which hydrogen bonds (*gray*) hold the residues in a helical configuration close to that of the "classical" alpha helix. Residues 41 through 45 and 50 through 54 (*left*) fold back against each other to form a "pleated sheet," also held together by hydrogen bonds. In addition the hydrogen bond between residues 1 and 40 ties the first 40 residues into a compact structure that may have been folded in this way before the molecule was fully synthesized (*see illustration at the bottom of these two pages*).

liquid. Such "polar" side chains are hydrophilic—attracted to water; they are found in aspartic acid and glutamic acid residues and in lysine, arginine and histidine residues, which have basic side groups. On the other hand, most of the markedly nonpolar and hydrophobic side chains (for example those found in leucine and isoleucine residues) are shielded from the surrounding liquid by more polar parts of the molecule. In fact, as was predicted by Sir Eric Rideal (who was at one time director of the Royal Institution) and Irving Langmuir, lysozyme, like myoglobin, is quite well described as an oil drop with a polar coat. Here it is important to note that the environment of each molecule in the crystalline state is not significantly different from its natural environment in the living cell. The crystals themselves include a large proportion (some 35 percent by weight) of mostly watery liquid of crystallization. The effect of the surrounding liquid on the protein conformation thus is likely to be much the same in the crystals as it is in solution.

It appears, then, that the observed conformation is preferred because in it the hydrophobic side chains are kept out of contact with the surrounding liquid whereas the polar side chains are generally exposed to it. In this way the system consisting of the protein and the solvent attains a minimum free energy, partly because of the large number of favorable interactions of like groups within the protein molecule and between it and the surrounding liquid, and partly because of the relatively high disorder of the water molecules that are in contact only with other polar groups of atoms.

Guided by these generalizations, many workers are now interested in the possibility of predicting the conforma-

side chains. Many of the other residues also were easily identifiable, but it was nevertheless most important for the rapid and reliable interpretation of the image that the results of the chemical analysis were already available. With their help more than 95 percent of the atoms in the molecule were readily identified and located within about .25 angstrom.

Further efforts at improving the accuracy with which the atoms have been located is in progress, but an almost complete description of the lysozyme molecule now exists [*see illustration on pages 362 and 363*]. By studying it and

the results of some further experiments we can begin to suggest answers to two important questions: How does a molecule such as this one attain its observed conformation? How does it function as an enzyme, or biological catalyst?

Inspection of the lysozyme molecule immediately suggests two generalizations about its conformation that agree well with those arrived at earlier in the study of myoglobin. It is obvious that certain residues with acidic and basic side chains that ionize, or dissociate, on contact with water are all on the surface of the molecule more or less readily accessible to the surrounding

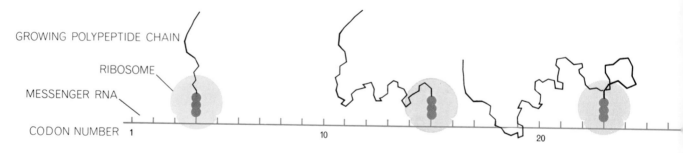

FOLDING OF PROTEIN MOLECULE may take place as the growing polypeptide chain is being synthesized by the intracellular particles called ribosomes. The genetic message specifying the amino acid sequence of each protein is coded in "messenger" ribonucleic acid (RNA). It is believed several ribosomes travel simultaneously along this long-chain molecule, reading the message as they go.

tion of a protein molecule from its chemical formula alone [see "Molecular Model-building by Computer," by Cyrus Levinthal; SCIENTIFIC AMERICAN Offprint 1043]. The task of exploring all possible conformations in the search for the one of lowest free energy seems likely, however, to remain beyond the power of any imaginable computer. On a conservative estimate it would be necessary to consider some 10^{129} different conformations for the lysozyme molecule in any general search for the one with minimum free energy. Since this number is far greater than the number of particles in the observable universe, it is clear that simplifying assumptions will have to be made if calculations of this kind are to succeed.

The Folding of Lysozyme

For some time Peter Dunnill and I have been trying to develop a model of protein-folding that promises to make practicable calculations of the minimum energy conformation and that is, at the same time, qualitatively consistent with the observed structure of myoglobin and lysozyme. This model makes use of our present knowledge of the way in which proteins are synthesized in the living cell. For example, it is well known, from experiments by Howard M. Dintzis and by Christian B. Anfinsen and Robert Canfield, that protein molecules are synthesized from the terminal amino end of their polypeptide chain. The nature of the synthetic mechanism, which involves the intracellular particles called ribosomes working in collaboration with two forms of ribonucleic acid ("messenger" RNA and "transfer" RNA), is increasingly well understood in principle, although the detailed environment of the growing protein chain remains unknown. Nevertheless,

it seems a reasonable assumption that, as the synthesis proceeds, the amino end of the chain becomes separated by an increasing distance from the point of attachment to the ribosome, and that the folding of the protein chain to its native conformation begins at this end even before the synthesis is complete. According to our present ideas, parts of the polypeptide chain, particularly those near the terminal amino end, may fold into stable conformations that can still be recognized in the finished molecule and that act as "internal templates," or centers, around which the rest of the chain is folded [see illustration at bottom of these two pages]. It may therefore be useful to look for the stable conformations of parts of the polypeptide chain and to avoid studying all the possible conformations of the whole molecule.

Inspection of the lysozyme molecule provides qualitative support for these ideas [see top illustration on opposite page]. The first 40 residues from the terminal amino end form a compact structure (residues 1 and 40 are linked by a hydrogen bond) with a hydrophobic interior and a relatively hydrophilic surface that seems likely to have been folded in this way, or in a simply related way, before the molecule was fully synthesized. It may also be important to observe that this part of the molecule includes more alpha helix than the remainder does.

These first 40 residues include a mixture of hydrophobic and hydrophilic side chains, but the next 14 residues in the sequence are all hydrophilic; it is interesting, and possibly significant, that these are the residues in the antiparallel pleated sheet, which lies out of contact with the globular submolecule formed by the earlier residues. In the light of our model of protein folding the obvious speculation is that there is no incentive to fold these hydrophilic residues in contact with the first part of the chain until the hydrophobic residues 55 (isoleucine) and 56 (leucine) have to be shielded from contact with the surrounding liquid. It seems reasonable to suppose that at this stage residues 41 through 54 fold back on themselves, forming the pleated-sheet structure and burying the hydrophobic side chains in the initial hydrophobic pocket.

Similar considerations appear to govern the folding of the rest of the molecule. In brief, residues 57 through 86 are folded in contact with the pleated-sheet structure so that at this stage of the process—if indeed it follows this course—the folded chain forms a structure with two wings lying at an angle to each other. Residues 86 through 96 form a length of alpha helix, one side of which is predominantly hydrophobic, because of an appropriate alternation of polar and nonpolar residues in that part of the sequence. This helix lies in the gap between the two wings formed by the earlier residues, with its hydrophobic side buried within the molecule. The gap between the two wings is not completely filled by the helix, however; it is transformed into a deep cleft running up one side of the molecule. As we shall see, this cleft forms the active site of the enzyme. The remaining residues are folded around the globular unit formed by the terminal amino end of the polypeptide chain.

This model of protein-folding can be tested in a number of ways, for example by studying the conformation of the first 40 residues in isolation both di-

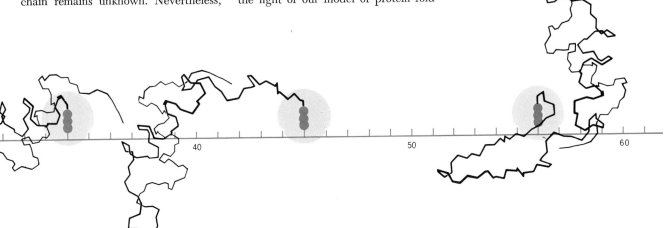

Presumably the messenger RNA for lysozyme contains 129 "codons," one for each amino acid. Amino acids are delivered to the site of synthesis by molecules of "transfer" RNA (*dark color*). The illustration shows how the lysozyme chain would lengthen as a ribosome travels along the messenger RNA molecule. Here, hypothetically, the polypeptide is shown folding directly into its final shape.

rectly (after removal of the rest of the molecule) and by computation. Ultimately, of course, the model will be regarded as satisfactory only if it helps us to predict how other protein molecules are folded from a knowledge of their chemical structure alone.

The Activity of Lysozyme

In order to understand how lysozyme brings about the dissolution of bacteria we must consider the structure of the bacterial cell wall in some detail. Through the pioneer and independent studies of Karl Meyer and E. B. Chain, followed up by M. R. J. Salton of the University of Manchester and many others, the structures of bacterial cell walls and the effect of lysozyme on them are now quite well known. The important part of the cell wall, as far as lysozyme is concerned, is made up of glucose-like amino-sugar molecules linked together into long polysaccharide chains, which are themselves cross-connected by short lengths of polypeptide chain. This part of each cell wall probably forms one enormous molecule—a "bag-shaped macromolecule," as W. Weidel and H. Pelzer have called it.

The amino-sugar molecules concerned in these polysaccharide structures are of two kinds; each contains an acetamido ($-NH \cdot CO \cdot CH_3$) side group, but one of them contains an additional major group, a lactyl side chain [see illustration below]. One of these amino sugars is known as N-acetylglucosamine (NAG) and the other as N-acetylmuramic acid (NAM). They occur alternately in the

polysaccharide chains, being connected by bridges that include an oxygen atom (glycosidic linkages) between carbon atoms 1 and 4 of consecutive sugar rings; this is the same linkage that joins glucose residues in cellulose. The polypeptide chains that cross-connect these polysaccharides are attached to the NAM residues through the lactyl side chain attached to carbon atom 3 in each NAM ring.

Lysozyme has been shown to break the linkages in which carbon 1 in NAM is linked to carbon 4 in NAG but not the other linkages. It has also been shown to break down chitin, another common natural polysaccharide that is found in lobster shell and that contains only NAG.

Ever since the work of Svante Arrhenius of Sweden in the late 19th century enzymes have been thought to work by forming intermediate compounds with their substrates: the substances whose chemical reactions they catalyze. A proper theory of the enzyme-substrate complex, which underlies all present thinking about enzyme activity, was clearly propounded by Leonor Michaelis and Maude Menten in a remarkable paper published in 1913. The idea, in its simplest form, is that an enzyme molecule provides a site on its surface to which its substrate molecule can bind in a quite precise way. Reactive groups of atoms in the enzyme then promote the required chemical reaction in the substrate. Our immediate objective, therefore, was to find the structure of a reactive complex between lysozyme and its polysaccha-

ride substrate, in the hope that we would then be able to recognize the active groups of atoms in the enzyme and understand how they function.

Our studies began with the observation by Martin Wenzel and his colleagues at the Free University of Berlin that the enzyme is prevented from functioning by the presence of NAG itself. This small molecule acts as a competitive inhibitor of the enzyme's activity and, since it is a part of the large substrate molecule normally acted on by the enzyme, it seems likely to do this by binding to the enzyme in the way that part of the substrate does. It prevents the enzyme from working by preventing the substrate from binding to the enzyme. Other simple amino-sugar molecules, including the trisaccharide made of three NAG units, behave in the same way. We therefore decided to study the binding of these sugar molecules to the lysozyme molecules in our crystals in the hope of learning something about the structure of the enzyme-substrate complex itself.

My colleague Louise Johnson soon found that crystals containing the sugar molecules bound to lysozyme can be prepared very simply by adding the sugar to the solution from which the lysozyme crystals have been grown and in which they are kept suspended. The small molecules diffuse into the protein crystals along the channels filled with water that run through the crystals. Fortunately the resulting change in the crystal structure can be studied quite simply. A useful image of the electron-density changes can be calculated from

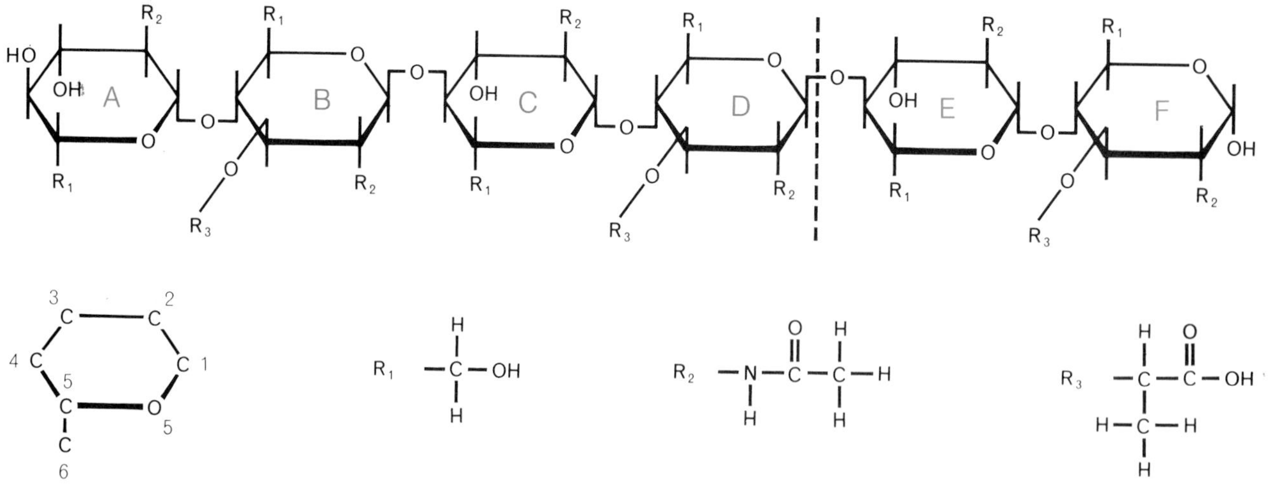

POLYSACCHARIDE MOLECULE found in the walls of certain bacterial cells is the substrate broken by the lysozyme molecule. The polysaccharide consists of alternating residues of two kinds of amino sugar: N-acetylglucosamine (NAG) and N-acetylmuramic acid (NAM). In the length of polysaccharide chain shown here

A, C and E are NAG residues; B, D and F are NAM residues. The inset at left shows the numbering scheme for identifying the principal atoms in each sugar ring. Six rings of the polysaccharide fit into the cleft of the lysozyme molecule, which effects a cleavage between rings D and E (see illustration on pages 366 and 367).

measurements of the changes in amplitude of the diffracted waves, on the assumption that their phase relations have not changed from those determined for the pure protein crystals. The image shows the difference in electron density between crystals that contain the added sugar molecules and those that do not.

In this way the binding to lysozyme of eight different amino sugars was studied at low resolution (that is, through the measurement of changes in the amplitude of 400 diffracted waves). The results showed that the sugars bind to lysozyme at a number of different places in the cleft of the enzyme. The investigation was hurried on to higher resolution in an attempt to discover the exact nature of the binding. Happily these studies at two-angstrom resolution (which required the measurement of 10,000 diffracted waves) have now shown in detail how the trisaccharide made of three NAG units is bound to the enzyme.

The trisaccharide fills the top half of the cleft and is bound to the enzyme by a number of interactions, which can be followed with the help of the illustration on pages 366 and 367. In this illustration six important hydrogen bonds, to be described presently, are identified by number. The most critical of these interactions appear to involve the acetamido group of sugar residue C [*third from top*], whose carbon atom 1 is not linked to another sugar residue. There are hydrogen bonds from the CO group of this side chain to the main-chain NH group of amino acid residue 59 in the enzyme molecule [*bond No. 1*] and from its NH group to the main-chain CO group of residue 107 (alanine) in the enzyme molecule [*bond No. 2*]. Its terminal CH_3 group makes contact with the side chain of residue 108 (tryptophan). Hydrogen bonds [*No. 3 and No. 4*] are also formed between two oxygen atoms adjacent to carbon atoms 6 and 3 of sugar residue C and the side chains of residues 62 and 63 (both tryptophan) respectively. Another hydrogen bond [*No. 5*] is formed between the acetamido side chain of sugar residue A and residue 101 (aspartic acid) in the enzyme molecule. From residue 101 there is a hydrogen bond [*No. 6*] to the oxygen adjacent to carbon atom 6 of sugar residue B. These polar interactions are supplemented by a large number of nonpolar interactions that are more difficult to summarize briefly. Among the more important nonpolar interactions, however, are those between sugar residue B and the ring system of residue

62; these deserve special mention because they are affected by a small change in the conformation of the enzyme molecule that occurs when the trisaccharide is bound to it. The electron-density map showing the change in electron density when tri-NAG is bound in the protein crystal reveals clearly that parts of the enzyme molecule have moved with respect to one another. These changes in conformation are largely restricted to the part of the enzyme structure to the left of the cleft, which appears to tilt more or less as a whole in such a way as to close the cleft slightly. As a result the side chain of residue 62 moves about .75 angstrom toward the position of sugar residue B. Such changes in enzyme conformation have been discussed for some time, notably by Daniel E. Koshland, Jr., of the University of California at Berkeley, whose "induced fit" theory of the enzyme-substrate interaction is supported in some degree by this observation in lysozyme.

The Enzyme-Substrate Complex

At this stage in the investigation excitement grew high. Could we tell how the enzyme works? I believe we can. Unfortunately, however, we cannot see this dynamic process in our X-ray images. We have to work out what must happen from our static pictures. First of all it is clear that the complex formed by tri-NAG and the enzyme is not the enzyme-substrate complex involved in catalysis because it is stable. At low concentrations tri-NAG is known to behave as an inhibitor rather than as a substrate that is broken down; clearly we have been looking at the way in which it binds as an inhibitor. It is noticeable, however, that tri-NAG fills only half of the cleft. The possibility emerges that more sugar residues, filling the remainder of the cleft, are required for the formation of a reactive enzyme-substrate complex. The assumption here is that the observed binding of tri-NAG as an inhibitor involves interactions with the enzyme molecule that also play a part in the formation of the functioning enzyme-substrate complex.

Accordingly we have built a model that shows that another three sugar residues can be added to the tri-NAG in such a way that there are satisfactory interactions of the atoms in the proposed substrate and the enzyme. There is only one difficulty: carbon atom 6 and its adjacent oxygen atom in sugar residue D make uncomfortably close contacts

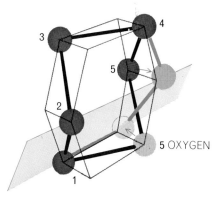

"CHAIR" CONFIGURATION (*gray*) is that normally assumed by the rings of amino sugar in the polysaccharide substrate. When bound against the lysozyme, however, the D ring is distorted (*color*) so that carbon atoms 1, 2 and 5 and oxygen atom 5 lie in a plane. The distortion evidently assists in breaking the substrate below the D ring.

with atoms in the enzyme molecule, unless this sugar residue is distorted a little out of its most stable "chair" conformation into a conformation in which carbon atoms 1, 2 and 5 and oxygen atom 5 all lie in a plane [*see illustration above*]. Otherwise satisfactory interactions immediately suggest themselves, and the model falls into place.

At this point it seemed reasonable to assume that the model shows the structure of the functioning complex between the enzyme and a hexasaccharide. The next problem was to decide which of the five glycosidic linkages would be broken under the influence of the enzyme. Fortunately evidence was at hand to suggest the answer. As we have seen, the cell-wall polysaccharide includes alternate sugar residues of two kinds, NAG and NAM, and the bond broken is between NAM and NAG. It was therefore important to decide which of the six sugar residues in our model could be NAM, which is the same as NAG except for the lactyl side chain appended to carbon atom 3. The answer was clear-cut. Sugar residue C cannot be NAM because there is no room for this additional group of atoms. Therefore the bond broken must be between sugar residues B and C or D and E. We already knew that the glycosidic linkage between residues B and C is stable when tri-NAG is bound. The conclusion was inescapable: the linkage that must be broken is the one between sugar residues D and E.

Now it was possible to search for the origin of the catalytic activity in the neighborhood of this linkage. Our task was made easier by the fact that John A.

Rupley of the University of Arizona had shown that the chemical bond broken under the influence of lysozyme is the one between carbon atom 1 and oxygen in the glycosidic link rather than the link between oxygen and carbon atom 4. The most reactive-looking group of atoms in the vicinity of this bond are the side chains of residue 52 (aspartic acid) and residue 35 (glutamic acid).

One of the oxygen atoms of residue 52 is about three angstroms from carbon atom 1 of sugar residue D as well as from the ring oxygen atom 5 of that residue. Residue 35, on the other hand, is about three angstroms from the oxygen in the glycosidic linkage. Furthermore, these two amino acid residues have markedly different environments. Residue 52 has a number of polar neighbors and appears to be involved in a network of hydrogen bonds linking it with residues 46 and 59 (both asparagine) and, through them, with residue 50 (serine). In this environment residue 52 seems likely to give up a terminal hydrogen atom and thus be negatively charged under most conditions, even when it is in a markedly acid solution, whereas residue 35, situated in a nonpolar environment, is likely to retain its terminal hydrogen atom.

A little reflection suggests that the concerted influence of these two amino acid residues, together with a contribution from the distortion to sugar residue D that has already been mentioned, is enough to explain the catalytic activity of lysozyme. The events leading to the rupture of a bacterial cell wall probably take the following course [see illustration on this page].

First, a lysozyme molecule attaches itself to the bacterial cell wall by interacting with six exposed amino-sugar residues. In the process sugar residue D is somewhat distorted from its usual conformation.

Second, residue 35 transfers its terminal hydrogen atom in the form of a hydrogen ion to the glycosidic oxygen, thus bringing about cleavage of the bond between that oxygen and carbon atom 1 of sugar residue D. This creates a positively charged carbonium ion (C^+) where the oxygen has been severed from carbon atom 1.

Third, this carbonium ion is stabilized by its interaction with the negatively charged aspartic acid side chain of residue 52 until it can combine with a hydroxyl ion (OH^-) that happens to diffuse into position from the surrounding water, thereby completing the reaction. The lysozyme molecule then falls away, leaving behind a punctured bacterial cell wall.

It is not clear from this description that the distortion of sugar residue D plays any part in the reaction, but in fact it probably does so for a very interesting reason. R. H. Lemieux and G. Huber of the National Research Council of Canada showed in 1955 that when a sugar molecule such as NAG incorporates a carbonium ion at the carbon-1 position, it tends to take up the same conformation that is forced on ring D by its interaction with the enzyme molecule. This seems to be an example, therefore, of activation of the substrate by distortion, which has long been a favorite idea of enzymologists. The binding of the substrate to the enzyme itself favors the formation of the carbonium ion in ring D that seems to play an important part in the reaction.

It will be clear from this account that although lysozyme has not been seen in action, we have succeeded in building up a detailed picture of how it may work. There is already a great deal of chemical evidence in agreement with this picture, and as the result of all the work now in progress we can be sure that the activity of Fleming's lysozyme will soon be fully understood. Best of all, it is clear that methods now exist for uncovering the secrets of enzyme action.

SPLITTING OF SUBSTRATE BY LYSOZYME is believed to involve the proximity and activity of two side chains, residue 35 (glutamic acid) and residue 52 (aspartic acid). It is proposed that a hydrogen ion (H^+) becomes detached from the OH group of residue 35 and attaches itself to the oxygen atom that joins rings D and E, thus breaking the bond between the two rings. This leaves carbon atom 1 of the D ring with a positive charge, in which form it is known as a carbonium ion. It is stabilized in this condition by the negatively charged side chain of residue 52. The surrounding water supplies an OH^- ion to combine with the carbonium ion and an H^+ ion to replace the one lost by residue 35. The two parts of the substrate then fall away, leaving the enzyme free to cleave another polysaccharide chain.

32

THE SYNTHESIS OF DNA

ARTHUR KORNBERG
October 1968

*Test-tube synthesis of the double helix that controls
heredity climaxes a half-century of effort by biochemists
to re-create biologically active giant molecules outside
the living cell*

My colleagues and I first undertook to synthesize nucleic acids outside the living cell, with the help of cellular enzymes, in 1954. A year earlier James Watson and Francis Crick had proposed their double-helix model of DNA, the nucleic acid that conveys genetic information from generation to generation in all organisms except certain viruses. We attained our goal within a year, but not until some months ago—14 years later—were we able to report a completely synthetic DNA, made with natural DNA as a template, that has the full biological activity of the native material.

Our starting point was an unusual single-strand form of DNA found in the bacterial virus designated ϕX174. The single strand is in the form of a closed loop. When ϕX174 infects cells of the bacterium *Escherichia coli*, the single-strand loop of DNA serves as the template that directs enzymes in the synthesis of a second loop of DNA. The two loops form a ring-shaped double helix similar to the DNA helixes found in bacterial cells and higher organisms. In our laboratory at the Stanford University School of Medicine we succeeded in reconstructing the synthesis of the single-strand DNA copies of viral DNA and finally in making a completely synthetic double helix. The way now seems open for the synthesis of DNA from other sources: viruses associated with human disease, bacteria, multicellular organisms and ultimately the DNA of vertebrates such as mammals.

An Earlier Beginning

The story of the cell-free synthesis of DNA does not start with the revelation of the structure of DNA in 1953. It begins around 1900 with the biochemical understanding of how the fermentation of fruit juices yields alcohol. Some 40 years earlier Louis Pasteur had convinced his contemporaries that the living yeast cell played an essential role in the fermentation process. Then Eduard Buchner observed in 1897 that a cell-free juice obtained from yeast was just as effective as intact cells for converting sugar to alcohol. This observation opened the era of modern biochemistry.

During the first half of this century biochemists resolved the overall conversion of sucrose to alcohol into a sequence of 14 reactions, each catalyzed by a specific enzyme. When this fermentation proceeds in the absence of air, each molecule of sucrose consumed gives rise to four molecules of adenosine triphosphate (ATP), the universal currency of energy exchange in living cells. The energy represented by the fourfold output of ATP per molecule of sucrose is sufficient to maintain the growth and multiplication of yeast cells. When the fermentation takes place in air, the oxidation of sucrose goes to completion, yielding carbon dioxide and water along with 18 times as much energy as the anaerobic process does. This understanding of how the combustion of sugar provides energy for cell metabolism was succeeded by similar explanations of how enzymes catalyze the oxidation of fatty acids, amino acids and the subunits of nucleic acids for the energy needs of the cell.

By 1950 the enzymatic dismantling of large molecules was well understood. Little thought or effort had yet been invested, however, in exploring how the cell makes large molecules out of small ones. In fact, many biochemists doubted that biosynthetic pathways could be suc-

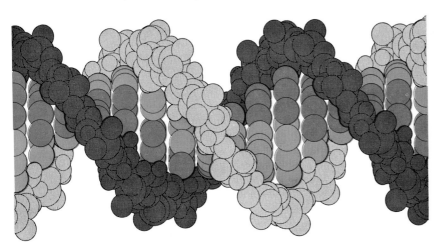

DOUBLE HELIX, the celebrated model of deoxyribonucleic acid (DNA) proposed in 1953 by James D. Watson and F. H. C. Crick, consists of two strands held together by crossties *(color)* that spell out a genetic message, unique for each organism. The Watson-Crick model explained for the first time how each crosstie consists of two subunits, called bases, that form obligatory pairs *(see illustrations on page 374)*. Thus each strand of the double helix and its associated sequence of bases is complementary to the other strand and its bases. Consequently each strand can serve as a template for the reconstruction of the other strand.

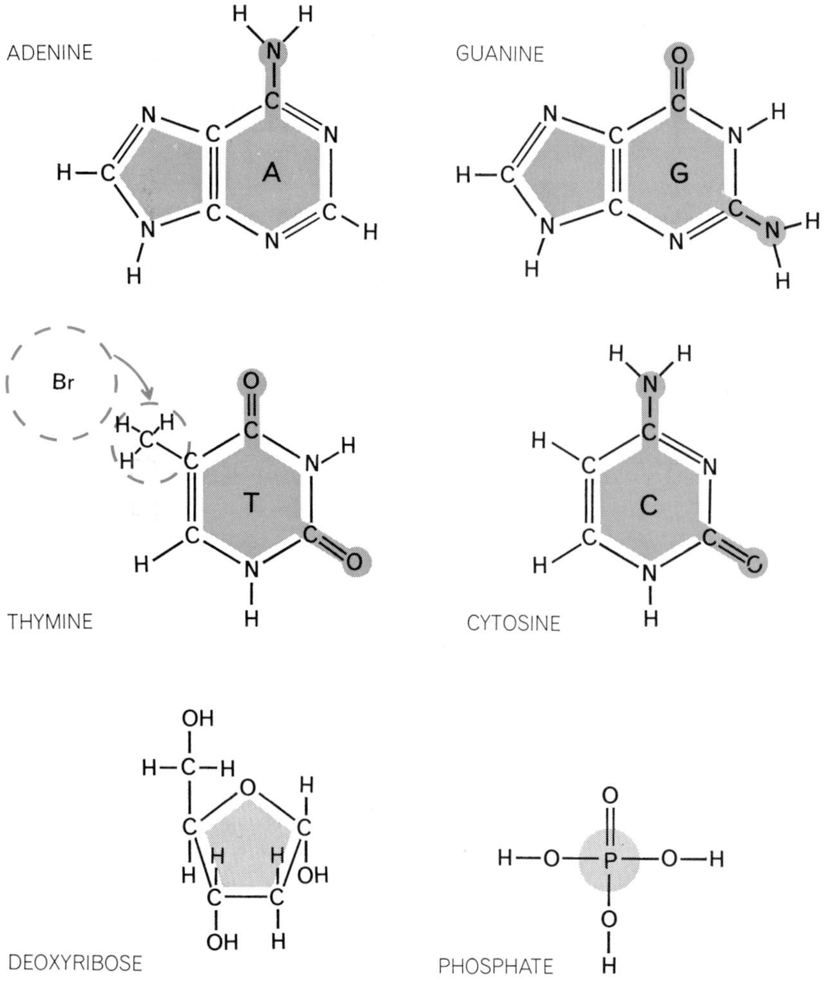

ADENINE

GUANINE

THYMINE

CYTOSINE

DEOXYRIBOSE

PHOSPHATE

DNA CONSTITUENTS are bases of four kinds, deoxyribose (a sugar) and a simple phosphate. The bases are adenine (*A*) and thymine (*T*), which form one obligatory pair, and guanine (*G*) and cytosine (*C*), which form another. Deoxyribose and phosphate form the backbone of each strand of the DNA molecule. The bases provide the code letters of the genetic message. For purposes of tagging synthetic DNA, thymine can be replaced by 5'-bromouracil, which contains a bromine atom where thymine contains a lighter CH_3 group.

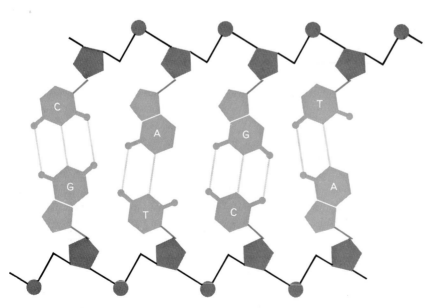

DNA STRUCTURE resembles a ladder in which the side pieces consist of alternating units of deoxyribose and phosphate. The rungs are formed by the bases paired in a special way, A with T and G with C, and held together respectively by two and three hydrogen bonds.

cessfully reconstructed in cell-free systems. Since then nearly two decades of intensive study have been devoted to the cell-free biosynthesis of large molecules. Two things above all have been made clear.

The first is that large molecules can be assembled in cell-free systems with the aid of purified enzymes and coenzymes. The second is that the routes of biosynthesis are different from those of degradation. Some biochemists had speculated that the routes of breakdown were really two-way streets whose flow might somehow be reversed. Now we know that the molecular traffic in cells flows on distinctive and divided highways. All cells have the enzymatic machinery to manufacture most of the subunits of large molecules from simple nutrients such as glucose, ammonia and carbon dioxide. Cells also have the capacity to salvage preformed subunits when they are available. On the basis of what has been learned the prospects are that in this century biochemists will assemble in the test tube complex viruses and major components of the cell. Perhaps the next century will bring the synthesis of a complete cell.

The Nucleotides

My co-workers and I were at Washington University in St. Louis when we made our first attempts to synthesize a nucleic acid in the test tube. By that time the constituents of nucleic acid were well known [*see illustrations at left*]. If one regards DNA as a chain made up of repeating links, the basic link is a structure known as a nucleotide [*see illustration on opposite page*]. It consists of a phosphate group attached to the five-carbon sugar deoxyribose, which is linked in turn to one of four different nitrogen-containing bases. The four bases are adenine (A), thymine (T), guanine (G) and cytosine (C). In the double helix of DNA the phosphate and deoxyribose units alternate to form the two sides of a twisted ladder. The rungs joining the sides consist of two bases: A is invariably linked to T and G is invariably linked to C. This particular pairing arrangement was the key insight of the Watson-Crick model. It means that if the two strands of the helix are separated, uncoupling the paired bases, each half can serve as a template for re-creating the missing half. Thus if the bases projecting from a single strand follow the sequence A, G, G, C, A, T..., one immediately knows that the complementary bases on the missing strand are T, C, C, G, T, A.... This base-pairing mechanism enables the cell to make accurate copies of the DNA molecule however many times the cell may divide.

When a strand of DNA is taken apart link by link (by treatment with acid or certain enzymes), the phosphate group of the nucleotide may be found attached to carbon No. 3 of the five-carbon deoxyribose sugar. Such a structure is called a 3′-nucleoside monophosphate. We judged, however, that better subunits for purposes of synthesis would be the 5′-nucleoside monophosphates, in which the phosphate linkage is to carbon No. 5 of deoxyribose.

This judgment was based on two lines of evidence. The first had just emerged from an understanding of how the cell itself made nucleotides from glucose, ammonia, carbon dioxide and amino acids. John M. Buchanan of the Massachusetts Institute of Technology had shown that nucleotides containing the bases A and G were naturally synthesized with a 5′ linkage. Our own work had shown the same thing for nucleotides containing T and C. The second line of evidence came from earlier studies my group had conducted at the National Institutes of Health. We had found that certain coenzymes, the simplest molecules formed from two nucleotides, were elaborated from 5′ nucleotide units. For the enzymatic linkage to take place the phosphate of the nucleotide had to be activated by an additional phosphate group [see illustration on next page]. Thus it seemed reasonable that activated 5′ nucleotides (nucleoside 5′ triphosphates) might combine with each other, under the proper enzymatic guidance, to form long chains of nucleic acid.

Our initial attempts at nucleic acid synthesis relied principally on two techniques. The first involved the use of radioactive atoms to label the nucleotide so that we could detect the incorporation of even minute amounts of it into nucleic acid. We sought the enzymatic machinery for synthesizing nucleic acids in the juices of the thymus gland, bone marrow and bacterial cells. Unfortunately such extracts also have a potent capacity for degrading nucleic acids. We added our labeled nucleotides to a pool of nucleic acids and hoped that a few synthesized molecules containing a labeled nucleotide would survive by being mixed into the pool. Even if there were net destruction of the pool of nucleic acids, the synthesis of a few molecules trapped in this pool might still be detected. The second technique exploited the fact that the nucleic acid could be precipitated by making the medium strongly acidic, whereas the nucleotide

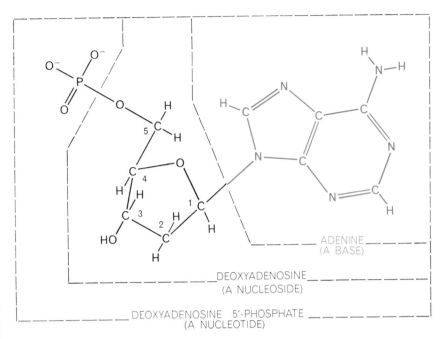

DNA BUILDING BLOCK, the monomer from which DNA polymers are constructed, is termed a nucleotide. There are four nucleotides, one for each of the four bases A, T, G and C. Deoxyadenosine 5′-phosphate, the nucleotide incorporating adenine, is shown here. If the phosphate group is replaced by a hydrogen atom, the structure is called a nucleoside.

precursors remained behind in solution.

Our first experiments with animal-cell extracts were uniformly negative. Therefore we turned to E. coli, which has the virtue of reproducing once every 20 minutes. Here we saw a glimmer. In samples to which we had added a quantity of labeled nucleotides whose radioactive atoms disintegrated at the rate of a million per minute we detected about 50 radioactive disintegrations per minute in the nucleic acid fraction that was precipitated by acid. Although the amount of nucleotide incorporated into nucleic acid was minuscule, it was nonetheless significantly above the level of background "noise." Through this tiny crack we tried to drive a wedge. The hammer was enzyme purification, a technique that had matured during the elucidation of alcoholic fermentation.

DNA Polymerase

In these experiments Uriel Littauer, a Fellow of the Weizmann Institute in Israel, and I observed the incorporation of adenylate (a nucleotide) from ATP into ribonucleic acid (RNA), in which the five-carbon sugar in the backbone of the chain is ribose rather than deoxyribose. Actually the first definitive demonstration of synthesis of an RNA-like molecule in a cell-free system had been achieved in the laboratory of Severo Ochoa in 1955. Working at the New York University School of Medicine, he

and Marianne Grunberg-Manago were investigating an aspect of energy metabolism and made the unexpected observation that one of the reactants, adenosine diphosphate (ADP), had been polymerized by cell juices into a chain of adenylates resembling RNA.

In our first attempts to achieve DNA synthesis in a cell-free system we used the deoxyribonucleoside called deoxythymidine. To Morris E. Friedkin, who was then at Washington University, we are grateful not only for supplying the radioactively labeled compound but also for the knowledge that the compound was readily incorporated into DNA by bone marrow cells and other animal cells. We were hopeful that extracts of E. coli would be able to incoporate deoxythymidine into nucleic acid by converting it first into the 5′ deoxynucleotide and then activating the deoxynucleotide to the triphosphate form. I found this to be the case. In subsequent months Ernest Simms and I were able to prepare separately deoxythymidine 5′-triphosphate and the other deoxynucleoside triphosphates, using enzymes or chemical synthetic routes. (In what follows the various deoxynucleosides in their 5′ triphosphate form will be designated simply by the initial of the base followed by an asterisk. Thus deoxythymidine 5′-triphosphate will be T*.)

In November, 1955, I. Robert Lehman, who is now at Stanford, started on the purification of the enzyme system in *E. coli* extracts that is responsible for converting T* into DNA. We were joined by Maurice J. Bessman some weeks later. Those were eventful days in which the enzyme, now given the name DNA polymerase, was progressively separated from other large molecules. With each step in purification the character of this DNA synthetic reaction became clearer. By June, 1956, when we participated at a conference on the chemical basis of heredity held at Johns Hopkins University, we could report two important facts about DNA synthesis in vitro, although we still lacked the answers to many important questions.

We reported first that preformed DNA had to be present along with DNA polymerase, and that all four of the de-oxynucleotides that occur in DNA (A, G, T and C) had to be furnished in the activated triphosphate form. We also reported that DNA from virtually any source—virus, bacterium or animal—could serve with the *E. coli* enzyme. What we still did not know was whether the synthetic DNA was a new molecule or an extension of a preexisting one. There were other questions. Did the synthetic DNA have the same chemical backbone and physical structure as natural DNA? Did it have a chemical composition typical of DNA, in which A equals T and G equals C, and in which, therefore, A plus G equals T plus C? Finally, and crucially: Did the chemical composition of the synthetic DNA reflect the composition of the particular natural DNA used to direct the reaction?

During the next three years these questions and related ones were resolved by the efforts of Julius Adler, Sylvy Kornberg and Steven B. Zimmerman. The synthetic DNA was shown to be a molecule with the chemical structure typical of DNA and the same ratio of A-T pairs to G-C pairs as the particular DNA used to prime, or direct, the reaction [*see illustration on page 378*]. The relative starting amounts of the four deoxynucleo-oside triphosphates had no influence whatever on the composition of the new DNA. The composition of the synthetic DNA was determined solely by the composition of the DNA that served as a template. An interesting illustration of this last fact justifies a slight digression.

Howard K. Schachman of the University of California at Berkeley spent his sabbatical year of 1957–1958 with us at Washington University examining the physical properties of the synthetic DNA. It had the high viscosity, the comparatively slow rate of sedimentation and other physical properties typical of natural DNA. The new DNA, like the natural one, was therefore a long, fibrous polymer molecule. Moreover, the longer the mixture of active ingredients was allowed to incubate, the greater the viscosity of the product was; this was direct evidence that the synthetic DNA was continuing to grow in length and in amount. However, we were startled to find one day that viscosity developed in a control test tube that lacked one of the essential triphosphates, G*. To be sure, no reaction was observed during the standard incubation period of one or two hours. On prolonging the incubation for several more hours, however, a viscous substance materialized!

Analysis proved this substance to be a DNA that contained only A and T nucleotides. They were arranged in a perfect alternating sequence. The isolated polymer, named dAT, behaved like any other DNA in directing DNA synthesis: it led to the immediate synthesis of more dAT polymer. Would any G* and C* be polymerized if these nucleotides were present in equal or even far greater amounts than A* and T* in a synthesis directed by dAT polymer? We found no detectable incorporation of G or C under conditions that would have measured the inclusion of even one G for every 100,-000 A or T nucleotides polymerized. Thus DNA polymerase rarely, if ever, made the mistake of matching G or C with A or T.

The DNA of a chromosome is a linear array of many genes. Each gene, in turn,

DEOXYADENOSINE 5'-PHOSPHATE

ATP

ATP

DEOXYADENOSINE 5'-TRIPHOSPHATE
(A*)

ACTIVATED BUILDING BLOCK is required when synthesizing DNA on a template of natural DNA with the aid of enzymes. The activated form of the nucleotide containing adenine is deoxyadenosine 5'-triphosphate, symbolized in this article by "A*." It is made from deoxyadenosine 5'-monophosphate by two different enzymes in two steps. Each step involves the donation of a terminal phosphate group from adenosine triphosphate (ATP).

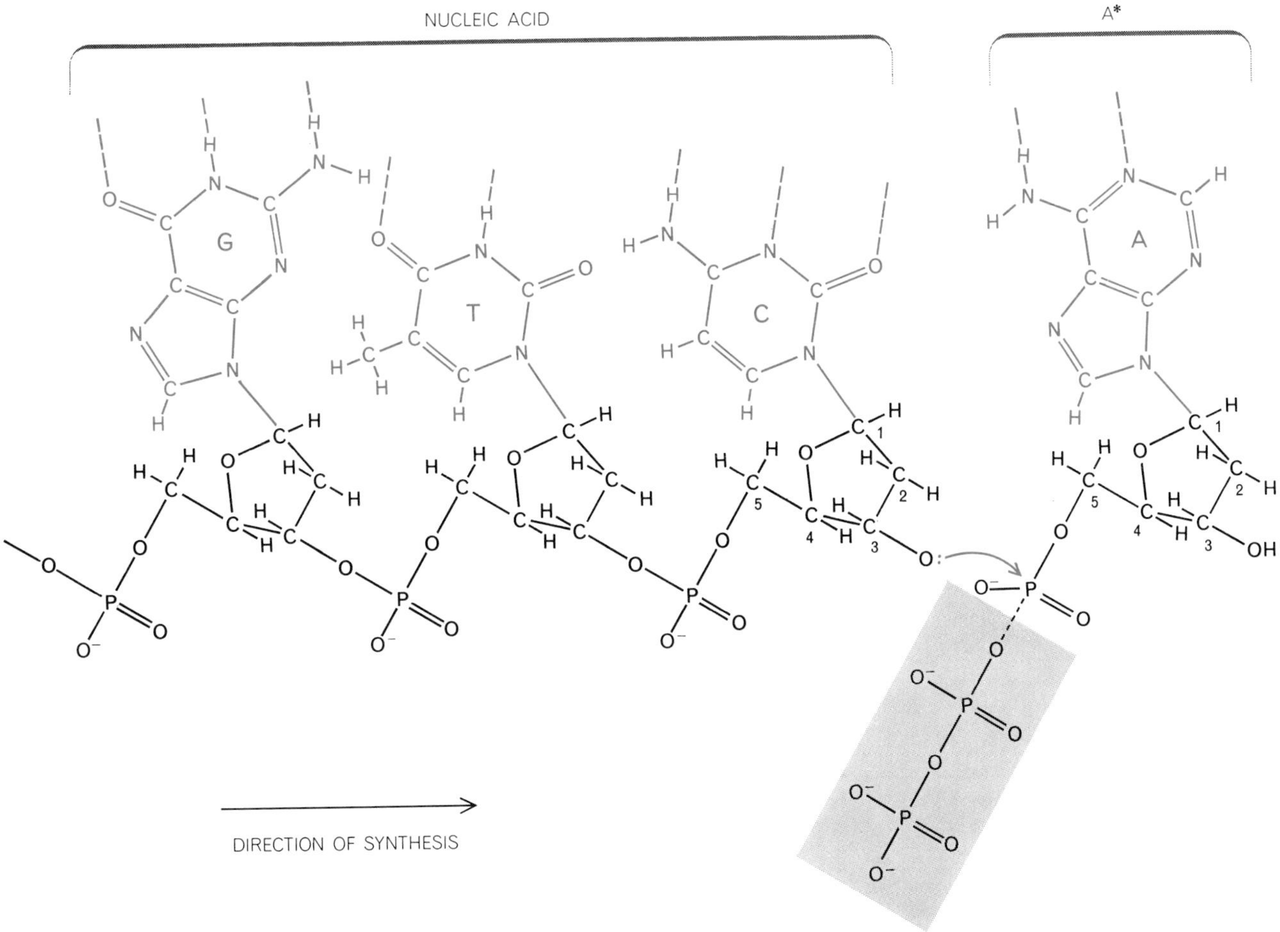

NUCLEIC ACID

A*

DIRECTION OF SYNTHESIS

SYNTHESIS OF DNA involves the stepwise addition of activated nucleotides to the growing polymer chain. In this illustration deoxyadenosine 5'-triphosphate (A*) is being coupled through a phosphodiester bond that links the 3' carbon in the deoxyribose portion of the last nucleotide in the growing chain to the 5' carbon in the deoxyribose portion of the newest member of the chain.

is a chain of about 1,000 nucleotides in a precisely defined sequence, which when translated into amino acids spells out a particular protein or enzyme. Does DNA polymerase in its test-tube synthesis of DNA accurately copy the sequential arrangement of nucleotides by base-pairing (A = T, G = C) without errors of mismatching, omission, commission or transposition? Unfortunately techniques are not available for determining the precise sequence of nucleotides of even short DNA chains. Because it is impossible to spell out the base sequence of natural DNA or any copy of it, we have resorted to two other techniques to test the fidelity with which DNA polymerase copies the template DNA. One is "nearest neighbor" analysis. The other is the duplication of genes with demonstrable biological activity.

Nearest-Neighbor Analysis

The nearest-neighbor analysis devised by John Josse, A. Dale Kaiser and myself in 1959 determines the relative frequency with which two nucleotides can end up side by side in a molecule of synthetic DNA. There are 16 possible combinations in all. There are four possible nearest-neighbor sequences of A (AA, AG, AT and AC), four for G (GA, GG, GT and GC) and similarly four for T and four for C. How can the frequency of these dinucleotide sequences be determined in a synthetic DNA chain? The procedure is to use a triphosphate labeled with a radioactive phosphorus atom in conducting the synthesis and to treat the synthesized DNA with a specific enzyme that cleaves the DNA and leaves the radioactive phosphorus atom attached to its nearest neighbor. For example, DNA synthesis is carried out with A* labeled in the innermost phosphate group, the group that will be included in the DNA product. This labeled phosphate group now forms a normal linkage (10^{16} times in a typical experiment!) with the nucleotide next to it in the chain—its nearest neighbor [see

illustration on page 379]. After the synthetic DNA is isolated it is subjected to degradation by an enzyme that cleaves every bond between the 5' carbon of deoxyribose and the phosphate, leaving the radioactive phosphorus atom attached to the neighboring nucleotide rather than to the one (A) to which it had originally been attached. The nucleotides of the degraded DNA are readily separated by electrophoresis or paper chromatography into the four types of which DNA is composed: A, G, T and C. Radioactive assay establishes the radioactive phosphorus content in each of these nucleotides and at once indicates the frequency with which A is next to A, to G, to T and to C.

The entire experiment is repeated, this time with the radioactive label in G* instead of A*. The second experiment yields the frequency of GA, GG, GT and GC dinucleotides. Two more experiments with radioactive T* and C* complete the analysis and establish the 16 possible nearest-neighbor frequencies.

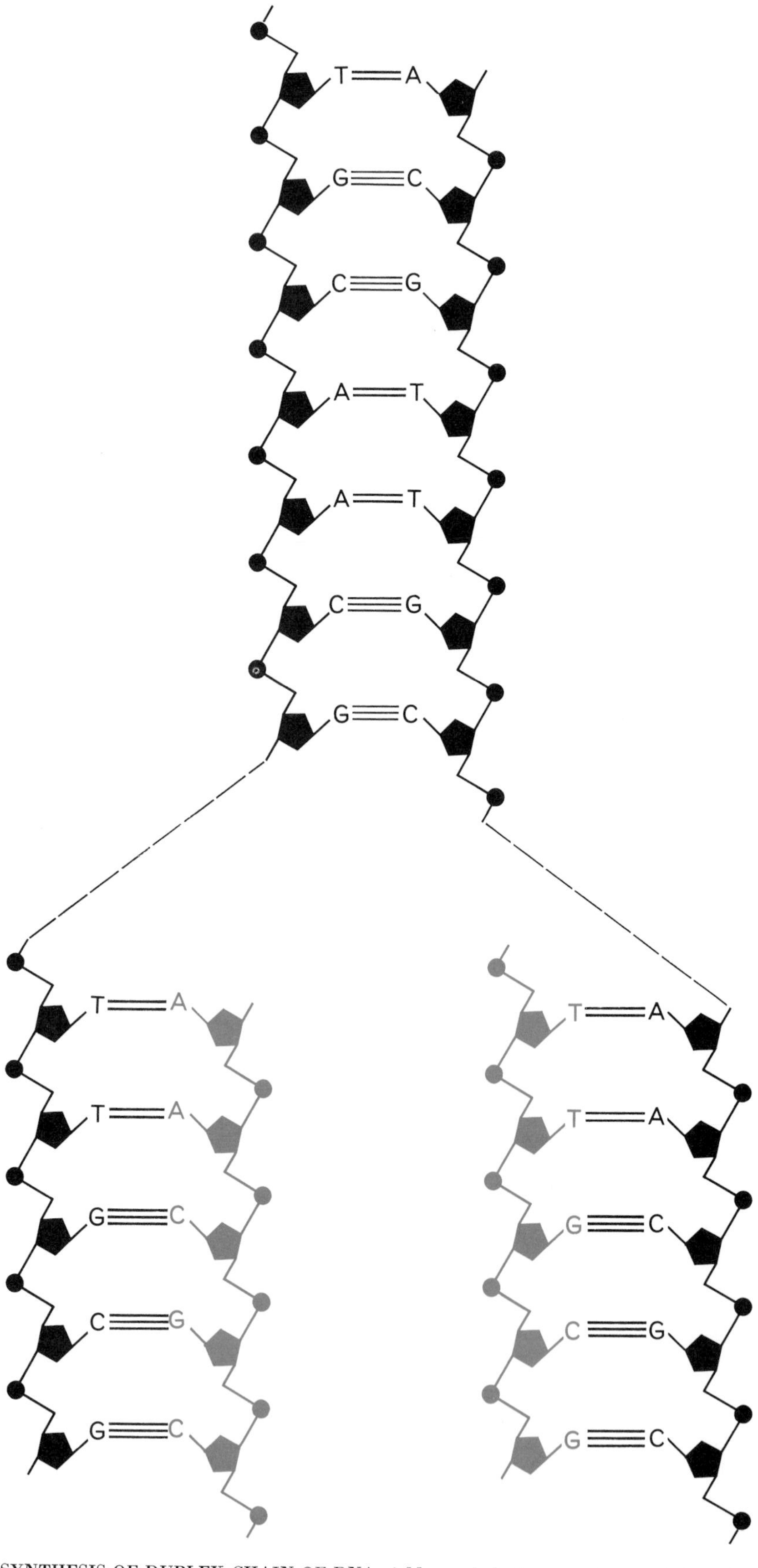

SYNTHESIS OF DUPLEX CHAIN OF DNA yields two hybrid molecules, consisting of a parental strand and a daughter strand, that are identical with each other and with the original duplex molecule. During the replicating process the parental duplex (*black*) separates into two strands, each of which then serves as the template for assembly of a daughter strand (*color*). The pairing of A with T and G with C guarantees faithful reproduction.

Many such experiments were performed with DNA templates obtained from viruses, bacteria, plants and animals. The DNA of each species guided the synthesis of DNA with what proved to be a distinctive assortment of nearest-neighbor frequencies. What is more, when a synthetic DNA was used as a template for a new round of replication, it gave rise to DNA with a nearest-neighbor frequency distribution identical with itself. Among the other insights obtained from these analyses was the recognition of a basic fact about the structure of the double helix. In replication the direction of the DNA chain being synthesized was found to run opposite to that of its template. By inference we can conclude that the chains of the double helix in natural DNA, as surmised by Watson and Crick, must also run in opposite directions.

Even with considerable care the accuracy of nearest-neighbor frequency analysis cannot be better than about 98 percent. Consequently we were still left with major uncertainties as to the precision of copying chains that contain 1,000 nucleotides or more, corresponding to the length of genes. An important question thus remained unanswered: Does DNA that is synthesized on a genetically or biologically active template duplicate the activity of that template?

One way to recognize the biological activity of bacterial DNA is to see if it can carry out "transformation," a process in which DNA from one species of bacteria alters the genetic endowment of a second species. For example, DNA from a strain of *Bacillus subtilis* resistant to streptomycin can be assimilated by a strain susceptible to the antibiotic, whereupon the recipient bacterium and all its descendants carry the trait of resistance to streptomycin. In other words, DNA molecules carrying the genes for a particular characteristic can be identified by their capacity for assimilation into the chromosome of a cell that previously lacked that trait. Yet when DNA was synthesized on a template of DNA that had transforming ability, the synthetic product invariably lacked that ability.

Part of the difficulty in synthesizing biologically active DNA lay in the persistence of trace quantities of nuclease enzymes in our DNA polymerase preparations. Nucleases are enzymes that degrade DNA. The introduction by a nuclease of one break in a long chain of DNA is enough to destroy its genetic activity. Further purification of DNA polymerase was indicated. Efforts over several years by Charles C. Richardson, Thomas Jovin, Paul T. Englund and LeRoy L. Bertsch resulted in a new procedure that was both simple and efficient. Finally, in April, 1967, with the assistance of the personnel and large-scale equipment of the New England Enzyme Center (sponsored by the National Institutes of Health at the Tufts University School of Medicine), we processed 100 kilograms of *E. coli* bacterial paste and obtained about half a gram of pure enzyme, free of the nuclease that puts random breaks in a DNA chain.

Unfortunately even this highly purified DNA polymerase has proved incapable of producing a biologically active DNA from a template of bacterial DNA. The difficulty, we believe, is that the DNA we extract from a bacterium such as *B. subtilis* provides the enzyme with a poor template. A proper template would be the natural chromosome, which is a double-strand loop about one millimeter in circumference. During its isolation from the bacterium the chromosome is broken, probably at random, into 100 or more fragments. The manner in which DNA polymerase and its related enzymes go about the replication of a DNA molecule as large and complex as the *B. subtilis* chromosome is the subject of current study in many laboratories.

The Virus φX174

It occurred to us in 1964 that the problem of synthesizing biologically active DNA might be solved by dealing with a simpler form of DNA that also has genetic activity. This is represented in viruses, such as φX174, whose DNA core is a single-strand loop. This "chromosome" not only is simpler in structure but also it is so small (about two microns in circumference) that it is fairly easy to extract without breakage. We also knew from the work of Robert L. Sinsheimer at the California Institute of Technology that when the DNA of φX174 invades *E. coli*, the first stage of infection involves the "subversion" of one of the host's enzymes to convert the single-strand loop into a double-strand helical loop. Sinsheimer called this first-stage product a "replicative form." Could the host enzyme that copies the viral DNA be the same DNA polymerase we had isolated from *E. coli*?

In undertaking the problem of copying a closed-loop DNA we could foresee some serious obstacles. Would it be possible for DNA polymerase to orient itself and start replication on a DNA

template if the template had no ends? Shashanka Mitra and later Peter Reichard succeeded in finding conditions under which the enzyme, as judged by electron microscope pictures, appeared to copy the single-strand loop. We then wondered if in spite of appearances in the electron micrographs, the DNA of φX174 was really just a simple loop. Perhaps, as had been suggested by other workers, it was really more like a necklace with a clasp, the clasp consisting of substances unrelated to the nucleotides we were supplying. Finally, we were aware from Sinsheimer's work that the DNA of φX174 had to be a completely closed loop in order to be infectious. We knew that our polymerase could only catalyze the synthesis of linear DNA molecules. How could we synthesize a genuinely closed loop? We were still missing either the clasplike component to insert into our product or, if the clasp was a mistaken hypothesis, a new kind of enzyme to close the loop.

Fortunately the missing factor was provided for us by work carried on independently in five different laboratories. The discovery in 1966 of a polynucleotide-joining enzyme was made almost simultaneously by Martin F. Gellert and his co-workers at the National Institutes of Health, by Richardson and Bernard Weiss at the Harvard Medical

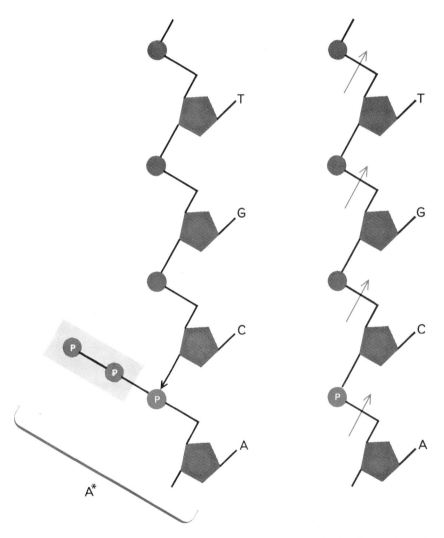

"NEAREST NEIGHBOR" ANALYSIS can reveal how often any of the four bases is located next to any other base in a single strand of synthetic DNA. Thus one can learn how often A is next to A, T, G or C, and so on. A radioactive phosphorus atom (*color*) is placed in the innermost position of one of the activated nucleotides, for example A*. The finished DNA molecule is then treated with an enzyme (*right*) that cleaves the chain between every phosphate and the 5′ carbon of the adjacent deoxyribose. Thus the phosphate is separated from the nucleotide on which it entered the chain and ends up attached to the nearest neighbor instead, C in the above example. The four kinds of nucleotide are separated by paper chromatography and the radioactivity associated with each is measured. The experiment is repeated with radioactive phosphorus linked to the other activated nucleotides.

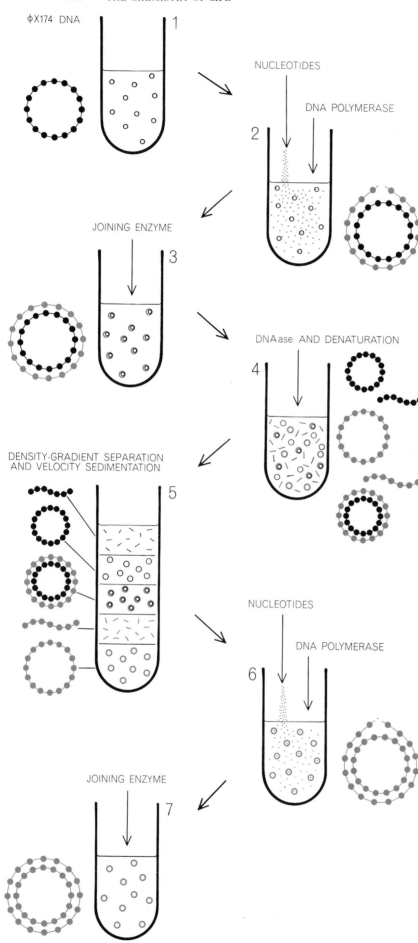

φX174 DNA

1

NUCLEOTIDES

DNA POLYMERASE

2

JOINING ENZYME

3

DNAase AND DENATURATION

4

DENSITY-GRADIENT SEPARATION
AND VELOCITY SEDIMENTATION

5

NUCLEOTIDES

DNA POLYMERASE

6

JOINING ENZYME

7

School, by Jerard Hurwitz and his colleagues at the Albert Einstein College of Medicine in New York, by Lehman and Baldomero M. Olivera at Stanford and by Nicholas R. Cozzarelli in my own group. It was the Lehman-Olivera preparation that we now employed in our experiments.

The polynucleotide-joining enzyme has the ability to repair "nicks" in the DNA strand. The nicks occur where there is a break in the sugar-phosphate backbone of one strand of the DNA molecule. The enzyme can repair a break only if all the nucleotides are intact and if what is missing is the covalent bond in the DNA backbone between a sugar and the neighboring phosphate. Provided with the joining enzyme, we were now in a position to find out whether it could work in conjunction with DNA polymerase to synthesize a completely circular and biologically active virus DNA.

By using the DNA of φX174 as a template we gained an important advantage over experiments based on transforming ability. Even if we were successful in synthesizing a DNA with transforming activity, this would still be of relatively limited significance. We could then say only that a restricted section of the DNA—a section as small as a part of a gene—had been assimilated by the recipient cell to replace a comparable section of its chromosome, substituting a proper sequence for a defective or incorrect one. However, Sinsheimer had demonstrated with the DNA of φX174 that a change in even one of its 5,500 nucleotides is sufficient to make the virus noninfective. . Therefore the demonstration of infectivity in a completely synthetic

SYNTHESIS OF φX174 DNA was accomplished by the following steps. Circular single-strand φX174 DNA, tagged with tritium, served as a template (1). Activated nucleotides containing A, G, C and 5′-bromouracil instead of T were added to the template, together with DNA polymerase. One of the activated nucleotides was tagged with radioactive phosphorus. The DNA synthesized on the template was complete but not yet joined in a loop (2). The loop was closed by the joining enzyme (3). Enough nuclease was now added to cut one strand in about half of all the duplex loops (4). This left a mixture of complete duplex loops, template loops, synthetic loops, linear template strands and linear synthetic strands. Since the synthetic strands contained 5′-bromouracil, they were heavier than the template strands and could be separated by centrifugation (5). The synthetic loops were then isolated and used as templates for making wholly synthetic duplex loops (6 and 7).

virus DNA would conclusively prove that we had carried out virtually error-free synthesis of this large number of nucleotides, comprising the five or six genes that carry out the virus's biological function.

In less than a year the test-tube synthesis of ϕX174 DNA was achieved. The steps can be summarized as follows. Template DNA was obtained from ϕX174 and labeled with tritium, the radioactive isotope of hydrogen. Tritium would thereafter provide a continuing label identifying the template. To the template were added DNA polymerase, purified joining enzyme and a cofactor (diphosphopyridine nucleotide), together with A^*, T^*, G^* and C^*. One of the nucleoside triphosphates was labeled with radioactive phosphorus. The radioactive phosphorus would thus provide a label for synthetic material analogous to the tritium label for the template. The interaction of the reagents then proceeded until the number of nucleotide units polymerized was exactly equal to the number of nucleotides in the template DNA. This equality was readily determined by comparing the radioactivity from the tritium in the template with the radioactivity from the phosphorus in the nucleotides provided for synthesis.

Such comparison showed that the experiments had progressed to an extent adequate for the formation of complementary loops of synthetic DNA. Complementary loops were designated ($-$) to distinguish them from the template loop ($+$). We had to demonstrate that the synthetic ($-$) loops were really loops. Had the polymerase made a full turn around the template and had the two ends of the chain been united by the joining enzyme? Several physical measurements, including electron microscopy, assured us that our product was a closed loop coiled tightly around the virus-DNA template and that it was identical in size and other details with the replicative form of DNA that appears in the infected cells. We could now exclude the possibility that some clasp material different from the nucleotide-containing compounds we had employed was involved in closing the virus-DNA loop.

The critical questions remaining were whether the synthetic ($-$) loops had biological activity—that is, infectivity—and whether the synthetic loops could in turn act as templates for the formation of a completely synthetic "duplex" DNA analogous to the replicative forms that were produced naturally inside infected

cells. In order to answer the first of these questions we had to isolate the synthetic DNA strands from the partially synthetic duplexes. For reasons that will be apparent below, we substituted bromouracil, a synthetic but biologically active analogue of thymine [see top illustration on page 374]. We then introduced just enough nuclease to produce a single nick in one strand of about half the population of molecules. The duplex loops that had been nicked would release a single linear strand of DNA; these single strands could be separated from their circular companions and from unnicked duplex loops by heating. Thus we were left with a mixture that contained ($+$) template loops, ($-$) synthetic loops, ($+$) template linear forms, ($-$) synthetic linear forms—all in about equal quantities—and full duplex loops.

It was at this point that the substitution of bromouracil for thymine became useful. Because bromouracil contains a bromine atom in place of the methyl group of thymine, it is heavier than thymine. Therefore a molecule containing bromouracil can be separated from one containing thymine by high-speed centrifugation in a heavy salt solution (the density-gradient technique perfected by Jerome R. Vinograd of Cal Tech). In this system the denser a substance is, the lower in the centrifuge tube it will settle. Thus from top to bottom of the centrifuge tube we obtained fractions containing the light single strands of thymine-containing ($+$) template DNA, the duplex hybrids of intermediate weight and finally the single-strand synthetic ($-$) DNA "weighted down" with bromouracil. The reliability of this fractionation was confirmed by three separate peaks of radioactivity corresponding to each of the fractions. We were further reassured by observations that the mean density of each fraction corresponded almost exactly to the mean density of standard samples of virus DNA containing bromouracil or thymine.

Still another physical technique involving density-gradient sedimentation was employed to separate the synthetic linear forms from the synthetic circular forms. The circular forms could then be used in tests of infectivity, by methods previously developed by Sinsheimer to demonstrate the infectivity of circular ϕX174 DNA. We tested our ($-$) loops by incubating them with E. coli cells whose walls had been removed by the action of the enzyme lysozyme. Infectivity is assayed by the ability of the virus to lyse, or dissolve, these cells when they are "plated" on a nutrient me-

dium. Our synthetic loops showed almost exactly the same patterns of infectivity as their natural counterparts had. Their biological activity was now demonstrated.

One further set of experiments remained in which the ($-$) synthetic loops were employed as the template to determine if we could produce completely synthetic duplex circular forms analogous to the replicative forms found in cells infected with natural ϕX174 virus. Because the synthetic ($-$) loops were labeled with radioactive phosphorus, this time we added tritium to one of the nucleotide-containing subunits (C^*). The remaining procedures were essentially the same as the ones described above, and we did produce fully synthetic duplex loops of ϕX174. The ($+$) loops were then separated and were found to be identical in all respects with the ($+$) loops of natural ϕX174 virus. Their infectivity could also be demonstrated. Sinsheimer had previously shown that, under these assay conditions, a change in a single nucleotide of the virus gave rise to a mutant of markedly decreased infectivity. Therefore the correspondence between the infectivity of our synthetic forms and their natural counterparts attested to the precision of the enzymatic operation.

Future Directions

The total synthesis of infective virus DNA by DNA polymerase with the four deoxynucleoside triphosphates not only demonstrates the capacity of this enzyme to copy a small chromosome (of five or six genes) without error but also shows that this chromosome, at least, is as simple and straightforward as a linear sequence of the standard four deoxynucleotide units. It is a long step to the human chromosome, some 10,000 times larger, yet we are encouraged to extrapolate our current conceptions of nucleotide composition and nucleotide linkage from the tiny ϕX174 chromosome to larger ones.

What are the major directions this research will take? I see at least three immediate and productive paths. One is the exploration of the physical and chemical nature of DNA polymerase in order to understand exactly how it performs its error-free replication of DNA. Without this knowledge of the structure of the enzyme and how it operates under defined conditions in the test tube, our understanding of the intracellular behavior of the enzyme will be incomplete.

A second direction is to clarify the

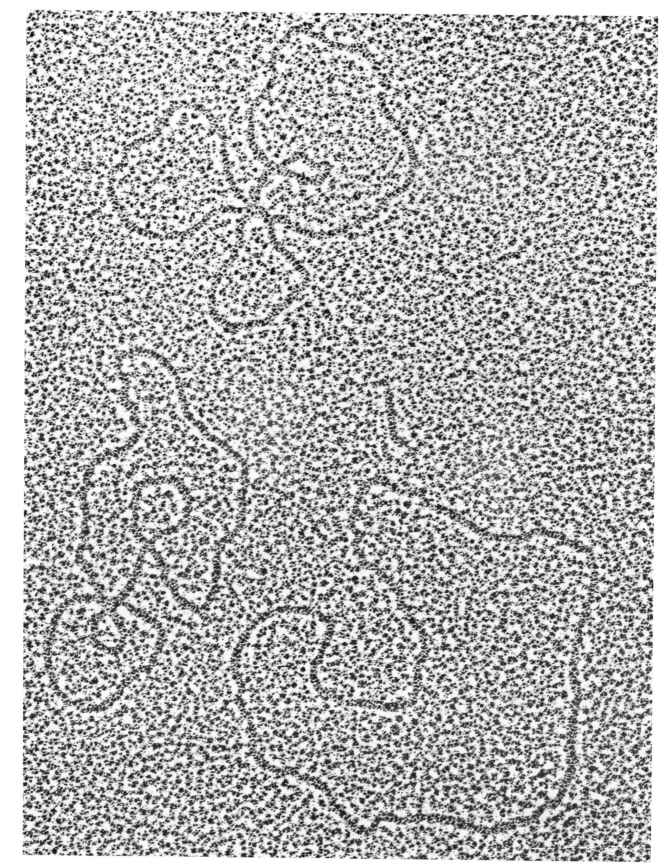

THREE CLOSED LOOPS OF DNA, each a complete double helix, are shown in this electron micrograph made in the author's laboratory at the Stanford University School of Medicine. One strand of each loop is the natural single-strand DNA of the bacterial virus φX174, which served as a template for the test-tube synthesis, carried out by enzymes, of a synthetic complementary strand. The hybrid molecules are biologically active. The enlargement is about 200,000 diameters. Each loop contains some 5,500 pairs of bases. If enlarged to the scale of the model on the opposite page, each loop of DNA would form a circle roughly 150 feet in circumference.

control of DNA replication in the cell and in the animal. Why is DNA synthesis arrested in a mature liver cell and what sets it in motion 24 hours after part of the liver is removed surgically? What determines the slow rate of DNA replication in adult cells compared with the rate in embryonic or cancer cells? The time is ripe for exploration of the factors that govern the initiation and rate of DNA synthesis in the intact cell and animal. Finally, there are now prospects of applying our knowledge of DNA structure and synthesis directly to human welfare. This is the realm of genetic engineering, and it is our collective responsibility to see that we exploit our great opportunities to improve the quality of human life.

An obvious area for investigation would be the synthesis of the polyoma virus, a virus known to induce a variety of malignant tumors in several species of rodents. Polyoma virus in its infective form is made up of duplex circular DNA and presumably replicates in this form on entering the cell. On the basis of our experience it would appear quite feasible to synthesize polyoma virus DNA. If this synthesis is accomplished, there would seem to be many opportunities for modifying the virus DNA and thus determining where in the chromosome its tumor-producing capacity lies. With this knowledge it might prove possible to modify the virus in order to control its tumor-producing potential.

Our speculations can extend even to large DNA molecules. For example, if a failure in the production of insulin were to be traced to a genetic deficit, then administration of the appropriate synthetic DNA might conceivably provide a cure for diabetes. Of course, a system for delivering the corrective DNA to the cells must be devised. Even this does not seem inconceivable. The extremely interesting work of Stanfield Rogers at the Oak Ridge National Laboratory suggests a possibility. Rogers has shown that the Shope papilloma virus, which is not pathogenic in man, is capable of inducing production of the enzyme arginase in rabbits at the same time that it induces tumors. Rogers found that in the blood of laboratory investigators working with the virus there is a significant reduction of the amino acid arginine, which is destroyed by arginase. This is apparently an expression of enhanced arginase activity. Might it not be possible, then, to use similar nonpathogenic viruses to carry into man pieces of DNA capable of replacing or repairing defective genes?

33

HYBRID NUCLEIC ACIDS

S. SPIEGELMAN

May 1964

One strand of nucleic acid will combine with another wherever the subunits of the two strands are complementary. Artificial combinations clarify the flow of information in the living cell

One of the most useful techniques for studying how genes work depends on the remarkable fact that certain chainlike molecules found in the living cell can "recognize" other chains whose molecular composition is complementary to their own. If one molecule is composed of subunits that can be symbolized by the sequence CATCATCAT..., it will recognize the complementary sequence GTAGTA- GTA... in a second molecule. As we shall see, these particular letters represent the chemical subunits that transmit the genetic information. When two such complementary chains are brought together under suitable conditions, they will "hybridize," or combine, to form a double-strand molecule in which the subunits C and G and A and T are linked by the weak chemical bond known as the hydrogen bond. This article will describe how hybridization has been exploited to study the cell's mechanism for manufacturing proteins.

A typical living cell synthesizes hundreds of different proteins, most of which serve as the enzymes, or biological catalysts, that mediate the myriad chemical reactions involved in growth and reproduction. Proteins are large chainlike molecules made out of some 20 different kinds of amino acids. According to current theory the sequence of amino acid units in a protein is specified by a single gene, and the genes are strung together in the chainlike molecules of deoxyribonucleic acid (DNA). The subunits of DNA that constitute the genetic code are four "bases": adenine (A), thymine (T), guanine (G) and cytosine (C). Normally DNA consists of two complementary chains linked by hydrogen bonds to form a double helix. Wherever A occurs in one chain, T occurs in the other; similarly,

G pairs with C. It is evident that each chain contains all the information needed to specify the complementary chain.

The flow of information in a cell begins with the base-pairing found in the double helix of DNA. Three principal modes of information transfer are distinguished by the end purposes they serve [*see illustrations on opposite page*]. The first is a duplication, which provides exact copies of the DNA molecule for transmission from one generation of cells to the next. The copying process utilizes the same "language" and the same "alphabet" that are present in the original material.

The second mode of transfer is a "transcription," which uses the same language but a slightly different alphabet. In this step DNA is transcribed into ribonucleic acid (RNA), a chainlike molecule that, like DNA, has four code units. Three are the same as those found in DNA: A, G and C. The fourth is uracil (U), which takes the place of thymine (T). One particular variety of RNA carries the actual program for protein synthesis. Although this variety of RNA is frequently called "messenger RNA," I prefer to speak of "translatable RNA" or "RNA messages." A "messenger" cannot be translated, but a message can.

The third mode of information transfer converts the information from the four-element language of translatable RNA to the 20-element language of the proteins. This step is properly regarded as a translation. Since every translation calls for a dictionary, it is not surprising that the cell uses one also. The cellular dictionary is made up of a collection of comparatively small RNA molecules known as transfer RNA (or soluble

RNA), which have the task of delivering specific amino acids to the site of protein synthesis. Each amino acid is attached to a transfer-RNA molecule by a specific activating enzyme.

The actual synthesis of protein molecules is accomplished with the help of ribosomes, which evidently serve to hold the translatable RNA "tape" in position while the message is being "read." Ribosomes are small spherical particles composed of protein and two kinds of RNA. One kind is about a million times heavier than a hydrogen atom; the other is about 600,000 times heavier. They are respectively called 23S RNA and 16S RNA, designations that refer to how fast they settle out of solution when they are spun at high speed in an ultracentrifuge.

Thus we see that cellular RNA is divided into two major categories: translatable and nontranslatable. The translatable variety (messenger RNA) constitutes only about 5 per cent of all the RNA in a cell; it is usually unstable and must be continuously resynthesized. The nontranslatable varieties of RNA (transfer RNA and the two kinds of ribosomal RNA) make up about 95 per cent of the RNA found in a cell and are extremely stable.

This picture of the genetic mechanism has arisen from the contributions of a large number of investigators using a wide variety of methods of analyzing gene function. I shall focus attention on some of the things that have been learned about the translatable and nontranslatable forms of RNA by exploiting the ability of RNA to hybridize with DNA of complementary composition. In effect this technique enables one to return an RNA molecule to the site of its synthesis on a particular stretch of DNA.

Early in 1958 my colleagues Masayasu Nomura and Benjamin D. Hall and I at the University of Illinois undertook to re-examine a remarkable experiment described in 1955 by Elliot Volkin and Lazarus Astrachan of the Oak Ridge National Laboratory. These workers had used radioactive isotopes to identify and study the RNA produced when the colon bacillus is infected with the bacterial virus designated T2. Infection occurs when T2 injects into the cell of the bacterium a double helix of DNA bearing all the information needed for the synthesis of new virus particles. Volkin and Astrachan had concluded that the RNA synthesized in the infected cells mimicked the composition of the T2 DNA.

At the time neither the experimenters nor anyone else thought that the RNA might represent a genetic message formed on a DNA template. It was suggested, rather, that this new kind of

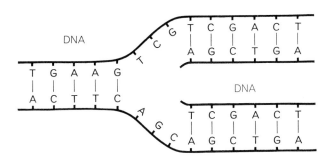

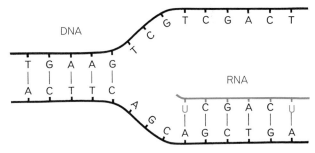

FLOW OF GENETIC INFORMATION involves duplication (*left*), transcription (*right*) and translation (*below*). Genetic information resides in giant chainlike molecules of deoxyribonucleic acid (DNA), in which the code "letters" are four bases: adenine (A), thymine (T), guanine (G) and cytosine (C). DNA normally consists of two complementary strands in which A pairs with T and G with C. During duplication, by an unknown mechanism, a new complementary strand is synthesized on each of the parent strands. In transcription only one strand of the DNA serves as a template and the new molecule formed is ribonucleic acid (RNA). In RNA the base uracil (U) takes the place of thymine as the partner of adenine. RNA molecules can be translatable or nontranslatable.

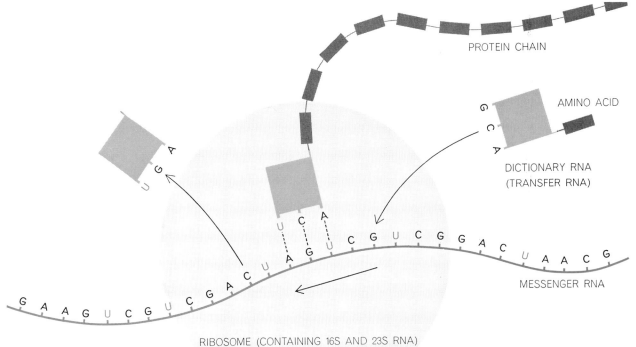

TRANSLATION PROCESS converts genetic information from the four-letter "language" of nucleic acids (DNA and RNA) into the 20-letter language of proteins. The letters of the protein language are the 20 amino acids that link together to form protein chains. If the DNA code is transcribed into translatable, or messenger, RNA, the RNA message becomes associated with one or more particles called ribosomes, which mediate the actual synthesis of protein. Ribosomes are made up of protein and two kinds of nontranslatable RNA, identified as 16S and 23S. Still another form of RNA called dictionary, or transfer, RNA delivers amino acids to the site of protein synthesis. It appears that a group of three bases in messenger RNA identifies each particular amino acid. According to one hypothesis the code group is "recognized" by a complementary set of bases in dictionary RNA. Evidently the ribosome serves as a "jig" for positioning amino acid subunits on the growing protein chain as the messenger RNA "tape" travels by.

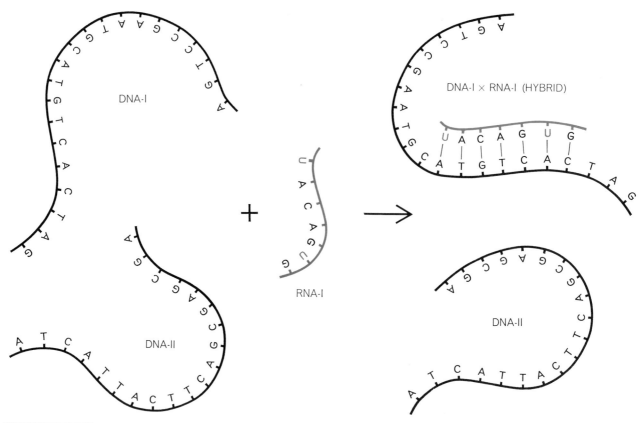

HYBRIDIZATION can occur when the base sequence in a strand of RNA matches up with that in single-strand ("denatured") DNA.

Here RNA-I is "challenged" with genetically related DNA-I and unrelated DNA-II. Only the genetically related strands hybridize.

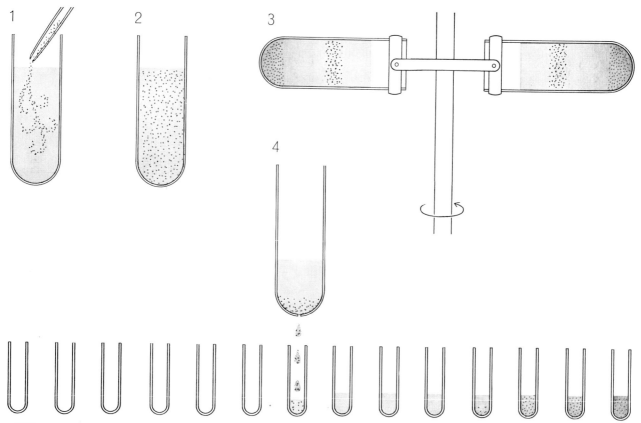

DENSITY-GRADIENT TECHNIQUE reveals if hybridization has taken place between RNA and DNA. The sample in question is added to a solution of cesium chloride (1 and 2). After centrifuga-

tion (3) the salt solution attains a smooth gradation in density. RNA (color), DNA (black) and RNA-DNA hybrids form layers according to their density. Fractions (4) can then be analyzed.

RNA was a precursor of the DNA needed to complete new virus particles. No doubt the experiment was misinterpreted and then neglected because it came so early in the modern history of DNA investigation. The helical model of DNA had been proposed only two years before by James D. Watson and F. H. C. Crick. Moreover, the experiment involved rather complex calculations and assumptions to support the view that the infected cells contained a distinctive new kind of RNA. It is clear in retrospect that this was the first experiment suggesting the existence of RNA copies of DNA.

It seemed to us that the Volkin-Astrachan observations were potentially so important that the design of an unequivocal experiment was well worth the effort. We set out, therefore, to see if bacterial cells infected with the T2 virus contained an RNA that could be specifically related to the T2 DNA. In our first experiments we sought evidence for this new type of RNA by physically isolating it from other RNA's. Two different procedures were successful. One (electrophoresis) measures the rate at which molecules migrate in an electric field; the other (sucrose-gradient centrifugation) measures their rate of migration when they are spun in a solution of smoothly varying density. Both of these methods showed that the RNA synthesized after virus infection was indeed a physically separable entity, differing in mobility and size from the bulk cellular RNA.

We found further that the ratio of the quantities of the bases (A, U, G and C) in the T2-specific RNA mimicked the ratio of the quantities of their counterparts (A, T, G and C) in the DNA of the virus. This suggested the possibility that the similarity might extend to a detailed correspondence of base sequence. A direct attack on this question by the complete determination of the sequences of bases was, and still is, too difficult.

Just at the right time, however, two groups of workers independently published experiments showing that if double-strand DNA was separated into single strands by heat (a process called denaturing), the two strands would re-form into a double-strand structure if the mixture was reheated and slowly cooled. This work was done by Julius Marmur and Dorothy Lane of Brandeis University and by Paul Doty and his colleagues at Harvard University. These investigators showed further that reconstitution of the double-strand molecule

occurs only between strands that originate from the same or closely related organisms. This suggested that double-strand hybrid structures could be formed from mixtures of single-strand DNA and RNA, and that the appearance of such hybrids could be accepted as evidence for a perfect, or near perfect, complementarity of their base sequences. It had already been shown by Alexander Rich of the Massachusetts Institute of Technology and by Doty that synthetic RNA molecules containing adenine as the only base would form hybrid structures with synthetic DNA molecules containing thymine as the only base.

With this work as background, we undertook to determine if T2 RNA would hybridize with T2 DNA. It was first necessary to solve certain technical problems. We had already devised methods for obtaining T2 RNA in a reasonable state of purity. The question was how to design the experiment so that if a hybrid structure formed, we could be certain of detecting it and identifying it as such.

All previous work on the reconstitution of two-strand DNA had involved sizable amounts of material that could form optically observable layers when it was spun in an ultracentrifuge. In our experiments the amount of hybrid material formed would probably be so small that it would escape detection by this method.

The detection method finally evolved combined several techniques. One depended on the fact that RNA has a slightly higher density than DNA; consequently RNA-DNA hybrids should have an intermediate density. Molecules of different densities can be readily separated by the density-gradient method developed by M. S. Meselson, Franklin W. Stahl and Jerome R. Vinograd at the California Institute of Technology. In this method the sample to be analyzed is added to a solution of a heavy salt, cesium chloride, and the mixture is centrifuged for about three days at more than 30,000 revolutions per minute. Under centrifugation the salt solution attains a smooth gradation in density, being most dense at the bottom of the sample tube and least dense at the top. The components of the sample migrate to layers at which their density

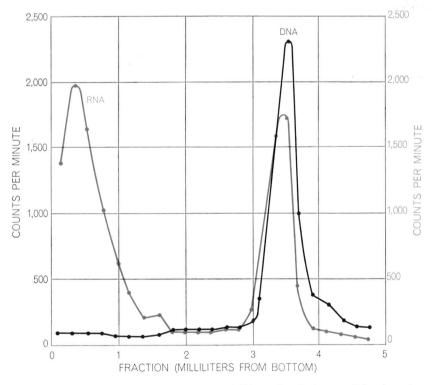

HYBRIDIZATION EXPERIMENT shows that RNA produced after a cell has been infected with the T2 virus is genetically related to the DNA of the virus. The RNA is labeled with radioactive phosphorus and the T2 DNA with radioactive hydrogen (tritium). The sample is subjected to density-gradient centrifugation (*see bottom illustration on opposite page*) and the radioactivity of the various fractions is determined. Although some of the RNA is driven to the bottom of the sample tube, much of it has hybridized with the lighter DNA fraction and thus appears between three and four milliliters above the bottom.

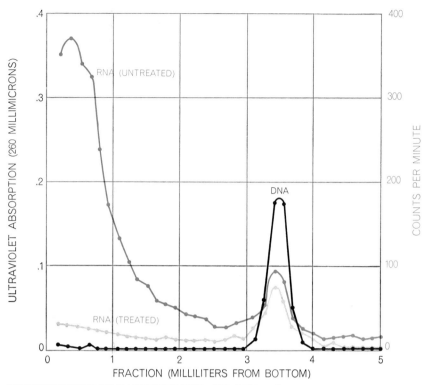

HYBRIDIZATION OF BACTERIAL RNA AND DNA is demonstrated for the bacterium *Pseudomonas aeruginosa*. Untreated RNA chiefly represents messenger RNA obtained by a special "step-down" procedure described in the text. In this experiment the presence of DNA in centrifuged fractions is determined by ultraviolet absorption. The coincident peaks in the two RNA curves represent RNA bound in RNA-DNA hybrids. "Treated RNA" refers to a portion of the sample that was treated before centrifugation with ribonuclease, an enzyme that normally destroys RNA. Although the enzyme has little or no effect on the hybridized RNA, it largely eliminates unhybridized RNA from the centrifuged sample.

exactly matches that of the salt solution. In place of the analytical ultracentrifuge we employed a centrifuge with swinging-bucket rotors, which permits actual isolation and analysis of various fractions. For this purpose the plastic sample tube is punctured at the bottom and the fractionated sample is withdrawn drop by drop for analysis [*see bottom illustration on page 386*].

To ensure a sensitive and unambiguous detection of the hybrid we labeled RNA with one radioactive isotope and DNA with another. The T2 RNA was labeled with radioactive phosphorus (P-32) and the T2 DNA with radioactive hydrogen (H-3). The beta particles emitted by P-32 have a characteristic energy different from those emitted by H-3; thus the isotopes can be assayed in each other's presence. The existence of hybrids in the centrifuged fractions would be signaled by the appearance of a layer containing the P-32 label of the RNA and the H-3 label of the DNA. Subsequently we observed that the layer of the hybrid fraction coincided closely with the layer of the unhybridized DNA. We could

therefore dispense with the radioactive label on DNA and establish its presence simply by its strong absorption of ultraviolet radiation at a wavelength of 260 millimicrons.

With these techniques we soon found that T2 RNA indeed hybridizes with T2 DNA. Furthermore, analysis of the hybrid confirmed that it was similar in overall base composition to T2 DNA. It was then necessary to show that hybrid formation occurs only between RNA and DNA that are genetically related. We exposed T2 RNA to a variety of unrelated DNA's from both bacteria and viruses. No hybrid formation could be detected, even with unrelated DNA's having an overall base composition indistinguishable from that of T2 DNA.

From these experiments one can conclude that T2 RNA has a base sequence complementary to that of at least one of the two strands in T2 DNA. Thus the similarity in base composition first noted by Volkin and Astrachan is a reflection of a more profound relatedness.

These experiments also tell us something about the events that take place when a virus invades a bacterial cell.

If precautions are taken to ensure that all the cells in a given sample are infected with the DNA virus, one finds that none of the RNA synthesized later can hybridize with the host DNA. This suggests that one of the first steps taken by a virulent virus in establishing infection is turning off production of the host's messenger RNA. Evidently RNA transcribed from the viral DNA provides the genetic messages needed for the formation of various proteins required to manufacture complete virus particles. Subsequent studies at the University of Cambridge by Sydney Brenner, François Jacob and Meselson have shown that the T2 messenger RNA is able to make use of ribosomes preexisting in the host cell for the synthesis of proteins.

We wondered next whether the transcription of the DNA code into RNA messages was a universal mechanism or whether it might be restricted to the simple mode of replication followed by viruses. The study of the flow of genetic information in normal cells is a problem of considerable difficulty. As noted above, about 95 per cent of the RNA present at any given moment is of the nontranslatable variety, consisting of ribosomal RNA and transfer, or dictionary, RNA. It is precisely because the translatable RNA molecules are so few—only about 5 per cent of the total amount of RNA—that they were overlooked for so long in normal cells. The detection of the RNA messages of T2 was made easy because the synthesis of ribosomal and transfer RNA is turned off in virus-infected cells.

We decided to look for a situation in normal cells that would imitate the advantages provided by infected ones. It had been known that the total RNA content of cells is positively correlated with rate of growth, and since most of the RNA is ribosomal RNA, a high growth rate implies a high content of ribosomes. What happens if cells are subjected to a "step-down" transfer, that is, a transfer from a rich nutrient medium to a poor one? The growth rate declines, usually by about half. More important, for a generation after they have been placed in a poorer medium the cells contain more ribosomes than they can usefully employ. We reasoned that in this period the synthesis of ribosomal RNA might stop. Since protein production continues at a low rate, however, some synthesis of RNA messages, which must be continuously replaced, should persist.

My colleague Masaki Hayashi undertook experiments to determine if this was the case. If it was, any RNA synthesized after step-down transition would be different from the ribosomal RNA. Hayashi selected three species of bacteria with DNA's of widely different base composition. In all three species the RNA synthesized after step-down transition possessed all the features that had characterized the RNA produced in virus-infected cells. These included instability, a base composition similar to that of the organisms' DNA's and a range of molecular sizes different from that of the ribosomal RNA.

Hybridization tests were carried out between the message-RNA fraction and genetically related DNA as well as with genetically unrelated DNA. The results were clear-cut. Hybrid structures were formed only when the mixture contained RNA and DNA of the same genetic origin. An experiment in hybrid formation that involved RNA and DNA from the bacterium *Pseudomonas aeruginosa* is summarized in the illustration on the opposite page.

This particular experiment illustrates an interesting and useful property of RNA-DNA hybrids. A portion of each sample of hybrid material was treated with the enzyme ribonuclease, which normally destroys RNA. One of the curves shows the amount of RNA in each fraction that was resistant to the enzyme. It can be seen that the RNA bound in the hybrid is quite resistant, whereas the free RNA is almost completely destroyed. This phenomenon turned out to be very useful for distinguishing between free and hybridized RNA. We can conclude from Hayashi's studies, and from those of others, that the flow of information from DNA to translatable RNA occurs normally in bacteria and is probably a universal mechanism in protein synthesis.

By the time these investigations were completed we were convinced that the RNA-DNA hybridization technique could be developed into an extremely powerful and versatile tool. Accordingly we decided to put it to a severe test. The problem we wanted to solve was this: Where do the nontranslatable molecules of RNA—ribosomal RNA and transfer RNA—come from?

Let us consider first the ribosomal variety. Two principal alternatives can be suggested for its mode of origin. Either it is formed on a DNA template or it is not. If it is formed on DNA, it should be complementary to some seg-

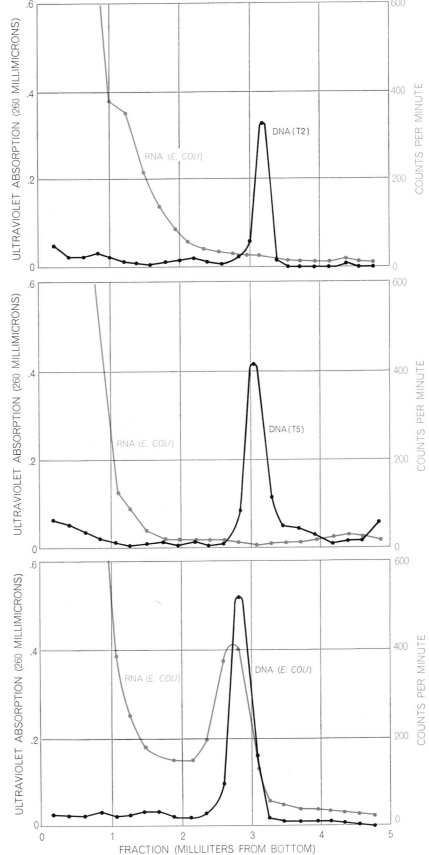

HYBRIDIZATION OF RIBOSOMAL RNA provides evidence that, like messenger RNA, it too is formed on a DNA template. In this experiment ribosomal RNA of the 23S variety was obtained from the colon bacillus (*Escherichia coli*). The top and middle curves show that no hybridization occurs when the RNA is challenged with single-strand DNA from the T2 and T5 viruses. When challenged with DNA from *E. coli*, however, hybridization is seen.

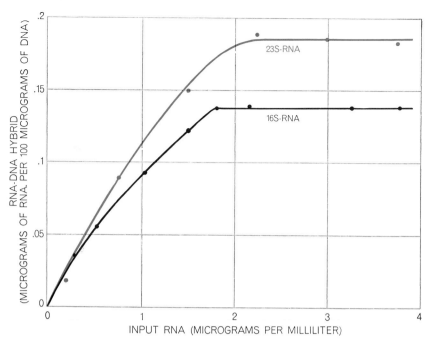

SATURATION CURVES indicate what fraction of the DNA molecule is set aside for producing the two forms of ribosomal RNA designated as 16S and 23S. The RNA and DNA samples were obtained from *Bacillus megaterium*. The results show that about .14 per cent of the DNA molecule is complementary to 16S and about .18 per cent to the 23S form.

ment of DNA and hence subject to hybridization.

It has been known for some time that the base composition of ribosomal RNA is not correlated with that of DNA found in the same cell. This, however, tells us nothing about the origin of the RNA; the DNA segment needed to serve as a template for ribosomal RNA might be so small as to constitute a nonrepresentative sample of the DNA's overall base composition.

Some three years ago one of my students, Saul A. Yankofsky, undertook the job of determining if hybridization could shed any light on this problem. The major complication was that a ribosomal RNA molecule appeared to be only about a ten-thousandth as long as the entire DNA molecule in a typical bacterial cell. We were faced, therefore, with the task of designing experiments that would detect hybridizations involving only a minute segment of DNA.

Theoretically the required sensitivity can be attained simply by labeling RNA so that it has a suitably high level of radioactivity. If no radioactivity was found in association with DNA, one could conclude that no hybrid had been formed. Experiments of this sort would require RNA labeled at a level of about one million counts per minute per microgram. The trouble with such high levels of radioactivity is that irrelevant "noise" can spoil the experiment. It is easy to detect 100 counts per minute

above the background level of radiation. Thus if as little as .0001 microgram of unhybridized RNA accidentally got into the DNA fraction, it would be detected and give a false reading. Such accidental contamination could occur in a number of ways. For example, the ribosomal RNA preparation might contain traces of radioactive translatable RNA that would hybridize with DNA. Small amounts of ribosomal RNA might be mechanically trapped by strands of DNA. Or there might be partial hybridization resulting from accidental coincidences of base complementarity over small regions.

By a variety of biological and technical stratagems it was possible to design a satisfactory experiment. Organisms were chosen with a DNA base composition far removed from that of ribosomal RNA, thereby making it possible to show that hybridized material actually contained ribosomal RNA. Contamination of the radioactive ribosomal RNA preparation by radioactive translatable RNA was eliminated by a simple trick. After the RNA in the cells was labeled with a suitable isotope the cells were transferred to a nonradioactive medium for a period long enough for the labeled RNA messages to disappear. Ribosomal RNA, being stable, retains its radioactive label. Finally, to avoid false readings from RNA that was either mechanically trapped or accidentally paired over short regions, all

suspected hybrids were treated with ribonuclease. The RNA in a genuine hybrid is resistant to this treatment.

It was noted earlier in this article that ribosomes contain two types of RNA, designated 23S RNA and 16S RNA. The outcome of a series of hybridizations between 23S RNA obtained from the colon bacillus and three different DNA preparations is presented in the illustration on the preceding page. A ribonuclease-resistant structure appears in the DNA-density region only when the DNA and the ribosomal RNA are from the same organism. These results clearly imply that ribosomal RNA is produced on a DNA template.

An extension of these studies gave us an answer to the following question: How much of the DNA molecule is set aside for turning out ribosomal RNA? To get the answer we simply add increasing amounts of ribosomal RNA to a fixed amount of DNA and determine the ratio of RNA to DNA in the hybrid at saturation. The illustration at the left shows the outcome of this experiment with the ribosomal RNA of *Bacillus megaterium*. The results indicate that approximately .18 per cent of the total DNA molecule is complementary to 23S RNA and .14 per cent to 16S RNA.

The difference in these two saturation values suggests that 23S RNA and 16S RNA are distinctively different molecules, but the evidence is not unequivocal. Although different in size, the two ribosomal RNA molecules have essentially the same base composition. There is still no direct way of telling whether they have the same or different base sequences. The similarity in base composition and the fact that the 23S RNA has about twice the weight of the 16S RNA had led, however, to the concept that the 23S-RNA molecule is a union of two 16S-RNA molecules.

To probe the matter further we designed an experiment to find out if the two kinds of ribosomal RNA compete for the same sites when they are hybridized with DNA. Hybridization mixtures were prepared that contained fixed amounts of DNA and saturating concentrations of 23S RNA labeled with P-32. To these we added increasing amounts of 16S RNA labeled with H-3, after which we determined the relative amounts of P-32 and H-3 in the hybrid structures. If the two kinds of RNA have an identical sequence, the entry of the H-3-labeled 16S RNA into the hybrid should displace an equivalent amount of P-32-labeled 23S RNA. If the sequences are different, the 16S RNA should hybridize as though the 23S material were not present. The experiment decisively supported the second alternative [*see illustration on this page*].

Following these experiments, there seemed little doubt that the third variety of RNA, transfer RNA, would also be found to originate on segments of DNA. The small size of transfer-RNA molecules made hybridization experiments even more difficult than the earlier ones. Nevertheless, the experiments were successfully carried out by Dario Giacomoni in our laboratory and by Howard M. Goodman in Rich's laboratory at M.I.T. Both workers obtained virtually identical results. They demonstrated by specific hybridization that the DNA of a cell contains sequences complementary to its molecules of transfer RNA. The amount of DNA set aside for the cell's genetic dictionary was found by both groups to be about .025 per cent, or less than a tenth of the combined space allotted to the two types of ribosomal RNA.

These experiments also ruled out an interesting possibility. The molecules of transfer RNA contain only about 80 bases (compared with about 2,000 for 16S RNA) and it was conceivable that the sequence of bases in transfer RNA's might be the same, or much the same, in the cells of different organisms. This possibility seemed more likely when Günter von Ehrenstein of Johns Hopkins University and Fritz A. Lipmann of the Rockefeller Institute showed, in a joint experiment, that transfer RNA's from the colon bacillus can serve as a dictionary in translating the RNA message for the synthesis of the protein hemoglobin from materials present in the red blood cells of the rabbit.

Giacomoni was able to show, however, that the base sequence in transfer-RNA molecules differs from organism to organism. In one such experiment a mixture of transfer-RNA molecules from two different organisms was challenged with DNA molecules obtained from one of them. For identification the genetically related transfer RNA was labeled with P-32 and the unrelated variety with H-3. Only the related RNA formed a hybrid; the genetically unrelated RNA did not [*see chart at left in illustration on next page*].

Instead of using one kind of DNA and two kinds of transfer RNA, one can reverse matters and also demonstrate specificity. For this purpose it is helpful to choose DNA preparations that migrate to different layers when they are subjected to density-gradient centrifuga-

tion. In such a mixture a hybrid will form only with radioactively labeled transfer RNA that is genetically related to one of the DNA's. In the experiment performed in our laboratory the DNA was obtained from two bacteria, *Pseudomonas aeruginosa* and *Bacillus megaterium,* and the transfer RNA was obtained only from the latter [*see chart at right in illustration on next page*].

These experiments reveal an interesting feature of the biological universe. It is assumed that only three of the 80-odd bases in a transfer-RNA molecule provide the means for "reading" the three-base code "words" in the RNA message. Although evidence is lacking on this point, it is possible that a temporary association between three bases in transfer RNA and three bases in the RNA message guarantees that the correct amino acid is deposited where it belongs in a growing protein chain [*see lower illustration on page 385*].

If this picture is accepted, what is

the role of the other 70-odd bases in transfer RNA? The function of the noncoding portion is unknown, but its presence provides an opportunity for biological individuality, from species to species, without disturbing the dictionary function of the molecule. The fact that the base sequences are different in the transfer RNA's of different organisms shows that this opportunity has not been neglected in the course of biological evolution.

We have now seen that all forms of RNA can be traced back to their point of origin on the DNA template. But the double-strand helix of DNA represents two templates, one the complement of the other. When any given segment of DNA is transcribed, two entirely different RNA molecules can be produced, depending on which strand of the DNA molecule serves as a template. Assuming that the entire length of the DNA molecule contains genetic

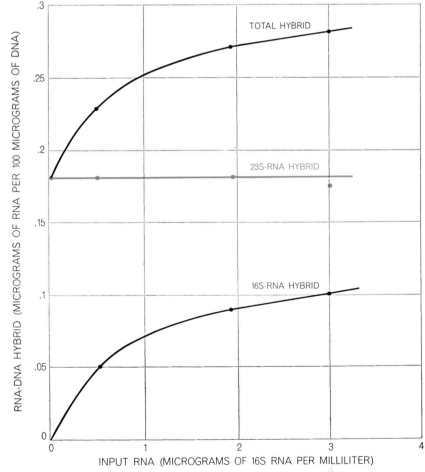

COMPETITION TEST shows that 16S and 23S ribosomal RNA form hybrids with different segments of the DNA molecule. The 16S RNA was labeled with tritium, the 23S RNA with radioactive phosphorus. Increasing amounts of 16S RNA were added to hybridization mixtures containing a saturating concentration of 23S RNA. Subsequently the relative amounts of tritium and radioactive phosphorus in the hybrids were determined. Since the two kinds of RNA hybridize without interference they must have different base sequences.

information that must be transcribed into RNA, there are three possibilities: (1) All of both strands are transcribed into complementary RNA; (2) both strands serve as templates, but in any given segment only one strand or the other is transcribed; (3) only one strand is transcribed.

Here again the hybridization test has supplied evidence to decide among the alternatives. Ideally what is required is a method of separating the two strands of the DNA molecule. If this could be done, one could test the various forms of RNA against each strand and determine if hybridization occurs.

Although the two strands of normal DNA can be separated, no way has yet been found to obtain a pure preparation containing strands of only one type. Fortunately nature provides a solution to the problem in the form of an organism that contains a single strand of DNA. The organism is the small DNA virus φX174, discovered in Parisian sewage about 30 years ago by French investigators. It is fairly easy to purify the virus particle and remove its DNA. Nature also provides a source of the complementary strand. When the virus infects a bacterial cell, the single strand of DNA serves as a template for the synthesis of a complementary strand, resulting in a normal double-strand DNA molecule. This molecule, known as the replicating form, can also be isolated for experimental purposes.

In order to run a hybridization test my co-workers Marie and Masaki Hayashi grew φX174 in infected cells in the presence of P-32 and extracted labeled molecules of translatable RNA. These molecules were then brought together with the single-strand DNA of φX174 and with a denatured sample of the double-strand form. The results obtained were satisfyingly clear. No hybrids were formed with the single-strand DNA, but excellent hybrids were produced with the DNA from the double-strand form. This implied that the RNA messages are complementary to the *other* strand in the two-strand DNA molecule, that is, the one not normally present in the φX174 particle. As a final confirmation we analyzed the base composition of the RNA that was hybridized. The results agreed with the expectation that it was complementary to only one of the two strands of the replicating form of φX174 DNA.

Using similar methods with other viruses, identical conclusions have now been drawn by two other groups: Glauco P. Tocchini-Valentini and his co-workers at the University of Chicago and Carol Greenspan and Marmur at Brandeis University. There seems little doubt that in all organisms only one strand of the DNA molecule serves as a template for RNA synthesis.

The original procedures of detecting hybrids involved lengthy high-speed centrifugations. Ekkehard K. F. Bautz of Rutgers University and Benjamin D. Hall of the University of Illinois have introduced the use of cellulose-acetate columns for hybridization experiments. Ellis T. Bolton and Brian J. McCarthy of the Carnegie Institution of Washington's Department of Terrestrial Magnetism have developed a convenient and rapid method using an agar column. Here the DNA is trapped on the agar gel and the RNA is hybridized with it. The RNA can then be removed by raising the temperature of the column and lowering the ionic strength of an eluting, or rinsing, solution.

The exploitation of the hybridization technique is still at an early stage, but it has already proved of great value in the analysis of gene function. It seems likely to play an increasingly important role in helping to illuminate many problems of molecular biology, including those pertinent to an understanding of the specialization of cells and biological evolution in general.

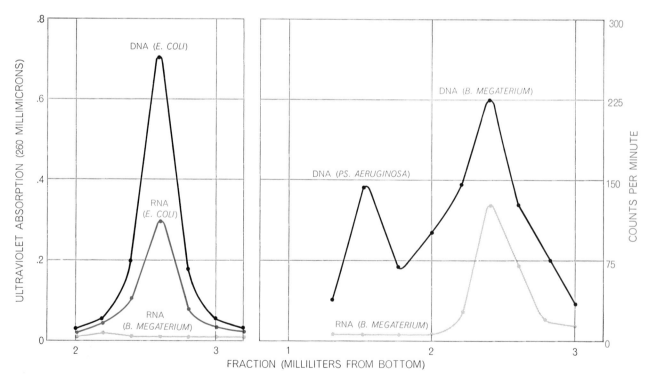

TESTS FOR GENETIC RELATIONSHIP can be carried out by challenging the RNA from two different organisms with the DNA from one of them. In one experiment (*left*) transfer RNA from *E. coli* was labeled with radioactive phosphorus; transfer RNA from *B. megaterium* was labeled with tritium. Only the former hybridizes with *E. coli* DNA. Conversely, in a second experiment (*right*), transfer RNA from *B. megaterium* hybridizes with genetically related DNA but not with DNA from *Ps. aeruginosa*.

34

THE GENETIC CODE: III

F.H.C. CRICK

October 1966

*The central theme of molecular biology is confirmed
by detailed knowledge of how the four-letter language
embodied in molecules of nucleic acid controls the
20-letter language of the proteins*

The hypothesis that the genes of the living cell contain all the information needed for the cell to reproduce itself is now more than 50 years old. Implicit in the hypothesis is the idea that the genes bear in coded form the detailed specifications for the thousands of kinds of protein molecules the cell requires for its moment-to-moment existence: for extracting energy from molecules assimilated as food and for repairing itself as well as for replication. It is only within the past 15 years, however, that insight has been gained into the chemical nature of the genetic material and how its molecular structure can embody coded instructions that can be "read" by the machinery in the cell responsible for synthesizing protein molecules. As the result of intensive work by many investigators the story

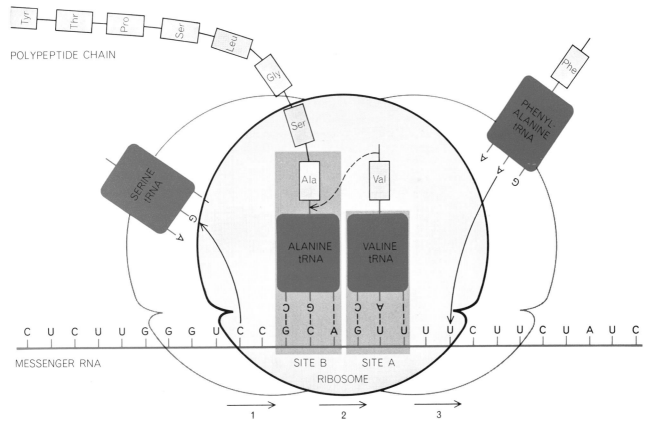

SYNTHESIS OF PROTEIN MOLECULES is accomplished by the intracellular particles called ribosomes. The coded instructions for making the protein molecule are carried to the ribosome by a form of ribonucleic acid (RNA) known as "messenger" RNA. The RNA code "letters" are four bases: uracil (U), cytosine (C), adenine (A) and guanine (G). A sequence of three bases, called a codon, is required to specify each of the 20 kinds of amino acid, identified here by their abbreviations. (A list of the 20 amino acids and their abbreviations appears on the next page.) When linked end to end, these amino acids form the polypeptide chains of which proteins are composed. Each type of amino acid is transported to the ribosome by a particular form of "transfer" RNA (tRNA), which carries an anticodon that can form a temporary bond with one of the codons in messenger RNA. Here the ribosome is shown moving along the chain of messenger RNA, "reading off" the codons in sequence. It appears that the ribosome has two binding sites for molecules of tRNA: one site (*A*) for positioning a newly arrived tRNA molecule and another (*B*) for holding the growing polypeptide chain.

AMINO ACID	ABBREVIATION
ALANINE	Ala
ARGININE	Arg
ASPARAGINE	AspN
ASPARTIC ACID	Asp
CYSTEINE	Cys
GLUTAMIC ACID	Glu.
GLUTAMINE	GluN
GLYCINE	Gly
HISTIDINE	His
ISOLEUCINE	Ileu
LEUCINE	Leu
LYSINE	Lys
METHIONINE	Met
PHENYLALANINE	Phe
PROLINE	Pro
SERINE	Ser
THREONINE	Thr
TRYPTOPHAN	Tryp
TYROSINE	Tyr
VALINE	Val

TWENTY AMINO ACIDS constitute the standard set found in all proteins. A few other amino acids occur infrequently in proteins but it is suspected in each case that they originate as one of the standard set and become chemically modified after they have been incorporated into a polypeptide chain.

of the genetic code is now essentially complete. One can trace the transmission of the coded message from its original site in the genetic material to the finished protein molecule.

The genetic material of the living cell is the chainlike molecule of deoxyribonucleic acid (DNA). The cells of many bacteria have only a single chain; the cells of mammals have dozens clustered together in chromosomes. The DNA molecules have a very long backbone made up of repeating groups of phosphate and a five-carbon sugar. To this backbone the side groups called bases are attached at regular intervals. There are four standard bases: adenine (A), guanine (G), thymine (T) and cytosine (C). They are the four "letters" used to spell out the genetic message. The exact sequence of bases along a length of the DNA molecule determines the structure of a particular protein molecule.

Proteins are synthesized from a standard set of 20 amino acids, uniform throughout nature, that are joined end to end to form the long polypeptide

chains of protein molecules [see illustration at left]. Each protein has its own characteristic sequence of amino acids. The number of amino acids in a polypeptide chain ranges typically from 100 to 300 or more.

The genetic code is not the message itself but the "dictionary" used by the cell to translate from the four-letter language of nucleic acid to the 20-letter language of protein. The machinery of the cell can translate in one direction only: from nucleic acid to protein but not from protein to nucleic acid. In making this translation the cell employs a variety of accessory molecules and mechanisms. The message contained in DNA is first transcribed into the similar molecule called "messenger" ribonucleic acid—messenger RNA. (In many viruses—the tobacco mosaic virus, for example—the genetic material is simply RNA.) RNA too has four kinds of bases as side groups; three are identical with those found in DNA (adenine, guanine and cytosine) but the fourth is uracil (U) instead of thymine. In this first transcription of the genetic message the code letters A, G, T and C in DNA give rise respectively to U, C, A and G. In other words, wherever A appears in DNA, U appears in the RNA transcription; wherever G appears in DNA, C appears in the transcription, and so on. As it is usually presented the dictionary of the genetic code employs the letters found in RNA (U, C, A, G) rather than those found in DNA (A, G, T, C).

The genetic code could be broken easily if one could determine both the amino acid sequence of a protein and the base sequence of the piece of nucleic acid that codes it. A simple comparison of the two sequences would yield the code. Unfortunately the determination of the base sequence of a long nucleic acid molecule is, for a variety of reasons, still extremely difficult. More indirect approaches must be used.

Most of the genetic code first became known early in 1965. Since then additional evidence has proved that almost all of it is correct, although a few features remain uncertain. This article describes how the code was discovered and some of the work that supports it.

Scientific American has already presented a number of articles on the genetic code. In one of them ["The Genetic Code," Offprint 123] I explained that the experimental evidence (mainly indirect) suggested that the code was a triplet code: that the bases on the messenger RNA were read three at a time and that each group corresponded to a

particular amino acid. Such a group is called a codon. Using four symbols in groups of three, one can form 64 distinct triplets. The evidence indicated that most of these stood for one amino acid or another, implying that an amino acid was usually represented by several codons. Adjacent amino acids were coded by adjacent codons, which did not overlap.

In a sequel to that article ["The Genetic Code: II," March, 1963] Marshall W. Nirenberg of the National Institutes of Health explained how the composition of many of the 64 triplets had been determined by actual experiment. The technique was to synthesize polypeptide chains in a cell-free system, which was made by breaking open cells of the colon bacillus (*Escherichia coli*) and extracting from them the machinery for protein synthesis. Then the system was provided with an energy supply, 20 amino acids and one or another of several types of synthetic RNA. Although the exact sequence of bases in each type was random, the proportion of bases was known. It was found that each type of synthetic messenger RNA directed the incorporation of certain amino acids only.

By means of this method, used in a quantitative way, the *composition* of many of the codons was obtained, but the *order* of bases in any triplet could not be determined. Codons rich in G were difficult to study, and in addition a few mistakes crept in. Of the 40 codon compositions listed by Nirenberg in his article we now know that 35 were correct.

The Triplet Code

The main outlines of the genetic code were elucidated by another technique invented by Nirenberg and Philip Leder. In this method no protein synthesis occurs. Instead one triplet at a time is used to bind together parts of the machinery of protein synthesis.

Protein synthesis takes place on the comparatively large intracellular structures known as ribosomes. These bodies travel along the chain of messenger RNA, reading off its triplets one after another and synthesizing the polypeptide chain of the protein, starting at the amino end (NH$_2$). The amino acids do not diffuse to the ribosomes by themselves. Each amino acid is joined chemically by a special enzyme to one of the codon-recognizing molecules known both as soluble RNA (sRNA) and transfer RNA (tRNA). (I prefer the latter designation.) Each tRNA molecule has its own triplet of bases, called an anticodon, that recognizes the relevant codon on the messenger RNA by pairing bases with it [*see illustration on page 393*].

Leder and Nirenberg studied which amino acid, joined to its tRNA molecules, was bound to the ribosomes in the presence of a particular triplet, that is, by a "message" with just three letters. They did so by the neat trick of passing the mixture over a nitrocellulose filter that retained the ribosomes. All the tRNA molecules passed through the filter except the ones specifically bound to the ribosomes by the triplet. Which they were could easily be decided by using mixtures of amino acids in which one kind of amino acid had been made artificially radioactive, and determining the amount of radioactivity absorbed by the filter.

For example, the triplet GUU retained the tRNA for the amino acid valine, whereas the triplets UGU and UUG did not. (Here GUU actually stands for the trinucleoside diphosphate GpUpU.) Further experiments showed that UGU coded for cysteine and UUG for leucine.

Nirenberg and his colleagues synthesized all 64 triplets and tested them for their coding properties. Similar results have been obtained by H. Gobind Khorana and his co-workers at the University of Wisconsin. Various other groups have checked a smaller number of codon assignments.

Close to 50 of the 64 triplets give a clearly unambiguous answer in the binding test. Of the remainder some evince only weak binding and some bind more than one kind of amino acid. Other results I shall describe later suggest that the multiple binding is often an artifact of the binding method. In short, the binding test gives the meaning of the majority of the triplets but it does not firmly establish all of them.

The genetic code obtained in this way, with a few additions secured by other methods, is shown in the table below. The 64 possible triplets are set out in a regular array, following a plan

FIRST LETTER	SECOND LETTER: U		C		A		G		THIRD LETTER
U	UUU	Phe	UCU	Ser	UAU	Tyr	UGU	Cys	U
	UUC		UCC		UAC		UGC		C
	UUA	Leu	UCA		UAA	OCHRE	UGA	?	A
	UUG		UCG		UAG	AMBER	UGG	Tryp	G
C	CUU	Leu	CCU	Pro	CAU	His	CGU	Arg	U
	CUC		CCC		CAC		CGC		C
	CUA		CCA		CAA	GluN	CGA		A
	CUG		CCG		CAG		CGG		G
A	AUU	Ileu	ACU	Thr	AAU	AspN	AGU	Ser	U
	AUC		ACC		AAC		AGC		C
	AUA		ACA		AAA	Lys	AGA	Arg	A
	AUG	Met	ACG		AAG		AGG		G
G	GUU	Val	GCU	Ala	GAU	Asp	GGU	Gly	U
	GUC		GCC		GAC		GGC		C
	GUA		GCA		GAA	Glu	GGA		A
	GUG		GCG		GAG		GGG		G

GENETIC CODE, consisting of 64 triplet combinations and their corresponding amino acids, is shown in its most likely version. The importance of the first two letters in each triplet is readily apparent. Some of the allocations are still not completely certain, particularly for organisms other than the colon bacillus (*Escherichia coli*). "Amber" and "ochre" are terms that referred originally to certain mutant strains of bacteria. They designate two triplets, UAA and UAG, that may act as signals for terminating polypeptide chains.

that clarifies the relations between them.

Inspection of the table will show that the triplets coding for the same amino acid are often rather similar. For example, all four of the triplets starting with the doublet AC code for threonine. This pattern also holds for seven of the other amino acids. In every case the triplets XYU and XYC code for the same amino acid, and in many cases XYA and XYG are the same (methionine and tryptophan may be exceptions). Thus an amino acid is largely selected by the first two bases of the triplet. Given that a triplet codes for, say, valine, we know that the first two bases are GU, whatever the third may be. This pattern is true for all but three of the amino acids. Leucine can start with UU or CU, serine with UC or AG and arginine with CG or AG. In all other cases the amino acid is uniquely related to the first two bases of the triplet. Of course, the converse is often not true. Given that a triplet starts with, say, CA, it may code for either histidine or glutamine.

Synthetic Messenger RNA's

Probably the most direct way to confirm the genetic code is to synthesize a messenger RNA molecule with a strictly defined base sequence and then find the amino acid sequence of the polypeptide produced under its influence. The most extensive work of this nature has been done by Khorana and his colleagues. By a brilliant combination of ordinary chemical synthesis and synthesis catalyzed by enzymes, they have made long RNA molecules with various repeating sequences of bases. As an example, one RNA molecule they have synthesized has the sequence UGUGUGUGUG.... When the biochemical machinery reads this as triplets the message is UGU–GUG–UGU–GUG.... Thus we expect that a polypeptide will be produced with an alternating sequence of two amino acids. In fact, it was found that the product is Cys–Val–Cys–Val.... This evidence alone would not tell us which triplet goes with which amino acid, but given the results of the binding test one has no hesitation in concluding that UGU codes for cysteine and GUG for valine.

In the same way Khorana has made chains with repeating sequences of the type XYZ... and also XXYZ.... The type XYZ...would be expected to give a "homopolypeptide" containing one amino acid corresponding to the triplet XYZ. Because the starting point is not clearly defined, however, the homopolypeptides corresponding to YZX... and ZXY... will also be produced. Thus poly-AUC makes polyisoleucine, polyserine and polyhistidine. This confirms that AUC codes for isoleucine, UCA for serine and CAU for histidine. A repeating sequence of four bases will yield a single type of polypeptide with a repeating sequence of four amino acids. The general patterns to be expected in each case are set forth in the table on this page. The results to date have amply demonstrated by a direct biochemical method that the code is indeed a triplet code.

Khorana and his colleagues have so far confirmed about 25 triplets by this method, including several that were quite doubtful on the basis of the binding test. They plan to synthesize other sequences, so that eventually most of the triplets will be checked in this way.

The Use of Mutations

The two methods described so far are open to the objection that since they do not involve intact cells there may be some danger of false results. This objection can be met by two other methods of checking the code in which the act of protein synthesis takes place inside the cell. Both involve the effects of genetic mutations on the amino acid sequence of a protein.

It is now known that small mutations are normally of two types: "base substitution" mutants and "phase shift" mutants. In the first type one base is changed into another base but the total number of bases remains the same. In the second, one or a small number of bases are added to the message or subtracted from it.

There are now extensive data on base-substitution mutants, mainly from studies of three rather convenient proteins: human hemoglobin, the protein of tobacco mosaic virus and the A protein of the enzyme tryptophan synthetase obtained from the colon bacillus. At least 36 abnormal types of human hemoglobin have now been investigated by many different workers. More than 40 mutant forms of the protein of the tobacco mosaic virus have been examined by Hans Wittmann of the Max Planck Institute for Molecular Genetics in Tübingen and by Akita Tsugita and Heinz Fraenkel-Conrat of the University of California at Berkeley [see "The Genetic Code of a Virus," by Heinz Fraenkel-Conrat; SCIENTIFIC AMERICAN Offprint 193]. Charles Yanofsky and his group at Stanford University have characterized about 25 different mutations of the A protein of tryptophan synthetase.

RNA BASE SEQUENCE	READ AS	AMINO ACID SEQUENCE EXPECTED
$(XY)_n$ ...	X Y X | Y X Y | X Y X | Y X Y ...	αβαβ
$(XYZ)_n$ ...	X Y Z | X Y Z | X Y Z ...	ααα
...	Y Z X | Y Z X | Y Z X ...	βββ
...	Z X Y | Z X Y | Z X Y ...	γγγ
$(XXYZ)_n$ ...	X X Y Z | X X Y Z | X X Y Z ...	αβγδαβγδ
$(XYXZ)_n$ ...	X Y X Z | X Y X Z | X Y X Z ...	αβγδαβγδ

VARIETY OF SYNTHETIC RNA's with repeating sequences of bases have been produced by H. Gobind Khorana and his colleagues at the University of Wisconsin. They contain two or three different bases (X, Y, Z) in groups of two, three or four. When introduced into cell-free systems containing the machinery for protein synthesis, the base sequences are read off as triplets (middle) and yield the amino acid sequences indicated at the right.

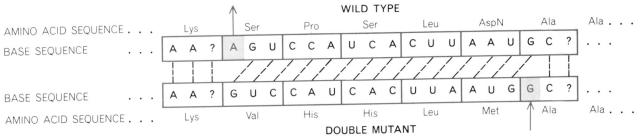

"PHASE SHIFT" MUTATIONS help to establish the actual codons used by organisms in the synthesis of protein. The two partial amino acid sequences shown here were determined by George Streisinger and his colleagues at the University of Oregon. The sequences are from a protein, a type of lysozyme, produced by the bacterial virus T4. A pair of phase-shift mutations evidently removed one base, A, and inserted another, G, about 15 bases farther on. The base sequence was deduced theoretically from the genetic code.

The remarkable fact has emerged that in every case but one the genetic code shows that the change of an amino acid in a polypeptide chain could have been caused by the alteration of a single base in the relevant nucleic acid. For example, the first observed change of an amino acid by mutation (in the hemoglobin of a person suffering from sickle-cell anemia) was from glutamic acid to valine. From the genetic code dictionary on page 395 we see that this could have resulted from a mutation that changed either GAA to GUA or GAG to GUG. In either case the change involved a single base in the several hundred needed to code for one of the two kinds of chain in hemoglobin.

The one exception so far to the rule that all amino acid changes could be caused by single base changes has been found by Yanofsky. In this one case glutamic acid was replaced by methionine. It can be seen from the genetic code dictionary that this can be accomplished only by a change of *two* bases, since glutamic acid is encoded by either GAA or GAG and methionine is encoded only by AUG. This mutation has occurred only once, however, and of all the mutations studied by Yanofsky it is the only one not to back-mutate, or revert to "wild type." It is thus almost certainly the rare case of a double change. All the other cases fit the hypothesis that base-substitution mutations are normally caused by a single base change. Examination of the code shows that only about 40 percent of all the possible amino acid interchanges can be brought about by single base substitutions, and it is only these changes that are found in experiments. Therefore the study of actual mutations has provided strong confirmation of many features of the genetic code.

Because in general several codons stand for one amino acid it is not possible, knowing the amino acid sequence, to write down the exact RNA base sequence that encoded it. This is unfortu-nate. If we know which amino acid is changed into another by mutation, however, we can often, given the code, work out what that base change must have been. As an example, glutamic acid can be encoded by GAA or GAG and valine by GUU, GUC, GUA or GUG. If a mutation substitutes valine for glutamic acid, one can assume that only a single base change was involved. The only such change that could lead to the desired result would be a change from A to U in the middle position, and this would be true whether GAA became GUA or GAG became GUG.

It is thus possible in many cases (not in all) to compare the nature of the base change with the chemical mutagen used to produce the change. If RNA is treated with nitrous acid, C is changed to U and A is effectively changed to G. On the other hand, if double-strand DNA is treated under the right conditions with hydroxylamine, the mutagen acts only on C. As a result some C's are changed to T's (the DNA equivalent of U's), and thus G's, which are normal-ly paired with C's in double-strand DNA, are replaced by A's.

If 2-aminopurine, a "base analogue" mutagen, is added when double-strand DNA is undergoing replication, it produces only "transitions." These are the same changes as those produced by hy-droxylamine—plus the reverse changes. In almost all these different cases (the exceptions are unimportant) the changes observed are those expected from our knowledge of the genetic code.

Note the remarkable fact that, al-though the code was deduced mainly from studies of the colon bacillus, it appears to apply equally to human beings and tobacco plants. This, to-gether with more fragmentary evidence, suggests that the genetic code is either the same or very similar in most or-ganisms.

The second method of checking the code using intact cells depends on phase-shift mutations such as the addi-tion of a single base to the message. Phase-shift mutations probably result from errors produced during genetic recombination or when the DNA mole-cule is being duplicated. Such errors have the effect of putting out of phase the reading of the message from that point on. This hypothesis leads to the prediction that the phase can be cor-rected if at some subsequent point a nucleotide is deleted. The pair of al-terations would be expected not only to change two amino acids but also to alter all those encoded by bases lying between the two affected sites. The reason is that the intervening bases would be read out of phase and there-fore grouped into triplets different from those contained in the normal message.

This expectation has recently been confirmed by George Streisinger and his colleagues at the University of Oregon. They have studied mutations in the protein lysozyme that were produced by the T4 virus, which infects the colon bacillus. One phase-shift mutation in-volved the amino acid sequence ...Lys—Ser—Pro—Ser—Leu—AspN—Ala—Ala—Lys.... They were then able to con-struct by genetic methods a double phase-shift mutant in which the cor-responding sequence was ...Lys—Val—His—His—Leu—Met—Ala—Ala—Lys....

Given these two sequences, the read-er should be able, using the genetic code dictionary on page 395, to decipher uniquely a short length of the nucleic acid message for both the original pro-tein and the double mutant and thus de-duce the changes produced by each of the phase-shift mutations. The correct result is presented in the illustration above. The result not only confirms sev-eral rather doubtful codons, such as UUA for leucine and AGU for serine, but also shows which codons are actual-ly involved in a genetic message. Since the technique is difficult, however, it may not find wide application.

Streisinger's work also demonstrates what has so far been only tacitly as-

ANTICODON	CODON
U	A G
C	G
A	U
G	U C
I	U C A

"WOBBLE" HYPOTHESIS has been proposed by the author to provide rules for the pairing of codon and anticodon at the *third* position of the codon. There is evidence, for example, that the anticodon base I, which stands for inosine, may pair with as many as three different bases: U, C and A. Inosine closely resembles the base guanine (G) and so would ordinarily be expected to pair with cytosine (C). Structural diagrams for standard base pairings and wobble base pairings are illustrated at the bottom of this page.

sumed: that the two languages, both of which are written down in a certain direction according to convention, are in fact translated by the cell in the same direction and not in opposite directions. This fact had previously been established, with more direct chemical methods, by Severo Ochoa and his colleagues at the New York University School of Medicine. In the convention, which was adopted by chance, proteins are written with the amino (NH_2) end on the left. Nucleic acids are written with the end of the molecule containing

a "5 prime" carbon atom at the left. (The "5 prime" refers to a particular carbon atom in the 5-carbon ring of ribose sugar or deoxyribose sugar.)

Finding the Anticodons

Still another method of checking the genetic code is to discover the three bases making up the anticodon in some particular variety of transfer RNA. The first tRNA to have its entire sequence worked out was alanine tRNA, a job done by Robert W. Holley and his collaborators at Cornell University [see the article "The Nucleotide Sequence of a Nucleic Acid," by Robert W. Holley, beginning on page 400]. Alanine tRNA, obtained from yeast, contains 77 bases. A possible anticodon found near the middle of the molecule has the sequence IGC, where I stands for inosine, a base closely resembling guanine. Since then Hans Zachau and his colleagues at the University of Cologne have established the sequences of two closely related serine tRNA's from yeast, and James Madison and his group at the U.S. Plant, Soil and Nutrition Laboratory at Ithaca, N.Y., have worked out the sequence of a tyrosine tRNA, also from yeast.

A detailed comparison of these three sequences makes it almost certain that the anticodons are alanine–IGC, serine–IGA and tyrosine–GΨA. (Ψ stands for pseudo-uridylic acid, which can form the same base pairs as the base uracil.) In addition there is preliminary evidence from other workers that an anticodon for valine is IAC and an anticodon for phenylalanine is GAA.

All these results would fit the rule that the codon and anticodon pair in an antiparallel manner, and that the pairing in the first two positions of the codon is of the standard type, that is, A pairs with U and G pairs with C. The pairing in the third position of the codon is more complicated. There is now good experimental evidence from both Nirenberg and Khorana and their co-workers that one tRNA can recognize several codons, provided that they differ only in the last place in the codon. Thus Holley's alanine tRNA appears to recognize GCU, GCC and GCA. If it recognizes GCG, it does so only very weakly.

The "Wobble" Hypothesis

I have suggested that this is because of a "wobble" in the pairing in the third place and have shown that a reasonable theoretical model will explain many of the observed results. The suggested rules for the pairing in the third position of the anticodon are presented in the table at the top of this page, but this theory is still speculative. The rules for the first two places of the codon seem reasonably secure, however, and can be used as partial confirmation of the genetic code. The likely codon-anticodon pairings for valine, serine, tyrosine, alanine and phenylalanine satisfy the standard base pairings in the first two places and the wobble hypothesis in the third place [see illustration on page 399].

Several points about the genetic code remain to be cleared up. For example, the triplet UGA has still to be allocated.

STANDARD AND WOBBLE BASE PAIRINGS both involve the formation of hydrogen bonds when certain bases are brought into close proximity. In the standard guanine-cytosine pairing (*left*) it is believed three hydrogen bonds are formed. The bases are shown as they exist in the RNA molecule, where they are attached to 5-carbon rings of ribose sugar. In the proposed wobble pairing (*right*) guanine is linked to uracil by only two hydrogen bonds. The base inosine (I) has a single hydrogen atom where guanine has an amino (NH_2) group (*broken circle*). In the author's wobble hypothesis inosine can pair with U as well as with C and A (*not shown*).

The punctuation marks—the signals for "begin chain" and "end chain"—are only partly understood. It seems likely that both the triplet UAA (called "ochre") and UAG (called "amber") can terminate the polypeptide chain, but which triplet is normally found at the end of a gene is still uncertain.

The picturesque terms for these two triplets originated when it was discovered in studies of the colon bacillus some years ago that mutations in other genes (mutations that in fact cause errors in chain termination) could "suppress" the action of certain mutant codons, now identified as either UAA or UAG. The terms "ochre" and "amber" are simply invented designations and have no reference to color.

A mechanism for chain initiation was discovered fairly recently. In the colon bacillus it seems certain that formylmethionine, carried by a special tRNA, can initiate chains, although it is not clear if all chains have to start in this way, or what the mechanism is in mammals and other species. The formyl group (CHO) is not normally found on finished proteins, suggesting that it is probably removed by a special enzyme. It seems likely that sometimes the methionine is removed as well.

It is unfortunately possible that a few codons may be ambiguous, that is, may code for more than one amino acid. This is certainly not true of most codons. The present evidence for a small amount of ambiguity is suggestive but not conclusive. It will make the code more difficult to establish correctly if ambiguity can occur.

Problems for the Future

From what has been said it is clear that, although the entire genetic code

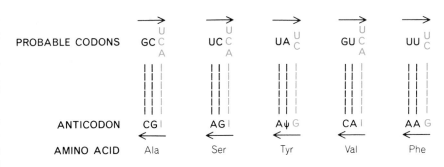

CODON-ANTICODON PAIRINGS take place in an antiparallel direction. Thus the anticodons are shown here written backward, as opposed to the way they appear in the text. The five anticodons are those tentatively identified in the transfer RNA's for alanine, serine, tyrosine, valine and phenylalanine. Color indicates where wobble pairings may occur.

is not known with complete certainty, it is highly likely that most of it is correct. Further work will surely clear up the doubtful codons, clarify the punctuation marks, delimit ambiguity and extend the code to many other species. Although the code lists the codons that *may* be used, we still have to determine if alternative codons are used equally. Some preliminary work suggests they may not be. There is also still much to be discovered about the machinery of protein synthesis. How many types of tRNA are there? What is the structure of the ribosome? How does it work, and why is it in two parts? In addition there are many questions concerning the control of the rate of protein synthesis that we are still a long way from answering.

When such questions have been answered, the major unsolved problem will be the structure of the genetic code. Is the present code merely the result of a series of evolutionary accidents, so that the allocations of triplets to amino acids is to some extent arbitrary? Or are there

profound structural reasons why phenylalanine has to be coded by UUU and UUC and by no other triplets? Such questions will be difficult to decide, since the genetic code originated at least three billion years ago, and it may be impossible to reconstruct the sequence of events that took place at such a remote period. The origin of the code is very close to the origin of life. Unless we are lucky it is likely that much of the evidence we should like to have has long since disappeared.

Nevertheless, the genetic code is a major milestone on the long road of molecular biology. In showing in detail how the four-letter language of nucleic acid controls the 20-letter language of protein it confirms the central theme of molecular biology that genetic information can be stored as a one-dimensional message on nucleic acid and be expressed as the one-dimensional amino acid sequence of a protein. Many problems remain, but this knowledge is now secure.

35

THE NUCLEOTIDE SEQUENCE OF A NUCLEIC ACID

ROBERT W. HOLLEY

February 1966

For the first time the specific order of subunits in one of the giant molecules that participate in the synthesis of protein has been determined. The task took seven years

Two major classes of chainlike molecules underlie the functioning of living organisms: the nucleic acids and the proteins. The former include deoxyribonucleic acid (DNA), which embodies the hereditary message of each organism, and ribonucleic acid (RNA), which helps to translate that message into the thousands of different proteins that activate the living cell. In the past dozen years biochemists have established the complete sequence of amino acid subunits in a number of different proteins. Much less is known about the nucleic acids.

Part of the reason for the slow progress with nucleic acids was the unavailability of pure material for analysis. Another factor was the large size of most nucleic acid molecules, which often contain thousands or even millions of nucleotide subunits. Several years ago, however, a family of small molecules was discovered among the ribonucleic

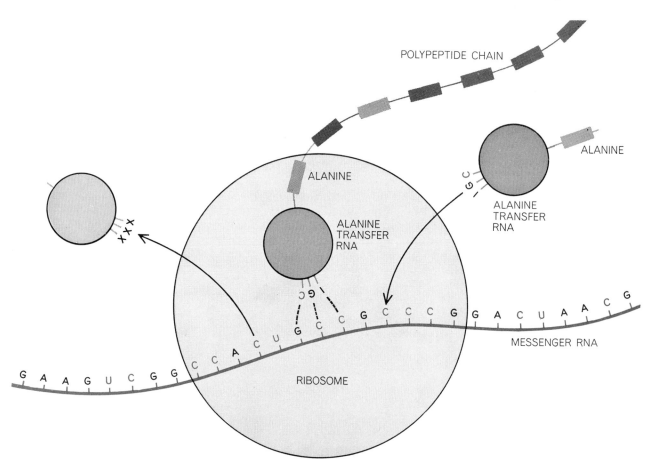

ROLE OF TRANSFER RNA is to deliver a specific amino acid to the site where "messenger" RNA and a ribosome (which also contains RNA) collaborate in the synthesis of a protein. As it is being synthesized a protein chain is usually described as a polypeptide. Each amino acid in the polypeptide chain is specified by a triplet code, or codon, in the molecular chain of messenger RNA.

The diagram shows how an "anticodon" (presumably I—G—C) in alanine transfer RNA may form a temporary bond with the codon for alanine (G—C—C) in the messenger RNA. While so bonded the transfer RNA also holds the polypeptide chain. Each transfer RNA is succeeded by another one, carrying its own amino acid, until the complete message in the messenger RNA has been "read."

HYPOTHETICAL MODELS of alanine transfer ribonucleic acid (RNA) show three of the many ways in which the molecule's linear chain might be folded. The various letters represent nucleotide subunits; their chemical structure is given at the top of the next two pages. In these models it is assumed that certain nucleotides, such as C—G and A—U, will pair off and tend to form short double-strand regions. Such "base-pairing" is a characteristic feature of nucleic acids. The arrangement at the lower left shows how two of the large "leaves" of the "clover leaf" model may be folded together. The triplet I—G—C is the presumed anticodon shown in the illustration on the opposite page. The region containing the sequence G—T—Ψ—C—G may be common to all transfer RNA's.

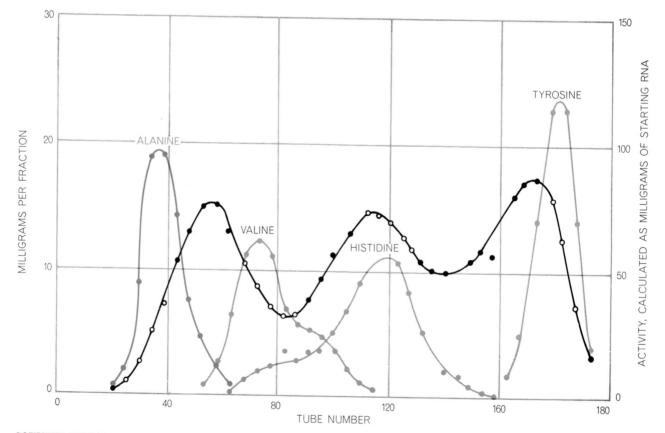

ADENYLIC ACID
p A

INOSINIC ACID
I

1-METHYLINOSINIC ACID
I^m

GUANYLIC ACID
G

1-METHYLG

NUCLEOTIDE SUBUNITS found in alanine transfer RNA include the four commonly present in RNA (A, G, C, U), plus seven others that are variations of the standard structures. Ten of these 11 different nucleotide subunits are assembled above as if they were linked together in a single RNA chain. The chain begins at the left with a phosphate group (*outlined by a small rectangle*) and is followed by a ribose sugar group (*large rectangle*); the two groups alternate to form the backbone of the chain. The chain ends at the right with

acids. My associates and I at the U.S. Plant, Soil and Nutrition Laboratory and Cornell University set ourselves the task of establishing the nucleotide sequence of one of these smaller RNA molecules—a molecule containing fewer than 100 nucleotide subunits. This work culminated recently in the first determination of the complete nucleotide sequence of a nucleic acid.

The object of our study belongs to a family of 20-odd molecules known as transfer RNA's. Each is capable of recognizing one of the 20 different amino acids and of transferring it to the site where it can be incorporated into a growing polypeptide chain. When such a chain assumes its final configuration, sometimes joining with other chains, it is called a protein.

At each step in the process of protein

COUNTERCURRENT DISTRIBUTION PATTERN shows two steps in the separation of alanine transfer RNA, as carried out in the author's laboratory. After the first step the RNA content in various collection tubes, measured by ultraviolet absorption, follows the black curve. Biological activity, indicated by the amount of a given amino acid incorporated into polypeptide chains, follows the colored curves. Pure transfer RNA's of four types can be obtained by reprocessing the tubes designated by open circles.

▲RIBOTHYMIDYLIC ACID   DIHYDROURIDYLIC ACID

T   U^h

MIXTURE OF URIDYLIC AND

-DIMETHYLGUANYLIC ACID   CYTIDYLIC ACID   URIDYLIC ACID   DIHYDROURIDYLIC ACIDS   PSEUDOURIDYLIC ACID

G^m   C   U   U*   ψ OH

a hydroxyl (OH) group. Each nucleotide subunit consists of a phosphate group, a ribose sugar group and a base. The base portion in the nucleotide at the far left, adenylic acid, is outlined by a large rectangle. In the succeeding bases the atomic variations are shown in color. The base structures without color are those commonly found in RNA. Black arrows show where RNA chains can be cleaved by the enzyme takadiastase ribonuclease T1. Colored arrows show where RNA chains can be cleaved by pancreatic ribonuclease.

synthesis a crucial role is played by the structure of the various RNA's. "Messenger" RNA transcribes the genetic message for each protein from its original storage site in DNA. Another kind of RNA—ribosomal RNA—forms part of the structure of the ribosome, which acts as a jig for holding the messenger RNA while the message is transcribed into a polypeptide chain [see illustration on page 400]. In view of the various roles played by RNA in protein synthesis, the structure of RNA molecules is of considerable interest and significance.

The particular nucleic acid we chose for study is known as alanine transfer RNA—the RNA that transports the amino acid alanine. It was isolated from commercial baker's yeast by methods I shall describe later. Preliminary analyses indicated that the alanine transfer RNA molecule consisted of a single chain of approximately 80 nucleotide subunits. Each nucleotide, in turn, consists of a ribose sugar, a phosphate group and a distinctive appendage termed a nitrogen base. The ribose sugars and phosphate groups link together to form the backbone of the molecule, from which the various bases protrude [see illustration at top of these two pages].

The problem of structural analysis is fundamentally one of identifying each base and determining its place in the sequence. In practice each base is usually isolated in combination with a unit of ribose sugar and a unit of phosphate, which together form a nucleotide. Formally the problem is analogous to de-

termining the sequence of letters in a sentence.

It would be convenient if there were a way to snip off the nucleotides one by one, starting at a known end of the chain and identifying each nucleotide as it appeared. Unfortunately procedures of this kind have such a small yield at each step that their use is limited. The alternative is to break the chain at particular chemical sites with the help of enzymes. This gives rise to small fragments whose nucleotide composition is amenable to analysis. If the chain can be broken up in various ways with different enzymes, one can determine how the fragments overlap and ultimately piece together the entire sequence.

One can visualize how this might work by imagining that the preceding sentence has been written out several times, in a continuous line, on different strips of paper. Imagine that each strip has been cut in a different way. In one case, for example, the first three words "If the chain" and the next three words "can be broken" might appear on separate strips of paper. In another case one might find that "chain" and "can" were together on a single strip. One would immediately conclude that the group of three words ending with "chain" and the group beginning with "can" form a continuous sequence of six words. The concept is simple; putting it into execution takes a little time.

For cleaving the RNA chain we used two principal enzymes: pancreatic ribonuclease and an enzyme called takadiastase ribonuclease T1, which was discovered by the Japanese workers K. Sato-Asano and F. Egami. The first

enzyme cleaves the RNA chain immediately to the right of pyrimidine nucleotides, as the molecular structure is conventionally written. Pyrimidine nucleotides are those nucleotides whose bases contain the six-member pyrimidine ring, consisting of four atoms of carbon and two atoms of nitrogen. The two pyrimidines commonly found in RNA are cytosine and uracil. Pancreatic ribonuclease therefore produces fragments that terminate in pyrimidine nucleotides such as cytidylic acid (C) or uridylic acid (U).

The second enzyme, ribonuclease T1, was employed separately to cleave the RNA chain specifically to the right of nucleotides containing a structure of the purine type, such as guanylic acid (G). This provided a set of short fragments distinctively different from those produced by the pancreatic enzyme.

The individual short fragments were isolated by passing them through a thin glass column packed with diethylaminoethyl cellulose—an adaptation of a chromatographic method devised by R. V. Tomlinson and G. M. Tener of the University of British Columbia. In general the short fragments migrate through the column more rapidly than the long fragments, but there are exceptions [see illustration on next page]. The conditions most favorable for this separation were developed in our laboratories by Mark Marquisee and Jean Apgar.

The nucleotides in each fragment were released by hydrolyzing the fragment with an alkali. The individual nucleotides could then be identified by paper chromatography, paper electrophoresis and spectrophotometric analysis. This procedure was sufficient to establish the sequence of each of the dinucleotides, because the right-hand member of the pair was determined by the particular enzyme that had been used to produce the fragment. To establish the sequence of nucleotides in larger fragments, however, required special techniques.

Methods particularly helpful in the separation and identification of the fragments had been previously described by Vernon M. Ingram of the Massachusetts Institute of Technology, M. Las-kowski, Sr., of the Marquette University School of Medicine, K. K. Reddi of Rockefeller University, G. W. Rushizky and Herbert A. Sober of the National Institutes of Health, the Swiss worker M. Staehelin and Tener.

For certain of the largest fragments, methods described in the scientific literature were inadequate and we had to develop new stratagems. One of these involved the use of an enzyme (a phosphodiesterase) obtained from snake venom. This enzyme removes nucleotides one by one from a fragment, leaving a mixture of smaller fragments of all possible intermediate lengths. The mixture can then be separated into fractions of homogeneous length by passing it through a column of diethylaminoethyl cellulose [see illustration on opposite page]. A simple method is available for determining the terminal nucleotide at the right end of each fraction of homogeneous length. With this knowledge, and knowing the length of each fragment, one can establish the sequence of nucleotides in the original large fragment.

A summary of all the nucleotide sequences found in the fragments of transfer RNA produced by pancreatic ribonuclease is shown in Table 1 on page 8. Determination of the structure of the fragments was primarily the work of James T. Madison and Ada Zamir, who were postdoctoral fellows in my laboratory. George A. Everett of the Plant, Soil and Nutrition Laboratory helped us in the identification of the nucleotides.

Much effort was spent in determining the structure of the largest fragments and in identifying unusual nucleotides not heretofore observed in RNA molecules. Two of the most difficult to identify were 1-methylinosinic acid and 5,6-dihydrouridylic acid. (In the illustrations these are symbolized respectively by I^m and U^h.)

Because a free 5'-phosphate group (p) is found at one end of the RNA molecule (the left end as the structure is conventionally written) and a free 3'-hydroxyl group (OH) is found at the other end, it is easy to pick out from Table 1 and Table 2 the two sequences that form the left and right ends of the alanine transfer RNA molecule. The left end has the structure pG—G—G—C— and the right end the structure U—C—C—A—C—COH. (It is known, however, that the active molecule ends in C—C—AOH.)

The presence of unusual nucleotides

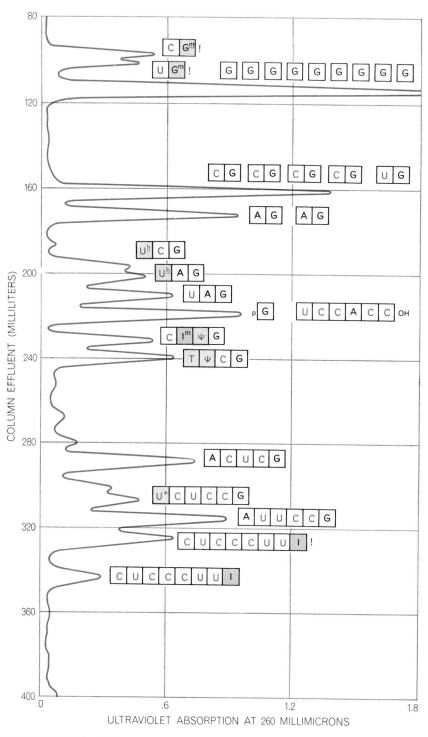

SEPARATION OF RNA FRAGMENTS is accomplished by chromatography carried out in a long glass column packed with diethylaminoethyl cellulose. The curve shows the separation achieved when the column input is a digest of alanine transfer RNA produced by takadiastase ribonuclease T1, an enzyme that cleaves the RNA into 29 fragments. The exclamation point indicates fragments whose terminal phosphate has a cyclical configuration. Such fragments travel faster than similar fragments that end in a noncyclical phosphate.

and unique short sequences made it clear that certain of the fragments found in Table 1 overlapped fragments found in Table 2. For example, there is only one inosinic acid nucleotide (I) in the molecule, and this appears in the sequence I–G–C– in Table 1 and in the sequence C–U–C–C–C–U–U–I– in Table 2. These two sequences must therefore overlap to produce the overall sequence C–U–C–C–C–U–U–I–G–C–. The information in Table 1 and Table 2 was combined in this way to draw up Table 3, which accounts for all 77 nucleotides in 16 sequences [see illustration on page 407].

With the knowledge that two of the 16 sequences were at the two ends, the structural problem became one of determining the positions of the intermediate 14 sequences. This was accomplished by isolating still larger fragments of the RNA.

In a crucial experiment John Robert Penswick, a graduate student at Cornell, found that a very brief treatment of the RNA with ribonuclease T1 at 0 degrees centigrade in the presence of magnesium ions splits the molecule at one position. The two halves of the molecule could be separated by chromatography. Analyses of the halves established that the sequences listed in the first column of Table 3 are in the left half of the molecule and that those in the second column are in the right half.

Using a somewhat more vigorous but still limited treatment of the RNA with ribonuclease T1, we then obtained and analyzed a number of additional large fragments. This work was done in collaboration with Jean Apgar and Everett. To determine the structure of a large fragment, the fragment was degraded completely with ribonuclease T1, which yielded two or more of the fragments previously identified in Table 2. These known sequences could be put together, with the help of various clues, to obtain the complete sequence of the large fragment. The process is similar to putting together a jigsaw puzzle [see illustrations on pages 408 and 409].

As an example of the approach that was used, the logical argument is given in detail for Fragment A. When Fragment A was completely degraded by ribonuclease T1, we obtained seven small fragments: three G–'s, C–G–, U–G–, U–G^m– and pG–. (G^m is used in the illustrations to represent 1-methylguanylic acid, another of the unusual nucleotides in alanine transfer RNA.) The presence of pG– shows that Frag-

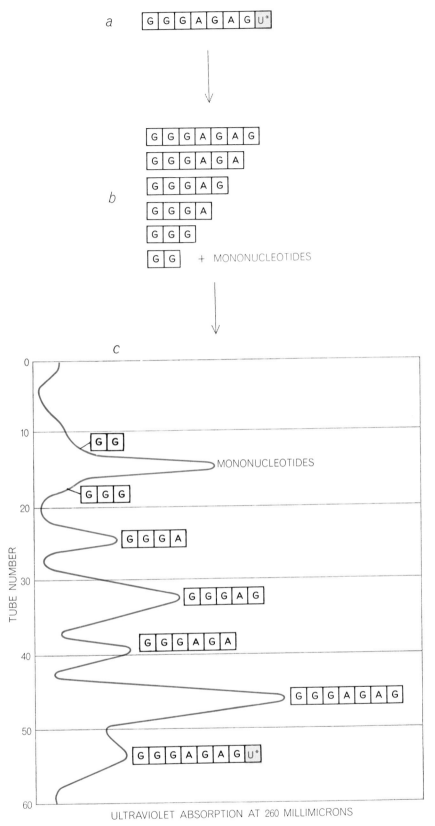

NEW DEGRADATION METHOD was developed in the author's laboratory to determine the sequence of nucleotides in fragments five to eight subunits in length. The example above begins with a fragment of eight subunits from which the terminal phosphate has been removed (a). When the fragment is treated with phosphodiesterase found in snake venom, the result is a mixture containing fragments from one to eight subunits in length (b). These are separated by chromatography (c). When the material from each peak is hydrolyzed, the last nucleoside (a nucleotide minus its phosphate) at the right end of the fragment is released and can be identified. Thus each nucleotide in the original fragment can be determined.

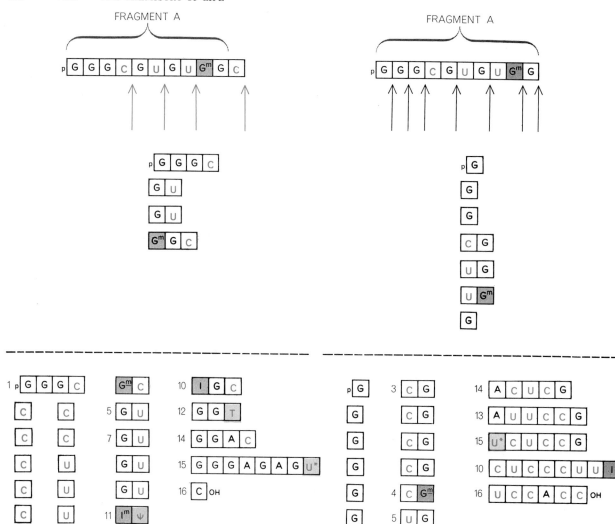

ACTION OF TWO DIFFERENT ENZYMES is reflected in these two tables. Table 1 shows the fragments produced when alanine transfer RNA is completely digested by pancreatic ribonuclease, which cleaves the molecule to the right of nucleotides containing bases with pyrimidine structures (C, U, U$\underline{h}$, ψ and T). The diagram at top left shows how pancreatic ribonuclease would cleave the first 11 nucleotides of alanine transfer RNA. The diagram at top right shows how the same region would be digested by takadiastase ribonuclease T1. Table 2 contains the fragments produced by this enzyme; they all end in nucleotides whose bases contain purine structures (G, G^m, G$\underline{m}$ and I). The numbers indicate which ones appear in the consolidated list in Table 3 on the opposite page.

ment A is from the left end of the molecule. Since it is already known from Table 3 that the left terminal sequence is pG–G–G–C–, the positions of two of the three G–'s and C–G– are known; the terminal five nucleotides must be pG–G–G–C–G–.

The positions of the remaining G–, U–G– and U–G^m– are established by the following information. Table 3 shows that the U–G^m– is present in the sequence U–G^m–G–C–. Since there is only one C in Fragment A, and its position is already known, Fragment A must terminate before the C of the U–G^m–G–C– sequence. Therefore the U–G– must be to the left of the U–G^m–, and the structure of Fragment A can be represented as pG–G–G–C–G–...U–G–...U–G^m–, with one G– remaining to be placed. If the G– is placed to the left or the right of the U–G– in this structure, it would create a G–G–U– sequence. If such a sequence existed in the molecule, it would have appeared as a fragment when the molecule was treated with pancreatic ribonuclease; Table 1 shows that it did not do so. Therefore the remaining G– must be to the right of the G^m–, and the sequence of Fragment *A* is pG–G–G–C–G–U–G–U–G^m–G–.

Using the same procedure, the entire structure of alanine transfer RNA was worked out. The complete nucleotide sequence of alanine transfer RNA is shown at the top of the next two pages.

The work on the structure of this molecule took us seven years from start to finish. Most of the time was consumed in developing procedures for the isolation of a single species of transfer RNA from the 20 or so different transfer RNA's present in the living cell. We finally selected a fractionation technique known as countercurrent distribution, developed in the 1940's by Lyman C. Craig of the Rockefeller Institute.

This method exploits the fact that similar molecules of different structure will exhibit slightly different solubilities if they are allowed to partition, or distribute themselves, between two nonmiscible liquids. The countercurrent technique can be mechanized so that the mixture of molecules is partitioned hundreds or thousands of times, while the nonmiscible solvents flow past each other in a countercurrent pattern. The solvent system we adopted was composed of formamide, isopropyl alcohol and a phosphate buffer, a modification of a system first described by Robert C. Warner and Paya Vaimberg of New York University. To make the method applicable for fractionating transfer RNA's required four years of work in collaboration with Jean Apgar, B. P. Doctor and Susan H. Merrill of the Plant, Soil and Nutrition Laboratory. Repeated countercurrent extractions of the transfer RNA mixture gave three of the RNA's in a reasonably homogeneous state: the RNA's that transfer the amino acids alanine, tyrosine and valine [*see bottom illustration on page 402*].

The starting material for the countercurrent distributions was crude transfer RNA extracted from yeast cells using phenol as a solvent. In the course of the structural work we used about 200 grams (slightly less than half a pound) of mixed transfer RNA's isolated from 300 pounds of yeast. The total amount of purified alanine transfer RNA we had to work with over a three-year period was one gram. This represented a practical compromise between the difficulty of scaling up the fractionation procedures and scaling down the techniques for structural analysis.

Once we knew the complete sequence, we could turn to general questions about the structure of transfer RNA's. Each transfer RNA presumably embodies a sequence of three subunits (an "anticodon") that forms a temporary bond with a complementary sequence of three subunits (the "codon") in messenger RNA. Each codon triplet identifies a specific amino acid [see "The Genetic Code: II," by Marshall W. Nirenberg; SCIENTIFIC AMERICAN, Offprint 153].

An important question, therefore, is which of the triplets in alanine transfer RNA might serve as the anticodon for the alanine codon in messenger RNA. There is reason to believe the anticodon is the sequence I–G–C, which is found in the middle of the RNA molecule. The codon corresponding to I–G–C could be the triplet G–C–C or perhaps G–C–U, both of which act as code words for alanine in messenger RNA. As shown in the illustration on page 2, the I–G–C in the alanine transfer RNA is upside down when it makes contact with the corresponding codon in messenger RNA. Therefore when alanine transfer RNA is delivering its amino acid cargo and is temporarily held by hydrogen bonds to messenger RNA, the I would pair with C (or U) in the messenger, G would pair with C, and C would pair with G.

We do not know the three-dimensional structure of the RNA. Presumably there is a specific form that interacts with the messenger RNA and ribosomes. The illustration on page 3 shows three hypothetical structures for alanine transfer RNA that take account of the propensity of certain bases to pair with other bases. Thus adenine pairs with uracil and cytosine with guanine. In the three hypothetical structures the I–G–C sequence is at an exposed position and could pair with messenger RNA.

The small diagram on page 401 indicates a possible three-dimensional folding of the RNA. Studies with atomic models suggest that single-strand regions of the structure are highly flexible. Thus in the "three-leaf-clover" configuration it is possible to fold one side leaf on top of the other, or any of the leaves back over the stem of the molecule.

One would also like to know whether or not the unusual nucleotides are concentrated in some particular region of the molecule. A glance at the sequence shows that they are scattered throughout the structure; in the three-leaf-clo-

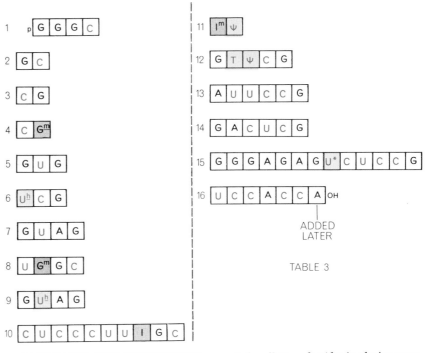

CONSOLIDATED LIST OF SEQUENCES accounts for all 77 nucleotides in alanine transfer RNA. The consolidated list is formed by selecting the largest fragments in Table 1 and Table 2 (*opposite page*) and by piecing together fragments that obviously overlap. Thus Fragment 15 has been formed by joining two smaller fragments, keyed by the number 15, in Table 1 and Table 2 on the opposite page. Since the entire molecule contains only one U*, the two fragments must overlap at that point. The origin of the other fragments in Table 3 can be traced in similar fashion. A separate experiment in which the molecule was cut into two parts helped to establish that the 10 fragments listed in the first column are in the left half of the molecule and that the six fragments in the second column are in the right half.

COMPLETE MOLECULE of alanine transfer RNA contains 77 nucleotides in the order shown. The final sequence required a care-

ful piecing together of many bits of information (*see illustration at bottom of these two pages*). The task was facilitated by degrada-

ver model, however, the unusual nucleotides are seen to be concentrated around the loops and bends.

Another question concerns the presence in the transfer RNA's of binding sites, that is, sites that may interact specifically with ribosomes and with

the enzymes involved in protein synthesis. We now know from the work of Zamir and Marquisee that a particular sequence containing pseudouridylic acid (Ψ), the sequence G–T–Ψ–C–G, is found not only in the alanine transfer RNA but also in the transfer RNA's for

tyrosine and valine. Other studies suggest that it may be present in all the transfer RNA's. One would expect such common sites to serve a common function; binding the transfer RNA's to the ribosome might be one of them.

Work that is being done in many

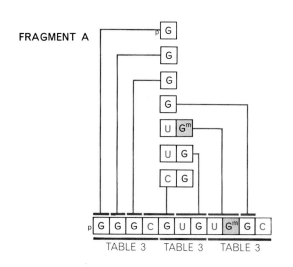

FRAGMENT A

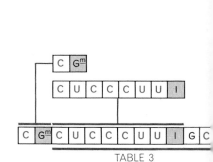

FRAGMENT B

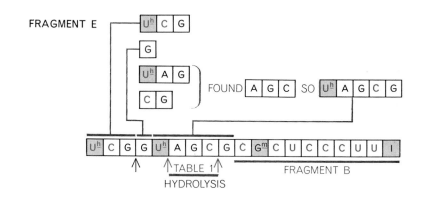

FRAGMENT E

REMAINDER OF LEFT HALF OF MOLECULE

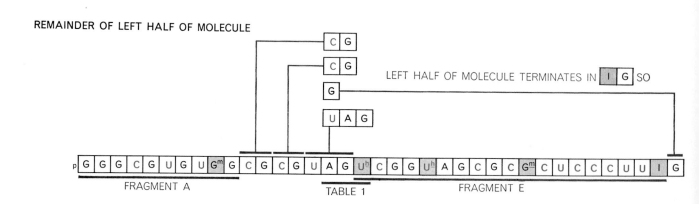

FRAGMENT C

FRAGMENT D

FRAGMENT F

FRAGMENT G

tion experiments that cleaved this molecule into several large fragments (*A, B, C, D, E, F, G*), and by the crucial discovery that

the molecule could be divided almost precisely into two halves. The division point is marked by the "gutter" between these two pages.

laboratories around the world indicates that alanine transfer RNA is only the first of many nucleic acids for which the nucleotide sequences will be known. In the near future it should be possible to identify those structural features that are common to various transfer RNA's,

and this should help greatly in defining the interactions of transfer RNA's with messenger RNA, ribosomes and enzymes involved in protein synthesis. Further in the future will be the description of the nucleotide sequences of the nucleic acids—both DNA and RNA—

that embody the genetic messages of the viruses that infect bacteria, plants and animals. Much further in the future lies the decoding of the genetic messages of higher organisms, including man. The work described in this article is a step toward that distant goal.

FRAGMENT C

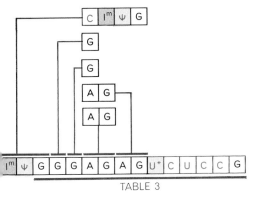

TABLE 3

FRAGMENT D

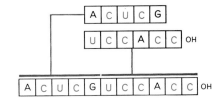

FRAGMENT F

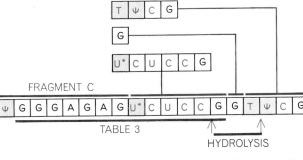

FRAGMENT C

TABLE 3

HYDROLYSIS

FRAGMENT G

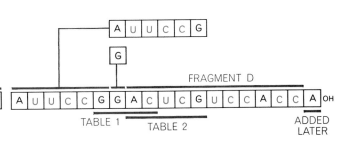

FRAGMENT D

TABLE 1   TABLE 2   ADDED LATER

ASSEMBLY OF FRAGMENTS resembled the solving of a jigsaw puzzle. The arguments that established the sequence of nucleotides in Fragment *A* are described in the text. Fragment *B* contains two subfragments. The larger is evidently Fragment 10 in Table 3, which ends in G—C—. This means that the C—Gᵐ— fragment must go to the left. Fragment *E* contains Fragment *B* plus four smaller fragments. It can be shown that *E* ends with I—, therefore the four small pieces are again to the left. A pancreatic digest yielded A—G—C—, thus serving to connect Uʰ—A—G— and C—G—. A partial digestion with ribonuclease T1 removed Uʰ—C—G—, showing it to be at the far left. The remaining G— must follow immediately or a pancreatic digest would have yielded a G—G—C— sequence, which it did not. Analyses of Fragments *A* and *E* accounted for everything in the left half of the molecule except for four small pieces. The left half of the molecule was shown to terminate in I—G—, thus the remaining three pieces are between *A* and *E*. Table 1 shows that one Uʰ is preceded by A—G—, therefore U—

A—G— must be next to *E*. The two remaining C—G—'s must then fall to the left of U—A—G—. Fragment *C* contains five pieces. Table 3 (Fragment 15) shows that the two A—G—'s are next to U* and that the two G—'s are to the left of them. It is also clear that C—Iᵐ—ψ—G— cannot follow U*, therefore it must be to the left. Fragment *D* contains two pieces; the OH group on one of them shows it to be to the right. Fragment *F* contains Fragment *C* plus three extra pieces. These must all lie to the right since hydrolysis with pancreatic ribonuclease gave G—G—T— and not G—T—, thus establishing that the single G— falls as shown. Fragment *G* gave *D* plus two pieces, which must both lie to the left (because of the terminal COH). Table 1 shows a G—G—A—C— sequence, which must overlap the A—C— in A—C—U—C—G— and the G— at the right end of the A—U—U—C—C—G—. Fragments *F* and *G* can join in only one way to form the right half of the molecule. The molecule is completed by the addition of a final AOH, which is missing as the alanine transfer RNA is separated from baker's yeast.

36

THE VISUALIZATION OF GENES IN ACTION

O. L. MILLER, JR.
March 1973

The electron microscope reveals individual genes being transcribed into RNA and their RNA being translated into protein. The pictures look remarkably like diagrams based on genetic and biochemical data

Since the middle 1950's a major objective in biology has been to document and add detail to what is called the central dogma of genetics: that DNA is the hereditary material, that its information is encoded in the sequences of its subunits that constitute the genes and that this information is transcribed into RNA and then translated into protein. By a variety of remarkable biochemical and genetic techniques many of the steps in this process of information transfer and the substances and cellular elements involved in them were identified; the DNA-protein dictionary was worked out, and the effect of different conditions and foreign agents on transcription and translation was determined. The increasingly detailed picture of gene action that emerged was necessarily based largely on indirect and collective evidence, however; there was little direct evidence of how individual genes function.

At the Oak Ridge National Laboratory in 1967 my colleagues and I began attempting to make electron micrographs of individual genes in action. The success we have had indicates that electron microscopy is potentially a valuable tool for the study of cell genetics at the molecular level. Meanwhile one of the gratifying aspects of our work has been that many of our pictures bear an almost uncanny resemblance to diagrams of transcription and translation that have been published over the years in technical journals and in magazines such as *Scientific American*. In other words, the pictures tend to confirm what had been proposed through consideration of painstakingly accumulated quantitative data.

The transfer of information from the deoxyribonucleic acid (DNA) of the genes into protein is accomplished in a series of steps involving the transcription of DNA into three kinds of ribonucleic acid (RNA). One is messenger RNA, which serves as the template on which amino acid subunits are assembled into proteins in the translation step. Another is ribosomal RNA, a constituent of the structures called ribosomes on which translation is accomplished. The third is transfer RNA, which carries the amino acid subunits of protein to their proper site along the messenger-RNA template.

The steps involved in the transcription of DNA into RNA and the translation of RNA into protein are essentially the same in all living cells, but the temporal and spatial relations of the steps in eukaryotic cells, whose DNA is confined within a membrane-bounded nucleus during most of the cell cycle, are different from the steps in prokaryotic cells, which lack such a nucleus. In prokaryotes, which include the many species of bacteria, transcription and translation of messenger RNA occur at the same time and place. In eukaryotic cells, which range from single-celled organisms such as the paramecium to the cells of higher vertebrates, the two processes are separated in time and space: transcription of DNA into the various RNA's that accomplish protein synthesis occurs in the nucleus; then the RNA's migrate through the membranous nuclear envelope into the cytoplasm, the main body of the cell, where the machinery for protein synthesis is located [*see illustration on page 412*].

We decided to begin by looking at the genetic machinery of a eukaryote: the oöcyte, or female reproductive cell, of amphibians such as frogs and salamanders. This type of oöcyte is a convenient cell with which to study genetic activity for a number of reasons; what was most important to us was that it would give us a good chance of seeing a great many identifiable genes, those that code for the two large molecules of RNA that are found in ribosomes. In most other eukaryotic cells these genes are found in clusters localized in nucleoli: dense bodies located at specific sites on certain chromosomes. In amphibian oöcytes large numbers of copies are made of the genes for ribosomal RNA, and these copies are in hundreds of nucleoli floating free and unattached to any chromosome. The result of this "amplification" of ribosomal-RNA genes is that a single oöcyte nucleus is equivalent to many hundreds of typical cell nuclei with respect to an identifiable genetic locus, and the amplified genes are not confusingly mingled with the other genes in the chromosomes.

There are two other advantages in studying amphibian oöcytes. One is that while they are growing they are actively synthesizing messenger RNA, and their chromosomes are in a greatly enlarged and uncoiled "lampbrush" stage where we might be able to see structural details of DNA being transcribed into messenger RNA. The final advantage is that these oöcytes are remarkably large. In some species they are as much as two millimeters in diameter at maturity and have nuclei as much as a millimeter in diameter. With a low-power microscope and jeweler's forceps one can readily isolate individual nuclei and manipulate their contents to prepare them for electron microscopy.

Barbara Beatty and I found that if the contents of an oöcyte nucleus were isolated and then put quickly into distilled water, the granular outer layers of the extrachromosomal nucleoli rapidly dispersed, allowing the compact fibrous

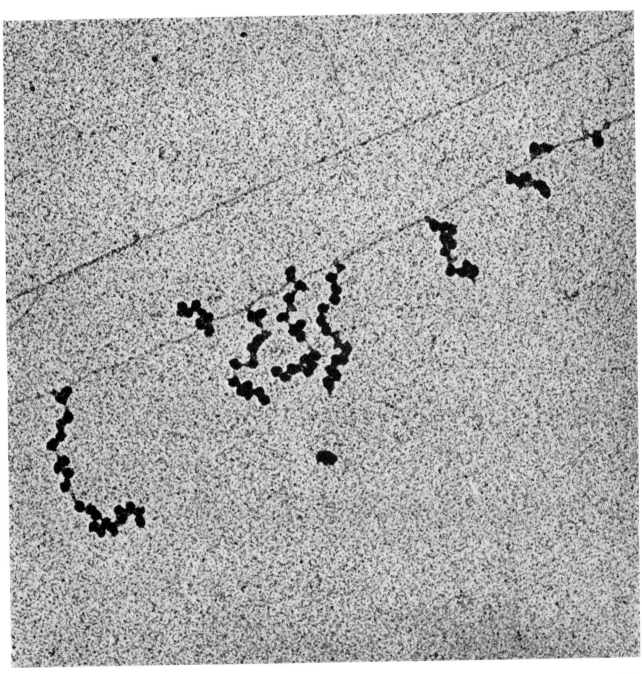

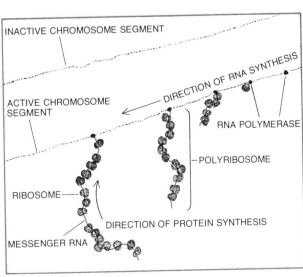

BACTERIAL GENE IN ACTION is enlarged 150,000 diameters in an electron micrograph made, like the others illustrating this article, by the author and his colleagues. The micrograph is interpreted in the somewhat simplified drawing at the right. One sees two segments of the chromosome of the bacterium *Escherichia coli*. The lower segment is active, that is, its DNA is being transcribed into messenger RNA and the RNA is being translated into protein. In the micrograph one can see molecules of RNA polymerase, the enzyme that catalyzes the transcription of DNA into RNA; one, at the far right, is at the approximate initiation site, where the transcription of each molecule of RNA begins. Successively transcribed messenger-RNA strands (*color in drawing*) in effect peel off toward the left; the longest one, now at the left of the micrograph, was the first to have been synthesized. As each RNA strand lengthens, ribosomes attach themselves and move along it, toward the chromosome, translating the RNA into protein (*not shown*).

INACTIVE CHROMOSOME SEGMENT

DIRECTION OF RNA SYNTHESIS

ACTIVE CHROMOSOME SEGMENT

RNA POLYMERASE

POLYRIBOSOME

RIBOSOME

DIRECTION OF PROTEIN SYNTHESIS

MESSENGER RNA

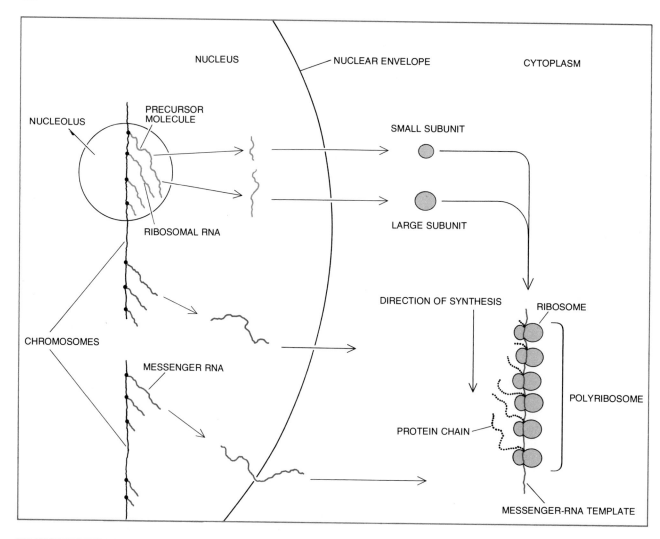

TRANSCRIPTION AND TRANSLATION occur respectively in the nucleus and the cytoplasm of eukaryotic cells. Ribosomal RNA (*light color*) transcribed from genes in the nucleolus forms a precursor molecule that is cleaved within the nucleolus to form the two large RNA molecules found in the two units of a ribosome. (RNA's from the same precursor are actually unlikely to end up in the same ribosome.) Messenger RNA's (*dark color*) transcribed from other genes serve as the templates on which polyribosomes assemble amino acid subunits into protein chains. Both kinds of RNA are complexed with protein (*not shown*) as they are being synthesized. In amphibian oöcytes a large number of copies of ribosomal-RNA genes are also present in extrachromosomal nucleoli.

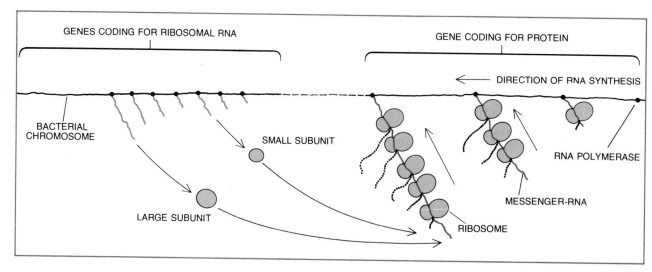

SIMULTANEOUS transcription and translation take place in prokaryotic cells such as bacteria, which have no true nucleus. Ribosomal RNA (*light color*) is transcribed from two genes, apparently contiguous, on certain segments of the single bacterial chromosome. Messenger RNA (*dark color*) is transcribed from genes at other sites and immediately translated into protein by ribosomes.

cores—which we hoped contained the genes we were looking for—to expand. We suspended such cores in a formalin solution to fix them and then spun them in a centrifuge tube where they were thrown against a thin carbon film on a wire-mesh specimen grid for an electron microscope. The problem was to spread the cores well without distorting them and destroying the pattern of gene activity we were looking for. We did that by treating the carbon films to make them hydrophilic, or water-attracting, and by rinsing the specimen grid in water containing an agent to reduce surface tension before drying it. Then we stained the specimen with a heavy metal to give it more contrast in the electron beam and made electron micrographs of the unwound cores.

Each unwound core appears as a long fiber in the form of a tangled, collapsed circle, along which there is a series of repeating shapes that we call matrix units [see illustration on next page]. Each of these units is about 2.4 microns long, and between each unit there is a segment of bare fiber that is usually somewhat shorter. Each matrix unit consists of a set of fibrils extending laterally from the long fiber. The length of the fibrils increases regularly along the fiber from one end of the matrix unit to the other, forming a structure with the outline of an arrowhead. All the matrix units on a single core fiber have the same polarity, that is, all the arrowheads point in the same direction. By treating the specimens with specific stains and enzymes we could identify the long axial fiber as a complex of DNA and protein and the lateral fibrils as a complex of RNA and protein. By means of autoradiography with radioactively labeled constituents of RNA we found that the matrix units were actively synthesizing RNA; the spacer segments between the matrix units were apparently not being transcribed. The nascent RNA molecules in the fibrils were complexed with protein that was made in the cell cytoplasm and that migrated into the nucleus and associated with the RNA molecules as they were being synthesized.

At about the same time we were making these observations some very pertinent biochemical information became available. Several investigators reported that the DNA in the extrachromosomal nucleoli of amphibian oöcytes contained subunit sequences that were complementary to the ribosomal RNA of ribosomes in oöcyte cytoplasm, thus confirming that the nucleolar DNA was the source of information for ribosomal RNA. The steps in the transfer of that

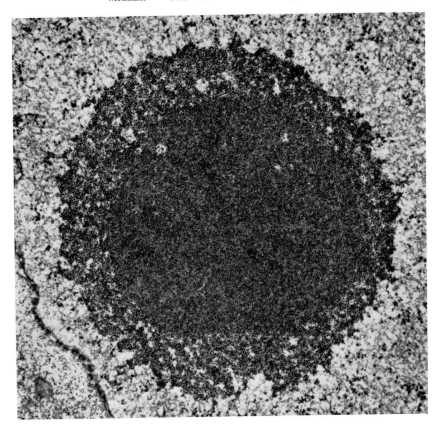

EXTRACHROMOSOMAL NUCLEOLUS in an oöcyte of the spotted newt, *Triturus viridescens*, is seen in thin section, enlarged 15,000 diameters. The nucleolus has a fibrous core and a granular cortex. Part of the nuclear envelope is visible (*lower left*), with the cytoplasm outside it. The core contains many copies of the genes for ribosomal RNA.

information were also being elucidated in many laboratories. It turned out that in both mammals and amphibians the two large RNA molecules that are found in ribosomes are not transcribed from DNA separately but are snipped out of a single precursor molecule. (Subsequent studies have shown that this is probably true in all eukaryotes.) The precursor molecule has a molecular weight of about 2.7 million in amphibians, and about 2.7 microns of double-strand DNA would be necessary to code for such a molecule. The matrix units we were observing were about 2.4 microns long, a figure reasonably close to the 2.7 microns of DNA. Since extrachromosomal nucleoli had been shown to contain DNA coding for ribosomal RNA, and since our observations localized RNA synthesis in the matrix units of nucleolar cores, we concluded that the DNA fiber in each matrix unit was indeed a gene coding for a precursor molecule of ribosomal RNA, and that the lateral fibrils contained such molecules in the process of being synthesized.

On some fibers that were slightly stretched during preparation we could see something more: we could resolve the individual molecules of RNA po-

lymerase, the enzyme that catalyzes the reading of DNA, that were active at each transcription site [see illustration on page 415]. Our success in visualizing nucleolar genes was due to the fact that when such genes are active, some 80 to 100 polymerase molecules are simultaneously transcribing each gene, forming fibrils of ribosomal-RNA precursor molecules (plus protein) in a graded set of lengths at successive stages of synthesis. The slight discrepancy between the observed 2.4-micron length of the matrix units and the 2.7-micron length expected for the genes can probably be explained by the fact that the DNA is unwound for transcription at each of the many transcription points; any bellying out of the DNA double helix would tend to shorten the overall length of the gene.

When we prepared amphibian oöcyte chromosomes in the same way that we had prepared the extrachromosomal nucleoli, we found that the fine detail of RNA synthesis could similarly be observed on the long lateral loops characteristic of the lampbrush stage [see illustrations on page 416]. Again we see the RNA polymerase molecules closely spaced along the active portions of the

414

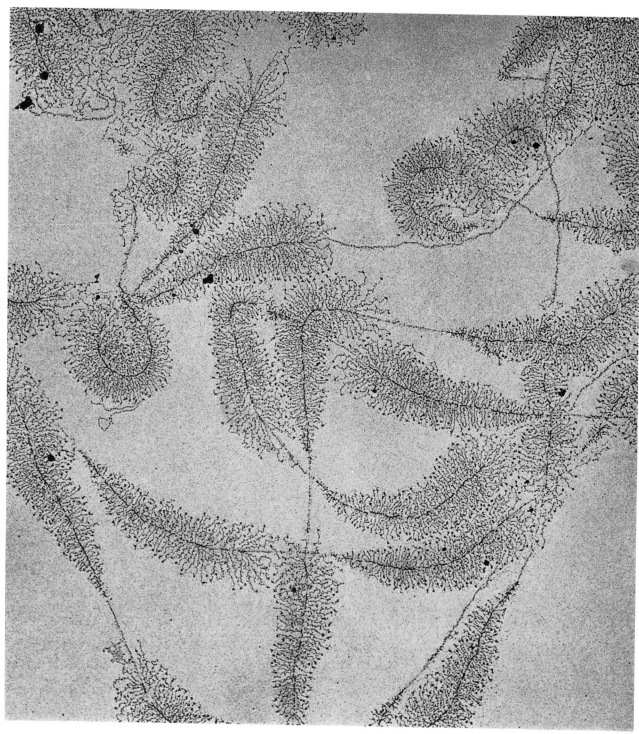

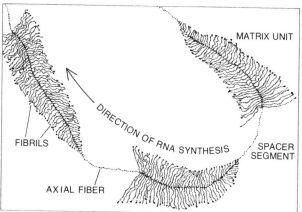

MATRIX UNIT

DIRECTION OF RNA SYNTHESIS

FIBRILS

SPACER
SEGMENT

AXIAL FIBER

ACTIVE GENES for ribosomal RNA are arrayed within an unwound nucleolar core that was stained with tungstic acid and enlarged 26,000 diameters in the electron micrograph, elements of
which are interpreted in the simplified drawing (*left*). One sees a
long axial fiber in the form of a collapsed circle, with a series of
arrowhead-shaped "matrix units" along it. Each unit is composed
of a set of fine lateral fibrils of graded lengths. By staining and
enzyme testing the axial fiber is identified as a complex of DNA
and protein, the fibrils as complexes of RNA and protein. The
segment of DNA within each matrix unit is a gene for ribosomal
RNA; each fibril is a protein-complexed ribosomal-RNA precursor
molecule that was being transcribed from the gene in the living
oöcyte. The stretches of the fiber between matrix units are inactive "spacer" segments the function of which is not yet known.

fiber, each at the base of one of the nascent RNA-protein fibrils that form a continuous gradient along the loop. The fibrils grow much longer than those on the nucleolar genes, however, and so they must contain very large RNA molecules. We do not know the subsequent function of the fibrils produced by the oöcyte chromosomes; presumably their RNA gives rise to messenger RNA's for the synthesis of specific proteins, but so far we cannot assign any specific messenger to any one loop.

With my help, Aimée H. Bakken has recently extended our work with eukaryotes to a mammalian cell: the human tumor cell line called HeLa, which has been established in laboratory cultures for 20 years and in which the synthesis of RNA and proteins has been intensively studied by chemical techniques. The fact that these cells are much smaller than oöcytes (only some 30 microns or less in diameter) called for different methods of isolating and dispersing the contents of their nuclei. In an attempt to avoid tedious micromanipulation we sought to develop a chemical method for disrupting cells and nuclei so that we could spread out the nucleoli and chromosomes without degrading the structures we hoped to observe. After many trials with various agents we attained success by treating the cells with low concentrations of Joy, an ordinary household dishwashing detergent.

We have isolated from HeLa cells (presumably from the nucleoli) active chromosome segments that we have identified tentatively as genes for ribosomal RNA [see top illustration on page 417]. The identification is based primarily on their striking structural similarity to the ribosomal matrix units of oöcytes, whose identity is on firm ground. In both cases the active segments of DNA are fully loaded with RNA polymerase molecules, with as many as 150 such polymerases per gene in HeLa cells; small, dense granules appear at the free ends of the growing fibrils, and neighboring active segments are separated by inactive spacer segments, which in the HeLa cells are about the same length as the putative ribosomal-RNA genes. The active segments are longer in HeLa cells than they are in the oöcyte but shorter than would be expected on the basis of the molecular weight of the ribosomal-RNA precursor in mammals: the segments are about 3.5 microns long instead of an expected 4.5 microns. Again the discrepancy is probably due to some unwinding of the DNA at each of the numerous closely spaced transcription sites.

In considering non-ribosomal-RNA

synthesis in HeLa cells, we knew from work done in other laboratories that 95 percent or so of the RNA synthesized in a HeLa cell at a given time does not survive to reach the cytoplasm but is degraded in the nucleus. This RNA is called heterogeneous nuclear RNA and its molecules range from about a tenth the size of the HeLa ribosomal-RNA-precursor molecule to many times that size. There is good evidence that small portions of the heterogeneous-nuclear-RNA molecules are not degraded and

are precursors of the messenger RNA's found in the HeLa cytoplasm.

In addition to the distinctive ribosomal-RNA-precursor fibrils we do see RNA-protein fibrils of various sizes at scattered sites along dispersed HeLa chromosomes, but none of these fibrils are arrayed in a graded set of lengths. This fact indicates that RNA synthesis is initiated much less frequently, and that the level of polymerase activity is therefore much lower, on active sites of HeLa chromosomes than it is on the active

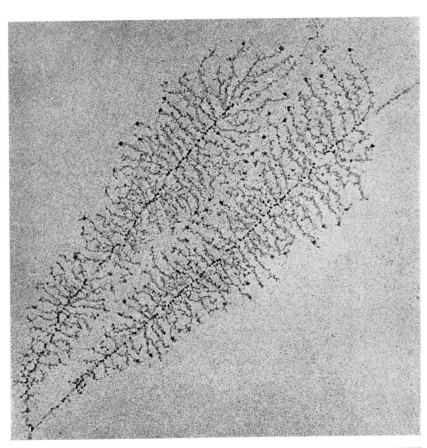

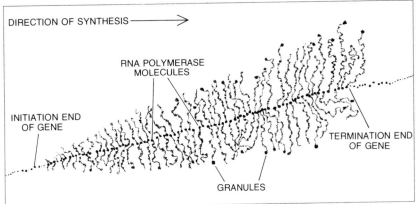

RNA POLYMERASE MOLECULES that are active at each transcription site are resolved on two nucleolar genes that were stretched in the process of preparation and are enlarged 48,000 diameters in the electron micrograph; one of the genes is reproduced in the drawing. The polymerase molecules are visible as dense granules at the base of each fibril; each polymerase was moving from left to right as it transcribed the gene. This micrograph also shows the granules, whose function is unknown, that are seen at the tip of maturing fibrils.

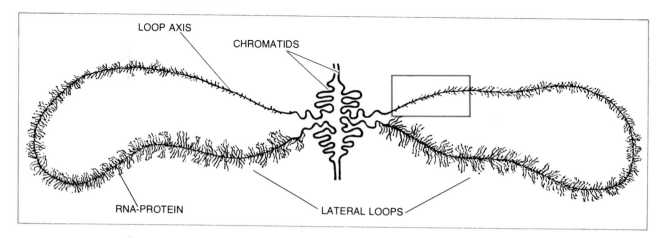

LAMPBRUSH CHROMOSOMES are a special form seen in oöcyte nuclei during such cells' long growth period. At the lampbrush stage the active portions of the sister chromatids that constitute each chromosome axis are greatly extended in lateral loops; the DNA axis of each loop is transcribed into RNA that is complexed with protein. The increasing length of the RNA molecules accounts for the increasing thickness of the loop from one point of juncture with the chromosome axis around to the next juncture. The colored rectangle shows approximately the segment of a lateral loop that is seen in the micrograph at the bottom of the page.

lateral loops of amphibian lampbrush chromosomes. Since heterogeneous nuclear RNA constitutes 95 percent or more of the RNA being synthesized in the HeLa nucleus, it seems likely that most of the dispersed RNA-protein fibrils we see attached to HeLa chromosomes contain nascent RNA of that kind. As in the case of the lampbrush chromosomes, however, no specific genetic role can yet be assigned to any of the observed RNA synthesis.

After our first success with amphibian oöcytes I collaborated with Charles A. Thomas, Jr., and Barbara A. Hamkalo at the Harvard Medical School in adapting the preparative techniques to a prokaryotic cell: the bacterium *Escherichia coli*. Later Hamkalo continued the work with me at Oak Ridge. Since there is a broad background of biochemical and genetic studies on the transcription and translation of various genes in *E. coli* and other bacteria, we could predict what active bacterial genes should look like if we could take their picture. The first problem, however, was to weaken the tough outer wall of the bacterial cells enough to open them up. We found that rapid cooling of growing *E. coli* followed by judicious application of the enzyme lysozyme removed enough of the wall to leave us with osmotically sensitive protoplasts: naked cells that could be burst open by osmotic shock when we diluted them in water. We then centrifuged the

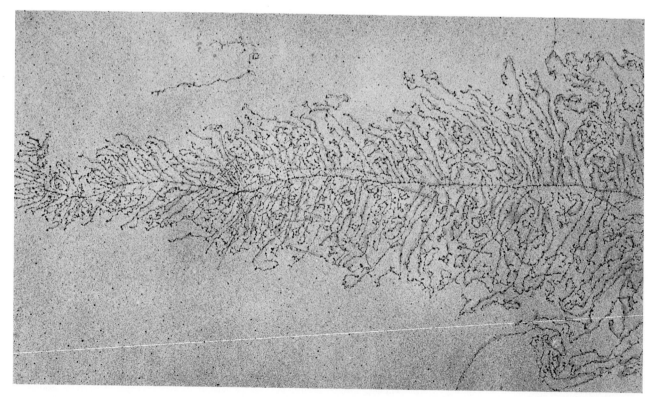

LATERAL LOOP of a lampbrush chromosome from a newt oöcyte is enlarged 22,000 diameters. The thin end, near the juncture with the chromosome axis, is at the left. RNA synthesis begins there, as shown by the increasing length of the fibrils from left to right. RNA polymerase molecules are visible at the base of each fibril. No specific genetic function can yet be assigned to these RNA's.

burst cells much as we had the oöcyte cores and made electron micrographs of their extruded contents.

We knew that in bacteria, unlike in eukaryotes, nascent messenger RNA is normally complexed with translating ribosomes. As soon as a newly initiated strand of messenger RNA is long enough, a ribosome attaches itself to the strand and translation of the RNA into protein begins. As the synthesis of the messenger RNA proceeds, giving rise to a longer strand of RNA, more ribosomes attach themselves to the strand. They form a string called a polyribosome, which continues the translation of the elongating messenger RNA. Any active genes coding for protein synthesis in our specimens should therefore have a graded set of strings of polyribosomes attached along their length.

Micrographs of the extruded *E. coli* contents show masses of thin fibers, some of them with attached strings of granules [*see bottom illustration at right*]. If we treat a specimen with the enzyme deoxyribonuclease, the fibers are destroyed; if we treat one with ribonuclease, the granular strings are removed from the fibers. We conclude that the fibers are portions of the bacterial chromosome and that the granules are ribosomes that were translating messenger RNA. The fact that there are no attached polyribosomes at all along large stretches of the chromosomes is consistent with reports that a large part of the *E. coli* chromosome either codes for rare species of RNA or is never transcribed. Also, our bacteria are grown under optimal conditions, so that many of the biosynthetic pathways may be inactive and the genes coding for their enzymes may be repressed.

At higher magnification the detailed configuration of the ribosome-coated regions—the genes coding for the synthesis of protein—is revealed [*see illustration on page 411*]. The gene is visible as a thin axial fiber. Along it there is a graded set of strings of ribosomes. At the junction point of each string of ribosomes with the gene we can resolve the active RNA polymerase molecule that is catalyzing the transcription. With negative staining we can see that the enzymes are smaller and more irregular in shape than they are in eukaryotic cells [*see top illustration on next page*]. Unfortunately the pictures do not show the growing protein chains that are presumably associated with each of the ribosomes; the amino acids that make up the chain are simply too small for our procedures to resolve.

As for ribosomal-RNA genes in *E. coli*,

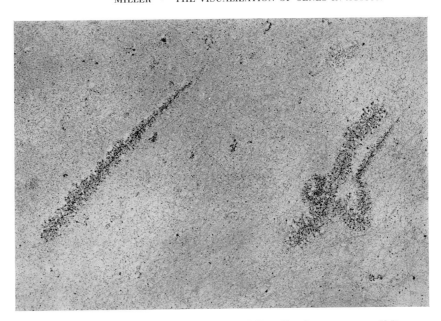

MATRIX UNITS isolated from the nucleus of a HeLa cell, a human tumor cell line, are identified as ribosomal-RNA genes on the basis of their similarity to the nucleolar genes of amphibian oöcytes: the graded set of closely spaced fibrils with a polymerase at the base of each, the dense granules at the tips of the fibrils and the occurrence of spacer segments (not resolved here) between the active genes. The enlargement is 16,000 diameters.

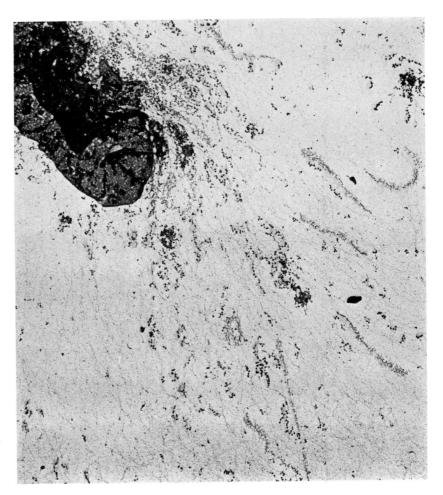

E. COLI CELL was exploded by osmotic shock. The micrograph shows the remains of the cell wall (*upper left*) and extruded cellular contents consisting of fine fibers with fibrillar segments and attached strings of granules. The fibers are portions of bacterial chromosome, the strings of granules are polyribosomes and the fibrillar segments are ribosomal genes. The enlargement is 22,000; the micrograph on page 411 further enlarges similar material.

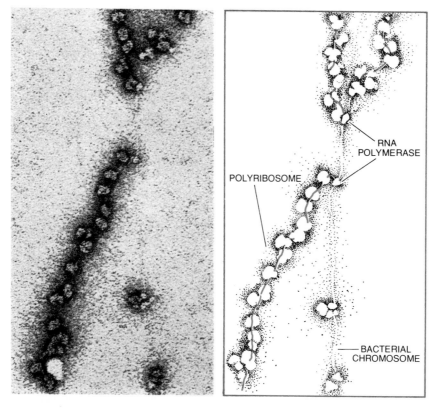

NEGATIVE STAINING (with uranyl acetate) outlines material from an *E. coli* cell, showing that the RNA polymerase molecules (*see drawing at right*) are smaller and more irregular than in eukaryotes. The staining reveals stretches of what is presumably messenger RNA but does not define the chromosome well. The enlargement is 150,000 diameters.

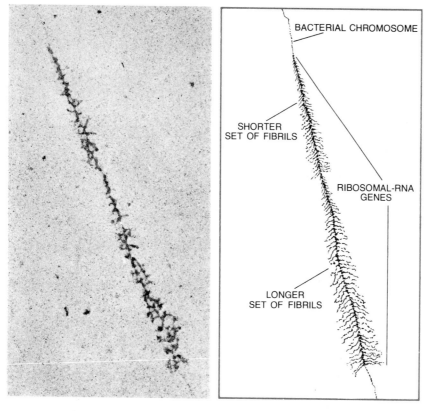

TWO CONTIGUOUS SETS of fibrils reflect transcription activity on genes for the two large RNA molecules of *E. coli* ribosomes. One matrix unit is about twice as long as the other and together they measure about 1.5 microns. The enlargement is 64,000 diameters.

we could predict that they should look very different from the genes for protein synthesis. Whereas the messenger RNA's of genes coding for proteins are contained within polyribosomes, the nascent ribosomal RNA should be complexed with some protein but not mature ribosomes. Quantitative studies of ribosomal-RNA synthesis in rapidly growing *E. coli*, combined with estimates that within each *E. coli* chromosome there are no more than seven segments coding for ribosomal-RNA synthesis, suggested that these segments should be relatively scarce in our preparations and should be densely packed with active RNA polymerase molecules. There was also evidence that the two large RNA molecules that are found in prokaryote ribosomes are synthesized separately rather than being cleaved from a single precursor as they are in eukaryotes; the genes for these two RNA's were thought to be contiguous, and so they should show up as two adjacent matrix units of fibrils. The molecular weights of the two bacterial ribosomal RNA's indicated that the matrix unit of one should be twice as long as the matrix unit of the other and that the total length of the two units should be close to 1.7 microns.

We have identified in our specimens segments of *E. coli* chromosome that meet these specifications rather precisely [*see bottom illustration at left*]. The segments have 60 to 70 attached fibrils arranged in two consecutive matrix units. As in the amphibian cells and the HeLa cells, these segments are somewhat shorter than had been predicted, again presumably because of the unwinding of DNA. The ribosomal-RNA genes of *E. coli* appear not to be close to one another on the chromosome as they are in eukaryotes. In fact, they are separated by chromosome segments that show the polyribosomal activity of protein-synthesis genes. We have not yet determined the minimum length of these intervening segments.

The fact that rather simple techniques of isolation and preparation have enabled us to see such fine structural details of genetic activity in several types of cell gives us confidence that electron microscopy will become an increasingly important tool for the study of cell genetics. Refinement of the techniques may make it possible to observe directly active genes from almost any kind of cell, and eventually to identify those genes with their products and learn how they function at different times in the cell's life cycle, how they are affected by varying conditions and how they relate to specific cell activities.

Nearly twenty-five years ago, I discussed the progress made in the science of chemistry during the first half of the twentieth century in an article in *Scientific American* that is reproduced in the first section of this book. I concluded my article by asking what the next fifty years would bring, and by making some predictions in answer to this question. Now that half of that fifty-year period has gone by, it may be interesting to examine my predictions, and to see to what extent they have already been fulfilled.

The first prediction was that by the year 2000 there would have been formulated a complete or nearly complete theory of the kinetics of chemical reactions, analogous to the now essentially completed system of chemical thermodynamics. Such a theory of chemical kinetics would enable a chemist to predict the rate of any chemical reaction with reasonable reliability. The theory would require a penetrating understanding of the nature of the forces between atoms and between molecules. Great progress had been made along these lines before 1950, and much progress has also been made since then (see "Chemistry by Computer," "The Nature of Aromatic Molecules," and some other articles in this book). One important feature of molecular structure, the restriction of rotation around single bonds, had been discovered by J. D. Kemp and K. S. Pitzer in 1937. Since then much detailed information about restricted rotation in the simpler molecules has been obtained, especially through the studies by microwave spectroscopy carried out by E. B. Wilson and his collaborators, and consideration of the effect of restricted rotation in determining molecular conformations has led to considerable progress in understanding the properties of moderately complex substances (see "The Shapes of Organic Molecules"). However, although it is known that the physiological properties of substances, such as their effect on the central nervous system, are determined by their molecular conformations, the understanding of the forces determining orientation around single bonds is not yet sufficient to permit reliable predictions to be made. I predict again that this goal will be largely reached before the end of the century—perhaps within a decade.

Considerable progress in studying very fast chemical reactions has been made by special techniques, such as with use of intersecting molecular beams, shock waves, and sudden increase in temperature (see "Relaxation Methods in Chemistry"). An especially valuable technique, introduced by Harden McConnell, involves the use of spin-labelled groups of atoms, containing an electron with unpaired spin, which can be studied by the methods of magnetic resonance spectroscopy. Application of these techniques, together with further quantum-mechanical studies, should lead to much progress.

The second prediction I made was that chemists would succeed in preparing catalysts to order. In fact, the preparation of specific catalysts remains essentially an empirical (trial-and-error) art, and the understanding of catalysis is still not profound enough to permit a catalyst to be made to order for a specific purpose.

Among the other predictions was that the second half of the twentieth century should witness the solution of the problem of the structure of proteins, nucleic acids, and other macromolecular constituents of living organisms, including enzymes and ultimately genes. In this field, which is now called molecular biology, progress has been astonishingly rapid. Following the pioneering work of

Fred Sanger with insulin, the amino-acid sequences of hundreds of proteins have already been determined, and many long polypeptide chains have been synthesized by the method developed by R. B. Merrifield (see "The Automatic Synthesis of Proteins"). The alpha helix and the beta-pleated sheet, which are the principal hydrogen-bonded structures of fibrous proteins, were discovered a quarter-century ago (see "Proteins"). Following the work of John Kendrew with myoglobin and Max Perutz with hemoglobin, structure determinations have been made by the X-ray diffraction methods for a number of other globular proteins, including cytochrome c (see "The Structure and History of an Ancient Protein"). Among these globular proteins are several enzymes, in whose molecules the region with catalytic activity can be identified. The molecule of one of these enzymes has a large hole in the center, and the region with catalytic activity is in this hole. The substrate molecules have access to this region through tunnels, and there is evidence that a feedback mechanism operates through the blocking of these tunnels by molecules produced by the catalysis when the concentration of these molecules becomes high.

A most astonishing event occurred in 1953, when the double helix of DNA was discovered by Watson and Crick. Nucleic acid had been identified earlier as the genetic substance, and in 1940 Max Delbrück and I had presented arguments to support the thesis that the gene consisted of two mutually complementary strands. The complementary structures were recognized by Watson and Crick as being hydrogen-bonded purine-pyrimidine pairs. This discovery led to the rapid elucidation of the genetic code (see "The Genetic Code: III") and to a tremendous amount of insight into the nature of the processes involved in heredity and in the functioning of living organisms.

Progress in molecular biology has been so rapid during the past twenty-five years that I now hesitate to make predictions about the future. I am convinced that many more important discoveries in molecular biology and molecular medicine will be made, that the science of chemistry itself will continue to flourish, and that through discoveries in chemistry scientists will be able to contribute to the struggle to diminish the amount of human suffering and to improve man's condition.

BIBLIOGRAPHIES

I CHEMISTRY: A PERSPECTIVE IN TIME

1. Chemistry

THE CHEMISTRY OF SILICONES. Eugene G. Rochow in *Scientific American*, Vol. 179, No. 4, pages 50–53, October, 1948.

FLUOROCARBONS. J. H. Simons in *Scientific American*, Vol. 181, No. 5, pages 44–47, November, 1949.

HOT ATOM CHEMISTRY. Willard F. Libby in *Scientific American*, Vol. 182, No. 3, pages 44–47, March, 1950.

PROTEINS. Joseph S. Fruton in *Scientific American*, Vol. 182, No. 6, pages 32–41, June, 1950.

THE SYNTHETIC ELEMENTS. I. Perlman and G. T. Sea-

borg in *Scientific American*, Vol. 182, No. 4, pages 38–47, April, 1950.

2. The Origin of Life

THE ORIGIN OF LIFE A. I. Oparin. Dover Publications, Inc., 1953.

A PRODUCTION OF AMINO ACIDS UNDER POSSIBLE PRIMITIVE EARTH CONDITIONS. Stanley L. Miller in *Science*, Vol. 117, No. 3046, page 528; May 15, 1953.

TIME'S ARROW AND EVOLUTION. Harold F. Blum. Princeton University Press, 1951.

II ATOMS AND THE CHEMICAL BOND

3. Atoms Visualized

FIELD IONIZATION OF GASES AT A METAL SURFACE AND THE RESOLUTION OF THE FIELD ION MICROSCOPE. Erwin W. Müller and Kanwar Bahadur in *The Physical Review*, Vol. 102, No. 3, pages 624–631; May 1, 1956.

A NEW MICROSCOPE. Erwin W. Müller in *Scientific American*, Vol. 186, No. 5, pages 58–62; May, 1952.

RESOLUTION OF THE ATOMIC STRUCTURE OF A METAL SURFACE BY THE FIELD ION MICROSCOPE. Erwin W. Müller in *Journal of Applied Physics*, Vol. 27, No. 5, pages 474–476; May, 1956.

STUDY OF ATOMIC STRUCTURE OF METAL SURFACES IN THE FIELD ION MICROSCOPE. Erwin W. Müller in *Journal of Applied Physics*, Vol. 28, No. 1, pages 1–6; January, 1957.

4. The Chemical Elements of Life

CERULOPLASMIN: A LINK BETWEEN COPPER AND IRON METABOLISM. Earl Frieden in *Bioinorganic Chemistry*, Advances in Chemistry Series 100. American Chemical Society, 1971.

CONTROL OF ENVIRONMENTAL CONDITIONS IN TRACE ELEMENT RESEARCH: AN EXPERIMENTAL AP-

PROACH TO UNRECOGNIZED TRACE ELEMENT REQUIREMENTS. Klaus Schwarz in *Trace Element Metabolism in Animals*, edited by C. F. Mills. E. and S. Livingstone, 1970.

THE PROTEINS: METALLOPROTEINS, VOL. V. Edited by Bert L. Vallee and Warren E. C. Wacker. Academic Press, 1970.

TRACE ELEMENTS IN BIOCHEMISTRY. H. J. M. Bowen. Academic Press, 1966.

TRACE ELEMENTS IN HUMAN AND ANIMAL NUTRITION. E. J. Underwood. Academic Press, 1971.

5. Chemistry by Computer

BISON: A NEW INSTRUMENT FOR THE EXPERIMENTALIST. Arnold C. Wahl, P. Bertoncini, K. Kaiser and P. Land in *International Journal of Quantum Chemistry*, Symposium No. 3, Part 2, pages 499–512; 1970.

CHEMISTRY FROM COMPUTERS. A. C. Wahl in *Argonne National Laboratory Reviews*, Vol. 5, No. 1, pages 43–69; April, 1969.

MOLECULAR THEORY OF GASES AND LIQUIDS. Joseph O. Hirschfelder, Charles F. Curtiss and R. Byron Bird. John Wiley and Sons, Incorporated, 1954.

VALENCE. Charles A. Coulson. Oxford University Press, 1961.

III MOLECULAR STRUCTURE AND BIOLOGICAL SPECIFICITY

6. The Stereochemical Theory of Odor

THE CHEMICAL SENSES. R. W. Moncrieff. Leonard Hill, Limited, 1951.

THE NATURE OF THE UNIVERSE. Lucretius. Translated by R. E. Latham. Penguin Books, 1951.

THE SENSES OF ANIMALS AND MEN. Lorus and Marjory Milne. Atheneum, 1962.

THE STEREOCHEMICAL THEORY OF OLFACTION. John E. Amoore, Martin Rubin and James W. Johnston, Jr., in Proceedings of the Scientific Section of the Toilet Goods Association, Special Supplement to No. 37, pages 1–47; October, 1962.

7. Pheromones

OLFACTORY STIMULI IN MAMMALIAN REPRODUCTION. A. S. Parkes and H. M. Bruce in Science, Vol. 134, No. 3485, pages 1049–1054; October, 1961.

PHEROMONES (ECTOHORMONES) IN INSECTS. Peter Karlson and Adolf Butenandt in Annual Review of Entomology, Vol. 4, pages 39–58; 1959.

THE SOCIAL BIOLOGY OF ANTS. Edward O. Wilson in Annual Review of Entomology, Vol. 8, pages 345–368; 1963.

8. The Structure of Cell Membranes

MEMBRANES OF MITOCHONDRIA AND CHLOROPLASTS. Edited by Efraim Racker. Van Nostrand Reinhold Company, 1969.

MEMBRANE MOLECULAR BIOLOGY. Edited by C. F. Fox and A. Keith. Sinauer Associates, 1972.

STRUCTURE AND FUNCTION OF BIOLOGICAL MEMBRANES. Edited by Lawrence I. Rothfield. Academic Press, 1971.

9. Molecular Isomers in Vision

THE CHEMISTRY OF VISUAL PHOTORECEPTION. Ruth Hubbard, Deric Bownds and Tôru Yoshizawa in Cold Spring Harbor Symposia on Quantitative Biology: Vol. XXX, Cold Spring Harbor Biological Laboratory, 1965.

CIS-TRANS ISOMERS OF RETINENE IN VISUAL PROCESSES. G. A. J. Pitt and R. A. Morton in Steric Aspects of the Chemistry and Biochemistry of Natural Products, edited by J. K. Grant and W. Klyne. Cambridge University Press, 1960.

MOLECULAR ASPECTS OF VISUAL EXCITATION. Ruth Hubbard and Allen Kropf in Annals of the New York Academy of Sciences, Vol. 81, Art. 2, pages 388–398; August 28, 1959.

VALENCE. Charles A. Coulson. Oxford University Press, 1961.

IV GASES, LIQUIDS, AND SOLIDS

10. Molecular Motions

EQUATION OF STATE CALCULATIONS BY FAST COMPUTING MACHINES. Nicholas Metropolis, Arianna W. Rosenbluth, Marshall N. Rosenbluth, Augusta H. Teller and Edward Teller in The Journal of Chemical Physics, Vol. 21, No. 6, pages 1,087–1,092; June, 1953.

FURTHER RESULTS ON MONTE CARLO EQUATIONS OF STATE. Marshall N. Rosenbluth and Arianna W. Rosenbluth in The Journal of Chemical Physics, Vol. 22, No. 5, pages 881–884; May, 1954.

MONTE CARLO EQUATION OF STATE OF MOLECULES INTERACTING WITH THE LENNARD-JONES POTENTIAL. I. A SUPERCRITICAL ISOTHERM AT ABOUT TWICE THE CRITICAL TEMPERATURE. W. W. Wood and F. R. Parker in The Journal of Chemical Physics, Vol. 27, No. 3, pages 720–733; September, 1957.

RADIAL DISTRIBUTION FUNCTION CALCULATED BY THE MONTE CARLO METHOD FOR A HARD SPHERE FLUID. B. J. Alder, S. P. Frankel and V. A. Lewinson in The Journal of Chemical Physics, Vol. 23, No. 3, pages 417–419; March, 1955.

11. The Solid State

ATOMIC STRUCTURE AND THE STRENGTH OF METALS. N. F. Mott. Pergamon Press, 1956.

ELEMENTS OF MATERIALS SCIENCE. L. H. Van Vlack. Addison-Wesley Publishing Company, Inc., 1966.

INTRODUCTION TO SOLID STATE PHYSICS. Charles Kittel. John Wiley and Sons, Inc., 1956.

PRINCIPLES OF THE THEORY OF SOLIDS. J. M. Ziman. Cambridge University Press, 1964.

SEVEN SOLID STATES. Walter J. Moore, W. A. Benjamin, Inc., 1967.

12. Ice

THE ELECTRICAL PROPERTIES OF ICE. L. Onsager and M. Dupuis in Electrolytes, edited by B. Pesce. Pergamon Press, 1962.

LATTICE STATISTICS OF HYDROGEN-BONDED CRYSTALS: I, THE RESIDUAL ENTROPY OF ICE. J. F. Nagle in The Journal of Mathematical Physics, Vol. 7, No. 8, pages 1484–1491; August, 1966.

MECHANISM FOR SELF-DIFFUSION IN ICE. L. Onsager

and L. K. Runnels in *Proceedings of the National Academy of Sciences,* Vol. 50, No. 2, pages 208–210; August, 1963.

RESIDUAL ENTROPY OF ICE. E. A. DiMarzio and F. H. Stillinger, Jr., in *The Journal of Chemical Physics,* Vol. 40, No. 6, pages 1577–1581; March, 1964.

SELF-DISSOCIATION AND PROTONIC CHARGE TRANSPORT IN WATER AND ICE. M. Eigen and L. De Maeyer in *Proceedings of the Royal Society,* Series A, Vol. 247, No. 1251, pages 505–533; October 21, 1958.

V   DYNAMICS OF CHEMICAL SYSTEMS

13. The Energy Resources of the Earth

ENERGY FOR MAN: WINDMILLS TO NUCLEAR POWER. Hans Thirring. Indiana University Press, 1958.

ENERGY RESOURCES. M. King Hubbert. National Academy of Sciences—National Research Council, Publication 1000-D, 1962.

ENVIRONMENT: RESOURCES, POLLUTION AND SOCIETY. Edited by William W. Murdoch. Sinauer Associates, 1971.

MAN AND ENERGY. A. R. Ubbelohde. Hutchinson's Scientific and Technical Publications, 1954.

RESOURCES AND MAN: A STUDY AND RECOMMENDATIONS. National Academy of Sciences–National Research Council, Committee on Resources and Man. W. H. Freeman and Company, 1969.

14. The Conversion of Energy

APPROACHES TO NONCONVENTIONAL ENERGY CONVERSION EDUCATION. Eric T. B. Gross in *IEEE Transactions on Education,* Vol. E-10, No. 2, pages 98–99; June, 1967.

EFFICIENCY OF THERMOELECTRIC DEVICES. Eric T. B. Gross in *American Journal of Physics,* Vol. 29, No. 1, pages 729–731; November, 1961.

ELECTRICAL ENERGY BY DIRECT CONVERSION. Claude M. Summers. Publication No. 147, The Office of Engineering Research, Oklahoma State University, March, 1966.

15. Relaxation Methods in Chemistry

FAST REACTIONS IN SOLUTION. E. F. Caldin. Blackwell Scientific Publications, 1964.

RELAXATION METHODS. M. Eigen and L. de Maeyer in *Technique of Organic Chemistry, Vol. VIII, Part II: Investigation of Rates and Mechanisms of Reactions,* edited by S. L. Friess, E. S. Lewis and A. Weissberger. Interscience Publishers Div., John Wiley and Sons, 1963.

RELAXATION SPECTROMETRY OF BIOLOGICAL SYSTEMS. Gordon G. Hammes in *Advances in Protein Chemistry,* Vol. 23, pages 1–57; 1968.

16. Catalysis

ADVANCES IN CATALYSIS AND RELATED SUBJECTS. Academic Press Inc., Publishers, 1948–1970.

CATALYTIC PROCESSES AND PROVEN CATALYSTS. Charles L. Thomas. Academic Press, Inc., 1970.

HETEROGENEOUS CATALYSIS. S. J. Thomson and G. Webb. John Wiley and Sons, Inc., 1968.

17. Why the Sea Is Salt

CHEMICAL OCEANOGRAPHY. Edited by J. P. Riley and G. Skirrow. Academic Press, 1965.

THE COMPOSITION OF SEA-WATER, SECTION I: CHEMISTRY, in *The Sea: Ideas and Observations on Progress in the Study of the Seas,* Vol. II, edited by M. N. Hill. Interscience Publishers, 1963.

MARINE CHEMISTRY: THE STRUCTURE OF WATER AND THE CHEMISTRY OF THE HYDROSPHERE. R. A. Horne. Wiley-Interscience, 1969.

THE OCEAN AS A CHEMICAL SYSTEM. Lars Gunnar Sillén in *Science,* Vol. 156, No. 3779, pages 1189–1197; June 2, 1967.

THE OCEANS: THEIR PHYSICS, CHEMISTRY, AND GENERAL BIOLOGY. H. U. Sverdrup, Martin W. Johnson and Richard H. Fleming. Prentice-Hall, Inc., 1961.

18. Fuel Cells

RECENT RESEARCH IN GREAT BRITAIN ON FUEL CELLS. F. T. Bacon and J. S. Forrest in *Transactions of the Fifth World Power Conference; Vienna, 1956,* Vol. 15, pages 5,397–5,412; 1957.

SYMPOSIUM ON FUEL CELLS. Ernst G. Baars, Chairman, in *Proceedings 12th Annual Battery Research and Development Conference,* pages 2–17; May 21–22, 1958.

TEXTBOOK OF ELECTROCHEMISTRY. G. Kortüm and J. O'M. Bockris. Elsevier Publishing Company, 1951.

19. The Oxygen Cycle

THE ATMOSPHERES OF THE PLANETS. Harold C. Urey in *Handbuch der Physik, Vol. LII, Astrophysics III: The Solar System,* edited by S. Flügge. Springer-Verlag, 1959.

ATMOSPHERIC AND HYDROSPHERIC EVOLUTION ON THE PRIMITIVE EARTH. Preston E. Cloud, Jr., in *Science,* Vol. 160, No. 3829, pages 729–736; May 17, 1968.

DISSOCIATION OF WATER VAPOR AND EVOLUTION OF OXYGEN IN THE TERRESTRIAL ATMOSPHERE. R. T. Brinkmann in *Journal of Geophysical Research,* Vol. 74, No. 23, pages 5355–5368; October 20, 1969.

THE EVOLUTION OF PHOTOSYNTHESIS. John M. Olson in *Science,* Vol. 168, No. 3930, pages 438–446; April 24, 1970.

HISTORY OF MAJOR ATMOSPHERIC COMPONENTS. L. V. Berkner and L. C. Marshall in *Proceedings of the National Academy of Sciences,* Vol. 53, No. 6, pages 1215–1226; June, 1965.

20. X-ray Crystallography

THE ARCHITECTURE OF MOLECULES. Linus Pauling and Roger Hayward. W. H. Freeman and Company, 1964.

THE CRYSTALLINE STATE, A GENERAL SURVEY: VOL. I. Sir Lawrence Bragg. Cornell University Press, 1962.

CRYSTALS: THEIR ROLE IN NATURE AND IN SCIENCE. Charles Bunn. Academic Press, 1964.

ORIGINS OF THE SCIENCE OF CRYSTALS. John G. Burke. University of California Press, 1966.

21. Chemical Analysis by Infrared

INFRARED AND RAMAN SPECTRA. Gordon B. B. M. Sutherland. Methuen and Company, 1935.

INFRARED AND RAMAN SPECTRA OF POLYATOMIC MOLECULES. Gerhard Herzberg. D. Van Nostrand Company, Inc., 1945.

INFRARED SPECTROSCOPY. R. Bowling Barnes, Robert C. Gore, Vurner Liddel, Van Zandt Williams. Rein-Hold Publishing Corporation, 1944.

22. Gas Chromatography

CHROMATOGRAPHY. Roy A. Keller, George H. Stewart and J. Calvin Giddings in *Annual Review of Physical Chemistry,* Vol. 11, pages 347–368; 1960.

PRINCIPLES AND PRACTICE OF GAS CHROMATOGRAPHY. Edited by Robert L. Pecsok. John Wiley and Sons, Inc., 1959.

23. Chemical Fossils

CHEMICAL EVOLUTION. M. Calvin in *Proceedings of the Royal Society,* Series A, Vol. 288, No. 1415, pages 441–466; November 30, 1965.

OCCURRENCE OF ISOPRENOID FATTY ACIDS IN THE GREEN RIVER SHALE. J. N. Ramsay, James R. Maxwell, A. G. Douglas and Geoffrey Eglinton in *Science,* Vol. 153, No. 3740, pages 1133–1134; September 2, 1966.

ORGANIC PIGMENTS: THEIR LONG-TERM FATE. Max Blumer in *Science,* Vol. 149, No. 3685, pages 722–726; August 13, 1965.

VII ORGANIC CHEMISTRY

24. Organic Chemical Reactions

MECHANISMS FOR LIQUID PHASE HYDROLYSES OF CHLOROBENZENE AND HALOTOLUENES. Albert T. Bottini and John D. Roberts in *Journal of the American Chemical Society,* Vol. 79, No. 6, pages 1,458–1,462; March 20, 1957.

THE STUDY OF ORGANIC REACTION MECHANISMS. Paul D. Bartlett in *Organic Chemistry: An Advanced Treatise,* Vol. 3, pages 1–121; 1953.

USES OF ISOTOPES IN ORGANIC CHEMISTRY. Dorothy A. Semenow and John D. Roberts in *Journal of Chemical Education,* Vol. 33, No. 1, pages 2–14; January, 1956.

25. The Nature of Aromatic Molecules

THE ANNULENES. Franz Sondheimer in *Accounts of Chemical Research,* Vol. 5, No. 3, pages 81–91; March, 1972.

AROMATICITY. P. J. Garratt. McGraw-Hill Book Company, 1971.

CARBOCYCLIC NON-BENZENOID AROMATIC COMPOUNDS. Douglas Lloyd. Elsevier Publishing Co., 1966.

MOLECULAR ORBITAL THEORY FOR ORGANIC CHEMISTS. A. Streitwieser, Jr. Wiley-Interscience, 1961.

ORGANIC REACTION MECHANISMS. Ronald Breslow. W. A. Benjamin, Inc., 1969.

26. The Shapes of Organic Molecules

CONFORMATION THEORY. Michael Hanack. Academic Press, 1965.

CONFORMATIONAL ANALYSIS. Ernest L. Eliel, Norman L. Allinger, Stephen J. Angyal and George A. Morrison. Interscience Publishers, 1965.

STEREOCHEMISTRY OF CARBON COMPOUNDS. Ernest L. Eliel. McGraw-Hill Book Company, Inc., 1962.

SYMPOSIUM: THREE-DIMENSIONAL CHEMISTRY. *Journal of Chemical Education,* Vol. 41, No. 2, pages 65–85; February, 1964.

27. Third-Generation Pesticides

THE EFFECTS OF JUVENILE HORMONE ANALOGUES ON THE EMBRYONIC DEVELOPMENT OF SILKWORMS. Lynn M. Riddiford and Carroll M. Williams in *Proceedings of the National Academy of Sciences,* Vol. 57, No. 3, pages 595–601; March, 1967.

THE HORMONAL REGULATION OF GROWTH AND REPRODUCTION IN INSECTS. V. B. Wigglesworth in *Advances in Insect Physiology: Vol. II,* edited by

J. W. L. Bement, J. E. Treherne and V. B. Wigglesworth. Academic Press Inc., 1964.

SYNTHESIS OF A MATERIAL WITH HIGH JUVENILE HORMONE ACITIVITY. John H. Law, Ching Yuan and

Carroll M. Williams in *Proceedings of the National Academy of Sciences*, Vol. 55, No. 3, pages 576–578; March, 1966.

VIII THE CHEMISTRY OF LIFE

28. Proteins

THE NATIVE AND DENATURED STATES OF SOLUBLE COLLAGEN. Helga Boedtker and Paul Doty in *Journal of the American Chemical Society*, Vol. 78, No. 17, pages 4,267–4,280; September 5, 1956.

THE OPTICAL ROTATORY DISPERSION OF POLYPEPTIDES AND PROTEINS IN RELATION TO CONFIGURATION. Jen Tsi Yang and Paul Doty in *Journal of the American Chemical Society*, Vol. 79, No. 4, pages 761–775; February 27, 1957.

POLYPEPTIDES, VIII: MOLECULAR CONFIGURATIONS OF POLY-L-GLUTAMIC ACID IN WATER-DIOXANE SOLUTION. Paul Doty, A. Wada, Jen Tsi Yang and E. R. Blout in *Journal of Polymer Science*, Vol. 23, No. 104, pages 851–861; February, 1957.

SYNTHETIC POLYPEPTIDES: PREPARATION, STRUCTURE, AND PROPERTIES. C. H. Bamford, A. Elliott and W. E. Hanby. Academic Press Inc., 1956.

29. The Automatic Synthesis of Proteins

THE CHEMISTRY AND BIOCHEMSITRY OF INSULIN. H. Klostermeyer and R. E. Humbel in *Angewandte Chemie: International Edition in English*, Vol. 5, No. 9, pages 807–822; September, 1966.

INSTRUMENT FOR AUTOMATED SYNTHESIS OF PEPTIDES. R. B. Merrifield, John Morrow Stewart and Nils Jernberg in *Analytical Chemistry*, Vol. 38, No. 13, pages 1905–1914; December, 1966.

SOLID PHASE PEPTIDE SYNTHESIS, I: THE SYNTHESIS OF A TETRAPEPTIDE. R. B. Merrifield in *Journal of the American Chemical Society*, Vol. 85, No. 14, pages 2149–2154; July 20, 1963.

THE SYNTHESIS OF BOVINE INSULIN BY THE SOLID PHASE METHOD. A. Marglin and R. B. Merrifield in *Journal of the American Chemical Society*, Vol. 88, No. 21, pages 5051–5052; November 5, 1966.

30. The Structure and History of an Ancient Protein

CONFORMATIONAL CHANGES UPON REDUCTION OF CYTOCHROME *c*. T. Takano, R. Swanson, O. B. Kallai and R. E. Dickerson in *Cold Spring Harbor Symposium on Quantitative Biology*.

FERRICYTOCHROME *c*: I, GENERAL FEATURES OF THE HORSE AND BONITO PROTEINS AT 2.8 A RESOLUTION. Richard E. Dickeron, Tsunehiro Takano, David Eisenberg, Olga B. Kallai, Lalli Samson, Angela Cooper and E. Margoliash in *The Journal of Biological Chemistry*, Vol. 246, No. 5, pages 1511–1535; March 10, 1971.

THE STRUCTURE AND ACTION OF PROTEINS. Richard E. Dickerson and Irving Geis. Harper and Row, Publishers, 1969.

THE STRUCTURE OF CYTOCHROME *c* AND THE RATES OF MOLECULAR EVOLUTION. Richard E. Dickerson in *Journal of Molecular Evolution*, Vol. 1, No. 1, pages 26–45; 1971.

31. The Three-dimensional Structure of an Enzyme Molecule

BIOSYNTHESIS OF MACROMOLECULES. Vernon M. Ingram. W. A. Benjamin, Inc., 1965.

INTRODUCTION TO MOLECULAR BIOLOGY. G. H. Haggis, D. Michie, A. R. Muir, K. B. Roberts and P. M. B. Walker. John Wiley and Sons, Inc., 1964.

THE MOLECULAR BIOLOGY OF THE GENE. J. D. Watson. W. A. Benjamin, Inc., 1965.

PROTEIN AND NUCLEIC ACIDS: STRUCTURE AND FUNCTION. M. F. Perutz. American Elsevier Publishing Company, Inc., 1962.

STRUCTURE OF HEN EGG-WHITE LYSOZYME: A THREE-DIMENSIONAL FOURIER SYNTHESIS AT 2 A. RESOLUTION. C. C. F. Blake, D. F. Koenig, G. A. Mair, A. C. T. North, D. C. Phillips and V. R. Sarma in *Nature*, Vol. 206, No. 4986, pages 757–763; May 22, 1965.

32. The Synthesis of DNA

ENZYMATIC SYNTHESIS OF DNA. Arthur Kornberg. John Wiley and Sons, Inc., 1961.

ENZYMATIC SYNTHESIS OF DEOXYRIBONUCLEIC ACID, VIII: FREQUENCIES OF NEAREST NEIGHBOR BASE SEQUENCES IN DEOXYRIBONUCLEIC ACID. John Josse, A. D. Kaiser and Arthur Kornberg in *The Journal of Biological Chemistry*, Vol. 236, No. 3, pages 864–875; March, 1961.

ENZYMATIC SYNTHESIS OF DNA, XXIV: SYNTHESIS OF INFECTIOUS PHAGE ϕX174 DNA. Mehran Goulian, Arthur Kornberg and Robert L. Sinsheimer in *Proceedings of the National Academy of Sciences*, Vol. 58, No. 6, pages 2321–2328; December 15, 1967.

33. Hybrid Nucleic Acids

DISTINCT CISTRONS FOR THE TWO RIBOSOMAL RNA COMPONENTS. S. A. Yankofsky and S. Spiegelman in *Proceedings of the National Academy of Sciences*, Vol. 49, No. 4, pages 538–544; April, 1963.

ORIGIN AND BIOLOGIC INDIVIDUALITY OF THE GENETIC DICTIONARY. Dario Giacomoni and S. Spiegelman in *Science*, Vol. 138, No. 3547, pages 1328–1331; December, 1962.

STRAND SEPARATION AND SPECIFIC RECOMBINATION IN DEOXYRIBONUCLEIC ACIDS: BIOLOGICAL STUDIES. J. Marmur and D. Lane in *Proceedings of the National Academy of Sciences*, Vol. 46, No. 4, pages 453–461; April, 1960.

THERMAL RENATURATION OF DEOXYRIBONUCLEIC ACIDS. Julius Marmur and Paul Doty in *Journal of Molecular Biology*, Vol. 3, No. 5, pages 585–594; October, 1961.

34. The Genetic Code: III

THE GENETIC CODE, VOL. XXXI: 1966 COLD SPRING HARBOR SYMPOSIA ON QUANTITATIVE BIOLOGY. Cold Spring Harbor Laboratory of Quantitative Biology.

MOLECULAR BIOLOGY OF THE GENE. James D. Watson. W. A. Benjamin, Inc., 1965.

RNA CODEWORDS AND PROTEIN SYNTHESIS, VII: ON THE GENERAL NATURE OF THE RNA CODE. M. Nirenberg, P. Leder, M. Bernfield, R. Brimacombe, J. Trupin, F. Rottman and C. O'Neal in *Proceedings of the National Academy of Sciences*, Vol. 53, No. 5, pages 1161–1168; May, 1965.

STUDIES ON POLYNUCLEOTIDES, LVI: FURTHER SYNTHESES, IN VITRO, OF COPOLYPEPTIDES CONTAINING TWO AMINO ACIDS IN ALTERNATING SEQUENCE DEPENDENT UPON DNA-LIKE POLYMERS CONTAINING TWO NUCLEOTIDES IN ALTERNATING SEQUENCE. D. S. Jones, S. Nishimura and H. G. Khorana in *Journal of Molecular Biology*, Vol. 16, No. 2, pages 454–472; April, 1966.

35. The Nucleotide Sequence of a Nucleic Acid

ISOLATION OF LARGE OLIGONUCLEOTIDE FRAGMENTS FROM THE ALANINE RNA. Jean Apgar, George A. Everett and Robert W. Holley in *Proceedings of the National Academy of Sciences*, Vol. 53, No. 3, pages 546–548; March, 1965.

LABORATORY EXTRACTION AND COUNTERCURRENT DISTRIBUTION. Lyman C. Craig and David Craig in *Technique of Organic Chemistry, Volume III, Part I: Separation and Purification*, edited by Arnold Weissberger. Interscience Publishers, Inc., 1956. See pages 149–332.

SPECIFIC CLEAVAGE OF THE YEAST ALANINE RNA INTO TWO LARGE FRAGMENTS. John Robert Penswick and Robert W. Holley in *Proceedings of the National Academy of Sciences*, Vol. 53, No. 3, pages 543–546; March, 1965.

STRUCTURE OF A RIBONUCLEIC ACID. Robert W. Holley, Jean Apgar, George A. Everett, James T. Madison, Mark Marquisee, Susan H. Merrill, John Robert Penswick and Ada Zamir in *Science*, Vol. 147, No. 3664, pages 1462–1465; March 19, 1965.

36. The Visualization of Genes in Action

ELECTRON MICROSCOPE VISUALIZATION OF TRANSCRIPTION. O. L. Miller, Jr., Barbara R. Beatty, Barbara A. Hamkalo and C. A. Thomas, Jr., in *Cold Spring Harbor Symposia on Quantitative Biology*, Vol. 35, pages 505–512; 1970.

MORPHOLOGICAL STUDIES OF TRANSCRIPTION. O. L. Miller, Jr., and Aimée H. Bakken in *Acta Endocrinologica*, Supplementum 168, pages 155–173; 1972.

PORTRAIT OF A GENE. Oscar L. Miller, Jr., and Barbara R. Beatty in *Journal of Cellular Physiology*, Vol. 74, No. 2, Part 2, Supplement 1, pages 225–232; October, 1969.

VISUALIZATION OF BACTERIAL GENES IN ACTION. O. L. Miller, Jr., Barbara A. Hamkalo and S. A. Thomas, Jr., in *Science*, Vol. 169, No. 3943, pages 392–395, July 24, 1970.

VISUALIZATION OF RNA SYNTHESIS ON CHROMOSOMES. O. L. Miller, Jr., and Barbara A. Hamkalo in *The International Review of Cytology*, Vol. 33, pages 1–25; 1972.